AF704095

CONF-870971

Proceedings of the Conference on

Geostatistical, Sensitivity, and Uncertainty Methods for Ground-Water Flow and Radionuclide Transport Modeling

San Francisco, California, September 15–17, 1987

Sponsored by U.S. Department of Energy and Atomic Energy of Canada Limited

Edited by

Bruce E. Buxton
Battelle Memorial Institute
Columbus Division
Columbus, Ohio

Associate Editors

Susanne M. Hogan
Battelle Memorial Institute
Office of Waste Technology Development
Willowbrook, Illinois

Lynn Copley-Graves
Battelle Memorial Institute
Project Management Division
Columbus, Ohio

Susan E. Brauning
Battelle Memorial Institute
Columbus Division
Columbus, Ohio

Organizing Committee

Edward S. Patera, Co-Chairman
Tin Chan, Co-Chairman
Albin Brandstetter
Bruce E. Buxton
Peter M. Clifton
William V. Harper
Susanne M. Hogan
Jane C. S. Long
Richard J. Schassburger
Claes Thegerström

BATTELLE PRESS

A portion of the funding for this conference was provided by the U.S. Department of Energy. However, the views and findings of the various papers are solely those of the authors and do not necessarily represent the policy of the agency. The agency encourages wide dissemination of the technical information contained herein, with due respect for the Publisher's rights regarding the complete volume.

Library of Congress Cataloging-in-Publication Data

Geostatistical, sensitivity, and uncertainty methods for ground-water flow and radionuclide transport modeling, CONF-870971

Papers presented at the DOE/AECL '87 Conference on Geostatistical, Sensitivity, and Uncertainty Methods for Ground-Water Flow and Radionuclide Transport Modeling held in San Francisco, California, on Sept. 15–17, 1987.

1. Nuclear power plants—Waste disposal—Environmental aspects—Mathematical models—Congresses. 2. Radioactive pollution of water—Mathematical models—Congresses. 3. Water, Underground—Pollution—Mathematical models—Congresses. 4. Groundwater flow—Statistical methods—Congresses. I. Buxton, Bruce E., 1953- . II. United States. Dept. of Energy. III. Atomic Energy of Canada Limited. IV. DOE/AECL '87 Conference on Geostatistical, Sensitivity, and Uncertainty Methods for Ground-Water Flow and Radionuclide Transport Modeling (1987: San Francisco, Calif.)

TD428.A86G64 1989 363.7'394 89-7376
ISBN 0-935470-45-X

Printed in the United States of America.

Copyright © 1989. Battelle Memorial Institute

All rights reserved. No part of this book may be reproduced or transmitted in any form or by any means, electronic or mechanical, including photocopying, recording or by any information storage and retrieval system, without permission of the publisher.

Battelle Press
505 King Avenue
Columbus, Ohio 43201-2693
614-424-6393; 1-800-451-3543

TD
428
.A86
G64
1989

Preface

This volume contains the proceedings of the DOE/AECL '87 Conference on Geostatistical, Sensitivity, and Uncertainty Methods for Ground-Water Flow and Radionuclide Transport Modeling held in San Francisco, California on September 15–17,1987. The conference was cosponsored by the United States Department of Energy (DOE) and Atomic Energy of Canada Limited (AECL).

Initial direction for the DOE/AECL '87 conference was provided by a cooperative research agreement in 1986 between DOE and AECL. Among other goals, that agreement called for a workshop on the statistical treatment of modeling data for ground-water flow and radionuclide transport studies to support the high-level nuclear waste disposal programs of the United States and Canada. The workshop gradually evolved into an international conference including the nuclear waste programs of those two countries along with as many other countries as possible. Ultimately, the conference was attended by almost 150 scientists from nine different countries.

The overall theme of the DOE/AECL '87 conference was the application of statistical methods for sensitivity and uncertainty analysis to nuclear waste disposal flow and transport modeling. The conference was organized into six technical sessions dealing with different topics in that theme:

- The role of statistical methods in the nuclear waste repository performance assessment plans of several waste disposal programs,
- Sensitivity and uncertainty methods for large computer codes,
- Stochastic hydrology,
- Statistical methods for modeling flow and transport in fractured media,
- Geostatistical methods, and
- The use of subjective information to supplement hard data in performance assessment studies.

Thirty-one presentations were made in the six sessions. Several presentations were invited from well-known authorities in the various technical fields. These invited presentations were supplemented by the contributed presentations of researchers from seven different countries. At the conclusion of the technical sessions, an open discussion session was held to address areas of concern to all conference attendees.

This volume includes papers for all of the thirty-one presentations along with introductory remarks from DOE, a transcript of the general discussion session, and a list of conference participants. All technical papers were subjected to a peer review after the conference. All papers were then edited and typeset for consistency of format into this volume of proceedings.

The technical program of the conference was established by an Organizing Committee composed of experts in the fields of nuclear waste, hydrology, and statistics. This Committee was responsible for the scope of the meeting, for the judging of abstracts to define the technical agenda, for chairing the technical sessions at the conference, and for soliciting the participation of an excellent group of conference participants. The members of the organizing committee were:

Edward S. Patera, U.S. Department of Energy,
Tin Chan, Atomic Energy of Canada Limited,
Albin Brandstetter, Battelle Office of Waste Technology Development,
Bruce E. Buxton, Battelle Columbus Division,
Peter M. Clifton, Rockwell Hanford Operations,
William V. Harper, Battelle Office of Nuclear Waste Isolation,
Susanne M. Hogan, Battelle Office of Waste Technology Development,
Jane C. S. Long, Lawrence Berkeley Laboratory,
Richard J. Schassburger, U.S. Department of Energy, and
Claes Thegerström, Nuclear Energy Agency.

Several individuals contributed fine reviews of the written contributions to this proceedings volume. The peer reviewers were:

Pradeep Aggarwal
Robert Andrews
Randall Barnes
Daniel Billaux
Albin Brandstetter
Katherine Campbell
Dwayne Chesnut
Richard Codell
Michael Grant
Amado Guzman
James Hanson
Bill Harper
Andreas Haug
Elizabeth Jacobson
André Journel
Muzaffer Kehnemuyi
Tony Knepp
Ram Kulkarni
Mark Larson
Jane C. S. Long
Gordon Luster
Donald Myers
Tom Naymik
Henning Omre
Ned Patera
Harry Perry
C. A. Rautman
Andrew Reisenauer
Brian Rutherford
Edward Taylor
Chin-Fu Tsang
Georges Verly
Joseph Wang

Susan K. Hall was responsible for the quality of the conference facilities and social program, as well as for ensuring that the conference ran smoothly.

Bruce E. Buxton
Battelle Memorial Institute
Columbus Division
Columbus, Ohio
September, 1988

Foreword

On September 15–17, 1987, the U.S. Department of Energy and Atomic Energy of Canada Limited co-sponsored the Conference on Geostatistical, Sensitivity, and Uncertainty Methods for Ground-Water Flow and Radionuclide Transport Modeling, with participation from nine other countries. This proceedings document contains the presentations and discussion at the conference.

Since that time, the United States Congress has significantly altered the national civilian high-level radioactive waste program, passing the Nuclear Waste Policy Amendments Act (NWPAA) on December 22, 1987. The primary impacts of this legislation are:

- Identification of one site, Yucca Mountain in Nevada, for site characterization, and termination of all other site-specific repository activities.
- Authorization to site, construct, and operate one monitored retrievable storage (MRS) facility. Results of an MRS Commission study of the need for an MRS facility are to be provided to Congress prior to commencement of this work. Site selection may proceed after the Secretary of Energy recommends a site for development as a repository.
- Direction to report to Congress between the year 2007 and 2010 on the need for a second repository.
- Appointment of a negotiator to seek a State or Tribe willing to host a repository or MRS facility.
- Broadening of financial assistance provisions.
- Formation of a Nuclear Waste Technical Review Board to evaluate the technical and scientific validity of DOE's high-level radioactive waste program.
- Study of subseabed disposal.
- Requirement to evaluate dry cask storage at civilian nuclear reactors until a permanent repository becomes available.
- Establishment of requirements for the transportation of nuclear wastes.

The content of these proceedings is, therefore, current as of prior to December 22, 1987. As a result of the NWPAA, the Salt Repository Project and Basalt Waste Isolation Project have been phased out. The results of performance assessment modeling studies done by these two projects are still relevant to the issue of safe disposal of high-level radioactive wastes, however, and articles about them are included in this document.

Table of Contents

Page

Geostatistical, Sensitivity, and Uncertainty Methods for Ground-Water Flow and Radionuclide Transport Modeling

Introductory Remarks to DOE/AECL '87 Conference on Geostatistical, Sensitivity, and Uncertainty Methods for Ground-Water Flow and Radionuclide Transport Modeling, San Francisco, California

RALPH STEIN
Director, Engineering and Geotechnology Division
Office of Civilian Radioactive Waste Management
Department of Energy

It is with genuine pleasure that I appear before you today because of the importance of this Conference, its cosponsorship with our colleagues from Canada, and the presence of other colleagues from abroad.

While we might find the denial of the obvious amusing, the serious business of this Conference is the pursuit of what is not obvious. How do we mark the boundaries of the unknown? How can we model the natural systems within those boundaries, and how do we define the dynamics of the natural systems in our model, now, as well as far into the future? Equally as important are questions related to how we validate our models. For without such validation we will be unable to license our nuclear waste repositories.

Analytical tools to deal with these problems, including those described at this Conference, have been developed that are at the cutting edge of technology. But the results of their application are subject to evaluation and challenge in the Congress as well as in the Courts. At the present time, nearly 40 items of legislation related to the high-level waste program have been submitted to Congress ranging from a moratorium on the Nuclear Waste Policy Act of 1982 (NWPA) to constraints on the selection of a repository site. In addition, there are over 35 lawsuits pending in the Courts against the Civilian Radioactive Waste Management (CRWM) Program of the U.S. Department of Energy (DOE). Perhaps the NWPA title should be changed to the Full Employment Act for Attorneys. In any event, one thing we can be certain of is that the surveillance of the CRWM Program, especially the validation of performance assessment models, will be intense and adversarial.

We recognize that some of the opposition to the program is rooted in the soil of political pressure, or a lack of understanding of the complex scientific and technical aspects of the program. Also, we know that not all of this opposition can be persuaded of the merits of the program on any grounds. For those who can be persuaded by facts and analysis, our approach is based on developing a sound scientific and technical foundation for the program, and the procedures we follow are fair, responsive, and open. I mention the knowledge and procedural aspects of our approach jointly because neither can be effective without the other.

A good example of the procedural part of this approach can be found in the plans for release of consultative drafts of the Site Characterization Plans (SCPs). The SCPs are large, technically complex documents that describe planned scientific and operational activities at each of the three candidate repository sites. The purposes of these activities are to:

- Determine the geologic, hydrologic, and geochemical conditions at a candidate site. A major use of the information collected during site characterization is to provide the basis for testing the adequacy of conceptual and statistical models on which performance assessment is dependent.
- Provide information needed for the design of the repository facility and to meet the licensing requirements of the U.S. Nuclear Regulatory Commission (NRC).

During discussions with the States and Indian Tribes, the NRC, and members of Congress, various concerns were raised that the 90-day statutory comment period would be insufficient time to allow for an adequate review and comment given the length and technical complexity of each SCP.

To be responsive to these concerns, DOE has concluded that the best interest of all the parties would be served by changing its procedure to a two-stage review process, beginning with an initial consultation process on the SCPs. To do this, DOE plans to simultaneously release all three SCPs as *consultation drafts* in early January 1988. After consultation, the SCPs would be prepared, to be followed by a 90-day public review with public hearings as called for in the NWPA. We hope that the change in procedure I have just described will help to alleviate the concerns and encourage full participation of the affected and interested parties.

While my discussion of the SCPs has focused on procedures, you will recall that the other component of our approach is the scientific and technical foundation of our program. This Conference is extremely important to that foundation because it has brought together the leaders of the performance assessment community. From the exchange of formal and informal views on geostatistical, sensitivity, and uncertainty analyses I expect that important progress will be made in assuring the safety of nuclear waste repositories. However, we all recognize that much work remains to be done to close the gap between the hope and the reality of siting a safe high-level waste repository.

In order for DOE to reach its programmatic goals to construct, operate, and close a repository, it is necessary for the repository to be licensed by the NRC, according to requirements and procedures spelled out in applicable U.S. laws and regulations. The requirements that must be met are summarized in the excellent Conference paper by Albin Brandstetter and Bruce Buxton of the Battelle organization. One of the requirements is for reasonable assurance that the repository will be safe for at least 10,000 years following repository closure with respect to radionuclide releases to the accessible environment.

The NRC has suggested that providing reasonable assurance may consist of DOE's analysis indicating that the likelihood of exceeding regulatory limits is low, that the NRC is satisfied that the performance assessment is sufficiently conservative, and its limitations well enough understood to ensure that repository performance will be within the predicted limits. The NRC has also suggested that "DOE use peer review to develop criteria for acceptability and applicability."

Because evaluating the safety of repositories involves predicting the extremely long-term behavior of a complex system of engineered and natural barriers, a unique scientific effort is needed. It is expected that validation of performance assessment models will be a significant, if not the key, topic addressed in both the NRC staff evaluation and public hearing processes leading to licensing. Surely, no search for the safety and protection of the public has ever been undertaken with such demands on scientific excellence.

Work that is responsive to these demands is well under way:

- The plans for the design and experimental work that will support the validation of performance assessment models will be contained in the site characterization study plans mentioned previously that are being developed for each candidate site.
- In addition, a generic system requirements document has been prepared, and associated requirements documents are being prepared by the regulatory projects that specify the requirements that must be satisfied by the disposal system at a specific location.
- At the international level, at the suggestion of the Swedish Nuclear Power Inspectorate, an ad hoc group with representatives from national and international organizations was formed to work on validation of radionuclide and groundwater transport models in the geosphere. To continue this work, a new project called INTRAVAL is being started with the first formal meeting to take place next month.

Based on these and other activities related to performance assessment models, I expect that performance assessment tools will undergo strong development in the near future. Different conceptual models for the geohydrologic structure of certain media and for different transport processes will be further developed. The description of repository systems will be further detailed. Also, processes of potential importance but little investigated, or couplings between processes that have not been incorporated will be investigated.

The ultimate aim of this and other work in this field is to achieve confidence in the models that are used for performance assessment. While the aim can be stated simply enough, agreement on what is *sufficient confidence* and how to achieve it is far more difficult. One solution to "how much is enough" is to do all the permutations and combinations of testing that the technical community and the participants in the program can think of regardless of how much its costs and the time that it takes. Our program is not structured that way. Instead it is designed to gather the information needed to show compliance with the regulations and to live within our budgetary constraints. But that is not the principal problem we have to deal with.

Our principal problems result from the fact that the traditional process of validation is not fully applicable to nuclear waste repository performance assessment modelling. This is due to the long time scales over which the repository must continue to isolate waste, so that direct experimental corroboration of models is not feasible. Also, the performance of the waste disposal system must be estimated for a variety of potential future environmental conditions. Other problems that we must consider are the time available for the development of performance assessment models; identification of the point where additional financial investment does not produce significant gains in the confidence that may be had in the safety assessment; and the potential adverse effects that site experimental investigations may have on the integrity of the site and its ability to isolate waste.

To resolve these problems, DOE must develop a validation strategy that provides the reasonable assurance specified by NRC regulations. The NRC recognizes the difficulties imposed by the time scales involved in high-level waste disposal, and accepts ''short-term testing supported by other avenues of inquiry such as peer review ... to obtain such assurance.'' The strategy I anticipate will go as far as one can go with the classical validation activities of system definition, scenario generation, consequence modeling, uncertainty analysis, and comparison with regulatory requirements. Beyond this we must try to extrapolate scientific information derived from laboratory experiments, field studies, in situ investigations, and observation of natural analogs to predict long-term repository performance. We have, with the assistance of the Pacific Northwest Laboratory Performance Assessment Scientific Support Program begun the process by developing a ''straw man'' for this validation effort. This was presented at the GEOVAL Symposium in Stockholm this past April.

Any such extrapolation must, of course, be subject to a detailed peer review. I envision the validation process of model extrapolation to begin with emphasis on technical features by specialists from each discipline involved in the performance assessment. To complete the process, I envision next a comprehensive peer review would take place by generalists with emphasis on completeness and consistency. Both levels of peer review would be interdisciplinary. Hopefully, this process will produce a consensus on the range of applicability of different modelling approaches and the important assumptions.

Establishing and implementing the process I have just described is a difficult task, and I feel that all of the members of the waste management community have a role and a responsibility in this undertaking. In particular, the attendees at this conference and the organizations they represent have an especially critical role in helping this process. As a first step, and within the next several weeks, I plan to initiate an effort that will start the development and describe a process for model validation. Like the development of the consultative draft stage for the SCPs, it will reflect the results of discussions with interested and affected parties.

When this undertaking is complete, I will bring the results to the scientific community for peer review. These results, because of their importance, must have the concurrence and consensus of the scientific community that I hope will be developed through a series of technical society meetings, NRC meetings, seminars, workshops, and conferences. Without such a supporting consensus of the scientific community, the credibility that reasonable assurance demands will be lacking. Without the skills and insight that you bring to the difficult task ahead we cannot hope to succeed.

After the peer review is complete and responses to the critique are incorporated into the final product, we will go to the NRC for a proposed rulemaking. You may ask why go to the NRC? What is there to gain, prior to the submittal of the license application in about 7 years? I believe that this issue of model validation is so important to successful licensing that, in my judgement, we should close with NRC, via rulemaking, on the validation process we have chosen. In short, show that the process is sound—and get it behind us. On the other hand, if the process is flawed, we should know this too, prior to submittal of a license application. Since rulemaking will take about 2 years to complete, proceeding now along this path provides the time needed to develop the process and proceed to rulemaking or correct any flaws prior to licensing.

I estimate that all these steps prior to rulemaking will take about 18 to 24 months. I hope that the attendees to this conference, as well as their colleagues and associates, will join me in making this effort work.

As you know from the number of proposed legislation and lawsuits directed against DOE, we are not free of suspicion by the affected States, Indian Tribes, and other parties. Recognizing this we have secured the oversight of technically competent and objective groups such as the National Academy of Sciences. Despite this the mistrust appears to continue.

The plan leading to rulemaking for model validation has no hidden agenda. It is directed toward facilitating the licensing process for an established and vital national objective. While we carry the burden of reasonable assurance, others have the burden of being responsible adversaries. If our models are faulty, let's correct them. If our procedures or processes need revision, as they did in the SCP review comment period, then we stand ready to do what is necessary. This is the best and only way to proceed. And it is what we intend to do, with what I hope will be your help and cooperation.

1 The Role of Geostatistical, Sensitivity, and Uncertainty Methods in Performance Assessment Studies

Performance Assessment, Site Characterization, and Sensitivity and Uncertainty Methods: Their Necessary Association for Licensing

NORMAN A. EISENBERG
U.S. Department of Energy
Washington, D.C.

LARRY D. RICKERTSEN
Weston Technical Support Team
Washington, D.C.

CHARLES VOSS
Pacific Northwest Laboratory
Washington, D.C.

Abstract

The regulatory requirements established by the U.S. Nuclear Regulatory Commission (NRC), the U.S. Environmental Protection Agency (EPA), and the U.S. Department of Energy (DOE) require that performance assessment analyses consider the uncertainty associated with the predictions of performance of a geologic repository for high-level nuclear waste. The uncertainty in performance predictions has several sources including:

1. Systematic and random error in measurement,
2. Spatial variations in geologic parameters,
3. Conceptual model uncertainties relative to geometrical configurations, major features, and boundary conditions,
4. Physicochemical process modeling, and
5. Future states of nature.

If the sources of uncertainty can be quantified, then the confidence in the performance assessment results can also be quantified. These types of uncertainty are not necessarily listed in order of importance, because the order may vary depending on the site. Some of these uncertainties may be reduced by taking appropriate measures. For example, use of more precise, better calibrated instruments can reduce measurement error. Other uncertainties, such as the spatial variations in geologic parameters, are inherent properties of a natural system. Such uncertainties cannot be reduced, but additional data may better specify spatial distributions of parameters.

This paper briefly reviews the sources of uncertainty in performance assessment analyses and discusses the role that sensitivity and uncertainty methods play in the DOE's effort to develop and license a nuclear waste repository in a geologic medium. In keeping with the focus of this conference, these topics are related primarily to groundwater movement and radionuclide transport analyses. Because methods for quantifying uncertainty exist and are applied primarily for uncertainties arising from spatial variability, and to a lesser degree alternative conceptual models, much discussion in this paper regards these uncertainties. Other types of uncertainty may be equally or more important, but less information is available.

Types of Uncertainty

Sensitivity and uncertainty analysis related to the development of a geologic repository for radioactive waste has been the subject of recent meetings and symposia (*1,2*). Contributors to these conferences have identified a variety of sources of uncertainty in describing the behavior of a geologic repository. For example, Cranwell and Helton (*3*) identify "process modeling uncertainty, model input data uncertainty and scenario uncertainty" as the primary components of uncertainty in analyzing the consequences and risks associated with geologic disposal of radioactive waste. Smith and Balderman (*4*) discuss natural geologic processes as sources of uncertainty in modeling a repository. Smith and Schwarz (*5*) discuss difficulties in characterizing hydrogeologic conditions at a potential repository site as a source of uncertainty in model predictions. Similarly, Channell, Wofsy, and Zand (*6*) discuss hydrologic parameter variation as a source of uncertainty in radionuclide transport predictions. Schalla and Leonhart (*7*) discuss data limitations as a source of uncertainty in formulating a conceptual hydrologic model of a potential repository site. In a rather comprehensive treatment, Kocher, Sjoreen, and Bard (*8*) identify insufficient site characterization, process model uncertainties, future evolution of the geologic environment, and near-field effects of the repository and waste emplacement as sources of uncertainty in estimating groundwater transport radionuclides.

In this paper we have selected five broad classes of sources of uncertainty, because of their important effect on performance assessment. It is the position of the DOE that site characterization should be guided in large measure by the need to resolve regulatory issues (*9*). Most of these issues will be resolved through a performance assessment. Thus site characterization may be viewed as an activity directed to reducing the uncertainties in performance assessments. Application of methods for sensitivity and uncertainty analysis is seen as a key element in evaluating site characterization activities, as planned and as executed, to determine the impact of site characterization on performance assessment.

The five types of uncertainty considered are (1) systematic and random error in measurement; (2) spatial variations in geologic parameters; (3) conceptual model uncertainties relative to geometrical configuration, major features, and boundary conditions; (4) physicochemical process modeling; and (5) future states of nature. Described briefly are these sources of uncertainty, methods for their treatment, and potential methods for their reduction.

Systematic and random error in measurement

This source of uncertainty includes inaccuracy in laboratory and field instruments. Also because so many parameters in the geosciences (typically geohydrologic parameters) are obtained by inferences from field measurements, this source of uncertainty includes uncertainties in the analyses performed to infer values of parameters. A typical example is estimating hydraulic conductivity from a slug test by matching discharge data to a theoretical curve. Another type of systematic measurement error is a result of samples being tested under a limited number of idealized conditions which may be convenient but not strictly representative of the actual conditions. Examples are uniaxial and triaxial tests to estimate the rock deformation modulus. These tests subject the rock specimen to a uniform stress field over the instrumented volume of rock while the in situ conditions are likely complex. Another source of error is a result of the sample being disturbed during sample collection, during preparation, or by the test apparatus. While new methods for obtaining samples, procedures for storage and handling, and new test equipment will minimize this source, they will still bias the test data to varying degrees. For example, cores of fractured rock obtained at depth may develop additional fractures or initial fractures may widen when the sample is brought to atmospheric pressure. Samples of groundwater obtained at depth may undergo significant chemical alteration if exposed to air, if temperature changes, or if contaminated by bacteria. These sources of uncertainty are typically treated by propagating the errors through the models in which uncertain values of parameters appear. Finally, the assumptions made in the interpretation of the test data will bias the results. Assumptions such as boundary and loading conditions are made to facilitate the analysis and interpretation of the test data. Parameters are often obtained based on some correlation with index properties or inferred from geophysical data. Typically, this is the smallest of all the sources of uncertainty. Standard error propagation techniques appear to be applicable and because the errors are small, perturbation methods are often applicable. Means for reducing these uncertainties include (1) careful calibration and use of instruments, (2) proper treatment of samples, (3) appropriate use of inferential analyses, and (4) vigorous application of quality assurance procedures to ensure implementation of 1–3.

Spatial variations in geologic parameters

All rock is heterogeneous with properties that exhibit considerable variation from point to point, not only between various strata but within a nominally homogeneous layer. In addition, it is not possible to obtain samples from every location within a rock mass and some uncertainty will be attributed to the limited availability of data. Typically, a property is expressed as an arithmetic average with the variability of the data indicated by some standard deviation. Higher moments of the statistical distribution may also be given or assumed, but the data to support fully such extrapolations are often lacking. In fact, a complete specification of the statistical distribution of parameters at every point in the space of interest would require measuring parameter values at a very large number of points in space. An additional complication is that it is very difficult to assess how much of the fluctuation from the average is due to natural variability and how much is introduced by measurement error. Further, many hydrologic tests, which provide inferential measures of hydraulic parameters, average properties over a subregion of space that frequently varies in size from one measurement point to another. For example, fixed-volume tests in higher conductivity zones are likely to average conductivity over a different size region than similar tests in a lower conductivity zone.

Heterogeneity of the geologic setting is an inherent property of any site chosen for waste disposal. The amount of data that can be gathered to characterize site heterogeneity is limited. Practical considerations of time, staff, and cost limit the data gathering at a given site. Furthermore, regulations require that site characterization be conducted in a fashion that does not significantly compromise the ability of the site to isolate waste. It is believed that noninvasive testing methods (e.g., radar and acoustic tomography) may provide some data otherwise unavailable because of this limitation; however, additional uncertainty is introduced in moving from such measurements to inferences about parameters of interest, such as hydraulic conductivity. Another limitation on data acquisition is imposed because many of the rocks of interest in waste isolation are relatively impermeable; because of this, any type of borehole or shaft may introduce a relatively large disturbance in the hydraulic system. This means that a very crude parallel of the Heisenberg uncertainty principle holds. Placing more boreholes in a relatively impermeable rock unit in order to reduce uncertainty regarding spatial variability of hydraulic parameters may in fact increase uncertainty by introducing significant changes into the hydraulic system.

Treatment of this class of uncertainty has received considerable attention. Methods of treatment include Monte Carlo schemes for sampling distributions of site parameters (*10,11*), the adjoint method approach for sensitivity and uncertainty analysis (*12*), kriging and other inverse problem approaches (*13*), and stochastic modeling in which parameter distributions are explicitly treated in deriving governing equations (*14*). Since a substantial part of this conference is devoted to discussing the relative merits of these different approaches and specific applications of them, we will not comment further. As discussed above, there are substantial barriers to reducing this source of uncertainty short of choosing a site with extraordinarily homogeneous geology.

Conceptual model uncertainties

This class of uncertainty includes uncertainty in the geometric relationships among features of the site, the definition of major features, and the geohydrologic boundary conditions. The conceptual model focuses on how many geohydrologic units adequately describe the site, the arrangement of the units, and the hydrologic boundary conditions of these units. From the point of view of a hydrologist, there may not be a clear distinction between uncertainties due to spatial variability of parameters and uncertainties due to the conceptual model. A fundamental problem in hydrology is specifying a sufficiently accurate model for the hydrology of a specified area when there is no uniquely correct model. Because the hydrologic structure of the site is extremely complex and heterogeneous, a common step in hydrologic modeling is the formulation of a conceptual model that provides a simplified description of the site. Because the site data always provide a limited representation of the true complexity of the site, many different simplified representations will fit the available data. Typically, a model is calibrated by adjusting model parameters until a desired goodness of fit is obtained. If additional data are obtained that are inconsistent with such a calibrated model, additional features may be added to provide additional degrees of freedom so that the model can accommodate both the earlier and later sets of data. In this sense conceptual model uncertainty can be viewed as a subset of the uncertainties due to spatial variability. For the purposes of this paper, however, it is convenient to differentiate between these uncertainties. The most important reason for this differentiation is that the sources of these types of uncertainty are substantially different. Parameter variability arises because the natural system with which we are dealing was not evolved in a uniform fashion; parametric variability describes the fundamental heterogeneity of the natural system. Conceptual model uncertainty arises from the desire of the analyst to simplify the treatment of the problem in order to make it more tractable and comprehensible. The conceptual model contains not just the simplifying assumptions by which the complex heterogeneity of the site is represented in some fashion, but is the broad analytical framework for assessing performance. Certainly some aspects of alternative conceptual models may be addressed by studies of parametric variability; however, the conceptual model must address a broader set of issues. For example, variations in hydraulic conductivity may address a range of alternative conceptual models for site hydrology. However, the conceptual model must also be consistent with other types of data, such as geophysical data and knowledge about the geologic origins of the site. Further, the more aspects of site performance that a conceptual model can treat, the higher will be its credibility. For example, an equivalent porous medium treatment of site hydrology may provide adequate estimates of groundwater travel time, but may not be suitable basis for estimating radionuclide transport, especially if phenomena such as matrix diffusion are important.

One approach to the treatment of this uncertainty is to derive estimates of performance using alternative conceptual models. Differences in these performance estimates then give a measure of the uncertainty resulting from the different conceptual models used. Such an approach, however, does not give a complete measure of this type of uncertainty, since many more reasonable conceptual models exist than those quantified. Furthermore, some conceptual models are more likely to be correct than others, so a need exists to somehow weight alternative models accordingly, in order to properly estimate these uncertainties. Another approach to conceptual model uncertainty is to use synthetic data sets to attempt to quantify the effects of undiscovered major features that may exist at a site. Conceptual model uncertainty may be reduced by increased site characterization aimed at discovering major features and conditions. It is also believed that the use of formalized statistical hypothesis testing methods to choose among alternative conceptual models can reduce the level of uncertainty and help to quantify it (*15*).

Physicochemical process modeling

This class of uncertainty arises from the choice of what processes to include in the description of the system, assumptions about the relationship among these processes and among the processes and the geometrical configuration, and how to describe the system physically and mathematically. This uncertainty arises because of the complexity of the physicochemical processes occurring in waste disposal configurations, the coupling of these various processes, and the irregular geometrical configurations in which they transpire. The dominant processes appear to be rather mundane and well understood when they occur in isolation; however, unusual couplings and configurations increase uncertainties. For example, the processes of fluid flow, diffusive and convective mass transport, and aqueous chemical reactions are all relatively well understood, when they take place under controlled conditions as might be experienced in a laboratory. The simultaneous occurrence of these processes in a complicated geometrical configuration such as a fractured rock raises considerable uncertainty about the correct model for mass transport. Another example of modeling uncertainty is the considerable differences over the correct paradigm for sorption and other chemical reactions between dissolved waste, groundwater, and rock.

Uncertainties due to physicochemical process modeling may be quantified to some degree by comparing performance estimates obtained from models that explicitly treat uncertain processes with estimates obtained from models not treating those processes. The INTRACOIN study did some of this sort of sensitivity analysis (*16*) by choosing extreme values of parameters in order to eliminate the effects of certain processes from transport models; for example, groundwater dispersivity was set to zero, thereby eliminating the effect of dispersion. Certain types of scoping calculations can also help to estimate the uncertainty due to process modeling. Values of variables obtained by solving sets of equations simplified by omitting terms describing certain processes, when substituted into the omitted terms, give a measure of the significance of those terms and hence the process not included. This type of uncertainty may be reduced by an adequate program of model validation, for example, as discussed at the recent GEOVAL symposium (*17,18,19*).

Future states of nature

This uncertainty arises because of the inability to predict precisely the conditions and environment in which the repository will be operating far into the future. It is clear that the multiple barrier approach implicit in the U.S. NRC regulations for nuclear waste disposal (10 CFR Part 60) is motivated by the desire to reduce this uncertainty (*20*). The U.S. EPA regulation for nuclear waste disposal (40 CFR Part 191) requires the portrayal of repository performance as a complementary cumulative distribution function (CCDF), which is itself a way of expressing the uncertainty due to future states of nature. Presentation of repository performance in a format that explicitly recognizes the relative probabilities of various possible futures provides a way of portraying this type of uncertainty. However, substantial difficulties attend substantiating the quantification of these probabilities. An approach to reducing uncertainties of this type is the vigorous pursuit of predictive modeling techniques for future states of nature, such as global climate modeling. Other approaches are the use of formalized expert opinion techniques to encode probabilities.

The Role of Sensitivity and Uncertainty Methods in Site Characterization

Preliminary performance assessment models have been developed to provide initial estimates of the groundwater travel time and the transport of radionuclides from the repository to the accessible environment. These analyses require many input parameters such as the properties of the system components and the conditions present. By applying sensitivity and uncertainty methods, these analyses are being used to assign priorities to the activities in the site characterization program. Sensitivity analyses are used to identify the relative importance of the input parameters to the predicted behavior. For those parameters that are relatively important, uncertainty methods are used to evaluate the sufficiency of the existing data base. If parameter variation has a significant effect on the predicted behavior and the associated parameter uncertainty is unacceptably high, tests to reduce the uncertainty to an acceptable level are needed and should be included in the site characterization activities. Conversely, if performance is insensitive to a parameter or a facet of the model in which it is contained, or if the range of and confidence in the parameter data are acceptable, the need for related testing during site characterization may be redirected or eliminated. Obviously, in developing the site characterization plan, the validity of the conceptual model and the significance of the parameter in performance analyses must also be considered.

A number of the performance requirements are stated as quantitative goals, e.g., groundwater travel time and radionuclide release rate to the accessible environment. For these requirements, performance assessment predictions can be used to evaluate the suitability of the repository site through direct comparison. The sources of uncertainty in the performance assessment model will result in a range of performance values with some upper bound of system performance as illustrated in Figure 1a. In these cases, it is possible to estimate the confidence or reasonable assurance that the system will perform within acceptable limits.

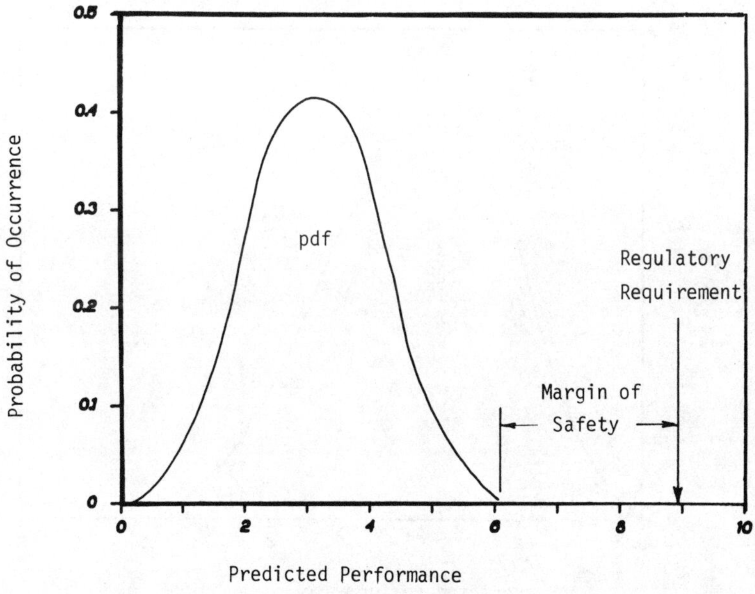

Figure 1a. **System clearly performing within regulatory limits.**

The degree to which the uncertainty in a predicted performance must be reduced depends on how close the predicted performance approaches regulatory limits. For repository systems where the predicted performance is considerably better than the regulatory limits (Figure 1a), thereby providing a large margin of safety, greater uncertainty is tolerable. For systems having a predicted performance near or above the regulatory requirement (Figure 1b), the sources of uncertainty will need to be reduced. Thus the level of effort in site characterization is also based on the performance estimates.

In addition to evaluating activities for site characterization, sensitivity and uncertainty methods are needed to evaluate the validity of the performance assessment prediction. The natural conditions are usually quite different from those assumed in an analysis. Rock masses are highly variable as compared to how they are normally modeled where an analysis may contain any number of assumptions and conceptual models to simplify and describe the system and processes involved. The assumptions and conceptual models are based on available data and related experience, a somewhat intuitive process. The subjective nature of this process is a source of uncertainty that must be quantified to the degree possible in order to estimate the margin of safety (assurance) for meeting Federal requirements.

One of the earliest examples of the potential difficulty in dealing with this type of uncertainty involved the hydrologic conceptual model for the Basalt Waste Isolation Project (BWIP). In 1982, the DOE submitted the BWIP Site Characterization Report (*21*) for technical review. The document was reviewed by the Nuclear Regulatory Commission (*22*), the State of Washington (*23*), and the U.S. Geological

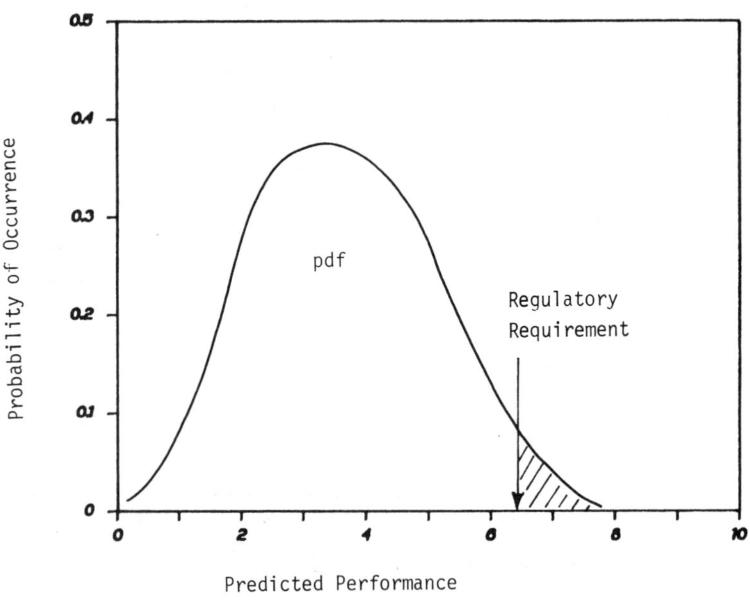

Figure 1b. **System with more uncertainty of compliance.**

Survey (*24*). The major source of controversy was the different interpretations given to the same data leading to different conceptual models and different modeling results. Clearly, methods are needed to reduce the subjectivity of data interpretation leading to the development of conceptual models.

One possible approach to resolving this problem is the use of hypothesis testing for analyzing the data and evaluating competing conceptual models. The approach forces a clear formulation of each component of a conceptual model or assumption which is then evaluated statistically. If the evaluation supports the conceptual model component, the analyst can have an increased confidence in that particular interpretation of the data. If the evaluation does not support the conceptual model component, either it must be rejected or additional data must be obtained in the hope that it will support the conceptual model component. The results are decisions that can be documented objectively and reviewed by decision makers in order to either achieve a consensus or provide the basis for further evaluation. An excellent example of this technique was developed by Doctor and Buchanan (*15*) who used geochemistry data from the BWIP site to test a groundwater hydrology conceptual model.

Examples of the Impact of Conceptual Model Uncertainty in Performance Assessments

Although site characterization has an impact on all five sources of uncertainty listed above, site characterization most strongly interacts with uncertainties related to spatial variability of parameters and alternative conceptual models. The following examples illustrate several cases where the use of sensitivity and uncertainty analysis methods have been employed, as an integral part of performance assessments, to evaluate the implication of parametric variability and to illustrate, in part, the uncertainties resulting from alternative conceptual models. These examples summarize information presented in the Environmental Assessments (*25a,b,c*); readers familiar with the analyses in those documents may wish to skip over these examples.

The preemplacement groundwater travel time for which a performance objective is set in 10 CFR 60.113(b) is not a precisely defined concept. The NRC conceptually defines this performance measure as the transit time of inert tracer particles in the groundwater. On this basis the preemplacement groundwater travel time could be defined as:

$$T = \int_S ds/V_p(s)$$

where S defines the pathway and V_p is the linear pore velocity along this pathway. This expression has been used in the Environmental Assessments to evaluate the groundwater travel time of candidate repository sites (*25a,b,c*).

Hanford Site—Pasco Basin. The deep basalts beneath the Hanford site form a layered sequence consisting of dense, fractured basalt flow interiors overlain by brecciated and vesicular flow tops. Some of the basalt flows are relatively thick (greater than 40 meters and continuous for many kilometers). The permeability of the basalt flow interiors appears to be lower than that of the flow tops, because of their smaller volume of interconnected fracture and pore space and because most of these fractures are filled or lined with secondary minerals. Thus, the flow interiors tend to act as confining units for groundwater contained in the flow tops. The permeability contrast between basalt flow tops and flow interiors promotes two-dimensional horizontal movement of groundwater along flow tops and one-dimensional vertical movement of groundwater through flow interiors.

The candidate repository horizon is in the Cohassett basalt flow about 950 m below the surface. The analysis in the BWIP Environmental Assessment (*25a*) gave results for a conceptual model of travel time in a layered sequence of basalt flow tops and flow interiors overlying the Cohassett flow interior (groundwater particles tracked from the base of the flow top immediately overlying the Cohassett flow interior).

The evaluation considered variations in the input parameters of these models to account for uncertainties due to limited knowledge of and spatial variability in the hydraulic characteristics of the flow interiors and flow tops. These parameters variations were propagated through the groundwater flow and travel time equations by means of a Monte Carlo technique (*26*). The parameter fields that were varied included the transmissivity and effective thickness of the flow tops and the vertical hydraulic conductivity and effective porosity of the flow interiors. Parameters defining the boundary conditions were not varied. Parameter probability distributions used in these analyses are given in Table 1. Figure 2, an example of typical results, shows the probability distribution of preemplacement groundwater travel time for the 4th (Concept D) of the conceptual models listed below. In general the median groundwater travel time in these analysis ranged between 61,000 years and 73,000 years, depending upon the range of parameters considered.

Other preliminary studies have been conducted. Table 2 summarizes the results of the most important of these studies. The variations in these results, in terms of both the computed groundwater travel time and the trends in the groundwater flow direction, reflect the differences in conceptual models used in the different studies.

These results indicate the parameters variations alone are not likely to address all uncertainties in groundwater travel time estimates. The need to address conceptual model uncertainty adequately is recognized by BWIP. The project has already identified alternative groundwater flow concepts that need to be investigated (*27*); these concepts are shown schematically in Figure 3 and are as follows:

> Concept A—This concept illustrates groundwater moving principally within heterogeneous, permeable flow tops separating flow interiors of relatively low vertical and horizontal hydraulic conductivity. Upward groundwater movement into shallower systems can occur as a result of (1) the positioning of flows where the front of one basalt flow of limited extent terminates atop of a more continuous flow, creating a direct fluid conduit between two flow tops or (2) groundwater leakage across low hydraulic conductivity flow interiors over large areas. In this concept, local features of relatively high hydraulic conductivity (e.g., thickening of flow-top breccia atop a spiracle) are not commonly juxtaposed. This concept depicts an anisotropic, heterogeneous flow system undisturbed by major folds and faults.

Parameter field	Probability distribution	Basis*
Flow-top transmissivity (T)	Lognormal Median = 0.15 m^2/d (1.6 ft^2/d) Standard deviation of $\log_{10}(T)$ = 1.83 Correlation range of $\log_{10}(T)$ = (0, 2, and 5 km (0, 1.2, and 3.1 mi))	Available data used to estimate median and log-standard deviation. Judgment used to estimate log-transmissivity correlation range.
Flow-top effective thickness	Uniform Range 10^{-3} to 10^{-1} m (10^{-3} to 10^{-1} ft)	Judgment, technical opinion surveys, and data from core samples.
Flow interior vertical hydraulic conductivity (K_v)	Lognormal Median = 5.0 x 10^{-13} m/s (1.4 x 10^{-7} ft/d), 1.5 x 10^{-12} m/s (4.2 x 10^{-7} ft/d), and 1.5 x 10^{-11} m/s (4.2 x 10^{-6} ft/d) Standard deviation of $\log_{10}(K_v)$ = 0.90	Derived from horizontal hydraulic conductivity data by assuming constant vertical to horizontal anisotropy ratios of 1, 3, and 30.
Flow interior effective porosity (n_e)	Lognormal Median = 10^{-4} Standard deviation of $\log_{10}(n_e)$ = 0.51 (95 percent of range between 10^{-5} and 10^{-3})	Judgment and data from core samples.

*See Clifton (1986) for detailed discussion.

Table 1. **Parameter field probability distributions for pre-waste-emplacement ground-water travel-time model (taken from 25a).**

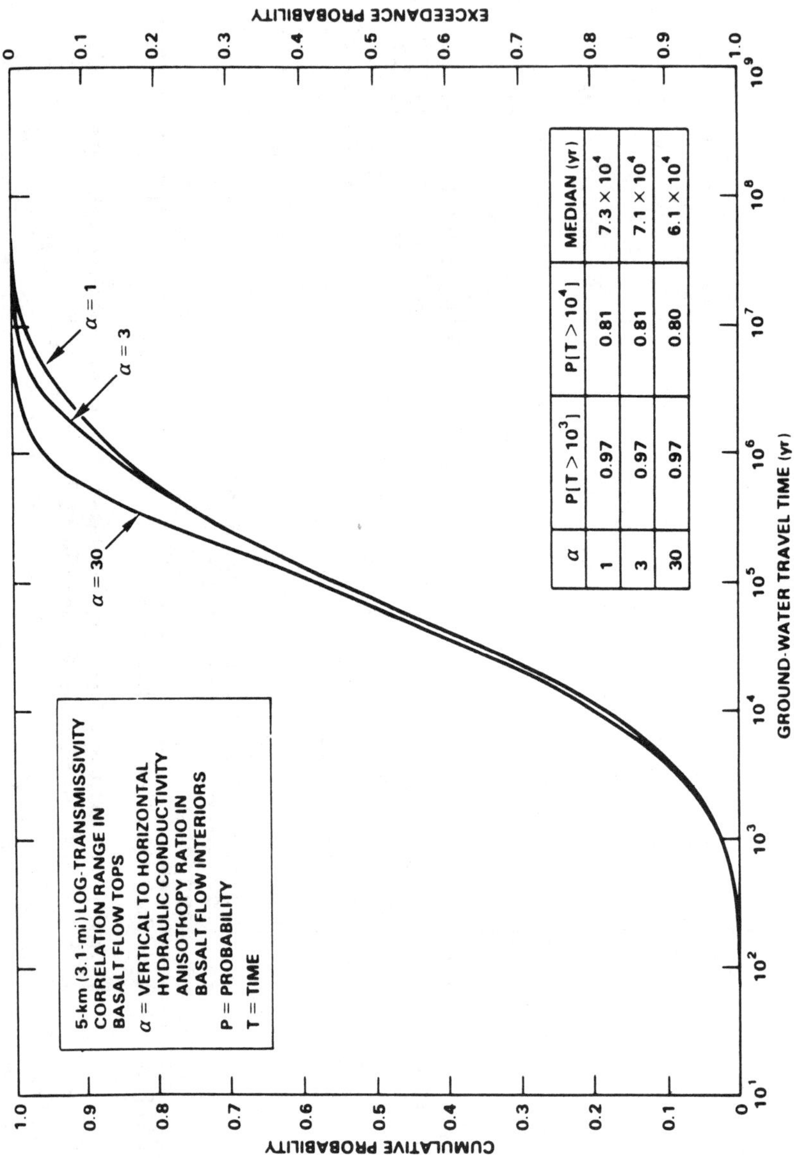

Figure 2. **Probability distribution of ground-water travel times (taken from 25a).**

Table 2. **Summary of ground-water travel-time estimates from modeling studies (taken from 25a).**

Study	Purpose of study	Ground-water travel-time distance and (or) direction	Ground-water travel time
Los Alamos Technical Associates, Inc. (LATA, 1981)	Initial estimates of ground-water movement from hypothetical repository	Northward, 12 km (7.5 mi) to the Columbia River	33,000 yr
Pacific Northwest Laboratory (Dove et al., 1981)	Demonstration of numerical modeling capability	Northward, 12 to 16 km (7.5 to 10 mi) to the Columbia River	13,000 to 17,000 yr
Rockwell Hanford Operations (Arnett et al., 1981)	Estimate ground-water travel times from reference repository location	Southwest, 32 km (20 mi) to beneath the Columbia River	30,000 yr
U.S. Nuclear Regulatory Commission (NRC, 1983b)	Sensitivity of ground-water travel times to variation of model inputs	10 km (6.2 mi) from hypothetical repository	20 yr to greater than 40,000 yr
Rockwell Hanford Operations (Clifton et al., 1983)	Stochastic analysis of ground-water travel times	10 km (6.2 mi) within flow top of host rock	Median ground-water travel time of 17,000 yr
Rockwell Hanford Operations (Clifton et al., 1984)	Stochastic analysis of ground-water travel times for draft environmental assessment	10 km (6.2 mi) within flow top of host rock	Median ground-water travel times of 17,000, 81,000, and 86,000 yr
Rockwell Hanford Operations (Arnett and Sagar, 1984)	Evaluation of deterministic approaches for ground-water travel-time calculations using a preliminary measured effective thickness value	10 km (6.2 mi) within flow top of host rock	3,600 to 142,000 yr

Concept B—This concept is similar to Concept A, except basalt flows are crossed by bedrock structural discontinuities having potentially larger vertical hydraulic conductivities than the confining aquitards. On a local scale, such discontinuities might represent individual tectonic fractures or shear zones. Regionally, these features could depict major fault or fold zones. If rock movement has occurred, such structures could depict zones where the lateral continuity of flow contacts is disrupted, causing flow contact to terminate against a flow interior of lower hydraulic conductivity. In this concept, structural discontinuities are heterogeneities having the potential for vertically connecting shallow and deep flow systems. Depending on the

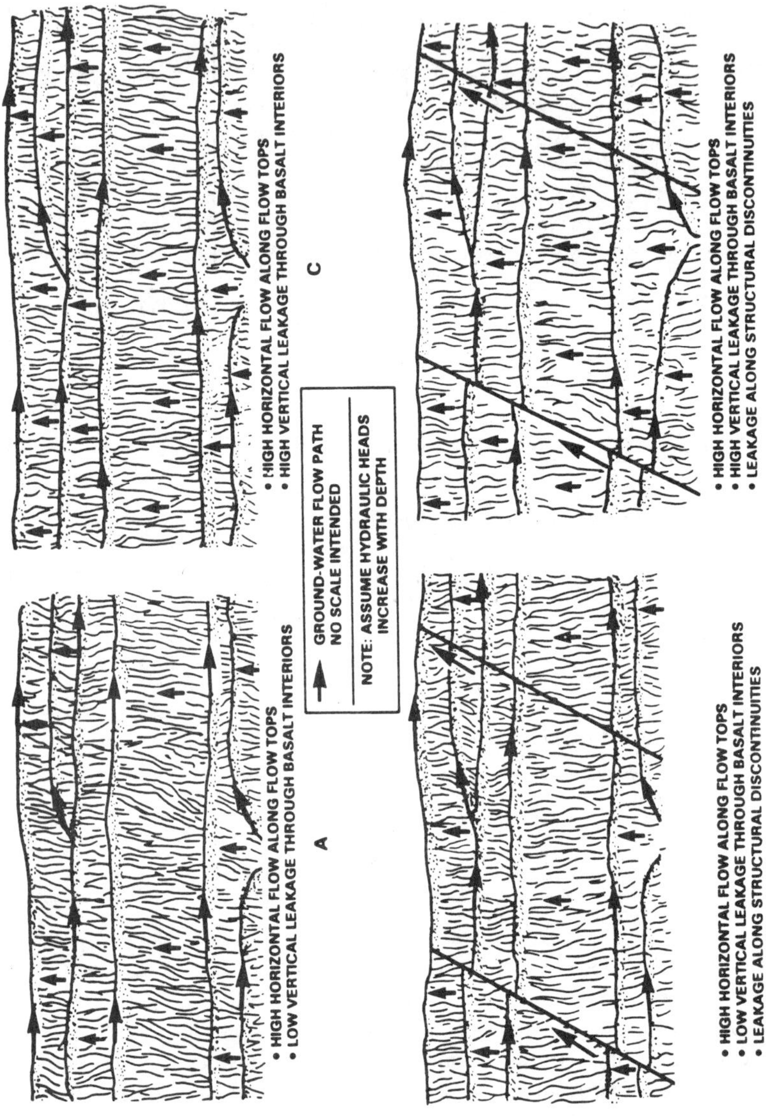

Figure 3. Alternative concepts for ground-water movement based on anisotropy contrasts and hypothetical structures (taken from 25a).

material filling the fractures these discontinuities could act as conduits of high hydraulic conductivity or as groundwater barriers. Overall, this concept depicts rock volumes of relatively low vertical leakage bounded by structural discontinuities.

Concept C—This concept represents a flow system characterized by lateral groundwater movement in flow tops bounded by basalt interiors of relatively high leakage. The anisotropy of flow-top and interior hydraulic conductivity is considerably less than in Concept A. In this concept, groundwater movement between deep and shallow systems occurs as a result of stratigraphic positioning or intersection of flow contacts, vertical leakage through unfilled or partially filled cooling fractures, and other primary features of relatively high hydraulic conductivity that are juxtaposed.

Concept D—This concept superimposes bedrock structural discontinuities on Concept C. As described in Concept B, such discontinuities might act as vertical conduits and (or) barriers of low hydraulic conductivity. This concept depicts rock zones of relatively high vertical leakage bounded by structural discontinuities.

These alternatives concepts have been investigated in a limited way through parameter variations (*26*). However, the variations taken into account so far do not yet address all uncertainties and geohydrologic characterization activities are currently under way to evaluate and reduce these uncertainties.

Deaf Smith County Site—Palo Duro Basin. For the Deaf Smith County site, the geohydrologic system relevant to the groundwater travel time evaluation can be thought of in terms of three different zones. The Ogallala and Dockum aquifers compose the first zone, here referred to as Hydrostratigraphic Unit A (HSU A). Underlying HSU A is a thick sequence of Permian evaporite rocks, characterized by extremely poor transmissive properties. This aquitard zone is referred to here as HSU B and is illustrated in Figure 4. Below HSU B is HSU C which includes the Wolfcamp and Pennsylvanian carbonate brine aquifer. The candidate repository horizon is in the Lower San Andres Unit 4 Salt in HSU B. The evaporite, interbed, and adjacent strata are depicted in Figure 4.

Although the groundwater models developed to date use different conceptualizations of the hydrogeologic setting, the following general conclusions for the Deaf Smith County site appear to be valid:

1. The regional vertical hydraulic gradient between HSU A and HSU C across HSU B is downward.
2. The regional horizontal gradient in HSU A and HSU C is approximately the same as the regional topographical gradient (about 0.004) from eastern New Mexico to southwestern Oklahoma.
3. The most important factor determining travel time estimates is the permeability of salt and other media of low permeability in HSU B.

The conceptual model of the regional flow system indicates that groundwater fluxes and flow velocities through the evaporite sequence in HSU B are very low. This model is consistently supported by independent lines of evidence from geochemical data, pressure and density distributions, permeability measurements, geologic and

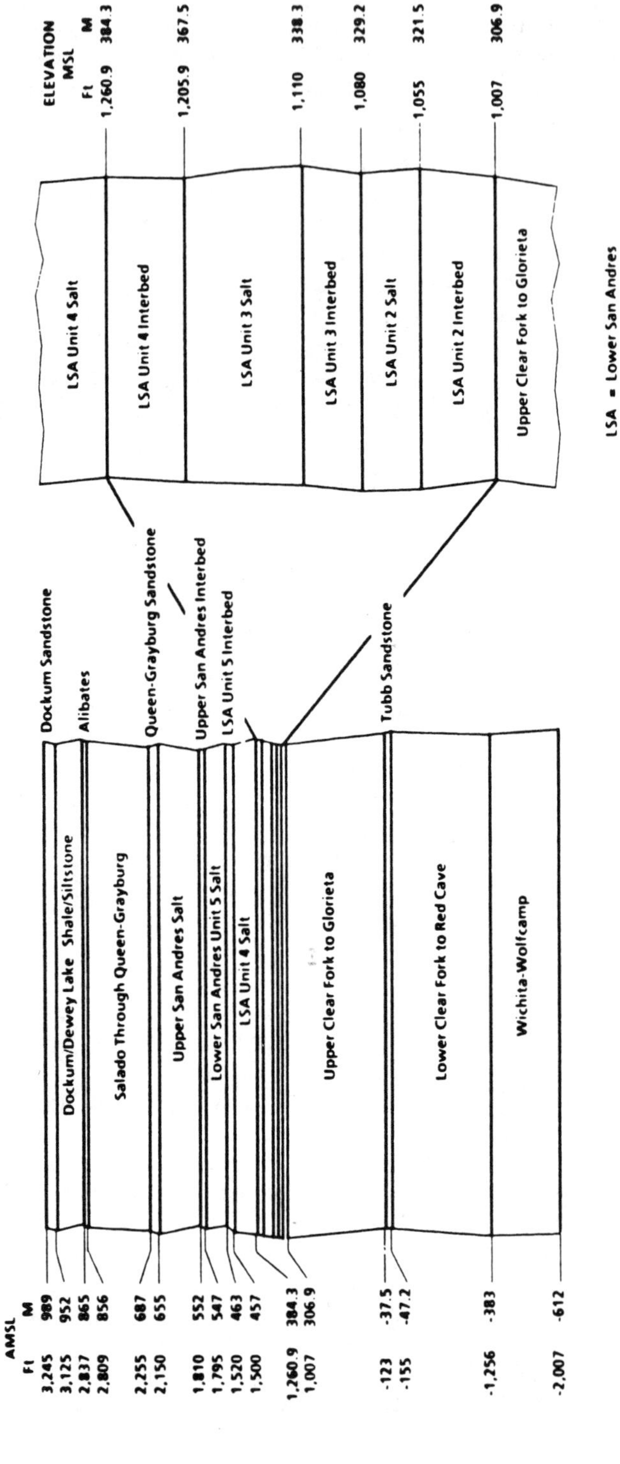

Figure 4. **Evaporite, interbed, and adjacent strata used for groundwater analysis in Deaf Smith County (taken from 25c).**

geophysical data, and modeling analyses. Although all the more brittle rocks in this system are fractured, the fractures are generally filled with secondary minerals, and the system appears to behave consistently as an extremely low-permeability, low-flux porous medium.

Currently available information suggests that the most applicable model for the local groundwater flow is porous-media flow through the bulk nonfractured, effective porosity of the stratigraphic units downward and laterally into the Wolfcamp series (in HSU C). A range of travel times can be calculated through a stochastic pathway analysis using known or estimated distributions of pressure, fluid density, permeability, and effective porosity.

Small-scale lithologic heterogeneities control the groundwater travel path from the edge of the disturbed zone to the accessible environment. In a stratified geologic environment consisting of heterogeneous layers of differing lithologies (salts, anhydrite, dolomite, siltstone, etc.), the groundwater travel path to the accessible environment is a function of the permeability of the varying lithologies, the horizontal and vertical hydraulic gradients within each layer, and the thickness and continuity of each layer. Because all of these values are uncertain, the expected travel path and travel time along this path to the accessible environment are also uncertain.

To address the uncertainties in the hydrogeologic parameters (permeability, porosity, vertical pressure distribution, horizontal hydraulic gradients, and total dissolved solids), a range of values for each parameter has been considered (*28*) as described below. In this analysis the ranges were defined by identifying one chance in 1,000 values which correspond to 3.1 standard deviations from the mean. These ranges are based on regional distributions of the uncertainty in the measurements conducted in the J. Friemel well located about 15 miles from the candidate site.

The pressures utilized in evaluating the vertical hydraulic gradients are based on observations made at the J. Friemel well within the Wolfcamp and Lower San Andres Unit 4 dolomite. The permeability distributions used for each layer are based on regional permeability distributions. In addition, the regional permeability distribution for the Lower San Andres was applied to all dolomite interbeds. The observed Wolfcamp permeability distribution was used for both the Wolfcamp and the porous Wichita dolomite. The regional sandstone distribution presented was applied to the Queen-Grayburg and Tubb sandstones. The permeability of each salt layer was assumed to be lognormally distributed with a range from 10^{-2} to 10^{-7} millidarcy. This range captures both the maximum observed laboratory values conducted at in situ confining pressures and the maximum observed in situ values. These values are believed to be conservative in that the vertical permeability through the evaporite aquitard suggested by modeling studies in the Palo Duro Basin is generally less than 3×10^{-4} millidarcy. The permeability distribution employed for the siltstone and shale layers was assumed to be lognormal with a range from 0.1 to 10^{-5} millidarcy based on Freeze and Cherry (*29*).

The porosity distributions used for each layer are based on either resistivity log interpretations or "textbook" ranges. The salt and dolomite interbed porosities are based on resistivity log interpretations and are considered to be normally distributed with a range of 0.3 percent to 1.64 percent and 0.4 percent to 8.0 percent, respectively.

The groundwater within evaporite and interbed layers is considered to be salt-saturated with a TDS content of 350,000 milligrams per liter. The TDS distribution in the Wolfcamp is considered to be triangular with a mode of 150,000 milligrams per liter plus or minus 50,000 milligrams per liter. The Dockum is considered to be freshwater while the Alibates, Salado through Queen-Grayburg, and the Queen-Grayburg sandstone are considered to be mixtures of fresh and saline groundwater with triangular distributions centered on 10,000 milligrams per liter (plus or minus 5,000 milligrams per liter), 175,000 milligrams per liter (plus or minus 25,000 milligrams per liter), and 250,000 milligrams per liter (plus or minus 50,000 milligrams per liter), respectively.

The horizontal hydraulic gradient used in the analyses of groundwater travel paths and travel times is considered to be approximately 0.004, based on observations in the Ogallala, Dockum, Lower San Andres Unit 4 carbonate, and the Wolfcamp. To account for uncertainty in this value, a triangular distribution has been assumed in the analysis. The lower end of the range in the San Andres salt and interbeds extends to 0.001, a value which has been generated based on the equivalent freshwater heads at the J. Friemel well.

Given the uncertainty, among others, in defining the extent of the disturbed zone, the travel paths and travel times presented below consider the disturbed zone to extend outside the Lower San Andres Unit 4 host salt to the base of the Lower San Andres Unit 5 dolomite and the top of the Lower San Andres Unit 4 dolomite. Therefore, travel times in the host salt are not added in the presented results.

The Latin Hypercube Sampling technique was used to define 1,000 realizations (representations of groundwater systems by selecting combinations of geohydrologic parameters) of the groundwater flow system. Typical results of these analyses are shown in Figures 5 and 6 in the form of a travel-time histogram and cumulative probability distribution for travel times.

These particular results refer to a net travel distance from the disturbed zone to the accessible environment of 1 km. The range of travel times is broad due to the uncertainty in the input parameters, particularly the permeability of the Lower San Andres Unit 4 carbonate. The calculated median travel time is about 88,000 years and the mean value is 370,000 years.

In this analysis correlations between hydrologic properties and spatial correlations of these properties were not incorporated. Furthermore, uncertainties in the conceptual model were not explicitly addressed, although the conservative assumptions used address these uncertainties to some extent indirectly.

There are indirect suggestions that subsurface faults or surface lineament features might penetrate all or most of the stratigraphic sequence from the basement below HSU C to the Dockum-Ogallala formation in HSU A in some locations. However, it should be noted that there is no direct evidence from the regional groundwater flow studies that the flow field in the area of Deaf Smith County is significantly affected by fractures or features which have permeability higher than the bulk porous-rock values. Nevertheless, because it is possible that these lineament factors could be associated with zones of higher permeability and/or lower porosity, their potential effects on travel time should be evaluated.

To evaluate the potential effects of possible hypothetical fracture zones on the travel-time distribution, it is assumed that surficial lineament features penetrate the

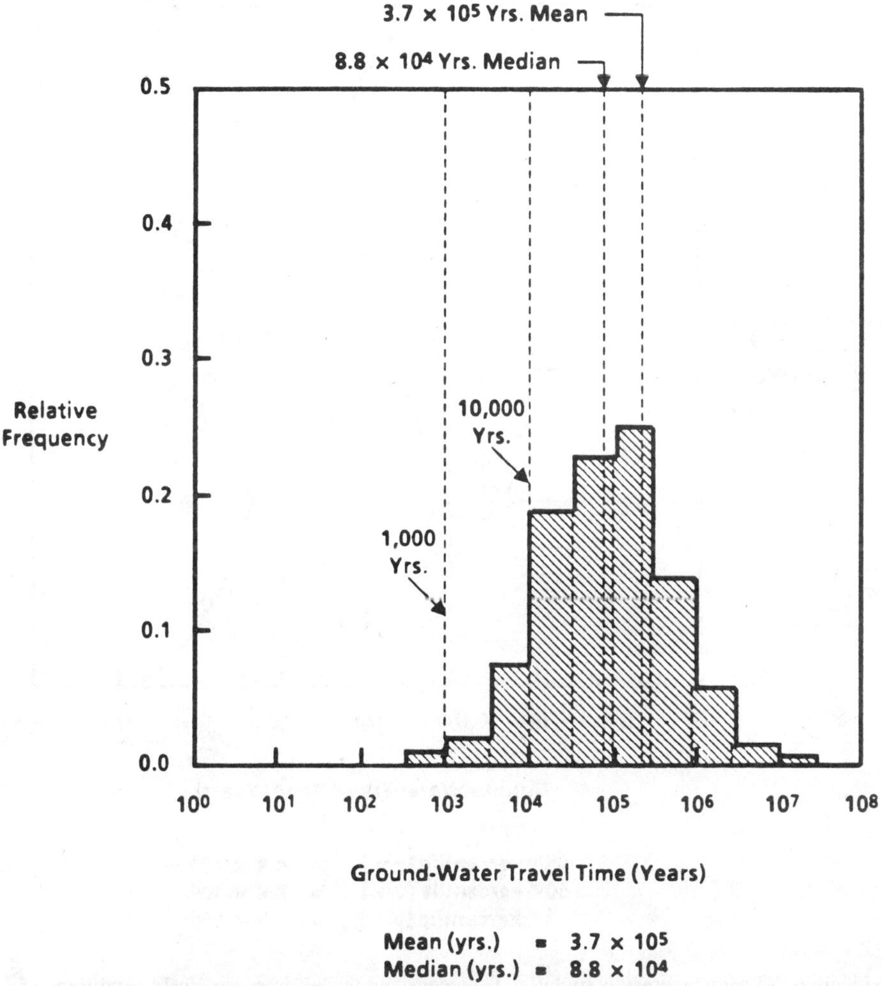

Figure 5. **Relative frequency of pre-waste-emplacement groundwater travel times for likely conditions of porous media flow in Deaf Smith County (taken from 25c).**

sedimentary strata and correspond to zones of lower effective porosity. No data on open fractures within brittle beds are available to quantify the probability of fracture zones intersecting likely particle travel paths. Therefore the following approach was used. Using the spatial distribution of surficial lineaments, 1,000 hypothetical fracture zones were randomly located within the region of interest. This information was used to define the distance a particle moves before intersecting a fracture zone. Within the fracture zone, the groundwater flux is considered to be unchanged and only the porosity is modified. When a particle intersects a fracture zone within a

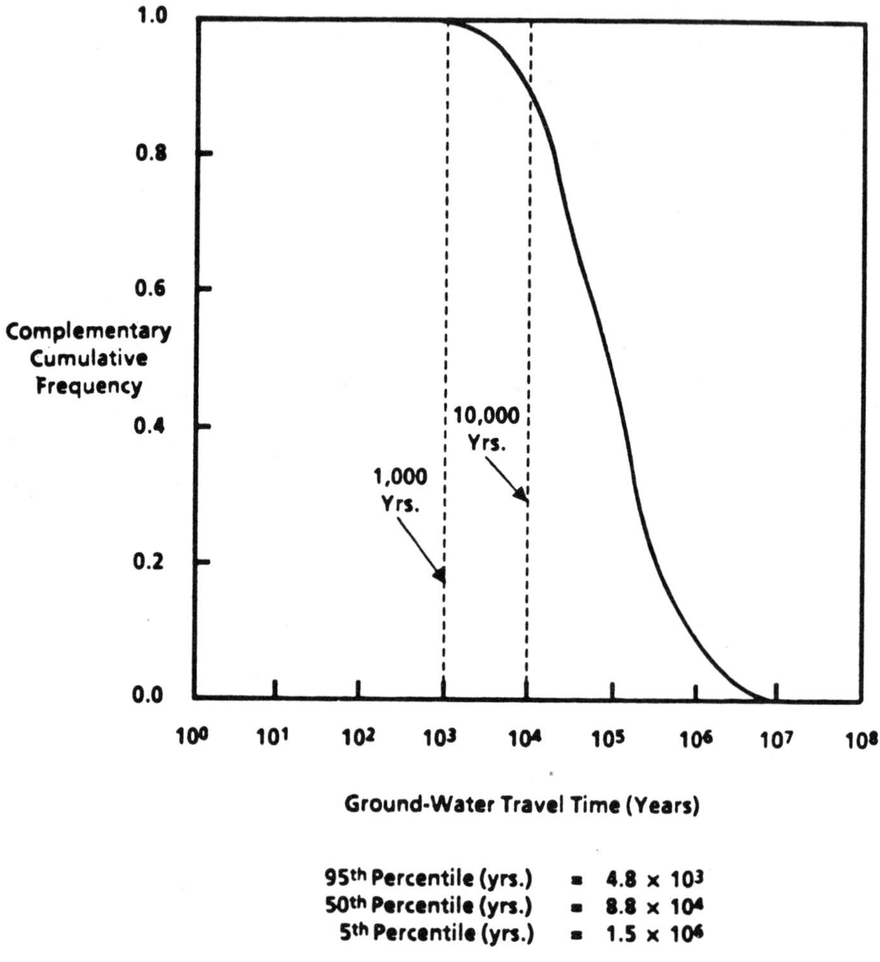

Figure 6. **Complementary cumulative frequency for travel times for likely conditions of porous media flow in Deaf Smith County (taken from 25c).**

brittle unit, the particle is assumed to stay within the fracture zone for the remaining distance to the accessible environment. The effective porosity of the fracture zone is decreased by two orders of magnitude, thus increasing the pore-water velocity in the hypothetical fracture zones.

Employing the above assumptions, the travel-time distributions for the unlikely fracture controlled pathway can be calculated. The median and mean travel times calculated, given the unlikely fracture controlled pathways, are 25,000 years and 310,000 years, respectively. The difference between the results for the likely and unlikely cases clearly illustrates the importance of explicitly addressing credible alternative conceptual models.

Yucca Mountain Site—Unsaturated Zone. The candidate repository horizon at the Yucca Mountain site is in the unsaturated zone, some 250 meters above the water table as shown in Figure 7. In this case the notions regarding groundwater travel time are somewhat different from the considerations made for groundwater movement under fully saturated conditions. The percolation of water occurs downward through fractures in the volcanic tuffs and through the porous rock matrix. Current information indicated that capillarity will generally cause the water to be retained within the pores in the rock matrix unless the percolation flux exceeds the capacity of the rock to transmit water. In this case, water can be transmitted downward through the fractures.

Because the thicknesses and hydraulic properties of the hydrogeologic units that make up the unsaturated zone at Yucca Mountain vary, an approach was developed to take this variability into account in calculation of travel times (*29*). The disturbed zone was assumed to extend to a distance 50 meters below the midplane of the

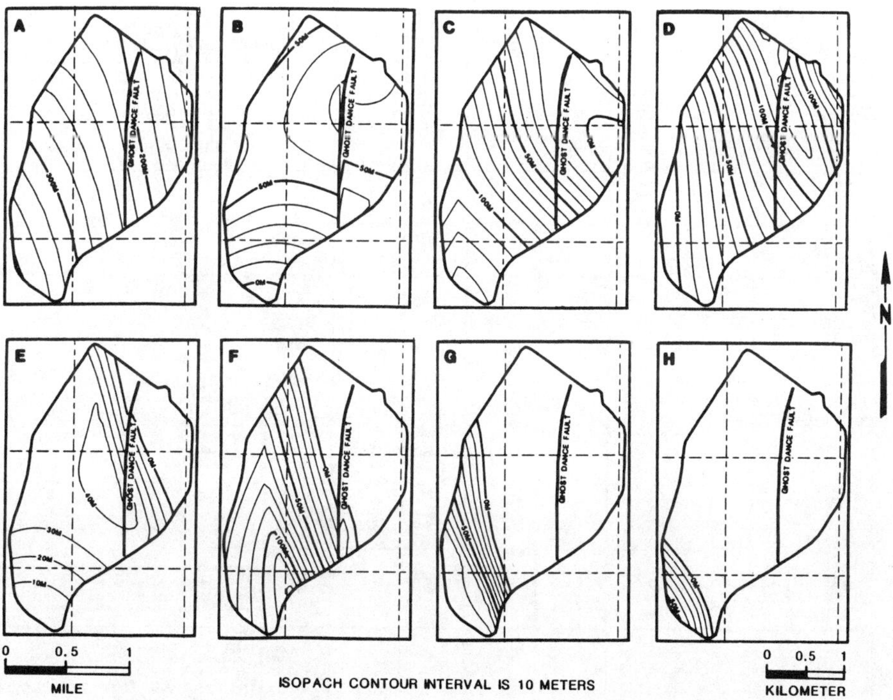

Figure 7. **Isopach contour maps: (a) total thickness from disturbed zone to the water table; (b) thickness of undisturbed Topopah Spring Welded Unit, TS_w; (c) thickness of the Calico Hills Vitric Unit, $CH_{n\text{-}v}$; (d) thickness of the Calico Hills Zeolitic Unit $CH_{n\text{-}z}$; (e) thickness of the Prow Pass Welded Unit, PP_w; (f) thickness of the Prow Pass Nonwelded Unit, PP_n; (g) thickness of the Bullfrog Welded Unit, BF_w; and (h) thickness of the Bullfrog Nonwelded Unit, BF_n (taken from 25b).**

repository. For purposes of analysis, the host rock within the perimeter of the repository was divided into 963 columns, extending from the disturbed zone boundary to the water table. Total unsaturated zone thickness and the thicknesses of hydrogeologic units between the disturbed zone and the water table are shown in Figure 7. The travel time to the water table through each of the 963 vertical columns was estimated. Table 3 shows the values for hydraulic parameters used for these calculations, including means and standard deviations for effective porosity and saturated matrix hydraulic conductivity for the seven hydrogeologic units considered in this calculation.

The three-dimensional volume of each hydrogeologic unit beneath the repository area was subdivided into 963, 76.2 meter square vertical columns, with each column

Table 3. **Parameters used in travel-time calculations for the unsaturated zone (from 25b).**

Parameter \ Hydrogeologic unit[a]	TS_w	CH_{n-v}	CH_{n-z}	PP_w	PP_n	BF_w	BF_n[b]	Remarks
Hydraulic gradient (i)	1.0	1.0	1.0	1.0	1.0	1.0	1.0	$i = \partial h/\partial l = \partial\psi/\partial l + \partial z/\partial l$ where $l = x,y,z$, $\partial\psi/\partial l \ll 1.0$ and $\partial z/\partial l = \partial z/\partial z = 1.0$, i.e. vertical gravity flow is assumed.
Mean saturated-matrix hydraulic conductivity K_s (mm/yr)[c]	0.7 (31)	107 (8)	0.5 (31)	88 (10)	22 (7)	118 (2)	22 (NA)	$\bar{K}_s = \ln^{-1}(\text{mean}[\ln(K_s)])$ Values in parentheses are the number of measurements.
$K_s - 1\sigma$ (mm/yr)	0.1	1.9	0.0	29	3	58	3	$\pm 1\sigma(K_s) = \bar{K}_s + \ln^{-1}(\ln\sigma[K_s])$
$\bar{K}_s + 1\sigma$ (mm/yr)	4.1	6,090	7.6	261	142	240	142	
Mean Effective Porosity $\bar{n}_e \pm 1\sigma$	0.11 ±.05 (138, 12)	0.32 ±.09 (23, 6)	0.27 ±.05 (65, 10)	0.24 ±.06 (27, 4)	0.25 ±.06 (75, 2)	0.22 ±.09 (120, 2)	0.25 ±.06 (NA)	$\bar{n}_e = n_b(1-S_r)$, where n_b is the mean bulk, dry porosity and S_r is residual saturation. Ordered pairs in parentheses are number of measurements of n_b and S_r, respectively.
Range of thicknesses (m)[d]	0-72 (98.5)	0-135 (95.3)	0-133 (94.5)	0-44 (83.2)	0-122 (63.1)	0-91 (25.6)	0-55 (7.5)	Thicknesses between disturbed zone and water table for area within the design repository boundaries. Values in parentheses are percentages of total repository are underlain by the units.
Particle velocity, V (mm/yr): q = 0.5 mm/yr	5.0	5.5	1.9	7.6	4.1	7.3	4.1	$V = (q/n_e)(q/K_s)^{-1/\epsilon}$, where n_e and k_s are mean values and q is flux.
Particle velocity, V (mm/yr): q = 1.0 mm/yr	8.9 (2,780)	9.4	3.4 (4,650)	12.8	7.2	12.5	7.2	Values in parentheses indicate fracture-flow velocity calculated as $(q-K_s)/0.0001$, where 0.0001 is the assumed effective porosity of fractures.
ϵ[e]	5.9	4.2	7.0	4.0	5.2	4.6	5.2	Empirical constant that represents the effects of the relationship between pore-size distribution and saturation on the amount of the effective porosity, n_e, available for flow; the effect of ϵ is to reduce flow area and thus increase particle velocity relative to values calculated using q/n_e.

[a] TS_w = Topopah Spring welded unit; CH_{n-v} = Calico Hills vitric unit; CH_{n-z} = Calico Hills zeolitic unit; PP_w = Prow Pass welded unit; PP_n = Prow Pass nonwelded unit; BF_w = Bullfrog welded unit; BF_n = Bullfrog nonwelded unit.

[b] Assumed to be hydrologically identical to PP_n.

[c] Saturated conductivity and effective-porosity data are from Sandia National Laboratories Tuff Data Base

[d] Range of thickness, Sandia National Laboratories Interactive Graphics Information System (IGIS)

[e] Values calculated from data

enclosing 5,086 square meters. Each column was divided into 3.05-meter-thick elements, giving a three-dimensional grid composed of 80,521 elements. See Figure 8 for a schematic drawing of the columns and elements used in the model. The particle velocity for each element within a particular hydrogeologic unit was calculated by using a value of saturated matrix hydraulic conductivity that was obtained by stan-

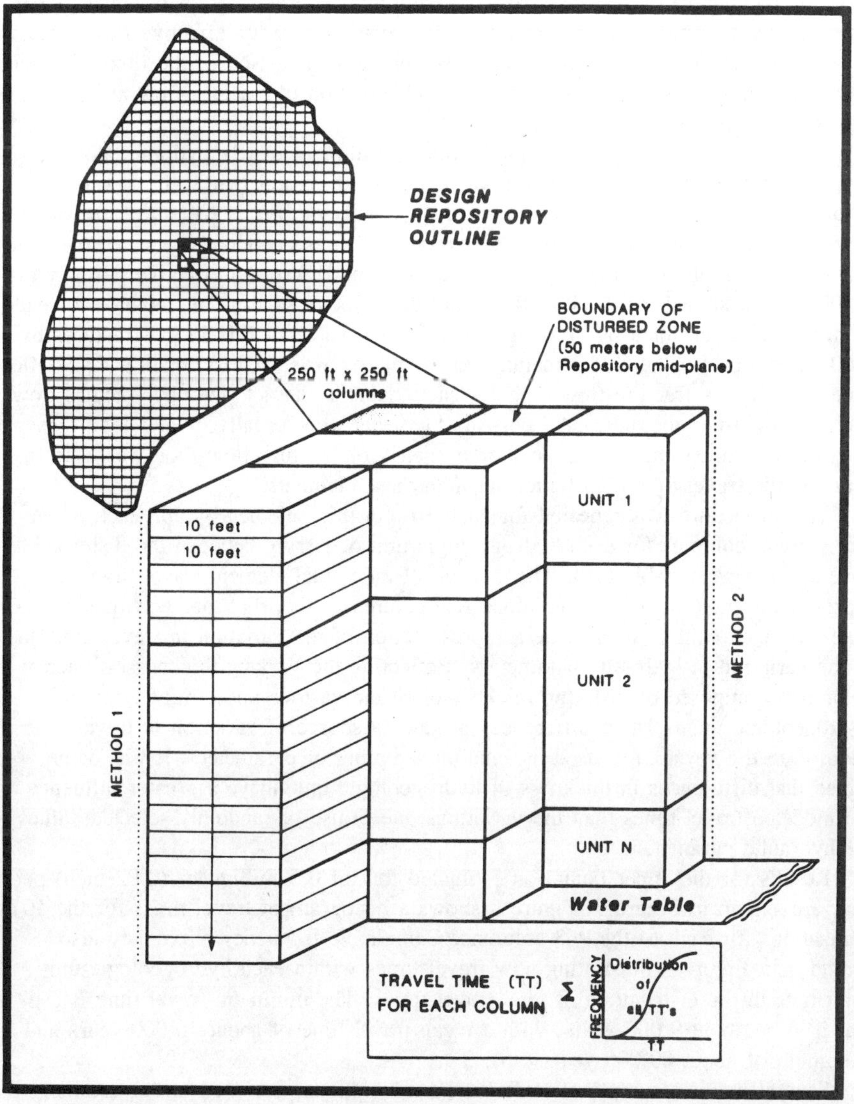

Figure 8. **Schematic diagram illustrating three-dimensional geometry for multiple elements 10 feet thick (taken from 25b).**

dard statistical sampling methods from the frequency distribution that best describes the conductivity data for the particular unit. If the flux value was less that 0.95 times the saturated matrix hydraulic conductivity (multiplied by a unit gradient), it was assumed that the flow within that element was entirely in the porous rock matrix, and a value of matrix effective porosity was then chosen by random sampling from the frequency distribution of porosity values for the appropriate hydrogeologic unit. The water particle velocity for each element was then calculated by dividing the flux value by the sampled effective porosity, assuming a hydraulic gradient equal to 1.0. The velocity was modified to account for the relation between effective flow area and saturation. The value of 0.95 was used for the comparison of the flux value with saturated conductivity to account for potential initiation of fracture flow at saturations less than 100 percent (*30*).

If the ratio of flux to the sampled value of saturated matrix hydraulic conductivity (multiplied by a unit gradient) was equal to or greater than 0.95, it was assumed that fracture flow occurred for that quantity of flux in excess of the value of saturated matrix conductivity (or in excess of 0.95 times conductivity when the ratio of flux to hydraulic conductivity is between 1.0 and 0.95). An effective porosity of 0.0001 was assumed for fracture flow, and the velocity of flow in fractures for each element was determined by dividing the calculated value of flux in the fractures by 0.0001. The portion of flux remaining in the matrix was used to obtain a matrix-flow time as well as a fracture-flow time for each element characterized by fracture flow. Matrix-flow times for these dual porosity elements were usually greater than the fracture-flow times; however, the shorter matrix or fracture flow time was used to generate the travel-time distribution through these elements.

This procedure was repeated for each 10-foot-thick element within each of the 963 vertical columns for each hydrogeologic unit occurring between the disturbed zone and the water table (*30*). The sum of all individual-element travel times represents a value of travel time along one column. This procedure was repeated 10 times for each column to provide a representation of the variation in travel time due to the variation in hydraulic parameters. Reflecting the varying thicknesses, each column is composed of differing thicknesses of the various unsaturated zone hydrogeologic units. These differences produce a source of variation in travel time in addition to the variation caused by random sampling of parameters. It was determined that differences in thickness of hydrogeologic units have a greater influence on the modeled travel times than the variations due to use of randomly selected values for hydraulic parameters.

Results for the upper bound on estimated flux of 0.5 millimeter (0.02 inch) per year are shown in Figure 9. Figure 9 shows a histogram of travel times for the 10 realizations for each of the 963 columns. Cumulative frequency curves are also provided in Figure 9 illustrating how travel times within each hydrogeologic unit contribute to the distribution of total travel times. The minimum travel time is estimated to be about 9,000 years, with a mean travel time of about 40,000 years and a maximum of about 80,000 years (*30*).

Several major assumptions underlie the calculations presented in this section; these assumptions represent important uncertainties in the conceptual and process models. The first assumption is that unsaturated zone flux below the disturbed zone

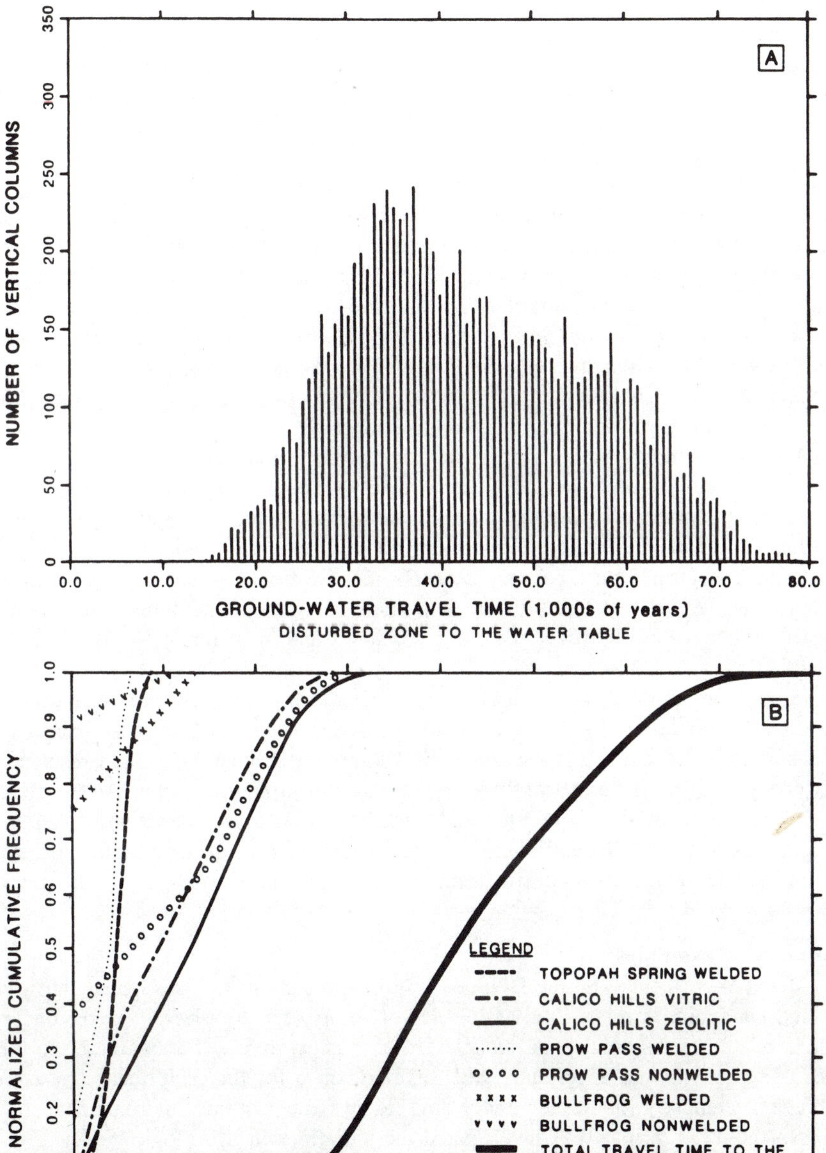

Figure 9. **Travel-time plots: (a) travel-time histogram for 9,630 realizations of travel-time model; flux = 0.5 millimeter (0.02 inch) per year; element thickness = 10 feet; (b) cumulative frequency of travel times for 9,630 realizations (normalized to a scale of 0.0 to 1.0) (taken from 25b).**

is vertical and uniformly distributed in time and space (i.e., the unsaturated zone at Yucca Mountain is assumed to have a vertical hydraulic gradient determined solely by gravity). The hydrologic properties of the rocks above the repository are probably able to smooth out episodic infiltration so that nearly steady-state flow occurs between the repository and the water table.

A second major assumption is that the effective hydraulic conductivity of the matrix of any given rock volume exactly equals the flux, because of the dependence of conductivity on saturation. This assumption is particularly important where the value of conductivity is less than the flux. In these cases, the matrix is assumed to conduct water up to an amount equal to the saturated conductivity, while the remainder of the flux travels in fractures. If the flux is less than the saturated conductivity for a particular interval, the relations among pore sizes, saturations, and capillary pressures are assumed to adjust to yield an effective conductivity just sufficient to pass the flux under a gravity-controlled gradient.

A third assumption underlying these calculations is that, at certain conditions of saturation, water probably does not move rapidly through fractures until flux approaches the saturated matrix hydraulic conductivity. This assumption precludes significant, sustainable fracture flow where the conductivity is more than the flux within the unsaturated tuffs beneath the repository. Some flow may take place in thin layers along the walls of fractures under partially saturated conditions. It is assumed that such flow would, however, be likely to have properties similar to flow in the matrix.

In addition as noted, these analyses do not address lateral diversion of the percolation that could occur, for example, at the interfaces of the tuff units. In particular, diversion of the flux to some feature such as the Ghost Dance fault could result in shorter travel times if the permeability of the fault is sufficiently high. At the present time, information needed to provide understanding of such an alternative conceptual model is not yet available and plans have been developed to provide such information in the site characterization program.

Summary of examples

Analyses performed by the DOE Repository Projects have investigated sensitivities and uncertainties related to parametric variability and, to a lesser extent, uncertainties related to the conceptual model. The examples discussed above describe the types of sensitivity and uncertainty analyses performed for the performance measure of preemplacement groundwater travel time, as presented in the Environmental Assessments (*25a,b,c*). BWIP investigated the variations in hydrologic model parameters due to limited knowledge of and spatial variability of hydraulic characteristics of the basalt flow interiors and basalt flow tops. These parametric studies explored, to some extent, alternative conceptual models for groundwater flow at the BWIP site, including anomalous connectivity between different stratigraphic units caused by spatial variation in the thickness of units and faulting and folding of the basalt flows. The Salt Repository Project Office (SRPO) investigated variations in hydrogeologic parameters (e.g., permeability, porosity, vertical pressure distribution, horizontal hydraulic gradients, and total dissolved solids) in the stratified geologic environment of the Deaf Smith County site. Additionally, the influence of hypothetical fracture zones was studied, to explore an alternative conceptual model. The Nevada Nuclear Waste Storage Investigations (NNWSI) analyzed variations in hydrologic properties and the thickness of units in the unsaturated tuff underlying the proposed repository location at Yucca Mountain.

These studies illustrate the substantial degree to which spatial variation of model parameters has been evaluated as part of the ongoing DOE program. The distributions of travel time derived by these analyses reflect the uncertainties due to those parameters included; however, not all sources of uncertainty have been quantitatively evaluated. It is expected that subsequent analyses will attempt to quantify major sources of uncertainty, but the degree to which this is accomplished depends on the relative importance of quantifying the particular uncertainty for demonstrating regulatory compliance and on the feasibility of such quantification.

Conclusions

Site characterization, performance assessment, and sensitivity and uncertainty analysis all have key, interrelated roles in obtaining a license for a nuclear waste repository. Performance assessment provides the quantitative analysis demonstrating compliance with regulatory requirements for long-term isolation of the waste. Site characterization provides a large body of the data needed (significant data also come from design activities) to conduct performance assessments. Sensitivity and uncertainty analyses may be used to (1) quantify certain types of uncertainties in the results of performance assessments, (2) evaluate the adequacy of performance assessment models, (3) evaluate the success of site characterization activities and identify the need for specific additional site characterization activities. In order to successfully license a repository it is necessary to reduce the uncertainties in performance assessments to the degree that a finding of reasonable assurance of compliance with regulatory requirements can be made. In this context, then, site characterization can be viewed as an activity directed to the reduction of uncertainties.

There are a number of different types of uncertainty associated with performance assessment computations. Sensitivity and uncertainty analysis methods have been extensively applied to studying and quantifying the uncertainties due to the spatial variability of parameters important in performance calculations. To a lesser, but significant, extent sensitivity and uncertainty analyses have been used to study uncertainties resulting from alternative conceptual models of repository sites. It is to these two sources of uncertainty that site characterization is most closely related. For this reason, this paper has focused on these. Given the need to develop the technical basis for licensing and site selection, within limited time and budgetary constraints, it is important to the DOE that site characterization activities be guided by a philosophy that is both comprehensive and efficient. The site characterization plans currently under preparation attempt to follow such a philosophy. The site characterization plans and the detailed study plans to follow address both parametric variability and conceptual model uncertainty. However, as the site characterization activities proceed and additional interim performance assessments are performed, there is a need to address the degree to which uncertainties are being successfully reduced. Furthermore, as site characterization provides more data, the ability to use performance assessments and uncertainty analyses to identify data needs is expected to increase. In order to address the degree to which site characterization reduces uncertainty, additional methodological developments in the area of uncertainty and sensitivity analysis may be warranted. Among the questions that arise in considering this need are:

1. What methods can be used to quantify modeling uncertainty?
   - How and to what degree can the effects of undiscovered site features be systematically quantified?
   - How and to what degree can the effects of unlikely physicochemical processes be quantified?
   - How and to what degree can the effects of various potential future environmental conditions be quantified?
2. How shall we treat uncertainties that cannot be quantified?
3. What methods can be used to combine uncertainties resulting from spatial variability, sparseness of measurement, measurement error, and modeling uncertainties?
4. What quantitative measures can be derived to indicate whether or not data gathered at a given site are sufficient to demonstrate regulatory compliance?
5. What quantitative measures can be derived to indicate what additional site data are needed and the locations of those data?

It is the intention of the DOE to coordinate performance assessment, site characterization, and uncertainty and sensitivity analysis methods development and use, so that these questions will be answered.

Literature Cited

1. U.S. Nuclear Regulatory Commission, 1981. *Proceedings of the Symposium on Uncertainties Associated with the Regulation of the Geologic Disposal of High-Level Radioactive Waste* (Gatlinburg, Tennessee, March 9–13, 1981), NUREG/CP-0022, U.S. Nuclear Regulatory Commission, Washington, D.C.
2. OECD Nuclear Energy Agency and U.S. Department of Energy, 1987. *Workshop on Uncertainty Analysis for System Performance Assessments* (Seattle, 24–26 February, 1987). To be published by OECD Nuclear Energy Agency, Paris, France.
3. Cranwell, Robert M. and Helton, Jon C., 1981. "Uncertainty analysis for geologic disposal of radioactive waste," in *Proceedings of the Symposium on Uncertainties Associated with the Regulation of the Geologic Disposal of High-Level Radioactive Waste*, U.S. Nuclear Regulatory Commission (Washington, D.C., March 9–13, 1981), NUREG/CP-0022, pp. 131–143.
4. Smith, Jay L. and Balderman, Morris A., 1981. "Natural geologic processes as sources of uncertainty in repository modeling," in *Proceedings of the Symposium on Uncertainties Associated with the Regulation of the Geologic Disposal of High-Level Radioactive Waste*, U.S. Nuclear Regulatory Commission (Washington, D.C., March 9–13, 1981), NUREG/CP-0022, pp. 247–256.
5. Smith, Leslie and Schwartz, Franklin W., 1981. "The role of hydraulic conductivity data in reducing uncertainty in radionuclide transport modeling," in *Proceedings of the Symposium on Uncertainties Associated with the Regulation of the Geologic Disposal of High-Level Radioactive Waste*, U.S. Nuclear Regulatory Commission (Washington, D.C., March 9–13, 1981), NUREG/CP-0022, pp. 305–312.

6. Channell, J. K., Wofsy, C., and Zand, S. M., 1981. "The effect of hydrologic parameter variation on radionuclide transport in the Rustler aquifer (southeastern New Mexico)," in *Proceedings of the Symposium on Uncertainties Associated with the Regulation of the Geologic Disposal of High-Level Radioactive Waste*, U.S. Nuclear Regulatory Commission (Washington, D.C., March 9–13, 1981), NUREG/CP-0022, pp. 365–372.
7. Schalla, R. and Leonhart, L. S., 1981. "Dealing with regional hydrologic data-base limitations case example: the Columbia river basalts," in *Proceedings of the Symposium on Uncertainties Associated with the Regulation of the Geologic Disposal of High-Level Radioactive Waste*, U.S. Nuclear Regulatory Commission (Washington, D.C., March 9–13, 1981), NUREG/CP-0022, pp. 415–423.
8. Kocher, D. C., Sjoreen, A. L., and Bard, C. S., 1983. *Uncertainties in Geologic Disposal of High-Level Wastes—Groundwater Transport of Radionuclides and Radiological Consequences*, U.S. Nuclear Regulatory Commission, Washington, D.C., July, 1983, NUREG/CP-2506.
9. U.S. Department of Energy, 1986. *Issues Hierarchy for a Mined Geologic Disposal System* (OGR/B-10), Washington, D.C., Sept. 1986, DOE/RW-0101.
10. Iman, R. L., J. C. Helton, and J. E. Campbell, 1978. *Risk Methodology for Geologic Disposal of Radioactive Waste: Sensitivity Analysis Techniques*, NUREG/CR-0394, U.S. Nuclear Regulatory Commission, October.
11. Iman, R. L., W. J. Conover, and J. E. Campbell, 1980. *Risk Methodology for Geologic Disposal of Radioactive Waste: Small Sample Sensitivity Analysis Techniques for Computer Models, With an Application to Risk Assessment*, NUREG/CR-1397, U.S. Nuclear Regulatory Commission, March.
12. Harper, William V., 1983. *Sensitivity/Uncertainty Analysis Techniques for Nonstochastic Computer Codes*, ONWI-444, Office of Nuclear Waste Isolation, Battelle Memorial Institute, Columbus, Ohio.
13. Delhomme, J. P., 1978. "Kriging in the hydrosciences," *Advances in Water Research, 1*(5): 251–266.
14. Gelhar, L. W., 1984. "Stochastic analysis of flow in heterogeneous porous media," in J. Bear and M. Y. Corapcioglu (editors), *Fundamentals of Transport Phenomena in Porous Materials*, Hingham, Massachusetts, Martinus-Nijhoff, p. 675.
15. Doctor, P. G. and J. A. Buchanan, 1986. The Role of Statistical Analysis in Development of Performance Assessment Conceptual Models (Draft), Pacific Northwest Laboratory, Richland, Washington.
16. INTRACOIN (International Nuclide Transport Code Intercomparison Study), May 1986. *Final Report Levels 2 and 3*, Swedish Nuclear Power Inspectorate, Stockholm.
17. Eisenberg, N. A., D. H. Alexander, and W. W.-L. Lee. "The role of validation activities in the U.S. Department of Energy's Office of Civilian Radioactive Waste Management Program," in GEOVAL-87 (April 7–9, 1987, Stockholm, Sweden). To be published.
18. Eisenberg, N. A., et al. "A proposed validation strategy for the U.S. DOE Office of Civilian Radioactive Waste Management Geologic Repository Projects," in GEOVAL-87 (April 7–9, 1987, Stockholm, Sweden). To be published.
19. Eisenberg, N. A. and A. E. Van Luik. "Validation activities addressing performance issues pertinent to the U.S. DOE Geologic Repository Projects," in GEOVAL-87 (April 7–9, 1987, Stockholm, Sweden). To be published.
20. Martin, J. B., M. J. Bell, and M. R. Knapp. "Uncertainties in the regulation and licensing of a high-level waste repository," in *Proceedings of the Symposium on Uncertainties Associated with the Regulation of the Geologic Disposal of High-Level Radioactive Waste*, U.S. Nuclear Regulatory Commission, Washington, D.C. (March 9–13, 1981), NUREG/CP-0022, pp. 3–16.

21. U.S. Department of Energy, 1982. *Site Characterization Report for Basalt Waste Isolation Project*, DOE/RL-82-3, Washington, D.C.
22. U.S. Nuclear Regulatory Commission, 1983. *Draft Site Characterization Analysis of Site Characterization Report for the Basalt Waste Isolation Project, Hanford, Washington*, NUREG-0960, 2 volumes, Washington, D.C.
23. Golder Associates, 1983. *Technical Review on the Site Characterization Report for the Basalt Waste Isolation Report*, 833-1007, D254/254A, State of Washington, Energy Facility Site Evaluation Council High-Level Nuclear Waste Management Task Force.
24. U.S. Geological Survey (USGS), 1983. *Review Comments on Site Characterization Report for Basalt Waste Isolation Project*, DOE/RL-82-3, Washington, D.C.

25a. U.S. Department of Energy, 1986. *Environmental Assessment, Reference Repository Location, Hanford Site, Washington,* DOE/RW-0070, Washington, D.C.

25b. U.S. Department of Energy, 1986. *Environmental Assessment: Yucca Mountain Site, Nevada Research and Development Area, Nevada*, DOE/RW-0073, Washington D.C.

25c. U.S. Department of Energy, 1986. *Environmental Assessment: Deaf Smith County, Texas*, DOE/RW-0069, Washington, D.C.

26. Clifton, R. M., 1986. *Groundwater Travel Time Analysis for the Reference Repository Location at the Hanford Site*, SD-BWI-7I-303, Rockwell Hanford Operations, Richland, Washington.
27. Gephart, R. C., S. M. Price, R. L. Jackson, and C. W. Myers, 1983. *Geohydrologic Factors and Current Concepts Relevant to Characterization of a Potential Nuclear Waste Repository Site in Columbia River Basalt, Hanford Site, Washington*, RHO-BW-SA-326P, Rockwell Hanford Operations, Richland, Washington.
28. Andrews, R. W., V. A. Kelley, J. A. McNerish, A. M. La Venny, and J. E. Campbell, 1985. *Travel Path/Travel Time Uncertainties of Salt Sites Proposed for High Level Waste Repositories*, INTERA Technologies, Austin, Texas.
29. Freeze, R. Allen and John A. Cherry, 1974. *Groundwater*, Prentice-Hall, Inc., Englewood Cliffs, New Jersey.
30. Sinnock, S., Y. T. Lin, and M. S. Tierney, 1986. *Preliminary Estimates of Groundwater Travel Time and Radionuclide Transport at the Yucca Mountain Repository Site*, SAND85-2701, Sandia National Laboratories, Albuquerque, New Mexico.

An Overview of Groundwater Flow and Radionuclide Transport Modeling in the Canadian Nuclear Fuel Waste Management Program

TIN CHAN
Atomic Energy of Canada Limited
Whiteshell Nuclear Research Establishment
Pinawa, Manitoba, Canada

Abstract

The Canadian Nuclear Fuel Waste Management Program is currently assessing the concept of disposing of nuclear fuel waste in a vault 500 m to 1000 m deep within plutonic rock in the Canadian Shield. This paper reviews the detailed geosphere modeling efforts at Atomic Energy of Canada Limited (AECL) and supporting fundamental research by AECL, Ontario Hydro, universities, and private consultants.

The paper presents (1) the mathematical and numerical formulations in the three-dimensional finite-element code, MOTIF, developed by AECL for modeling coupled ground-water flow, heat transport, and contaminant transport in fractured porous media; (2) verification and validation studies; (3) application of the code to simulate the migration of inert contaminants from a hypothetical nuclear waste vault in a conceptual geosphere model constructed from extrapolation of field data; and (4) the connection between the detailed geosphere model and a simplified one-dimensional network submodel for the system variability analysis code (SYVAC).

Introduction

The Canadian Nuclear Fuel Waste Management Program is currently investigating the concept of disposing of immobilized nuclear fuel waste in a vault deep within plutonic rock. A systems variability analysis approach (*1,2*) is being used to assess the risk to humans from the decaying radionuclides. In the evaluation, the system is divided into three subsystems: the vault, the geosphere, and the biosphere. The behavior of each component is represented by a submodel in the overall systems assessment computer code, SYVAC (*2*). The geosphere submodel describes the transport of radionuclides through the groundwater from the vault to the biosphere. While the overall system assessment requires a simple submodel that can be economically executed a large number of times in a Monte Carlo simulation, it is fundamental that it be based on phenomena observed in relevant hydrogeological settings. Atomic Energy of Canada Limited (AECL) is using data obtained from investigations of a specific research area on the Canadian Shield to develop representative geosphere submodels.

Figure 1 illustrates the steps involved in developing a SYVAC submodel of a research area. Development of a submodel of a research area begins with the assimilation of all relevant geological, geophysical, geochemical, and hydrogeological data for the region. Based on this information, a detailed conceptual model of the hydrogeological regime is formulated. A hypothetical disposal vault is located within the model, and the coupled equations describing fluid flow and heat transport are solved numerically to predict groundwater velocity distributions. Contaminant transport models are then used to determine a suitable form for the SYVAC geosphere submodel and estimate characteristics of potential pathways to the biosphere.

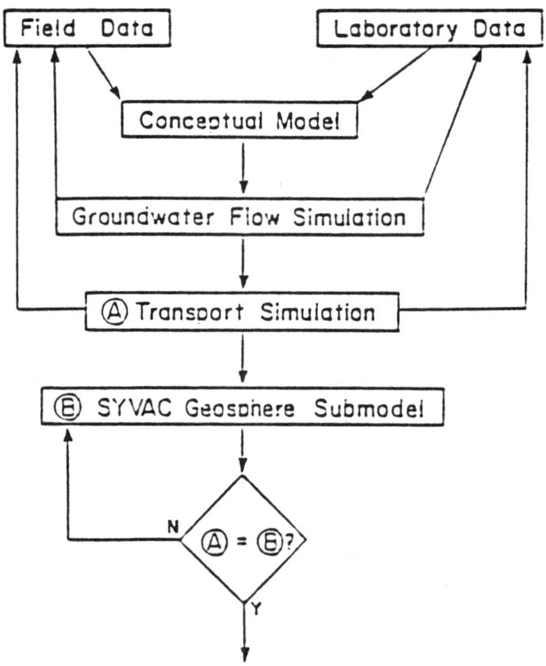

Figure 1. **Strategy for developing the geosphere submodel in SYVAC. Modified from (*7*).**

In order to model the detailed processes of groundwater flow, heat transport, and contaminant transport within plutonic rock formations, AECL has developed the finite-element code, MOTIF (*3*). The numerical accuracy of the code has been verified by comparing analytical and independent numerical solutions (*4*). Its ability to correctly model groundwater flow in a fractured rock mass has been validated by comparing the model results with field measurements of drawdown and groundwater seepage associated with the sinking of the access shaft for the Underground Research Laboratory (URL) in Manitoba, Canada (*5,6*). The MOTIF finite-element code is currently being applied to simulate a hypothetical reactor waste vault in a conceptual geosphere model based on data from a research area (*7,8,9*).

Fundamental supporting research is also being conducted to complement the approach described above.

The MOTIF Code

The computer code MOTIF (**M**odel **O**f **T**ransport **I**n **F**ractured/Porous Media) has been developed by AECL to model the coupled processes of saturated or unsaturated fluid flow, conductive and convective heat transport, brine transport, and single-species radionuclide transport in a rock mass of low permeability intersected by a few major fracture zones of relatively high permeability. Using the Galerkin finite-element method, the MOTIF code solves the three partial differential equations governing these physical phenomena, i.e., the fluid mass balance equation, the contaminant mass balance equation, and the heat energy balance equation.

Governing equations

The governing equations for groundwater flow, heat transport, and solute transport can be derived from the conservation of mass, momentum, and energy (*10*). Additional coupling effects are described by equations of state, which give density and viscosity as functions of pressure, temperature, and concentration. Here, we first present the governing equations for porous media, and then discuss how these equations are modified for application to fractures.

The condition of continuity (for water) in unsaturated porous media is written in tensor form as

$$\frac{\partial}{\partial t}(S_w \rho \theta) + \frac{\partial}{\partial x_i}(S_w \rho \theta u_i) - I = 0 \quad (1)$$

where S_w = degree of saturation, θ = porosity, ρ = density, t = time, x_i = i-th cartesian coordinate, u_i = i-th component of the average interstitial fluid velocity, and I = the specific production rate. For application to fracture flow, based on Darcy's law generalized for unsaturated flow,

$$u_i = -\frac{k_r k_{ij}}{S_w \rho \theta \mu} \frac{\partial}{\partial x_j}(p + \rho g x_3) \quad (2)$$

where k_r = relative permeability, k_{ij} = permeability tensor, μ = dynamic viscosity, p = fluid pressure, and g = gravitational acceleration. Both S_w and k_r are functions of p. Consequently, k_r is a function of S_w regardless of whether the medium is isotropic or anisotropic. In using the MOTIF code for simulations, the user has to supply the characteristic curves $S_w(p)$, and $k_r(p)$.

$$\rho\,(p, T, C) = \rho_{ref}\,[1 + \Delta T\,(A + \Delta TB) + \alpha_f\, p + \varepsilon C] \quad (3)$$

where

ρ_{ref} = reference fluid density for Equation (3)
T = temperature, C
T_{ref} = reference temperature, C
$\Delta T = T - T_{ref}$
A, B = coefficients
p = pressure above atmospheric pressure
α_f = fluid compressibility at atmospheric pressure

$$\varepsilon = \frac{\rho_{max} - \rho_{min}}{\rho_{ref}\,(C_{max} - C_{min})}$$

ρ_{max}, ρ_{min} = maximum and minimum densities corresponding to maximum and minimum concentrations, C_{max} and C_{min}, respectively;

and

$$\mu(T) = A_1 \exp\,[B_1/(273 + T)] \tag{4}$$

where A_1 and B_1 are coefficients. Although viscosity only varies with temperatures in the current version of MOTIF, the formulation in MOTIF allows the use of a more general form

$$\mu = \mu\,(\rho, C, T)$$

with only a minor modification to the code.

By manipulating Equations (1) through (4), one obtains

$$\rho\rho_o g \left\{ \theta c_\psi + S_w \left[\theta\alpha_f + (1 - \theta)\alpha_s \right] \right\} \frac{\partial h}{\partial t} + S_w\theta \left[\frac{\partial \rho}{\partial C}\frac{\partial C}{\partial t} + \frac{\partial \rho}{\partial T}\frac{\partial T}{\partial t} \right] - \frac{\partial}{\partial x_i}\left[\rho k_r \frac{k_{ij}\,\rho_o g}{\mu} \frac{\partial}{\partial x_j}(h + e x_3) \right] - I = 0 \tag{5}$$

where

$e = (\rho_o - \rho)/\rho_o$, ρ_o = reference density, $h = (p/\rho_o g) + x_3$ = reference hydraulic head,
α_f = fluid compressiblity, α_s = bulk solid compressibility, and
$c_\psi = \partial S_w/\partial p$ = moisture capacity.

The equation of solute (or one-species radionuclide) mass balance is written as

$$\frac{\partial}{\partial t}\left\{\left[S_w\theta + (1-\theta)a_1\right]C\right\} + \frac{\partial}{\partial x_i}(\theta S_w u_i C)$$

$$-\frac{\partial}{\partial x_i}\left(S_w D_{ij}\,\theta\,\frac{\partial C}{\partial x_j}\right) - I_c$$

$$+ [\theta S_w + (1-\theta)a_1]\,\lambda C = 0 \qquad (6)$$

where C = concentration of solute (or radionuclide), a_1 = adsorption constant, D_{ij} = tensor of hydrodynamic dispersion coefficients = $a_{ijkl}\, u_k\, u_l/u + D_d\, T^*_{ij}$, a_{ijkl} =dispersivity tensor, u = magnitude of interstitial velocity, D_d = molecular diffusion coefficient, T^*_{ij} = tortuosity tensor, λ = radioactive decay constant, and I_c = specific solute production rate.

For isotropic media, the tensor of hydrodynamic dispersion coefficients takes the form

$$D_{ij} = D_T\,\delta_{ij} + (D_L - D_T)\,\frac{u_i u_j}{u^2} + T^* D_d \delta_{ij} \qquad (7)$$

where

D_L = longitudinal convective dispersion coefficient
= $a_L u$
a_L = longitudinal dispersivity
D_T = transverse convective dispersion coefficient
= $a_T u$
a_T = transverse dispersivity
T^* = scalar tortuosity
δ = Kronecker delta.

The energy conservation condition yields

$$\frac{\partial}{\partial t}\left[(\rho c)_e T\right] + \frac{\partial}{\partial x_i}(S_w\theta\rho_f c_f\, u_i T) - \frac{\partial}{\partial x_i}\left[E_{ij}\,\frac{\partial T}{\partial x_j}\right] - I_H = 0 \qquad (8)$$

where $(\rho c)_e = S_w\theta\rho c_f + (1-\theta)\rho_s c_s$, c_f = specific heat capacity of fluid, ρ_s = density of solid skeleton, c_s = specific heat capacity of solid skeleton, $E_{ij} = S_w\theta\rho c_f\, a_{ijkl}\, u_k\, u_l/u + \lambda^T_{ij}$ where λ^T_{ij} = effective bulk thermal conductivity tensor, and I_H = specific heat production rate.

The conceptual model inherent to the formulation employed in the MOTIF code for application to fractured/porous media is that the rock body being simulated either contains a large enough number of fractures to be represented on average as heterogeneous porous media, or else there are so few fractures that they can be either neglected or simulated explicitly by special finite elements. Dominant fractures or fracture zones that cannot be lumped into the porous media representation are simulated by planar fracture elements as discussed below. This, of course, assumes that the geometry of these discrete fractures or fracture zones, as well as their hydraulic properties and heat and solute transport properties can be determined through site characterization.

For application to a fracture or fracture zone, Equation (5) for fluid flow, Equation (6) for solute transport, and Equation (8) for heat transport are integrated over the thickness of the fracture or fracture zone. The following assumptions are required for this procedure to be valid for fluid flow:

1. The hydraulic conductivity in the fracture or fracture zone is large compared with that in the background rock.
2. The hydraulic gradient across the fracture or fracture zone is relatively small so that the heads on opposite sides of the fracture or fracture zone are not much different from the averaged head across the fracture thickness.
3. The flow is predominately along the axis of the fracture or the fracture zone.
4. The local variation (within a finite element) in fracture or fracture zone thickness is negligible.

For solute transport these assumptions are required:

1. The solute is always instantaneously and thoroughly mixed across the fracture or fracture zone.
2. The concentration in the background rock immediately outside the fracture or fracture zone is approximately equal to the concentration averaged over the fracture thickness.

For heat transport these assumptions are required:

1. The fluid and solid phases in the fracture or fracture zone are always in thermodynamic equilibrium.
2. The fluid is always instantaneously and thoroughly mixed across the fracture or fracture zone.
3. The temperature in the background rock immediately outside the fracture or fracture zone is approximately equal to the temperature averaged over the fracture thickness.

The current model of the MOTIF code does not include a dual-porosity formulation.

Numerical Methods

In the MOTIF code, the Galerkin weighted residual method, in conjunction with the finite-element approximation, is used for spatial discretization. This renders each governing equation into a system of ordinary differential equations. Finite-difference approximation for the temporal derivatives further reduces each into a sparse set of simultaneous algebraic equations. The time-integration scheme incorporated in MOTIF enables the user to specify fully explicit Crank-Nicholson, or fully implicit finite difference, or any scheme in between. The three sets of equations for flow, solute transport, and heat transport are solved sequentially, as illustrated in Figure 2, using a numerical algorithm known as "skyline profiling" (*11*), based on Gaussian elimination. Solution of the flow equation provides the spatial distribution and time

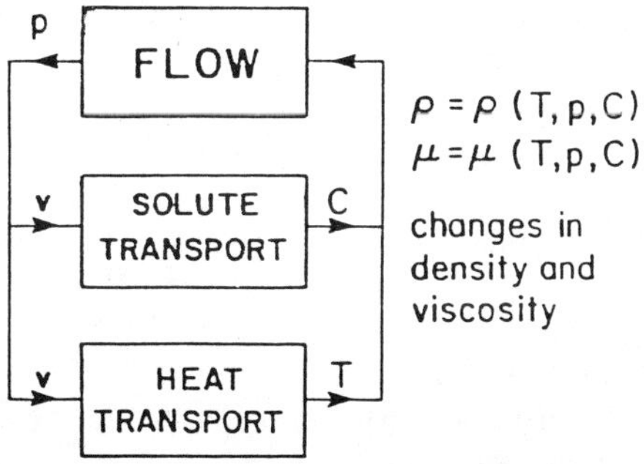

Figure 2. **Schematic representation of the MOTIF algorithm for solving the coupled processes of groundwater flow, solute transport, and heat transport. Reprinted with permission from (*7*).**

evolution of hydraulic head or, equivalently, pressure and velocity. The latter is fed into the convection terms in the solute transport and heat transport equations. The pressure, p, temperature, T, and concentration, C, solutions are substituted into the constitutive equations to update the values of fluid density, ρ, and viscosity, μ, before advancing to the next time step.

The equations to be solved are generally nonlinear. In MOTIF this nonlinearity is handled by Picard iteration (*12*). To make it practical to perform 3-D simulations, a version of MOTIF has been optimized to take advantage of the parallel processing and pipeline architecture of the Floating Point Systems FPS-364/MAX computer. Benchmark testing has indicated computational speeds exceeding 50 Mflops (mega floating operations per second).

Three types of isoparametric elements are available in MOTIF: a hexahedron, a 2-D quadrilateral element, and a 1-D line element. These elements are all defined in a 3-D space; thus the hexahedron element can be used to represent porous media in a 3-D model, while the quadrilateral element can be used to represent either porous media in a 2-D model or planar fractures or fracture zones in a 3-D model. Similarly the line element can be used to represent either porous media in a 1-D model, or planar fractures or fracture zones in a 2-D model, or narrow channels and pipes in a 3-D model. A combination of these can be employed in a single model. It should be noted that in applying the 2-D element in MOTIF to simulate a planar fracture or fracture zone, or the 1-D element to simulate channel flow, it has been assumed that these features can be approximated, respectively, by 2-D or 1-D porous media.

Verification

For quality assurance the MOTIF code has been verified by comparison with analytical solutions for the following problems:

1. a semi-infinite strip recharge basin,
2. a pumping well intersected by a vertical fracture plane,
3. a finite, confined circular aquifer with an injection well intersected by a horizontal fracture plane,

4. a one-dimensional radionuclide transport with hydrodynamic dispersion, linear absorption, and radioactive decay,
5. one-dimensional heat transport, and
6. hydrothermal convection around an exponentially decaying spherical heat source.

In addition, the MOTIF solutions have been compared with other numerical solutions for

1. thermal buoyancy flow,
2. concentration-driven flow and
3. topographically driven steady-state flow in a porous medium traversed by two intersecting fracture zones.

Most of the verification cases listed above have been discussed by Chan et al. (*4*).

Validation

The ability of the MOTIF code to correctly simulate groundwater flow in a plutonic rock formation was assessed by using the code to predict the drawdown and water inflow caused by excavation of the shaft at the Underground Research Laboratory (URL) in a granitic pluton on the Canadian Shield.

The access shaft for the URL has been constructed to a depth of 255 m in a previously undisturbed, saturated, granitic pluton near the town of Lac du Bonnet, Manitoba. Prior to the excavation, detailed site evaluation was carried out to provide the data to construct conceptual models of the hydrogeology of the URL vicinity. The geological and hydrogeological investigations revealed that three major, low-dipping fracture zones largely controlled the patterns of groundwater movement and groundwater chemistry within the rock mass (Figure 3). Other permeable, subvertical fracture zones correspond to the locations of surface topography lineaments; however, their depths have only been inferred from observations made during some of

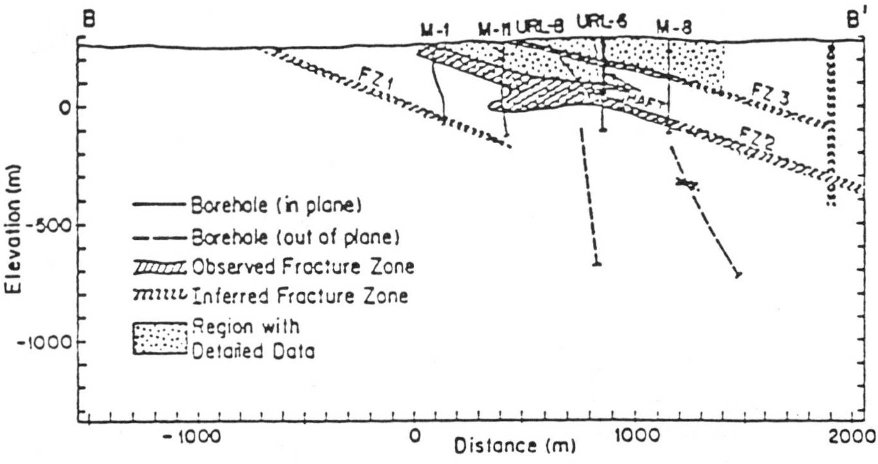

Figure 3. **Section BB′ (see Figure 8) showing observed and inferred geology near the URL. Reprinted with permission from (*7*).**

the large-scale groundwater pumping tests. Except for these distinctive fracture zones, the rock mass was found to be relatively unfractured, with fracturing being more pronounced near ground surface. This work also revealed a complex spatial pattern of permeability within each of the fracture zones. This pattern strongly controlled both the groundwater movement and chemistry that existed within each zone (*13,14*).

Before shaft excavation began in 1984 May, 3-D models of the URL vicinity were developed using the MOTIF code and used to predict the hydrogeological responses during and after shaft construction (*15*). These predictions are now being compared with the responses measured at the site.

Two models of different scales were generated: a steady-state, regional-scale (tens of kilometers) model and a transient, local-scale (a few kilometers) model. Plan views of the finite-element meshes for these two models are shown, respectively, in Figures 4 and 5. The regional-scale model was employed to study the regional flow regimes and to help define boundary conditions for the local-scale model. In the MOTIF models, three types of elements were employed: 3-D continuum elements to represent the background rock mass, planar elements to represent a thin veneer of surficial deposits as well as major fracture zones, and line elements to represent shafts and drifts. Time-dependent seepage boundary conditions were used to simulate sequential excavation.

Hydraulic conductivities were determined by interpreting a variety of single-borehole or inter-borehole straddle packer tests. The hydraulic conductivities of the background rock mass were assumed to be horizontally uniform and to vary from 10^{-12} m/s at 1000 m below mean sea level to 10^{-8} m/s at the ground surface. For the fracture zones, an average value of 10^{-6} m/s was used.

In the models using the MOTIF code, the top boundary had a prescribed head condition for surface water bodies and an infiltration condition elsewhere. Infiltration rates through the top surface of the models were adjusted so that the computed water levels corresponded to the observed long-term data, or surface topography in the absence of data. The bottom boundary was impermeable and the vertical boundaries were constant-head boundaries. For the local model, the boundary head values were obtained from the steady-state regional model. The values of the hydrogeological parameters for the major fracture zones in the local model were adjusted until calculated drawdowns matched those recorded in the field during various pumping tests.

This calibrated local model was used to predict time-variant piezometric responses that would occur at 171 monitoring locations in the rock mass around the excavation due to the sinking of the main access shaft. The rate of seepage into the shaft was also calculated.

A comparison between the recorded inflow rates to the shaft and the predicted values of Guvanasen et al. (*15*) is illustrated in Figure 6. The model predicted that the inflow rate would reach a maximum when Fracture Zone 3 was intersected by the shaft. The model also predicted that subsequently the inflow rate would gradually decline to a constant value. This overall trend agrees reasonably well with the general trend of the observed data. However, the maximum inflow actually occurred when the shaft excavation was deeper than Fracture Zone 3, because unanticipated seepages were encountered in subvertical fractures located below Fracture

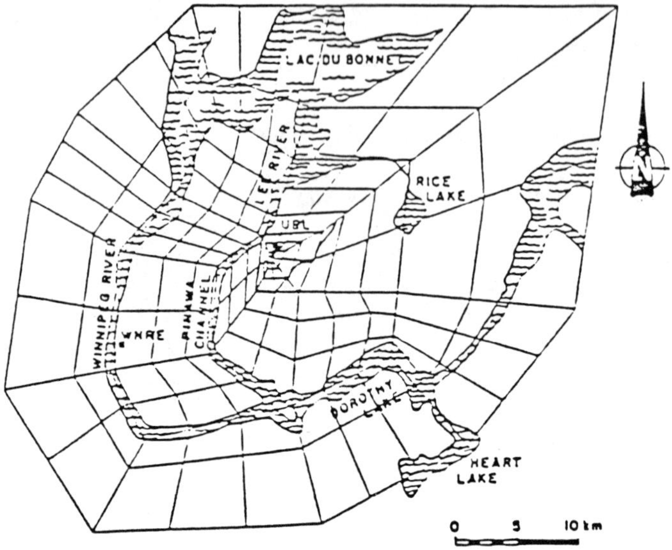

Figure 4. **Plan view of regional-scale, finite-element mesh for URL drawdown modeling.**

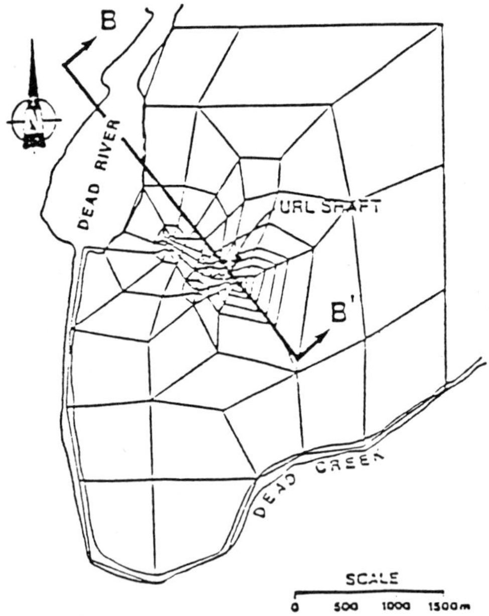

Figure 5. **Plan view of local-scale, finite-element mesh for URL drawdown modeling.**

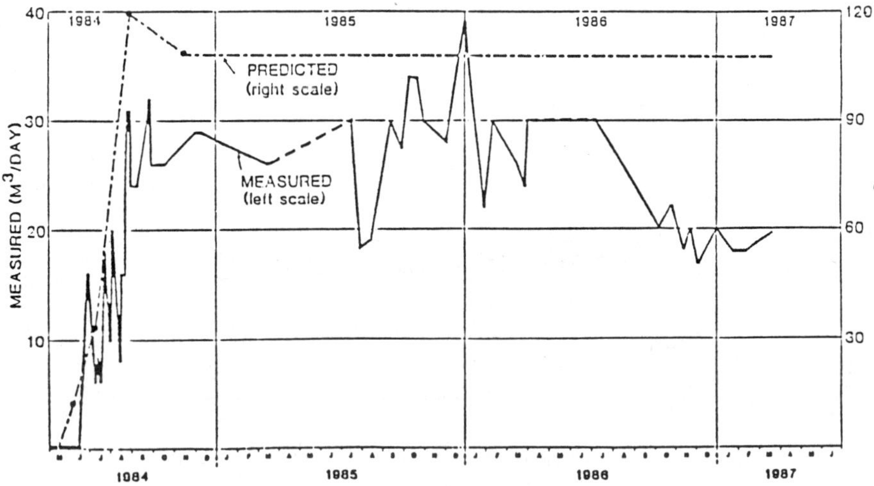

Figure 6. **Total groundwater inflow to the URL facility.**

Zone 3. Furthermore, the predicted inflow rate was approximately four times the measured rate.

For most of the 171 piezometer locations the magnitude and the temporal variation of the drawdown are fairly accurately predicted by the model. An example of the agreement between predicted and measured piezometric response within fracture zone 3 is presented in Figure 7. Further details of the comparison can be found in reference 6.

The results of the URL drawdown experiment give confidence in the ability of the MOTIF code to simulate regional groundwater flow problems, at least for potential disposal sites with hydrogeological characteristics similar to those found at the Whiteshell Research Area.

The reason for the discrepancies between the predicted and measured discharges has not yet been unambiguously identified, but there are some plausible suggestions:

1. Because of the limited computational speed of the available computer hardware at the time this simulation was performed, the finite-element mesh immediately surrounding the shaft was not sufficiently dense.
2. The processes of hydraulic testing and model calibration might not have determined the hydraulic properties accurately enough in the near field of the shaft.
3. The changes in hydraulic conductivity in the fracture zones in the vicinity of the shaft in response to excavation were not included in the model.

Near-field straddle-packer tests during the excavation of the shaft and one of the drifts have indicated localized decreases in fracture hydraulic conductivity by a factor of two or so following excavation. A preliminary analysis showed that stress redistribution alone could not fully explain these decreases.

It is recognized that errors in groundwater velocity calculations would lead to errors in solute transport calculations. Accordingly further theoretical, numerical, and

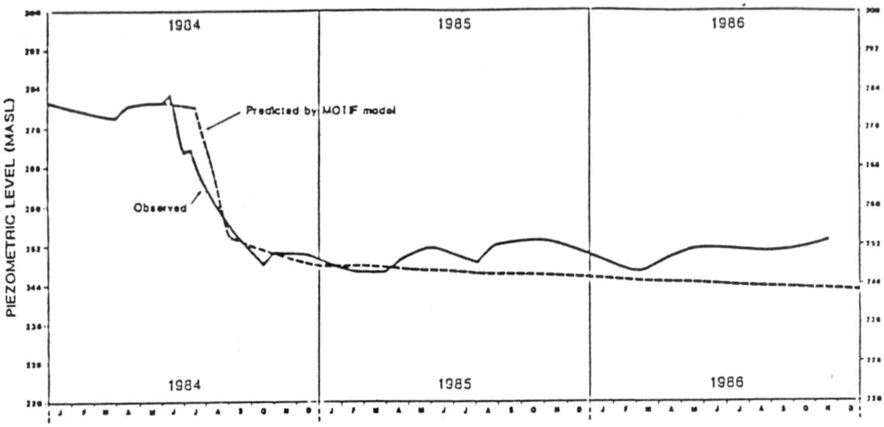

Figure 7. **Piezometric drawdown in fracture zone 3 at URL-1 (93.4- to 131.0-m interval).**

field investigations are necessary to resolve the problem of discharge calculations. However, our field hydrogeologists consider the discrepancies mentioned above to be small in comparison with the uncertainty and variability in hydraulic parameters in a typical plutonic rock mass in the Canadian Shield. Furthermore, the general agreement between the predicted and measured head distributions means that the far-field hydraulic gradients and, therefore, possibly also the far-field velocities, were predicted with reasonable accuracy. For the purpose of disposal system performance assessment, the travel time and pathline calculations as well as the transport calculations are influenced predominantly by the far-field velocities.

Tracer tests are currently being performed in the Whiteshell Research Area. Prior to the tracer tests, hydraulic interference tests were performed and the drawdown and recovery data were used to estimate the distribution of hydraulic conductivity and porosity. The estimated hydraulic conductivity distribution was then used along with a range of values for the dispersivities and porosity in MOTIF transport models to predict the breakthrough curves in various boreholes. Concentration data are still being collected. Comparison of predicted and observed breakthrough curves would allow the in situ transport parameters to be estimated and provide some kind of model validation.

As discussed near the end of this paper, uncertainties and variabilities in the disposal system are treated by means of Monte Carlo simulations.

Whiteshell Research Area Geosphere Modeling

Conceptual regional hydrogeological model

A detailed conceptual hydrogeological model has recently been constructed based on field data from the Whiteshell Research Area. This research area has been selected for the modeling study because of the large amount of relevant field data that are currently available and can be used to develop a representative and physically based conceptual model. Although no nuclear waste will be disposed of in the research area, the modeling study considers various scenarios involving a hypothetical disposal vault located at a depth of 500 m.

The Whiteshell Research Area covers about 750 km^2 in southeastern Manitoba (Figure 8), including the sites of the URL and Whiteshell Nuclear Research Establishment (WNRE). A major portion of the area is part of the Lac du Bonnet Batholith, a large granitic pluton. The Winnipeg River system is assumed to provide a stable hydrogeological boundary nearly surrounding the area. In general, the topography slopes from southeast to northwest and the water table closely follows the topography.

Until recently, subsurface investigations in boreholes have been limited primarily to the WNRE and URL sites. However, the area that must be modeled to analyze the impact of the hypothetical disposal vault is larger than that for which detailed data are currently available. Therefore, the conceptual model of the region has been developed from detailed studies at the WNRE and URL sites and from geological features interpreted from air photography and regional geophysical surveys (*16*). The model incorporates more recent data than those used to construct the regional flow model for the URL drawdown experiment.

The linear features in Figure 8 are assumed to be subvertical fracture zones, the major ones being at least 4 km deep and the minor ones 1 km deep. Figure 9, which

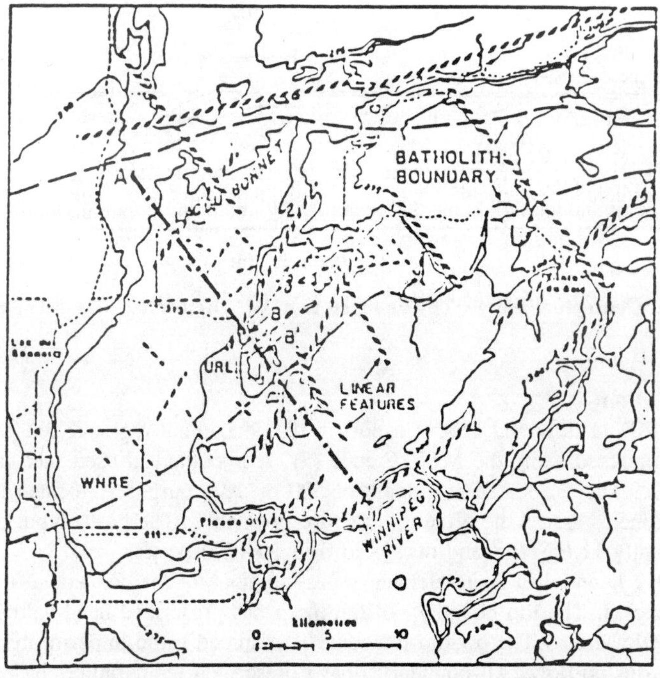

Figure 8. **Whiteshell Research Area. Reprinted with permission (*7*).**

is an extension of the previous cross section (Figure 3) through the URL site, shows the conceptualized geology. Low-dipping fracture zones are assumed to intersect the surface at vertical zones and most of them are extended to intersect another vertical zone at depth. This is to permit the analysis of a regionally continuous fracture zone system, which is a conservative assumption with regards to radionuclide transport. The fracture zones are considered to be porous media and uniform in thickness. The rock mass is divided into five layers of porous media with permeability and porosity decreasing with depth. The permeability of the fracture zones is two to four orders of magnitude higher than that of the rock mass.

It is important to note that the conceptual model is one possible interpretation based on available data. It has been constructed to allow sensitivity analyses with respect to assumptions about features and their properties. Recently, the field investigations have been expanded to encompass the entire region of the conceptual model. These investigations will help to refine the conceptual model and will also provide specific input data for calibration.

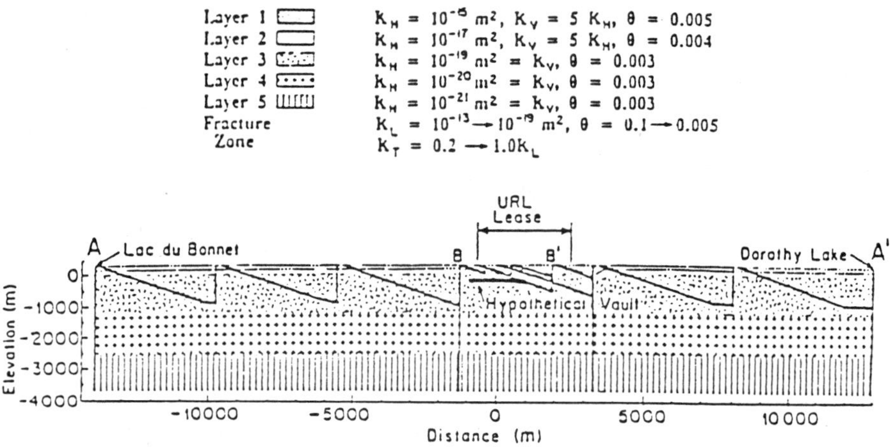

Figure 9. **Conceptualized geology along section AA′. Reprinted with permission (*7*).**

Flow simulation

Coupled fluid flow and heat transport in the conceptual regional geosphere model are being simulated using the MOTIF code (*7*). A hypothetical used fuel vault, 1.9 km × 1.9 km, is located at a depth of 500 m. It is purposely located to intersect a fracture zone to assess the effect of fracture proximity. The heat output of the vault is initially 11.6 W/m^2 and decays to 0 in 100,000 years.

Several 2-D and 3-D finite-element representations of the conceptual model have been constructed. The top boundary of the mesh has prescribed head values equal to water-table elevations. The bottom boundary is assumed to be impermeable and the side boundaries no-flow. All boundaries have prescribed temperatures based on geothermal measurements at the URL (*17*).

Initially, a 2-D model corresponding to the cross section of Figure 9 was constructed to help define an adequate 3-D representation, select a suitable form for the SYVAC geosphere submodel, and assist in sensitivity analysis. This 27-km by 4-km section is essentially along a flow line, i.e., there is little flow perpendicular to the plane. Element properties, such as permeability and porosity, have been selected as average values for the units in the conceptual model but are adjusted for sensitivity analysis.

To check numerical convergence of predictions, two meshes and three time-stepping sequences of differing refinement were used in this analysis. All elements were 2-D. Figure 10 illustrates the coarse-mesh option for the vault region. The coarse mesh consists of a total of 914 quadrilateral elements and 985 nodes. The fine mesh has been constructed by dividing each element of the coarse mesh in four.

Transient simulations were conducted for a period of 1 million years using geometrically increasing time steps. Example predictions of temperature and velocity components near the vault are shown in Figure 11. Locations M and L are about 100 m above the vault in the rock mass and in the low-dipping fracture zone that intersects the vault, respectively (see Figure 10). Location V is about 200 m above the vault in a vertical fracture zone 325 m from the edge of the vault.

For this simulation, the velocity in fracture zones near the vault is on the order of 1 m/a. In the rock mass outside the fracture zones, velocities are typically four orders of magnitude lower. In general, the perturbations in both temperature and velocity last about 100,000 years. The temperature rise extends about 2000 m from the vault, with convective heat transport being negligible. The temperature at the vault centroid reaches a peak value of 70 °C after 60 years and then rapidly

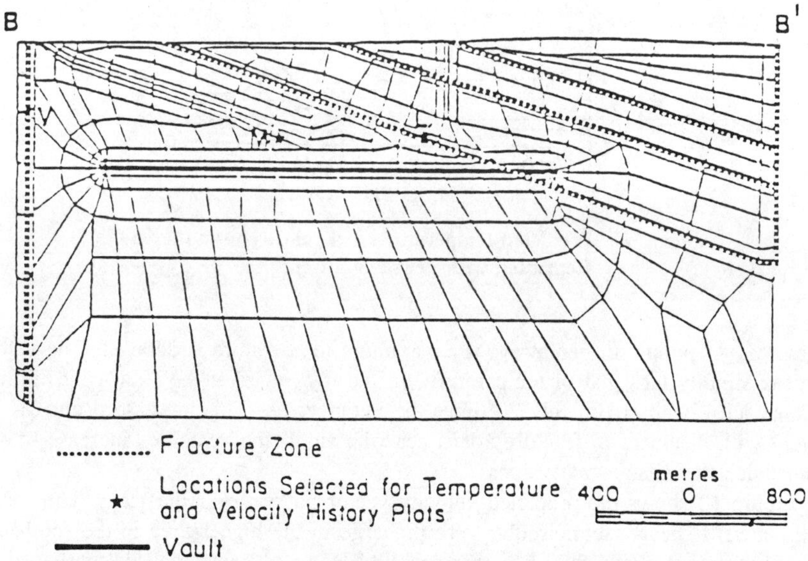

Figure 10. **Two-dimensional, finite-element mesh for vault region. Reprinted with permission (7).**

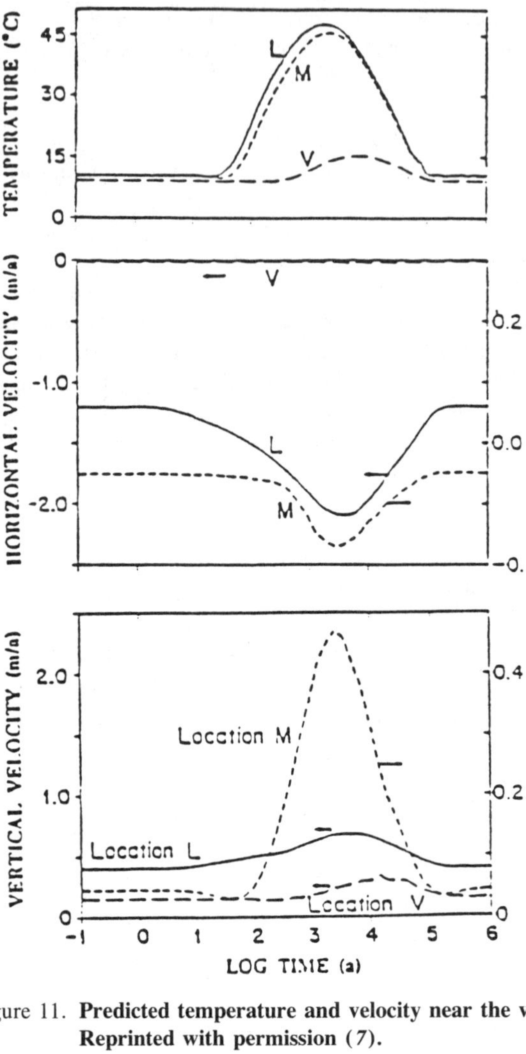

Figure 11. **Predicted temperature and velocity near the vault. Reprinted with permission (7).**

decreases. At points further away, the maximum temperature is delayed. The velocity peak slightly lags that of temperature. In the rock mass at the vault centroid, the peak velocity is approximately 20 times the initial value. This increase drops off rapidly with distance. In fracture zones near the vault, the velocities increase to about twice the steady-state values.

Figure 12 shows the predicted velocity vector pattern near the vault. The steady-state pattern is nearly semicircular with recharge in the higher area in the southeast, lateral movement above the lower permeability layers at depth, and discharge up through the left side of the vault to the low-lying area in the northwest. The local surface topography (as reflected in the water table) drives the flow. Flow is up the

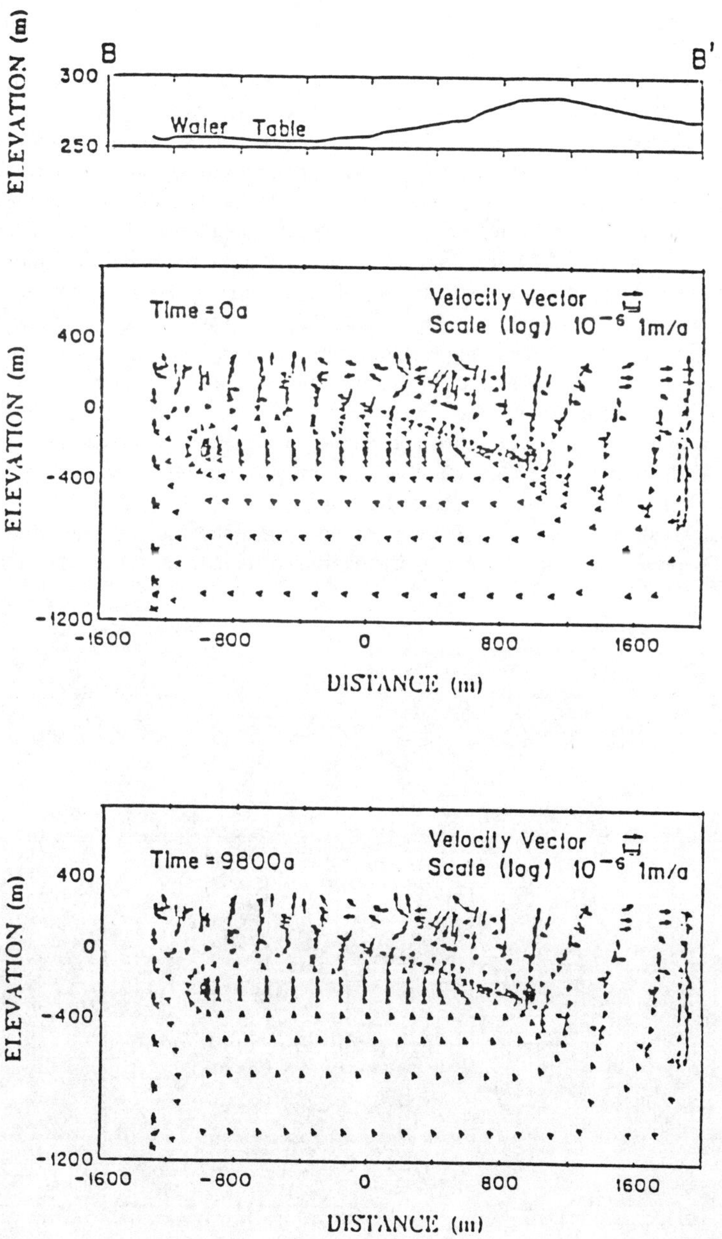

Figure 12. **Water table profile and predicted velocity distribution near the vault. Reprinted with permission (*7*).**

low-dipping fracture zone intersecting the vault. This general pattern is consistent with field observations. At 9800 years, the overall perturbation in velocity field is near its maximum; the flow pattern is now nearly vertical through almost the entire vault because of the buoyancy effect of the waste heat.

Two 3-D transient models have been constructed: a coarse-mesh, regional-scale model and a fine-mesh, local-scale model. The regional model is bounded by the Winnipeg River on the east, south, and west and by Lac du Bonnet on the north (see Figure 8). It extends to a depth of 4 km and will be used to study the regional flow regimes and help define boundary conditions for the local model. A plan view of the finite-element mesh for the regional model is shown in Figure 13. The background rock is represented by 3-D elements while the fracture zones are represented by 2-D elements. The local model covers an area of approximately 80 km^2 centered on the URL lease area and extends to a depth of 1.5 km. A NW-SE section (corresponding to AA′ in Figure 13) of the finite-element mesh for the local geosphere model is shown in Figure 14. Only 3-D elements are employed in the local model.

Preliminary comparison between the 2-D and 3-D geosphere models has indicated that there are both qualitative and quantitative differences in the groundwater velocity distributions and, consequently, contaminant transport predicted by these models. Details are reported by Reid and Chan (*9*).

Analysis of the sensitivity of predicted flows near the vault is being done using both 2-D and 3-D models. Factors being examined include numerical discretization;

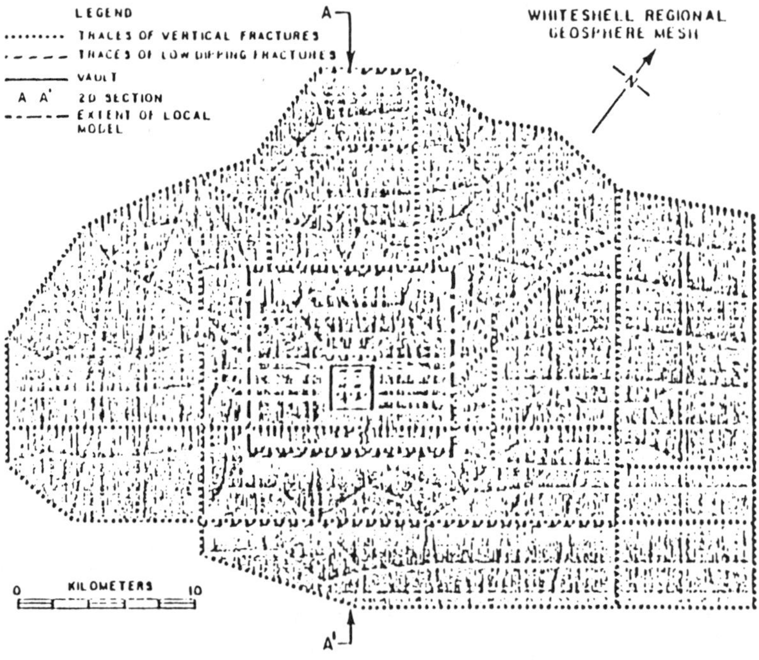

Figure 13. **Plan view of the finite-element mesh for the 3-D regional geosphere model. Reprinted with permission (*7*).**

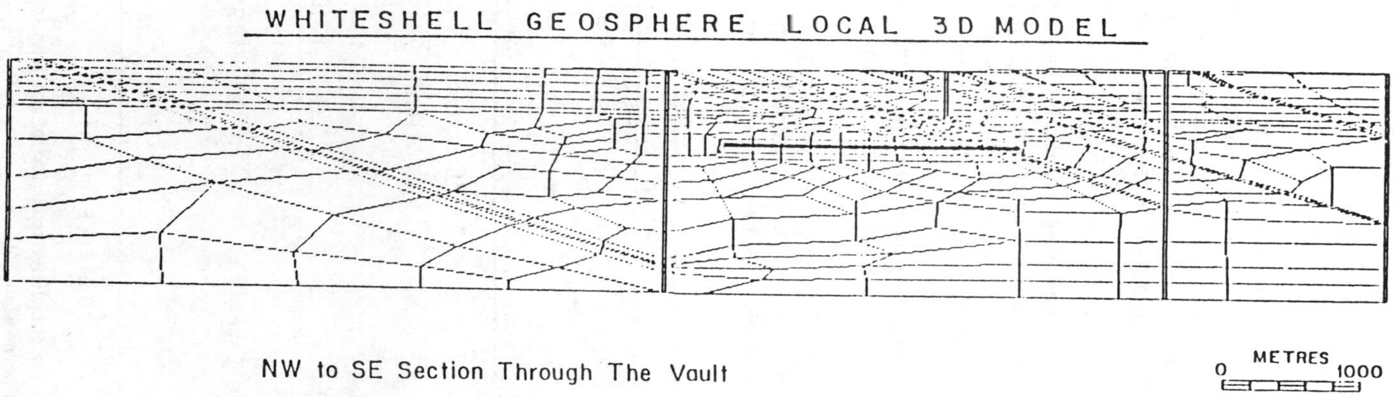

Figure 14. **NW to SE vertical section of the finite-element mesh for the 3-D local geosphere model. Elements representing the vault are indicated by the dark area in the central portion of the mesh. Reprinted with permission (7).**

variation in the location and nature of boundary conditions; variations in the fracture zone conceptualization; values of parameters, such as hydraulic conductivity and porosity, assigned to features in the conceptual model; and altered properties resulting from vault construction.

Interesting examples of sensitivity analyses using the 2-D finite element geosphere model have been reported by Chan and Scheier (*8*). These analyses indicate that fracture-zone avoidance is a very effective method to retard convective transport. However, grouting of a major fracture zone intersection may increase or decrease the travel time of an inert contaminant particle depending on the location of the source.

Transport simulation

As an initial step in transport modeling, particle tracking codes are being used to calculate radionuclide pathlines and travel times based only on groundwater velocity distributions predicted by MOTIF. Figure 15 illustrates travel predictions using the previously described 2-D flow analysis results. The path lengths from the vault to the surface do not vary a great deal, being two to three times the depth of the vault. However, the travel times in this simulation vary from 1200 years to 26 million years, the fastest path, D, being up the low-dipping fracture zone that intersects the vault.

These results indicate that a wide range of radionuclide transport scenarios have to be considered. One extreme involves rapid convection in major fracture zones with transient thermal effects being important. In the other extreme, convective trans-

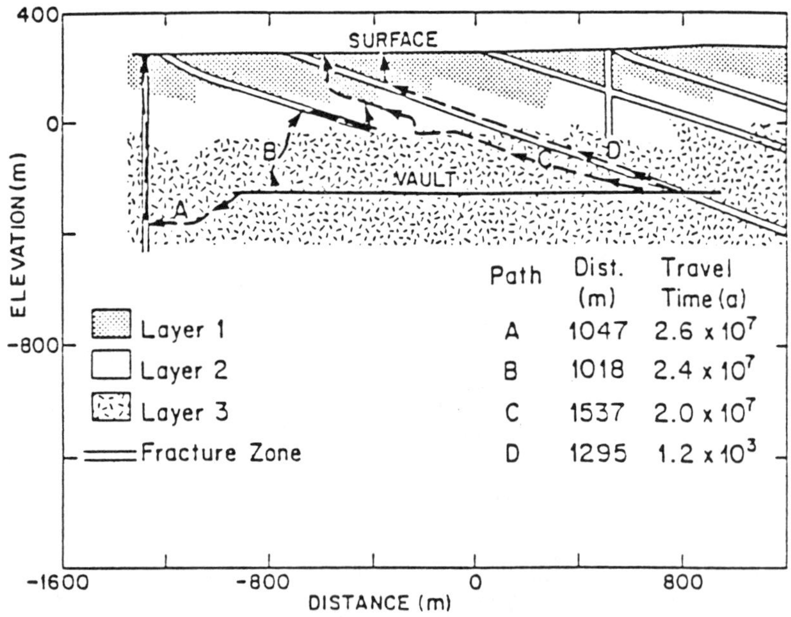

Figure 15. **Predicted paths for convective transport of particles released from the vault. Reprinted with permission (*7*).**

port through the unfractured bulk rock may be negligible compared to diffusion, dispersion, and retardation mechanisms. The relative importance is very sensitive to the proximity of major fracture zones to the vault.

A detailed 2-D, finite-element, convective-dispersive transport model for the vault region is now being analyzed using MOTIF to ensure adequate representation of diffusion and dispersion in the SYVAC submodel. Preliminary results indicate a rapid convection of contaminants up the low-dipping fracture zone and then a significant dispersion in the near-surface layers. Dispersion and diffusion significantly accelerate the first arrival of contaminants along the slow paths but have only minor influence on transport along the faster paths.

The SYVAC Geosphere Submodel

From the results of MOTIF flow simulations, particle tracking, and MOTIF transport simulations, a simplified SYVAC geosphere submodel has been constructed. The current form of the submodel is a network of 1-D linear segments (*18*). An example is shown in Figure 16.

The geometry of the network and the groundwater flow velocity in each segment are determined by interpretation of the output of the MOTIF finite-element models. Two options are available: (1) the average groundwater flow velocity for each segment can be calculated from the appropriate element velocities output by the MOTIF runs and read into the SYVAC submodel; or (2) the nodal head values obtained from the MOTIF runs can be input into the submodel, which then calculates the average velocity for each segment. Within each segment the 1-D contaminant mass-balance equations (including chemical sorption) for a radionuclide decay chain are solved analytically (*19*) assuming constant velocities, zero initial concentrations, and simple boundary conditions. Other input parameters to the network submodel include dispersion coefficients, diffusion coefficients, and retardation factors for the various radio-

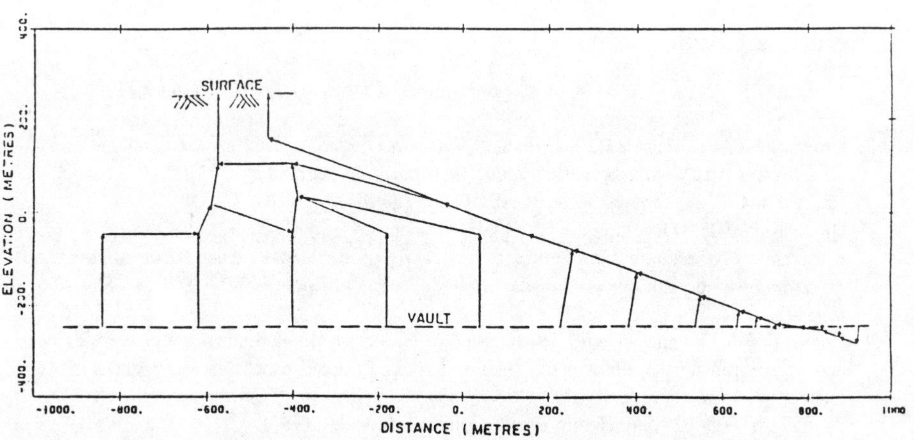

Figure 16. **Example network for SYVAC-geosphere submodel.**

nuclides. These have to be consistent with the input parameters in the MOTIF simulations and laboratory or field data. Because SYVAC uses a Monte Carlo simulation approach to accommodate uncertainty and variability in the system, probability distribution functions have to be derived for the input parameters of the geosphere submodel. Where data are lacking, the distribution functions would have to be derived from expert opinion.

SYVAC selects a realization, defined to be a possible state of the system, by randomly sampling a value for each parameter from its specified distribution. This set of parameter values is then used within the submodels to estimate the transport of contaminant from the vault to the biosphere, and leads to a predicted impact for that realization. The selection and prediction process is repeated, typically 1000 times or more, and the results are summarized in, for example, a histogram of predicted consequences versus their frequency of occurrence.

If the groundwater geometric and flow velocity or head data from the MOTIF flow model are transferred to the SYVAC geosphere submodel as described above, then the hydraulic parameters cannot be sampled within the submodel. To overcome this deficiency a finite-element steady-state groundwater flow code has been assembled using a selected subset of subroutines of the MOTIF code. This can be used for flow simulations using networks of 1-D elements in 3-D space. Because of the computational efficiency of 1-D elements, it is feasible to incorporate this 1-D code into SYVAC. This approach would facilitate the consistent sampling of hydraulic and transport parameters in the Monte Carlo simulations.

The SYVAC geosphere submodel requires many simplifying approximations; for example, transport for the network is not determined by solving the system of equations for the entire network but rather by solving the transport equations for each of the segments sequentially. A necessary step in the development of the SYVAC geosphere submodel is to compare its transport calculations for a nonsorbing, nondecaying tracer with detailed 1-D network, 2-D, and 3-D MOTIF transport simulations. This comparison is currently in progress.

Literature Cited

1. Dormuth, K. W. and R. D. Quick, *International Journal of Energy Systems,* Vol. 1, pp. 125–127 (1981).
2. Dormuth, K. W. and G. R. Sherman, *SYVAC—a computer program for assessment of nuclear fuel waste management systems, incorporating parameter variability,* Atomic Energy of Canada Limited Report, AECL-6814 (1981).
3. Guvanasen, V., "Development of a finite-element code and its application to geoscience research," *Proceedings 17th Information Meeting of the Nuclear Fuel Waste Management Program,* Atomic Energy of Canada Limited Technical Record, TR-199*, pp. 554–566 (1984).
4. Chan, T., V. Guvanasen, and J.A.K. Reid, "Numerical modelling of coupled fluid, heat and solute transport in deformable fractured rock. Coupled processes associated with nuclear waste repositories," *Proc. Int. Symposium* (Berkeley, California), Ed. C. F. Tsang, pp. 605–625, Academic Press, Orlando, Florida (1987).
5. Davison, C. C. and V. Guvanasen, "Hydrogeological characterization, modeling and monitoring of the site of Canada's Underground Research Laboratory," presented at Inter-

national Association of Hydrogeologists 17th International Congress on Hydrogeology of Rocks of Low Permeability, Tucson, January 7–12 (1985).

6. Davison, C. C., T. Chan and N. W. Scheier, "Experimental activities of the Canadian Nuclear Fuel Waste Management Program to validate geosphere models," to appear in Proceedings, GEOVAL 87, Stockholm, April (1987).
7. Chan, T., N. V. Scheier and J.A.K. Reid, "Finite-element thermohydrogeological modelling for Canadian nuclear fuel waste management," *Proceedings of the Second International Conference on Radioactive Waste Management,* Toronto, Canadian Nuclear Society, pp. 653–660 (1986).
8. Chan, T. and N. W. Scheier, "Finite-element simulation of groundwater flow and heat and radionuclide transport in a plutonic rock mass," *Proceedings, 6th International Congress on Rock Mechanics,* Montreal, August 30–September 3, pp. 41–46, Balkema Press (1987).
9. Reid, J.A.K. and T. Chan, "Sensitivity of Whiteshell geosphere modeling results to dimensionality (2-D/3-D) of the computer simulations employed," these Proceedings (1989).
10. Bear, J., *Dynamics of fluids in porous media*, American Elsevier, New York (1972).
11. Bathe, K. J. and E. L. Wilson, *Numerical methods in finite element analysis*, Prentice-Hall, Englewood Cliffs, N.J. (1976).
12. Huyakorn, P. S. and G. F. Pinder, *Computational methods in subsurface flow*, Academic Press, New York (1983).
13. Davison, C. C., "Hydrogeological characterization of the site of Canada's Underground Research Laboratory," Proceedings of the IAH International Symposium on Groundwater Resource Utilization and Contaminant Hydrogeology (Montreal, Canada), pp. 310–335, May 21–23 (1984).
14. Davison, C. C., *Ground Water Monitoring Review,* Vol. 4, No. 4, pp. 95–103, Fall (1984).
15. Guvanasen, V., J.A.K. Reid and B. W. Nakka, "Predictions of hydrogeological perturbations due to the construction of the Underground Research Laboratory," Atomic Energy of Canada Limited Technical Record, TR-344* (1985).
16. Scheier, N. W. and S. H. Whitaker, "Development of a geosphere model for use in assessing the performance of a nuclear fuel waste disposal vault," Proceedings of the Joint Annual Meeting of the Geological Association of Canada, Mineralogical Association of Canada and the Canadian Geophysical Union, Ottawa (1986).
17. Drury, M. J., "Geothermal logging of URL boreholes," Earth Physics Branch, Energy, Mines and Resources Canada Internal Report, 82-2 (1981).
18. Goodwin, B. W., T. H. Andres, P. A. Davies, D. M. LeNeveu, T. W. Melnyk, G. R. Sherman and D. M. Wuschke, "Post-closure environmental assessment of the disposal of nuclear fuel waste for the Canadian Fuel Waste Management Program," *Radioactive Waste Management and the Nuclear Fuel Cycle,* Vol. 8, pp. 241–272 (1987).
19. Heinrich, W. F. and T. Andres, "Response functions of the convection-dispersion equations describing radionuclide migration in a semi-infinite medium," *Ann. Nucl. Energy,* Vol. 12, pp. 685–691 (1985).

---

*Unrestricted, unpublished report available from SDDO, Atomic Energy of Canada Limited Research Company, Chalk River, Ontario KOJ 1JO.

Treatment of Uncertainties in Ground-Water Flow Modeling in the Swiss Radioactive Waste Program

PETER HUFSCHMIED
National Cooperative for the Storage of Radioactive Waste
Baden, Switzerland

Abstract

The most plausible scenario for release of radionuclides from a geologic repository is by transport in moving ground-water. Accordingly, modeling of ground-water flow plays an important role in Swiss performance assessment work. Modeling takes place at different scales, beginning at a regional scale which gives boundary conditions for a smaller local-scale model. At the repository scale, a third model is being used to study the influence of major hydraulic discontinuities near the repository. A fourth model finally describes the fine-scale flow in individual water-bearing features which are most important for radionuclide transport in the geosphere. Sensitivity of model parameters and approaches to treat uncertainties are different for the various scales. Main sources of uncertainty are (1) incomplete available data about the spatial distribution of hydrogeologic parameters and (2) measurement/interpretation errors. Significant effort has already been involved in understanding and quantifying uncertainty in results of hydraulic testing, and increasing attention is being focused upon the problem of the incompleteness of data. Efforts are under way to better quantify both types of uncertainties. Approaches to treat uncertainty in derived model parameters are discussed. Ultimately the methods proposed should allow one to link field data uncertainty and incomplete information to the uncertainty of hydrodynamic input parameters for radionuclide transport calculations. The methodology is being tested within the international project HYDROCOIN. The impact of hydrologic uncertainty considerations upon the ongoing and planned site characterization work as well as associated R&D needs of the Swiss program are discussed.

1 Introduction

Since 1985, when Nagra submitted a report on technical feasibility and achievable safety of final repositories for radioactive waste in Switzerland (so-called Project

Gewähr 1985, Nagra, 1985b) to the Federal Government, Nagra's work has shifted from feasibility studies towards site characterization, site selection, and repository design. Highest priority is given to the repository for low- and intermediate-level waste, where four potential sites are currently being investigated. The high-level waste program continues with the completion of the regional investigation of crystalline bedrock in Northern Switzerland. In addition, storage of high-level and intermediate-level waste in sedimentary rock is under consideration.

For Project Gewähr 1985 it was adequate to apply a bounding-type safety analysis based on scenario and consequence analysis for a limited number of scenarios and parameter sets. This approach has the advantage of being transparent and flexible, but for site characterization, site selection, and repository design, some of its limitations are obvious. Worst-case models are not necessarily suitable to develop a sound understanding of reality. Furthermore, they can lead to overemphasis on predicted high individual consequences resulting from combinations of very unlikely conservative parameters or particular scenarios of low probability or small overall consequences (Zuidema, 1986). In recognition of these limitations, efforts are currently being made to develop methodologies which can

1. better utilize the experimental evidence derived from field and laboratory investigations during the site characterization phase, in order to calibrate and validate performance assessment models, and
2. account for uncertainties in model predictions in a more objective and quantitative way.

This development is further supported by the fact that increasing emphasis is begin placed upon estimating radiological risk rather than individual dose.

In this paper only ground-water flow models are discussed. Such models are usually some kind of submodels in the performance assessment process. Input data for ground-water flow models depend on geologic information as well as on field and laboratory evidence. This information usually is uncertain and incomplete. The output of ground-water flow models (e.g., ground-water travel time) defines input to other submodels such as geosphere transport codes. Probabilistic performance assessment ultimately requires that uncertainties in input parameters are propagated through uncertainties on model parameters to the relevant output. Such an approach, currently being tested by Nagra within the scope of the international project HYDROCOIN, is discussed in Chapter 5.

Parameter variations are the most generally applicable approach to probabilistic modeling. However, this approach requires that probabilities are assigned to different parameter sets. With only a few parameter values varied over a wide range, only bounding-type analyses such as the ones conducted for Project Gewähr 1985 are possible. This type of analysis will continue to play a role in the early phases of a site investigation program when only few data are available. An example for such an approach is discussed in Chapter 4.

It should be noticed that ground-water flow models not only may be used in safety analysis. They might also provide valuable input to design of site characterization programs (exploration strategies) and repository layout. An example for site characterization is briefly presented in Section 2.2.

2 Role of Ground-Water Modeling in Safety Analysis and Site Characterization

2.1 Safety analysis

For safety analysis of repositories for high-level as well as for low- and intermediate-level wastes, release of nuclides into ground-water and transport by ground-water to the biosphere were identified as being the most likely (base case) scenario for most host rocks and potential sites considered in the Swiss program. The model chain for this base case scenario defined in Project Gewähr 1985 (Nagra, 1985b) is given in Figure 2.1.

Ground-water flow models based on regional and local hydrogeology as well as on near-field hydrology provide input to subsequent safety models. The most relevant output of ground-water flow models may be summarized as follows:

1. Flow through the repository and distribution of flow among different water-conducting features potentially intersecting the repository form a basis for near-field chemistry, performance of engineered barriers, and radionuclide transport in the near field.
2. Modeling of flow paths through the geosphere and the distribution of ground-water among water-conducting features in the geosphere define input important to radionuclide transport calculations in the geosphere.
3. Modeling of ground-water flow in aquifers surrounding the host rock is used to quantify radionuclide dilution in these aquifers.
4. Discharge areas to the biosphere are identified in order to support selection of appropriate biosphere scenarios and identification of biosphere compartments (including quantification of dilution processes).

2.2 Site characterization

Ground-water flow models can be utilized in the site characterization process for the planning of exploration programs. A recent example, taken from the Swiss low and intermediate waste program, concerns the planning of site investigations at the Wellenberg site, Canton Nidwalden (Nagra, 1987a). Figure 2.2 shows an isometric sketch of the three-dimensional model with the different hydrogeologic units. The host rock consists of Valanginien Marl embedded between two folded limestone layers, the Axen Nappe and the Drusberg Nappe. At present, it is unknown whether the underlying rock below the marl consists of Tertiary shales or limestone formations of the Axen or Drusberg Nappes. For Tertiary rock, a significantly lower permeability is expected than for limestone.

A preliminary three-dimensional ground-water flow model has been established (Nagra, 1987b) based on geologic surface mapping. Hydraulic parameters were estimated from a nearby highway tunnel or taken from the literature. Present uncertainty in hydraulic conductivity is reflected by the wide parameter range, which usually comprises two to three orders of magnitude. Uncertainty in boundary conditions (head in bounding limestone formations) is also significant. At present no drillholes are available at the site.

Parameter variation methods were used to treat uncertainties in boundary conditions and hydraulic conductivity values. Based on expert judgment, parameter sets were subdivided into cases with higher and cases with lower probability. Modeling of 57 parame-

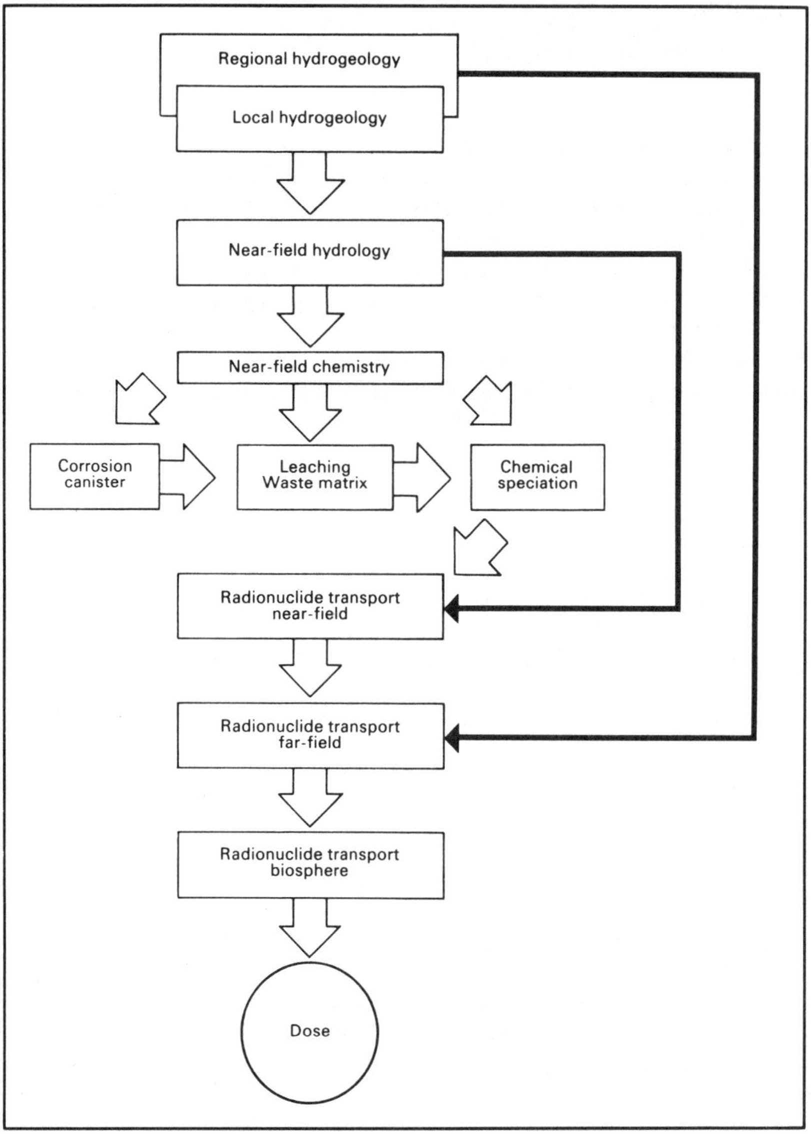

Figure 2.1. **Model chain for safety analysis for base case ground-water scenario in Project Gewähr 1985.**

ter sets clearly demonstrated that hydraulic head in the Axen Nappe and permeability of the underlying are the most sensitive parameters. Results of the modeling were used for three purposes:

1. to support selection of optimum borehole locations
2. to plan borehole testing programs and define priorities

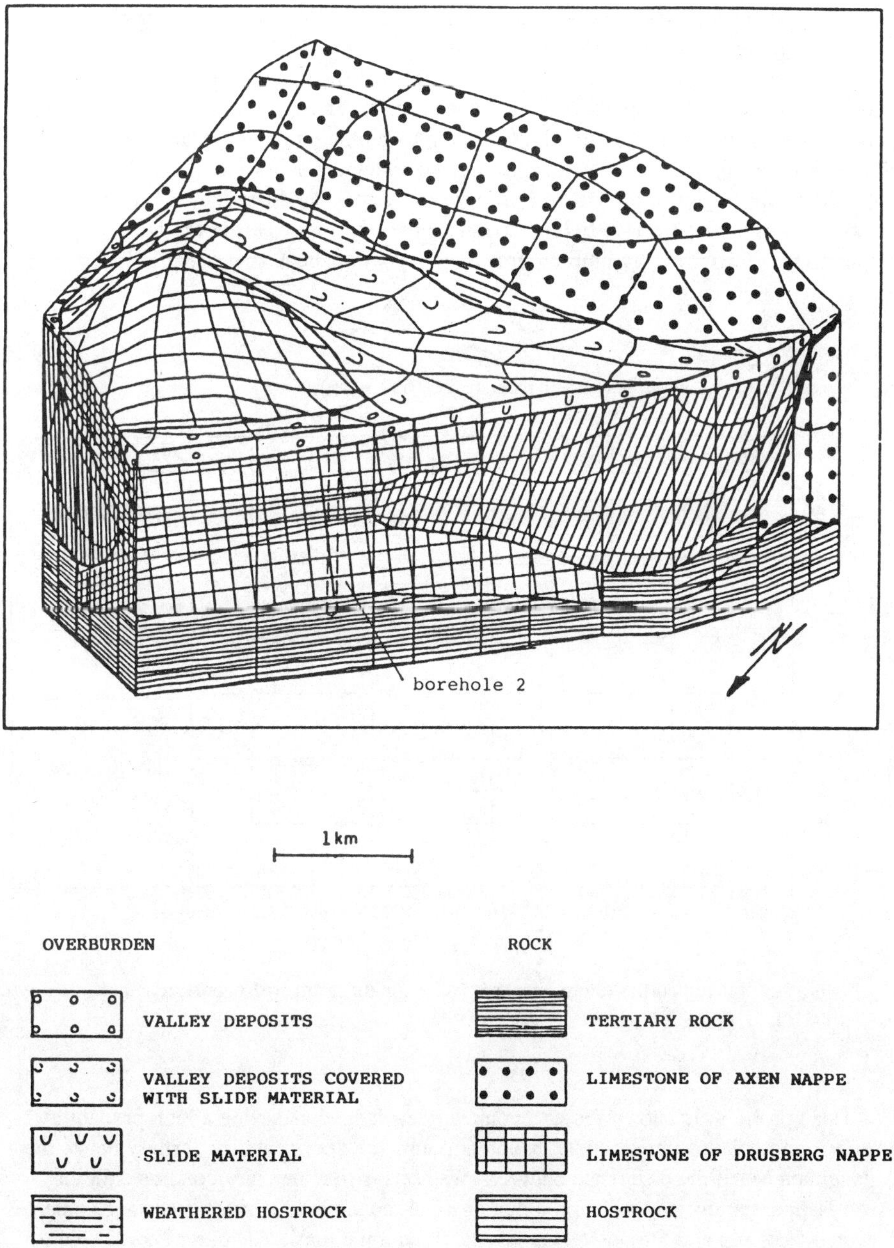

Figure 2.2. **Isometric sketch of the preliminary three-dimensional ground-water flow model at the Wellenberg site.**

3. to establish "pre-test" predictions of the head profile prior to drilling of the boreholes as a help to guide decisions during field work.

To illustrate the type of model results available for the purposes mentioned above, Figure 2.3 is shown. It displays predicted head profiles along a borehole that is planned to be drilled as shown in Figure 2.2. According to geologic prediction, the borehole should first hit relatively shallow unconsolidated valley deposits, then it should pass through the upper part of the Drusberg Nappe (limestone) and later enter into the Valanginien Marl (host rock). Below the marl either the lower part of the Drusberg Nappe (limestone of inverse limb of Drusberg Nappe) or the Tertiary should be hit.

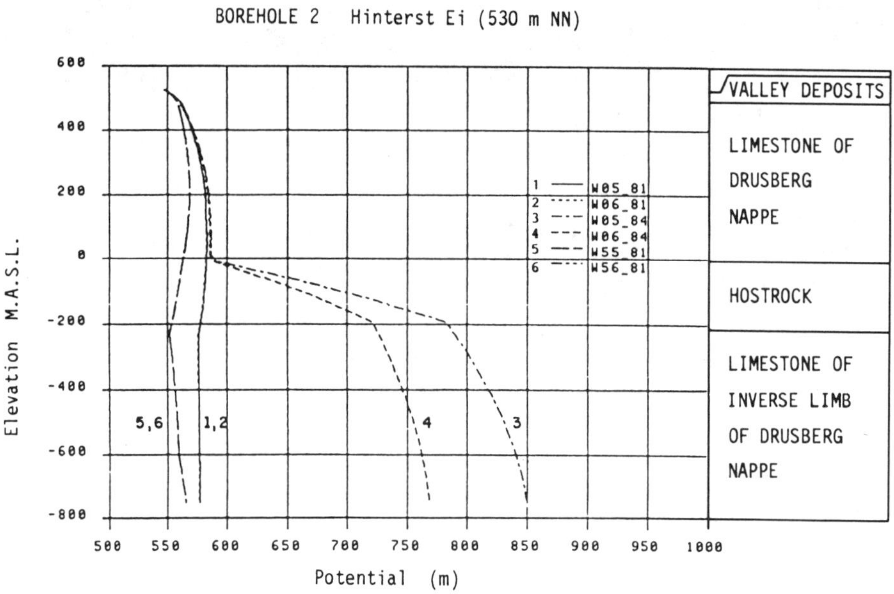

Figure 2.3. **Head profiles along borehole No. 2 for different hydrogeologic conditions.**

Head profiles are shown for six parameter sets judged as having a high probability of occurrence. Parameter sets 1, 2, 5, and 6 assume low permeability Tertiary below the Valanginien Marl; the difference between these sets is the boundary conditions in the Axen Nappe. Parameter sets 3 and 4 demonstrate the effect of various limestone contacts between the Axen and Drusberg Nappes underneath the marl. The latter is predicted to cause a marked increase of hydraulic head in the lower part of the borehole with artesian conditions up to 300 m above wellhead. It is obvious that information such as that presented in Figure 2.3 is highly valuable for planning purposes as well as for continuous reassessment of the hydrogeologic conditions in a later stage during drilling and testing activities.

3 Ground-Water Flow Models Used in the Swiss Program

3.1 FEM301

The computer program FEM301, originally developed at the University of Neuchâtel (Kiraly, 1985), is a finite-element code which solves for steady-state ground-water flow with constant density in an equivalent three-dimensional anisotropic porous medium. The three-dimensional ground-water flow field may contain linear or planar hydraulic discontinuities such as thin aquifers between aquitards or permeable shear zones. Such features are modeled with one- or two-dimensional elements embedded in the three-dimensional element network. Input for the program consists of horizontal and vertical geometry of the relevant hydrogeologic features, hydraulic conductivity tensors, and surface and lateral boundary conditions. Boundary conditions may be defined by either specified hydraulic head or flux. The output of the model consists of nodal point head or flux values. Extensive pre- and postprocessing codes have recently been developed in order to simplify input generation and to further evaluate and graphically display output data. Color graphics proved to be an efficient way to display three-dimensional output.

A postprocessing routine for pathline and travel time calculations within the three-dimensional network has recently been added to the code.

Significant effort has been put into verification of FEM301 against analytical solutions of two- and three-dimensional flow problems (Kiraly, 1985), and detailed comparison has been made against numerical solutions of other ground-water flow codes within the international project HYDROCOIN (OECD, 1988; Kimmeier et al., 1986).

3.2 FREESURF

FREESURF (Nagra, 1985a) is a two-dimensional finite-element code for steady-state ground-water flow within an equivalent anisotropic porous medium. The code allows one to tackle free surface and seepage face problems. In addition, pathlines and stream functions may be calculated. Verification of the code is reported in OECD (1988).

4 Modeling of Ground-Water Flow in Northern Switzerland Crystalline Rock for Project Gewähr 1985

Due to the high topographic relief and the relatively complex tectonic and hydrogeologic situation presented by the folding and the overthrust of the Jura and the sequence of sedimentary aquifers and aquitards, and the expected deep-reaching flow in the crystalline rock, it was necessary to start modeling at a large regional scale. The regional model gave boundary conditions to a smaller local-scale model.

4.1 Regional model

The regional model (23,000 km^2) extends from the Aare Massif in the Central Alps in the south to the Black Forest in the north and from Lake Constance on the east to the Aare River on the west (Kimmeier et al., 1985). The hydrogeologic situation is rather complicated: the Mesozoic sedimentary cover over the crystalline basement contains some important limestone aquifers separated by impermeable formations of clay, anhydrite, and rock salt. The Mesozoic sediments are folded and overthrust in some areas, allowing potential water flow through fault zones and thrust zones connecting different sedimentary aquifers. The area is traversed by a Permocarboniferous trough with a sedimentary filling which is, in general, of fairly low permeability, although some open fractures have been observed. For more details, the reader is referred to Nagra (1985b).

A top view of the finite-element network is presented in Figure 4.1. Vertically the model comprises up to 15 layers, 12 of them representing sedimentary units; the remaining three layers are used to model the crystalline basement underneath the sedimentary cover.

Results from the regional model, Figure 4.1, indicate that the principal recharge to the crystalline basement occurs where it outcrops in the Aare Massif and Black Forest. Significant vertical leakage through the Tertiary Molasse in Central Switzerland can also recharge the crystalline. The principal discharge of the crystalline is to the Rhine River along the border between Switzerland and Germany. The regional flow system in the crystalline basement is controlled by the regional confinement provided by the salt and anhydrite strata of the Middle Muschelkalk aquitard. If the permeability of this confining layer is increased, then more local flow systems are predicted to extend into the crystalline basement and crystalline ground-water may also discharge to the Aare, Reuss, and Limmat Rivers.

In order to validate the regional model, comparisons with observed heads in the crystalline basement as well as hydrochemical and geothermal evidence were used. For details, the reader is referred to Andrews et al. (1985) and McCombie et al. (1987).

4.2 Local model

Although the regional model adequately defines the overall ground-water flow regime in Northern Switzerland, it is less suitable for detailed description of ground-water flow directions in the Nagra investigation area. In order to better approximate flow in the crystalline basement as well as to incorporate a greater degree of structural, stratigraphic, and topographic complexity, a more fine-scale local model has been constructed (Kimmeier et al., 1985). The local model (900 km^2) comprises the Folded Jura and the Table Jura between the Aare River in the south and the Rhine River in the north. To the west it extends to the Ergolz River, while to the east the model boundary follows an arbitrary south to north line east of Baden.

The structural geologic features modeled with more detail in the local model are fault zones and aquifer outcrops. In addition the finer discretization allowed the large variation in surficial topography to be more precisely modeled. A top view of the element network is shown in Figure 4.2. Hydraulic head and general water flow directions in the more permeable upper crystalline layer are presented in the same figure.

As already mentioned, the approach to sensitivity and uncertainty analysis for Project Gewähr 1985 consisted of parameter variations. Literature survey and borehole hydraulic tests were used to compile a data base on hydraulic conductivities of the different aquifers and aquitards. Based on this information, modeled hydraulic conductivities were subjected to variations of approximately one to four orders of magnitude. No lateral variation of hydraulic conductivity within a particular hydrogeologic unit was applied. Figure 4.3 shows some results of the parameter variations with respect to gradients and flow directions at three locations within the model. Gradients and flow directions were

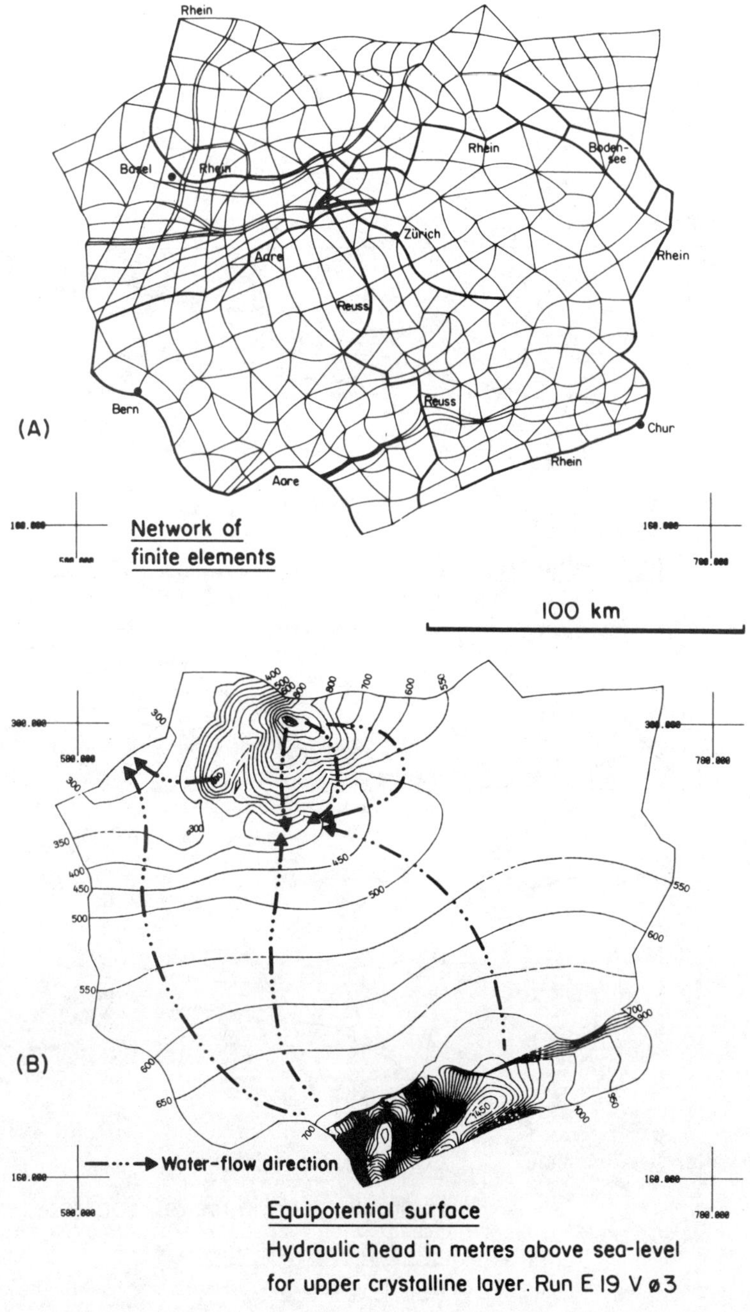

Figure 4.1. **Regional hydrodynamic model: network of finite elements, equipotential surfaces, and ground-water flow direction in the upper crystalline.**

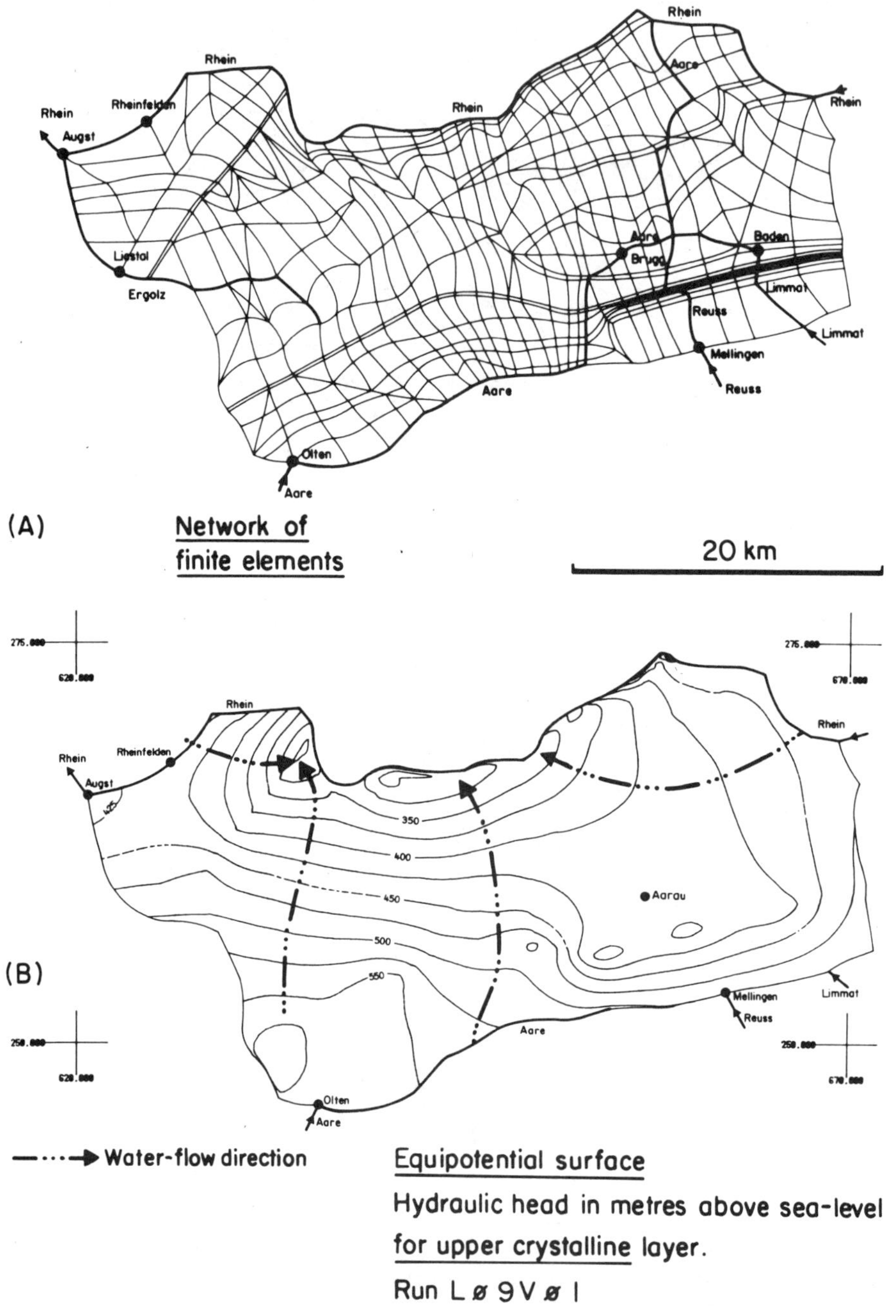

Figure 4.2. **Local hydrodynamic model: network of finite elements, equipotential surfaces, and ground-water flow direction in the upper crystalline.**

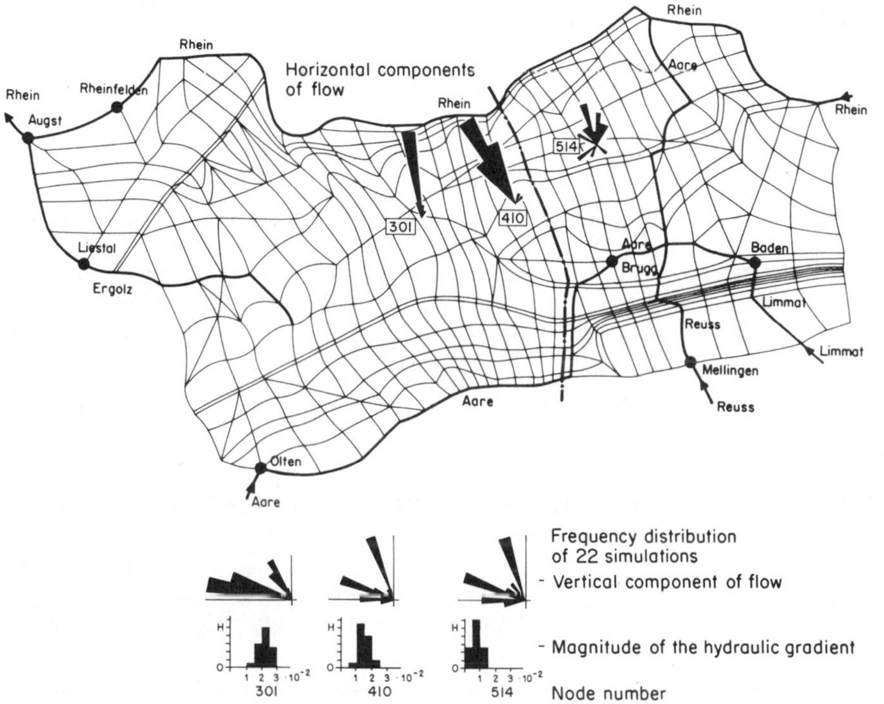

Figure 4.3. **Horizontal and vertical components of ground-water flow vector and hydraulic gradient at three locations in the middle crystalline layer. Results of 22 simulations with the local model.**

subsequently used in a smaller repository-scale model and to define fine-scale flow through water-bearing features.

4.3 Repository-scale model

For safety analysis, the generic question remained to be solved about the possible influence of highly conductive shear zones near the repository. FREESURF was used to model ground-water flow and travel paths between the repository and such conductive features. It was assumed that site exploration at depth allows one to identify such a zone up to a distance of 200 m from the repository. Thus the minimum distance between the repository and the zone was assumed to be 200 m. As part of a sensitivity analysis a series of cases (case A–case D), as shown in Figure 4.4, were investigated. Figure 4.5 shows equipotential lines and flow paths for case D assuming vertical flow upwards through the repository to the shear zone. Compared to the base case assuming upward flow through 500 m of crystalline rock without a high permeability shear zone, for case D, water travel time decreased from 1060 years to 288 years. For details about other cases and consequences to radionuclide transport through the host rock, the reader is referred to Nagra (1986).

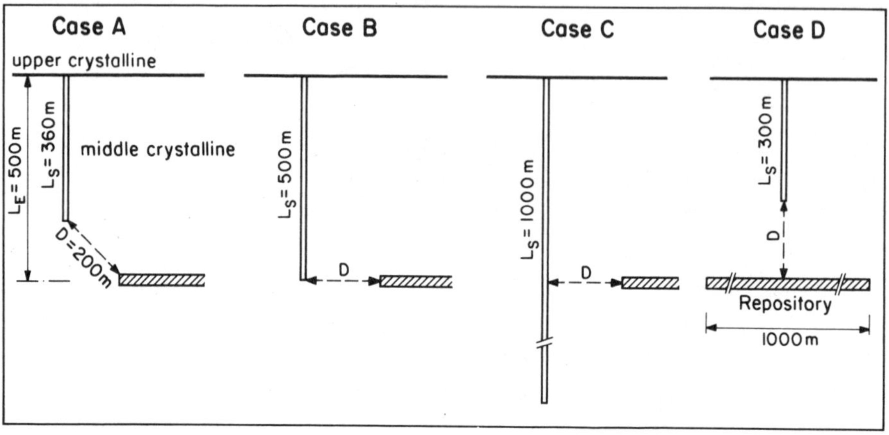

Figure 4.4. **Influence of vertical hydraulic discontinuities near a repository in the middle crystalline layer. Cases A to D.**

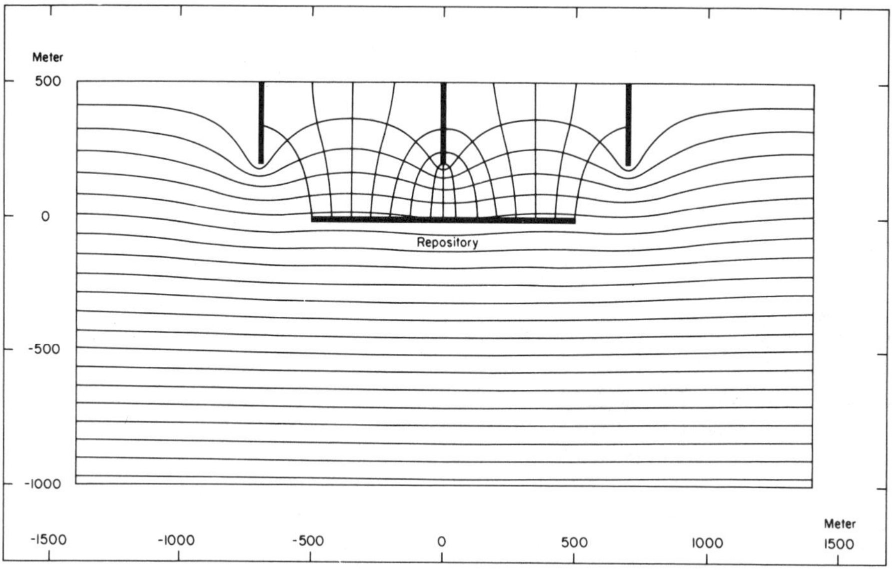

Figure 4.5. **Equipotential lines for hydraulic head and travel paths from a repository in the middle crystalline layer under presence of highly permeable discontinuities. Case D, horizontal spacing of discontinuities 700 m.**

4.4 Flow through water-bearing features

Safety analyses for Project Gewähr 1985 have shown that for radionuclide transport through the geosphere the fine-scale water flow systems are of paramount importance (Nagra, 1985b). The following two flow systems could be identified within permeable zones ($K > 10^{-11}$ m/s) of the Böttstein borehole:

— System I: Kakirite with quartz veins

Vugular to porous water-bearing quartz veins embedded in variably developed impermeable kakirite zones form water-bearing systems with $K = 10^{-7}$ to 10^{-9} m/s, average vertical spacing around 150 to 200 m, and horizontal separation around 60 to 80 m.

— System II: Strongly fissured dykes

Larger (over 0.3 m) aplite and pegmatite dykes with frequent fissuring (non-healed fissures and shear zones) form water-bearing systems with $K = 10^{-7}$ to 10^{-11} m/s, average vertical spacing around 400 m, and horizontal separation around 250 m.

Ambiguity in identification of the appropriate conceptual model for flow and radionuclide transport through the two systems is one of the main sources of uncertainty in the safety analysis for Project Gewähr 1985.

For the identification and hydraulic characterization of the fine-scale water flow systems, core inspection, geophysical logging, and hydraulic testing were used. A rather new method, fluid logging, has proven to be especially valuable to identify the location of water-bearing features. The method is currently being developed into a quantitative tool that should allow the determination of very low inflow rates on the order of liters/day from individual features.

5 Chalk River Test Case

A small ground-water flow system near Chalk River Nuclear Laboratories (CRNL), Canada, has been selected among other test cases by the international interdisciplinary project HYDROCOIN as a test case for level 2 and level 3. Level 2 deals with validation of ground-water flow models using field and laboratory experiments; level 3 is concerned with sensitivity and uncertainty analysis. For more details the reader is referred to HYDROCOIN progress reports (SKI, 1986, 1987a, b).

Nagra has adopted the Chalk River case to test strategies for calibration and validation of ground-water flow codes and further develop methods for sensitivity and uncertainty analysis.

5.1 Site description

The Chalk River test case involves a small ground-water flow system in fractured monzonitic gneiss. An extensive experimental program has been conducted within an area of about 200 × 150 × 50 m. The investigations comprise core logging, fracture orientation statistics, single-hole and cross-hole hydraulic packer tests, monitoring of hydraulic heads in boreholes, and hydraulic conductivity tensor determinations. Figure 5.1 gives a schematic view of the site, indicating locations of 17 boreholes, four major fracture zones, and a low-permeability diabase dyke. Detailed information about the site and the experimental data has been presented by Raven (1985).

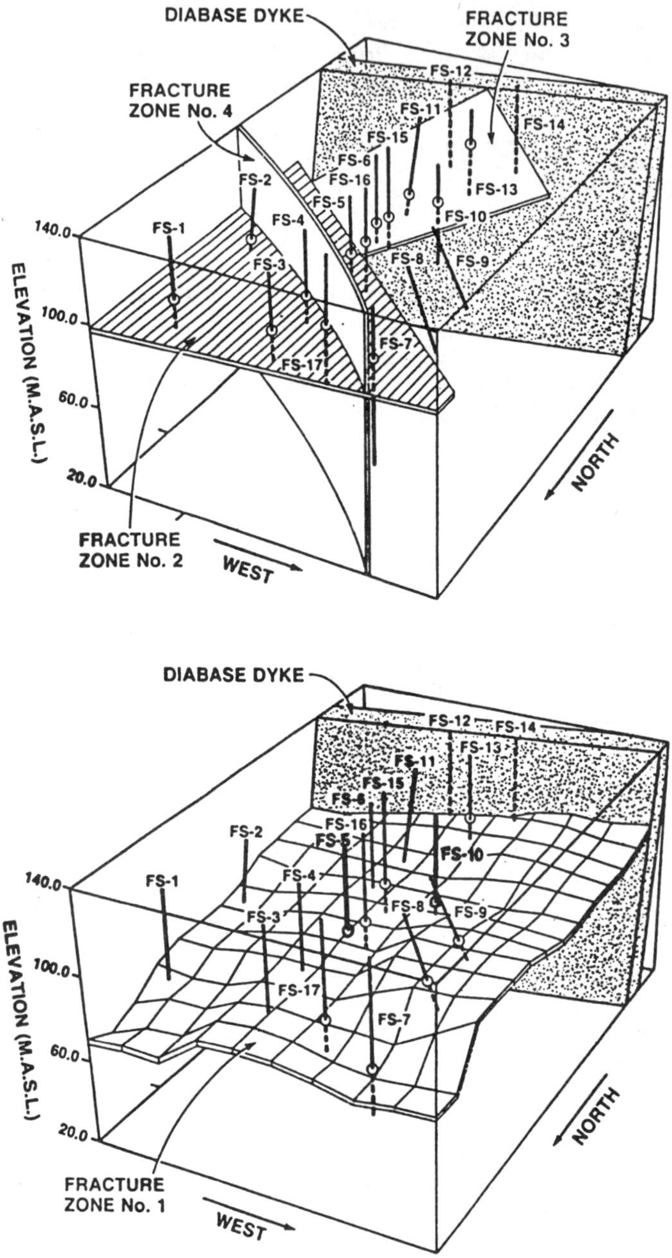

Figure 5.1. **Isometric sketches of CRNL ground-water flow study site showing the location of boreholes, fracture zones, and diabase dyke. From Raven (1985).**

5.2 Strategy for calibration, validation, and sensitivity/ uncertainty analysis for ground-water flow models

The methodology for calibration, validation, and sensitivity/uncertainty analysis discussed below follows an approach based on maximum likelihood estimation theory recently presented by Carrera and Neuman (1986a, b, c). The rather general method may provide a suitable framework for the calibration and validation of ground-water flow models as well as for other models such as the hydraulic test or tracer test interpretation models. An extension to sensitivity and uncertainty analysis has recently been proposed by Wacker et al. (1987). The strategy for calibration, validation, and sensitivity/ uncertainty analysis is outlined in Figure 5.2. Details are presented in Resele et al. (1986) and Job et al. (1987).

5.3 Conceptual model

5.3.1 Hydrodynamic model and discretization. Taking advantage of the mixed dimension capabilities of FEM301, a three-dimensional finite-element mesh, Figure 5.3, with embedded two-dimensional fractures, Figure 5.4, was generated. The rock mass between the ground surface and the bottom of the model is subdivided into seven horizontal layers. Fracture zones 3 and 2 are embedded between the rock matrix layers. The bottom boundary is defined by assuming no flow below the highly permeable fracture zone 1. Lateral no-flow boundaries follow the diabase dyke and assumed vertical flow symmetry lines. Fixed-head boundaries are adopted at ground surface. All calculations were done for steady-state flow conditions.

Hydraulic conductivities within each rock layer and transmissivities in each fracture zone are assumed constant. Fixed-head boundaries are treated as known constant values. The model comprises the following unknown parameters:

1. one hydraulic conductivity tensor for each of the seven rock mass layers. Isotropy in the horizontal plane was assumed, leaving two unknown coefficients (horizontal and vertical hydraulic conductivity) to be determined for each layer.
2. four isotropic transmissivity values, one for each fracture zone.

All free model parameters amount to a total of 18 unknowns to be determined.

5.3.2 Prior information on model parameters. Cross-hole interference tests were conducted in 90 packer-isolated intervals. From this data base, Raven (1985) derived equivalent single-fracture aperture and storativity values for each fracture zone and the vertical hydraulic diffusivity of the rock mass. In addition, based on Snow's model (Snow, 1965), fracture logs and injection test data were combined to generate a tensorial representation of hydraulic conductivity for each interval. Resele et al. (1986) used this data base to derive the following prior estimates for the above-mentioned model parameters:

1. the "best-guess" (prior) mean model parameters (log-hydraulic conductivities and log-transmissivities)
2. confidence intervals for the model parameters (standard deviations of log-conductivity and transmissivity values).

Best-guess prior estimates are given in Table 5.1. In addition, standard deviations for the head values were estimated based on the stochastic relationship between hydraulic conductivity and head variance (Bakr et al., 1978).

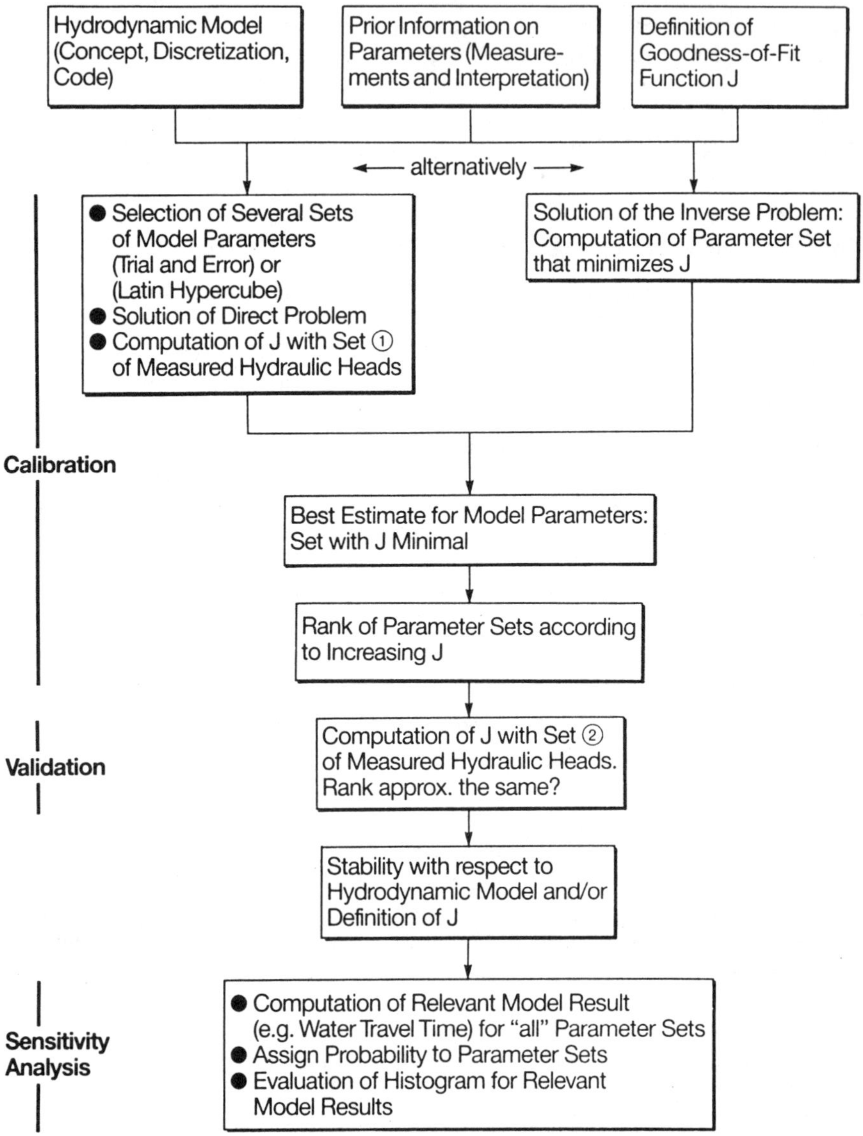

Figure 5.2. **Strategy for calibration, validation, and sensitivity/uncertainty analysis for ground-water flow models. From Job et al. (1987).**

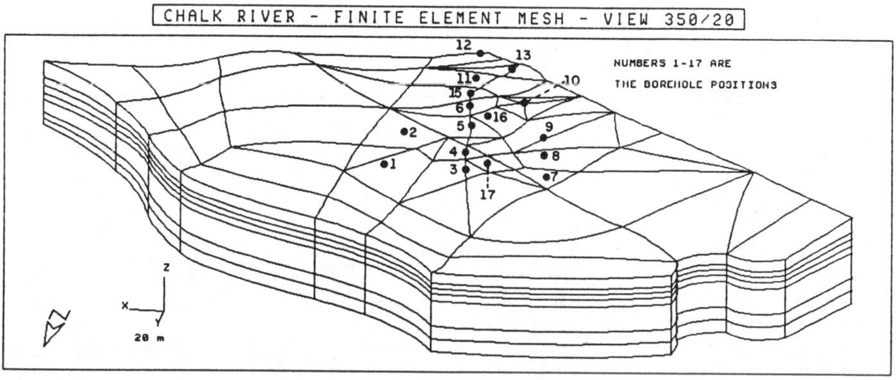

Figure 5.3. **Finite element mesh.**

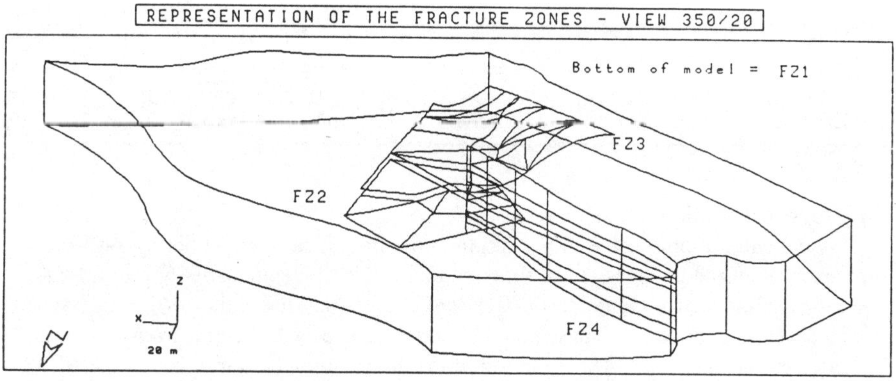

Figure 5.4. **Two-dimensional fracture zones.**

5.3.3 Goodness-of-fit function. This step involves the definition of a goodness-of-fit function J. Also called the objective function, J is a measure of discrepancy between measured and computed quantities of both model parameters (hydraulic conductivities, fluxes) and dependent variables (hydraulic head). Based on Carrera and Neuman (1986a), the following J-function was chosen:

$$J = J_h + J_p = (h^* - h)^T Vh^{-1} (h^* - h) + (p_i^* - p)^T V_i^{-1} (p_i^* - p)$$

where J_h measures the goodness-of-fit of head values and J_p is the goodness-of-fit measure for the parameter values. Further, $h^* - h$ is the difference between observed head h^* and computed head h while $p_i^* - p$ is the difference between prior estimates p_i^* of model parameters and computed model parameters p. V_h and V_i are weighting matrices that depend on the standard deviations of prior estimates of head and model parameters, respectively.

Table 5.1. **Comparison between prior estimates of model parameters and best-fit model parameters for run version 16.**

Parameter		Prior estimate	Best fit
Rock mass hydraulic conductivity (m/s)			
Layers 1, 4–7	horizontal	2.8 E−09	1.0 E−10
Layers 1, 4–7	vertical	4.8 E−10	4.0 E−11
Layers 2, 3	horizontal	2.8 E−09	1.0 E−09
Layers 2, 3	vertical	4.8 E−10	4.0 E−10
Fracture zone transmissivities (m^2/s)			
Zone 1		2.3 E−06	1.0 E−07
Zone 2		7.8 E−08	1.0 E−07
Zone 3		1.3 E−07	1.0 E−08
Zone 4		1.0 E−07	1.0 E−07

5.4 Model calibration

Model calibration involves the solution of the inverse problem either by automatic minimization of the goodness-of-fit function (Carrera and Neuman, 1986b) or alternatively by trial-and-error techniques (Wacker et al., 1986). Trial and error involves selection of several parameter sets based on the modeler's experience and learning curve or, e.g., by random sampling. After solution of the direct problem, J may be computed. With both approaches the parameter set minimizing J is called the best-estimate parameter set. It bears noting that trial and error will lead only to an approximation of the best parameter set and, in a multidimensional parameter space, it might be difficult to manually identify the best solution. However, trial and error has the advantage of being easy to implement because it requires only repeated solution of the direct problem and subsequent computation of J. Trial and error is also best suited for evaluating conceptual model uncertainty. In addition, it provides a good understanding of system behavior.

Figure 5.5 shows the values of the goodness-of-fit function for 23 different sets of parameters subsequently selected by the modelers in the trial-and-error process. J is generally decreasing with increasing run version number, which demonstrates that the modelers were following a learning curve.

Figure 5.6 shows a comparison between computed and observed head values for nine boreholes for the optimal run, i.e., the one that minimizes J. The general trend of decreasing head with depth, which is controlled by the presence of the flat-lying, highly permeable fracture zone 1 with low head, is well reproduced by the model for each borehole. No systematic error seems to exist.

Prior estimates and best-fit solutions for rock mass hydraulic conductivities and fracture zone transmissivities are given in Table 5.1 for run version 16. Comparison

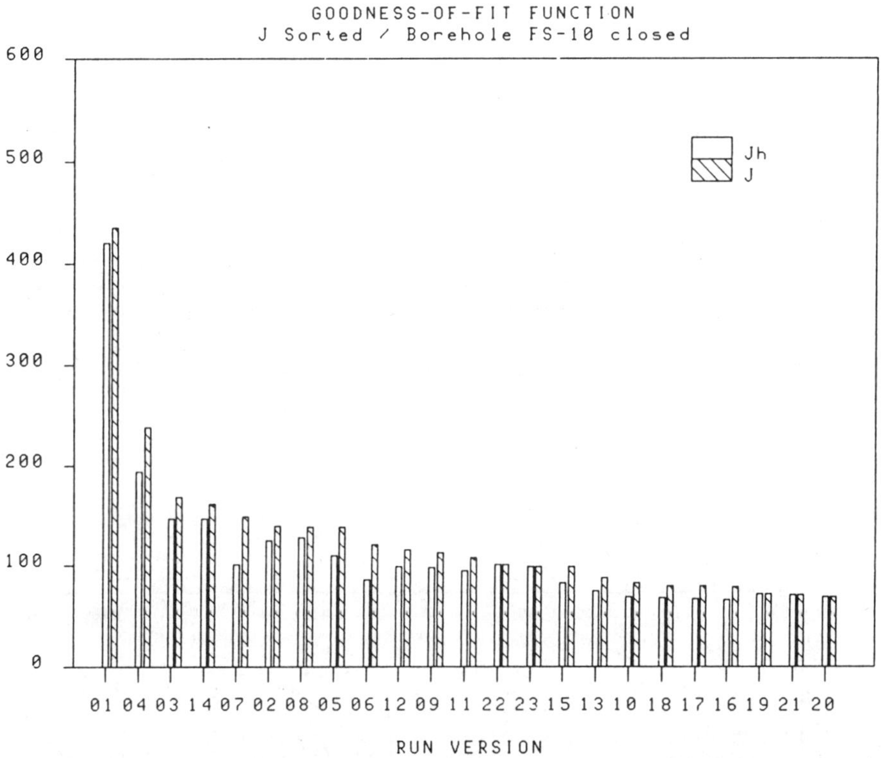

Figure 5.5. **Model calibration by trial and error based on the goodness-of-fit function for measured head values from the campaign of December 15, 1982 (borehole FS-10 closed).**

reveals a difference in the order of one magnitude. This is not surprising, given the relatively low weight that has been assigned to the prior estimates by the modelers. Figure 5.7 demonstrates that the objective function J is dominated by the measure of agreement between observed and predicted head J_h. It should be noticed that the difference in J for the latter run versions is rather small, despite the fact that some significant changes in model parameters were made. This may indicate the difficulty of trial and error to progress beyond a certain goodness-of-fit. On the other hand, it might indicate a "non-uniqueness" situation (weak identifiability of model parameters based on head values only). This would stress the need to incorporate reliable prior information for model parameters. In order to further investigate these questions, work is currently under way to automatically solve the inverse problem.

5.5 Model validation

The aim of validating a site-specific model is to build up confidence that the model can correctly predict the site-specific physical phenomena for which it is intended. This can be done, for example, by demonstrating that the model can predict quantities other

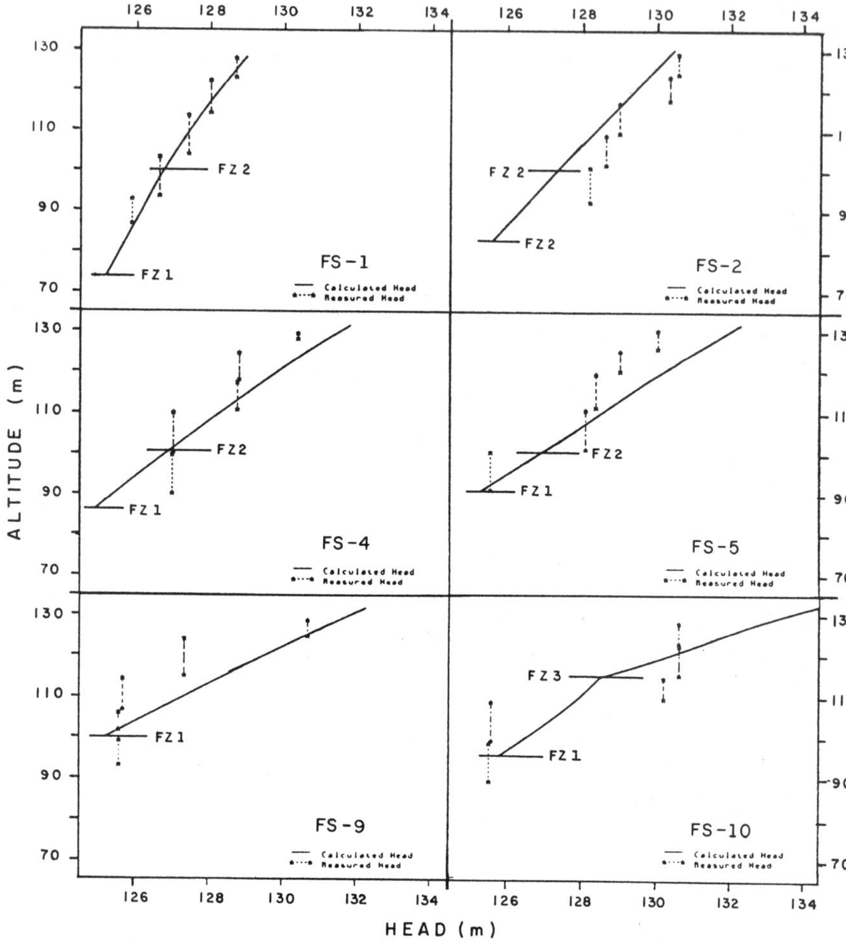

Figure 5.6. **Comparison between measured and calculated head for nine boreholes from optimum run for the flow state of December 15, 1982 (borehole FS-10 closed).**

than those used to calibrate the model. In our case, the model was validated by calculating the goodness-of-fit function for a set of independent measurements representing a different flow state at November 1, 1984. During this flow state, borehole FS-10 was kept open, providing a short-circuit between fracture zones 3 and 1. The short-circuit significantly altered the head distribution in the system (Raven, 1985). The goodness-of-fit for the new flow state is again measured with J. Comparison between Figures 5.5 and 5.7 demonstrates that the ranking of runs is very similar for both flow states. With the data set used for calibration, run No. 20 is the best, while the same run is second best with the validation data set. Further, J of run No. 20 is only slightly higher for the validation set. This might be expected, given the generally higher head gradients observed in fracture zones 1 and 3 with the validation flow state.

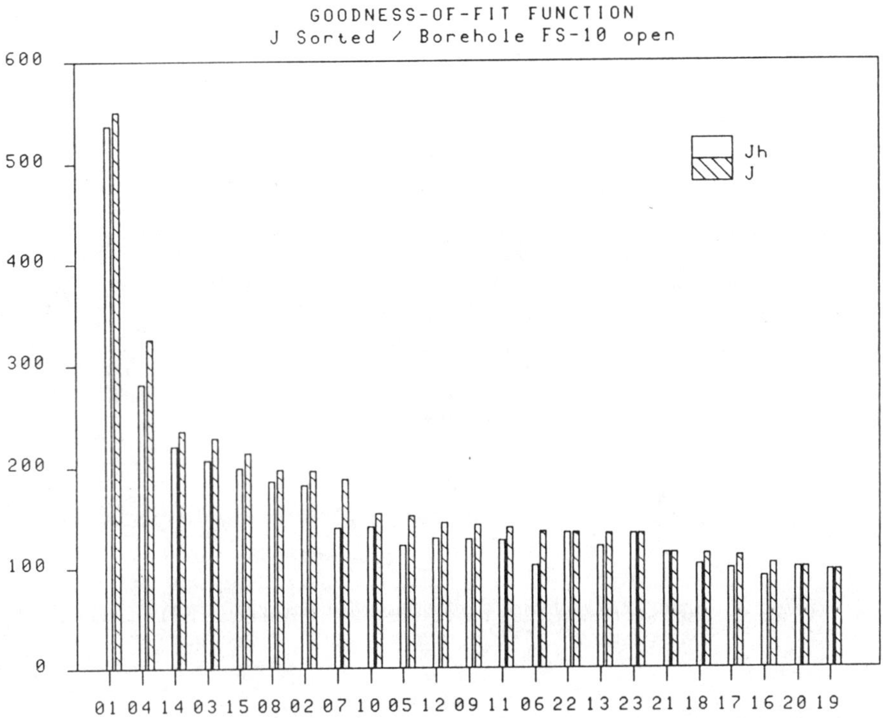

Figure 5.7. **Model validation by trial and error based on the goodness-of-fit function for measured head values from the campaign of November 1, 1984 (borehole FS-10 open).**

5.6 Sensitivity and uncertainty analysis

5.6.1 Definition of a performance criterion. Sensitivity and uncertainty analysis requires definition of an appropriate performance criterion. Here we will discuss water travel time between a starting point (top of borehole FS-11 was selected) and a boundary plane (northern model boundary was selected). Figure 5.8 shows the pathlines for the different runs for the flow state of December 15, 1982.

Travel times for water were calculated as shown in Table 5.2, assuming a flow porosity of 3 E−05 in the rock matrix and a value of 5 E−04 in the fracture zones. For most of the parameter sets, particles first pass through the rock and are then captured by the vertical fracture zone 4 in which they flow downwards until they encounter the horizontal fracture zone 1. In this zone they are transported in a north direction until the model boundary is hit. As can be seen in Figure 5.8 also, some particles (run versions 09 and 11) get trapped in the model. This is most likely caused by insufficient discretization. Discretization problems associated with particle tracking have been identified in the HYDROCOIN project as being a major source of uncertainty (SKI, 1986). Ideally,

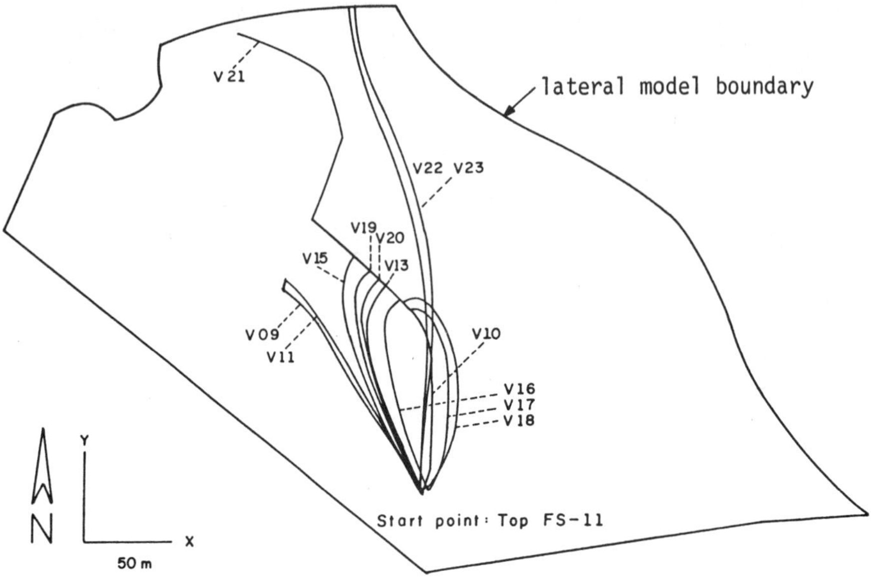

Figure 5.8. **Top view of water flow trajectories starting from top FS-11 location.**

Table 5.2. **Water travel time and probability for runs 1 to 20 for flow state December 15, 1982. Ranking according to goodness-of-fit function J.**

Run version	Goodness-of-fit, m^2	Water travel time, years	Probability of parameter set
20	70.1	0.20	0.58
21	71.8	53.9	0.58
19	72.5	12.	0.58
16	80.0	17.4	0.41
17	80.2	10.5	0.41
18	80.6	10.5	0.41
10	83.7	27.0	0.01
13	88.6	10.8	E−04
15	98.5	41.2	E−09
23	99.4	140.	E−09
22	101.1	146.	E−09
11	108.2	(304.)	E−17
09	112.7	(349.)	E−17

( ) = trapped particles.

stability of the solution should always be checked by comparing tracking results based on a different degree of discretization.

5.6.2 Assignment of probabilities to parameter sets. Assignment of probabilities to each parameter set is one of the most difficult and crucial steps in sensitivity and uncertainty analysis; however it is inevitable if probabilistic performance assessment methods are to be applied. Most of the probabilistic performance assessment methods used to date (Monte Carlo simulations) assign probabilities to parameter sets based on a priori information on the parameter pdf (see, for example, Andrews et al., 1987). Information on dependent variables (e.g., head) is generally neglected in such approaches. Wacker et al. (1987) have proposed to assign probabilities to parameter sets based on posteriori parameter pdf derived from the goodness-of-fit function J, which includes both prior parameter estimates as well as dependent-variable information (head, etc.). A detailed discussion of this approach is beyond the scope of this paper. Relative probabilities of a parameter set are presented in Table 5.2 together with associated travel times. Cumulative probability that water flow time is exceeded is presented in Figure 5.9. What should be noticed is the fact that the three parameter sets with the highest likelihood of occurrence (runs 20, 21, and 19) have a rather broad travel time distribution between 0.20 and 53.9 years. This indicates that in a fractured system water travel time can be very sensitive to small changes in model parameters.

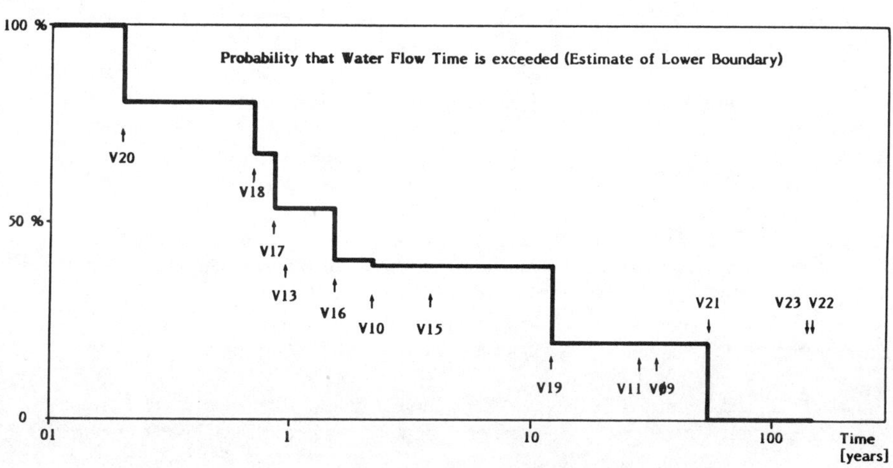

Figure 5.9. **Probability that water flow time is exceeded.**

6 Summary and Conclusions

The development of the Swiss radioactive waste program from feasibility studies towards site characterization, site selection, and repository design calls for more objective methods for calibration and validation of performance assessment models, including quantitative approaches for sensitivity and uncertainty analysis. For the example of ground-water flow models, utilized at different scales, beginning at a large regional scale down to the fine scale of water-bearing discontinuities in fractured crystalline rock, various approaches for modeling, calibration, validation, and sensitivity/uncertainty analysis were briefly discussed. The examples indicate that results of ground-water flow models may be utilized for different purposes such as performance assessment, site characterization, and repository design.

The proposed maximum likelihood method based on prior information about model parameters and dependent variables seems to be a promising tool that can be applied to model calibration, model validation, as well as sensitivity and uncertainty analysis. At Nagra, the method is currently being applied in HYDROCOIN level 2 and 3 and in regional ground-water modeling. In addition applications are being explored in the interpretation of hydraulic tests and tracer tests. Further developments are needed towards a more automatic procedure for model calibration and the quantification of confidence bounds on derived model parameters and model results.

Literature Cited

1. Andrews, R. W., Kimmeier, F., Perrochet, P., and Kiraly, L. (1985): "Validation of Hydrogeologic Models to Describe Groundwater Flow in the Crystalline Basement of Northern Switzerland," *Proceedings of Materials Research Society Symposia 1985*, Pittsburgh, Pennsylvania, Vol. 50, pp. 107–115, MRS.
2. Andrews, R. W., LaVenue, A. M., and McNeish, J. A. (1987): "Evaluating the Effect of Sampling and Spatial Correlation on Ground-Water Travel Time Uncertainty Coupling Geostatistical, Stochastic, and First Order, Second Moment Methods," Paper presented at DOE/AECL '87 Conference on Geostatistical Sensitivity and Uncertainty Methods for Ground-Water Flow and Radionuclide Transport Modeling, September 15–17, 1987, San Francisco, California (this volume).
3. Bakr, A. A., Gelhar, L. W., Gutjahr, A. L., and MacMillan, J.R. (1978): "Stochastic Analysis of Spatial Variability in Subsurface Flows, 1. Comparison of One- and Three-Dimensional Flows," *Water Resources Research*, Vol. 14, pp. 263–171.
4. Carrera, J., and Neuman, S. P. (1986a): "Estimation of Aquifer Parameters under Transient and Steady-State Conditions—1. Maximum Likelihood Method Incorporating Prior Information," *Water Resources Research*, Vol. 22, No. 2, pp. 199–210.
5. Carrera, J., and Neuman, S. P. (1986b): "Estimation of Aquifer Parameters under Transient and Steady-State Conditions—2. Uniqueness, Stability and Solution Algorithms," *Water Resources Research*, Vol. 22, No. 2, pp. 211–227.
6. Carrera, J., and Neuman S. P. (1986c): "Estimation of Aquifer Parameters under Transient and Steady-State Conditions—3. Application of Synthetic and Field Data," *Water Resources Research*, Vol. 22, No. 2, pp. 228–242.
7. Job, D., Resele, G., and Wacker, C. (1987): *Hydrodynamic Modelling of the Chalk River Block, Calibration, Validation and Sensitivity Analysis*, Report prepared for Nagra by Motor Columbus Consulting Engineers, Baden, Switzerland.
8. Kimmeier, F., Perrochet, P., Kiraly, L., and Andrews R. W. (1985): *Simulation par modèle* mathématique des écoulements souterrains entre les Alpes at la Forêt Noire, Nagra Technical Report NTB 84-50, Nagra, Baden, Switzerland.
9. Kimmeier, F., Wacker, C., Perrochet, P., Hufschmied, P., and Kiraly L. (1986): *Code Verification of the Groundwater Flow Model FEM301—HYDROCOIN— Level 1—Case 2 and Case 6*, Nagra Technical Report NTB 86-22, Nagra, Baden, Switzerland.
10. Kiraly, L. (1985): *FEM301—A Three-Dimensional Model for Groundwater Flow Simulation*, Nagra Technical Report NTB 84-49, Nagra, Baden, Switzerland.

11. McCombie, C., Gautschi, A., Thury, M., Hufschmied, P., and Hadermann, J. (1987): "Swiss Programmes for Validation of Performance Assessment Models by Comparison with Experimental Data," Paper presented at GEOVAL-87 Symposium, April 7–9, Stockholm, Switzerland.
12. Nagra (1985a): *Project Gewähr 1985—Endlager für schwach- und mittelaktive Abfälle: Sicherheitsbericht*, Nagra Projekt Bericht, NGB 85-08, Baden, Switzerland.
13. Nagra (1985b): *Project Gewähr 1985—Nuclear Waste Management in Switzerland: Feasibility Studies and Safety Analyses*, Nagra Project Report, NGB 85-09, June 1985, Baden, Switzerland.
14. Nagra (1986): *Ueberlegungen zur Ausdehnung kristalliner Wirtgesteine für die Endlagerung radioaktiver Abfälle in der Nord-schweiz*, Nagra Technischer Bericht NTB 86-26, Baden, Switzerland.
15. Nagra (1987a): *Sondiergesuch Wellenberg*, Nagra Sondiergesuch NSG 18, Nagra, Baden, Switzerland.
16. Nagra (1987b): *Hydrodynamische Modellierung Wellenberg*, Report prepared for Nagra by Motor Columbus Consulting Engineers, Baden, Switzerland (to be published).
17. Raven, K. G. (1985): *Hydraulic Characterization of a Small Groundwater Flow System in Fractured Monzonitic Gneiss*, Report on hydrogeologic research activities, conducted by Contaminant Hydrogeology Section National Hydrology Research Institute, Inland Water Directorate, Environmental Canada for AECL, April 1985.
18. Resele, G., Wacker, C., and Job, D. (1986): *Application of Goodness-of-Fit Measures to the Calibration of Chalk River Block Flow System*, Report prepared for Nagra by Motor Columbus Consulting Engineers, Baden, Switzerland.
19. SKI (1986): *HYDROCOIN Progress Report No. 4, January–August 1986*, Swedish Nuclear Power Inspectorate, Stockholm, Sweden.
20. SKI (1987a): *HYDROCOIN Progress Report No. 5, September 1986–March 1987*, Swedish Nuclear Power Inspectorate, Stockholm, Sweden.
21. SKI (1987b): *HYDROCOIN Progress Report No. 6, April–September 1987*, Swedish Nuclear Power Inspectorate, Stockholm, Sweden.
22. OECD (1988): *The International HYDROCOIN Project, Groundwater Hydrology Modelling Strategies for Performance Assessment of Nuclear Waste Disposal, Level 1: Code Verification*, Report prepared by the international project HYDROCOIN, coordinated by Swedish Nuclear Power Inspectorate, published by Nuclear Energy Agency, Organisation for Economic Cooperation and Development, Paris, France.
23. Snow, D. T. (1965): "A Parallel Plate Model of Fractured Permeable Media," Ph.D. Thesis, University of California, Berkeley.
24. Wacker, C., Job, D., and Resele G. (1987): "The Application of a Goodness-of-Fit Measure to the Calibration, Validation and Sensitivity Analysis of a 3-D Groundwater Flow Model," Paper presented at special poster symposium, Evaluation of Groundwater Flow and Transport Modeling Strategies, AGU Spring Meeting, May 18–22, 1987, Baltimore, Maryland.
25. Zuidema, P. (1986): "Comments on the Application of Probabilistic Assessment Codes to Performance Assessment," Paper presented at NEA/OECD Meeting on Probabilistic Safety Assessment Codes, London, November 11–14, 1986.

The Role of Geostatistical, Sensitivity, and Uncertainty Analysis in Performance Assessment

ALBIN BRANDSTETTER
Battelle Office of Waste Technology Development
Willowbrook, Illinois

BRUCE E. BUXTON
Battelle Columbus Division
Columbus, Ohio

Abstract

The overall U.S. Civilian Radioactive Waste Management Program is briefly described for the benefit of participants not familiar with the program or recent developments. The Nuclear Waste Policy Act of 1982 is summarized. A status report is given on the selection of the Texas, Nevada, and State of Washington sites for site characterization leading to the recommendation of a site for a first repository. The current activities of the second repository project are then described, and the relationship between the first and second repository projects is defined. The regulatory basis for performance assessment is identified, covering descriptions of the numerical standards imbedded in 40 CFR Part 191, 10 CFR Part 60, and 10 CFR Part 960. The requirements for considering uncertainties stated in these regulations and draft technical positions of the U.S. Nuclear Regulatory Commission are summarized. The importance of geostatistical, sensitivity, and uncertainty analyses and their relative merits are pointed out. The sources of uncertainties are identified, and current methods for geostatistical, sensitivity, and uncertainty analyses are reviewed.

Introduction

This conference was organized in support of the national programs for the disposal of high-level nuclear wastes in Canada and the United States of America, managed by Atomic Energy of Canada Ltd. (AECL) and the U.S. Department of Energy (DOE), respectively. The need for this conference was recognized in the cooperative agreement signed by AECL and the DOE in 1986 for waste isolation research and development of common interest to both countries. The principal wastes considered by these programs are spent nuclear fuel from civilian nuclear power reactors and high-level nuclear waste from the reprocessing of spent nuclear fuel, if this option is exercised in the future. Both countries have decided that the preferred option for this disposal will be the isolation of the wastes in geologic formations at depths between approximately 300 and 1,000 m below the land surface. The U.S. decision for geologic disposal, made in 1981, is based on a comprehensive evaluation of disposal alternatives documented in the Final Environmental Impact Statement, Management of Commercially Generated Radioactive Waste (*1*). The decision itself is documented in the *Federal Register* (*2*).

The U.S. plan is to construct mined shafts and tunnels, much like a conventional mine (Figure 1), and to emplace the waste in metallic containers in separate emplacement holes in the tunnel floors or walls (Figure 2) (*3*). The surface facilities are expected to require an area of about 200 hectares, while the underground facilities will encompass an area of about 1,000 hectares (for emplacement of the wastes at a single level). Emplacement of the wastes would occur over a period of 30 to 50 years, after which the surface facilities will be dismantled and the underground openings will be backfilled. For safety assessment and related regulatory purposes, the period up to the end of backfilling is called the preclosure phase, while the period thereafter is called the postclosure phase.

The principal concern for this method of disposal is the potential contamination of ground-water supplies by the radioactive materials and their potential release to the ground surface and to surface waters during the postclosure phase. Consequently, the occurrence and movement of ground water, and the interactions between the

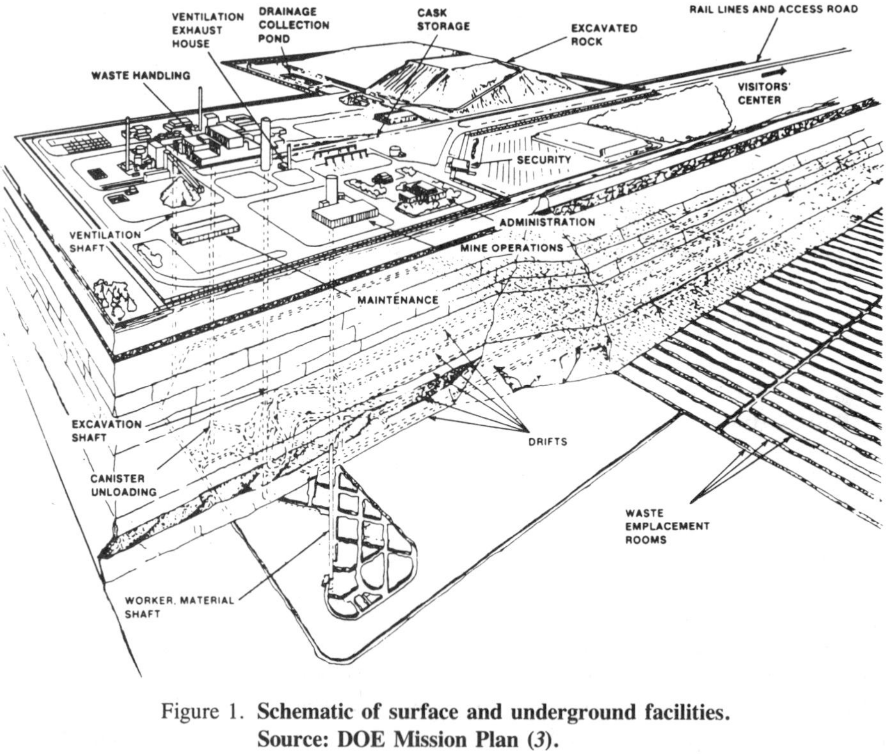

Figure 1. **Schematic of surface and underground facilities. Source: DOE Mission Plan (*3*).**

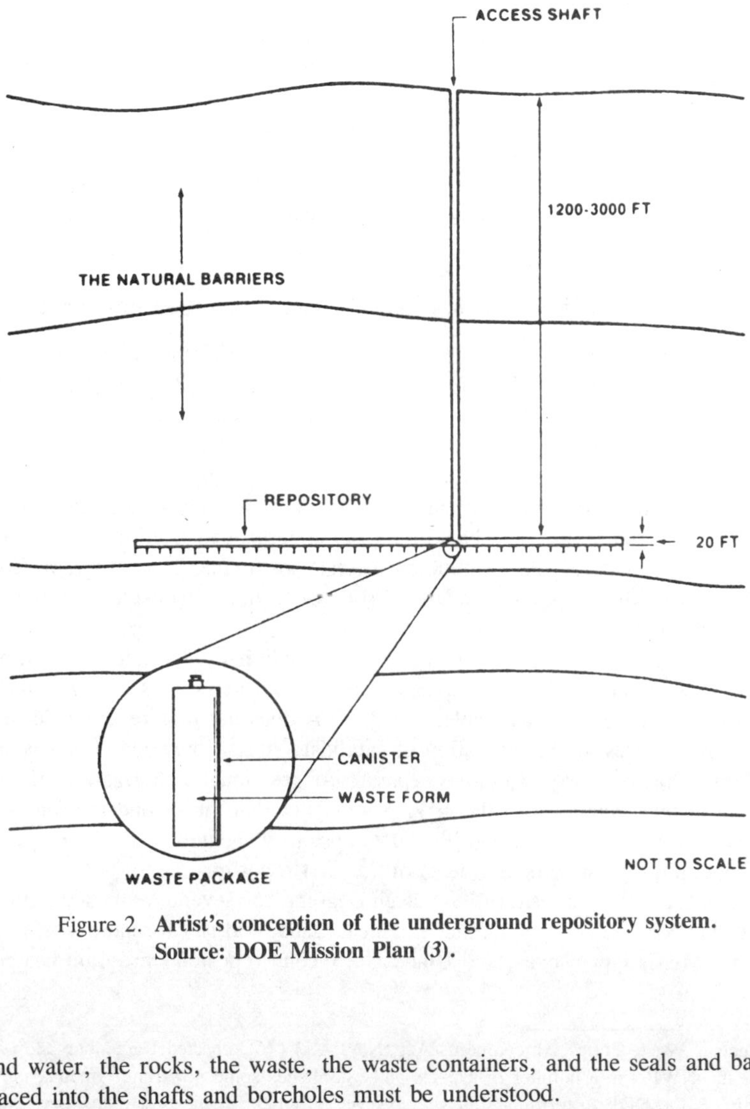

Figure 2. **Artist's conception of the underground repository system. Source: DOE Mission Plan (*3*).**

ground water, the rocks, the waste, the waste containers, and the seals and backfills emplaced into the shafts and boreholes must be understood.

Due to the long half-lives of some of the radionuclides, the safety of a repository must be predicted over thousands of years. Single-value (''best'' or ''conservative'') estimates will not suffice because of the great uncertainties in characterizing heterogeneous geologic formations at great depths, which are compounded by the uncertainties inherent in long-term predictions. It will be imperative, therefore, to estimate the uncertainties in predicted repository performance in order to establish levels of confidence in the predictions.

Due to the high cost of site investigations and engineering studies, it is also very important to determine the relative importance of parameters and processes in order to establish priorities for research and development, site studies, and engineering design. This can be accomplished through sensitivity analyses that determine the effects of variations in one parameter or process on other parameters or processes.

And last but not least, statistical techniques will be needed to assist in defining the uncertainties in the observed data and the computed results. Geostatistical techniques will be needed to estimate the spatial distribution and uncertainties of parameter values derived from point measurements. Thus, geostatistical, sensitivity, and uncertainty analyses are crucial not only to overall performance assessment of a nuclear waste repository, but to overall waste isolation program planning.

Status of U.S. Civilian Radioactive Waste Management Program

The current policy and legal basis for the U.S. program for the disposal of civilian high-level nuclear waste and spent fuel was established by the Nuclear Waste Policy Act of 1982 (NWPA)[1] (*4*). This act places the responsibility of developing and operating a nuclear waste repository with the DOE. This responsibility includes research and development, site selection, engineering design, construction, operation (waste transportation and emplacement), and decommissioning (dismantling of the surface facilities and sealing and backfilling the underground facilities). The NWPA also establishes schedules and the funding mechanism for the U.S. program. The program is currently managed by DOE's Office of Civilian Radioactive Waste Management (OCRWM).

The NWPA authorizes investigations for two regionally distributed repositories. It limits the inventory at the first repository to 70,000 metric tons of heavy metal until a second repository is available. The NWPA does not require a second repository; it only requires the identification of candidate sites. Construction and operation of a second repository would require separate congressional authorization. If authorized, the schedule contained in the NWPA stipulates that the second repository should be operated in parallel with the first repository, beginning five years after the start of operation (waste emplacement) of the first repository.

Studies for a first repository have been ongoing for several years, concentrating on basalt in the State of Washington; tuff in Nevada; and salt formations in Louisiana, Mississippi, Texas, and Utah. The second repository program has concen-

[1]The Nuclear Waste Policy Amendments Act (NWPAA) (*26*), enacted December 22, 1987, amends the NWPA in a number of aspects and establishes some new requirements, namely: (1) it directs the DOE to terminate all site-specific activities for the first repository, other than at the Yucca Mountain site in Nevada; (2) it prohibits site-specific activities for a second repository and defers consideration of the need for a second repository until 2007; (3) it requires the evaluation of the need for monitored retrievable storage; (4) it directs the DOE to negotiate with States and Indian tribes willing to host a repository or monitored retrievable storage facility; (5) it requires the evaluation of subseabed disposal; (6) it requires the evaluation of dry cask storage at civilian nuclear reactors until a permanent repository becomes available; (7) it establishes requirements for the transportation of nuclear wastes; (8) it includes provisions for technical program reviews; and (9) it includes provisions with respect to benefits to repository host States and Indian tribes. The NWPAA does not change the schedule established by the NWPA for the first repository.

trated on national surveys of crystalline rocks (*5*), leading to the recommendation for preliminary field investigations at twelve areas in seven eastern States in early 1986 (*6*). In May of 1986, the DOE recommended to the President and the President concurred with the selection for extensive site characterization of three sites as candidates for a first repository (Figure 3), namely the Deaf Smith County bedded salt site in Texas, the Hanford basalt site in Washington, and the Yucca Mountain tuff site in Nevada (*7*). Figure 3 also shows the regions considered for a second repository in crystalline rock formations.

At the same time, the DOE deferred all siting-related activities for the second repository program (*8*). This decision was based on revised nuclear waste generation projections, which indicated that a second repository may not be needed as soon as indicated by the NWPA. Since May 1986, research and development of technology for a potential second repository has continued, but on a non-site-specific basis. This includes the evaluation of other than crystalline rock types.

In June 1987, the DOE proposed to the Congress revised schedules for both the first and second repository programs. If approved by the Congress, these schedules will result in delays of the schedules mandated by the NWPA (*9*). The proposed delays reflect revised time estimates for the needed site characterization, including exploratory shaft construction and underground testing, and for public involvement in the siting and licensing process, including consultation and concurrence with the affected States and Indian Tribes, as well as recent reductions in the funds appropriated by the Congress.

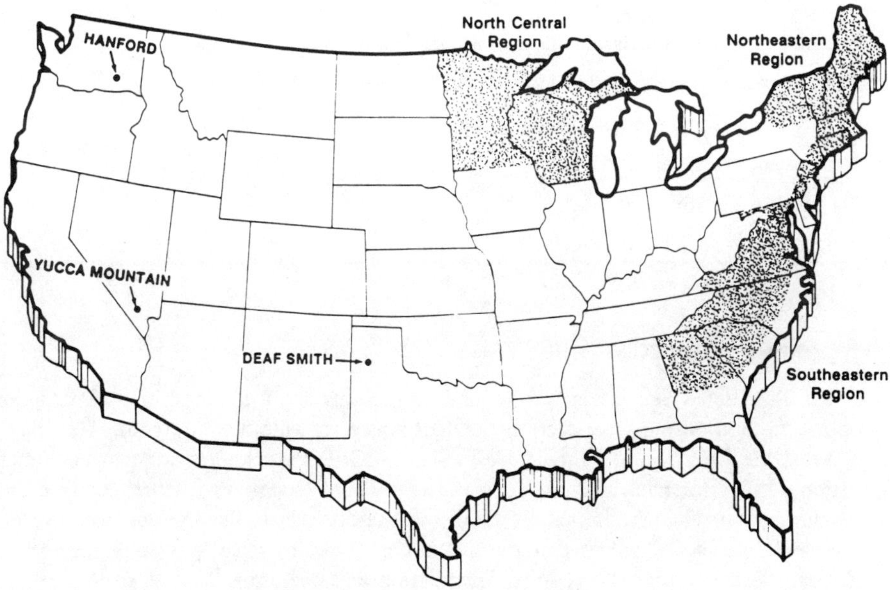

Figure 3. **Location of the three candidates sites for the first repository and of the crystalline regions considered for the second repository.**
Source: DOE Mission Plan and Amendment (*3,7*).

The key milestones proposed by the DOE for the first repository program are the submittal of a license application to the U.S. Nuclear Regulatory Commission (NRC) in 1995, the start of construction in 1998, and the start of waste emplacement in 2003 (Table 1). The key milestones proposed by the DOE for the second repository program are the start of a national survey of potential host rocks in 1995, the recommendation of sites for site characterization in 2007, the submittal of a license application to the NRC in 2015, and the start of waste emplacement in 2023 (Table 2). Actual schedules have not yet been established; they depend on congressional action and perhaps on court decisions arising from litigation over the DOE's siting decisions.

Table 1. **Proposed schedule for first repository (*7*).**

Start of exploratory shaft construction	
tuff	4th quarter 1988
basalt	2nd quarter 1989
salt	4th quarter 1989
Start of in situ testing	
tuff	2nd quarter 1990
basalt	4th quarter 1991
salt	4th quarter 1991
DOE issues draft environmental impact statement	1993
DOE issues final environmental impact statement	1994
DOE recommends site to the President	1994
DOE submits license application to the NRC	1995
NRC issues construction authorization	1998
Start of construction	1998
Start of operation (waste emplacement)	2003

Regulatory Aspects

In the United States, several Federal regulations apply to the siting, construction, operation, and postclosure safety of nuclear waste repositories. As required by the NWPA, the DOE issued siting guidelines in 10 CFR Part 960 (*8*) for the recommendation of sites for nuclear waste repositories.[2] These guidelines list the technical and institutional considerations that the DOE must address in evaluating and recommending candidate sites. Nuclear facilities, including civilian nuclear power plants and nuclear waste repositories, require construction and operating licenses from the

[2]CFR = Code of Federal Regulations.

Table 2. **Proposed schedule for second repository *(7)*.**

Begin national survey	1995
Complete national survey	1997
DOE issues draft area recommendation report	2000
DOE issues final area recommendation report	2002
DOE identifies potentially acceptable sites	2002
DOE recommends sites for characterization	2007
President recommends site to the Congress	2015
DOE submits license application to the NRC	2015
NRC issues construction authorization	2018
Start of construction	2018
Start of operation (waste emplacement)	2023

NRC. The NRC has issued regulations for the licensing of nuclear waste repositories in 10 CFR Part 60 (*9*). Responsibility for establishing standards for the radiological health and safety of the public rests with the U.S. Environmental Protection Agency (EPA), which has issued radiological standards for the protection of the public health and safety for both the preclosure and postclosure phases of nuclear waste repositories in 40 CFR Part 191 (*10*). These EPA regulations are incorporated in the NRC regulations by reference. In addition, 10 CFR Part 20 (*11*,*12*), issued by the NRC for nuclear facilities in general, applies to radiological protection of the work force during the preclosure phase. The DOE siting guidelines are compatible with the regulatory requirements of the NRC and the EPA.

The key stipulations contained in these regulations relevant to the subject of this conference pertain to the following numerical standards for postclosure performance:

1. Limits on cumulative releases of radionuclides to the accessible environment resulting from significant processes and events over the first 10,000 years following repository closure (40 CFR 191.13 and Appendix A, Table 1) (*10*). Different levels of compliance are required for different probabilities of occurrence of the disruptive processes and events (40 CFR 191.13 and Appendix B) (*10*).
2. Limits on radionuclide concentrations for significant sources of ground water for the first 1000 years following repository closure, and limits on radiation dose equivalents to individuals from all potential pathways, including from drinking water (40 CFR 191.15 and 191.16) (*10*).
3. Disqualification of a site if the pre-waste-emplacement ground-water travel time along the fastest path of likely radionuclide travel from the disturbed zone to the accessible environment is less than 1000 years, unless a different travel time is approved by the NRC (10 CFR 60.113) (*9*). It is a favorable

condition if this ground-water travel time exceeds 10,000 years (10 CFR 960.4-2-1) (*8*).

4. Substantially complete containment of the waste within the waste packages for the first 300 to 1000 years following repository closure (10 CFR 60.113) (*9*).
5. Following waste container failure, limits on annual radionuclide release rates from the repository into the surrounding geologic formations (10 CFR 60.113) (*9*).
6. Reasonable assurance that the repository will be safe for at least 100,000 years following repository closure with respect to radionuclide releases to the accessible environment (10 CFR 960.3-1-5), and considering geohydrologic (960.4-2-1), geochemical (960.4-2-2), and climatic changes (960.4-2-4) (*8*).
7. Reasonable assurance that waste exhumation would not occur by erosion during the first one million years following repository closure (10 CFR 960.4-2-5) (*8*).

All of these regulations recognize the great uncertainties in predicting repository performance over thousands of years and, consequently, require only reasonable assurance that the requirements can be met (*9*). The EPA regulations (*10*) assume that the DOE will, whenever practicable, assemble the results of all performance assessments that determine compliance with the EPA release limits into a complementary cumulative probability distribution that indicates the probability of exceeding various levels of cumulative release. This distribution function should encompass all sources of uncertainties from all potentially significant disruptive processes and events, considering also the probabilities of occurrence of these processes and events. The EPA will assume that the disposal system will be in compliance with its release limits if this can be demonstrated by a single distribution function.

The above indicates that the need for uncertainty analyses is not only recognized, but required by the regulations governing the siting and licensing of nuclear waste repositories. And, as previously indicated, sensitivity analyses are required for overall program planning in order to set priorities for research and development, site studies, engineering design, and as a part of the uncertainty analyses. Statistical, including geostatistical, analyses are then necessary as part of the performance assessments of nuclear waste repositories to interpret and interpolate input data and computed results.

Definitions

An *uncertainty analysis* estimates the uncertainty in a system's performance resulting from the uncertainty of one or more factors associated with the system (*13*).

A *sensitivity analysis* determines the effects of variation(s) in one or more parameters on the performance of a system or some part of it.

Both uncertainty and sensitivity analyses require definition of the system, the ranges of parameters over which the system is to be investigated, and a description of the characteristics of the system that is to be evaluated (*13*).

Geostatistical techniques are a set of statistical methods designed for geologic or, more generally, earth science applications (*14,15,16*). These techniques first quantify the intrinsic correlation structure of the physical or chemical process being studied, and then use this structure to tailor more precise and risk-qualified estimates of performance. Geostatistical methods can be applied to both uncertainty and sensitivity studies.

The following three definitions have been offered for the term *performance assessment* as applied to nuclear waste repositories:

1. *DOE (in 10 CFR 960.2).* Any analysis that predicts the behavior of a system or system component under a given set of constant and/or transient conditions. Performance assessments will include estimates of the effects of uncertainties in data and modeling (*8*).
2. *NRC (in NUREG-0960, volume 1, page 9-1).* The process of quantitatively evaluating component and system behavior, relative to containment and isolation of radioactive wastes, to support development of a high-level waste repository and to determine compliance with the numerical criteria associated with the regulation 10 CFR Part 60 (*13*).
3. *EPA (in 40 CFR 191.12).* An analysis that (1) identifies the processes and events that might affect the disposal system; (2) examines the effects of these processes and events on the performance of the disposal system; and (3) estimates the cumulative releases of radionuclides, considering the associated uncertainties, caused by all significant processes and events. These estimates shall be incorporated into an overall probability distribution of cumulative release to the extent practicable (*10*).

Sources of Uncertainty

There are four major sources of uncertainty in performance assessments:

1. *Scenario specification.* All significant processes and events that could lead to the release of radionuclides to the accessible environment must be identified, along with their probability of occurrence.
2. *Model definition.* For each radionuclide release scenario, a specific set of conceptual site and engineering design models and mathematical performance models must be defined. These models define the system to be modeled and stipulate the particular physical and chemical processes that are expected to affect postclosure performance.
3. *Parameter specification.* Each conceptual and mathematical model is expressed in terms of various operational parameters. Performance assessment analyses require the assignment of values to these parameters based on design-specific and site-specific data.
4. *Model calculations.* Detailed performance assessment models are evaluated with complex computer codes. These codes are subject to numerical inaccuracies from various sources such as miscoding and round-off errors.

In a repository license application, the Federal regulations require the applicant to take into account performance assessment uncertainties in the demonstration of compliance (*8,9,10*). And the positions of the NRC both on a licensing and on a performance assessment strategy strongly endorse the evaluation of uncertainty (*17,18*). However, the decisions are left to the individual applicant concerning the specific analyses that will be performed to support the license application.

It is not necessary to establish complete assurance that the regulatory requirements, including numerical standards, will be met. The NRC recognizes this and states in 10 CFR 60.101 the following (*9*):

> "While these performance objectives and criteria are generally stated in unqualified terms, it is not expected that complete assurance that they will be met can be presented. A reasonable assurance, on the basis of the record before the Commission, that the objectives and criteria will be met is the general standard that is required."

This indicates, indirectly, the importance of uncertainty analyses for estimating the level of assurance that can be associated with the performance predictions.

In the following sections, some performance assessment analysis needs and existing methods are outlined in terms of the four sources of uncertainty. Each section will address applicable uncertainty, sensitivity, and geostatistical methods.

Scenario Specification

The first step in a performance assessment is to define the scenarios, that is, all significant processes and events that could result in the release of radionuclides to the accessible environment. The uncertainty introduced to performance assessment estimates from the incorrect specification of scenarios is not easy to quantify with formal statistical methods. For the most part, this source of error must be controlled by peer review and expert judgment.

Analysis needs

The EPA regulations (*10*) assume that it will be possible to formulate a single complementary cumulative probability distribution function for each disposal system that will indicate the probability of exceeding various levels of cumulative radionuclide release to the accessible environment, and that will incorporate all significant release scenarios and sources of uncertainty. In practice, it will be difficult to construct such a probability distribution for at least three reasons:

1. All significant radionuclide release scenarios must be completely and correctly specified.
2. The probability of occurrence for each scenario must be accurately estimated.
3. The correlation, or redundancy, between the various scenarios must be quantified.

Over the thousands of years for which the repository is designed to contain the nuclear waste, many events and processes will occur. If the performance assessment incorrectly specifies these events and processes, then the performance estimates can be quite different from the true performance that will be realized by the repository. An uncertainty analysis attempts to quantify the likely magnitude of these differences.

Because not all possible scenarios can be anticipated, the uncertainty analysis should attempt to bound the set of possible scenarios with a wide range of scenarios. This range should be carefully selected and sufficiently wide to ensure that if the estimated repository performance shows compliance under the selected range of scenarios, then the actual repository performance will also be in compliance under the particular scenario that will develop over time. To help identify the conditions that define the bounds of repository performance, a detailed sensitivity analysis should be conducted to find the particular events and processes that most significantly affect performance. This analysis will help in selecting the range of scenarios that should be considered.

To calculate the expected repository performance aggregated over all possible scenarios, the performance in each scenario must be combined with the correct probability of occurrence for each scenario. For scenarios that share similar events and processes, the probabilities of occurrence may be highly correlated. Since the probabilities and correlations cannot be exactly known for events so uncertain and so far into the future, the assumed probabilities and correlations must be systematically varied in a sensitivity and uncertainty analysis to judge the impact of the probability and correlation assumptions on the aggregated performance estimates.

Existing methods

There are probably few formal sensitivity and uncertainty tools that can be used to assess the errors in performance estimates due to the incorrect specification of scenarios. Incorrect specification of scenarios, however, could potentially be the most significant source of error, particularly in the early stages of a repository project when few data are available. The following list gives some methods that can be used:

1. Event-tree analysis can help define the set of anticipated and unanticipated processes that should be considered in the scenario analysis.
2. Peer review by a panel of experts can be used to judge the completeness and accuracy of the scenarios, probabilities, and correlations considered.
3. Monte Carlo sampling methods can be used to define a range of occurrence probabilities and correlations for evaluating scenarios.

Model Definition

For each scenario considered, the release and movement of radionuclides in the environment must be evaluated in the context of a specific conceptual model for the site and for the engineering design. The effects of various physical and chemical processes within the conceptual model are in turn calculated with specific mathematical models. All of these conceptual and mathematical models simplify reality with a series of explicit and implicit assumptions about the nature of the physical and chemical systems and processes. These simplifications are necessary to make complex natural systems and processes tractable for mathematical analyses. However, the simplifications introduce errors into the performance estimates. This is a problem that is common to all components of the repository system, from the engineered barriers to the long-term geologic processes. In the early stages of a repository project, model definition errors can be the most significant source of uncertainty. Even for licensing, these errors can be substantial because long-term processes are being modeled.

Analysis needs

Uncertainty due to the conceptual models is introduced to the performance estimates whenever assumptions are made to simplify the actual systems or processes. An uncertainty analysis needs to evaluate these assumptions and simplifications as to their completeness and validity. The models must take into account all of the significant processes affecting performance, and they must correctly reflect the effects of the processes.

A sensitivity analysis for the conceptual models should determine which conceptual and mathematical components most significantly affect the performance measures. This determination should consider not only which models most affect the performance calculations, but also which models are most uncertain. The sensitivity information then helps allocate funds for the validation of various models. A particular model should be thoroughly investigated if it dramatically affects the performance calculations and it has not been adequately validated. For example, if a homogeneous, porous medium model is replaced by a discrete fracture model with highly conductive fractures, there might be dramatic changes in the resulting estimates of ground-water flow. Also, most fracture flow models are newer and less tested than most porous flow models.

Geostatistical analyses can be performed to help support some underlying model assumptions. These analyses examine the available data from various perspectives to see whether the assumptions seem to be supported by the data. They can help identify features such as major geologic structures for the conceptual models and correlation structures for the mathematical models.

Model validation procedures are designed to investigate the extent of the errors in the performance estimates that result from the approximations that the computer codes must employ to represent complex systems and processes. For example, the flow of ground water might be calculated with a finite-element computer code that replaces heterogeneous rock properties, such as hydraulic conductivity, which can vary significantly within a distance of a few meters, by uniform hydraulic conductivities within elements that might represent thousands of cubic meters. Approximations are unavoidable when dealing with natural systems; it is important, however, to quantify the extent of the approximations with validation studies to help judge the accuracy of the performance estimates.

Validation studies cannot completely control errors introduced by modeling simplifications and approximations; in most cases, however, they can help reduce those errors. In a validation study, the estimates that are produced by a performance model are compared with data from field or laboratory sampling programs that measure the same processes in the modeled systems. The degree to which a computer code can be validated depends upon the type of performance process it models. In some cases, it is impossible to obtain data on the naturally occurring process. For example, changes in a ground-water flow system at a potential repository site caused by future long-term climatic changes and tectonic events cannot be measured during the short time of repository development. In other cases, it may be possible to obtain validation data that approximate the natural process being modeled. For example, accelerated tests of waste canister degradation, performed in the few years of repository development, may provide data that approximate the degradation of a canister in a natural system over hundreds of years. For still other processes, reliable validation data can be obtained during the time of repository development. For example, predicted ground-water flow rates in the near-field environment for the first few postclosure years can be validated against measurements of the ground-water flow during the site characterization, construction, and operation of the repository. Note that because so many postclosure performance processes extend for hundreds or thousands of years into the future, it is in many cases difficult to obtain the appropriate data needed for validation studies.

Performance assessment models for uncertainty analysis also need to be validated to the extent possible. Uncertainty is generally expressed by generating a probability distribution for each performance measure. This distribution gives various ranges for the possible values of the performance measure, along with the probability that the measure will fall into each range. For example, for a particular site an uncertainty analysis for ground-water travel time might indicate that the true travel time has a 70 percent probability of being within an interval from 9,000 to 11,000 years, and a 30 percent probability of being outside of that interval. The probability distribution, that is, the intervals along with their corresponding probabilities, is calculated from a mathematical model, just as are the other performance measures. The probability model must be validated to support the findings of the uncertainty analysis.

Existing methods

At least three types of analyses can be conducted to assess performance uncertainty due to the definition of conceptual and mathematical models:

1. Write down a complete listing of all of the underlying assumptions, both explicit and implicit, for all of the models used. These lists should be well documented and they should be subjected to a critical peer review by a qualified panel of experts.
2. Perform validation studies to confirm the assumptions whenever possible. Some assumptions, for example about major geologic structures, can be validated by field investigations. Other assumptions, particularly about long-term processes, cannot be validated by field or laboratory studies.
3. Develop models that define the range, instead of just single values, for each performance measure. The difference between the computed and measured performance should bound the error due to model definition.

Several geostatistical techniques can be used to evaluate the validity of models. Some of these techniques are used to support conceptual models, while others are used to support mathematical models. For the most part, geostatistical methods are used for far-field studies rather than for the very near-field and engineered systems. Usually the geostatistical analyses are conducted in an iterative fashion with the geologists, hydrologists, and engineers. The modelers first propose a model. The characteristics of the model are then examined against trends and structures seen in the data. The data structures in turn suggest modifications to the conceptual models.

The following geostatistical methods can be applied to performance assessment studies:

1. Variogram analyses can quantify the spatial autocorrelation structure of individual parameters, such as hydraulic conductivity and potentiometric head.

The autocorrelation structure corresponds closely to the geologic structure of the site. For example, anisotropies seen in a hydraulic-conductivity variogram should correspond to horizontal versus vertical hydraulic conductivity differences at the site.

2. Similarly, cross-variogram analyses can quantify the spatial correlation structure between various parameters. These analyses might help support or refute independence assumptions used in a model. For example, a fracture network simulation model might assume that fracture densities and fracture apertures are spatially independent.
3. Cross-validation studies can help support stationarity assumptions. A repository site might be subdivided into locally homogeneous zones for performance assessment analyses. Cross-validation methods can help define the subdivisions by looking for homogeneities in the available data.
4. Statistical distribution checks can support various probabilistic models needed for performance assessment studies. These models might be used to demonstrate compliance with the environmental regulations, or to simulate various processes. For example, hydraulic conductivities might be simulated using a lognormal distribution assumption.

Sensitivity studies for model definitions are mostly nonstatistical in nature. These studies amount to changing the underlying assumptions, and hence the models, and evaluating how the performance estimates are affected. For example, a simulation of rock joint patterns in a fractured rock might be accomplished with a random Poisson process, or with a spatially autocorrelated stochastic process.

The validation of mathematical models for performance uncertainty analyses is somewhat different from the validation of other performance models, such as groundwater flow models. The validation of probability models must confirm the probability distribution calculations for each performance measure. The calculated probabilities predict the behavior of the performance measures that would result if the governing parameters, such as hydraulic conductivity and head, are allowed to vary within a range of plausible values. For instance, in the ground-water travel time example cited above, if the travel time were to be recalculated 100 times with equally probable parameter values, the uncertainty analysis would predict that 70 of the travel times will fall between 9,000 and 11,000 years and 30 of the travel times will fall outside that interval. The range of possible values for the governing parameters is determined from the uncertainty seen in the measured data.

To validate the probability models, resampling studies must be conducted that repeatedly and randomly select parameter values from the range of plausible values, and then the resulting performance must be calculated with each new set of parameters. At least two types of resampling studies can be conducted to help validate probability models:

1. Conditional simulation can help validate the probability distribution calculations made for performance measure uncertainty. Conditional simulation generates a dense set of thousands of parameter values that exhaustively describe the rock properties of a potential repository site. These values define a simulated rock body and are constructed with the same statistical attributes, such as correlation structure, as seen in the available measured data. Once the

simulated rock body has been created, repeated field sampling programs can be performed by extracting data from the set of simulated parameters. The performance models can be evaluated with each repeated set of data, and the results can be compared with the predictions that were made with the probability models. The probability models are validated when the results from the resampling study match the predicted results from the probability calculations.

2. Cross-validation can help validate the uncertainty calculations for the input parameters. Parameter uncertainty calculations are an input to the performance uncertainty calculations, and they help allocate resources for additional field and laboratory data-gathering activities. In cross-validation studies, each datum is excluded from the set of parameter data, and the value of the excluded datum is estimated from the remaining data. After all data have been excluded, estimated, and returned to the data set, the magnitude of the estimation errors is compared to the magnitude predicted by the parameter uncertainty calculations (e.g., made with a procedure such as kriging).

Parameter Specification

The performance assessment analyses require the assignment of specific values to the operational parameters of the various conceptual and mathematical models used. These parameter values will be based upon design and site data that are either measured, the result of design analyses, or assigned through expert judgment.

Analysis needs

The estimation of model parameters is subject to greater errors when few measured data are available to form the estimates. Also, the performance measures are influenced more by some parameters than by others. Uncertainty, sensitivity, and geostatistical studies can help determine how many data are required for each parameter to reduce the estimation errors to acceptable levels.

Uncertainty analysis. To date, the evaluation of uncertainty in performance estimates has focused more on parameter specification than on the other three sources of uncertainty. There are several analytical tools from various scientific disciplines that can be applied to this problem. And this is an area where repository developers have the most control over uncertainty. Since parameter specification errors are primarily due to a lack of data, they can be significantly reduced through the developmental phases of the project by simply collecting more data. Of course, the data collection activities should be designed to reduce uncertainty as much as possible for each dollar spent. And at licensing, some residual uncertainty will still exist due to the complexity of natural systems and the long-term processes that must be modeled for repository performance.

Uncertainty due to parameter specification comes from two sources, measurement inaccuracy and sampling variability. These sources of uncertainty are common to all models in the performance assessment from the engineered barrier system to the far-field ground-water flow system. Measurement inaccuracy is due to an inability to exactly measure many natural systems and processes. For example, the hydraulic conductivity of rock must be inferred from flow tests, it cannot be directly measured. Also, the scale of measurement is often inappropriate for the scale of the performance assessment analyses. For example, in many ground-water flow calculations, measurements of rock properties are required for large blocks of rock (e.g., a block might be many thousands of cubic meters in size), while field and laboratory measurements are made on small volumes of rock (e.g., analyzed drill cores are usually only several cubic centimeters in size). And some parameters cannot be measured at all and must be assigned by expert judgment. For example, the changes in potentiometric head over several thousand years cannot be measured.

Even if natural processes could be measured exactly, a source of parameter specification uncertainty would remain due to sampling variability. Rock properties are not constant across a site; there is an intrinsic heterogeneity, that is, a natural spatial variability, of the rock properties. The only way in which this heterogeneity might ever be exactly described is by an exhaustive sampling campaign with millions of samples taken at very close intervals. The fact that such a sampling campaign is never possible results in errors between the inferred rock properties from the sampling and the true rock properties of the site.

From a regulatory point of view, uncertainty analysis is required primarily by the total system radionuclide release standard through the complementary cumulative distribution function (*10*). However, in order to support the license application and to design research and data collection programs to reduce uncertainty, a performance assessment uncertainty analysis should quantify the likely errors in all of the models contributing to the performance estimates due to measurement inaccuracies and sampling variability.

Sensitivity analysis. A performance assessment sensitivity analysis should assess the magnitude of the changes in performance estimates due to incremental changes in each of the operational parameters of the performance models. When these sensitivity measures are combined with the measures of uncertainty for each of the parameters, decisions can be made as to where future sampling and research activities can most reduce performance assessment uncertainties. However, these decisions will not be automatic. There are several different performance measures that will each have a different set of sensitivities. And there are several submodels contributing to the performance calculations. Also each submodel may have dozens of parameters that must be specified. In fact, when the parameters are spatially varying, such as the rock properties, there may be literally hundreds of parameters.

Besides guiding future research and data collection needs, sensitivity studies can be used to help reduce the complexity of the performance assessment models used. If changes in a particular parameter can be shown to have relatively little effect on a performance measure, then the performance model can be simplified by dropping that parameter. These simplifications can be made in both generic and site-specific studies.

Geostatistical analysis. Geostatistical analyses can contribute to both the performance assessment uncertainty and sensitivity studies. Analyses are often necessary to interpolate between data locations for spatially varying parameters such as hydraulic conductivity. Geostatistical techniques can assess the sampling variability and estimation errors for spatially varying parameters. Geostatistical analyses can quantify the geologic and correlation structures. And simulation studies can examine the sensitivity of performance measures to the spatial distribution of geologic parameters.

Existing methods

Several methods can be drawn from different scientific disciplines for uncertainty, sensitivity, and geostatistical analyses. Some of the methods available for repository research have been compared by authors such as McKay, Beckman, and Conover (*19*); Thomas (*20*); Harper (*21*); Harper and Gupta (*22*); Iman and Helton (*23*); and Porn and Akerlund (*24*).

Uncertainty methods. The methods available for performance assessment uncertainty analysis can be categorized into one of five groups:

1. Statistical sampling methods can evaluate the range of likely performance estimates by defining a representative set of values for the operational parameters, which are then used as inputs to the performance models. The values for the parameters are typically chosen by some random sampling scheme, taking into account the probability distributions and correlations of the parameters. Methods that fall into this broad grouping are Monte Carlo sampling, Latin hypercube sampling, and discrete event simulation.
2. Some performance models have uncertainty components built directly into their underlying equations. For example, the methods of stochastic hydrology incorporate variance components for the parameters defining ground-water flow, thus leading to variability in the estimates of flow.
3. In some cases, first-order variance approximations can be made for the performance measures. These approximations are based on Taylor-series expansions of the performance function around nominal values for the operational parameters. They require knowledge of the sensitivity coefficient and covariance structure for each of the parameters.
4. Classical statistical variance component methods can be used to assess measurement accuracy for the operational parameters. These methods generally require that replicate measurements be made for the parameters under identical conditions. For example, when analyzing a water sample for chemical composition, the sample should be split into subsamples and each subsample should be analyzed to assess the accuracy of the analysis procedure.
5. When parameter values can only be specified by expert judgment, parameter uncertainty can be quantified with the use of Bayesian methods. These methods pool any available hard data with the expert opinions to form a performance estimate that is more accurate than either source could have provided independently.

Sensitivity methods. The methods available for performance assessment sensitivity analysis for parameter estimation can be summarized into three groups:

1. Statistical experimental design methods can be used to screen a large number of operational parameters down to a smaller number of parameters for detailed sensitivity and uncertainty analyses. These methods establish a

limited number of values of each parameter for processing in the performance models. The resulting performance estimates are then used in standard regression and analysis-of-variance procedures to compare the effects of the various parameters on the performance measure.

2. Statistical sampling methods such as Monte Carlo and Latin hypercube sampling can be used in conjunction with regression techniques to judge the effects of various input parameters on the performance measure output. In contrast to the fixed parameter values established by experimental design methods, sampling methods allow a component of randomness to enter into the establishment of the parameter values. The resulting performance estimates are analyzed with partial correlation and standardized regression coefficients to judge which parameters have the greatest effect on performance.
3. The adjoint method critically examines every computational step in a performance model and calculates the partial derivatives of the performance measure with respect to every parameter in the model, thereby generating an adjoint version of the model (*25*). The adjoint model is then evaluated with nominal values of the parameters. The resulting sensitivity coefficients provide very detailed information with which to judge the effects of the various parameters.

Generally speaking, the experimental design methods are used to screen a large number of input parameters down to a smaller number that will be further analyzed with Monte Carlo sampling methods. The sampling methods require many evaluations (perhaps hundreds of runs) of each performance model, and are therefore better suited to situations where there are several performance models but each has only a few input parameters. In contrast, the adjoint method calculates sensitivity coefficients for any number of parameters with only one evaluation of the adjoint model. However, generating the adjoint model can be expensive in terms of time and computer resources, and it must be repeated for each performance model. It should also be noted that none of these techniques is well suited to analyzing the changes in sensitivity that may occur with respect to different locations at the repository site. If the performance measure is evaluated at one location at the site (e.g., ground-water travel time from a location at the center of the disturbed zone), then the calculated sensitivities are also specific to that location.

Geostatistical methods. Several geostatistical methods are available for uncertainty and sensitivity analyses in connection with performance assessment parameter estimation. Those methods can be summarized into three groups:

1. Variogram analyses can be used to estimate several parameters of both conceptual and mathematical models. The variogram function quantifies the autocorrelation (for one parameter) or cross-correlation (for two parameters) structure of a site by calculating experimental correlation coefficients from the available data. These correlation coefficients, along with data means and variances, are parameters of various mathematical models used in the performance assessment analyses. The variogram function also quantifies parameters of geologic structure such as the range of correlation and anisotropy ratios. In addition, an upper bound can be established for the measurement error variability with the nugget variance of the variogram. And for parameters that

exhibit extreme variability, the correlation analysis can be refined to separately examine various ranges of the parameters (e.g., the high conductivity data separately from the low conductivity data) with an indicator variogram analysis.

2. Kriging is a geostatistical interpolation method for parameters showing spatial variability. Kriging uses the variogram's quantification of the correlation structure to efficiently interpolate values of the parameters at locations away from known data locations. The method also calculates the corresponding estimation variance for the interpolated values. That variance is a measure of the lack of data for specifying the unknown parameter values. Depending on the particular geology of the repository site, several varieties of kriging are available for either single parameters or several parameters. Some of these are simple, ordinary, universal, indicator, probability, and multigaussian kriging.
3. In ways similar to the Monte Carlo sampling methods previously described, the geostatistical conditional simulation method can be used for uncertainty and sensitivity studies. This method generates a random field of parameter values (e.g., a field of hydraulic conductivities covering the potential repository site) that has the same spatial correlation structure and exactly matches the available data from the real site. The advantage of simulating the parameter values is that data can be generated at a very close spacing, much closer than the density of the available data. By simulating various random fields and calculating performance estimates with each, a range of estimates is established with which to judge the uncertainty in the performance assessment measure. Also, by examining the changes in the performance assessment estimates with the changes in the simulated parameter inputs, a sensitivity study can be conducted to isolate the most important parameters for each performance measure.

Model Calculations

In most cases the conceptual and mathematical performance models are too complex to be routinely evaluated with hand calculations. In those cases, the performance estimates must be calculated with complex computer codes. The accuracy of the performance estimates is therefore affected by round-off errors and any errors in the coding itself. Calculational errors can also be introduced during hand calculations, when using calculators, and during application of commercial software such as spreadsheet programs. In addition, errors can be introduced during the transfer of data from their original sources into the format required for the calculations.

Analysis needs

Calculation errors can be controlled by the use of quality assurance procedures for code verification and calculations.

All performance assessment computer codes must be verified according to well-established quality assurance guidelines. Code verification procedures are designed to check that the executable statements in the computer program make the exact calculations that are required in the mathematical performance models. For example, if an executable statement has a misplaced sign or digit, it could still execute but would give erroneous results. Code verification seeks to eliminate all errors introduced from sources such as miscoding and rounding off.

Calculation procedures are designed to check that no errors are introduced during the transfer of data from original sources to the analysis input, that rounding errors are acceptable based on the expected and justified accuracy of the data and calculational results, and that no errors are introduced during the calculations.

Existing methods

Quality assurance procedures exist in all projects of the U.S. Civilian Radioactive Waste Management Program to help guide and document code verification and calculations in order to eliminate coding and calculation errors. Code verification generally entails one or more of the following: comparing code results with (1) hand calculations, (2) verified analytical solutions or approximations, and (3) verified numerical computer codes (benchmarking), all designed to perform the same type of analysis. Errors in the calculations are controlled by having the entire or key portions of the calculations repeated by investigators who are not involved in the original calculations. Calculation errors can be almost completely controlled if adequate resources are applied to verifying the computer codes and checking the data and calculations.

Research and Development Needs

The Federal Regulations clearly mandate uncertainty analyses in the performance assessment studies for nuclear repository research. However, those regulations do not attempt to define the specific uncertainty analyses and methods that should be used. For some analyses, adequate methods already exist. But in many areas, several important needs should be addressed in the future research and development plans for uncertainty, sensitivity, and geostatistical methods:

1. Methods should be developed to quantify the uncertainty in performance estimates due to inadequacies in the conceptual and mathematical models.
2. Methods for showing compliance with the EPA standards (*10*) must be developed. At this time, it is not clear that a single complementary cumulative probability distribution function, incorporating all sources of uncertainty, can be reliably constructed, or that such a distribution function is the correct tool with which to judge compliance.
3. More experience is needed on the application of geostatistical methods to parameters important to repository performance, particularly to fractured rock properties. These studies should address both autocorrelation for single parameters and cross-correlation for pairs of parameters.
4. Flexible, user-friendly software systems need to be developed for uncertainty, sensitivity, and geostatistical analyses on mainframe and/or microcomputers.
5. Methods should be developed to validate the statistical models for performance uncertainty analyses.

6. The role in repository design of qualitative uncertainty measures, assigned by expert judgment, as opposed to quantitative uncertainty measures calculated with mathematical models, should be investigated.
7. A formal peer review process should be developed, along with panels of experts for peer reviews. Also, methods are needed for quantifying the uncertainty in expert opinions.
8. Methods need to be formalized for using the uncertainty and sensitivity information to guide repository development.

Conclusion

This conference, as stated in the conference announcement, is intended to present the state of the art in geostatistical, sensitivity, and uncertainty analysis methods needed for performance assessments of potential nuclear waste repositories in deep geologic formations, with particular emphasis on their relevance to groundwater flow and radionuclide transport modeling. The conference is intended to guide future research and stimulate cooperation between nuclear waste disposal programs on an international basis. We believe that this conference, which features internationally known experts, will accomplish this and will benefit not only the Canadian and U.S. nuclear waste management programs, but the conference participants and other interested scientists as well.

Literature Cited

1. U.S. Department of Energy, *Final Environmental Impact Statement, Management of Commercially Generated Radioactive Waste*, DOE/EIS-0046F (October 1980).
2. *Federal Register*, *46*, 26677 (May 14, 1981).
3. U.S. Department of Energy, *Mission Plan for the Civilian Radioactive Waste Management Program*, DOE/RW-0005 (June 1985).
4. Nuclear Waste Policy Act of 1982, 97th Congress, Public Law 97-425 (January 7, 1983).
5. Office of Crystalline Repository Development, *A National Survey of Crystalline Rocks and Recommendations of Regions to Be Explored for High-Level Radioactive Waste Repository Sites*, OCRD-1 (April 1983).
6. U.S. Department of Energy, *Draft Area Recommendation Report for the Crystalline Repository Project: Overview*, DOE/CH-15(0) (January 1986).
7. U.S. Department of Energy, *OCRWM Mission Plan Amendment*, DOE/RW-0128 (June 1987).
8. 10 CFR Part 960, "Nuclear Waste Policy Act of 1982; General Guidelines for the Recommendation of Sites for the Nuclear Waste Repositories; Final Siting Guidelines" (December 6, 1984).
9. 10 CFR Part 60, "Disposal of High-Level Radioactive Wastes in Geologic Repositories, Technical Criteria" (July 31, 1986).
10. 40 CFR Part 191, "Environmental Standards for the Management and Disposal of Spent Nuclear Fuel, High-Level and Transuranic Radioactive Wastes; Final Rule" (September 19, 1985).
11. 10 CFR Part 20, "Standards for Protection Against Radiation" (March 5, 1980).
12. 10 CFR Part 20, "Standards for Protection Against Radiation, Proposed Rule" (January 9, 1986).

13. U.S. Nuclear Regulatory Commission, *Draft Site Characterization Analysis of the Site Characterization Report for the Basalt Waste Isolation Project, Hanford, Washington Site*, NUREG-0960 (March 1983).
14. Journel, A. G., and Ch. J. Huijbregts, *Mining Geostatistics*, Academic Press, London, reprinted with corrections (1981).
15. Clark, Isobel, *Practical Geostatistics*, Applied Science Publishers Ltd., London (1979).
16. Verly, Georges, Michel David, Andre G. Journel, and Alain Marechal, *Geostatistics for Natural Resources Characterization, Parts 1 and 2*, D. Reidel Publishing, Dordrecht, Holland (1984).
17. U.S. Nuclear Regulatory Commission, *Draft Generic Technical Position on Licensing Assessment Methodology for High-Level Waste Geologic Repositories* (July 1984).
18. U.S. Nuclear Regulatory Commission, *Revised Modeling Strategy for HLW Performance Assessment* (July 23, 1984).
19. McKay, M. D., R. J. Beckman, and W. J. Conover, "A Comparison of Three Methods for Selecting Values of Input Variables in the Analysis of Output from a Computer Code," *Technometrics*, *21* (2) (May 1979).
20. Thomas, R. E., *Uncertainty Analysis*, ONWI-380, Office of Nuclear Waste Isolation, Battelle Memorial Institute, Columbus, OH (1982).
21. Harper, W. V., *Sensitivity/Uncertainty Analysis Techniques for Nonstochastic Computer Codes*, BMI/ONWI-444, Office of Nuclear Waste Isolation, Battelle Memorial Institute, Columbus, OH (1983).
22. Harper, W. V., and S. K. Gupta, *Sensitivity/Uncertainty Analysis of a Borehole Scenario Comparing Latin Hypercube Sampling and Deterministic Sensitivity Approaches*, BMI/ONWI-516, Office of Nuclear Waste Isolation, Battelle Memorial Institute, Columbus, OH (1983).
23. Iman, R. L., and J. C. Helton, *A Comparison of Uncertainty and Sensitivity Analysis Techniques for Computer Models*, NUREG/CR-3904, SAND84-1461, Sandia National Laboratories, Albuquerque, NM (1985).
24. Porn, Kurt, and Ove Akerlund, *Procedures for Uncertainty and Sensitivity Analysis in Repository Performance Assessment*, TR 85-18, Swedish Nuclear Fuel and Waste Management Co., Stockholm, Sweden (October 1985).
25. Oblow, E. M., *An Automated Procedure for Sensitivity Analysis Using Computer Calculus*, ORNL/TM-8776, Oak Ridge National Laboratory, Oak Ridge, TN (1983).
26. Nuclear Waste Policy Amendments Act of 1987. 100th Congress, Public Law 100–203 (December 22, 1987).

Regulatory Perspectives of Concept Assessment

PETER A. FLAVELLE
Atomic Energy Control Board
Ottawa, Ontario, Canada

Abstract

The Atomic Energy Control Board is the lead agency for the regulatory review of the Assessment of the Canadian Concept for Nuclear Fuel Waste Disposal being done by Atomic Energy of Canada Limited and Ontario Hydro. This paper describes the regulatory perspective of how the Concept Assessment could demonstrate the feasibility of a disposal system conforming to regulatory requirements.

The long-term aspects of Concept Assessment encourage the use of various predictive techniques for different time scales. Each technique will have a different potential for establishing confidence in the predictions. The predicted performance of a facility during operation should have a very high confidence, as it can be based on standard engineering calculations and the predictions can be validated later by monitoring during operation. The predictions of the transient period following closure of the facility should achieve a medium level of confidence, since they can be based on extrapolations of predictions of operational performance, using models that can be calibrated with monitoring data and with averaged input data derived from natural analogue studies. Predictions based on fundamental processes will have a medium level of confidence when made to intermediate times after closure. Long-term predictions using generic or typical input data or Monte Carlo calculations of simplified models will have the least confidence, and yet they can still contribute to the confidence that the disposal concept will conform to regulatory requirements.

Regulatory Initiatives

The announcement of the inception of the Nuclear Fuel Waste Management Program by the federal government of Canada and the provincial government of Ontario in June of 1978 (*1*) assigned the responsibility of performing research into the storage and transportation of irradiated fuel to Ontario Hydro, a provincial power utility, and the responsibility of performing research into the immobilization and disposal of the nuclear fuel wastes to Atomic Energy of Canada Limited (AECL), a federal crown corporation. AECL's overall task "to verify that permanent disposal in a deep underground repository in intrusive igneous rock is a safe, secure, and desirable method of disposing of radioactive waste" (*1*) is known as Concept Assessment.

In August of 1981 a second Canada-Ontario joint statement (*2*) identified the Atomic Energy Control Board (AECB), which regulates the nuclear industry in Canada, as the "lead agency for the regulatory and environmental review of the disposal concept."

In keeping with its designated role, the AECB has been developing regulatory requirements, criteria by which to judge the acceptability of the concept and the Concept Assessment documentation, and guidelines addressing how these requirements and criteria could be met. Policy statements published by the AECB as Regulatory Documents have undergone extensive review by the nuclear industry and by the public before being finalized.

The AECB Regulatory Document R-104 (*3*) presents objectives, requirements, and guidelines dealing with the long-term aspects of all radioactive waste disposal, including nuclear fuel wastes. This document sets the individual radiological risk limit at 10^{-6} fatal cancers and serious genetic defects in a year and recommends a risk conversion factor of 0.02 per sievert. This conversion factor reflects the probability of a health effect from a given exposure, but does not include the probability of that exposure occurring. The document also identifies 10,000 years as the period for which compliance must be demonstrated. Guidelines to meet the general requirement of compliance to the radiological risk limit without reliance on institutional controls are presented in Regulatory Document R-104, and are discussed in greater detail in Information Report Info-0217 (*4*).

Requirements more specific to deep geological disposal of nuclear fuel wastes are published in Regulatory Document R-71 (*5*). This policy statement presents requirements for the disposal concept, for assessing the disposal concept, and for documenting that assessment. It also contains discussions on the nature of Concept Assessment. In these discussions the life of a geologic repository is described in terms of preclosure and postclosure phases. "Closure" is defined as the sealing and decommissioning of the disposal facility. Preclosure, then, comprises site selection, facility installation, facility operation, and decommissioning. After closure, the facility goes through a period of transient conditions until quasi-steady-state conditions develop. For the purposes of this discussion, the postclosure phase is arbitrarily divided into three time scales: short-, intermediate-, and long-term postclosure periods, corresponding to different rates of change of the transient conditions, and roughly equivalent to decades, centuries, and millennia, respectively.

The geologic characteristics required for an acceptable disposal site are presented and discussed in Regulatory Document R-72 (*6*). Preliminary recommendations to the AECB concerning the hydrogeologic factors of regulatory importance have been made by consulting experts in Information Report Info-0223 (*7*).

Regulatory View of Concept Assessment

Concept Assessment is a complex undertaking. To be tractable, it requires a structured approach, beginning with a definition of the problem. The problem definition should specify explicitly what questions need to be answered and what form the answers should take. These specifications then determine what tools and methods are appropriate to use. The problem definition can then be the basis for clear, explicit statements of what is to be accomplished and how. This approach is typical of large-scale, complicated engineering problems.

The discussion in Regulatory Document R-71 (*5*) states, "The concept will be refined until it can be shown with an acceptable level of confidence that the performance requirements established by the regulatory agencies will likely be met. At the completion of Concept Assessment, there must be little doubt that the concept which is finally accepted can be undertaken successfully in Canada." When predicting the performance of a disposal system, Concept Assessment must examine all pertinent conditions and processes that may affect the disposal system, and the predictions made must have a demonstrably acceptable level of confidence.

Thus, one general problem definition for Concept Assessment is that AECL must demonstrate with an acceptable level of confidence that its disposal concept is both theoretically possible and achievable (Figure 1). For the purposes of this discussion, "theoretically possible" is defined as the existence of a set of conditions or circumstances under which the disposal concept will meet regulatory requirements. By "achievable," it is meant that those conditions or circumstances can be duplicated by a disposal facility, i.e., that the concept is technically feasible.

PROBLEM DEFINITION

DEMONSTRATE CONDITIONS (OR CIRCUMSTANCES) UNDER WHICH THE CONCEPT MEETS REGULATORY REQUIREMENTS (THEORETICALLY POSSIBLE)

DEMONSTRATE THAT THOSE CONDITIONS CAN BE DUPLICATED IN A DISPOSAL FACILITY (ACHIEVABLE)

ASSESS THE CONFIDENCE THAT CAN BE PLACED IN BOTH DEMONSTRATIONS

Figure 1. **Example of a problem definition for Concept Assessment.**

The confidence referred to in R-71 is subjective in nature. Hence, it cannot be quantified a priori, and an "acceptable" level of confidence cannot be prescribed in advance. Although criteria by which to judge the confidence in Concept Assessment have not been derived, it is recognized that evaluating the confidence in Concept Assessment will become an important regulatory issue and will not be a trivial task. The degree of confidence that can be achieved is dependent in part on the tools and techniques used and the confidence that can be placed in each.

Methods for solving a problem can be classified into three categories: quantitative, semi-quantitative, and qualitative (Figure 2). Quantitative methods, such as those commonly used for engineering problems, calculate specific values for unknown variables and result in statements like "The groundwater velocity will be 500 centimetres per year." Uncertainty in these calculated values can arise from inaccuracies in the method of calculation (often an equation or a computer model) and inaccuracies in the input data. Quantitative methods allow the uncertainty in the

METHODS OF SOLUTION

QUANTITATIVE : Engineering Calculations
(CALCULATES A VALUE)

SEMI-QUANTITATIVE : Limiting/Bounding Calculations
(IT IS NOT POSSIBLE TO EXCEED THIS VALUE)
: Probabilistic Calculations
(THIS VALUE IS NOT LIKELY TO BE EXCEEDED)

QUALITATIVE : Natural Analogues
(IT HAPPENS THIS WAY IN NATURE)

Figure 2. **Methods of problem solution for Concept Assessment.**

calculations to be evaluated explicitly by comparisons between predictions and observations or between results based on different input data or different calculational methods. The confidence that can be placed in conclusions based on these methods of analysis is in general inversely related to the uncertainty in the calculated values.

Although the same calculations can be used in both quantitative and semi-quantitative methods of solving problems, the interpretation of calculated results is different. Semi-quantitative methods calculate limiting values, including maxima, minima, or asymptotes. These methods allow qualitative statements to be made about the range of unknown quantities of interest, such as "It is not possible to exceed this value" and "It is not likely this value will be exceeded."

There are two basic approaches to semi-quantitative methods. The first uses the set of input data that will calculate a limiting value, such as "The equilibrium aqueous concentration of calcium is 70 milligrams per litre." The conclusion that can be drawn from this result is "The long-term calcium concentration will approach 70 milligrams per litre." The second approach calculates several specific values of the unknown from several sets of input values to evaluate a bounding value. This approach is useful when the input data are very uncertain, and the conclusion that can be drawn is based on the maximum (or minimum) value predicted. Such a conclusion might be "The maximum temperature predicted is 87 degrees, so the highest temperature should not exceed 100 degrees." Conclusions should be drawn from the results of semi-quantitative analyses only when the calculational methods are used within their limits of applicability. The confidence that can be placed in such conclusions will be determined by the uncertainty in the calculated values.

Qualitative methods are essentially logical arguments, with little or no reference to quantitative values. Such arguments are usually based on observation, experience, or expert opinion. When reasoned estimates are used (e.g., expert opinion) the confidence in the argument depends on the quality of the reasoning used. Natural analogues can be used in qualitative arguments, leading to the conclusion "It happens this way in nature." The confidence placed in such arguments by the technical community depends on how closely the natural analogue matches the conditions of the problem of interest. This comparison can be facilitated, and the determination of the confidence simplified, by defining appropriate problem specifications.

Demonstration of Theoretical Possibility

There are many methods available to demonstrate that a disposal concept is theoretically possible (Figure 3). Logical arguments based on natural analogues can be used to qualitatively demonstrate isolation capabilities of the geosphere. To make full use of arguments based on natural analogues, the conditions at the analogues must be compared critically to the expected conditions at a disposal facility, and the significance of any discrepancies between the two should be assessed. The greater the similarity between a natural analogue and a disposal system, the greater the confidence that can be placed in the argument. The extent and detail of the comparison between the conditions at a natural analogue and those expected in a disposal facility depend on the level of detail needed for the logical argument, which should be defined as part of the problem specification.

One example of a natural analogue is the Cigar Lake uranium ore deposit in Saskatchewan, Canada. At Cigar Lake a high-grade uranium ore is confined by a clay halo, even though the host sandstone is highly fractured and hydraulically very conductive (*8*). This ore body demonstrates geochemical containment in the presence of an open flow-field and is a strong natural analogue argument in support of the theoretical possibility of the concept. AECL's current disposal concept places used fuel, which is about 99% uranium, in an emplacement canister that is embedded in a clay buffer. Although the conditions at the Cigar Lake ore body are not identical to those expected in this disposal concept and the discrepancies should be evaluated, the similarity between the uranium ore body in a clay halo and the uranium fuel in a

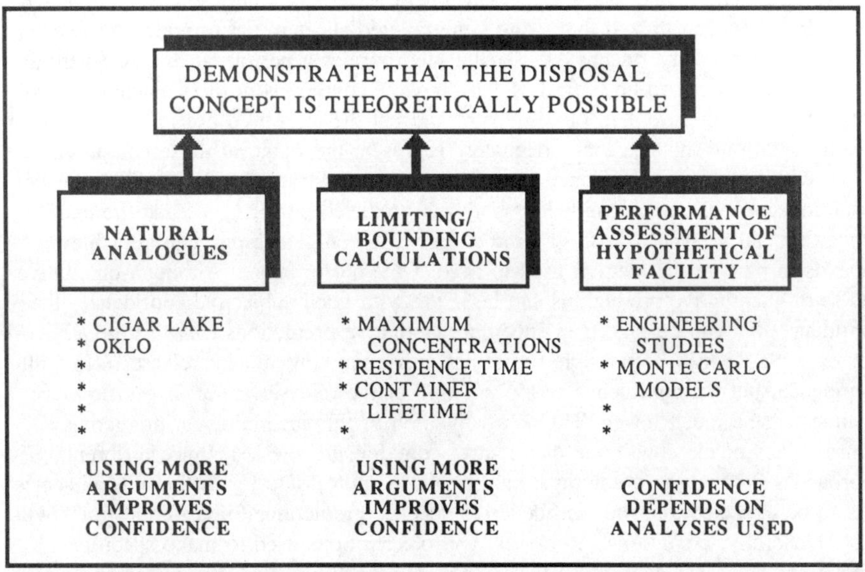

Figure 3. **Methods to demonstrate the conditions or circumstances under which the disposal concept will meet regulatory requirements and to improve confidence in the demonstration.**

clay buffer should inspire confidence in the conclusions drawn from this natural analogue.

Another natural analogue is the natural reactors at Oklo, in Gabon, Africa. There, fission products have not migrated any significant distance from the location of the reactors, although this system is chemically open (*9*). This analogue also demonstrates the capacity of the geosphere to isolate radioactive waste. Although the differences between the conditions at Oklo and those expected in AECL's disposal concept need to be critically evaluated, the fact that fission products are retained in the open chemical system improves the confidence that there are conditions under which a disposal system would perform to meet regulatory requirements.

The theoretical possibility of a disposal system performing adequately can also be shown by limiting or bounding calculations. These can be empirical or mechanistic (process-controlled) models. One such model would be to use solubility constraints to estimate the maximum possible contaminant concentration under steady-state conditions. Other examples might include the calculation of limiting values of residence time in the geosphere using various structural models, or the calculation of limiting values of container lifetime based on various corrosion models and various groundwater chemistry models. The confidence that can be placed in such calculations can be very high if they are based on models that are amenable to validation.

The theoretical possibility of a disposal concept can also be demonstrated using a performance assessment of a hypothetical disposal facility, as AECL is doing. This approach should be able to define a set of conditions within which a disposal facility will meet regulatory requirements. Initially, a performance assessment of a disposal facility on a generic site in the Canadian Shield area of Northern Ontario was undertaken. It is now recognized that using generic site data may not provide a usable result, so the current Concept Assessment incorporates a performance assessment at several sites claimed to be typical of the range of conditions in the Canadian Shield. This introduces the problem of demonstrating that the sites used are indeed typical of Shield conditions and that they adequately represent the range of those conditions.

The ability of a performance assessment to demonstrate the plausibility of a disposal concept will be determined by both the predicted performance and the confidence that can be placed in those predictions. The confidence that can be achieved depends in part on the method used to predict the performance. Although qualitative and semi-quantitative predictions can be made with good subjective confidence, the confidence that can be placed in specific quantitative predictions can vary greatly. For example, it may be possible to predict maximum contaminant release rates with confidence, but the confidence in a prediction of the release rate at a specific place or time will be much lower. This is a reflection of the uncertainty in predicting future values of the conditions that control contaminant releases. Since different approaches for making predictions may be more appropriate for different time scales, it is to be expected that the confidence in specific predictions for different times will vary. Generally, conditions, processes, and observations used to make specific predictions have less confidence the further into the future they are extrapolated.

Demonstration of Achievability

Concept Assessment must demonstrate not only the set of conditions under which a disposal system will meet regulatory requirements, but also that those conditions can be duplicated in a disposal facility. That is, the concept must be shown to be technically feasible (Figure 4). This means that the performance assessment and any other analyses of a disposal concept must result in the derivation of a set of specifications for the siting, construction and operation of a disposal facility. It must also demonstrate that it is possible to site, construct, and operate a disposal facility to those specifications.

For siting, the analyses used in Concept Assessment should demonstrate the capability to distinguish between an acceptable and an unacceptable site. This implies that the analyses should identify the important parameters by which to judge a site and derive a pass/fail criterion for each of them individually or as interrelated subsets. The important parameters would be used to guide site characterization studies, and the pass/fail criteria would be the specifications used in site selection.

For facility construction and operation, Concept Assessment should identify the parameters that are important to the predictions as performance indicators and should also specify operational targets and normal operating ranges for each of them. Associated with this is the need to demonstrate that the technology is available to monitor these parameters. If it is not technically possible to measure an important parameter, then it is not useful as a performance indicator.

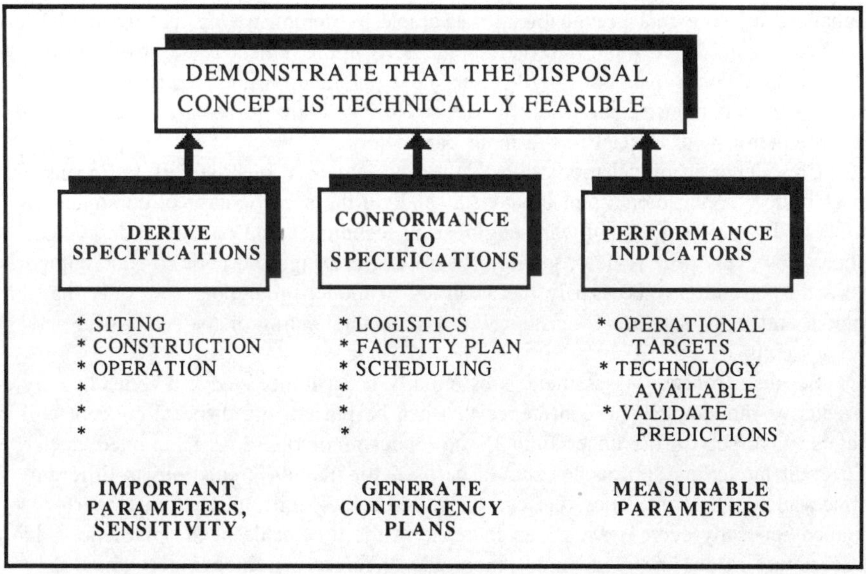

Figure 4. **Methods to demonstrate that conditions or circumstances under which the disposal concept will meet regulatory requirements can be duplicated by a disposal system.**

Sensitivity analyses will be vital in identifying the important parameters for siting, in identifying performance indicators for construction and operation, and for deriving pass/fail criteria and operational targets.

The demonstration of technical feasibility should also include plans and schedules to show the logistics of siting, constructing, operating, and decommissioning a disposal facility. These are necessary to show that it would be possible to operate a real facility and to generate contingency plans in anticipation of abnormal or accident conditions.

Development of Confidence in Predictions

One of the regulatory requirements is that models be validated where possible. Validation is accomplished by comparing observed behavior to model predictions based on real input conditions. Validation should provide a measure of the uncertainty in the model predictions based on discrepancies between the predicted and observed data. Thus, for a given range of input values, a validated model will have a quantified uncertainty. This should provide greater confidence in model predictions based on valid input data than if the uncertainty were not measured.

A "partial validation" is when the range of real input data used to validate the model is smaller than the range of input data for the problem of interest. A partial validation will also improve confidence in model predictions, but not to the extent that a full validation would.

Where validation is not possible, confidence in the model predictions should be enhanced to the extent practicable, for example by demonstrating that the model has realistic sensitivity to input data variations. This improvement in the level of confidence applies only over the range of the input parameter values for which realistic sensitivity is demonstrated, where model sensitivity is the local partial derivative of the prediction with respect to the input parameters.

Confidence in model predictions should be increased qualitatively by means of peer review of the model and input data, and, in the specific case of computer models, by use of good software engineering techniques and code-to-code comparisons. However, peer review, good software engineering, and code-to-code comparisons do not address necessarily the accuracy of model predictions, but only that the model and input data are consistent with our understanding of the real world and processes in it.

Because Concept Assessment is essentially a feasibility study, it relies heavily on predictive modelling. The confidence that can be placed in a disposal concept will be related inversely to the uncertainty in the model predictions. As mentioned earlier, different modelling approaches should be used for making predictions at different time scales. The confidence that can be placed in a specific prediction of performance generally decreases with an increase in the time scale of the prediction, due to uncertainty in the occurrence and magnitude of future conditions upon which the model predictions are based. Table 1 presents some examples of analyses that can be used in a performance assessment of a disposal facility and how confidence in the assessment can be developed.

Table 1. **Analyses that could be used to build confidence in Performance Assessment.**

Phase	Preclosure		Postclosure	
Period	Operation	Short-Term	Intermediate-Term	Long-Term
Time Scale		Decades	Centuries	Millennia
Methods of Assessment	Engineering calculations based on design data Finite element Finite difference Systems analysis	Extrapolation of engineering calculations Mechanistic and process-based calculations	Extrapolation of engineering calculations using data from analogues Mechanistic calculations Detailed probabilistic calculations Qualitative arguments based on analogues	Extrapolation of engineering and mechanistic calculations Probabilistic calculations based on derived and empirical relationships Qualitative arguments based on natural analogues
Methods of Developing Confidence	Analogous to other facilities Potential to calibrate models using data from operation Potential to validate predictions with operational data	Predictions can be based on models that have the potential to be calibrated during operation using real data Predictions can be based on models that have the potential to be validated for operational conditions Should be concordance of predicted values from different methods and for common times		
Relative Confidence Level	High	Medium	Medium–Low	Low

The preclosure phase of a disposal concept includes siting, facility construction, waste packaging and emplacement, and facility decommissioning. The predictions made for the preclosure phase should be able to achieve a low uncertainty, resulting in a high degree of confidence. Most of them will be based on common engineering calculations, employing such methods as conservative estimates, numerical methods of analysis such as finite difference or finite element computer programs, limiting conditions, reliability analysis, and sound engineering judgement. These calculations should be similar to those performed for other engineering projects such as mines, dams, or generating stations.

When a disposal facility's life is over, it will not be allowed to close until there is substantial evidence that the facility is performing adequately as predicted, and will continue to do so. Thus, one aspect of an acceptable concept is that some of the predictions that will be made in support of licensing can be validated using monitoring data from construction and operation of the facility. This will provide confidence in predictions of continued acceptable performance, which is a precondition for closure.

Predictions for decades after closure, when the transient conditions will likely be changing most rapidly, should be based on calculations similar to those used for the preclosure phase. In fact, short-term postclosure predictions will probably be extrapolations of some preclosure calculations. During operation of the facility these postclosure predictions would be refined, using models validated or at least calibrated with monitoring data. Concept Assessment, then, should be able to justify a moderate degree of confidence in the predictions for the short-term postclosure period. There is also the potential to greatly enhance the confidence in short-term postclosure predictions using models validated by monitoring once the facility is in operation.

Predictions for the intermediate postclosure period will have a somewhat lower level of confidence, since they are being made for centuries into the future and there is less certainty about the conditions that will exist then. These predictions should be made with the same methods of extrapolation as for the short-term, as well as using data averaged over centuries (for example, climatic data). Detailed models based on fundamental processes should be used, as should qualitative arguments based on natural analogues, to improve the confidence in the predictions for the intermediate term.

Specific predictions of behavior in the long-term postclosure period, based on conditions that may exist millennia after closure, are expected to have the greatest uncertainty. The long time scale makes extrapolation of detailed calculations not only highly uncertain but prohibitively expensive. Calculations at these longer time scales are more amenable to the use of assessment models using simplified relationships and limiting assumptions. These assessment models include Monte Carlo type computer models, where distributed input data reflect the uncertainty in future conditions. The algorithms used in Monte Carlo calculations are usually simplifications of more detailed models, so the algorithms themselves include the uncertainties of both the underlying detailed models and of the simplifications. The confidence, then, that can be placed in a single or deterministic prediction using a simplified algorithm and uncertain input data is not very high.

This lack of confidence in a single prediction is balanced by making a large number of predictions using the full range of the distributed input data. The set of predictions is interpreted in a semi-quantitative manner to arrive at a bounding value for the predictions. The confidence, then, that can be placed in a set of Monte Carlo calculations is derived from the use of a large number of different input conditions to explore the domain of possible output conditions, but this confidence is constrained by the appropriateness of the input data and the applicability of the algorithms used. The confidence in these predictions cannot be improved beyond a certain point simply by increasing the number of Monte Carlo calculations.

There is another general approach for demonstrating the feasibility of the concept to improve confidence in the predictions. When there are several different methods being used at several different time scales, there should be a concordance among predictions for common times (Figure 5). Early predictions for the long-term postclosure period should show reasonable agreement with predictions made for intermediate- and short-term time scales, even though different models may be used. Similarly, early predictions for the intermediate time scale should show reasonable agreement with short-term predictions. This would link the more uncertain long-term predictions to the less uncertain short-term predictions, thus enhancing confidence in the overall predicted performance of the disposal facility.

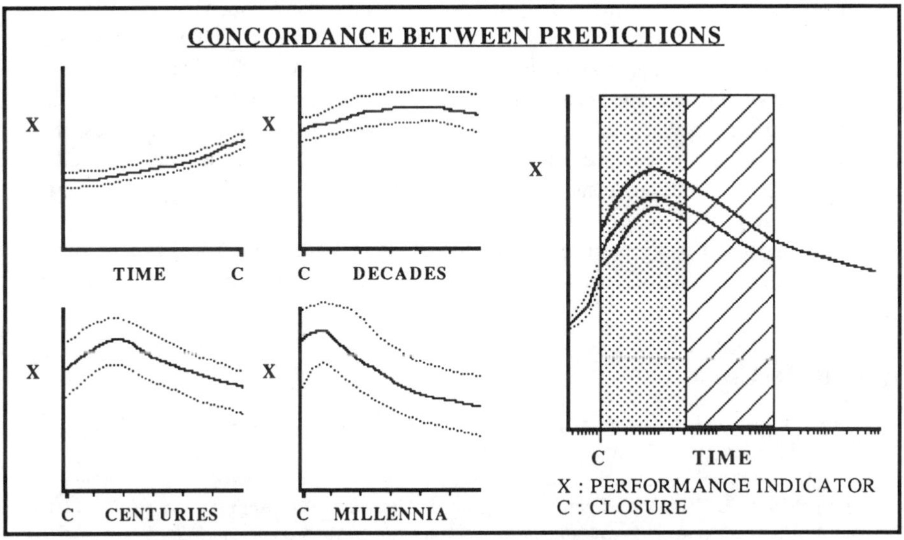

Figure 5. **Concordance at common times between predictions made for different time scales.**

Summary

The Atomic Energy Control Board (AECB) is the lead agency for the regulatory review and evaluation of the Canadian Nuclear Fuel Waste Management Program being developed by Atomic Energy of Canada Limited (AECL) and Ontario Hydro. AECL is responsible for performing and documenting the Concept Assessment, a program to demonstrate that deep geological disposal in crystalline rock is "a safe, secure and desirable method of disposing of radioactive wastes." The AECB has developed regulatory requirements, guidelines, and criteria by which to evaluate Concept Assessment.

Concept Assessment is viewed by the AECB as essentially a feasibility study to show that the disposal concept is theoretically and technically possible. There must be an acceptable level of confidence in the conclusions that AECL will draw concerning the feasibility of the concept. The level of confidence is considered to be directly related to the number and quality of arguments presented in support of conclusions concerning the concept, and inversely related to the uncertainty in calculations performed in support of those conclusions.

Concept Assessment is defined as a demonstration that a set of conditions exists within which a disposal system would meet regulatory requirements, and a demonstration that that set of conditions can be duplicated by the proposed disposal system. The specification of the confidence in the demonstrations requires quantification of calculational uncertainties, the use of several different types of supporting arguments including semi-quantitative and qualitative approaches, the use of models which are validated or have the potential to be validated, and the concordance of predictions from various sources.

Several arguments are identified that may be useful for Concept Assessment, ranging from logical arguments based on natural analogues to a performance assessment of a hypothetical disposal facility on a typical site. Methods of accomplishing a performance assessment are presented in terms of how they should contribute to the confidence in the results. The level of confidence will vary with different types of analyses and, for some analyses, will deteriorate the further into the future a prediction is made. No single analysis or technique is seen to provide the necessary confidence in the demonstration of the feasibility of the disposal concept. Rather, an integrated approach using a variety of methods to generate a suite of different supporting arguments for different time scales is seen as being appropriate for Concept Assessment.

Literature Cited

1. Energy, Mines and Resources Canada, News Release, "Canada/Ontario Radioactive Waste Management Program," Ottawa (1978, June).
2. Energy, Mines and Resources Canada, News Release, "Canada-Ontario Joint Statement on the Nuclear Fuel Waste Management Program," Ottawa (1981, August).
3. Atomic Energy Control Board, *Regulatory Objectives, Requirements and Guidelines for the Disposal of Radioactive Wastes—Long-Term Aspects*, Regulatory Document R-104, Regulatory Policy Statement, Ottawa (1987, June).
4. Wagstaff, K.P., "Regulatory Criteria for the Disposal of Radioactive Wastes," *Proceedings of the Second International Conference on Radioactive Waste Management, 1986*, September 7–11, Winnipeg, Manitoba, Canadian Nuclear Society, Toronto (1986); also as Atomic Energy Control Board, Information Document Info-0217, Ottawa (1986, September).
5. Atomic Energy Control Board, *Deep Geological Disposal of Nuclear Fuel Waste: Background Information and Regulatory Requirements Regarding the Concept Assessment Phase*, Regulatory Document R-71, Regulatory Policy Statement, Ottawa (1985, January).
6. Atomic Energy Control Board, *Geological Considerations in Siting a Repository for Underground Disposal of High-Level Radioactive Waste,* Regulatory Document R-72, Regulatory Guide, Ottawa (1987, September).
7. Lafleur, D.W. and K.G. Raven, *Hydrogeologic Factors to be Addressed in Disposal Guidelines,* Info-0223, Atomic Energy Control Board Information Document, Ottawa (1986, December).
8. Cramer, J.J., "A Natural Analogue for a Fuel Waste Disposal Vault," *Proceedings of the Second International Conference on Radioactive Waste Management, 1986*, September 7–11, Winnipeg, Manitoba, Canadian Nuclear Society, Toronto (1986).
9. Jakubick, A.T. and W. Church, *Oklo Natural Reactors: Geological and Geochemical Conditions—A Review,* Info-0179, Atomic Energy Control Board Information Document, Ottawa (1986, February).

Role of Uncertainty in the Basalt Waste Isolation Project

ANTHONY J. KNEPP and DAVID H. DAHLEM
U.S. Department of Energy
Richland, Washington

Abstract

The current national Civilian Radioactive Waste Management (CRWM) Program to select a mined geologic repository will likely require the extensive use of probabilistic techniques to quantify uncertainty in predictions of repository isolation performance. Performance of nonhomogeneous, geologic, hydrologic, and chemical systems must be predicted over time frames of thousands of years and therefore will likely contain significant uncertainty. A qualitative assessment of our limited ability to interrogate the site in a nondestructive manner coupled with the early stage of development in the pertinent geosciences support this statement. The success of the approach to incorporate what currently appears to be an appreciable element of uncertainty into the predictions of repository performance will play an important role in acquiring a license to operate and in establishing the level of safety associated with the concept of long-term geologic storage of nuclear waste.

This paper presents a brief background on the Hanford Site and the repository program, references the sources that establish the legislative requirements to quantify uncertainties in performance predictions, and summarizes the present and future program at the Hanford Site in this area. The decision to quantify significant sources of uncertainties has had a major impact on the direction of the site characterization program here at Hanford. The paper concludes with a number of observations on the impacts of this decision.

Background

By the beginning of the next century, the United States plans to begin operation of the first geologic repository for the permanent disposal of spent commercial nuclear fuel and high-level radioactive waste. Public Law 97-425, the Nuclear Waste Policy Act (NWPA) of 1982, establishes (1) the process for selecting a site and constructing the repository; (2) criteria for evaluating the isolation potential of the site; and (3) the process for operating, closing, and decommissioning the facility. Under the NWPA, nine sites were identified as potentially acceptable, and three of the nine were recommended for site characterization. After completion of site characterization, one of the recommended three will house the nation's first geologic repository.

The three sites recommended for site characterization are Hanford, Washington; Yucca Mountain, Nevada; and Deaf Smith County, Texas. Each site is currently preparing plans for site characterization. This paper concentrates on work at the Hanford, Washington site by the Basalt Waste Isolation Project (BWIP); however, the discussion applies equally well throughout the CRWM Program.

Hanford, Washington

The Hanford Site (Figure 1) is in an agricultural region in southeastern Washington drained by the Columbia River. The U.S. Department of Energy's (DOE's) 560-square-mile nuclear reservation completely surrounds the site.

The geologic setting is characterized by the presence of layered flood basalts. Erupting at varying intervals between 10 to 30 million years ago, the basalt covered an area of over 100,000 square miles in parts of Washington, Idaho, and Oregon. After the eruptions ended, the lava fields were tilted as a unit to the southwest.

Anticlinal ridges bound the Hanford Site to the north and south (Figure 2): the Saddle Mountains to the north and the Rattlesnake Hills to the south. Hydraulic head and geophysical data indicate the existence of a nearly impermeable impediment to flow which defines the western side of the basin (DOE, 1986). As indicated by the

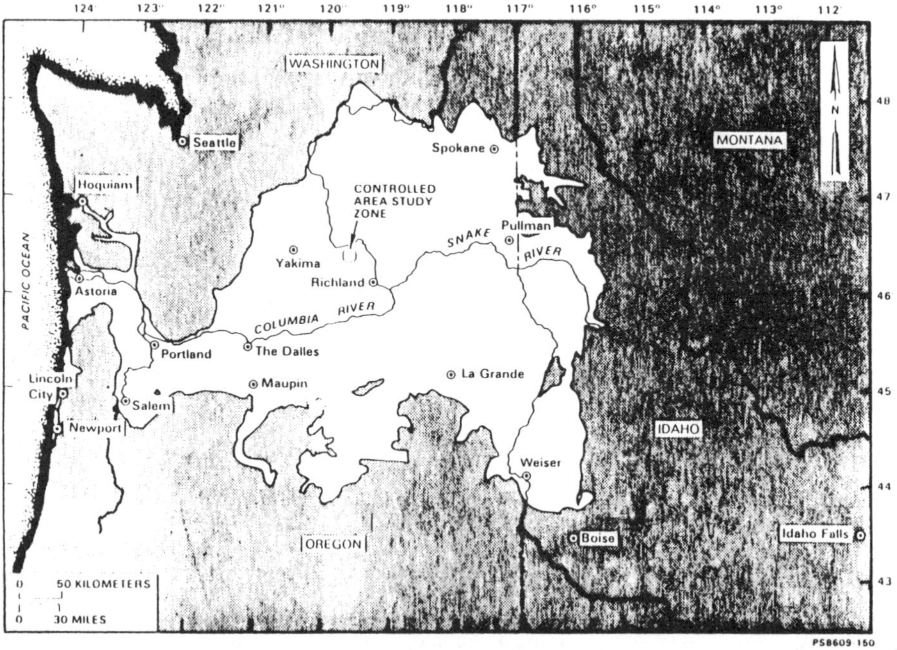

Figure 1. **Location map.**

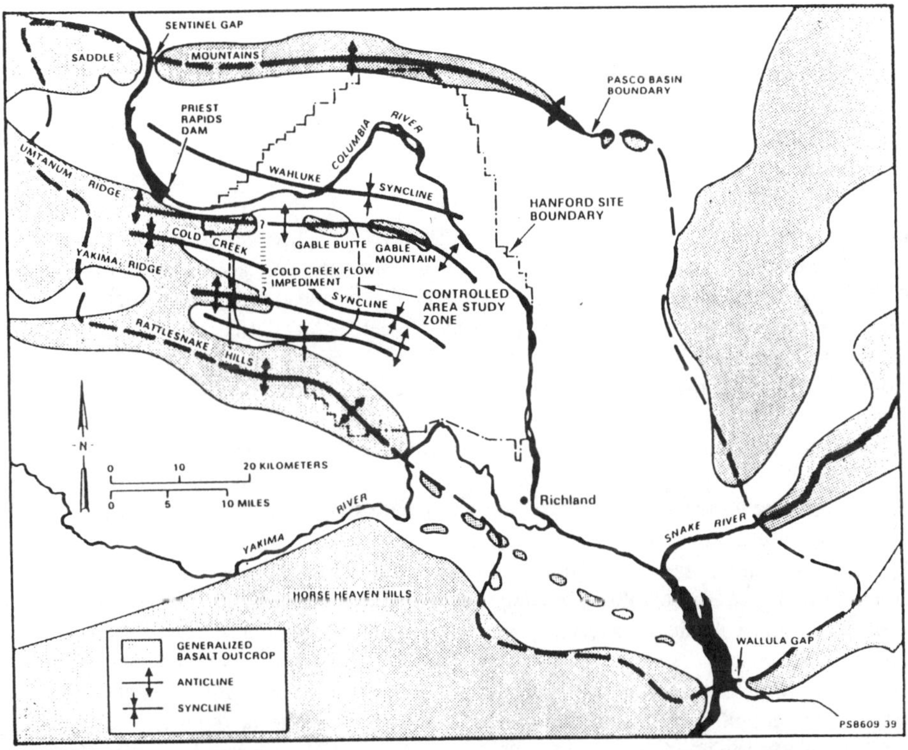

Figure 2. **Structural elements in the region around the proposed basalt repository.**

east-west trending ridges, the site is under considerable north-south compression (Davis, 1977).

The basalt flows collectively form the Columbia River Basalt Group which consists of three formations: Grande Ronde, Wanapum, and the Saddle Mountains. The total basalt section at Hanford has a thickness of over 10,000 feet with some individual lava flows reaching thicknesses of over 200 feet. Flows of the Grande Ronde sequence account for nearly 85 percent of the total volume of basalt flows in the group. Overlying the basalts are semiconsolidated sediments of the Ringold Formation and finally the Quaternary-age glaciofluvial deposits of the Hanford Formation (Tolan et al., 1987).

Seismicity records and seismic monitoring of the region indicate the site is geologically stable, disturbed primarily by low-magnitude earthquake ''swarms'' (U.S. Nuclear Regulatory Commission, 1982).

Structural features include the east-west Yakima fold belt and regional faults trending north-south and northwest-southeast. Regional lineaments are conjectured traces of these features. Folds of the Yakima complex are characterized by sharp, partially overturned anticlines separated by broad structurally simple synclines. The Cold Creek Syncline is the host environment of the proposed repository.

Groundwater at repository depth is dilute, alkaline (pH = 9), and reducing (Eh = − 400 mv). Ambient temperature is approximately 52 °C (Early et al., 1986).

Legislative Requirement to Evaluate Uncertainty

The NWPA and ancillary legislation clearly establish the requirement to explicitly quantify the major uncertainties in the prediction of long-term repository performance. The requirement to evaluate uncertainty in repository system performance is traceable to legislation and guidelines promulgated by the U.S. Nuclear Regulatory Commission (NRC) and the U.S. Environmental Protection Agency (EPA). Three passages lay the foundation for this effort. Certainly the most explicit statement comes from the EPA 40 CFR Part 191, "Environmental Standards for the Management and Disposal of Spent Nuclear Fuel." This regulation states that the disposal system must have a reasonable expectation that the cumulative releases of radionuclides to the accessible environment for 10,000 years after disposal shall have a likelihood of less than 1 chance in 10 of exceeding certain prescribed limits and also have a likelihood of less than 1 chance in 1000 of exceeding 10 times these same limits. It is noteworthy that the legislation recognizes that there will "inevitably be substantial uncertainties in projecting disposal system performance."

Elsewhere in the same legislation, the term "performance assessment" is described and given probabilistic connotation. Paraphrasing 40 CFR Part 191, performance assessment means an analysis that identifies the process, the events, and their effects on the disposal system and estimates the cumulative release of radionuclides, *considering the associated uncertainties,* caused by all significant processes and events. This definition has been interpreted at the Hanford Site to require an explicit handling of uncertainties.

And finally, a third source of the requirement to explicitly incorporate uncertainty comes from the NRC's "Draft Generic Position on Groundwater Travel Time" (Codell, 1986). Groundwater travel time (GWTT) is described as a distributed variable rather than an exact quantity and is properly quantified through the development of a cumulative density function. Compliance with the GWTT objective would be demonstrated "if it could be shown that any particle ... has a (100-x)% or greater probability of arrival at the accessible environment in a time greater than 1000 years." It should be recognized that NRC Technical Positions are not requirements in the strictest sense, but do define an approach that if followed should facilitate licensing should a site be selected.

It should also be recognized that a fairly strong position can be made that it would clearly help the applicant for an NRC repository construction license to consider in the site characterization program the scope of the effort that should be undertaken to define and quantify significant uncertainties in the prediction of repository and site performance.

The scope of this effort at BWIP is currently being defined and will be published after the release of the Site Characterization Plan in a series of documents called "Study Plans." In most instances the type and quantity of data required are the result of professional judgment, probabilistic modeling results, and recognition of budget and schedule constraints. Modeling studies, which helped define the scope of the data collection program, are summarized in the following section.

Status of Major Performance Analyses

Various stochastic techniques were evaluated for use at Hanford (Sagar et al., 1987); however, the most commonly used is that of Monte Carlo simulations. In this method, the uncertain parameters are treated as either random variables with assigned univariate probability density functions or as spatially variable stochastic fields with multivariate probability density functions. Sufficient sampling of the parameter field distributions is conducted to converge on the statistics of the governing distributions. For each realization of the uncertain parameters, a model is executed to produce a suite of output realizations.

We have conducted three large modeling efforts in which parametric uncertainty has been explicitly modeled. They are:

1. Source term modeling,
2. Groundwater travel time modeling, and
3. Far-field radionuclide transport modeling.

Detailed results for all three modeling efforts can be found in the Hanford Environmental Assessment (DOE, 1986) and are summarized below.

Source term modeling included an analysis of the release of radionuclides from a waste container and their migration through surrounding barriers, i.e., bentonite-basalt packing and fractured basalt rock. The primary transport mechanism was assumed to be molecular diffusion. The approach included the simplification of system geometry and processes, the development of a stochastic container lifetime model, and the development of a second model to integrate the stochastic functions obtained from individual waste containers to estimate total release. Radionuclide solubility and the sorption coefficient were the main stochastic parameters.

In the second modeling effort, a preliminary estimate of groundwater travel time was made using 2- and 3-dimensional applications of convective steady state groundwater transport equations. The groundwater analysis is used to compute seepage velocity fields before tracing pathlines and calculating travel times. Studies have accounted for uncertainties in transmissivity and porosity and in some cases boundary conditions.

The third major analysis resulted in predictions of the magnitude of releases from the repository to the accessible environment (defined to be 5 kilometers away [40 CFR Part 191]). An analytic solution was developed for the 1-dimensional transient convection-dispersion equation in a flow tube. A stochastic, time-dependent radionuclide source term and velocity field were obtained from the previously described analyses. Additionally, the transport properties of the site (sorption and dispersion coefficient) were assumed to be random.

Future plans

In the near future, emphasis will be placed on improving the conceptual basis for some of the less developed models and on verifying, benchmarking, and in some cases validating a number of the better established ones. Analytical effort will concentrate on sensitivity and uncertainty studies useful for establishing priorities in the data collection effort. When applicable, predictions of field test results will be made.

In the area of uncertainty, we are considering a trial application of the adjoint technique to develop a complete suite of sensitivities for a near-field transport model. The value of these sensitivities will be evaluated as to their usefulness in helping define the field program as compared to other approaches and sources of information.

There are also plans to develop a stochastic systems performance assessment model similar to the Canadian SYVAC code. This model is currently conceptualized to consist of three parts:

1. A time-dependent stochastic source term for radionuclide release from the waste container. The transport mechanism would be primarily diffusional.
2. A 2-dimensional far-field geosphere model that will handle heterogeneous, anisotropic transport, flow in fractures, thermally induced buoyancy, radionuclide decay, and sorption processes.
3. A biosphere model that would be capable of computing cumulative releases over the period of simulation and individual dose rates.

Programmatic Perspective of Uncertainty Analyses

It is still quite early in the program to evaluate the value of uncertainty analyses in meeting the project's objectives mentioned earlier, i.e., resolving issues, focusing data collection, and facilitating design. The fairly obvious benefit that has accrued is an explicit quantification of parametric uncertainty. The use of uncertainty techniques has permitted BWIP to conduct a more intellectually satisfying analysis that has gained at least partial acceptance from the regulatory community. Replacing a "worst case" approach to assessing the system over 10,000 years has been very important. Postulating a "worst case" for a geologic system with performance criteria that extend for 10,000 years is by its nature subjective and thus open to a wide variance in definition.

A number of difficulties have also surfaced. Large-scale uncertainty analyses place considerable demands on computer and personnel resources. Analytical efforts are greatly expanded and, if care is not taken, the complexity of the analytical effort can easily become an end in itself. Focus is thus inadvertently lost.

An attempt to quantify spatial or intrinsic variability, or statistical relationships between parameters such as correlation, will require the collection of extensive data bases. However, a number of forces are at work that would tend to limit the potential size of the data base. The depth of the repository makes drilling, instrumentation, and testing both time consuming and expensive. Many of the zones to be tested are hydrologically tight and require long periods to equilibrate and additionally make the collection of representative data difficult to obtain. The need to not compromise isolation capability also tends to limit the number of boreholes drilled for the program. It is increasingly obvious that the quantification of parameter variability and parameter correlation will be accomplished with techniques other than the simple collection of field data. These techniques will include formal methods to encode probabilities in ways that they are reproducible and traceable, and that deal with natural human biases. Numerical experimentation will be used to evaluate the importance of selected distributions or model structure to predicted repository performance. The use of such techniques is predicated on the expected robustness of the ability of the site to isolate wastes to very conservative modeling assumptions.

Validation of probabilistic models will likely compound the problems that already exist in validating deterministic geoscience models. Validation of geoscience models is currently a poorly defined area. Work is currently required to define just what is meant by validation of a geoscience model, what is being validated, and how the process that leads to a validated model should work. The lack of clarity in this area leads to the conclusion that the unique features of probabilistic models will require effort beyond that needed for deterministic ones and will lead to additional issues that must be resolved before experimental programs can be proposed and implemented.

Obviously only a select number of problem areas can be approached with the concept of directly quantifying the major sources of uncertainties affecting their respective solutions. Ideally, sufficient sampling should eliminate all but intrinsic sources of uncertainty, but to find oneself in this fortuitous situation is as rare as it is ideal. As we examine our future plans, we note that a number of criteria have been proposed to evaluate and guide future work in this area. Figure 3 lists those for consideration.

The first criterion reflects the concept that if a deterministic approach can be used to provide appropriate solutions to a problem, such an approach is generally preferable over techniques that attempt to explicitly quantify uncertainties. If there appears to be an "Inability to effectively decide based on design or worst case analy-

CRITERIA FOR SELECTIVE APPLICATION

OF UNCERTAINTY ANALYSIS TECHNIQUES

1. INABILITY TO EFFECTIVELY DECIDE BASED ON DESIGN OR WORST CASE ANALYSES.

2. KNOWLEDGE BY DECISION MAKERS OF UNCERTAINTY METHODS.

3. WELL DEFINED PROBLEM.

4. CLEAR STATEMENT OF THE UNCERTAINTIES TO BE INCLUDED.

Figure 3. **Criteria for the selective application of uncertainty analysis techniques.**

ses'' (Figure 3, Criterion 1), then an analysis incorporating uncertainty techniques can be justified. The second criterion, ''Knowledge by decision makers of uncertainty methods,'' should serve as a warning that difficult decisions remain even though the sophistication of the analysis is much greater than the usual design event or deterministic methods. Decision makers should recognize that a subjective decision reflecting a level of uncertainty that is willing to be accepted will always remain regardless of the analytical approach. Large efforts are required to acquire sufficient data, build the appropriate software, and conduct the analyses. The acceptance of the results as a valid assessment of the uncertainty inherent in the problem is often dependent on the background of the regulator or reviewer. Hopefully, the uncertainty techniques will lead to a better understanding of the uncertainties in a given isolation system in this program. The final two criteria, ''Well defined problem'' and ''A clear statement of the uncertainties to be included,'' both refer to the problems usually encountered in maintaining a clear, well-defined direction to an analysis that will directly quantify uncertain parameters. Limited studies should be completed that define the sensitivity of various decision options to the modeled parameters, their distributions, and the modeling assumptions before committing to large-scale uncertainty analyses.

Conclusions

The following summarizes the direction of the evaluation of uncertainty at Hanford, the approach being followed, and potential problems:

1. BWIP is committed to quantifying as much of the uncertainty in long-term predictions of performance as is practicable. There is legislative precedent to do so, and the nature of the questions needing resolution compel such an approach.
2. Future expansion of this portion of the analytical effort beyond the current level will be limited.
3. The expected field and laboratory data base will not be sufficient to completely satisfy the analytical needs of model validation. Some novel solutions are needed.
4. Formal probability encoding techniques will be needed to ''quantify'' expert judgement and to fill the gap between available field and laboratory data and analytical needs.

We expect the use of uncertainty techniques to be an asset in examining the uncertainties found in the evaluation of long-term system performance. However, it is important that focus is maintained on project goals and that reasonable expectations of data availability temper future analytical plans.

Literature Cited

1. Benjamin, J. R., and C. A. Cornell, *Probability, Statistics and Decision for Civil Engineers,* McGraw-Hill Book Company, 1970.
2. Codell, R., ''Draft Generic Position on Groundwater Travel Time,'' Office of Nuclear Material Safety and Safeguards, U.S. Nuclear Regulatory Commission, June 1986.

3. Davis, G. A., *Tectonic Evaluation of the Pacific Northwest: Precambrian to Present,* Preliminary Safety Analysis Report, Amendment 23, Subappendix 2RC, WPPSS Nuclear Project No. 2, Richland, WA, 1977.
4. Early, T. O., M. D. Mitchell, and G. D. Spice, *A Hydrochemical Data Base for the Hanford Site, Washington,* SD-BWI-DP-061, Rev. 1, Rockwell Hanford Operations, Richland, WA, 1986.
5. Sagar, B., P. M. Clifton, and A. J. Knepp, ''Application of Groundwater Models to Geologic Disposal of High-Level Nuclear Wastes,'' National Waterwell Association, 1987.
6. Tolan, T. L., S. P. Reidel, M. H. Benson, J. L. Anderson, K. R. Fecht, and D. A. Swarson, ''Revision to the Extent and Volume of the Columbia River Basalt Group,'' Vol. 19(6), p. 83, *Proceedings of Annual Meeting of a Cordilleran Section,* Geologic Society of America, Hilo, Hawaii, May 20–22, 1987.
7. U.S. Department of Energy, *Environmental Assessment, Reference Repository Location, Hanford Site, Washington,* Vol. 1, DOE/RW-0070, Office of Civilian Radioactive Waste Management, May 1986.
8. U.S. Department of Energy, *OCRWM Mission Plan Amendment,* DOE/RW-0128, Office of Civilian Radioactive Waste Management, June 1987.
9. U.S. Environmental Protection Agency, ''Environmental Standards for the Management and Disposal of Spent Nuclear Fuel, High-Level and Transuranic Radioactive Wastes; Final Rule,'' 40 CFR Part 191, Sept. 19, 1985.
10. U.S. Nuclear Regulatory Commission, *Safety Evaluation Report Related to the Operation of WPPSS Nuclear Project No. 2,* NUREG-0892, Supplement No. 1, 1982.

2 Sensitivity and Uncertainty Methods for Large Computer Codes

Deterministic Methods for Sensitivity and Uncertainty Analysis in Large-Scale Computer Models*

B. A. WORLEY, E. M. OBLOW,
F. G. PIN, and R. E. MAERKER
Engineering Physics and Mathematics Division
Oak Ridge National Laboratory
Oak Ridge, Tennessee

J. E. HORWEDEL, R. Q. WRIGHT,
and J. L. LUCIUS
Computing and Telecommunications Division
Oak Ridge National Laboratory
Oak Ridge, Tennessee

Abstract

The fields of sensitivity and uncertainty analysis have traditionally been dominated by statistical techniques when large-scale modeling codes are being analyzed. These methods are able to estimate sensitivities, generate response surfaces, and estimate response probability distributions given the input parameter probability distributions. Because the statistical methods are computationally costly, they are usually applied only to problems with relatively small parameter sets. Deterministic methods, on the other hand, are very efficient and can handle large data sets, but generally require simpler models because of the considerable programming effort required for their implementation. The first part of this paper reports on the development and availability of two systems, GRESS and ADGEN, that make use of computer calculus compilers to automate the implementation of deterministic sensitivity analysis capability into existing computer models. This automation removes the traditional limitation of deterministic sensitivity methods. The second part of the paper describes a Deterministic Uncertainty Analysis (DUA) method that uses derivative information as a basis to propagate parameter probability distributions to obtain result probability distributions. The last part of the paper demonstrates the deterministic approach to sensitivity and uncertainty analysis as applied to a sample problem that models the flow of water through a borehole. The sample problem is used as a basis to compare the cumulative distribution function of the flow rate as calculated by the standard statistical methods and the DUA method. The DUA method gives a more accurate result based upon only two model executions compared to fifty executions in the statistical case.

*Research sponsored by National High-Level Waste Management Program of the U.S. Department of Energy, under contract No. DE-AC05-840R21400 with Martin Marietta Energy Systems, Inc.

Introduction

The Office of Nuclear Waste Isolation (ONWI) performed sensitivity and uncertainty studies as part of its performance assessment of a high-level nuclear waste repository in salt [*1,2*]. The role of the sensitivity analysis is to provide a means to limit the scope of the more complicated problem of quantifying uncertainties. Uncertainty analyses will be performed to support design reliability studies, to produce a cost-benefit analysis in conjunction with cost estimates, to ensure compliance with regulatory criteria, and to help identify important research and development needs.

The sensitivity analysis of computer-generated results consists of determining the effect of model data upon the calculated results of interest. Because computer model equations can be differentiated analytically, sensitivities can be precisely defined and calculated in a deterministic fashion using both direct and adjoint methods [*3–10*]. The deterministic approach is particularly suited to large-scale problems for which direct perturbation of the model data becomes impractical from a cost standpoint. The main drawback to the deterministic approach has been the initial manpower investment to add the computational capability for calculating the necessary derivatives into existing computer models.

For quantification of uncertainties in computer-generated results, the problem can be expressed more precisely as the propagation of input uncertainties through models by the laws of probability to obtain output uncertainties. (The uncertainty associated with whether the computer model accurately reflects the physical phenomena is a problem of model validation and is not addressed in this paper.) Uncertainties of computer results are of primary interest in applications such as repository performance assessment in which experimental validation is not possible or practical. Because of the complicated nature of the computational structure of large computer models, and because of the large number of input and data parameters associated with such models, to date almost all uncertainty analysis of computer results has been performed using a statistical approach [*11–14*].

This paper presents a comprehensive approach to sensitivity and uncertainty analysis of large-scale computer models that is analytic (deterministic) in principle and that is firmly based on the model equations. The theory and application of two systems based upon computer calculus, GRESS [*15,16*] and ADGEN [*17*], are discussed relative to their role in calculating model derivatives and sensitivities without a prohibitive initial manpower investment. A Deterministic Uncertainty Analysis (DUA) method [*18*] that retains the characteristics of analytically computing result uncertainties based upon parameter probability distributions is then introduced. Finally, the role of sensitivity analysis and the DUA method are demonstrated on a sample problem describing the flow of water through a borehole.

Deterministic Sensitivity Analysis

A brief description of general sensitivity theory is given here as an aid to understanding the problem of applying this theory to computer models. The example to be discussed will be that of a general set of nonlinear equations given by

$$y = F(y,c) \quad , \tag{1}$$

where y represents the dependent variable being solved for, c represents the user-specified model data or parameter set, and F defines the model equations. The particular form chosen in Eq. (1) is one that can be used generally to represent equations coded in the FORTRAN programming language. The left side of the equation can represent the stored value of the variable calculated from the functional formula on the right side.

Because the number of components of the vector y calculated in any typical large-scale modeling problem is large, it is useful to define a generic result for such a calculation that is of particular interest to the model user. Typically many results will be needed for analysis, but in most cases they form a much smaller set than the actual set of y component values. A typical result will be defined as

$$R = h(y) \ , \tag{2}$$

where R is a single number that is a function of the solution to Eq. (1). For notational ease, the generic parameter α_i will be used to denote any individual parameter. The total number of parameters in the problem will be assumed to be M so that the index on α_i will run from 1 to M.

The basic problem in any sensitivity study is to find the rate of change in the result R arising from changes in any model parameters. For the generic parameter α_i, then, the quantity of interest is the numerical value of $dR/d\alpha_i$ given analytically by

$$\frac{dR}{d\alpha_i} = \frac{\partial h}{\partial y}\frac{dy}{d\alpha_i} \tag{3}$$

Because the functional dependence of R on y through h(y) is defined analytically by the model user, only $dy/d\alpha_i$ needs to be generated in order to evaluate Eq. (3). The procedure needed to get $dy/d\alpha_i$ is to differentiate Eq. (1) as follows:

$$\frac{dy}{d\alpha_i} = \frac{\partial F}{\partial y}\frac{dy}{d\alpha_i} + \frac{\partial F}{\partial c}\frac{dc}{d\alpha_i} \ . \tag{4}$$

Rearranging Eq. (4) yields the following set of coupled equations to solve for $dy/d\alpha_i$,

$$\left(I - \frac{\partial F}{\partial y} \right) \frac{dy}{d\alpha_i} = \frac{\partial F}{\partial c}\frac{dc}{d\alpha_i} \ , \tag{5}$$

or in more compact form,

$$Ay_i' = s_i \quad , \quad i = 1, \ldots, M \quad , \tag{6}$$

where I is the identity matrix and A, y_i', and s_i are given by

$$A = I - \frac{\partial F}{\partial y} \ , \tag{7}$$

$$y_i' = \frac{dy}{d\alpha_i} \quad , \tag{8}$$

and

$$s_i = \frac{\partial F}{\partial c} \frac{dc}{d\alpha_i} \quad . \tag{9}$$

If Eq. (6) were solved directly for y_i', the result could be used in Eq. (3) to evaluate $dR/d\alpha_i$. This method of sensitivity analysis is called the "direct" approach and is a classical methodology that has received a great deal of attention in the literature [*3,7*]. Because Eq. (6) must be solved each time a new α_i is defined, the direct approach is most suitable for problems with relatively few input parameters of interest, for problems in which the solution of Eq. (6) is very inexpensive compared to the solution of the model itself, or for analytical problems in which the inverse of A can be explicitly determined.

For large-scale models with a large data base in which the ultimate objective is still the evaluation of $dR/d\alpha_i$, the intermediary step of solving for $dy/d\alpha_i$ and its inherent computational inefficiency can be avoided. For such problems the "adjoint" approach is far more applicable. In this methodology, use is made of the fact that Eq. (6) is linear in y_i', and appropriate adjoint equations can therefore be developed specifically to evaluate Eq. (3).

Defining the matrix adjoint of A as A* and using the usual definition of this adjoint give the identity

$$u^{tr}Av = v^{tr}A^*u \quad , \tag{10}$$

where u and v are arbitrary vectors and A* is defined as

$$A^* = A^{tr} \tag{11}$$

Here the tr superscript represents the transpose of the vector or matrix.

If specific vectors for the problem at hand are chosen for u and v, the problem-specific adjoint equation can be set up as follows:

$$A^*y^* = s^* \quad , \tag{12}$$

where

$$A^* = A^{tr} = \left(I - \frac{\partial F}{\partial y} \right)^{tr} \quad . \tag{13}$$

Choosing s* as

$$s^* = (dh/dy)^{tr} \quad , \tag{14}$$

Eq. (3) can now be evaluated as follows:

$$\frac{dR}{d\alpha_i} = y^{*tr} \frac{\partial F}{\partial c} \frac{dc}{d\alpha_i}, \quad i = 1, \ldots, M, \tag{15}$$

where y* is now the solution to

$$\left(I - \frac{\partial F}{\partial y} \right)^{tr} y^* = \left(\frac{dh}{dy} \right)^{tr} \tag{16}$$

The simplicity of the adjoint approach lies in the fact that Eq. (16) needs to be solved only once to get any and all sensitivities in the problem. This is a result of Eq. (16) being independent of the definition of α_i. The particular choice of α_i is reflected only in the evaluation of Eq. (15), which involves simple vector products. In essence, the adjoint approach reduces the computational effort needed to evaluate $dR/d\alpha_i$ from solving many coupled linear equations to the evaluation of several vector products. For large-scale systems with many thousands or even millions of parameters, this represents orders of magnitude in computational efficiency.

It should be noted here that both the direct and adjoint equations [i.e., Eqs. (6) and (16)] are in any case far easier to solve than the original model [Eq. (1)]. Both Eqs. (6) and (16) are linear, while Eq. (1) is nonlinear. The direct and adjoint approaches, however, require the results of the original model equations to be available in order to set up Eqs. (6) and (16), because the A matrix and the vectors s and s* depend on y.

In order to solve either the direct or adjoint sensitivity analysis, then, the model user must first generate the matrices $\partial F/\partial y$ and $\partial F/\partial c$ from the original nonlinear computer model. For large-scale problems this generally requires a great deal of painstaking human effort. First, the model equations must be extracted from the computer coding. They must then be differentiated with respect to all parameters of interest, and finally direct or adjoint sets of equations must be set up for computational solution. Successful automation of this procedure greatly reduces the human effort involved, potentially by orders of magnitude. The advantage of automation of sensitivity model development is therefore great indeed. The next two sections discuss two automated systems that use calculus precompilers to add capability to existing FORTRAN computer models for solving the direct and adjoint equation procedures.

GRESS: An Automated System for Solving the Direct Sensitivity Problem

For large-scale computer models, the equations are usually very complex and tied closely to and embedded in complex model logic and data-handling routines. In addition, for nonlinear problems, the numerical solution procedure often precludes an easy separation of the modeling equations from other parts of the model coding structure. For these reasons, a general system was developed to automate the application of computer calculus in existing codes. The system first developed for solving the direct sensitivity problem was the GRadient-Enhanced Software System (GRESS). Details of the GRESS system are given in Refs. 15 and 16, and the underlying ideas are briefly summarized herein.

The basic principle of GRESS is to read the model source program and search for model equations. These are identified uniquely by the appearance in the FORTRAN source program of the "=" symbol. Because all FORTRAN "equations" so identified occur in the form of Eq. (1) (i.e., with a single dependent variable on the left side of such an expression), GRESS can search for and analyze each equation in terms of its functional dependence on y and c. The basic computer calculus operations of GRESS are then used to compute the successive elements of $\partial F/\partial c$ and $\partial F/\partial y$ as each expression is encountered. The differentiation is carried out analytically using calculus software for all permissible FORTRAN functions and operators, and the results are computed and stored numerically using the local (current) values of the independent and dependent variables. GRESS takes advantage of the fact that in solving Eq. (5), the matrix $(I - \partial F/\partial y)$ is lower triangular and the y vector can be computed by forward substitution. The important point is that the components of y are solved successively as each equation is differentiated and that the $(I - \partial F/\partial y)$ matrix does not have to be stored. (The adjoint problem requires the storage of this matrix, as will be discussed in the next section.)

GRESS only recognizes real-variable store operations as valid equations (i.e., the left side variable in a FORTRAN equation must be real), because continuous derivatives are to be calculated. Also, the left-hand side of an equation is treated as a separate component of y each time it is executed (including each execution in a DO LOOP). The calculation of $\partial F/\partial y$ and $\partial F/\partial c$ in effect means that GRESS can be used to calculate the derivative of any real variable in the model with respect to any other real variable in the model. All derivatives are available for both internal and external use. For example, the derivatives $dR/d\alpha_i$ are used in the DUA method to be described later.

The application of GRESS to an existing FORTRAN model is illustrated in Figure 1. In the preliminary preprocessing step, the FORTRAN model is separated into two subsets. The first, which contains all main sequence computations, is hand-modified for submission to the GRESS precompiler. The amount and nature of these modifications depend on the particular application at hand and the limitations of the current version of GRESS. GRESS presently recognizes all FORTRAN 77 functions with the exception of complex functions. The second subset (possibly null) is composed of subroutines whose only communication with the first subset is through the arguments in their calling sequence. These subroutines are usually associated with input, output, and peripheral program-analysis functions. They do not require GRESS compilation and may usually be submitted unchanged to the FORTRAN compiler.

The next step is the GRESS precompilation in which the automated code translation necessary to compute derivatives is performed using automated computer calculus. This step consists primarily of a rearrangement of the program data structure and a substitution of calls to GRESS interpretive software in place of all arithmetic lines of coding. All arithmetic operations of the original model are precompiled into a pseudomachine code (the GRESS P-code) for use during program execution. The

ORNL–DWG 85-18000

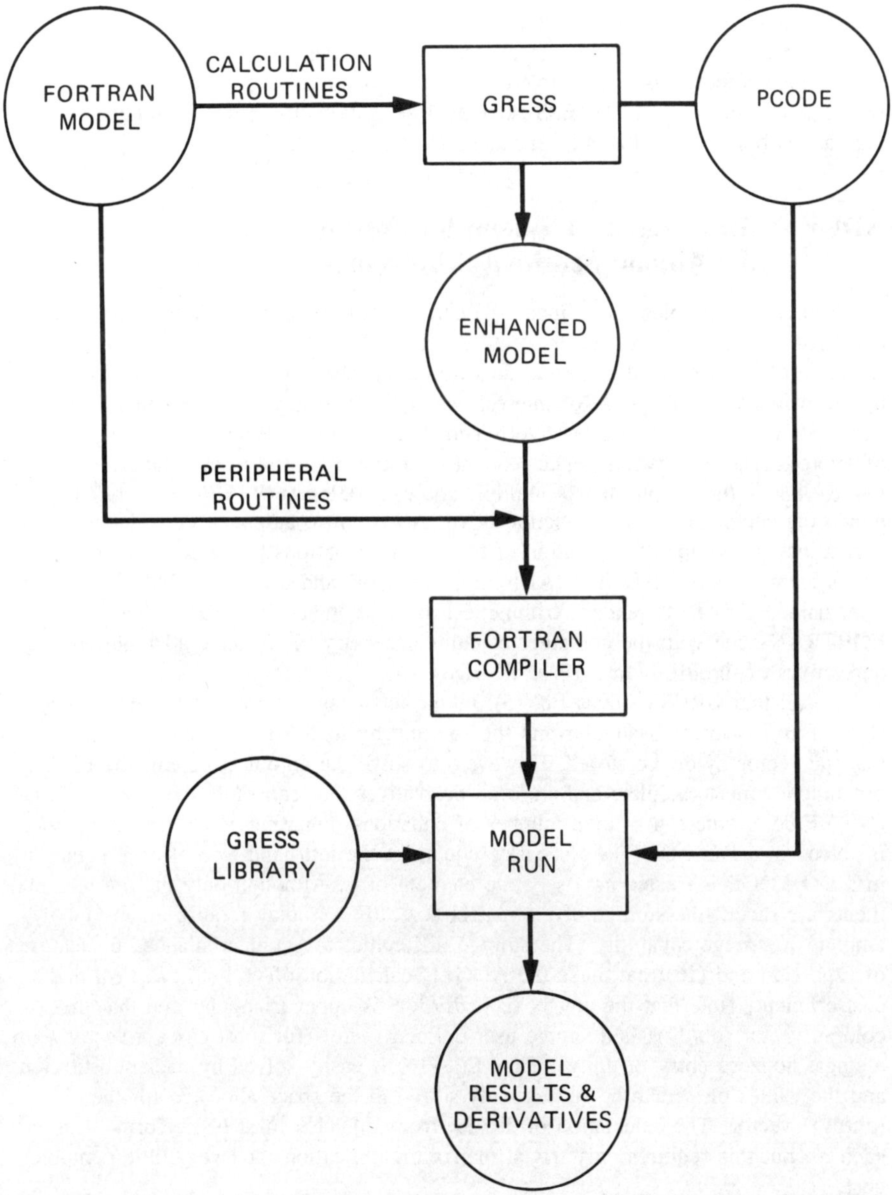

Figure 1. **GRESS flowchart.**

two output files of this step are the enhanced model and the binary P-code file. These two files and a set of GRESS software subroutines supporting the enhanced model are compiled and run as a normal FORTRAN program to produce both the reference model results and gradient information. The gradients and reference results are used to calculate the sensitivities.

GRESS has undergone extensive verification during its development. To date, five major computer models of interest to ONWI have been enhanced using GRESS, with direct comparison of GRESS-calculated derivatives to perturbation-derived derivatives being made for each enhanced model [*19–23*].

ADGEN: An Automated System for Solving the Adjoint Sensitivity Problem

The adjoint problem is defined by Eqs. (12–16). As previously mentioned, the calculation of the adjoint solution vector y^* from Eq. (16) is not a function of the selection of input parameter α_i and thus needs only be performed once to determine the derivatives of a response of interest with respect to any parameter of interest. The matrix $\partial F/\partial c$ must also be determined, but it too is independent of the parameter of interest. The only parameter-dependent operation required to calculate the derivative $dR/d\alpha_i$ is the simple matrix multiplication operation $(y^{*tr})(\partial F/\partial c)(dc/d\alpha_i)$ in which the vector $dc/d\alpha_i$ is a function of α_i. A system to automate the calculation of derivatives based upon the solution of the adjoint equations has been developed [*17*]. The system is named ADGEN (ADjoint GENerator) and uses a GRESS-like precompiler named EXAP (Extended Arithmetic Processor) that will enhance any FORTRAN code with the computer calculus necessary to calculate all required derivatives of the $\partial F/\partial y$ and $\partial F/\partial c$ matrices.

Recall that GRESS solves Eq. (5), taking advantage of the fact that the matrix $(I - \partial F/\partial y)$ is lower triangular and the solution by forward substitution requires only that the vector $dy/d\alpha$ be stored. However, to solve the adjoint problem, the EXAP precompiler must calculate and store all derivatives that constitute the $n \times n$ matrix $(I - \partial F/\partial c)^{tr}$, where n = total number of equations, counting each time an equation is solved in a DO LOOP as a separate equation; the left-hand side of each equation in a DO LOOP is treated as a separate element of y. Although only the nonzero elements are saved, the storage of the matrix $(I - \partial F/\partial c)^{tr}$ may require a substantial amount of storage capability. The storage difficulties are counterbalanced by features of Eqs. (15) and (16) that make the ADGEN calculation of y^* both practical and cost efficient. Note that the matrix $(I - \partial F/\partial c)^{tr}$ is upper triangular and that the column vector $(dh/dy)^{tr}$ is a simple user-defined vector (for most cases a vector with a single nonzero entry of unity). Thus Eq. (16) is easily solved by back substitution, and the values of y^* can be successively stored in the space allocated for the $(dh/dy)^{tr}$ vector. The calculation of $dR/d\alpha_i$ from Eq. (15) must be performed for each α_i, but this requires only trivial matrix multiplications and very little computer cost.

The ADGEN system is shown schematically in Figure 2. Like GRESS, the ADGEN precompiler EXAP produces a binary file that essentially contains all the arithmetic operations of the code being enhanced. The enhanced code has CALL

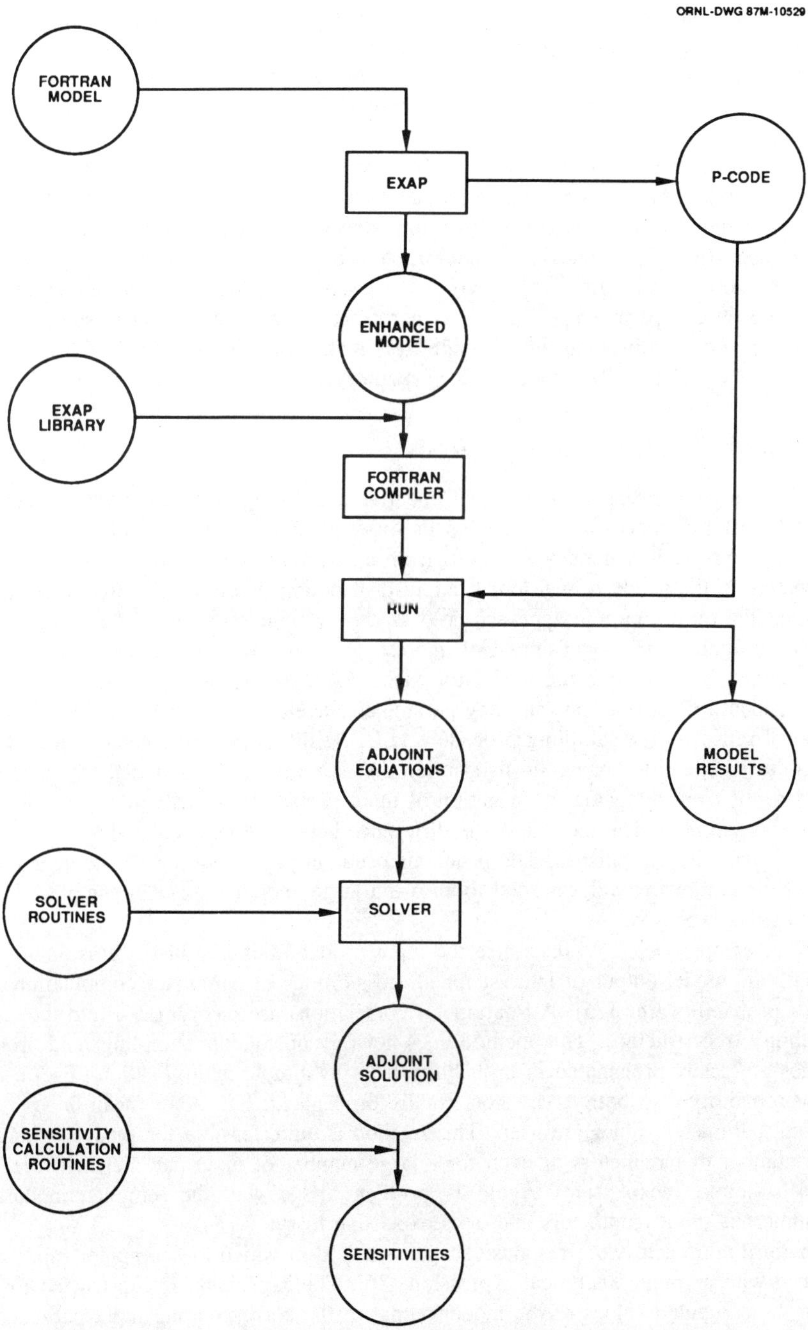

Figure 2. **ADGEN flowchart**.

statements that access the P-CODE, and it calculates the normal model results as well as the derivatives making up the $\partial F/\partial y$ and $\partial F/\partial c$ matrices. One major difference from the GRESS schematic (Figure 1) is the requirement to store the $(I - \partial F/\partial c)^{tr}$ matrix and the need for a postprocessor solver routine to calculate the adjoint solution.

The testing and verification of ADGEN have been performed on the PRESTO-II computer model [*24*], which has approximately 6,900 lines of coding. The sample problem included in Ref. 24 calculates a time-dependent radiation dose to humans from transport of 42 radionuclides. The problem has a large data base of approximately 69,000 input parameters. ADGEN was used to calculate the derivatives of the dose to all 69,000 parameters, and direct perturbation was used to verify selected derivatives. For this problem the $(I - \partial F/\partial y)^{tr}$ matrix had approximately 870,000 rows, requiring about 19,700 kilobytes of memory.

Deterministic Uncertainty Analysis

The analytical propagation of input uncertainties through a calculational model is unfeasible, if not impossible, for all but the most simple models. The difficulty lies in mapping probability density functions from an M-dimensional space of input parameters to the singly dimensioned output distribution function. To circumvent this problem, the most common approach is to randomly sample the input distributions and then calculate the model output of interest, constructing a probability distribution of the output by rerunning the model for each sample set of input parameters. The input probability distributions and any parameter correlations are handled, in a statistical sense, in the sampling procedure [*13,14*]. But hopefully the sampling procedure will lead to an output distribution that is representative of that which would result from the actual propagation of input probability distributions. As the number of sampling sets increases, the difference between the calculated and "true" output distribution diminishes. The problems occur in practice when the number of runs of the computer model needed to ensure a large enough statistical sample becomes too expensive.

Another approach is to discretize the input probabilities into histograms and evaluate the model output of interest for all possibilities of parameter combinations to form a probability tree [*25*]. All parameter correlations are incorporated into the probability tree structure. This method does not rely on random sampling, and probabilities are easily propagated in probability trees by simple multiplication. The histogram probability distributions are not actually propagated, but rather mean or endpoint parameter values are used. This method is quite feasible for models with a small number of parameters or even for a large number of input parameters if the model is simple (inexpensive). Again the problem arises when the computer model has numerous input parameters and/or is expensive to run.

A third approach is the response surface method in which the computer model is replaced with a simple analytical expression [*26*]. The expression is constructed by fitting the computed values of the model output to the corresponding input parameters, or more generally, to chosen functions of the input parameters. Traditionally, statistical techniques are used to choose the set of computer model runs to be made for this fitting. The uncertainty in the response is then determined by a second, more extensive, statistical sampling and evaluation of the response surface. The advantage of replacing the model with the response surface is the drastically reduced computational time to compute the expression result compared to running the computer model. The disadvantage is the introduction of error in the calculated output by replacement of the model with a simple expression.

The DUA method [*18*] combines the characteristics of the response surface method and probability trees. Statistical sampling is not required, and probabilities are propagated analytically within discretized numerical meshes that encompass the parameter space. The approach underlying the deterministic calculation of uncertainties in the DUA method relies upon (1) a replacement of the computer model with an analytical function relating the responses of interest to the parameters of interest, and (2) discretizing the parameter space and calculating the expected value of the response within each discrete parameter space "mesh." The parameters of interest are chosen to be those that are "uncertain," meaning that they have known or assumed probability distributions. The parameters of interest may often include the entire set of data used by the computer model.

This deterministic approach differs from the response surface methods in two ways. First, the analytical function is constructed based on the response value of interest as well as the partial derivatives of the response with respect to each of the parameters. The classical response surface method constructs the surface (analytical expression) based on only the response value at each parameter space point. Thus the degrees of freedom with which to fit the response to the parameter values are much greater in the DUA method than in the response surface methods. There is, of course, no reason to distinguish this aspect of the DUA method from response surface methods if the response surfaces are constructed using derivative data; however, in the classical response surface methods these derivatives are assumed to be unavailable, most likely because the response surface methods grew out of experimental design fields in which only set points of the control variables (parameters in our terminology) and the experimental measured values (responses) are known. As the response surface methods came to be used for replacement of large, complex computer models, again the derivative information was not used because of the difficulty of calculating partial derivatives chained through complex computational paths. However, the development of efficient methods for calculating derivatives and sensitivities for large-scale computer model results has progressed steadily based upon a firm theoretical foundation. Moreover, with the availability of GRESS and ADGEN, the calculation of derivative information for the purpose of improving the formation of response surfaces is both practical and cost-effective. This availability of derivative information is a key component in the DUA method.

The second feature that distinguishes the DUA method from response surface methods, and a feature that it has in common with probability tree methods, is that the entire parameter space is spanned. In the response surface methods, the construction of response surfaces has been used primarily to dramatically increase the number of sampling points in a statistical determination of response probability distributions, because the evaluation of the analytical response surface expression is much less expensive to obtain than the corresponding computer model result. However, only by spanning the entire parameter space can probabilities be propagated, either through a computer model or through an analytical expression. Spanning the entire parameter space is practical only if the discretization of parameter probability distributions is performed over a reasonably large mesh. In probability tree methods, for example, the probability distributions are typically replaced with the high and low values of the distribution. The DUA method extends the probability tree methods into a more rigorous propagation of probabilities in two ways:

1. Because an analytical expression relates the response to the parameters, the expected value of the response over each discretized mesh can be calculated analytically and thus gives a more meaningful value than just a single sampling point within the mesh.
2. Because the computer model is replaced with an analytical expression, a finer mesh size can be constructed over the parameter space and a more accurate representation of the parameter probability distributions of most interest can be obtained.

Another point to be made in favor of propagating probabilities through an expression that can only approximate the original computer model, and one that makes the DUA method possible for computer models with a large number of parameters, is the integral nature of the probability distribution of the response. The probability distribution of the response of interest is an integral quantity, and errors introduced by replacement of the computer model with an analytical expression are most often washed out when the parameter space is completely spanned.

Formation of response surface using derivative information

The DUA method replaces the computer model with a response surface by relating the response of interest as calculated by the computer model to the parameter values by techniques that incorporate knowledge of the partial derivatives of the response with respect to the parameters of interest. The simplest form of a response surface is one formed by linear extrapolation from reference space points to each mesh of the discretized parameter space. Within each mesh the response surface is linear with respect to the parameters, and the calculation of the expected value of the response within the mesh, given parameter probability functions, is straightforward. An extrapolation scheme that makes use of the sensitivities is outlined in Ref. 18.

A more general approach for construction of a response surface is a least-squares fitting technique. Possible schemes for incorporating derivative information into the standard fitting technique are presented in Ref. 18. Basically, to construct a response surface to a given order of expansion, the use of derivative information reduces the number of computer runs required to uniquely determine the expansion coefficients by a factor of approximately 1/M, where M is the number of parameters. One can either construct a global response surface or define local response surfaces over subregions of the parameter space. A local fit of the response values and derivatives using a low-order function may be more desirable than a global fit using a higher order function to fit a large portion of the data because a higher order fit involving many response points may result in a very radically behaved function in the parameter space not near the fitted points. For this reason most of our research to date has focused upon either local fitting or linear extrapolation from reference parameter space points. By careful selection of these parameter space points for which model results will be obtained (using Latin Hypercube Sampling, for example), the number of computer runs can be held to a small fraction ($<< 1/M$) of the number required for the conventional construction of a response surface.

Finally, sensitivity analysis plays an important role in the formation of the response surface by eliminating those parameters that have a negligible effect on the result of interest based on their sensitivities and uncertainty ranges. Also, the derivative information from the reference model runs can be used to identify the parameters that occur exclusively in a given combination. Such identification reduces the parameter space by replacement of the individual parameters with the particular combination. For example, if the derivative of the response with respect to each of two parameters is the same at each reference space point sampled, the two parameters most likely appear in the model as a sum of each other, and a single parameter representing the sum of the two can be used in the formation of the response surface in place of the two individual parameters. The sample problem exemplifies these uses of sensitivity and derivative data in the formation of the response surface.

Propagation of probabilities

The propagation of parameter probability distributions from the multidimensional parameter space to the singly dimensioned result space is determined by the governing system of equations and the input variable probability density functions (PDFs). In theory, this propagation can be performed analytically by convolution of the integral of the parameter space into a discrete number of integrals of the singly dimensioned response space, in which each integral is over a monotonically changing function representing the result. However, because the identification of the convolution integrals, in particular the limits of the integrals, is virtually impossible for all but the simplest problems, and because the model equations are nonlinear and complexly intertwined in general, the propagation of probability distributions through computer models cannot be treated analytically in the strictest sense.

The propagation of parameter probability distributions in the DUA approach is performed by discretizing the M-dimensional parameter space (M = number of parameters) into L meshes, with each mesh denoted by m_ℓ. The probability of mesh m_ℓ occurring within the entire parameter space, $p(m_\ell)$, is calculated as well as the expected value of the response function within the mesh, $E(R_\ell)$, where R_ℓ represents the response function within m_ℓ. The probability $p(m_\ell)$ is assigned to $E(R_\ell)$ to obtain the probability of $E(R_\ell)$ within the discrete space of expected values. The pairs of $p(m_\ell)$ and $E(R_\ell)$ are reordered such that $E(R_1) < E(R_2) < \ldots E(R_L)$ and as such constitute the probability density function of the response R over the parameter space. The cumulative distribution function (CDF) of R, C(R), is the running sum of the reordered $p(m_\ell)$ paired with the corresponding value of $E(R_\ell)$. In the limit as $L \rightarrow \infty$, C(R) approaches the true CDF of R as calculated using the response function.

Let the functional form of the response within m_ℓ be given by

$$R_\ell = g_\ell(c) \tag{17}$$

where $g_\ell(c)$ is the response surface function within m_ℓ resulting either from a fitting procedure or from a linear expansion from one or more reference space points. The vector c is the M-dimensional parameter vector given by $c = \{\alpha_1, \alpha_2, \ldots, \alpha_M\}_{tr}$. Given the joint probability function of c as $P(c) = P(\alpha_1, \alpha_2, \ldots \alpha_M)$, the probability that $c \subset m_\ell$ is given by

$$p(m_\ell) = P(c \subset m_\ell) = \int_{m_\ell} P(c)dc \quad , \tag{18}$$

and the expected value of the response R within m_ℓ, $E(R_\ell)$, is

$$E(R_\ell) = \int_{m_\ell} g(c)\, P(c)\, dc/p\,(m_\ell) \quad . \tag{19}$$

The values of $p(m_\ell)$ and $E(R_\ell)$ as calculated by Eqs. (18) and (19) are used to construct the PDF and CDF of the response R.

Sample Problem

Reference 27 describes a sample problem that exemplifies the use of uncertainty analysis in high-level waste applications. The sample problem consists of three coupled equations with eight input parameters and three dependent variables. The analysis focuses on one of the three dependent variables, the flow rate, as the response of interest, and statistical techniques are used to calculate the cumulative distribution of the flow rate given probability distributions for the eight input parameters.

The governing equations describe the downward flow of water through a borehole that is drilled from the ground surface through two aquifers. For a fully penetrating well and no ground-water gradient, the steady-state flow through the upper aquifer into a borehole is given by

$$Q = \frac{2\pi\,(H_u - H_{uw})\,T_u}{\ell n\,(r/r_w)} \tag{20}$$

where Q = flow, m^3/yr
T_u = transmissivity of upper aquifer, m^2/yr
H_u = potentiometric head of upper aquifer, m
H_{uw} = steady-state potentiometric head in borehole at upper aquifer, m
r = radius of influence, m
r_w = radius of borehole, m.

Similarly, the steady-state flow from the borehole to the lower aquifer is given by

$$Q = \frac{-\,2\pi\,(H_\ell - H_{w\ell})\,T_\ell}{\ell n\,(r/r_w)} \tag{21}$$

where T_ℓ = transmissivity of lower aquifer, m^2/yr
H_ℓ = potentiometric head of lower aquifer, m

$H_{w\ell}$ = steady-state potentiometric head in borehole at lower aquifer, m.

The flow of water through the borehole is assumed to be laminar and isothermal and is given by

$$Q = \frac{\pi r_w^2 (H_{wu} - H_{w\ell}) K_w}{L} \qquad (22)$$

where K_w = hydraulic conductivity of borehole, m/yr

L = length of borehole, m.

In Eqs. (20–22), Q, H_{wu}, and $H_{w\ell}$ are dependent variables; the flow rate of water, Q, is the response of interest. The uncertainty problem is to calculate the cumulative distribution function of Q, C(Q), given the probability density functions of the eight input parameters r_w, r, T_u, T_ℓ, H_u, H_ℓ, K_w, and L. The probability density functions of these eight parameters are given in Ref. 27.

The standard statistical approach for calculating C(Q) is to define a design matrix based upon the PDFs of the parameters. Several sampling procedures are available for determining a suitable design matrix. For this problem, Ref. 27 investigates the formation of design matrices based upon the Latin Hypercube Sampling (LHS) procedure using 10 and 50 design points. The choice of the sets of input parameters in a design matrix hopefully removes as much bias as possible from the selection procedure so that each calculated value of the response is of equal probability. Thus, the probability of a calculated response is 1/N, where N is the number of input sets in the design matrix and formation of C(Q) is performed by ranking the values of Q from lowest to highest and apportioning a probability of 1/N to each value. Figure 3 shows the plots of C(Q) resulting from the use of the 10-point and 50-point LHS design matrices. Clearly, the 10-point LHS design matrix does not result in a curve for C(Q) that closely matches that of the 50-point set. It is important to keep in mind that a design matrix based on N input sets requires that the computer model be run N times to determine C(Q). For this sample problem, the "computer model" consists of Eqs. (20–22), and these were solved 10 times for the 10-point design matrix and 50 times for the 50-point design matrix.

Deterministic sensitivity and uncertainty analysis of the sample problem

The DUA method was applied to this sample problem and the results compared to the published statistical results in Ref. 27. The choice of N reference points from which the response surface is formed in the application of DUA to this problem was chosen to be a subset of the 10-point LHS design matrix. For each reference point i, i=1, . . . ,N, defined by the 8-dimensional parameter vector $c_i = \{r_w^i, r^i, T_u^i, T_\ell^i, H_u^i, H_\ell^i, K_w^i, L^i\}^{tr}$, the derivative vector $dR/dc = \{(\partial Q/\partial r_w)_i, (\partial Q/\partial r)_i, (\partial Q/\partial T_u)_i, (\partial Q/\partial T_\ell)_i, (\partial Q/\partial H_u)_i, (\partial Q/\partial H_\ell), (\partial Q/\partial K_w)_i, (\partial Q/\alpha L)_i\}^{tr}$, and the response $Q(c_i)$ were calculated deterministically. Sensitivities of Q with respect to each parameter, defined by $(\alpha_k/Q(c)_i)/(\partial Q/\partial \alpha_k)_i$, were also computed. Parameters with sensitivities and uncertainty ranges such that their influence on Q was negligible were dropped from the parameter space for the purpose of forming the response surface. As a result, the parameters T_u, r, and T_ℓ were not used in the formation of the response surface. In

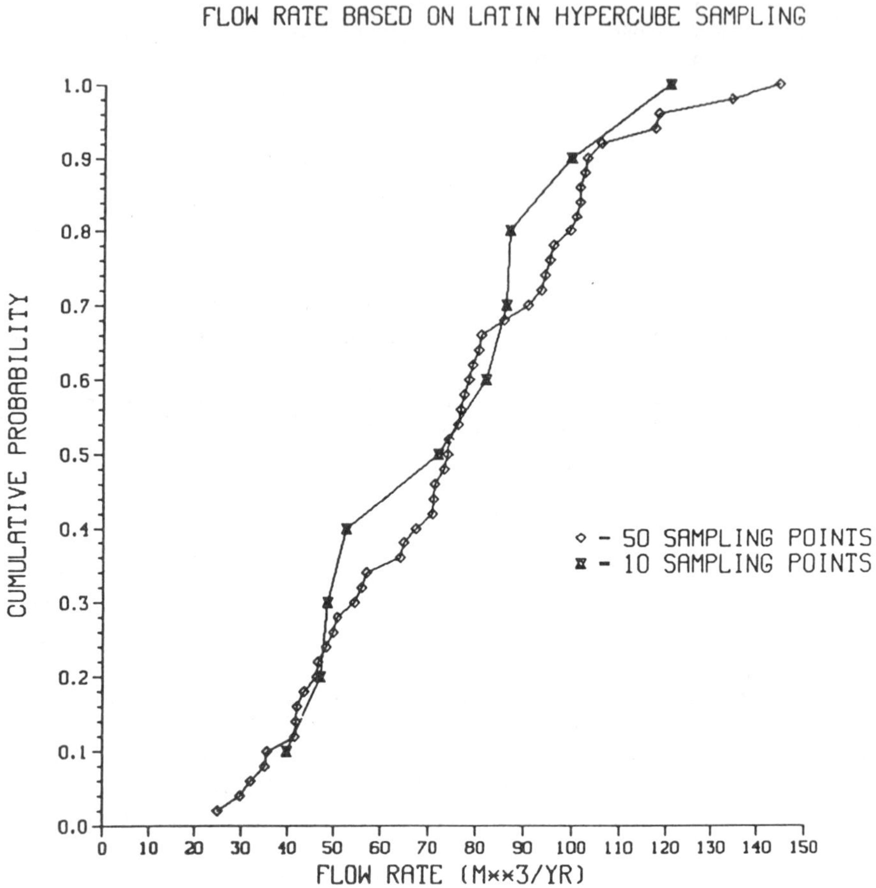

Figure 3. **Cumulative distribution functions for 10-point and 50-point LHS design matrices (from data published in Ref. 27)**.

addition, the derivatives of H_u and H_ℓ are equal but opposite in sign, indicating that the two parameters occur in the model in the combination of $(H_u - H_\ell)$. Therefore these two parameters were replaced by the single parameter $(H_u - H_\ell)$. The response surface was then constructed based on only the four significant parameters $(H_u - H_\ell)$, r_w, K_w, and L. This reduction in the number of significant parameters and the combination of parameters illustrate the role of sensitivity analysis in the DUA method as applied to this sample problem.

In this application, the response surface was formed by linear extrapolations of Q from the reference points. The entire parameter space of significant parameters as identified in the sensitivity analysis (r_w, $H_u - H_\ell$, K_w, and L) was divided into L discrete, nonoverlapping meshes, m_ℓ, $\ell = 1 \ldots L$. The expected value of Q within each mesh, $E(Q_\ell)$, was determined by replacing Q_ℓ for g(c) in Eq. (19). Here, $Q_\ell = Q(c)$, $c \subset m_\ell$, where within m_ℓ, Q(c) was calculated by extrapolation. The mesh probability $p(m_\ell)$ was calculated from Eq. (18) using the parameter probability distributions from Ref. 27; $p(H_u - H_\ell)$ had to be calculated separately using the individual distributions. The probability assigned to each $E(Q_\ell)$, $\ell = 1 \ldots L$, was the corresponding value of $p(m_\ell)$. As discussed previously, in the DUA method the number of meshes, L, is chosen such that the entire reduced parameter space is covered by nonoverlapping discrete meshes and therefore the entire probability space is complete.

As a benchmark against which a comparison of the DUA method and the statistical results from Ref. 27 could be compared, the sample problem model was executed 2304 times to approximate the "true" CDF of Q for this problem. A comparison of this benchmark 2304-point CDF to the statistical 50-point CDF from Ref. 27 is shown in Figure 4. The CDF based on the 50-point LHS design matrix is a fairly accurate representation of the true CDF of Q. DUA method results were obtained by forming a response surface by extrapolation from two reference model runs and propagating parameter PDFs over a discrete mesh consisting of 2304 meshes. As shown in Figure 5, the CDF of Q calculated deterministically based on the DUA

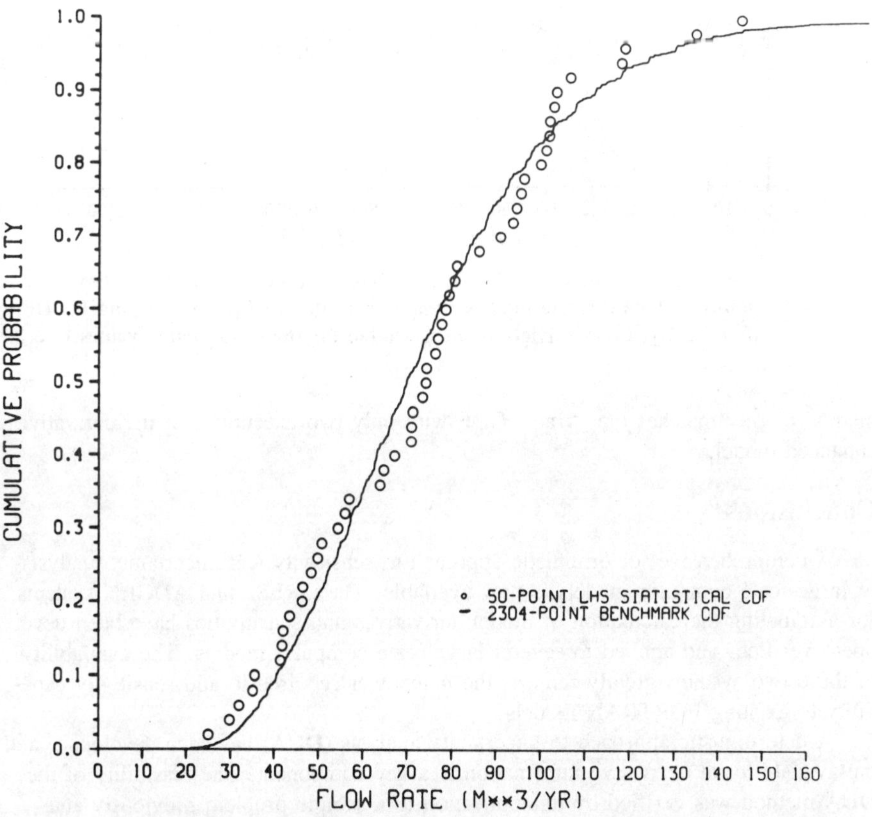

Figure 4. **Comparison of the 50-point LHS CDF to the 2304-point benchmark CDF.**

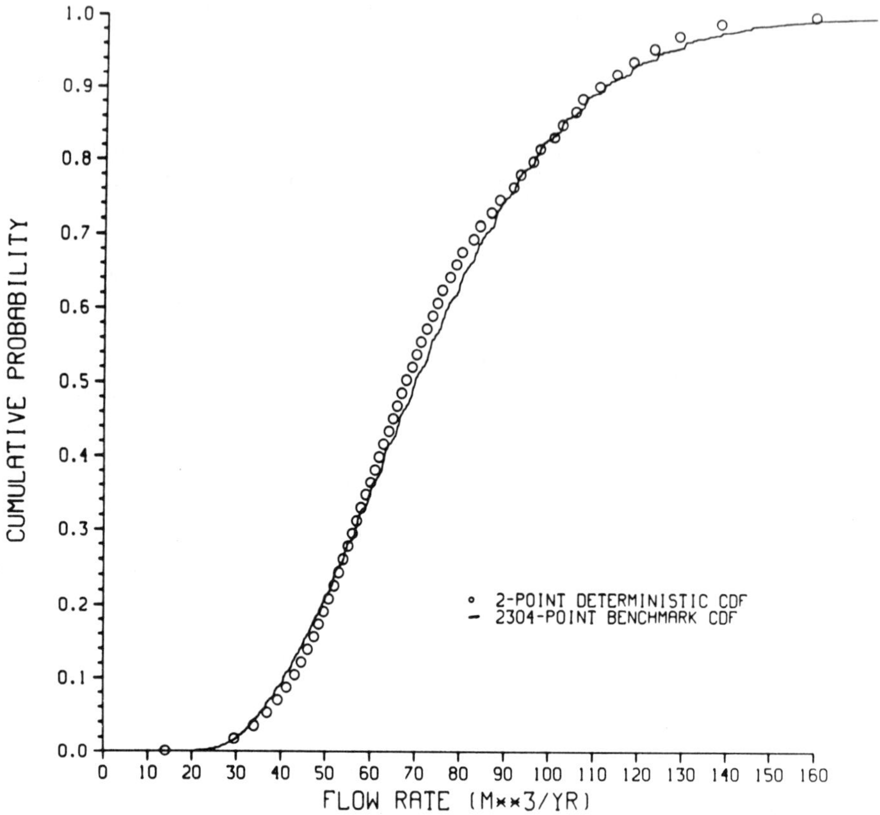

Figure 5. **Comparison of the deterministic CDF based on an extrapolation from two points to 2304 discrete meshes compared to the 2304-point benchmark CDF. (For clarity, every fortieth point is plotted for the extrapolated values.)**

method closely matches the "true" CDF with only two executions of the derivative-enhanced model.

Conclusions

A comprehensive, deterministic approach to sensitivity and uncertainty analysis of large-scale computer models is now available. The GRESS and ADGEN systems for automating the calculation of model derivatives and sensitivities have been developed, verified, and applied to several large-scale computer models. The availability of these two systems greatly reduces the manpower required to add sensitivity capability to existing FORTRAN models.

A deterministic approach to uncertainty analysis (DUA) has been developed, and the availability of derivative information is a key component. The feasibility of the DUA method was verified by its application to a sample problem previously analyzed using a statistical approach. The sample problem results show that simple linear extrapolation from two space points produces a CDF of the response of interest that more closely matches a benchmark 2304-point CDF than does the CDF based on a 50-point LHS design matrix. The reduction in model runs by a factor of 25 and the increased accuracy in calculating the CDF of the response of interest provide strong evidence that a substantial savings in computational cost is possible. This reduction is offset by the additional cost of calculating derivatives, but the deterministic calculation of model derivatives has been shown in the published literature to be both feasible and fairly cost efficient—certainly much less than a factor of 25. The availability of the GRESS and ADGEN systems for adding derivative-taking capability to existing models makes the DUA approach even more practical. The strong analytical foundation of propagating probabilities deterministically is another desirable feature of the DUA approach.

Literature Cited

1. Thomas, R. E., *Uncertainty Analysis,* ONWI-380, Office of Nuclear Waste Isolation, Battelle Memorial Institute, Columbus, Ohio (1982).
2. Harper, W. V., *Sensitivity/Uncertainty Analysis Techniques for Nonstochastic Computer Codes,* ONWI-444, Office of Nuclear Waste Isolation, Battelle Memorial Institute, Columbus, Ohio (1983).
3. Tomović, R., and M. Vukobratović, *General Sensitivity Theory,* Elsevier North-Holland, Inc., New York (1972).
4. Stacey, W. M., *Variational Methods in Nuclear Reactor Physics*, Academic Press, New York (1974).
5. Greenspan, E., "Developments in Perturbation Theory," *Advances in Nuclear Science and Technology*, Vol. 9, Academic Press, New York (1976).
6. Oblow, E. M., "Sensitivity Theory for Reactor Thermal-Hydraulics Problems," *Nucl. Sci. Eng., 68*, 322 (1978).
7. Frank, P. M., *Introduction to System Sensitivity Theory*, Academic Press, New York (1978).
8. Cacuci, D. G., *J. Math. Phys.* (NY), *22*, 2794 (1981).
9. Cacuci, D. G., *J. Math. Phys.* (NY), *22*, 2803 (1981).
10. Weisbin, C. R., et al., "Sensitivity and Uncertainty Analysis of Reactor Performance Parameters," *Advances in Nuclear Science and Technology*, Vol. 14, Plenum Press, New York (1982).
11. McKay, M. D., R. J. Beckman, and W. J. Conover, "A Comparison of Three Methods for Selecting Values of Input Variables in the Analysis of Output from a Computer Code," *Technometrics, 21*, 239–245 (1979).
12. Iman, R. L., J. C. Helton, and J. E. Campbell, *J. Quality Technol., 13*, 3, 174 (1981).
13. Iman, R. L., J. C. Helton, and J. E. Campbell, *J. Quality Technol., 13*, 4, 232 (1981).
14. Iman, R. L., and W. J. Conover, "A Distribution-Free Approach to Inducing Rank Correlation Among Input Variables," *Communications in Statistics, B11*(3), 311–334, (1982).
15. Oblow, E. M., *An Automated Procedure for Sensitivity Analysis Using Computer Calculus*, ORNL/TM-8776, Oak Ridge National Laboratory, Oak Ridge, Tennessee (1983).
16. Oblow, E. M., *GRESS, Gradient-Enhanced Software System*, ORNL/TM-9658, Oak Ridge National Laboratory, Oak Ridge, Tennessee (1985).

17. Pin, F. G., E. M. Oblow, J. E. Horwedel, and J. L. Lucius, "ADGEN: An Automated Adjoint Code Generator for Large-Scale Sensitivity Analysis," paper presented at Am. Nuc. Soc. Winter Mtg., Los Angeles, California (November 1987).
18. Worley, B. A., *Deterministic Uncertainty Analysis,* ORNL-6428, Oak Ridge National Laboratory (1987).
19. Oblow, E. M., F. G. Pin, and R. Q. Wright, "Sensitivity Analysis Using Computer Calculus: A Nuclear Waste Isolation Application," *Nucl. Sci., 94,* 46 (1986).
20. Worley, B. A., R. Q. Wright, F. G. Pin, and W. V. Harper, "Application of an Automated Procedure for Adding a Comprehensive Sensitivity Calculation Capability to the ORIGEN2 Point Depletion and Radioactivity Decay Code," *Nucl. Sci. Eng., 94,* 180 (1986).
21. Pin, F. G., B. A. Worley, E. M. Oblow, R. Q. Wright, and W. V. Harper, "An Automated Sensitivity Analysis Procedure for the Performance Assessment of Nuclear Waste Isolation Systems," *Nucl. and Chem. Waste Man., 6,* 255 (1986).
22. Worley, B. A., R. Q. Wright, and F. G. Pin, *A Finite-Line Heat Transfer Code with Automated Sensitivity-Calculation Capability,* ORNL/TM-9975, Oak Ridge, Tennessee (1986).
23. Worley, B. A., and J. E. Horwedel, *A Waste Package Performance Assessment Code with Automated Sensitivity-Calculation Capability*, ORNL/TM-9976, Oak Ridge National Laboratory, Oak Ridge, Tennessee (1986).
24. Fields, D. E., et al., *PRESTO-II: A Low-Level Waste Environmental Transport and Risk Assessment Code*, ORNL-5970, Oak Ridge National Laboratory, Oak Ridge, Tennessee (1986).
25. Sutcliffe, W. G., "Uncertainty Analysis: An Illustration from Nuclear Waste Package Development," *Nucl. and Chem. Waste Manag., 5,* 131 (1984).
26. Myers, R. H., *Response Surface Methodology*, Allyn and Bacon, Inc., Boston, Massachusetts (1971).
27. Harper, W. V., and Gupta, S. K., *Sensitivity/Uncertainty Analysis of a Borehole Scenario Comparing Latin Hypercube Sampling and Deterministic Sensitivity Approaches*, BMI/ONWI-516, Office of Nuclear Waste Isolation, Battelle Memorial Institute, Columbus, Ohio (1983).

Improving the Efficiency of Monte Carlo Simulation for Groundwater Transport Models

D. VENEZIANO
Civil Engineering Department
Massachusetts Institute of Technology
Cambridge, Massachusetts

R. KULKARNI, G. LUSTER
G. RAO, and A. SALHOTRA
Woodward-Clyde Consultants
Oakland, California

Abstract

When applied to complicated groundwater models or when used to estimate the probability of exceeding high pollutant concentrations, the Monte Carlo method becomes either imprecise or expensive. Two methods are proposed here that substantially reduce the computational requirements in comparison to Monte Carlo (MC) simulation when the objective is to estimate a given fractile of the output concentration. The methods are general with respect to the probability distribution of the uncertain source and model parameters and with respect to the algorithm that estimates the desired concentration, given such parameters. One of the methods uses a combination of several uncertainty-propagation techniques; the other is an adaptation of the Rackwitz-Fiessler algorithm of system reliability analysis. Both methods employ recursive linearization of the mass transport model.

The methods have been implemented in a FORTRAN computer code and used to evaluate the 0.9 fractile of downgradient concentration using a semianalytical three-dimensional mass transport model. Results are typically close to direct MC estimates, and the CPU time is about two orders of magnitude less than that required by the MC method. This is an important feature of the proposed methods, especially for the evaluation of exposure through complicated numerical algorithms.

Of the two methods considered, the one based on the Rackwitz-Fiessler method is recommended for two reasons: (1) it is numerically more stable, and (2) it produces results that do not depend on parameters to be selected by the user, such as the sample size and the random-number seed.

An important feature of the methods, which is not available in standard Monte Carlo, is the analytical evaluation of sensitivity of the desired fractile to the input random variables. Knowledge of the parameters that are more influential is obviously important to refine their treatment in the transport model and their probabilistic representation. Although not implemented at present in the FORTRAN program, a method is described to include the effect of parameters that vary randomly in time or space.

Introduction

The U.S. Environmental Protection Agency (EPA) is required to regulate the disposal of toxic, hazardous substances in landfills such that they do not pose a risk to human health or the environment. The long-term risk at a land-disposal facility is evaluated in terms of the probability that the concentration of a given chemical at a downgradient, downwind, or downstream monitoring point exceeds a chemical-specific threshold level. The movement of a chemical from the point of disposal to the monitoring point depends on a number of environmental, chemical, and medium-specific parameters that typically exhibit spatial and temporal variability. Superimposed on this is the uncertainty arising from lack of sufficient information about some of the parameters. Thus, the system under consideration is best described by a probabilistic representation of the input vector $X = (X_1, \ldots, X_K)^T$. Depending on the objectives of the analysis, uncertainty on the vector X may reflect a site-specific, a class-specific, or a generic state of knowledge.

Given this probabilistic framework, a problem of interest is that of finding the probability distribution of the concentration, Y, at the receptor location. Regulatory decisions (whether to ban the land disposal of a chemical, whether a landfill is adequate, etc.) are then based on a comparison of a specific fractile of this concentration distribution with some preselected threshold values. A specific approach suggested by the EPA in 1986 (*Federal Register 51*:9, p. 1602 ff) is to estimate the 90th percentile of the receptor concentration Y, assuming the release of a unit concentration of contaminant at the disposal site, and then to "back-calculate" the maximum source concentration that would be allowed at the disposal site in order to ensure that the resulting receptor concentration would not exceed a specified threshold value with greater than 10% probability. (The back-calculation is straightforward, because in the transport model that is used [Salhotra et al., *1*], the receptor concentration is proportional to the source concentration.) The estimation of an upper fractile of a model output distribution is thus a problem of great interest to the EPA. The methods presented in this paper were developed primarily in response to this problem.

The standard approach to calculating quantities such as the P-fractile Y_P or the probability Prob[$Y > y^*$] that Y exceeds a set value y^* is Monte Carlo (MC) simulation. Although conceptually simple, general in its applicability, and asymptotically exact, the MC method has the drawback of being computationally intensive. This is a serious limitation, especially when accurate modeling of the transport problem requires the use of detailed numerical procedures, based for example on two- or three-dimensional finite-element codes. Another case where MC is inefficient is the estimation of extreme fractiles (Y_P with P very close to 1) or very small probabilities (Prob[$Y > y^*$] with y^* very large). For such cases, we propose two alternatives to direct MC. The first alternative combines features of existing uncertainty-propagation techniques (Monte Carlo, response surface, Rackwitz-Fiessler) but is otherwise new. The second alternative is a more direct adaptation of the Rackwitz-Fiessler procedure for system reliability. Although the focus here is on the estimation of fractiles Y_P, both procedures can be easily modified to produce exceedance probabilities.

The components of X are allowed in the first method to have any joint probability distribution, provided that Monte Carlo simulation from that distribution is possible. In the second method, the probability distribution of X is also unrestricted, although the condition of independence among nonnormal variables makes computation easier. In both cases, the dependence of Y on X is defined algorithmically through a computer code. The type and complexity of the underlying groundwater model is therefore unrestricted, except that it should be feasible to use the code about 30 to 50 times with different input vectors X. This represents a significant saving with respect to direct Monte Carlo, for which the number of simulations would typically be of the order of 1000 or greater.

Basis of Proposed Methods

Suppose for the moment that the components of X are random variables (the case with random functions will be considered later) and denote by D_P the region of X-space where $Y(X)$ exceeds the P-fractile Y_P. This region and the boundary B_P on which $Y = Y_P$ are illustrated in Figure 1 for the case of two input variables. Clearly, the probability content of D_P is $(1-P)$, so that

$$\text{Prob}[D_P] = \int_{D_P} f_X(x)\, dx = 1 - P \tag{1}$$

where f_X is the joint probability density function of X.

The idea behind the first, so-called Modified Monte Carlo (MMC) method is that a good approximation to the probability content of D_P is obtained by linearizing the boundary B_P around its maximum-density point X'_P, i.e., around

$$X'_P = \{x : f_X(x) = \max, \text{ subject to } Y(x) = Y_P\} \tag{2}$$

The approximating half-space D'_P is shown in Figure 1, bounded by a dashed straight line.

If one could find D'_P, one could also find the point X'_P and calculate Y_P exactly, as $Y_P = Y(X'_P)$. Unfortunately, D'_P is difficult to obtain. An alternative approximation to D_P, which also is easier to obtain than D'_P, is one that is designed to preserve the probability content of D_P (D'_P does not). This is the half-space D''_P such that $\text{Prob}[D''_P] = 1-P$. This second approximation is the one used by the MMC method. Once D''_P has been found, Y_P is estimated as

$$Y''_P = Y(X''_P) \tag{3}$$

in which X''_P is the point on the boundary of D''_P where f_X is maximum. The region D''_P and the point X''_P are shown in Figure 2. In the MMC method, X''_P is further replaced with an estimate $\hat{X}''_P$, obtained through Monte Carlo sampling (explained below).

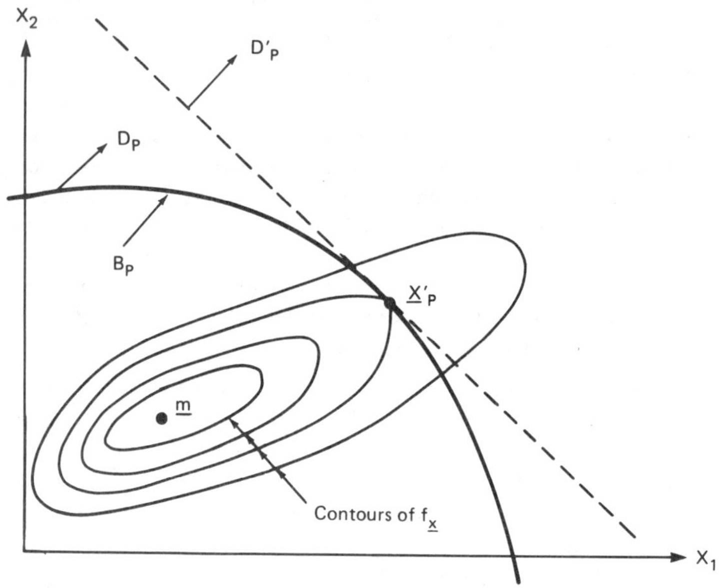

Figure 1. **Definition of regions D_P and D'_P.**

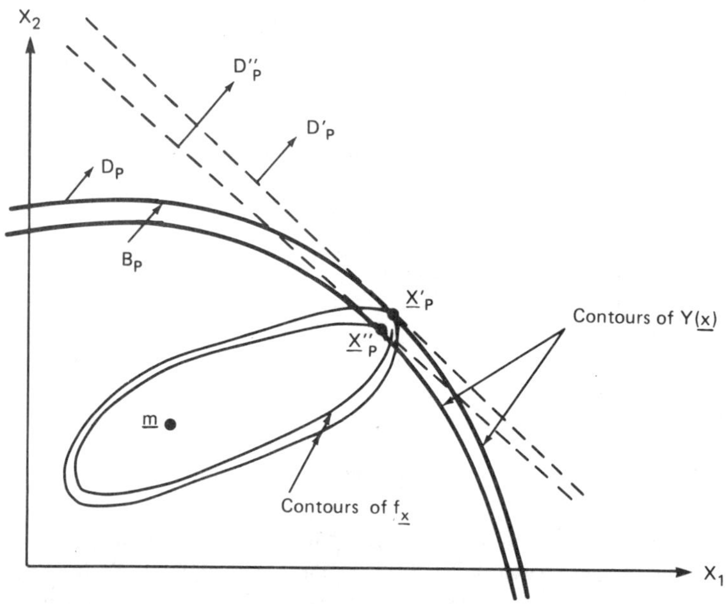

Figure 2. **Definition of regions D_P and D''_P.**

The second, Rackwitz-Fiessler (RF) method uses a similar approximation to D_P, in which however the point X_P'' is found analytically after the actual distribution of X is replaced with a multivariate normal distribution. The approximating normal distribution is such that the values of the true marginal probability densities and cumulative distributions of X are reproduced at the point X_P''.

Local-linearization procedures similar to the approximation in Equation 3 (with X_P'' calculated by either the MMC or the RF method) have been used extensively in engineering reliability, where they have been found to produce consistently good results (Ellingwood et al., *2*; Ditlevsen, *3*).

Modified Monte Carlo (MMC) Method

A procedure to implement the MMC method numerically, and in particular to estimate the vector X_P'' in Equation (3), is described below. Frequently one is interested not only in estimating Y_P but also in measuring the sensitivity of Y_P to changes in the distribution of the input variables $X_1, \ldots, X_K$. A method to perform sensitivity analyses of this type, which are difficult in standard Monte Carlo applications, is also described.

Estimation of Y_P

The main problem in the calculation of Y_P'' in Equation (3) is of course the estimation of the vector X_P''. The MMC method produces estimates of X_P'' by an iterative procedure: At the beginning of iteration "i," an approximation X_i to X_P'' is available. During that iteration, two operations are performed:

1. The region D_{P_i}'' is defined by linearizing $Y(X)$ around X_i, giving

$$Y_i''(X) = a_{o_i} + \sum_{k=1}^{K} a_{k_i} X_k \tag{4}$$

2. A new approximation X_{i+1} to X_P'' is generated and convergence is checked. These operations are described next in greater detail.

Step 1: Linearization of Y(*X*). Linearization is through Taylor expansion. The coefficients a_{o_i} and a_{k_i} in Equation (4) are obtained from (K+1) responses calculated by the computer code for the exact calculation of Y:

$Y(X_i)$ (already available from iteration i−1; see below), and

$$Y(X_{i_k}),\ k=1, \ldots, K \tag{5}$$

where X_{i_k} is X_i perturbed in the kth component.

Step 2: Calculation of X_{i+1} and check of convergence. The idea is to find X_{i+1} as the point of maximum density on the boundary of D_{P_i}'':

$$X_{i+1} = \left\{x{:}f_X(x) = \text{max, subject to } Y_i''(x) = Y_{P_i}''\right\} \tag{6}$$

where Y_{P_i}'' is the P-fractile of Y_i'' in Equation (4).

For particular distributions of X, X_{i+1} can be found analytically or semianalytically. An important such case is where X has a multivariate normal distribution $N(m,\underline{\Sigma})$. In this case, the point X_{i+1} is given by

$$X_{i+1} = m + Z_p\, \alpha_i \tag{7}$$

where Z_P is the P-fractile of the standard normal distribution and the vector α_i is calculated as

$$\alpha_i = \frac{\underline{\Sigma}\, a_i}{\left[a_i^T\, \underline{\Sigma}\, a_i\right]^{1/2}} \tag{8}$$

The vector $a_i = (a_1, \ldots, a_{K_i})^T$ has as components the coefficients in the summation of Equation (4).

In the case where X has a generic distribution, the vector X_{i+1} can be found through Monte Carlo simulation: a set of vectors x_j, $j=1, \ldots, J$ is simulated from the distribution F_X, and probability density $f_j = f_X(x_j)$ and linearized response $Y_j = Y_i(x_j)$ from Equation (4) are evaluated at each x_j. (Notice that significant saving of time and improvement in convergence can be realized by simulating the vectors x_j and calculating the densities f_j only once, in Iteration 1. The same vectors and densities are then used in subsequent iterations, for which only the linearized responses Y_j have to be recalculated.) With x_j, f_j, and Y_j known, the vector X_{i+1} is obtained as follows:

1. Select Y_P'' as $Y_{P_i}'' = [(P)(n+1)]^{th}$ ranking value of Y_j, rounding or interpolating if $(P)(n+1)$ is not an integer. This estimator of Y_P replaces the estimator of Equation (3) for the present case of a generic distribution.
2. Let j^* be the index j for which f_j is maximum among the simulations for which $Y_j \geq Y_{P_i}''$. Then

$$X_{i+1} = x_j^* \tag{9}$$

Convergence is checked by comparing Y_{P_i} with $Y_{P_{i-1}}$, where

$$Y_{P_i} = Y(X_{i+1}) \tag{10}$$

Special provisions must be made in the first iteration: one might set X_1 equal to the mean vector m and initialize the estimate Y_{P_0}'' to be compared with Y_{P_1}'' in Equation (10) by using $Y_{P_0}'' = Y(X_1)$.

An illustration for the case of only one input variable X is shown in Figure 3. Notice that in this case the exact fractile Y_P is found in just two iterations, irrespective of the starting value X_1. This is because D_P'' does not depend on the linearization of $Y(X)$. In higher dimensions, D_P'' depends on the linearization; this is why additional iterations are needed.

Sensitivity analysis

Several measures of sensitivity of Y_P to input conditions may be defined. The one proposed here utilizes the linearization of $Y(X)$ at convergence, say

$$Y^*(X) = a_o^* + \sum_{k=1}^{K} a_k^* X_k \tag{11}$$

Let X have mean m and covariance matrix $\underline{\Sigma}$. (If these quantities are not known analytically, they can be calculated numerically from F_X or estimated from the Monte Carlo sample $x_1, \ldots, x_J$.) Then a measure of the sensitivity of Y_P to uncertainty on the input variable X_k is obtained by comparing the actual second-moment reliability index β_o with the conditional reliability index β_k that results from setting X_k equal to its mean value m_k. These two indices are given by

$$\beta_o = \frac{|Y_P'' - m^*|}{\sigma^*} \qquad \beta_k = \frac{|Y_P'' - m^*|}{\sigma_k^*}$$

where $m^* = a_o^* + \sum_{k=1}^{K} a_k^* m_k$

$$\sigma^{*2} = \sum_{q=1}^{K} \sum_{r=1}^{K} a_q^* a_r^* \mathrm{Cov}[X_q, X_r]$$

$$\sigma_k^{*2} = \sigma^{*2} - \frac{1}{\mathrm{Var}[X_k]} \sum_{q=1}^{K} \sum_{r=1}^{K} a_q^* a_r^* \mathrm{Cov}[X_k, X_q] \mathrm{Cov}[X_k, X_r] \tag{12}$$

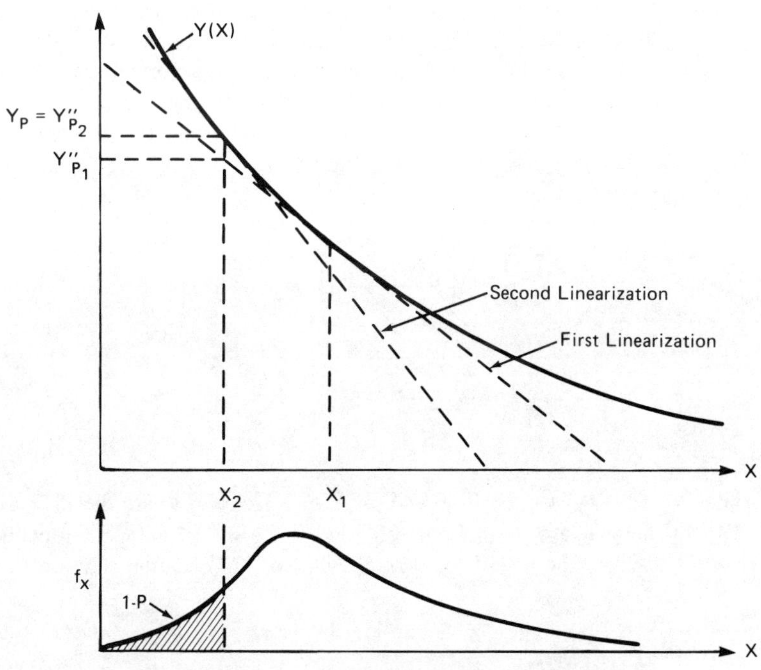

Figure 3. **Proposed method for a single input variable.**

In the above expressions, m^* and σ^{*2} are the mean and the variance of $Y^*(X)$, which is $Y''(X)$ at final convergence. σ_k^{*2} is the conditional variance of $Y^*(X)$ obtained if X_k is fixed at m_k. Because σ_k^{*2} is smaller than σ^{*2}, $\beta_k \geq \beta_o$. The expression for σ_k^{*2} is exact in the case of jointly normal vectors X. The larger the difference $(\beta_k - \beta_o)$ is, the greater the influence of uncertainty in X_k is on Y_P.

Rackwitz-Fiessler (RF) Method

In 1978, Rackwitz and Fiessler (*4*) proposed a method of system reliability analysis which, with various extensions and modifications, has been used very successfully in the evaluation of structural safety (Ditlevsen, *3*; Ellingwood et al., *2*). The method aims at estimating the probability with which a random vector X with given distribution F_X belongs to a region D of X space. In application to system reliability, D is the so-called safe region, i.e., the set of all points x for which the system survives. A special case is when survival depends on the value of a response quantity $Y(X)$ and occurs for $Y \leq y^*$, where y^* is a given constant. Then the region D is given by

$$D = \{x : Y(x) \leq y^*\} \tag{13}$$

The Rackwitz-Fiessler (RF) method is an iterative procedure, which in the ith iteration performs the following three steps:

1. Tangent linearization of the boundary of D. In the case when D is given by Equation (13), the method linearizes $Y(x)$ about a point x_i that satisfies $Y(x_i) = y^*$. Then the result is formally identical to Equation (4).
2. Replacement of the actual distribution of X, F_X, with a normal distribution. For the case in which X has independent components, i.e., when $F_X = \Pi_k F_{X_k}$, the approximating distribution is the product of independent normal distributions $N(m_k, \sigma_k^2)$. The parameters m_k and σ_k^2 are found by imposing the conditions

$$\frac{1}{\sigma_k} \phi \left(\frac{X_{ik} - m_k}{\sigma_k} \right) = f_{X_k} (X_{ik}) \tag{14}$$

$$\Phi \left(\frac{X_{ik} - m_k}{\sigma_k} \right) = f_{X_k} (X_{ik})$$

where f_{X_k} and F_{X_k} are the PDF and CDF of X_k, ϕ and Φ are the PDF and CDF for the standard normal variable, and X_{ik} is the kth component of the vector X_i used in the linearization step. Equation (14) has the solution

$$\sigma_k = \phi(\gamma_k)/f_{X_k} (X_{ik}) \tag{15}$$

$$m_k = X_{ik} - \gamma_k \, \sigma_k$$

where $\gamma_k = \Phi^{-1} [F_{X_k} (X_{ik})]$

Notice that, by virtue of Equations (14), the approximating normal distribution $N(m_k, \sigma_k^2)$ has the same probability density and cumulative probability as the true distribution F_{X_k}, for $X_k = X_{ik}$.

3. A new linearization point X_{i+1} is found, as

$$X_{i+1} = m + \delta \underline{\Sigma} a \tag{16}$$

where m and $\underline{\Sigma}$ are the mean vector and covariance matrix of the multivariate normal approximation in Step 2, a is the vector of coefficients a_{k_i} in Equation (4) (from Step 1), and δ is a constant such that $Y(X_{i+1}) = y^*$.

Notice that the RF method calculates the probability with which the response Y exceeds a given value y*. In the practical problem to be solved in this paper, the situation is reversed, in the sense that one requires the value of Y that is exceeded with a given probability P. In order to make use of the RF algorithm, one needs only to modify the selection of the linearization point X_{i+1}. Because this point is found under the condition that X has a normal distribution, one can calculate X_{i+1} using Equations (7) and (8).

The RF method differs from the MMC method in two respects:

1. The linearization point X_{i+1} is always found through Equation (7). Hence, no Monte Carlo simulation is necessary and the final estimate of Y_P is unique and independent of the user;
2. The actual distribution of X is replaced with a normal distribution, using Equations (15). This approximation is commonly a good one.

Sensitivity of Y_P to the input distribution can be analyzed using the method described previously with the parameters m and $\underline{\Sigma}$ of the approximating normal distribution at convergence. Startup provisions and convergence criteria are the same as for the MMC method.

For application to system reliability, the RF method has been extended to cases when the vector X has dependent components (Hohenbihler and Rackwitz, 5). One can use that same extension in the calculation of Y_P. Notice, however, that if the dependent components have a joint normal distribution, then no modification is necessary to the procedure described above, because Equations (7) and (8) hold also when the covariance matrix $\underline{\Sigma}$ is nondiagonal.

Improvement of Efficiency and Extensions

Differential treatment of influential and less influential variables

In some cases, it may be known from experience or from physical reasoning that some of the input parameters are not influential on the outcome Y. In other cases, the less influential parameters are not intially known but can be identified in the course of the analysis. This does not mean that uncertainty on such parameters should be ignored; rather, it means that the noninfluential variables can be treated in less detail, with savings in computational cost but no significant reduction in accuracy. The method described assumes that the less influential variables cannot be discriminated a priori. Additional savings are realized if at least partial discrimination is possible before any calculation is made.

Identification of the less influential variables can be made during Iteration 1 (or at any subsequent iteration of choice), based on the changes in concentration ΔY associated with changes in the input variables proportional to their respective standard deviations. For example, if ΔY_{max} is the maximum change over all the variables, then one might consider as noninfluential a variable X_k for which the ratio $\Delta Y_k/\Delta Y_{max}$ is smaller than a given value, say 0.1 or 0.05. The less influential variables are tagged, set to current values in the vector X_i, and their influence coefficients a_{k_i} are held constant in all subsequent iterations. At convergence, such variables are reactivated (by making sensitivity runs also with respect to them) and the final estimate of Y_P is obtained using the full linearized model.

For the MMC method, I iterations require (I)(K+1) response calculations using the exact groundwater model, J Monte Carlo simulations of the input vector X, and IJ calculations of the linearized response. For realistic I (e.g., 10), K (e.g., 20), and J (e.g., 1000), most of the time is spent in using the groundwater model. The variant described above reduces the number of response calculations to at most $2(K+1) + (I-2)(K_1+1)$, where K_1 is the number of variables that are more influential on Y. For the above values of I and K and for $K_1 = 4$, the number of response calculations is reduced from 210 to 82. A similar reduction in the number of response calculations is attained in the RF method.

Input random functions

In multidimensional groundwater models, variables such as hydraulic conductivity, dispersivity coefficients, and groundwater seepage velocity might be represented as uncertain functions of space. Other parameters, for example the leachate concentration at the source, might be regarded in transient models as random functions of time. Because a random function includes an infinite number of random variables and the previous methods can deal only with a small number of variables, a different approach must be followed in this case. A convenient procedure is as follows:

Step 1: The method as previously described is used with all random functions set to mean values.

Step 2: With the vector X set equal to X* (this is the vector X_i at convergence), the random functions are simulated M times (e.g., 10 to 20 times), and the associated concentrations $Y_{(m)}$, m = 1, . . . ,M, are calculated using the exact algorithm. Let Y* be the concentration when the random functions are set to mean values and X is set to X^*. Then one might represent the effect of the random functions on Y through either an additive random variable with equal probabilities at the points $(Y_{(m)} - Y^*)$ or a multiplicative random variable with equal probabilities at the points $(Y_{(m)}/Y^*)$.

Step 3: Suppose that an additive model is chosen. Then, in the MMC method, each linearized response Y_j(j = 1, . . . ,j) is replaced with M responses

$$Y_{jm} = Y_j + Y_{(m)} - Y^*, \; m = 1, \ldots, M \tag{17}$$

and Y_P is estimated as the value exceeded by $(1-P)$ (JM) responses Y_{jm}. In the RF method, a random input variable X_{K+1} is added to the vector *X*. The distribution of X_{K+1} is inferred from the quantities $(Y_{(m)} - Y^*)$ or $(Y_{(m)}/Y^*)$, depending on whether the effect on Y is additive or multiplicative. Iteration continues with K+1 input variables, until final convergence.

The above treatment of random functions has two limitations:

1. The treatment is adequate if the random-fluctuation components have a small influence on Y, but is rough if such fluctuations are highly influential.
2. In the case of several random functions, only their aggregated influence on Y_p can be assessed. In order to discern the contribution from each random function one should use in Step 2 a controlled experimental design, with consequent increase of computational costs.

Variable transformations

A convenient feature of the fractile Y_P is that it is invariant with respect to one-to-one transformations of the response variable Y. Hence, for example, if Y is replaced with lnY or Y^3, then Y_P becomes $\ln Y_P$ or $Y_P{}^3$, respectively. In some situations, transformations of *X* may also be appropriate.

Although the fractile Y_P enjoys this invariance property, the estimator Y_P'' in Equation (3) does not, and this is why a transformation is of interest here. Ideally, one should use a transformation such that the surface B_P in Figure 1 is a plane, because in this case $Y_P'' = Y_P$. Such a transformation may make the dependence of Y on *X* almost linear and may therefore also reduce the number of iterations to convergence.

A number of techniques have been suggested for the transformation of variables in regression models (e.g., Box and Tidwell, *6*; Box and Cox, *7*; Friedman and Stuetzle, *8*; Breiman and Friedman, *9*). For our case, the ACE method of Breiman and Friedman is particularly attractive, because its objective is to find functions g(Y) and $h_k(X_k)$ such that the regression of g(Y) against the transformed variables $h_k(X_k)$ is linear. Finding suitable functions g and h_k requires repeated simulation of vectors *X* and calculation of the associated responses Y(*X*). However, it should be necessary to make such simulations only once, using a distribution F_X appropriate for a generic site. (The form of the functions g and h_k ordinarily would not vary from site to site.) Using g(Y) in place of Y and $h_k(X_k)$ in place of X_k would linearize the boundary B_P in Figure 1. Therefore, the boundaries B_P and B_P'' in Figure 2 would coincide and the linearization errors of the present methods would be eliminated.

Computer Program

A FORTRAN program has been written to implement the MMC and RF methods, as well as standard Monte Carlo simulation (MC method). The program can handle uncertain vectors *X* with arbitrary probability distributions; however, in its present form, it allows probabilistic dependence to exist only among the components of *X* with normal distributions. Features that are included are sensitivity analysis and automatic screening with differential treatment of the less influential variables. An additional feature, which is available when the MMC or MC methods are used, is the calculation of a confidence interval for the fractile Y_P. In the MMC case, this interval relates to the linearized function $Y^*(X)$ as defined in Equation (11). The method uses the property that order statistics are beta distributed (Johnson and Kotz, *10*, p. 38).

Confidence intervals at level 80% will be reported in the numerical examples.

Numerical Examples

In order to test the proposed methods, we have used a semi-analytical groundwater transport model, so that the procedures could be evaluated under a variety of conditions at low computational cost. The model selected for this purpose is the saturated-zone model described by Salhotra et al. (*1*). None of the inputs to that model is a random function. Except for this feature, all aspects of the RF and MMC methods can be tested. In particular, one can check the algorithm for different distributions of the vector *X* and for different conditions of probabilistic dependence among its components. Results of several runs are presented and compared with direct MC simulations. In addition to conducting these tests, we have considered simpler made-up problems of uncertainty propagation for which the exact value of Y_P is known. Results for these simpler problems are reported first. In all calculations, P is set to 0.9, which is the value used by Salhotra et al. (*1*).

Cases with known solution

Four cases are considered, all with independent random variables X_k:

Case 1: $Y = X_1 + X_2 + X_3$ Exact 0.9-fractile: $Y_{0.9} = 2.157$

$X_k \sim U[0,1]$

Case 2: $Y = \exp\{X_1 + X_2 + X_3 + X_4 + X_5\}$ $Y_{0.9} = 3566$

$X_1 \sim N(0,1)$

$X_2 \sim N(1,4)$

$X_3 \sim N(0,9)$

$X_4 \sim N(-1,9)$

$X_5 \sim N(2,0.25)$

Case 3: $Y = \exp\{X_1 + X_2\}$ $Y_{0.9} = 48.89$

$X_k \sim EX(1)$

Case 4: $Y = X_1^2 + X_2^2 + X_3^2 + X_4^2 + X_5^2$ $Y_{0.9} \cong 17.02$

$X_k \sim N(1,1)$

Notation for the probability distributions is as follows:

U[a,b] = uniform distributionn between a and b
$N(m,\sigma^2)$ = normal distribution with mean value m and variance σ^2
EX(m) = exponential distribution with mean value m

The values reported above for $Y_{0.9}$ are exact for the first three cases. For Case 4, Y has a noncentral χ^2 distribution and the reported value of $Y_{0.9}$ is the result of 2000 Monte Carlo simulations.

Estimates of $Y_{0.9}$ from various runs are given in Table 1. From this table and from other cases not reported here, one can conclude as follows:

1. The MMC method is more robust than the RF method with respect to the distribution of the random variables X_k. In particular, the estimate of $Y_{0.9}$ from the RF method deteriorates when the variables X_k have high skewness (Case 3).
2. The RF method is exact in Case 2, for which the variables X_k have normal distributions and the contour surfaces of Y in *X*-space are flat. Notice, however, that in Case 2, the response Y is a highly nonlinear function of the

Table 1. **Summary results for four problems with known fractile $Y_{0.9}$.**

Case (known $Y_{0.9}$)	Method	NSIM[(1)]	NCODE[(2)]	$Y_{0.9}$	Central 80% Confid. Int.
1 (2.157)	RF	—	9	2.311	—
	MMC	500	9	2.122	(2.078,2.229)
	MC	500	500	2.122	(2.078,2.229)
2 (3566)	RF	—	19	3566	—
	MMC	500	13	2220	(−564,5067)
	MMC	1000	19	2643	(1214,4239)
	MMC	2000	19	2957	(2166,4149)
	MC	500	500	2819	(1691,5067)
	MC	1000	1000	2670	(1770,4424)
	MC	2000	2000	2977	(2356,4316)
3 (48.89)	RF	—	10	30.08	—
	MMC	500	7	54.10	(39.50,65.97)
	MMC	1000	7	58.34	(43.61,69.66)
	MMC	2000	7	48.29	(41.14,55.71)
	MC	500	500	53.30	(41.41,67.23)
	MC	1000	1000	55.08	(45.44,70.70)
	MC	2000	2000	48.21	(41.70,56.23)
4[(3)] (~17.02)	RF	—	13	12.38	—
	MMC	500	13	11.60	(10.79,12.51)
	MMC	2000	13	11.78	(11.46,12.34)
	MC	500	500	16.19	(15.33,17.42)
	MC	2000	2000	17.02	(16.62,17.49)

(1) Number of Monte Carlo simulations.
(2) Number of times exact value of Y is calculated.
(3) "Known" fractile based on 2000 MC simulations.

Table 2. **Definition of input variables for Groundwater Transport Model. Distributions of X_1 through X_7 are provided in Table 3.**

		Random Variables		
X_1	=	DD	(mm)	particle diameter
X_2	=	BAQFR	(m)	aquifer thickness
X_3	=	T	(°C)	temperature
X_4	=	VW	(m/yr)	seepage velocity
X_5	=	QC	(m/yr)	infiltration rate
X_6	=	AW	(m^2)	area of disposal site
X_7	=	α_L	(m)	longitudinal dispersivity
		Constant or Functionally Dependent Variables		
X_8	=	$\theta = 0.261 - 0.0385 \ln X_1$		porosity
X_9	=	$\sigma = \sqrt{X_6}/6$	(m)	width of Gaussian source term
X_{10}	=	$\alpha_T = X_7/8$	(m)	transverse dispersivity
X_{11}	=	$\alpha_Z = X_7/160$	(m)	vertical dispersivity
X_{12}	=	PH = 6.2		pH
X_{13}	=	FOC = 0.01		organic carbon content
X_{14}	=	DKAO = 0	(1/mole/yr)	acid catalyzed rate constant
X_{15}	=	DKNO = 0	(1/yr)	neutral hydrolysis rate constant
X_{16}	=	DKBO = 0	(1/mole/yr)	base catalyzed rate constant
X_{17}	=	TR = 25	(°C)	reference temperature
X_{18}	=	DKOC = 0		normalized distribution coefficient
X_{19}	=	BLAM = 0.0693	(1/yr)	biodegradation coefficient
X_{20}	=	XW = 152.4	(m)	distance to well

input variables. As a consequence, both MMC and MC perform poorly. (Notice the large widths of the confidence intervals, even for 2000 simulations.)

3. Both MMC and RF are biased in Case 4. This is the only case in which the contour surfaces of Y in *X*-space are curved, so that tangent linearization introduces errors.

Table 3. **Distributions of input random variables for the Groundwater Transport Model.**

Variable	Distribution Information(1) Case 1	Case 2	Case 3	Case 4
X_1 = Particle Diameter (DD,mm)	U(0.0004,0.1)	LN(−3,0.25)	LN(−3,0.25)	U(0.0004,0.1)
X_2 = Aquifer Thickness −3m (BAGFR,m)	EXP(75.6)	EXP(75.6)	EXP(75.6)	EXP(75.6)
X_3 = Temperature (T, °C)	N(14.4,28.0)	N(14.4,28.0)	N(14.4,28.0)	N(14.4,28.0)
X_4 = Seepage Velocity (VW,m/yr)	N(300,2500)	LN(4.6,1)	LN(3,1)	N(300,2500)
X_5 = Infiltration Rate (QC,m/yr)	U(0.1,0.5)	LN(−1.5,0.25)	LN(−1.5,0.25)	U(0.1,0.5)
X_6 = Area of Facility (AW,m^2)	LN(5,0.25)	LN(5,0.25)	LN(5,0.25)	N(180,2500)
X_7 = Longit. Dispersivity (α_L,m)	HISTOGRAM(2)	LN(2.5,1)	LN(2.5,1)	HISTOGRAM(2)

(1) Distribution notation: U(range), EXP(mean), N(mean, variance), LN(logarithmic mean, logarithmic variance). All variables are independent, except for (X_3, X_4) and (X_4, X_6) in Case 4, which have correlation coefficients 0.5 and 0.4, respectively. (These correlations were not intended to be physically realistic.)

(2) $f_{X_7}(x) = 0.11111$, for $0.1 \le x < 1$
0.06667, for $1 \le x < 10$
0.003334, for $10 \le x < 100$

4. In all cases, RF and MMC require only a few evaluations of the exact response (NCODE in Table 1). The confidence widths for MMC are comparable to those for the MC method, which requires many more responses; however, note that the MMC confidence intervals apply to the linearized responses, which may be biased.

As a general conclusion, the RF method is more accurate if the contour surfaces of Y are nearly flat (this condition does *not* imply linearity of Y as a function of X) and the distributions of the most influential random variables have low skewness. The MMC method is appropriate when Y is nearly linear in X. This emphasis on linearity in the relationship between Y and X reinforces the conclusion stated previously that, where strong nonlinearities are present, Y and/or the variables X_k should be transformed so that the relationship becomes approximately linear.

Table 4. **Summary result for the four groundwater transport problems of Table 3.**

Case	Method	NSIM(1)	NCODE(2)	$Y_{0.9}$	Central 80% Confid. Int.	CONV(3)
1	MC	1000	1000	0.0151	(0.0147,0.0157)	—
	MMC	1000	21	0.0186	(0.0178,0.0190)	Y
	MMC	2000	39	0.0189	—	N
	RF	—	33	0.0172	—	Y
2	MC	1000	1000	0.0622	(0.0564,0.0665)	—
	MMC	1000	18	0.0782	(0.0744,0.0822)	Y
	MMC	2000	18	0.0779	(0.0744,0.0812)	Y
	RF	—	35	0.0600	—	Y
3	MC	1000	1000	0.0940	(0.0906,0.0978)	—
	MMC	1000	41	0.2190	—	N
	MMC	2000	27	0.0386	—	N
	RF	—	52	0.1140	—	Y
4	MC	1000	1000	0.0150	(0.0146,0.0153)	—
	MMC	1000	20	0.0153	(0.0145,0.0163)	Y
	MMC	2000	18	0.0160	(0.0153,0.0167)	Y
	RF	—	33	0.0178	—	Y

(1) Number of Monte Carlo simulations.
(2) Number of times groundwater model used.
(3) Convergence of algorithm.

Examples using a mass-transport model

Next we show results from calculations using the saturated-zone transport model described by Salhotra et al. (*1*). The input includes the 20 variables listed in Table 2, of which 7 are considered uncertain. Four cases are examined, with the distributions specified in Table 3. Notice that Table 3 gives only the distributions of variables X_1 through X_7: the remaining input parameters, X_8 through X_{20}, are set to constant values or are calculated from $(X_1, \ldots, X_7)$, as indicated in Table 2. Also notice that all cases except Case 4 consider the variables $X_1, \ldots, X_7$ to be independent. Cases 2 and 3 are identical, except for the logarithmic mean of the seepage velocity, X_4. This parameter was found in preliminary sensitivity analyses to be quite important: the downstream concentration Y increases rapidly with increasing X_4 and reaches a maximum for $X_4 \cong 10$ m/yr ($\ln X_4 \cong 2.3$). For higher seepage velocities, the concentration decreases steadily.

Table 4 summarizes the results of several runs and shows that, except for Case 3 where the MMC method did not converge, the MMC and RF methods produced results in close agreement with direct MC simulation. Lack of convergence occurs in the MMC method when the linearization point alternates between two (more rarely,

Table 5. **Example of Sensitivity Analysis: Case 2 in Tables 3 and 4, using the RF method.**

	β_o = **1.283**	
Var. K	β_K	$\beta_K - \beta_o$
1	1.285	0.002
2	1.283	0.000
3	1.283	0.000
4	1.900	0.617
5	1.555	0.272
6	1.335	0.052
7	1.321	0.038

among several) simulated values of *X*. If such a condition occurs, a warning is issued and calculations continue using the first duplicated linearization point. The RF method converged in all cases analyzed, although on average it required a larger number of evaluations of the downstream concentration than the MMC method did.

As an example of sensitivity analysis, Table 5 reports the values of β_o and β_k (k = 1, . . . ,7) from Equation (12), for Case 2, RF method. Because sensitivity is with respect to the parameters of the fitted normal distributions, β_o has the value 1.283 that corresponds to the 0.9 fractile of the standard normal variable. The differences (β_k and β_o) indicate that seepage velocity (X_4) has a dominating influence on uncertainty of Y, followed in importance by the infiltration rate (X_5).

Acknowledgment

This work was sponsored by U.S. Environmental Protection Agency under Contract No. 68-03-6304.

Literature Cited

1. Salhotra, A. M., Gaboury, D. R., Huyakorn, P. S., and Mulkey, L. A., "A Multimedia Exposure Assessment Model for Evaluating Land Disposal of Hazardous Waste," *Proceedings of the National Conference on Hazardous Wastes and Hazardous Materials*, March 16–18, 1987, Washington, D.C., Hazardous Materials Control Research Institute, Silver Spring, MD (1987).

2. Ellingwood, B., Galambos, T. V., MacGregor, J. G., and Cornell, C. A., ''Development of a Probability Based Load Criterion for American National Standard A58,'' U.S. Department of Commerce, National Bureau of Standards, Washington, D.C. (1980).
3. Ditlevsen, O., ''Basic Reliability Concepts,'' *Proceedings,* NATO Advanced Study Institute on Reliability Theory and Its Applications in Structural and Soil Mechanics, Bornholm, Denmark (1982).
4. Rackwitz, R. and Fiessler, B., ''Structural Reliability under Combined Random Load Sequences,'' *Computers and Structures, 9,* 489–494 (1978).
5. Hohenbihler, M. and Rackwitz, R., ''Non-normal Dependent Vectors in Structural Safety,'' *J. Engrg. Mech. Div., ASCE, 107* (EM6), 1227–1238 (1981).
6. Box, G.E.P. and Tidwell, P. W., ''Transformation of the Independent Variables,'' *Technometrics, 4,* 531–550 (1962).
7. Box, G.E.P. and Cox, D. R., ''An Analysis of Transformations,'' *J. Royal Statist. Soc., B 26,* 211–243 (1964).
8. Friedman, J. H. and Stuetzle, W., ''Projection Pursuit Regression,'' *J. Am. Statist. Assoc., 76,* 376, 817–823 (1981).
9. Breiman, L. and Friedman, J. H., ''Estimating Optimal Transformations for Multiple Regression and Correlation,'' *J. Am. Statist. Assoc., 80,* 391, 580–619 (1985).
10. Johnson, N. L. and Kotz, S., *Distributions in Statistics: Continuous Univariate Distributions—2,* Houghton Mifflin Company, Boston (1970).

Sensitivity of Whiteshell Geosphere Modeling Results to Dimensionality (2-D/3-D) of the Computer Simulations Employed

J. A. KEITH REID and TIN CHAN
Atomic Energy of Canada Limited
Whiteshell Nuclear Research Establishment
Pinawa, Manitoba, Canada

Abstract

The Whiteshell Geosphere Conceptual Model has been assembled as a basis for two- and three-dimensional simulations to predict water flows from a hypothetical nuclear waste vault. The conceptual model is consistent with surface and subsurface geological, geophysical, and hydrogeological data from the granite batholith and surrounding country rock that comprise the Whiteshell Research Area (WRA) in southeastern Manitoba, Canada.

The MOTIF finite-element code has been applied to simulate the coupled processes of groundwater flow and heat transport. Convective transport of inert contaminants from a hypothetical nuclear fuel waste vault, through the geosphere, to the biosphere was simulated using a particle-tracking technique. Hydraulic properties of the elements of the intersection of a major fracture zone with the hypothetical vault were changed to represent the effect of grouting. Results in the form of heads, water velocities, and particle tracks from two- and three-dimensional finite element simulations are compared.

Some three-dimensional (3-D) velocity patterns and fracture-grouting effects cannot be represented in two-dimensional (2-D) models.

1. Introduction

The Canadian Nuclear Fuel Waste Management Program is currently assessing the concept of disposing of nuclear fuel waste deep within plutonic rock. A Monte Carlo simulation code, SYVAC (SYstem Variability Analysis Code), is being applied to assess the radiological impact of this geological disposal. Due to the large number of sampled runs required for Monte Carlo simulation, the geological system has to be represented by a highly idealized model. This precludes a direct comparison of SYVAC predictions with field data. It is still possible, however, to make an indirect connection by first constructing detailed 2-D and 3-D models based on laboratory and field data to simulate the transport of contaminants and water from the nuclear fuel waste vault, through the geosphere, to the biosphere. The results of the detailed models are then used to construct a simplified SYVAC model (*1*,*3*).

This paper describes the detailed 3-D geosphere model, the results obtained using the 3-D model, how the results differ from 2-D modeling results, and how dimensionality of the simulation causes these differences.

2. Three-Dimensional Modeling

The modeled area in the Whiteshell Research Area is first introduced. An outline of the 3-D local model is presented along with boundary conditions and the simulation methodology used. Some results of the 3-D model are then presented showing 3-D heads, predicted discharge times, locations, and surface exit zones, along with the detailed effects of grouting.

2.1 Modeled area

Figure 1 shows the Whiteshell Research Area bounded by the Winnipeg River system. The Winnipeg River flows south, through a chain of lakes, over a hydro

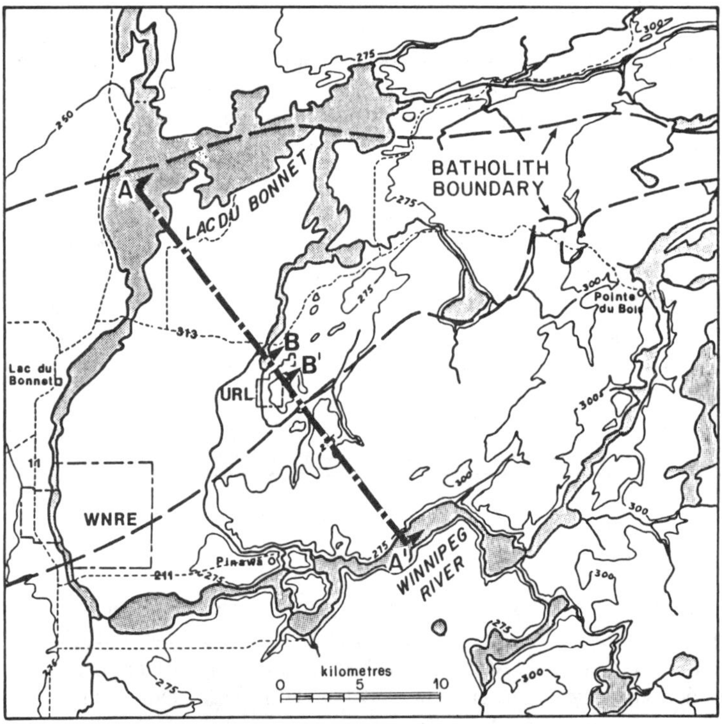

Figure 1. **Whiteshell Research Area outline map indicating the Winnipeg River System/Lac du Bonnet water bodies. Reprinted with permission (*2*).**

dam, and then into Lac du Bonnet. The 3-D regional model represents the area bounded by the Winnipeg River system, except in the northeast where it is bounded by a fault scarp. No-flow boundaries are used for the perimeter of the model and the bottom (at 4-km depth). The surface of the model is a specified head boundary using the topography and lake/river levels as estimators. The temperature boundaries are taken as 6 °C at surface and increasing with depth according to the gradient observed in field investigations. The section used in the 2-D regional modeling falls on the 27-km-long line A-A′.

The conceptualized geology has intact rock and fracture zone permeabilities that decrease with depth in layers, with the permeability of the fracture zones being two to four orders of magnitude higher than that of the intact rock. The conceptual model is fully described by Chan et al. (*3*).

A plan view of the finite-element mesh used in the 3-D regional model is shown in Figure 13 of Chan (*1*). The local area model is an 8-km × 10-km × 1.5-km thick block within the regional model covering the area of main interest for water flow passing through the approximately 2-km × 2-km area of the hypothetical vault. The purpose of the regional model is to estimate boundary conditions for the local model. In this paper, only preliminary results of the local model are reported.

A topographic map, Figure 2, showing the local model area and vault surface projection, indicates some important features of the topography. The Pinawa Channel flows northeast, Boggy Creek flows north, and the Lee River flows northeast. The Underground Research Laboratory (URL) is situated at the knoll southeast of Boggy Creek. An important topographic feature is a small peninsula at the junction of the Pinawa Channel and Boggy Creek protruding across the 2-D section.

True steady-state simulations were produced by first estimating the temperatures and pressure distributions using a constant fluid properties calculation. This system state was then employed as the initial condition for a transient thermal simulation of several steps, and fluid properties were allowed to vary with temperature and pressure until the calculated temperatures and pressures did not change further with time. No vault heat transient simulations are reported here.

2.2 Three-dimensional local model

Figure 3 is a plot of the topology generated by a contouring program using the 3-D mesh surface nodal z-coordinates, which are the main features of the topography represented in the model. The knoll at the URL is evident, along with the ridge to the northeast of the URL. The main water bodies are approximated by the shaded area in Figure 3. The model is built up of 11 vertical slices of nodal planes. The slice illustrated in Figure 4 shows the vertical fractures, V1 through V3, and low-dipping fractures, LD1 through LD4. The vault is indicated as a dark line in this diagram. For the results of rapid water flows through the hypothetical vault, situated at 500 m below surface, our area of main interest is between LD1 and V2 on this section.

Temperature boundary conditions are specified for the 2028 nodes distributed over the six surfaces of the model. The surface boundary temperature is 6 °C. The side boundary temperature increases linearly with depth according to the natural gradient (11.6 °C/km) observed in the area. Specified head boundary conditions are

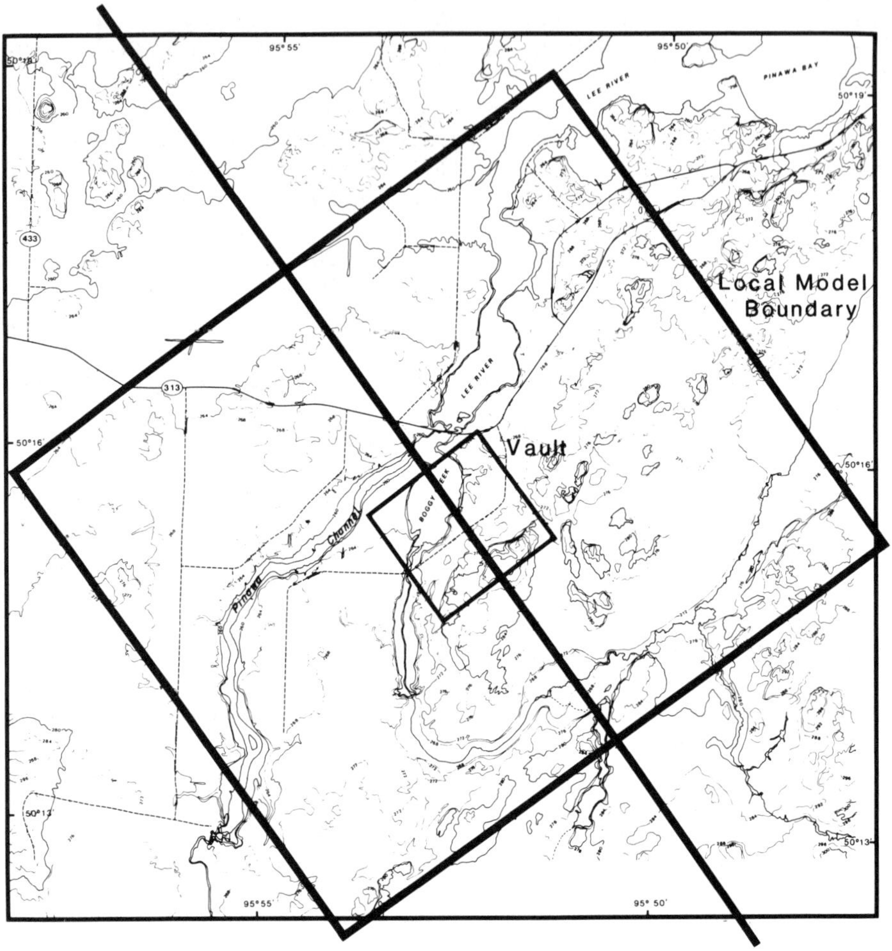

Figure 2. **Topographic map showing the local model area, Lee River, Boggy Creek, and the Pinawa Channel. (The 2-D section, indicated by the central line, runs through the centre of the local area. The URL knoll lies on the 2-D section SE of the Lee River/Boggy Creek junction.)**

applied to the upper surface nodes and, as a first approximation, no-flow boundaries were applied to the other five surfaces.

2.3 Three-dimensional model results

Results from the model are presented in three forms: (1) 3-D head plots, (2) particle transit-time contours and exit locations by particle starting point in the vault, and (3) surface discharge locations shown on surface topology maps.

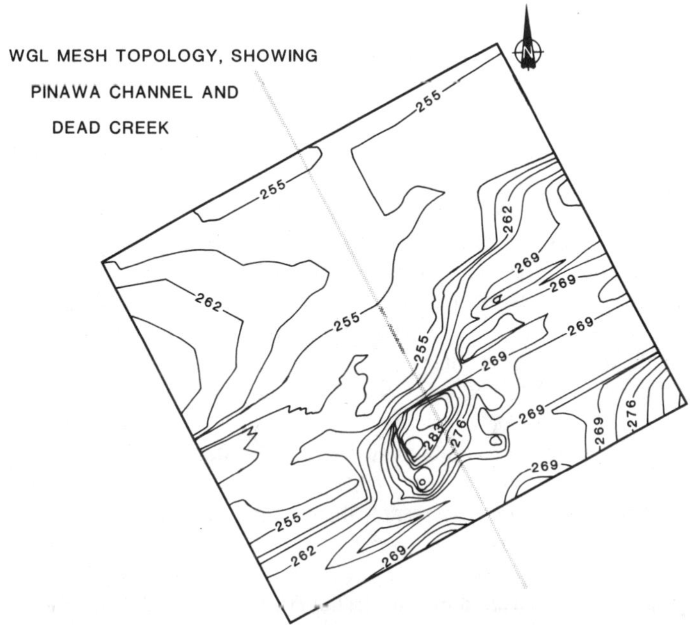

Figure 3. **The mesh topology contours generated from the Z components (elevations) at the surface nodes in the 3-D local model. (Major surface features, including the URL knoll, are evident.) Adapted from Chan and Scheier (*2*).**

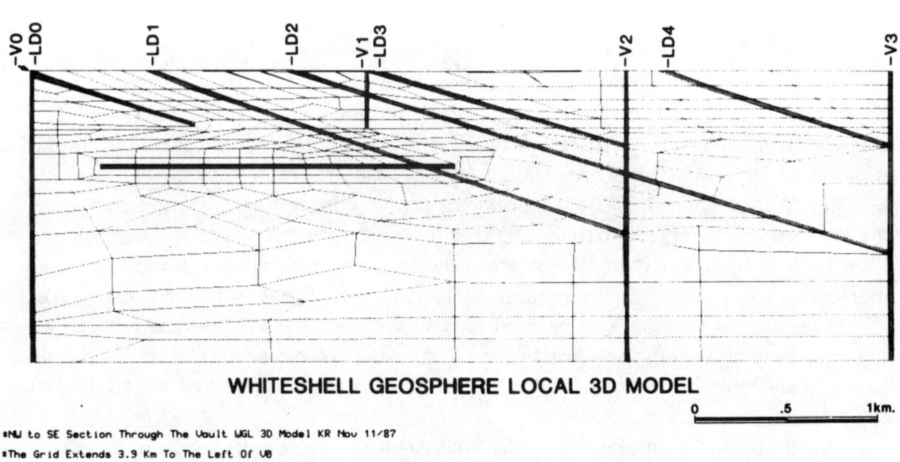

Figure 4. **Northwest to southeast vertical section through part of the local model showing the orientation of various vertical (V0) and low dipping (LD0) fractures. (The model grid extends 3.9 km to the left of fracture zone V0.)**

2.3.1 Three-dimensional head plots. Figure 5 shows head contours for six surfaces. By making only four folds, the reader may assemble these surfaces into a 3-D model that illustrates the controlling factors for water flows from the hypothetical vault to the surface.

> Fold a photocopy of Figure 5 in half, along the dotted line A-A′, such that the dotted line remains visible. (All four folds will be made this way.) Keeping the A-A′ fold, fold at the dashed lines B-B′ and C-C′. This will form a backward "Z" with the base labeled "vault from above" and the top labeled "surface from above." The slanting sheet "LD1 from west" shows the fracture zone, as in Figure 4, dipping at 18.5° to the southeast. A final slight fold along D-D′ brings the LD1-vault junction into contact.

This approximately 2-km × 2-km × 0.5-km part of the 8-km × 10-km × 1.5-km Whiteshell Geosphere Local (WGL) Model shows the features having most influence on rapid water flows from the vault to surface. The water movements are gravity-driven by higher elevation of the water table under the knoll shown on the surface contour plot. These heads are transferred, through the surface elements, and appear in modified form on the surface elements from below plot. The heads are also transferred through the fracture zones and intact rock, to produce elevated heads at the LD1-vault junction. The heads diminish up the fracture toward the topographic and hydraulic low at Boggy Creek. This head decreasing up the fracture zone causes upward flow, normal to the head contours, and at an angle of about 45° to the dip of LD1.

In the vault, heads also diminish from the LD1-vault intersection toward the west causing slow water flows through the intact rock. More significantly, heads in the hypothetical vault increase to the east of the LD1-vault intersection, causing water flows toward the LD1 fracture zone.

2.3.2 Vault plan views. Figure 6a shows the plan view of the vault with logarithmic contours of the discharge times for particles initiated at locations throughout the vault area. A total of 135 tracks were run for each case, concentrated in the areas of greatest interest near the fracture zone (LD1) and vault intersection.

The contour lines show transit times for purely convective flows, initiated at points in the vault. Figure 6a indicates that the time required to reach the surface is low (0.8 Ka [1000 years] to 10 Ka) when the particles are initiated in the LD1-vault intersection. The area of low discharge time corresponds exactly to the fracture zone LD1-vault intersection and can be used to delineate the intersection. The transit time increases gradually to the southeast of the LD1-vault intersection. The transit time increases suddenly when the particles are initiated to the northwest of the LD1-vault intersection.

Figure 6b shows, in plan view, the vault with emergence discharge locations indicated. These are strongly related to the transit times. The long-time discharges surface at the Pinawa Channel or the Boggy Creek-Pinawa Channel Confluence. In contrast, the short-time discharges all come to surface in the Boggy Creek Discharge. Figures 6c and 6d are discussed under grouting results.

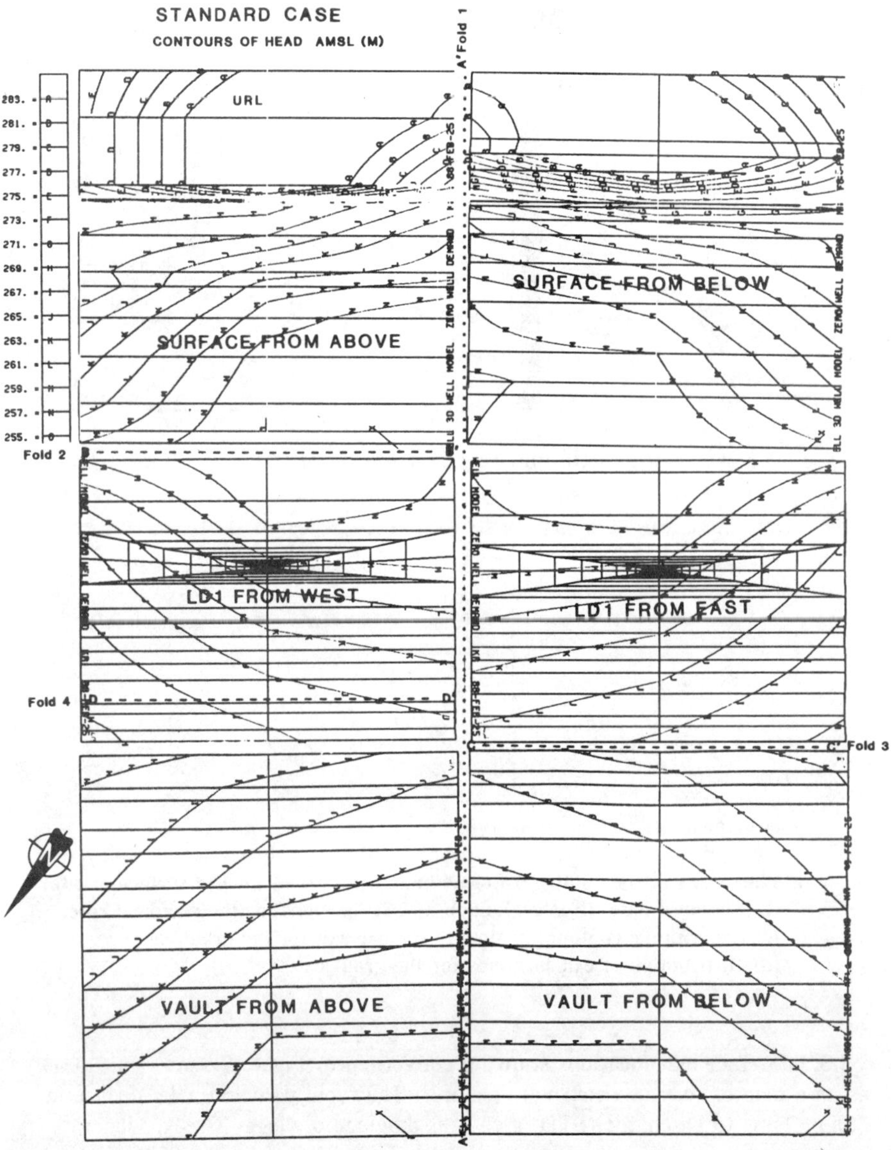

Figure 5. **3-D representation of head contours for the standard 3-D local model. (Fold a photocopy of Figure 5 in half, along the dotted line A-A′, so that the dotted line remains visible. (All four folds will be made this way.) Keeping the A-A′ fold, then fold at the dashed lines B-B′ and C-C′. This will form a backward "Z" with the base labeled "vault from above" and the top labeled "surface from above." The slanting sheet "LD1 from west" shows the fracture zone, as in Figure 4, dipping at 18.5 degrees to the southeast. A final slight fold along D-D′ brings the LD1-vault intersection into contact.)**

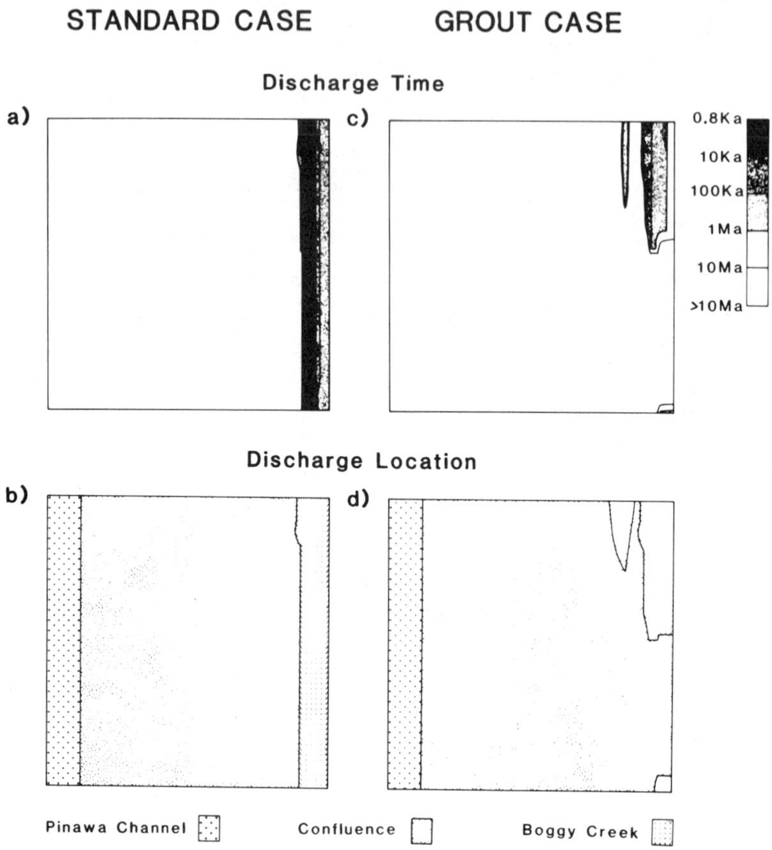

Figure 6. **Plan views of the vault: (a) Logarithmic contours of time to surface exit for the standard case. (b) Distribution of exit locations for the standard case. (c) Logarithmic contours of time to surface exit for the grout case. (d) Distribution of exit locations for the grout case.**

2.3.3 Surface exit locations showing convergence. Figure 7 shows exit locations on a map of surface water body features. The rapid particle tracks surface in the Short Time to Discharge (STD) area near the edge of Boggy Creek. Under purely convective, topographically driven flow, the total area for the STD discharge is only about 40% of the vault particle track initiation area for these tracks. The particle track initiation area is, in turn, a small fraction of the total vault area, as illustrated in Figure 6b.

For the Long Time to Discharge (LTD) tracks to the Pinawa Channel and Confluence, the surface area covered is less than 5% of the initiation area in the vault. However, for the LTD discharges, part of the transport would be diffusion- rather than convection-dominated. The convective flow would spread out these LTD pathways both in time and in exit location.

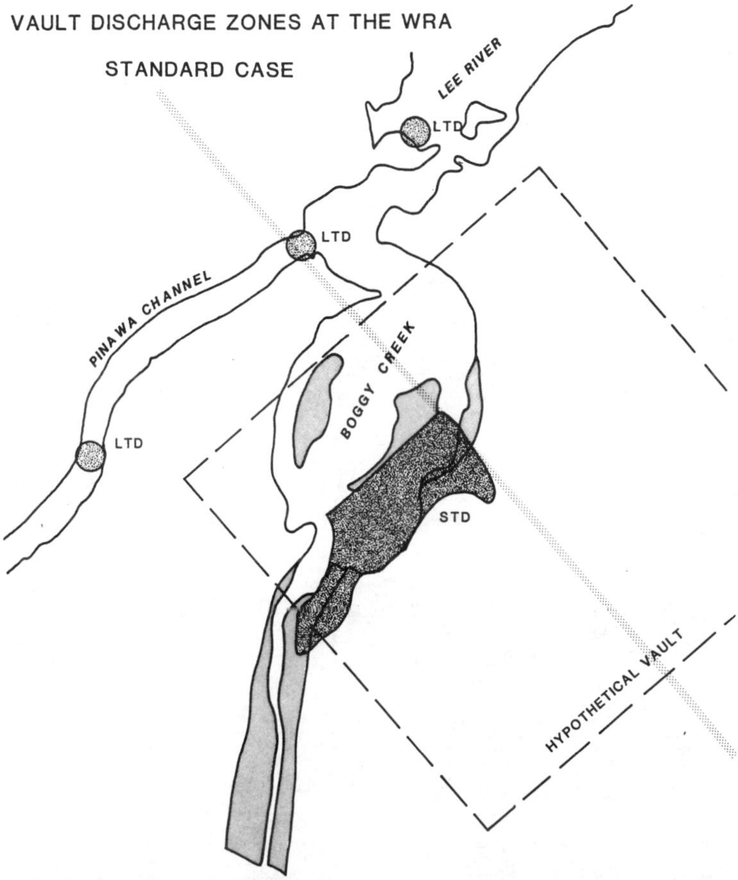

Figure 7. **Surface topographic feature map showing standard case Short Time to Discharge (STD) and Long Time to Discharge (LTD) locations in relation to the hypothetical vault surface projection.**

For purely convective flow, the total surface discharge area is less than 10% of the vault plan area.

2.4 Effect of grouting

Grouting was simulated by decreasing the permeability of the elements that correspond to the intersection of the fracture zone (LD1) and the vault, to correspond to the vault rather than the fracture zone properties. This change decreased the permeability by five orders of magnitude. The influence of grouting is seen in (1) raising the heads upstream of the intersection and lowering the heads downstream of the intersection, (2) increasing transit times, and (3) increasing convergence of the particle tracks leading to a smaller area of surface discharge.

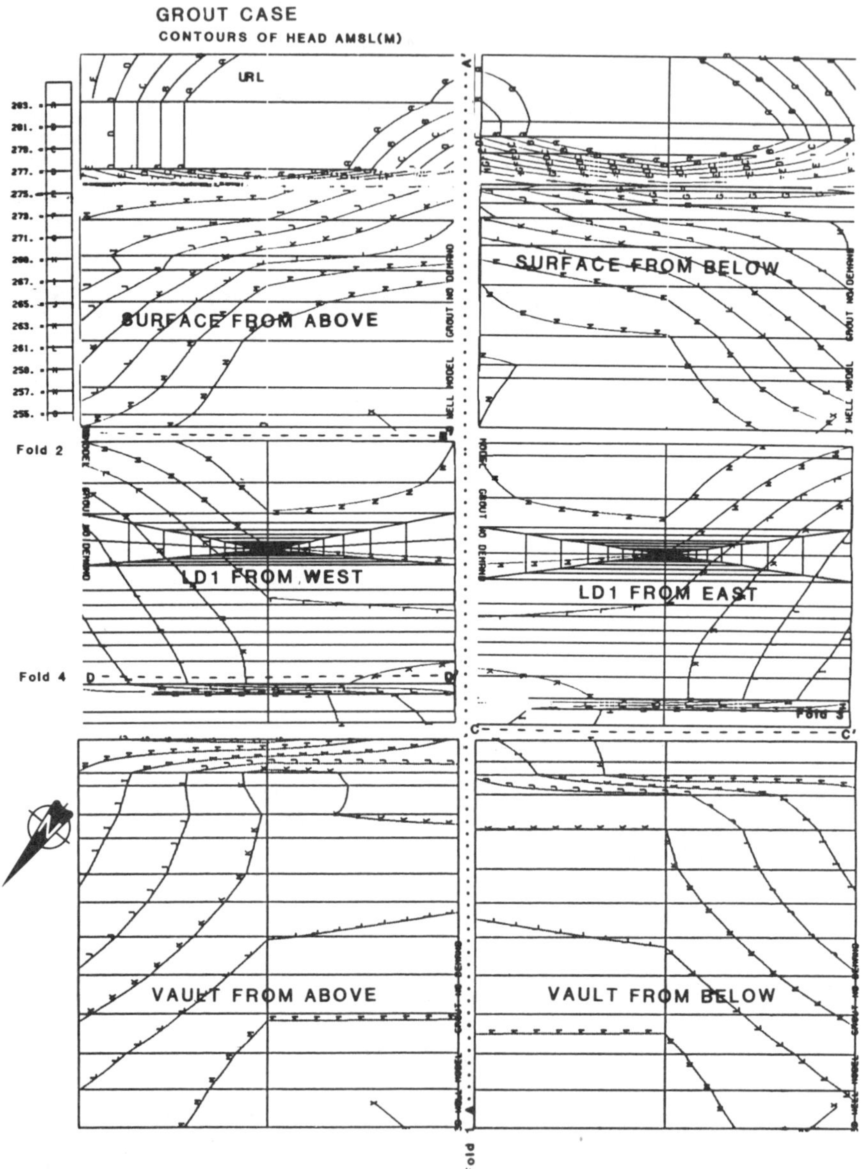

Figure 8. **3-D representation of head contours for the grout case of the local model. (To assemble, follow the instructions for Figure 5.)**

2.4.1 Modification of heads in three dimensions. Fold Figure 8 following the instructions in Section 2.3.1 for Figure 5. A comparison with the standard case, Figure 5, shows heads are elevated east of the LD1-vault intersection and lowered to the west of the LD1-vault intersection leading to a flow of water around the grouted zone. As in the 2-D modeling reported by Chan and Scheier (*2*), particles that would otherwise not enter the fracture zone may now be swept into it as water flows around the grouted area in the fracture zone and rock matrix.

The 3-D head difference plot, shown in Figure 9, again folded according to the instructions in Section 2.3.1, shows the hydraulic head differences between the standard and the grouted cases. This figure highlights the elevation of head upstream from the grout and the depression downstream from the grout. This effect in the LD1 fracture zone is also seen in the vault and is significant to the movement of water and radionuclides from the vault to the surface.

2.4.2 Vault plan views. Comparison of Figures 6b and 6d indicates that discharge locations are changed in the grout case for particles originating both in the south half of the LD1-vault intersection and to the southwest of the intersection. Tracks from these locations, in the standard case, move through the fracture zone to the Boggy Creek Discharge. In the grout case, elevated heads upstream of the grout cause these tracks to move out of the LD1-vault intersection along the vault, through the rock mass, and to discharge at much greater distances in the confluence discharge area.

Because of the sweep effects, two particle tracks, which moved to the confluence discharge in the standard case, move to the Boggy Creek Discharge in the grout case. The travel time is reduced (Figure 6c), but remains longer than 100 Ka.

The relocated pathways produce extended transit times. Figure 6c indicates that tracks moving to Boggy Creek Discharge have order-of-magnitude increases in transit time. Tracks in the south section of the LD1-vault intersection and to the southwest of this area switch from the Boggy Creek Discharge to the Confluence Discharge and have transit times extended by several orders of magnitude.

2.4.3 Surface exit locations. Figure 10 shows surface exit locations of particle tracks from the grouted case. The STD location in Boggy Creek is reduced in plan area from the standard case. The convergence effect for particle tracks in the Boggy Creek Basin is accentuated by the grouting. In this case the discharge area is less than 10% of the particle-track initiation area in the vault.

The added LTD location near the north end of Boggy Creek is the exit location for the two swept tracks seen in Figure 6c. The added LTD location near the south end of Boggy Creek is the exit location for the two particle tracks initiated at the south corner of the vault and shown as Boggy Creek Discharge in Figure 6c.

The major area of the vault has similar exit locations in the standard and grout cases. However, for the STD start points, changes in exit location and increases in transit time are caused by the presence of grout in the LD1-vault intersection.

3. Differences Between the Two- and Three-Dimensional Results

Here 2-D results, presented in Chan et al. (*3*), are compared with 3-D results presented in this paper.

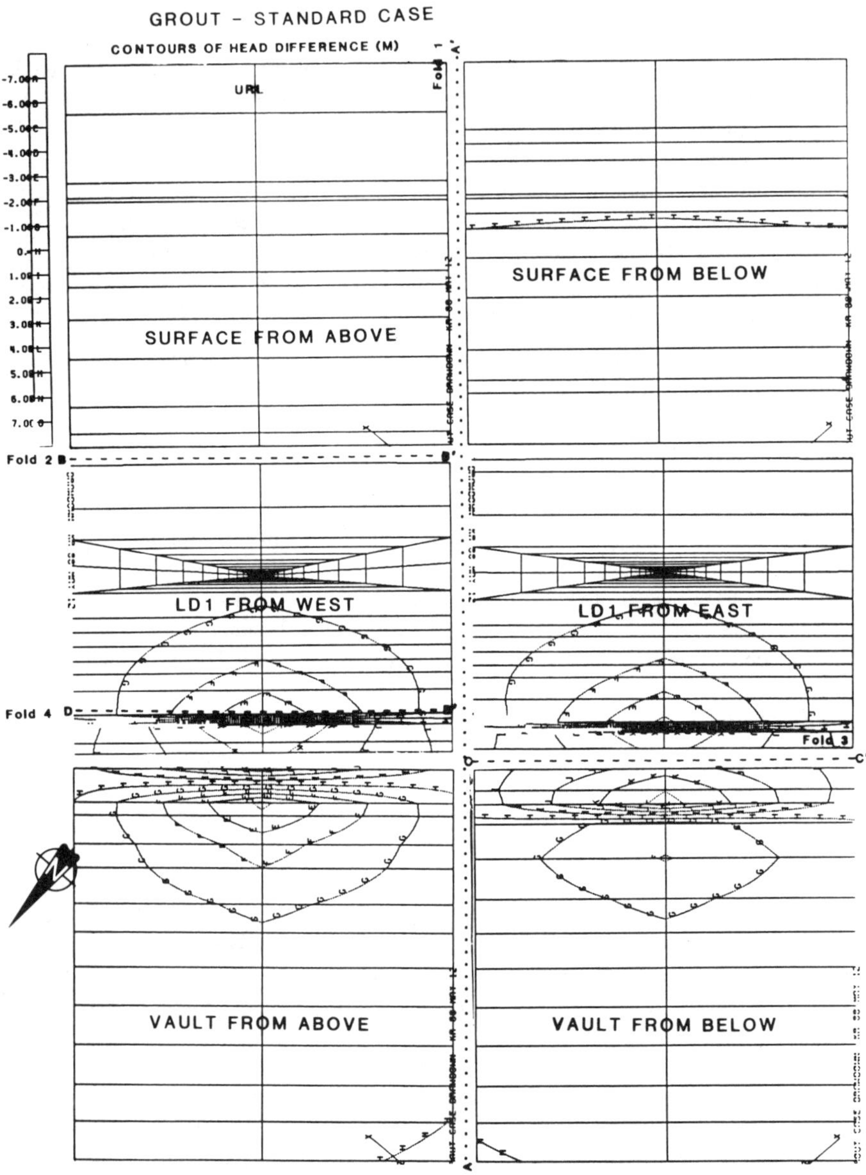

Figure 9. **3-D representation of head difference contours calculated as steady-state grouted head minus steady-state standard case head. (For assembly instructions see Figure 5.)**

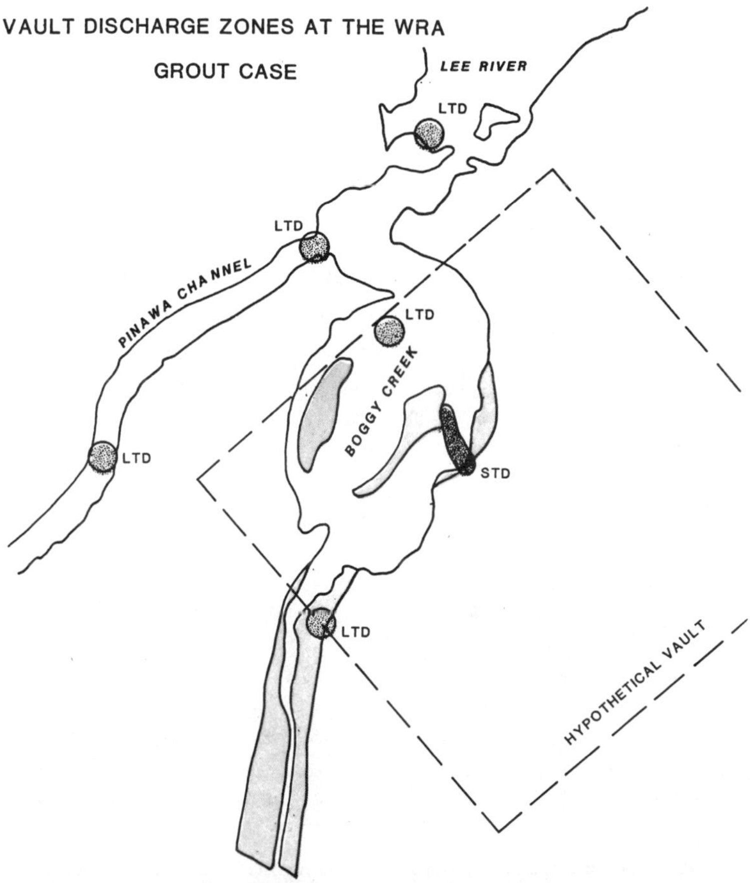

Figure 10. **Surface topographic feature map showing the grout case Short Time to Discharge (STD) and Long Time to Discharge (LTD) locations in relation to the hypothetical vault surface projection. (Intermediately shaded areas signify wetland.)**

In fracture zone LD0 in the 2-D simulation (*3*), the flow was downward, caused by the small peninsula between Boggy Creek and the Lee River. This peninsula crosses the 2-D section. In the 3-D local model, flow in the LD0 fracture zone is upward and across the zone into the section of the 2-D model. This contributes to the converging flows toward the Pinawa Channel/Boggy Creek Confluence. Because of this LD0 effect, there is no 3-D particle track corresponding to the 2-D path that flows up from the vault and then down the LD0 fracture zone before reaching the surface (path B in the 2-D simulation).

In both the 2-D and 3-D simulations when grout is present, some particles move more quickly due to the sweep effect, where grout causes increased water flow velocities in the intact rock near the grouted zone. See Figure 11 for the 2-D results. However, in the 3-D simulation, the effect is seen to be very localized to the north-

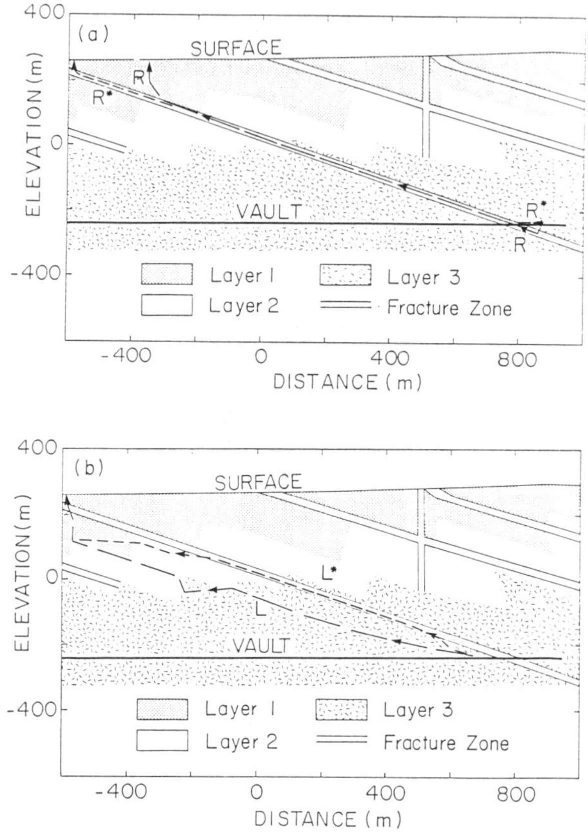

Figure 11. **2-D model results showing impact of grouting the fracture zone/vault intersection on predicted paths for convective transport of particles released from the vault: Paths R and L—no grouting; Paths R* and L*—with grouting. Reprinted with permission (*2*).**

east edge of the vault. Velocities in the area around the grouted material at the edge of the vault increase in the case of a grouted fracture/vault intersection and the particle is "swept" past the grout.

There are 3-D tracks corresponding to "D" in the 2-D modeling. These behave similarly in the 2-D and 3-D simulations in that the grouting causes them to switch to an extremely long path time and length. However, in the 3-D standard case, tracks in the central area of the intersection of LD1 and the vault behave dissimilarly in that they come to the surface in under 1000 a, somewhat less than the time predicted for particle track D in the 2-D model (*3*).

4. Causes of the Two- and Three-Dimensional Sensitivity

To examine the causes of the differences between 2-D and 3-D modeling, Figure 12 depicts cause "1" as the peninsula in the 3-D model and in the 2-D model

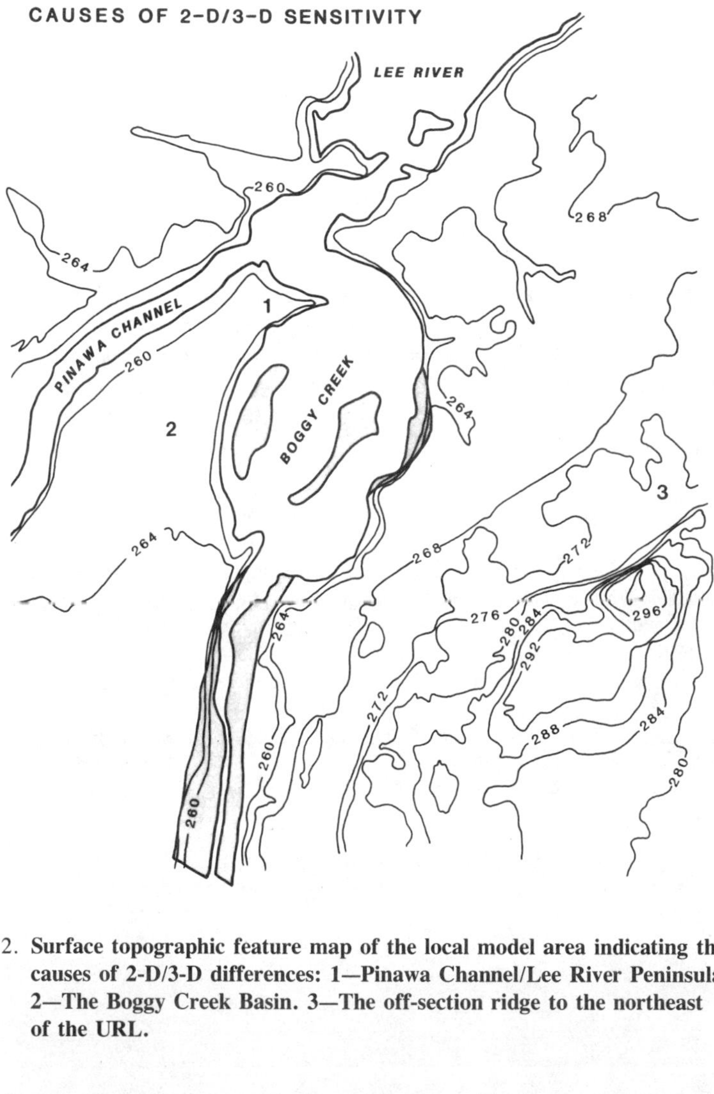

Figure 12. **Surface topographic feature map of the local model area indicating the main causes of 2-D/3-D differences: 1—Pinawa Channel/Lee River Peninsula. 2—The Boggy Creek Basin. 3—The off-section ridge to the northeast of the URL.**

that projects out across the 2-D section between Boggy Creek and the Lee River. This leads to 3-D results contradicting the 2-D prediction of flow downward in LD0. Cause "2" is the Boggy Creek depression where particle track convergence is caused by a basin that is both outside and within the 2-D section. Cause "3" is the ridge, off the 2-D section, to the northeast of the URL. This produces flows in the third dimension and accentuates the convergence effects of Boggy Creek Basin, increasing water flow velocity up the LD1 fracture zone and lowering transit times to the surface.

Cause "4" is that the complete grouting of the fracture in 2-D cannot simulate the increased effect of convergence of groundwater within the fracture zone around the grouted area. If there is grout in only part of the fracture, water can flow around the ends of the grout in the fracture, thereby increasing the convergence of the particle paths and decreasing the extent of the sweep effect.

5. Conclusion

There are both qualitative and quantitative differences between the results of 2-D and 3-D models. In a 2-D model, topological or topographical features that are outside the section cannot be properly represented. The convergence effect where groundwater converges in a fracture around the grout area requires study in three dimensions. Finally, discharge areas can be calculated only by using 3-D models.

When a large number of runs are necessary, such as in parameter sensitivity and system variability studies, 2-D models are useful because they are cheaper to run and simpler to interpret. However, crucial results need to be checked by a selected number of runs with 3-D models.

Literature Cited

1. Chan, T., "An Overview of Groundwater Flow and Radionuclide Transport Modeling in the Canadian Nuclear Fuel Waste Management Program," in these Proceedings (1988).
2. Chan, T., and N. W. Scheier, "Finite-Element Simulation of Groundwater Flow and Heat and Radionuclide Transport in a Plutonic Rock Mass," *Proceedings, 6th International Congress on Rock Mechanics,* Montreal, August 30–September 3, pp. 41–46, Balkema Press (1987).
3. Chan, T., N. W. Scheier, and J. A. Keith Reid, "Finite-Element Thermohydrogeological Model for Canadian Nuclear Fuel Waste Management," *Proceedings of the Second International Conference on Radioactive Waste Management*, Toronto, Canadian Nuclear Society, pp. 653–660 (1986).

Development of a Methodology of Geochemical Sensitivity Analysis for Performance Assessment

MALCOLM D. SIEGEL
Waste Management Systems Division
Sandia National Laboratories
Albuquerque, New Mexico

SIDNEY L. PHILLIPS
Information and Computing Sciences Division
Lawrence Berkeley Laboratory
University of California
Berkeley, California

JAMES O. LECKIE
Environmental Engineering and Science Division
Department of Civil Engineering
Stanford University

WALTON R. KELLY
Division of High-Level Waste Management
Office of Nuclear Material Safety and Safeguards
U.S. Nuclear Regulatory Commission
Washington, D.C.

Abstract

The objective of this geochemical sensitivity analysis is to formulate criteria for valid simplification of radionuclide transport models used in performance assessment. In the past, limitations in available models, data, and computer codes have necessitated the use of simplified representations of radionuclide chemistry and water/rock interactions in analyses of high-level waste (HLW) repositories. The validity of these simplifying approximations is a crucial element in the design of performance assessment methods and must be assessed relative to the appropriate performance measure and ranges of physicochemical parameter values.

In this paper, we illustrate the complementary roles played by

1. simple transport models designed to bound the radionuclide discharge over wide ranges of physicochemical conditions,
2. detailed geochemical models that provide a structured method to obtain reliable information about solute/rock/water interactions, and
3. sensitivity and uncertainty analysis techniques used to assess the significance of errors in both data and models.

In this multiyear project, progress has been made in the following areas:

1. compilation of computerized data bases of critically evaluated thermodynamic data and empirical sorption (batch K_d) data for waste elements,
2. assessment of the potential use of surface complexation models in existing performance assessment methodologies, and
3. simulation of radionuclide discharge for generic and site-specific systems over 10,000-year periods.

Introduction

Sensitivity analyses and performance assessment calculations of radionuclide transport from HLW repositories rely on simplified representations of complex chemical behaviors of aqueous radionuclides and hydrologic flow fields. The structure of these calculations is strongly influenced by

1. current theories describing the transport and reaction of chemical species in natural systems,
2. the availability of data from both empirical and mechanistic studies of solute/rock/water interactions in the specific geochemical environments of interest, and
3. the costs associated with computer simulations of large systems.

Limitations in these three areas prevent the inclusion of all possible phenomena in any single analysis. A crucial element in the design of the models to be used in these studies is the validity of the simplifying assumptions that are used.

Sandia National Laboratories (SNLA) is assisting the U.S. Nuclear Regulatory Commission (NRC) in the development of a performance assessment methodology for HLW repositories. For the past several years, an effort has been made to develop a method of geochemical sensitivity analysis. This work has been a collaborative effort among staff at SNLA, Lawrence Berkeley Laboratory (LBL), and the NRC. The objective of this analysis is to assess the significance of errors in calculations of radionuclide discharge due to various approximations used to describe geochemical processes. The purpose of this paper is to describe the approach of this work by evaluating some questionable assumptions inherent in performance assessment models and associated sensitivity analyses.

Model Validity and Simplification

General principles

Zeigler (*1*) provides a useful framework for defining criteria for valid model simplification. He identifies the following five distinct elements of a model/simulation:

1. the real system, which is the source of observations,
2. experimental frames, which are selected observations from the real system,
3. the base model, a hypothetical model that is valid in all experimental frames,
4. lumped models, which are simple models designed for particular experimental frames, and
5. computer codes, which carry out the simulations.

Various methods of simplification have been discussed by Henize (*2*). He suggests that a weakness in many sensitivity studies of large systems is the failure to include all significant feedback mechanisms between the individual components (i.e., submodels) of the system. A common error in sensitivity analyses involves holding the value of one or more variables constant while varying the value of a dependent variable during a parametric analysis of system behavior. If significant interactions between the variables have been ignored, then the results of the analysis can be misleading.

Henize also notes that another common error in sensitivity analyses and simulations is the failure to direct the calculations toward the appropriate performance measure. The number of independent and dependent variables that could be used to describe even a simple system is potentially very large. If the most useful subset of parameters is not identified early in the analysis, then the simulations and sensitivity analyses can be inefficient or irrelevant.

Criteria for valid simplification of models used in performance assessment should be defined in light of these basic concepts. The appropriate experimental frame must be defined and the significance of potential interactions must be assessed relative to the correct performance measure. The application of these general concepts is applied to specific aspects of performance assessment models in the following section.

Simplification of geochemical models for performance assessment calculations and sensitivity analysis

In the performance assessment methodology developed by SNLA for the NRC, the Latin Hypercube sampling (LHS) procedure (*3*) is used to create combinations of values of input parameters for large numbers of calculations in sensitivity analyses. As illustrated in Figure 1, ranges and distributions are assigned to each input variable, values of each variable are selected, and combinations of parameter values are used for calculations with the NEFTRAN (*4*) or SWIFT-II (*5*) codes. These computer codes calculate radionuclide travel times and discharge rates at the end point of a hydrologic flow path. The results of many such calculations produce a distribution of values of the performance measure that can be used in sensitivity analyses or to assess compliance of a site with the EPA Standard 40 CFR Part 191 (*6,7*). If the results of these calculations are to be valid, then the following must be shown:

1. It is possible to assign realistic distributions of the input parameters using available data, and
2. The correlations between the input variables as represented by the LHS procedure reflect the true relationships in nature.

When these assumptions are not valid, then the analysis may suffer from the corresponding two errors:

1. use of an inappropriate experimental frame and
2. use of an incomplete or incorrect set of relationships among the variables to represent the system.

The Latin Hypercube sampling procedure allows for the assignment of linear correlations between variables. Generally, however, the nature of the relationships among variables is not known and many parameters are assumed to be uncorrelated.

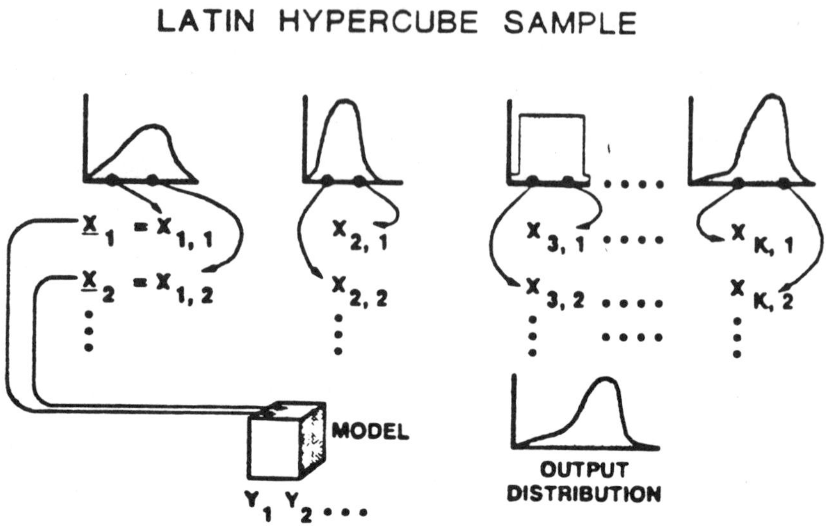

Figure 1. **Use of the Latin Hypercube sampling procedure in sensitivity analysis. Distributions and correlations are assigned to each variable X; values $X_{n,k}$ of the variables are sampled and combined into data sets $\underline{X}_k$ for model input. Multiple calculations produce a distribution of output values Y_k.**

For example, the radionuclide distribution coefficient K_d, matrix porosity, and ground-water velocity are commonly assumed to be independent. In a later section of this paper, the adequacy of this assumption will be examined.

Figure 2 illustrates the general structure of radionuclide transport calculations carried out by the NEFTRAN and SWIFT-II computer codes in performance assessment studies (*4,5*). Interactions among radionuclides, water, and rock are represented by retention factors whose values are determined using the following simplifying assumptions:

1. A single chemical species exists for each radionuclide.
2. The composition of the ground water is constant in each transport leg.
3. The rock composition is constant in each transport leg.
4. Reversible equilibrium sorption that can be described by a linear or Freundlich isotherm is the only chemical process occurring.
5. Chemical exchange between fractures and the rock matrix can be described by a bulk retardation factor, a linear-driving-force (mass transfer) approximation, or an effective diffusion coefficient.
6. Radionuclide transport by colloids or organic complexes is negligible.

Figure 3 contrasts the lumped model of chemical interactions (3a) used in these performance assessment models with a hypothetical base model for the real system (3b). The lumped model includes only reversible sorption and diffusion of radionuclides into dead-end pores of the rock matrix. The base model includes a number of other processes such as metal-organic complexation, oxidation-reduction, inorganic

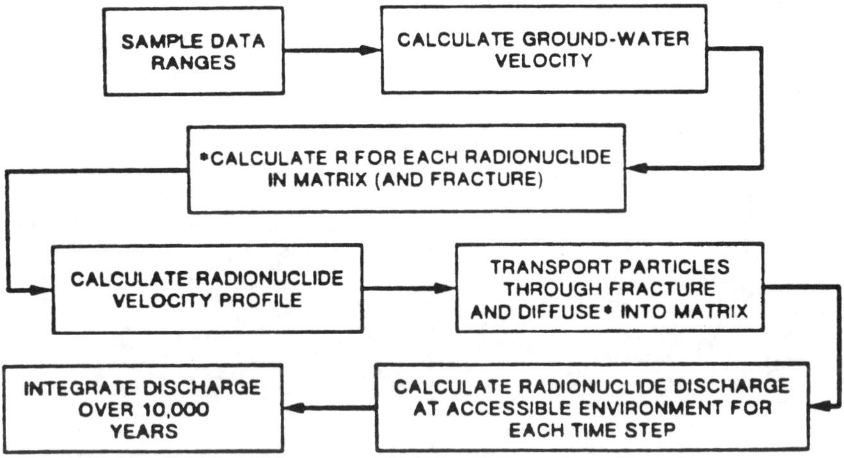

Figure 2. **Outline of radionuclide transport calculations as carried out by the NEFTRAN computer code.**

complexation, irreversible sorption, ion filtration, and colloid formation. It is generally accepted that the base model depicted in Figure 3b is more predictively valid than the lumped model in Figure 3a. It has been suggested that more sophisticated computer codes that include all of these phenomena be used in performance assessment calculations (*8*). In later sections of this paper, however, it will be shown that limitations in available data and computer codes currently prevent the use of such comprehensive models in a performance assessment methodology.

One alternative to the use of either the inadequate lumped model described in Figure 3a or the impractical base model in Figure 3b is to give no credit to geochemical retardation in assessments of compliance with the EPA Standard. In a later section of this paper, it will be shown that uncertainties in hydrologic data for potential sites preclude reliance solely on hydrologic barriers to radionuclide release.

Development of a Method for Geochemical Sensitivity Analysis

Role of geochemical sensitivity analysis in performance assessment

Although it is anticipated that advances in geochemistry, hydrology, and computer science will allow for comprehensive descriptions of radionuclide transport at the repository sites at some future time, it is prudent to design an interim approach to address the aforementioned weaknesses in the transport models currently used in performance assessment. Such an approach should identify site-specific conditions under which the approximations used in the current performance assessment models significantly hinder our ability to assess compliance with the EPA Standard. Another objective of this analysis would be to determine if the available solubility and sorption data are adequate to define the experimental frame for performance assessment models.

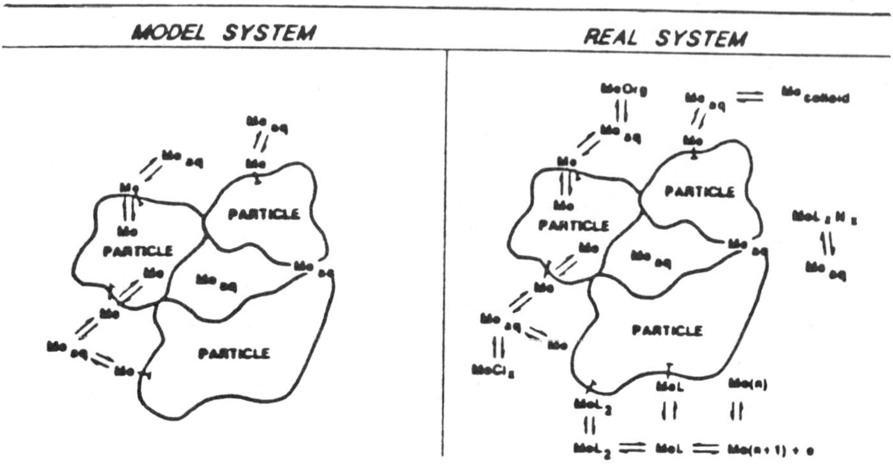

Figure 3. **Comparison of 'model' system as represented in transport codes that use a K_d or retardation factor with a 'real' system containing many potentially important chemical processes. The processes illustrated in the figure are described in the text.**

The components of the analysis are identified in Figure 4. The probabilistic risk assessment (PRA) codes NEFTRAN, CDFPLT, and LHS are being used to assess the amount of geochemical retardation needed to ensure compliance with the EPA Standard 40 CFR Part 191 at a hypothetical basalt repository site. The Sandia Sorption Data Management System (SSDMS) and the Aqueous Solutions Database (ASD) are being developed to store and facilitate retrieval of K_d values and thermodynamic property values, respectively. These parameter values are used either in the PRA

COMPONENTS OF ANALYSIS

Data Bases
SSDMS
ASD

PRA Codes
NEFTRAN
LHS
CDFPLT

Speciation Codes
MINEQL
PHRQPITZ

Coupled Codes
TRANQL

Figure 4. **Components of a method of geochemical sensitivity analysis. The method described in the text involves complementary use of probabilistic risk assessment (PRA) codes, data bases of sorption and thermochemical data, and computer codes that can model chemical speciation and transport.**

codes or in speciation codes such as PHREEQE (*9*) and MINEQL (*10*). Coupled reaction/transport codes such as TRANQL (*11*) do not use a retardation factor to approximate chemical behavior of the radionuclides. With the proper modifications, it is hoped that codes like these can be used to carry out simulations in which many of the assumptions associated with retardation factors can be relaxed. In the sections that follow, examples of the use of each of the components with site-specific data are given.

Performance assessment calculations for a hypothetical basalt site

The radionuclide flow path assumed in these calculations is shown schematically in Figure 5. The flow path is similar to that assumed in performance assessment studies of the proposed repository site at the Hanford Reservation in Washington State (*12,13,14*). The repository is located in the interior of a basalt flow. Radionuclides flow vertically from the repository through the dense interior to a porous flow top and then horizontally 5 km to the accessible environment. Data from several sources were used to calculate cumulative distribution functions for ground-water travel times in the horizontal leg of this flow path (*15,16,17*). Curves 1 to 3 in Figure 6 were calculated from hydraulic head and conductivity data assuming 1-dimensional Darcian flow. These curves illustrate the large uncertainty associated with the current estimates for ground-water travel time at the basalt site.

Curve 4 in Figure 6 shows a distribution in which 15% of the ground-water travel times are less than 1000 years and 99.9% of the travel times are less than 10,000 years. This distribution is more conservative than one that had been proposed by the NRC (*18*) to satisfy the preclosure ground-water travel time requirement of

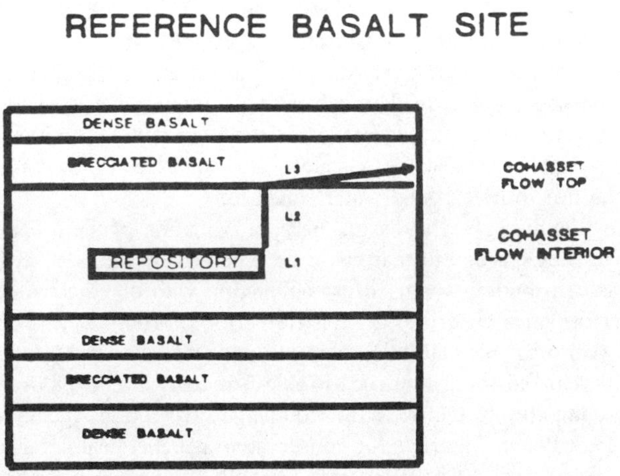

Figure 5. **Hypothetical reference basalt site used in geochemical sensitivity analysis. Data from the Basalt Waste Isolation Plant (BWIP) site in Washington State are used for many of the system variables; however, it is not claimed that the reference site accurately represents any real site. The flow path used in the radionuclide discharge calculations is indicated along L1, L2, and L3.**

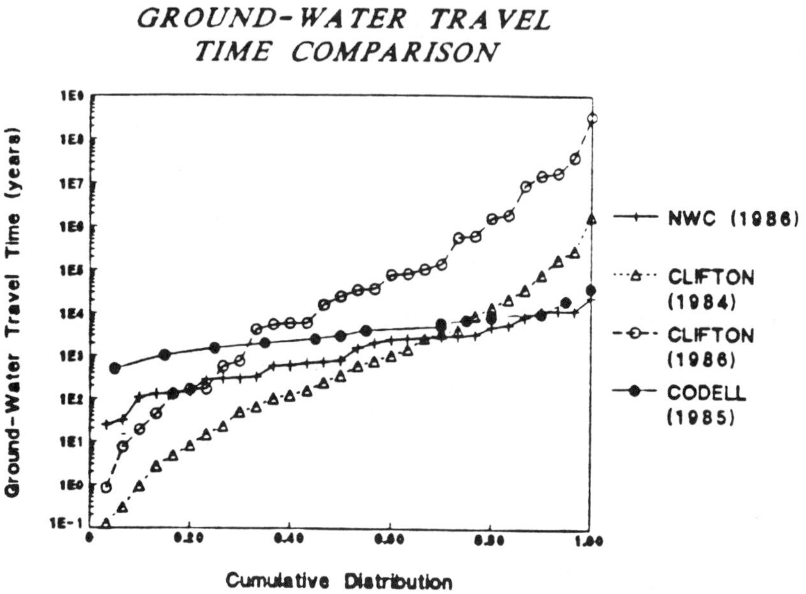

Figure 6. **Comparison of cumulative distribution functions of ground-water travel times for basalt repository site. See text for discussion of method and assumptions used to calculate each curve. Curve identified as Codell (1985) (*18*) was used in radionuclide discharge calculations.**

the NRC regulation for high-level waste disposal 10 CFR Part 60 (*19*). (Note that this interpretation of the requirement was not retained in a more recent NRC draft technical position paper [*37*].) In the analysis that follows, it was assumed that the ground-water travel time in the vertical leg of the flow path (L2 within the dense interior) was negligible and that the postclosure travel time within the horizontal leg was equal to the preclosure ground-water travel time.

Curve 4 in Figure 6 gives a reasonable approximation of the most conservative (worst-case) hydrologic scenario that could be accepted by the NRC for a licensable HLW site. This distribution was used in conjunction with nominal values for other NRC HLW performance criteria (i.e., release period of 100,000 years and isolation period of 300 years) to calculate integrated radionuclide discharges for a repository inventory of 10-year-old spent fuel rods (Table 3 in *20*). The 10,000-year integrated discharges are compared to the U.S. Environmental Protection Agency (EPA) Standard 40 CFR Part 191 in Figure 7 for conservative geochemical conditions (i.e., $K_d = 0$). The distribution curve lies above the EPA limit indicating that this scenario would lead to violations of the EPA Standard. The construction of complementary cumulative distribution functions (CCDFs) and their use in assessing compliance of a repository with the EPA Standard are discussed elsewhere (*21*). These calculations show that if the performance of a repository site were only as good as the minimum requirements of the three numerical performance criteria described above, the site

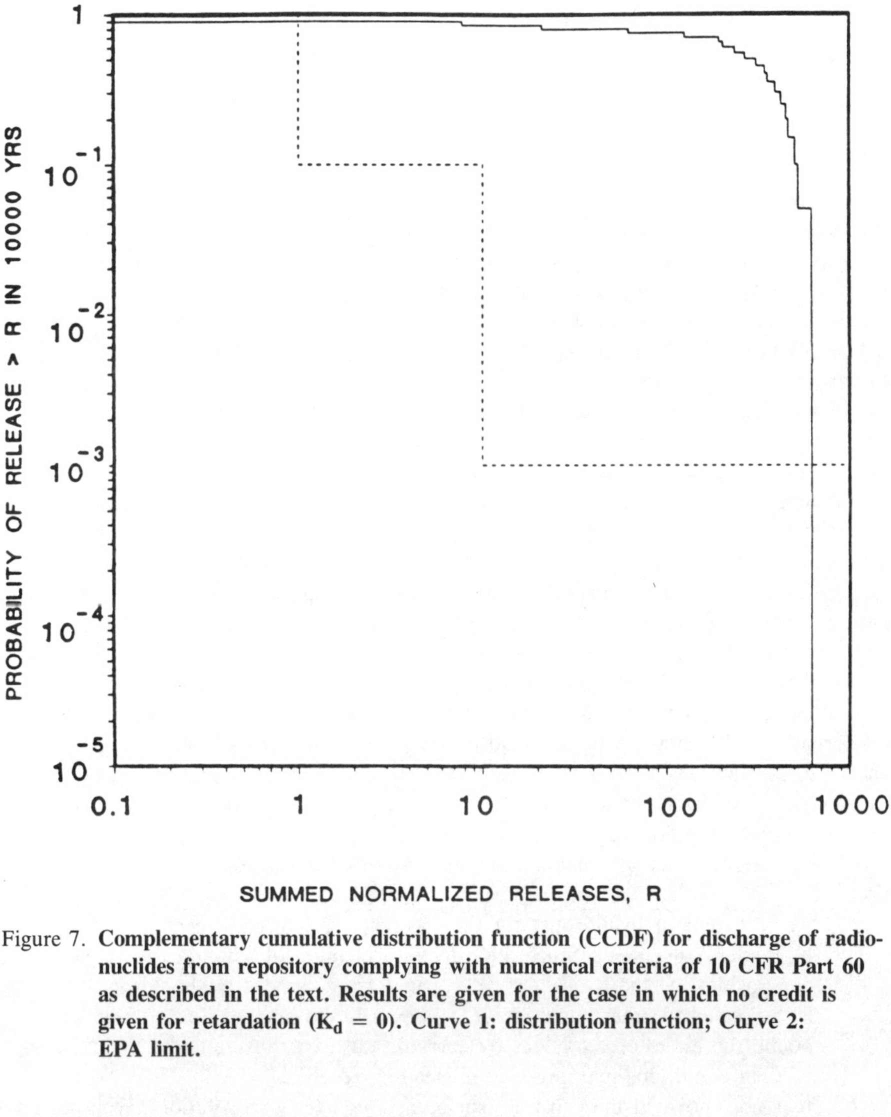

Figure 7. **Complementary cumulative distribution function (CCDF) for discharge of radionuclides from repository complying with numerical criteria of 10 CFR Part 60 as described in the text. Results are given for the case in which no credit is given for retardation (K_d = 0). Curve 1: distribution function; Curve 2: EPA limit.**

would violate the EPA Standard 40 CFR Part 191, if no geochemical processes retard radionuclide migration.

Review of available sorption data relevant to a basalt repository site

An estimate of the radionuclide retardation due to geochemical processes is required to determine if the basalt site previously described will comply with the EPA Standard. A first step in this process was to choose conservative values of retardation factors and sorption ratios (K_d's or R_d's) which can be used to calculate realistic upper bounds to radionuclide discharges. Before this could be done, however, it was necessary to critically evaluate the available data to determine if such conservative values could be selected in a technically defendable way.

The Sandia Sorption Data Management System (SSDMS) has been developed to store and organize data obtained from batch and column radionuclide sorption experiments (*22*). All available batch (K_d) data relevant to the Basalt Waste Isolation Plant (BWIP) site were compiled in the data base (*23*). These data have been collected for rock-water combinations that are considered representative for the proposed repository site. Extensive descriptions of experimental conditions and procedures are included in the data base to allow examination of the correlations between experimental parameters and K_d values. Statistical and graphical utilities are available to facilitate these analyses. Based on this evaluation, each datum is assigned a quality index to facilitate subsetting and manipulation of the data. Detailed descriptions of the data fields, programs, and scope of the available data can be found in (*22,23*).

A major concern emerged from our critical evaluation of the available batch sorption data. In general, the data have been acquired without an understanding of the properties of the substrates or identification of the species present in solution. The potential for the introduction of spurious effects is large in these experiments. Many of the K_d values obtained may include the effects of processes other than reversible sorption (cf. Figure 3b) and cannot be used to calculate valid retardation factors for use in solute transport equations. Without a prior model for the chemical system, critical evaluation of these data has proceeded in an empirical manner. In the past, incorrect laboratory procedures and faulty data often have been identified by trial and error (see *23* and cited works). Observations of the effects of experimental procedures on the measured K_d values have been used to determine if the data are conservative. The data can be accepted on only a provisional basis until they are shown to be flawed. There is no way to show that the data are valid due to the incomplete descriptions of the solutions, solids, and experimental procedures, and the lack of theoretical guidance.

These specific areas of uncertainty were identified during our review of the sorption data:

1. At best, data are valid only for the specific physicochemical conditions of a particular experiment. Since the effects of changes of solution and substrate composition on the K_d values are not understood, the data should not be extrapolated to other conditions. As shown in the next section, attempts to bound the range of K_d values by carrying out experiments under the extrema of field conditions may produce misleading results.
2. It is not known if the substrate surfaces in contact with solution during the experiments are representative of in situ mineral surfaces. Methods used to prepare samples for batch experiments often include grinding the samples. This procedure may create new sites that bind radionuclides by processes other than sorption (e.g., electron transfer and coprecipitation). Such processes would give nonconservative (high) K_d values.
3. The nature and rates of chemical reactions that may control the speciation of sorbed and aqueous species are not known. Although an apparent steady state is reached in many batch experiments, it is not known if true equilibrium has been achieved. In many cases the observation of a sorption/desorption hyster-

esis suggests that the binding process is not reversible under the conditions of the experiment.

Figure 8 summarizes the results of the review of available K_d data for neptunium in basalt/water systems (*23*). In this figure, average K_d values for several rock/water combinations are plotted. It can be seen that the K_d values depend on parameters that can be related to field conditions such as rock type, ground-water type, and temperature. In addition, a significant dependence on experimental parameters that are not easily related to in situ conditions is observed. These include the solution: solid ratio and redox conditions of the experiment and a variety of sample preparation techniques that are lumped together under the 'Quality' index indicated in the figure.

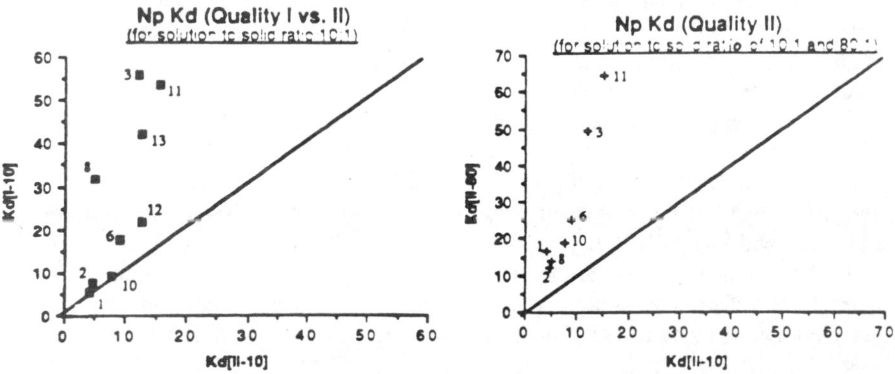

Key	Sorption Interval	Ground Water	Redox	Temp. (°C)	Sorption Category
1	McCoy Canyon	GR-1	Oxic	25	basalt
2	Pomona	GR-1	Oxic	25	basalt
3	Pomona	GR-1	Oxic	25	smectite
4	Pomona	GR-2	Oxic	25	basalt
5	Pomona	GR-2	Oxic	25	smectite
6	Umtanum	GR-1	Oxic	25	basalt
7	Umtanum	GR-2	Oxic	25	basalt
8	McCoy Canyon	GR-1	Oxic	60	basalt
9	McCoy Canyon	GR-2	Reducing	60	basalt
10	Pomona	GR-1	Oxic	60	basalt
11	Pomona	GR-1	Oxic	60	smectite
12	Umtanum	GR-1	Oxic	60	basalt
13	Umtanum	GR-2	Oxic	60	basalt

Figure 8. **Summary of K_d (ml/gm) values for neptunium in different combinations of rocks and waters found at the BWIP site. After evaluation of experimental method used to obtain K_d value, a quality index of I is assigned to "good" data, and an index of II is assigned to "less reliable" data. Axis labels identify quality index and solution:solid ratio. Table identifies each point plotted.**

These data can be sorted by type and quality, and distributions of K_d values can be obtained as shown in Figure 9. The setting of ranges for valid sensitivity analyses, however, is complicated by the lack of an underlying model for radionuclide behavior. Such a model would enable us to determine what part of the range of K_d values is due to spurious effects that are not relevant to field conditions. In Figure 9, it is clear that subsetting the data on the basis of solution and solid composition produces a narrow distribution compared to the range for all data. However, the limits of the range can be set only by subjective judgement based on these empirical, "process-blind" observations.

Use of detailed chemical models to predict radionuclide behavior

An alternative and more robust method to describe radionuclide/solid reactions is provided by surface complexation models such as the triple-layer model or the constant capacitance model (*24,25*). These approaches allow for a complete accounting for all species in solution, solids, or sorbed onto solid/solution interfaces. They are based on electrostatic models for the structure of the solid/solution interface. Surface reaction sites compete with aqueous ligands in complexation reactions involving free cations or anions.

Calculations of the speciation of neptunium under oxidizing conditions in water with major solute compositions similar to those reported at the BWIP site have been carried out with the MINEQL code (*14,25*). Figures 10 to 12 present the results of these calculations and illustrate the power of the surface complexation models. Figure 10 is a distribution diagram for neptunium in a model system illustrating features common to many of the systems that were studied. The system parameters are

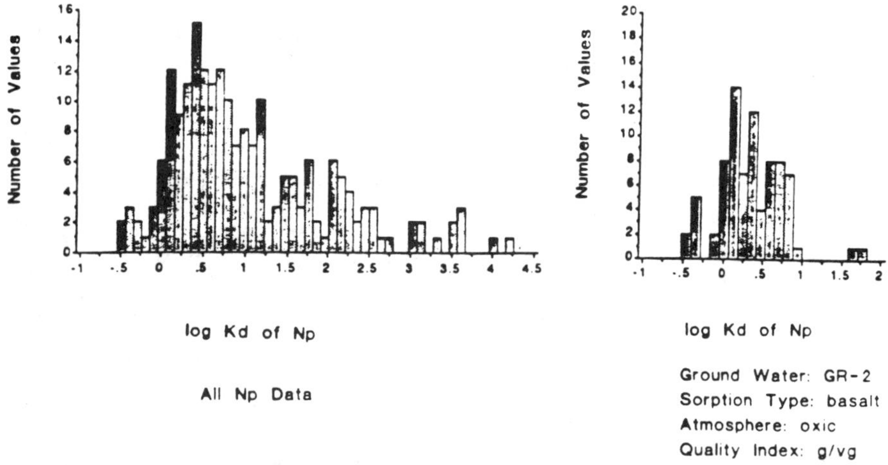

Figure 9. **Histograms of K_d values for neptunium. (a) All available sorption data for basalt environments; (b) Highest quality data for unaltered basalt in GR-2 synthetic water, under oxic conditions (quality index g/vg, i.e., 'good' or 'very good').**

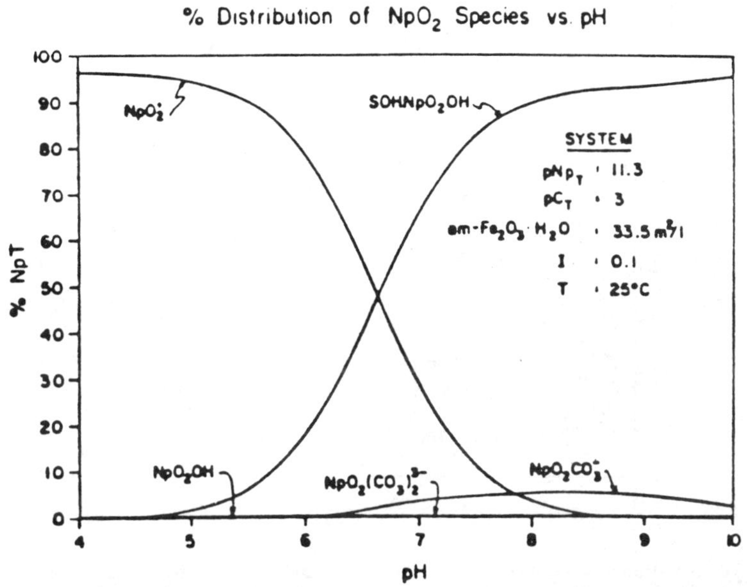

Figure 10. **Distribution diagram for neptunium species under oxic conditions. % Np_T refers to percent total neptunium present as a particular species. Only Np(V) species were assumed present.**

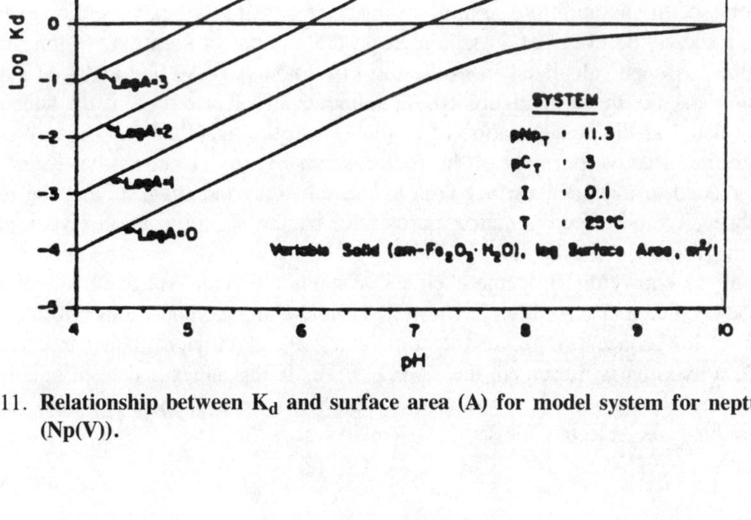

Figure 11. **Relationship between K_d and surface area (A) for model system for neptunium (Np(V)).**

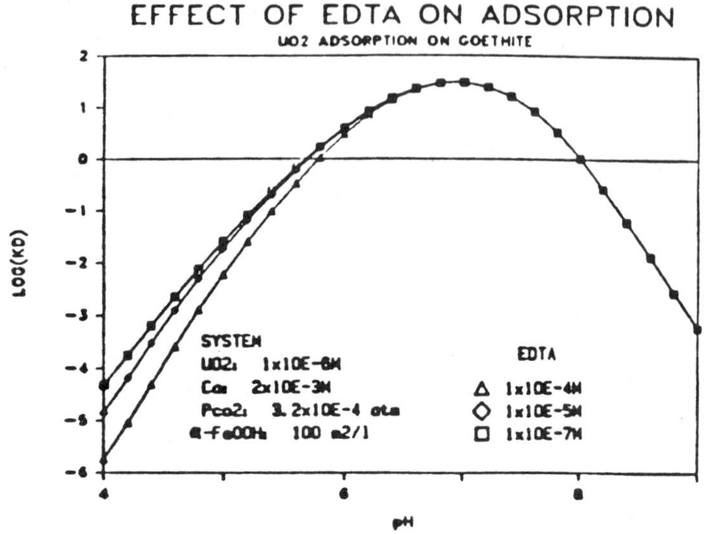

Figure 12. **Relationship between K_d for uranium (UO_2^{++}) and pH in the presence of a complexing ligand, EDTA, and goethite (FeOOH).**

described in the figure and figure caption. It can be seen that at low pH values, most of the neptunium is present as the aqueous neptunyl ion NpO_2^+. At pH values above 6.5, most of the neptunium is sorbed onto the amorphous iron oxyhydroxide substrate (indicated as $SOHNpO_2OH$). These calculations suggest that under oxic conditions at pH values similar to those measured at the BWIP site (pH = 8 to 11), neptunium will sorb strongly onto iron oxyhydroxide present in water-bearing fractures.

Calculations like those illustrated in Figure 10 can be used to evaluate the validity of some of the assumptions commonly made in sensitivity analyses. For example, Figure 11 shows the effect of surface area on the amount of sorption. In this figure, a K_d value has been calculated from the ratio of amounts of sorbed and free neptunium from data like those in Figure 10. The amount of surface area of the substrate is directly related to the concentration of available sorption sites and also may be related to the effective porosity of the rock matrix. Figure 11 shows that K_d is directly related to available surface area of the substrate and suggests that the use of uncorrelated K_d and effective matrix porosity distributions in the Latin Hypercube sample may not be realistic.

Figure 12 shows that in some chemical systems the actual range of K_d values cannot be predicted from a few "extreme" ground-water compositions. In the presence of low concentrations of the complexing agent, ethylenadiametetracetic acid (EDTA), a maxima is present in the K_d-pH curve. If K_d values had been obtained only from measurements at pH values of 6 and 8, an overly conservative K_d range would probably be selected for sensitivity analyses.

Uncertainties in speciation and sorption calculations

Figures 10 to 12 do not consider the uncertainties in the thermochemical data used to calculate the K_d values. These uncertainties may be significant and should be addressed in a method of geochemical sensitivity analysis. Two main types of uncertainty should be considered. These are

1. questions of the existence of certain species, and
2. numerical uncertainties in the values of complexation constants (log K).

An example of the first type of uncertainty is the controversy concerning the existence of the $U(OH)_5^-$ species. In their model for the hydrolysis of tetravalent uranium, Baes and Mesmer (*26*) assumed that the $U(OH)_5^-$ species was a dominant species in alkaline solutions in contact with the solid UO_2. Based on this assumption and assuming a regular progression of hydrolysis constants, they calculated formation constants for $U(OH)_2^{2+}$, $U(OH)_3^+$, and $U(OH)_4$. Recently, however, Ryan and Rai (*27*) and Bruno (*28*) have challenged the evidence supporting the existence of the $U(OH)_5$ species and thereby the values of the other hydrolysis constants. The controversy over the aqueous model for uranium is a source of uncertainty that should be included in speciation and solubility calculations.

Numerical uncertainties in the values of equilibrium constants can be significant even in systems for which there is general agreement on the aqueous model. For example, estimates for the value of the formation constant of $PuCO_3^{2+}$ range from 10^{13} (*29*) to 10^{47} (*30*). The calculated solubility and identity of the dominant species depend on the value of the constant that is used in speciation calculations.

The Aqueous Solutions Database (ASD) has been developed to store and process thermodynamic property values and associated uncertainties for sensitivity analysis. The data base contains a well-documented, critically evaluated, internally consistent set of values for ΔG_f°, ΔH_f°, C_p°, and S° for elements of interest in HLW disposal. The features of the data base are described in Figure 13; the procedure used to critically evaluate the data is discussed in detail in Phillips et al. (*31,32*). The data base is accessible to any interested researchers over telecommunication networks such as MILNET.

The ASD has been designed to facilitate examination of the sensitivity of calculated solubility limits and speciation to uncertainties in the thermodynamic data. One kind of analysis, illustrated in Figure 14, involves randomized sampling of log K values from user-defined distributions. The extrema and shape of the distributions can be determined from actual experimental uncertainties or can be calculated from the free energies and propagated uncertainties stored in the ASD. By sampling repeatedly from log K distributions and carrying out replicate speciation calculations, uncertainties could be calculated for the estimated concentrations of aqueous species. Examination of the resulting response surface by a variety of statistical techniques could identify the particular constants that dominate the uncertainty in the calculated solubility, speciation, and sorption.

Use of detailed geochemical models in performance assessment calculations

The inclusion of chemical phenomena described in Figure 3b in performance assessment calculations will require computer codes that couple chemical speciation calculations like those carried out by the MINEQL code with equations for advec-

THERMODYNAMIC DATABASE FOR GSA

- ASD FORMAT 1: GHS and Cp data
- ASD FORMAT 2: log K for specific reaction
- ADDREACT: add new reaction to FORMAT 2
- COMPLOGK: compute log K $\pm$ σ for reaction
- BRIDGE: produce new MINEQL database from ASD data with log K $\pm$ σ and $\Delta H_r \pm \sigma$
- LHSMIN: prepare file for multiple runs of MINEQL for different log K
- MINQLMLT: carry out multiple MINEQL runs and write specified results to a file for post-processing

Figure 13. **Components of the Aqueous Solutions Database for use in sensitivity analysis.**

FUNCTIONS OF ASD PREPROCESSOR IN SPECIATION SENSITIVITY ANALYSIS

1. Reads ΔG values and σ from ASD for basis species

2. Calculates log K's and uncertainties for input to LHS

3. Uses LHS to produce nxm array of log K values

4.1 Reads vector of log K's input for a single calculation

4.2 Performs speciation calculation and writes results to a file

4.3 Chooses important species and writes their concentrations or other parameters to a file

4.4 Repeats for each vector

5. Prepares input for sensitivity analysis on results by graphical or other methods

Figure 14. **Example of use of the Aqueous Solutions Database in geochemical sensitivity analysis.**

tive/diffusive transport. Several such codes have been described in the literature (*11,33,34*). The potential use of the TRANQL code (*11*) in performance assessment studies has been evaluated as part of this study. The evaluation has considered both the availability of data needed by the code as well as the computer resources required for multiple, 10,000-year, 5-km simulations. As already discussed, large numbers of simulations are needed to construct the CCDFs used to assess compliance of HLW sites with the EPA Standard 40 CFR Part 91.

A critical evaluation of radionuclide sorption data was carried out to identify data sets that could be processed with the triple-layer model (TLM) (*25*). Figure 15 summarizes the chemical systems for which high-quality data were found. The criteria used in the review are discussed in (*25*). It can be seen that only data for simple hydrous oxides in simple electrolyte systems are available. Present formulations of the triple-layer model are not applicable to carbonates, salts, or materials with fixed charges such as clays. In addition, the model cannot be applied rigorously in solutions more concentrated than an ionic strength of 0.1 molal. These limitations mean that, at present, the TLM cannot be applied directly to many of the substrates or solutions that must be modeled in HLW performance assessment.

Examination of the TRANQL code (*11*) revealed that the available version of the code was not useful for performance assessment calculations. The code cannot simulate many important chemical phenomena such as oxidation/reduction, radioactive production, or precipitation. The code is inefficient; estimates of the cost of running a single simulation of transport of a single element over a 5-km distance for a 10,000-year time period range from $10,000 to $40,000 (*14,35*).

Based on the information discussed above, it has been concluded that without major improvements in both data bases and computer codes, detailed chemical models cannot be used in place of codes such as NEFTRAN or SWIFT-II to assess the compliance of a repository site with the EPA Standard 40 CFR Part 191. At

RESULTS OF STANFORD REVIEW OF AVAILABLE TLM DATA

Substrate	Electrolyte	Metals
γ-Al_2O_3	NaCl	-
α-Fe_2O_3	KCl	-
α-FeOOH	KNO_3, NaCl, Na_2SO_4, KCl, $MgCl_2$, $CaCl_2$, $CaCO_3$	UO_2^{2+}
Am-$Fe_2O_3H_2O$	$NaNO_3$, K_2SO_4, $NaHCO_3$	UO_2^{2+}, Cd^{2+} NpO_2^+, SeO_4^{2-}
δMnO_2	$NaNO_3$, NaCl	-
Am-SiO_2	NaCl, KCl	-

Figure 15. **Summary of chemical systems for which high-quality data are available for the triple-layer model.**

present, detailed chemical speciation and chemical transport models are best used to provide insights into the nature of significant interactions and correlations among system variables. The codes should be used to assess the validity of simplifying assumptions used in performance assessment tools and to extend sensitivity and uncertainty analysis of system behavior to detailed geochemical processes.

The complementary roles of several modeling approaches and computer codes in this kind of sensitivity analysis are illustrated in Figure 16. Computer codes such as PHREEQE (*9*) or MINEQL (*10*) can be used to define possible ranges in groundwater compositions and corresponding radionuclide speciation under site-specific conditions. Such information can be used to assess, if available sorption data (K_d's or retardation factors) can be used to bound radionuclide sorption behavior at the HLW sites, and to examine the nature of interactions among geochemical parameters. The performance assessment codes NEFTRAN (*4*) and SWIFT (*5*) represent radionuclide/rock interactions in terms of retardation factors and mass transfer (exchange) coefficients. Site-specific hydrologic data for fractured and porous media, experimental sorption data, and the results of the speciation calculations could be used to estimate ranges for these parameters and to assign meaningful correlations among system variables. Coupled reaction/transport codes like TRANQL (*11*) could be used to carry out transport simulations for a limited number of conditions in order to deter-

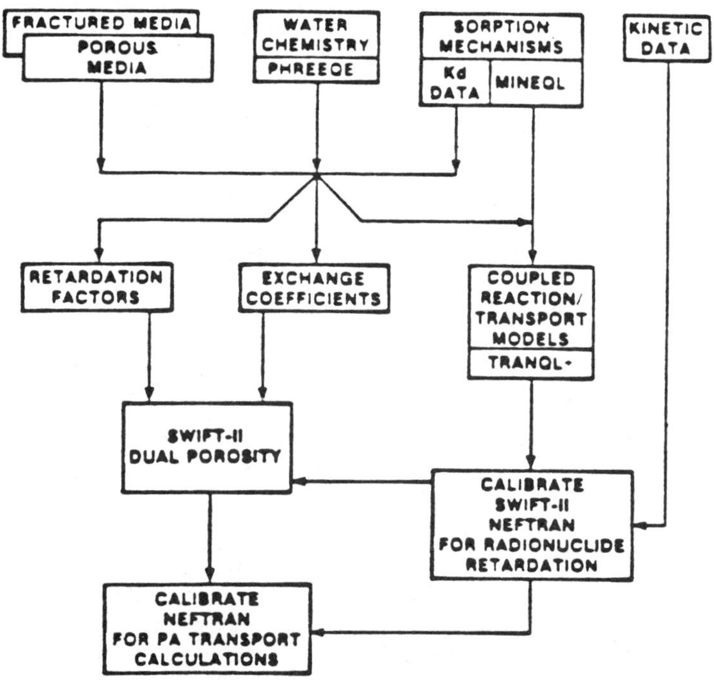

Figure 16. **Complementary roles of different models discussed in this paper in transport sensitivity analysis.**

mine the range of discharge rates that result when interactions between solution composition and sorption behavior are included in the model. The results of these calculations could be compared to simulations carried out with the SWIFT and NEFTRAN codes to determine if the simple geochemical approximations in those codes provide conservative results (i.e., overestimates of radionuclide discharge).

Kinetic effects

The assumptions of homogeneous and heterogeneous chemical equilibria are implicit in chemical speciation codes such as MINEQL and PHREEQE and in transport codes such as SWIFT and NEFTRAN. Previously (*36*), criteria for the validity of three approximate methods to calculate radionuclide discharges in fractured porous rock were derived:

1. an equivalent-porous-medium approximation in which the bulk rock matrix is in equilibrium with the fracture fluid,
2. a linear-driving-force approximation in which local chemical equilibrium exists between the pore fluids and the rock matrix and radionuclide diffusion rates are proportional to the difference in average radionuclide concentrations in the rock matrix and fracture fluids, and
3. an approximation where radionuclide diffusion into the matrix is calculated assuming a semi-infinite matrix.

The first of the above approximations is implicit in the use of both a bulk retardation factor to calculate radionuclide travel times and the coupled reaction/transport code TRANQL. The second approximation is used by the NEFTRAN code to model matrix diffusion. Analytical solutions based on the third assumption appear commonly in the literature and in texts.

These criteria can be seen as a subset of a set of more general criteria for the validity of the assumption of local equilibrium assumption (LEA). Criteria for the validity of this assumption could be based on the identification of conditions that allow for chemical reaction rates to proceed to equilibrium at rates substantially greater than the advective flow of ground water. Additional theoretical and experimental work is required to obtain the data needed to define criteria for the LEA for radionuclides under conditions relevant to HLW disposal.

Summary and Conclusion

The need for and form of a method of geochemical sensitivity analysis can be seen to emerge from the above discussion. The objective of this analysis is to define criteria for the valid simplification of transport models to be used in performance assessment. An underlying principle of the work is that the validity must be assessed relative to the proper performance measure, in the relevant experimental frame, and considering all significant interactions among system variables. The performance measures of interest are the 10,000-year integrated discharges of each radionuclide at the boundary of the accessible environment as defined by the EPA Standard. The experimental frames are the ranges of physicochemical parameters expected at the candidate repository sites.

In this analysis the probabilistic risk assessment (PRA) codes NEFTRAN, CDFPLT, and LHS are being used to assess the amount of geochemical retardation needed to ensure compliance with the EPA Standard 40 CFR Part 191 at a hypothetical basalt repository site. The Sandia Sorption Data Management System (SSDMS) and the Aqueous Solutions Database (ASD) are being developed to store and facilitate retrieval of K_d values and thermodynamic property values, respectively. These parameter values are used either in the PRA codes or in speciation codes such as PHREEQE (*9*) and MINEQL (*10*). Coupled reaction/transport codes such as TRANQL (*11*) do not use a K_d value to approximate chemical behavior of the radionuclides. With the proper modifications, it is hoped that codes like these can be used to carry out simulations in which many of the assumptions associated with K_d's and retardation factors can be relaxed.

Commonly used techniques in batch sorption experiments do not allow for an explicit accounting of the fate of all of the radionuclides introduced into the solution. In general, the available K_d data are inadequate to show that the sorption of radionuclides is the process responsible for the observed K_d value. With the available data, it cannot be shown that a K_d value of zero would not be realistic under in situ conditions. Therefore, a K_d value of zero should be included in sensitivity analyses.

A critical review of data for use in the more robust triple-layer sorption model was carried out. In addition, an evaluation of the suitability of the TRANQL code for use in performance assessment was made based on code flexibility and computing efficiency. Based on these reviews, it was concluded that without major improvements in both data bases and computer codes, detailed chemical models cannot be used in place of codes such as NEFTRAN or SWIFT-II to assess the compliance of a repository site with the EPA Standard 40 CFR Part 191. At present, detailed chemical speciation and chemical transport models are best used to provide insights into the nature of significant interactions and correlations among system variables. The codes can be used to assess the validity of simplifying assumptions used in performance assessment tools and to extend sensitivity and uncertainty analyses of system behavior to detailed geochemical processes.

Acknowledgements

This report is an integration of several subprojects that have been carried at Sandia National Laboratories, Lawrence Berkeley Laboratory, and Stanford University. The authors wish to acknowledge the contributions of several persons who have made significant contributions to the program. K. L. Erickson has assisted the design of the sensitivity analysis project and has been responsible for much of the work related to kinetics and transport theory. P. Davis, R. Guzowski, and K. Brinster provided valuable assistance in formulating a conceptual model for a hypothetical basalt site. E. Sheppard carried out calculations to construct the CCDF in Figure 7. M. Moore wrote the software for the SSDMS, and S. Faith assisted in the design of the SSDMS and helped compile and review the sorption data for basaltic rocks. D. Kent and V. S. Tripathi assisted in the evaluation of experimental data for use in the triple-layer model and evaluated the potential use of TRANQL in performance assessment studies. F. Hale implemented the utility programs for the ASD and assisted in verification of the thermodynamic data.

This work was supported by the U.S. Nuclear Regulatory Commission and performed at Sandia National Laboratories, operated for the U.S. Department of Energy under contract number DE-AC04-76DP00789; at Stanford University, and at Lawrence Berkeley Laboratory.

Literature Cited

1. Zeigler, B. P., *Theory of Modeling and Simulation,* Wiley Interscience, New York (1976).
2. Henize, John, *Large Scale Systems, 1,* 89 (1980).
3. Iman, R. L., J. M. Davenport, and K. Ziegler, *Latin Hypercube Sampling (Program User's Guide),* SAND79-1473, Sandia National Laboratories, Albuquerque, NM (1980).
4. Longsine, Dennis E., Evaristo J. Bonano, and Charlene P. Harlan, *User's Manual for the NEFTRAN Computer Code,* NUREG/CR-4766, (SAND86-2405), Sandia National Laboratories, Albuquerque, NM (1987).
5. Reeves, Mark, D. S. Ward, N. D. Johns, and R. M. Cranwell, *Theory and Implementation for SWIFT-II, The Sandia Waste-Isolation Flow and Transport Model for Fractured Media,* NUREG/CR-3328, (SAND83-1159), Sandia National Laboratories, Albuquerque, NM (1986).
6. Siegel, M. D., M. S. Chu, and R. E. Pepping, "Compliance assessments of hypothetical geological nuclear waste isolation systems with the Draft EPA Standard," in *Scientific Basis for Nuclear Waste Management VI,* D. G. Brookins (Ed.), *Mat. Res. Soc. Symp. Proc., 15,* Elsevier Science Publishing Co., Inc., Amsterdam (1983).
7. U.S. Environmental Protection Agency, "Environmental Standards for the Management and Disposal of Spent Nuclear Fuel, High-Level and Transuranic Radioactive Wastes; Final Rule," 40 CFR Part 191," *Federal Register, 50,* 38066 (1985).
8. Muller, A. B., D. Langmuir, and L. E. Duda, "The formulation of an integrated physicochemical-hydrologic model for predicting waste nuclide retardation in geologic media," in *Scientific Basis for Nuclear Waste Management VI,* D. G. Brookins (Ed.), *Mat. Res. Soc. Symp. Proc., 15,* Elsevier Science Publishing Co., Inc., Amsterdam (1983).
9. Parkhurst, D. L., D. C. Thornstenson, and L. N. Plummer, *PHREEQE—a Computer Program for Geochemical Calculations,* U.S. Geological Survey, Reston, VA (1980).
10. Westall, J. C., J. L. Zachary, and F.M.M. Morel, *MINEQL: A Computer Program for the Calculation of Aqueous Systems,* Tech. Note 18, Water Control Laboratory, Mass. Inst. of Technol., Boston, MA (1976).
11. Cederberg, G., "TRANQL: A Ground-Water Mass-Transport and Equilibrium Chemistry Model for Multicomponent Systems," Ph.D. Dissertation, Stanford University, Stanford, CA (1985).
12. U.S. Department of Energy, *Environmental Assessment, Reference Repository Location, Hanford Site, Washington,* DOE/RW-0070, Office of Civilian Radioactive Waste Management, Washington, DC (1986).
13. Bonano, E. J., M.S.Y. Chu, R. M. Cranwell, and P. A. Davis, "Development of Performance Assessment Methodology for Nuclear Waste Isolation in Geologic Media," in *The Symposium on Groundwater Flow and Transport Modeling for Performance Assessment of Deep Geologic Disposal of Radioactive Waste: A Critical Evaluation of the State of the Art,* NUREG/CP-0079, (PNL-SA-13796, CONF-8505180), Pacific Northwest Laboratory, Richland, WA (1986).

14. Siegel, M. D. (Ed.), *Geochemical Sensitivity Analysis, Interim Progress Report, FY86-FY87,* SAND87-2633, Sandia National Laboratories, Albuquerque, NM (1987).
15. Clifton, P. M., and R. C. Arnett, *Preliminary Uncertainty Analysis of Pre-Waste-Emplacement Groundwater Travel Times for a Proposed Repository in Basalt,* SD-BWI-TA-013, Rockwell Hanford Operations, Richland, WA (1984).
16. Clifton, P. M., *Groundwater Travel Time Analysis for the Reference Repository Location at the Hanford Site,* SD-BWI-7I-303, Rockwell Hanford Operations, Richland, WA (1986).
17. Nuclear Waste Consultants, *Review of "Groundwater Travel Time Analysis for Reference Repository Location at the Hanford Site,"* SD-BWI-TI-303, letter report to the Nuclear Regulatory Commission (June 13, 1986).
18. Codell, R., *Draft Generic Technical Position on Groundwater Travel Time,* U.S. Nuclear Regulatory Commission, Washington, DC (1985).
19. U.S. Nuclear Regulatory Commission, "Disposal of High-Level Radioactive Wastes in Geologic Repositories; Technical Criteria," 10 CFR Part 60, Washington, DC (June 30, 1983).
20. Siegel, M. D., and M. S. Chu, Technical Assistance for Regulatory Development: Review and Evaluation of the Draft EPA Standard 40 CFR 191 for Disposal of High-Level Waste, Volume 3: A Simplified Analysis of a Hypothetical Repository in a Tuff Formation, NUREG/CR-3235, Sandia National Laboratories, Albuquerque, NM (1983).
21. Ortiz, N. R., and R. M. Cranwell, *Risk Assessment Methodology for High Level Waste: Assessing Compliance with the EPA Draft Standard Including Uncertainties,* SAND82-0596, Sandia National Laboratories, NM (1982).
22. Faith, Stuart, Malcolm Moore, and Malcolm Siegel, *Sandia Sorption Data Management System (SSDMS):* Volume 1. *User's Guide,* SAND87-2634, Sandia National Laboratories, Albuquerque, NM (1987).
23. Siegel, Malcolm D., and Stuart Faith, *Sandia Sorption Data Management System (SSDMS):* Volume 2. *Evaluation of Sorption Data for a Repository in Basalt,* SAND87-2635, Sandia National Laboratories, Albuquerque, NM (1987).
24. Westall, John, and Herbert Hohl, *Advances in Colloid and Interface Science* (Amsterdam) *12,* 265 (1980).
25. Kent, Douglas B., V. S. Tripathi, N. B. Ball, and J. O. Leckie, *Surface-Complexation Modeling of Radionuclide Adsorption in Sub-Surface Environments,* NUREG/CR-4807, (SAND86-7175), Sandia National Laboratories, Albuquerque, NM (1987).
26. Baes, C. F., and R. E. Mesmer, *The Hydrolysis of Cations,* John Wiley and Sons, Inc., New York (1976).
27. Ryan, J. L., and D. Rai, *Polyhedron, 2,* 947 (1983).
28. Bruno, J., D. Ferri, and I. Grenthe, *Acta Chem. Scand., A40,* 428 (1986).
29. Silva, R. J., "Carbonate Complexation of Pu(IV) in Aqueous Solutions," Paper No. 44, Div. of Nucl. Chem. and Tech., 189th Amer. Chem. Soc. Mtg., Miami Beach, FL (1985).
30. Moskvin, A. I., and A. D. Gel'man, *Jour. Inorg. Chem. (USSR) 3,* 198 (1958).
31. Phillips, S. L., F. Hale and M. D. Siegel, "A Thermodynamic Data Management System for Nuclear Waste Isolation Performance Assessment," presented at the International Conference on Thermodynamics of Aqueous Systems with Industrial Applications, Airlie House, Warrenton, VA (May 10–14, 1987).
32. Phillips, S. L., F. V. Hale, L. F. Silvester, and M. D. Siegel, *Thermodynamic Tables for Nuclear Waste Isolation: Volume 1. Aqueous Solutions Database,* NUREG/CR-4864, (LBL-22860, SAND87-0323), Sandia National Laboratories, Albuquerque, NM (1987).
33. Miller, C. W., "Toward a Comprehensive Model of Chemical Transport in Porous

Media,'' in *Scientific Basis for Nuclear Waste Management VI,* in D. G. Brookins (Ed.), *Mat. Res. Soc. Symp. Proc., 15,* 481 (1983).
34. Yeh, G. T., and V. S. Tripathi, *HYDROGEOCHEM —A Coupled Model of HYDROlogical Transport and GEOCHEMical Equilibria in Multicomponent Systems,* ORNL-6371, Oak Ridge National Laboratory, Oak Ridge, TN (1987).
35. Siegel, M. D. (Ed.), *Geochemical Sensitivity Analysis, Progress Report FY84 to FY85,* SAND85-1644, Sandia National Laboratories, Albuquerque, NM (1985).
36. Erickson, K. L., M.S.Y. Chu, M. D. Siegel, and W. Beyeler, ''Approximate Methods to Calculate Radionuclide Discharges for Performance Assessment of HLW Repositories in Fractured Rock,'' in *Waste Management 86,* 2, 337, Post, R. G. (Ed.), University of Arizona, Tucson, AZ (1986).

Representative Elements: A Step to Large-Scale Fracture System Simulation

T. M. CLEMO
Idaho National Engineering Laboratory
Idaho Falls, Idaho

Abstract

Large-scale simulation of flow and transport in fractured media requires the development of a technique to represent the effect of a large number of fractures. Representative elements are used as a tool to model a subset of a fracture system as a single distributed entity. Representative elements are part of a modeling concept called dual permeability. Dual permeability modeling combines discrete fracture simulation of the most important fractures with the distributed modeling of the less important fractures of a fracture system.

This study investigates the use of stochastic analysis to determine properties of representative elements. Given an assumption of fully developed laminar flow, the net fracture conductivities and hence flow velocities can be determined from descriptive statistics of fracture spacing, orientation, aperture, and extent. The distribution of physical characteristics about their mean leads to a distribution of the associated conductivities. The variance of hydraulic conductivity induces dispersion into the transport process.

Simple fracture systems are treated to demonstrate the usefulness of stochastic analysis. Explicit equations for conductivity of an element are developed and the dispersion characteristics are shown. Explicit formulation of the hydraulic conductivity and transport dispersion reveals the dependence of these important characteristics on the parameters used to describe the fracture system. Understanding these dependencies will help to focus efforts to identify the characteristics of fracture systems. Simulations of stochastically generated fracture sets do not provide this explicit functional dependence on the fracture system parameters.

Successful simulation of flow and transport in fractured reservoirs depends on overcoming three major obstacles. These are (1) a description of the reservoir, which includes the important characteristics of the fracture system; (2) a model of single fractures that realistically represents flow, transport, and interactions with the matrix material; and (3) a method to combine the fracture system description and the fracture model into a usable tool. The final obstacle is the focus of the dual permeability technique being developed at the Idaho National Engineering Laboratory.

Dual permeability refers to two levels of simulating fractures. The larger and hence most important fractures are modeled discretely, and the smaller more numerous fractures are modeled as a continuum with representative properties. The continuum blocks that surround and are bounded by discrete fractures are called representative elements. Dual permeability modeling focuses on the effects of the dominant flow carriers and yet retains the influence of smaller fractures without having to treat them in detail. The approach lies between the concepts of discrete fracture modeling and the continuum or representative elementary volume (REV) approach.

Dual permeability modeling should not be confused with the dual porosity approach. The dual porosity approach treats both the fracture network and the matrix material as different continua that are everywhere in contact. The term distinguishes the high-porosity fracture system from the low-porosity matrix. The dual permeability concept treats high permeability fractures discretely and lower permeability fractures as a continuum. Rather than the domains being everywhere in contact, the dual permeability approach relies on a well-defined interface between the discrete fractures and the continuum representation.

The representative element is the key to the dual permeability concept. These features must faithfully duplicate the influence of a fracture subsystem. The properties of an element can be determined either analytically or numerically through small-scale simulations. The amount of complexity that is incorporated into a representative element is a controllable parameter because the difficulty in obtaining the properties of elements can be adjusted by the level of detail included in the discrete portion of the dual permeability model.

In this paper, discrete and continuum modeling are reviewed and compared to dual permeability modeling. The important characteristics of representative elements are discussed and contrasted to the REV. The paper includes the development of representative elements for a simplistic two-dimensional fracture system at two levels of complexity.

Modeling Concepts

The discrete approach incorporates the most detail of any of the modeling techniques. All fractures which are considered relevant are modeled as individual entities. The fractures can be described from knowledge of the individual fractures in the system or be a stochastically generated realization based upon a statistical description of the fracture system. To model all fractures as individual entities presents a large burden to the computational process. Presently, the computational burden limits discrete fracture flow and transport simulations to reservoirs with few relevant fractures or small portions of a fracture system which may not include enough of the relevant fractures.

Continuum modeling is on the opposite end of the spectrum from discrete fracture modeling. The continuum approach is based on a lumped parameter model of the fracture system, where the continuum is composed of REVs. These volumes model the hydraulic and transport behavior of a large number of fractures. For simulations of real fracture systems, the hydraulic and transport properties are determined from a statistical description of the fracture system. A requirement is that they represent a fracture system that is sufficiently large, such that effects of individual fractures cannot be distinguished in the response of the model. The scale must be large enough so that the fractured rock can be treated as if it were homogeneous. Reducing a large number of fractures to a single representation has become an active area of research. Discrete fracture simulations on a subreservoir scale are a promising approach to determining properties for REVs [Dershowitz (*1*), Long (*2*), Schwartz et al. (*3*)].

A homogeneous porous media approximation has been the most common method of representing a fracture network as a continuum. A major assumption of the porous media model is that transport dispersion can be modeled as a Gaussian random process using a dispersivity coefficient to determine the variance of transport about the mean movement. A number of studies have called into question the validity of the porous media approximation [Simmons (*4*), Dagan (*5*), Schwartz et al. (*3*)].

The concept of dual permeability modeling is to use the best of each of these approaches. Where needed, the high detail of discrete modeling is used, but where the detail is not required, the computational burden is reduced. The homogeneity and indistinguishable fracture requirements of REVs do not apply to representative elements. Dual permeability can be used to model systems that do not meet these requirements. Systems that are influenced by large flow paths or are not large enough to meet homogeneity can be modeled using the dual permeability approach.

Representative Elements

Representative elements (RE) model the connection between the discrete fractures for both flow and transport purposes. They are not a model of the fundamental nature of the fracture system. The hydraulic properties of an RE are conductivity, both along and perpendicular to the head gradient, porosity, and transit time distribution. Transit time is the time for solute to cross the element. The dispersion process causes the transit time to be distributed rather than a fixed value. Transit time distributions are commonly described by a breakthrough curve. The hydraulic properties of the elements will normally be a function of the size and shape of the element. This is in contrast to the REVs, which are required to be sufficiently large such that they are independent of changes in size. A critical aspect of the elements is that the discrete fractures which bound the element allow some simple approximations of the boundary conditions. Because the discrete fractures are the larger ones in the system, they generally have either a small or a linear pressure drop. Imposing a fixed or linear head boundary condition on an element makes it much easier to describe the properties of the element.

Representative element properties are conductivity, both along and perpendicular to the head gradient; porosity; transport direction; and transit time distribution. While it would be easier to generate representative elements from known fracture geometries, reservoir fracture systems will most likely be described in terms of probability distributions of size, spacing, and other pertinent characteristics. The properties of an RE should be described probabilistically. Either the mean values or the complete distribution could then be integrated with the discrete fracture in the dual permeability model.

Representative element properties can be found two ways. One is stochastic analysis, which is the analytical condensation of the physical fracture system description into permeability and transport parameters. The use of stochastic analysis is demonstrated in this paper. An alternative technique, one that may ultimately prove more powerful, is through stochastic simulations of the elements. Investigations of this technique have been conducted in modeling flow in a three-dimensional fracture system [Miller and Allman (*6*)], and flow and transport in a two-dimensional fracture system [Miller and Clemo (*7*)].

Example Fracture System

The two-dimensional fracture system used as the basis for this example (Figure 1) was derived from a study of the Raft River geothermal field [Miller et al. (*8*)]. In Figure 1, two dominant fracture sets are composed of a few large and widely spaced fractures. A third set of small, closely spaced fractures completes the flow system. The first observation to be made from Figure 1 is the sparse nature of the large fractures. To model this system with a continuum model would require a much larger REV than depicted in the figure. The second observation to note is the large number of small fractures. A discrete simulation of this system would have to ignore most of

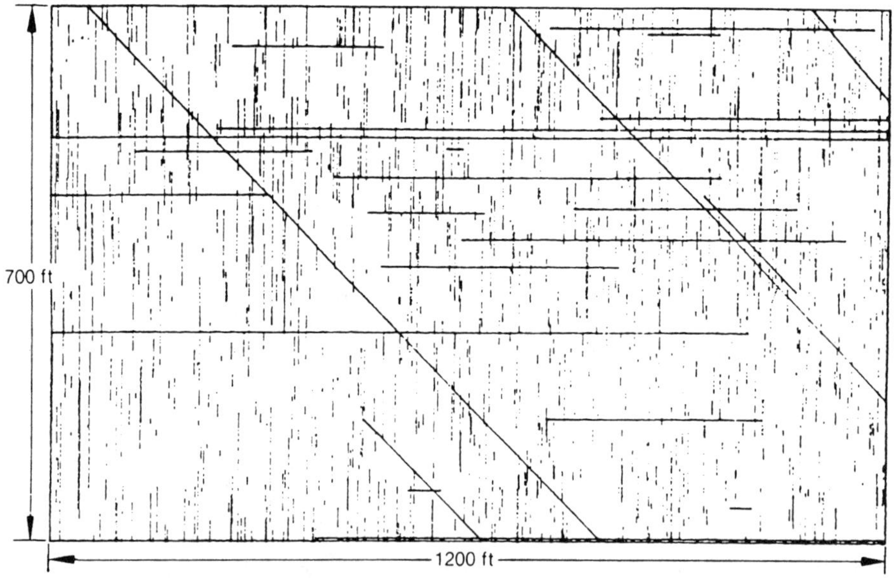

Figure 1. **Two-dimensional representation of a fracture system. Reprinted with permission (*13*).**

these fractures. Unfortunately, these lesser fractures may provide a significant hydraulic connection between the larger fractures.

In the Raft River study statistical distributions of fracture spacing for three fracture sets were developed from acoustic televiewer data. For the purposes of this example the mean fracture lengths are taken as 2.15 times the mean spacing. This assumption was developed from fracture spacing and length data of Mahtab et al. (*9*) and Gale et al. (*10*), which showed spacing-to-length ratios of 3.16 and 1.15, respectively. Fracture centers are assumed to be uniformly distributed in space. The fracture spacing is then exponentially distributed (*11*). Fracture length is also assumed to be exponentially distributed and independent of the space between fractures. A fixed fracture orientation is used to reduce the complexity of the analysis. A lognormal aperture distribution is assumed. The mean of the aperture distribution is assumed either linearly correlated to fracture length or independent of length.

Equation 1 is the exponential probability density function for the fracture length. Equation 2 is the lognormal probability density function for the aperture. Dershowitz (*1*) includes a good discussion of these probability density functions.

$$P(\ell) = 1/\lambda_\ell \, e^{-1/\lambda_\ell} \tag{1}$$

$$P(b) = \frac{1}{(2\pi\,\sigma_b^2)^{1/2}} \, \frac{e^{-[\ln^2(b/\bar{b})/2\sigma_b^2]}}{b} \tag{2}$$

Fracture Flow Model

Fractures are assumed to behave as infinite parallel plates. Flow in a fracture is modeled using the cubic law

$$Q/J = cb^3 \tag{3}$$

Velocities are assumed to be low enough that a plug flow model is applicable for transport. The matrix material is treated as impermeable to both flow and diffusion. Matrix interactions have been neglected to simplify the following analysis. Matrix interactions need to be considered in a functional model.

Element Development

Two types of elements are developed in this paper. In each case the dual permeability model treats all but the smallest of the fractures in the two widely spaced sets discretely. Only the connection provided by the closely spaced set to these discrete fractures is modeled. The first type of RE (Type 1) ignores the members of the widely spaced set that are not included as discrete fractures. The second RE (Type 2) includes these smaller fractures in a limited sense. In this treatment the added fractures provide connections within the element between the other members of the element but are not allowed to penetrate the sides of the element. In both cases, the

REs have discrete fractures bounding the sides perpendicular to the closely spaced set and are impermeable along the parallel sides.

Figure 2 depicts a typical element of the second type with the boundary conditions of its sides identified. The two boundaries of Figure 2 that are labeled constant head boundaries represent the intersection of this element with the discrete fractures. These two boundaries are connected by the numerous vertical fractures. The thicker lines represent the smaller fractures of the horizontal fracture set. Inclusion of the horizontal fractures distinguishes this element as Type 2. Two no-flow boundaries are parallel to the vertical fracture set and hence cannot be crossed by fractures of this set. Fractures of the horizontal set may cross these boundaries. For Type 2 elements the no-flow boundary condition is an approximation. The element width must be large enough to ensure the approximation is appropriate.

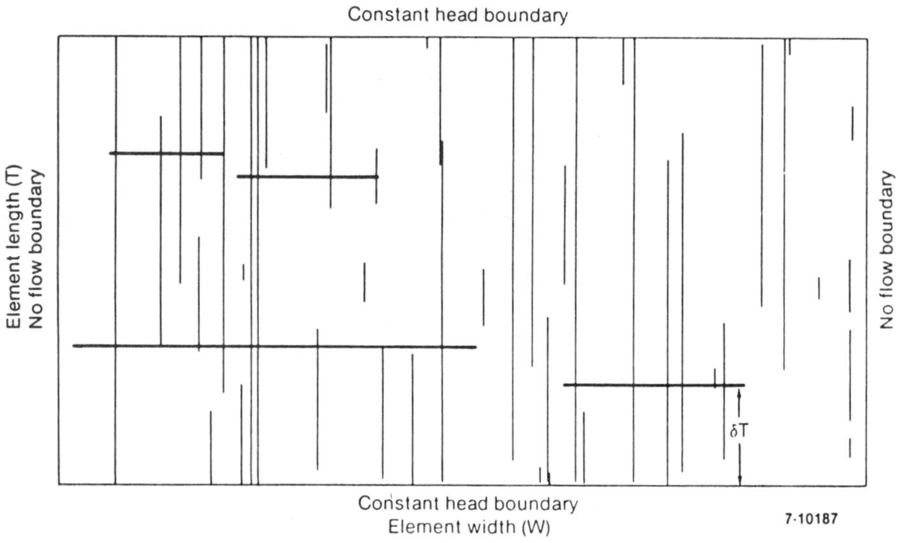

Figure 2. **Sample representative element. Modified from (*13*).**

Conductivity

If the aperture is independent of all other factors and is lognormally distributed, then the expected conductivity of a single fracture is

$$<k> = \frac{1}{(2\pi\sigma_b^2)^{1/2}} \int_0^\infty cb^2 \, e^{-[\ln^2(b/\bar{b})/2\sigma b^2]} db \tag{4}$$

$$<k> = c\bar{b}^3 \, e^{9\sigma b^2/2} \tag{5}$$

Equation (5) reveals a strong dependence of the average conductivity upon the variance of the logarithm of aperture.

The distribution of fracture length also affects the element conductivity. Only fractures that connect both sides of the element contribute to flow in this parallel system. The fracture will connect the two sides only if the length, ℓ, is such that

$$\ell > 2Z \text{ if } Z > T/2 \text{ or } \ell > 2(T-Z) \text{ if } Z < T/2$$

where Z is the distance from the bottom of the element to the fracture center and T is the element thickness. This criterion is symmetric about T/2.

For a negative exponential distribution of fracture length, the probability distribution function of connecting fractures becomes

$$P(\ell) = \begin{cases} \dfrac{\ell - T}{\lambda_\ell^2} e^{-\ell/\lambda_\ell} & \text{for } \ell > T \\ 0 & \text{for } \ell < T \end{cases} \tag{6}$$

For fractures with aperture unrelated to length and expected fracture spacing, λ_s, the conductivity per unit width of the element is

$$<k> = \frac{c\bar{b}^3}{\lambda_s} e^{9\sigma_b^2/2} e^{-T/\lambda_\ell} \tag{7}$$

The expression for conductivity in Equation (7) has the units of a material property (i.e., conductivity per unit area normal to the direction of flow). The conductivity of the element is not a material property. It is a measure of the effective hydraulic conductance of a specific element. Equation (7) reveals that the effective conductivity of the element decreases exponentially with increasing thickness, T, due to the exponential length distribution of fractures.

Next assume that the aperture is a direct function of the length as

$$b = \alpha\ell \tag{8}$$

The expected conductivity of an element is then

$$<k> = \frac{c\alpha^3}{\lambda_s} (T^3 + 6T^2\lambda_\ell + 18T\lambda_\ell^2 + 24\lambda_\ell^3) e^{-T/\lambda_\ell} \tag{9}$$

Figure 3 presents a comparison of the sensitivity of conductivity to element thickness for apertures correlated and uncorrelated to fracture length. Figure 3 makes it clear

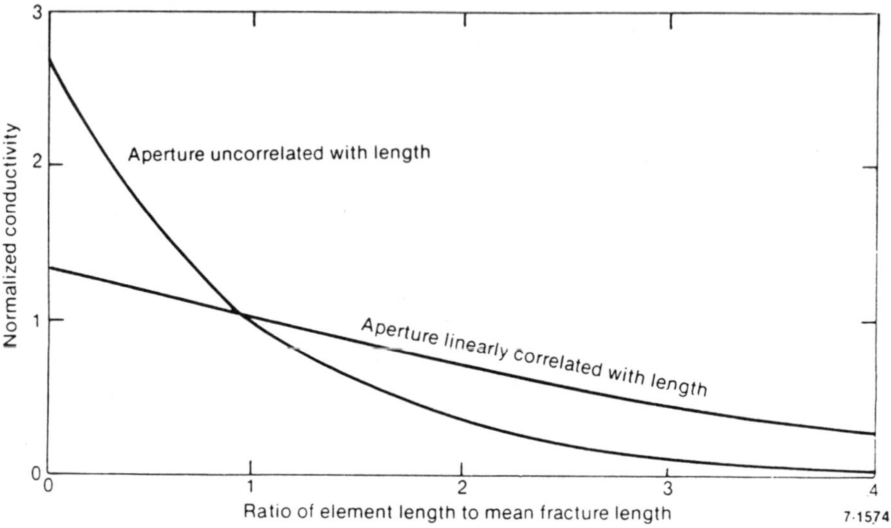

Figure 3. **Conductivity sensitivity to element thickness. Reprinted with permission (*13*).**

that more attention needs to be paid to mean fracture length if apertures are assumed to be uncorrelated with length.

If aperture is related to fracture size by a lognormal distribution where $\bar{b} = \beta\ell$ and σ_b^2 = constant, then

$$<k> = \frac{c\beta^3}{\lambda_s} e^{9\sigma_b^2/2} (T^3 + 6T^2\lambda_\ell + 18T\lambda_\ell^2 + 24\lambda_\ell^3)\, e^{-T/\lambda_\ell} \tag{10}$$

which is identical to the direct proportionality case with $\alpha = \beta e^{3\sigma_b^2/2}$.

To extend this case to nonparallel element faces (Figure 4) the conductance is averaged over the thickness. Conductance is the flow per unit pressure drop, Q/dP, rather than flow per unit pressure gradient. For an element varying from T_a on one side, to T_b on the other, the net conductivity of the element is

$$<k> = \frac{(T_b + T_a)}{2(T_b - T_a)} \frac{\lambda_\ell c\beta^3 e^{9\sigma_b^2/2}}{\lambda_s} \Bigg\{ (T_b^2 + 8T_b\lambda_\ell + 26\lambda_\ell^2)\, e^{-T_b/\lambda_\ell}$$

$$-(T_a^2 + 8T_a\lambda_\ell + 26\lambda_\ell^2)\, e^{-T_a/\lambda_\ell} \tag{11}$$

$$+ 24\lambda_\ell^2 \left[\ln\left(\frac{T_b}{T_a}\right) + \sum_{n=1}^{\infty} \frac{(T_b - T_a)^n}{\lambda_\ell^n\, n \cdot n!} \right] \Bigg\}$$

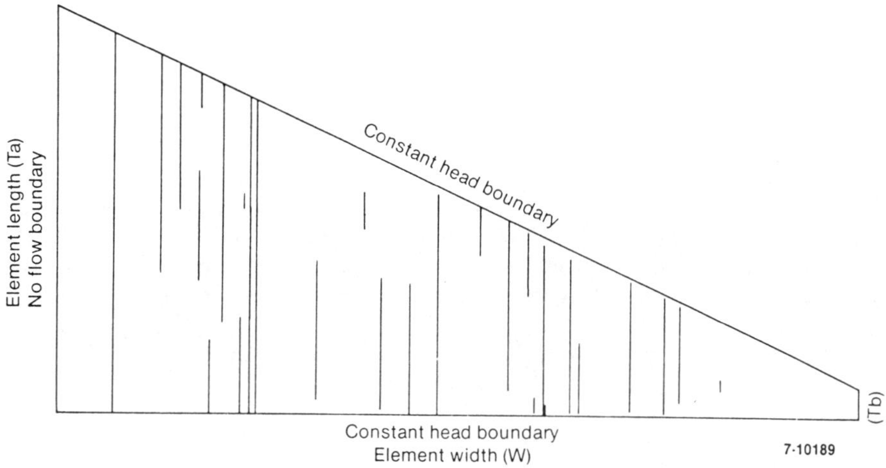

Figure 4. **Representative element between nonparallel fractures.**

If the element contains cross fractures (Type 2), then the conductivity of the element needs to be generated by a recursive development of subelement conductivity and laws governing the combination of the conductivities of the elements. As an example, consider the region below the cross fracture in the lower right of Figure 2. The thickness of one subelement is the distance between the element boundary and the cross fracture, δT. The upper subelement then has a thickness of $T - \delta T$. From previous results, the conductivity of the lower subelement with uncorrelated aperture is

$$<k> = \frac{c\bar{b}^3}{\lambda_s} e^{9\sigma b^2/2} e^{-\delta T/\lambda_\ell} \tag{12}$$

and for the upper subelement

$$<k> = \frac{c\bar{b}^3}{\lambda_s} e^{9\sigma b^2/2} e^{-(T - \delta T)/\lambda_\ell} \tag{13}$$

These subelements must be combined to form a subelement that spans the thickness of the element. Two approximations are made at this point. First, the fractures that span the element do not interact with the cross fracture. Second, the fluid that travels from one fracture to another through the cross fracture travels a distance equal to the largest mean spacing of the connecting fractures of the subelements. The conductivity of the spanning subelement is found from

$$k_{new} = T \bullet \left[\frac{\delta T}{k_{\delta T}} + \frac{T - \delta T}{k_{T - \delta T}} + \frac{T_{cross}}{k_{cross}} \right]^{-1} - k_T \tag{14}$$

where

$$T_{cross} = \lambda_s e^{\delta T/\lambda_\ell}$$

$$k_{cross} = c\bar{b}_2^{\,3} e^{9\sigma b_2^2/2}$$

This new conductivity introduced to the RE by the cross fracture is added to the previously determined conductivity to find that of an RE with a single cross fracture. This procedure is repeated for each crossing fracture. The expected conductivity for each element is found by taking the expectation over dT and ℓ_2 of the set of small cross fractures.

Consider a single crossing fracture located randomly in the element.

$$\text{Assuming } T_{cross}/k_{cross} << \frac{\delta T}{k_{\delta T}} + \frac{T - \delta T}{k_{T - \delta T}}$$

The expected value of k_{new} for a uniformly distributed δT is

$$<k>_{new} = \frac{c\bar{b}^3}{\lambda_s} e^{9\sigma b^2/2} \left[\frac{T^2}{\lambda_\ell[(T - \lambda_\ell) + \lambda_\ell e^{-T/\lambda_\ell}]} \right]^{-1} e^{-T/\lambda_\ell} \qquad (15)$$

This is the additional conductivity per unit width of the spanning element above the representative element without a cross fracture. The expected conductivity for a representative element with one crossing fracture that does not intersect the sides of the element is

$$<k>_{T_2} = <k>_T + \lambda_{\ell_2} (1 - e^{-\ell_2/\lambda_{\ell_2}}) <k>_{new} \qquad (16)$$

where ℓ_2' is the upper cutoff length of the crossing fracture set.

This procedure is easily extended to a multiple randomly placed crossing fracture with one serious limitation. Crossing fractures cannot cross subelement boundaries as illustrated in the situation depicted in Figure 5. The additional resistance to flow introduced by t_{cross}/k_{cross} would not substantially change the analysis.

The conductivities developed above are the expected values for elements of width, W. It is important to consider the effect of the width of the element on the variance of the conductivity. The variance of the conductivity of the fractures increases proportionately to the expected number of fractures, W/λ_s, contained by the element. The variance in the number of fractures represented by the element is also W/λ_s for an exponential spacing of fractures [Parzen (*12*)]. These variances of the element conductivity are independent and therefore additive. The conductivity of the elements also increases linearly with W/λ_s, which means the relative variance decreases linearly with the ℓ element width.

As mentioned earlier, the conductivity determined in this analysis has the units of a material property, but is a measure of the expected conductivity of a specific

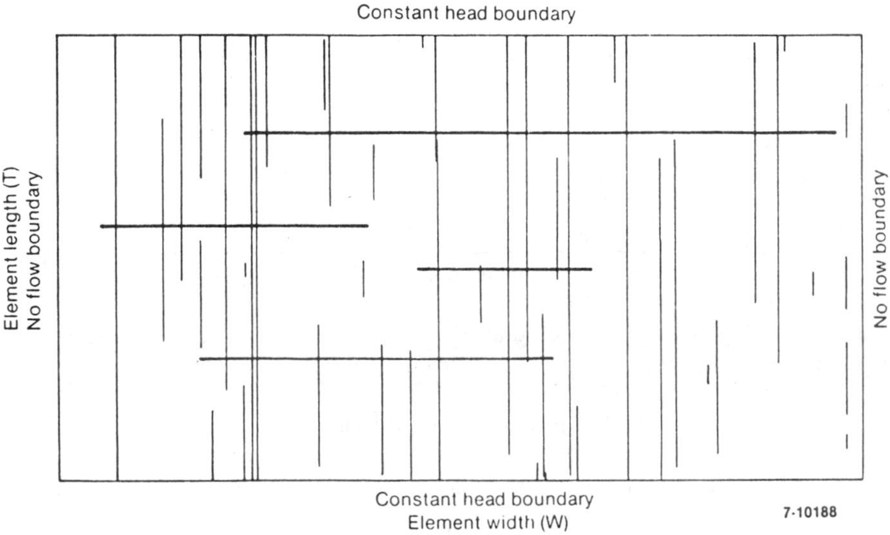

Figure 5. **Untreated fracture network.**

geometry. The representative element properties are strictly modeling tools and should not be construed as a measure of physical properties.

Transport

Solute transport simulation requires consideration of the variation of fluid velocities about a mean since this variation introduces dispersive effects into a flow system. The dispersive nature of REs can be determined from the probability distribution of the conductivity. For the parallel plate plug flow model the velocity per unit pressure gradient in a fracture is

$$v = cb^2 \tag{17}$$

The transit time is then

$$t = T/cb^2 \tag{18}$$

Equation (18) reveals that the transit time is uniquely determined by the aperture, b, of the fracture. The conductivity of a single fracture is a measure of the amount of fluid flux traveling through that fracture for an applied pressure gradient. Hence, the probability distribution of transit times is the probability distributionn of the conductivity transformed by the relation between T and b as defined by Equation (18). The three combinations of aperture and length correlations used for conductivity are also studied for the transit time of Type 1 elements. For aperture uncorrelated with length

$$P(t) = \frac{1}{(2\pi\sigma_b{}^2)^{1/2}} \frac{\bar{t}^{3/2}}{2t^{5/2}} e^{-[\ln^2(t/\bar{t})/8\sigma_b{}^2]} e^{-9\sigma b^2/8} \tag{19}$$

where $\bar{t} = T/c\bar{b}^2$.

If aperture varies linearly with length, $b = \alpha\ell$, and length is exponentially distributed, then

$$P(t) = \frac{T^2}{2\alpha^4\lambda_\ell^2} \frac{\left[\left(\frac{T}{ct}\right)^{1/2} - \alpha T\right] e^{-\left[\left(\frac{T}{ct}\right)^{1/2} - \alpha T\right]/\alpha\lambda_\ell}}{(T^3 + 6T\lambda_\ell + 18T\lambda_\ell^2 + 24\lambda_\ell^3)} \tag{20}$$

If aperture is lognormally distributed about a mean given by $\bar{b} = \beta\ell$, and length exponentially distributed, then the transit time probability density function is

$$P(t) = P(t{:}\bar{b}) \bullet P(\bar{b})d\bar{b} \tag{21}$$

where $P(t{:}\bar{b})$ is the probability of t given $\bar{b}$. $P(t{:}\bar{b})$ is defined by Equation (19) and

$$P(\bar{b}) = \frac{\bar{b} - \beta T}{\beta^2\lambda_\ell^2} e^{-(\bar{b} - \beta T)/\beta\lambda_\ell} \tag{22}$$

Figure 6 is a plot of the three transit time distributions described by Equations (19–21). The figure also includes the Gaussian probability distribution for comparison. These density distributions all have similar means and standard deviations. The Gaussian distribution was included because it is used to model dispersion in a porous media approximation. The Gaussian probability density differs in character from the other distributions. It rises much earlier and in fact always has a nonzero probability at zero time. This means that the initial tracer breakthrough will occur earlier with a porous approximation. It also lacks the long tail of the other distributions. These factors illustrate the inappropriateness of a porous media approximation for representative elements.

Summary

Representative elements are a major component of the concept of dual permeability modeling. A dual permeability model treats the most important fractures discretely and incorporates the rest of the fractures in representative elements. Representative elements are a continuum representation of these less important fractures. The elements model the flow and transport between the major fractures and are not a fundamental description of inherent properties of the less important fractures.

Dual permeability has the potential of simulating fracture systems that cannot be treated by either of the two conventional approaches. Dual permeability modeling can simulate flow systems that contain more fractures than can be treated discretely. The restricting conditions of homogeneity that limit the usefulness of continuum modeling do not apply to the dual permeability approach.

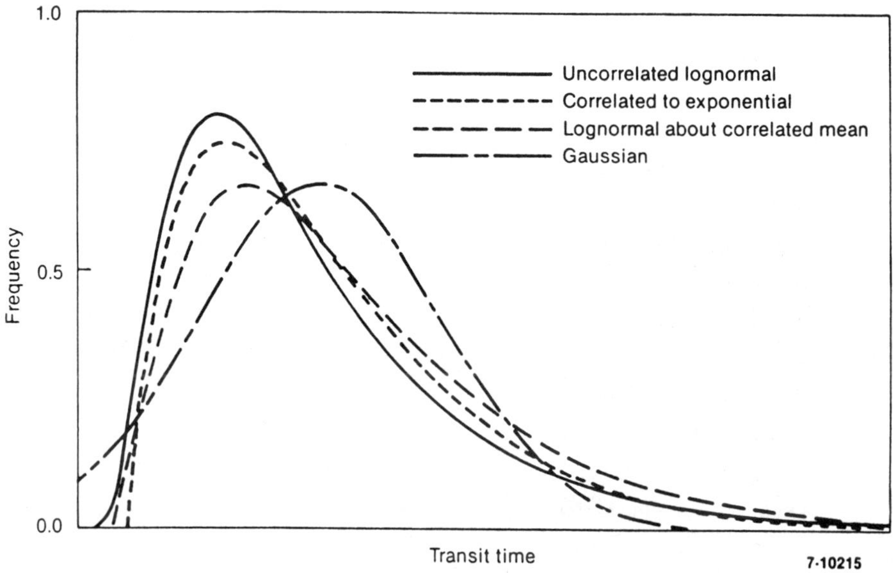

Figure 6. **Transit time probability distributions.**

The key to a practical dual permeability model is the development of the properties of the representative element. This work has shown that for simple systems these properties can be determined analytically through the use of stochastic analysis. The important properties are conductivity and transport time, which are closely related. These properties can only be determined in a statistical sense because the underlying fracture parameters are statistical. Even these simple elements would greatly reduce the number of fractures that would have to be modeled discretely.

More complex fracture systems or flow models will require that representative properties be determined through stochastic numerical studies and modeling of fracture subsystems. Studies are being performed at the Idaho National Engineering Laboratory on this topic.

Acknowledgment

This work was performed by EG&G Idaho, Inc. for the U.S. Department of Energy under Contract No. DE-AC07-76ID01570.

Notations

- b fracture hydraulic aperture
- $\overline{b}$ average (expected) aperture
- c constant incorporating the effects of viscosity and density on flow for parallel plate fractures

dP	pressure drop across the element
k	hydraulic conductivity[a]
ℓ, ℓ_2	length of a fracture[b]
λ_ℓ, λ_{ℓ_2}	expected length of a fracture[b]
λ_s, λ_{s_2}	average spacing of fracture centers normal to the fracture faces[b]
$< >$	expectation operator
p( )	probability density function
Q	flow rate
J	pressure gradient
t	time
T	thickness of the element
δT	thickness of a subelement
v	average velocity of fluid in a fracture
W	width of the element, normal to fracture faces
σ_b^2	variance of ln(b)

(a) Subscript δT, $T-\delta T$, T refers to Type 1 elements of that thickness. New refers to composite subelement. T_2 refers to Type 2 elements.

(b) Subscript 2 refers to the crossing fracture set.

Literature Cited

1. Dershowitz, W. S., "Rock Joint Systems," Ph.D. thesis, Massachusetts Institute of Technology, Boston (1984).
2. Long, J.C.S., *Verification and Characterization of Continuum Behavior of Fractured Rock at AECL Underground Research Laboratory,* BMI/OCRD-17, LBL-14975, University of California, prepared for the Office of Crystalline Repository Development, Columbus, Ohio (1984).
3. Schwartz, F. W., L. Smith, and A. S. Crowe, *Water Resources Research, 19*(5), 1253–1265 (1983).
4. Simmons, C. S., *Water Resources Research, 18*(4), 1193–1214 (1982).
5. Dagan, G., *Water Resources Research, 18*(4), 835–848 (1982).
6. Miller, J. D. and D. W. Allman, "Dual Permeability Modeling of Flow in a Fractured Geothermal Reservoir," *Proceedings of the Eleventh Workshop on Geothermal Reservoir Engineering,* Stanford (1986).
7. Miller, J. D. and T. M. Clemo, "Dual Permeability Modeling of Solute Transport in Discrete Fracture Systems," *Proceedings of the Thirteenth Annual Workshop on Geothermal Reservoir Engineering,* Stanford (1988).
8. Miller, J. D., D. W. Allman, and D. Wulf, *Raft River Fracture Characterization and Reservoir Simulation,* SE-PB-84-58, Idaho National Engineering Laboratory (1984).
9. Mahtab, M. A., D. D. Bolstad, and F. S. Kendorski, *Analysis of the Geometry of Fractures in San Manuel Copper Mine, Arizona,* Bureau of Mines, RI 7715 (1973).

10. Gale, J. E., A. Rouleau, and L. C. Atkinson, "Hydraulic Properties of Fractures," *Proceedings IAH Symposium on Hydrogeology of Rocks of Low Permeability,* Tucson, 1–6 (1985).
11. Hudson, J. A. and S. D. Priest, *International Journal of Rock Mechanics, Mineral Science, and Geomechanics, 16,* 339–362 (1979).
12. Parzen, E., *Stochastic Processes,* Holden-Day, p. 30 (1962).
13. Clemo, T., "Representative Element Modeling of Fracture Systems Based on Stochastic Analysis," *Proceedings of the Eleventh Workshop on Geothermal Reservoir Engineering,* Stanford, 71–76 (1986).

3 Stochastic Hydrology

Some Uncertainties About Uncertainty

R. ALLAN FREEZE
The University of British Columbia
Vancouver, British Columbia, Canada

GHISLAIN DE MARSILY
Ecole Superieure des Mines de Paris
Fontainebleau, France

LESLIE SMITH
The University of British Columbia
Vancouver, British Columbia, Canada

JOEL MASSMANN
Michigan Technological University
Houghton, Michigan, U.S.A.

Abstract

Regulatory and engineering performance criteria for the siting and design of underground nuclear waste repositories in North America require the use of probabilistic concepts to take into account uncertainties associated with the hydrogeologic environment. Probabilistic calculations are needed at the siting stage to estimate pre-emplacement groundwater travel times and potential radionuclide flux rates, and at the design stage to assess the risk associated with alternative design options. To carry out these calculations, hydrogeologists are applying geostatistical models and stochastic simulation methods originally developed to assess piezometric response in near-surface unconsolidated aquifers over limited spatial distances and short time frames with relatively abundant data. This paper will highlight the uncertainties associated with the transfer of this technology to the assessment of advective transport in deep rock formations over large distances and long time frames under conditions of sparse data availability. Attention will be directed primarily to the 1,000-year groundwater travel time regulatory criterion. A Bayesian geostatistical methodology for calculating the probability of meeting this criterion is summarized and assessed, in terms of both its conceptual strengths and its operational weaknesses. Issues of special concern revolve around (1) the appropriateness of the Bayesian approach, (2) problems associated with inverse calibration procedures, (3) identification of trends and zonations in hydraulic conductivity, and (4) problems associated with small-sample statistics. The last of these may affect (1) testing the hypothesis of a lognormal distribution for hydraulic conductivity, (2) testing for stationarity, (3) determining the form and parameter values for the autocorrelation function, (4) identifying the existence of embedded structures, (5) obtaining a significant degree of uncertainty reduction during Bayesian updating, and (6) applying inverse procedures.

Introduction

The hydrogeologic assessment of nuclear waste sites in North America is being carried out with numerical mathematical models that solve boundary-value problems based on the partial differential equations that describe groundwater flow and radionuclide transport in heterogeneous geologic media. In the past decade it has been recognized by hydrogeologic modelers that the innate heterogeneity of the geologic system creates uncertainty as to the correct parameter values to employ in the models at all locations other than those few locations where measurements are available. A flood of recent research results has put in place (1) a geostatistical methodology, based on the theory of spatial stochastic processes, that can be used to quantify the uncertainty in input parameters, and (2) methods of stochastic simulation that can be used to assess the impact of uncertainties in input parameters on the uncertainties associated with predicted output variables. These techniques have been greeted with enthusiasm by the agencies responsible for the development and regulation of deep repositories for high-level nuclear waste. The stochastic concepts are ideally suited for incorporation into the risk assessments that must accompany the siting, design, and performance assessment of a nuclear waste repository.

The authors of this article have been among those researchers who have encouraged the application of stochastic methods. However, we now wish to place on record some words of caution. We have two basic concerns. First, we note that these techniques were originally developed to assess piezometric response in near-surface unconsolidated aquifers, over limited spatial distances and short time frames, with relatively abundant data. They may not be as valuable when applied to the assessment of radionuclide transport in deep rock formations, over large distances and long time frames, under conditions of sparse data availability. Second, we note that the lag time that has traditionally existed between the presentation of research results and their application in engineering projects of societal importance has become vanishingly small; we worry that the telescoping of this sequence has not allowed for sufficient investigation of the possibilities and limitations of the stochastic methodologies.

Our purpose in this paper is to summarize the steps in what we consider to be the best geostatistical approach to hydrogeologic prediction; and to identify for each step the limitations and assumptions that deserve greater attention before routine application of these methods in nuclear waste programs can be considered defensible. It is not our intent to introduce into the nuclear waste literature another set of red herrings; we will try to identify those issues that should engender real concern and those that should not.

The siting and design of nuclear repositories in North America, especially in the United States, are being driven by performance criteria laid down by regulatory agencies. Current performance criteria include specifications related to (1) canister life, (2) radionuclide release rates from the disturbed zone, (3) radionuclide arrival rates at the accessible environment, and (4) pre-waste-emplacement groundwater travel times from the disturbed zone to the accessible environment. In order to present our concerns about geostatistical methodology in the clearest light, we will limit discussion in this paper to consideration of the pre-waste-emplacement travel-time criterion. A more complete statement of the U.S. criterion can be gleaned from the Federal Register [10 CFR 60.113(a)(2)]: "The geologic repository shall be located so that the pre-waste-emplacement groundwater travel time along the fastest path of likely radionuclide travel from the disturbed zone to the accessible environment shall be at least 1,000 years . . . . " The U.S. Nuclear Regulatory Commission has subsequently clarified the criterion through a series of Draft Generic Position papers. In one of these (Codell, *1*), the desirability of a probablistic approach is clearly enunciated. It is stated that "the estimation of groundwater travel time must accommodate spatial variability, temporal variability, and uncertainty." Further, "compliance with the 1000-year objective would be demonstrated if it could be shown that any tracer particle leaving the disturbed zone has a (100-X)% or greater probability of arriving at the accessible environment in a time greater than 1000 years, where X is a small number." An acceptable value of X is not specified, but it is suggested that "a value greater than a few percent and considerably less than 50%" would be desirable. Preliminary probabilistic assessments of groundwater travel times at repository sites have already appeared (Andrews et al., *2*; Clifton, *3*; Lin et al., *4*).

All four North American sites (the three U.S. sites currently under investigation and the as-yet unchosen Canadian site on the Precambrian Shield of northern Ontario) will involve the assessment of groundwater flow and transport through fractured rock. There has been, and continues to be, considerable research into the question of whether flow and transport in fractured rocks is best treated in terms of discrete fracture zones or as an equivalent porous medium. We do not address this issue in this paper. In the porous-medium approach, the geostatistical method is generally applied directly to hydraulic conductivity data. In the discrete-fracture approach, it is generally applied to the geometric properties of the discrete fractures (orientation, density, length, aperture, etc.). Most of the issues we discuss are relevant to both modes of attack, but to avoid continual caveats and asides, we will limit our discussions to the porous-medium approach.

For the same reason, we emphasize travel time calculations that are based on steady-state, saturated flow regimes. Most regional flow systems of interest are thought to be in steady state, although there are recent studies suggesting that some sites may be experiencing long-term transient decays from earlier climates or topographies on a geologic time scale. The limitation to saturated flow places the Yucca Mountain site outside the framework of discussion, although similar concepts could be applied there.

Uncertainty in pre-waste-emplacement groundwater travel time may be due to (1) model uncertainty or (2) parameter uncertainty. In the first category are uncertainties related to such issues as the validity of Darcy's law, the appropriateness of an equivalent porous-medium approach, the use of a steady-state flow field, the use of a one-dimensional or two-dimensional model to represent three-dimensional flow systems, and the presence of isothermal conditions. In the second category are uncertainties related to the spatial variability and values of the hydraulic properties of the media, and the type and values of the boundary conditions on the boundaries of the flow field. Uncertainties in the first category are less amenable to quantitative stochastic analysis than those in the second category.

For steady, saturated flow, the hydraulic properties of interest are the hydraulic conductivity and the porosity. Of these two, it is generally held that the hydraulic conductivity is the parameter of greater uncertainty, and most of the stochastic simulations to date have emphasized this fact by treating hydraulic conductivity as the only uncertain parameter. We shall continue this precedent, but we draw attention to this decision as the first stochastic assumption worthy of note. We are replacing what may possibly be a complex multivariate spatial stochastic process by a simpler univariate one.

Some authors argue for a third category of uncertainty in addition to model uncertainty and parameter uncertainty. It would be called calibration uncertainty and it relates to the uncertainty in the spatial distribution of hydraulic heads, which constitute the output variable in direct predictive applications of hydrogeologic models, but which may be used as input variables in inverse simulations to estimate hydraulic conductivity. We do not follow that approach, considering that this category is included in the estimation error associated with the determination of a parameter.

One of the primary goals of the investigative programs at the waste sites is to evaluate the probability density function of groundwater travel time in order to determine whether the value of X is likely to be acceptable to regulatory agencies. This evaluation is achieved by taking measurements of both hydraulic conductivity and hydraulic head. Measurements may raise or lower the prior estimate of the mean value of travel time, which may in turn raise or lower the value of X. Measurements also reduce uncertainty, tightening the pdf for travel time, thus reducing X. The reduction in uncertainty in travel time results from the reduction in uncertainty in parameter values. Measurements of hydraulic conductivity reduce parameter uncertainty directly; measurements of hydraulic head reduce it indirectly through inverse simulation. The iterative process by which sites are measured and modeled constitutes both a calibration and a validation. Viewed as calibration, the process provides quantitative reduction in parameter uncertainty; viewed as validation, it provides qualitative reduction in model uncertainty. Because measurements reduce uncertainty, they have worth; but they also have a cost. Application of geostatistical methodology in a risk-cost-benefit framework allows us to place data worth into the context of decision theory as it may apply to siting and design at nuclear waste sites.

A decision-theory approach is best carried out in a Bayesian statistical framework. It is therefore necessary to distinguish between Bayesian statistics and classical statistics. The differences are philosophical, but they have operational implications. With the classical statistical approach, one must await the existence of data before determining the probability density functions for the input parameters. The moments of the pdf are estimated from the data, and confidence intervals on these estimates are dependent on the number of data points available. With the Bayesian approach, one always has a prior distribution in hand and the arrival of data allows one to use Bayes theorem to update the estimates of the moments of the prior distribution to those of a posterior distribution. It is possible to set up a prior distribution before taking any data at all. Such a pdf is subjective; it is based on measurements from other similar sites, and on the analyst's experience and common sense. After an initial set of measurements is taken at a site, and assuming that such a data network is relatively sparse, the posterior distribution determined with the Bayesian approach will still be strongly influenced by the subjective prior. With a classical statistical approach, no such influence will be felt. As the number of data points becomes large, the estimates provided by the two approaches converge. For the sparse data networks anticipated at nuclear waste sites, probabilities calculated with the Bayesian approach will differ from those calculated with the classical approach.

To summarize then, we will assess geostatistical methodology (1) in a Bayesian decision-theory framework, (2) as it applies to the probability of achieving the 1,000-year-travel-time regulatory criteria, (3) using a steady-state model of saturated flow, (4) at fractured-rock sites that can be treated as equivalent porous media, and (5) at sites where the uncertain hydraulic property of greatest importance is the hydraulic conductivity. We therefore consider only advective transport, because the focus on travel time does not require consideration of the chemical processes. The methodology is equally applicable to more complex advective-dispersive systems.

Summary of the Proposed Geostatistical Approach

Many questions arise in the course of site investigation that can be addressed with geostatistical analysis applied in a decision-theory framework. Among those related to the travel-time criterion are the following:

1. What is the probability, X, of failing to meet the 1,000-year criterion at a given site?
2. Given two alternative sites, which one is "best" with respect to the 1,000-year criterion?
3. For a given site, which of various repository locations is the "best," with respect to the 1,000-year criterion?
4. Given two or more alternative measurement programs, which one is most likely to produce the greatest reduction in uncertainty with respect to the 1,000-year criterion?
5. How much data is "enough"? When can data measurement programs be terminated?

Let us assume that we are going to try to answer these types of questions using a stochastic-simulation approach with a 2D or 3D, steady-state, saturated, finite-element model. In this section, we provide a sequential list of the steps required to carry out such an analysis.

Seven of the eight steps address the first question; one step (Step 7) addresses questions 2 through 5. In this section, we present the steps in a straightforward manner without discussion of potential limitations. In the following section, we provide a table that critically assesses the steps in terms of assumptions and limitations. In the final section, we discuss, in more detail, the specific issues of greatest concern.

Step 1. Select form of spatial stochastic process for hydraulic conductivity

It is usual to assume that hydraulic conductivity is a stationary process that is lognormally distributed; that is, we can define a parameter, Y, where:

$$Y = \ln K \tag{1}$$

such that a probability density function for Y is normal. The first two moments of the normal pdf for Y are given by the mean, μ, and the variance, σ.

Further, it is necessary to assume a variogram or an autovariance structure for Y. Three forms of the covariance function are in common use. The first is an exponential model:

$$C(x) = \sigma^2 e^{-|x|/\lambda} \tag{2}$$

where x is distance and λ is an exponential decay parameter, known as the correlation length. This is a one-parameter representation of the autocorrelation properties of Y. It is fully specified by the correlation length, λ, which is the distance over which C decays to a value of e^{-1}. It is commonly recognized to be anisotropic, leading to a three-parameter representation in three dimensions, given by λ_x, λ_y, λ_z.

A second type of autocovariance structure is the spherical model:

$$C(x) = \sigma^2 - c^2[1-\delta(x)] - (\sigma^2 - c^2)\,[3/2(x/a) - 1/2(x/a)^3] \qquad x < a$$

$$C(x) = 0 \qquad x \geq a \tag{3}$$

where $\delta(x)$ is the Kronecker delta ($\delta = 1$ if $x = 0$, $\delta = 0$ if $x \neq 0$). This model has two parameters, c^2, the so-called nugget effect which represents small-scale variability or measurement noise, and the range, a, which is the distance after which the autocovariance function is zero. In comparing these first two models, one finds that the exponential model reaches zero (within 5%) at a distance of 3λ, which is thus close to the range, a, of the spherical model.

The third type of autocovariance structure is the linear model in two dimensions:

$$C(x,y) = \sigma^2 \left(1 - \frac{|x|}{b_1}\right)\left(1 - \frac{|x|}{b_2}\right) \tag{4}$$

There are two parameters, b_1 and b_2.

Step 2. Determine Bayesian prior distribution

Let us assume that we are going to carry out our travel-time calculations with a finite-element network that uses a single-subscript numbering notation and finite-element blocks: i = 1, 2, 3 ....n. It will be necessary to assign values of Y_i for all i. Prior to making any measurements of Y, we will have uncertainty as to the values of Y_i; for any i there is an infinite number of possible values. We wish to estimate Y_i based on the assumption that these possible values are normally distributed with expected value, μ_i, and uncertainty, σ_{ii}^2. For the entire set of n blocks, there will be a vector of expected values $\{\mu_i\}$, of the estimates Y_i, and a covariance matrix $[\sigma_{ij}^2]$, of the estimates of Y_i, in which the diagonal terms, σ_{ii}^2 represent the uncertainty of Y_i, and the off-diagonal terms reflect the autocovariance structure of the estimates.

To illustrate, consider a one-dimensional flow system (Figure 1a) with a sequence of Y values, $Y_i = Y(x_i) = \{Y_1, Y_2, \ldots\ldots\ldots Y_n\}$, each being defined for a small block of porous medium around $x = x_i$. The vector of expected values will be of length n; the autocovariance matrix n by n. Figure 1b shows a 12-node example.

(a)

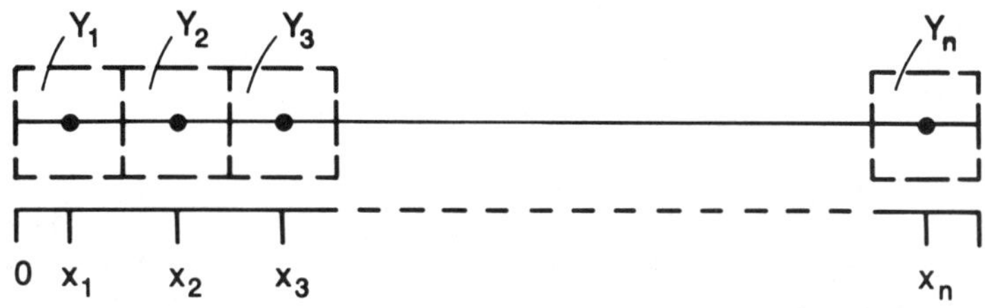

(b)

$$\{\mu_i\} = \{0.6\ 0.6\ 0.6\ 0.6\ 0.6\ 0.6\ 0.6\ 0.6\ 0.6\ 0.6\ 0.6\ 0.6\}$$

$$\left[\sigma^2_{ij}\right] = \begin{bmatrix} 1.10 & 0.83 & 0.55 & 0.28 & & & 0 \\ 0.83 & 1.10 & \ddots & \ddots & \ddots & & \\ 0.55 & \ddots & 1.10 & \ddots & \ddots & \ddots & \\ 0.28 & \ddots & \ddots & \ddots & \ddots & \ddots & 0.28 \\ & \ddots & \ddots & \ddots & 1.10 & \ddots & 0.55 \\ & & \ddots & \ddots & \ddots & 1.10 & 0.83 \\ 0 & & & 0.28 & 0.55 & 0.83 & 1.10 \end{bmatrix}$$

(c) Measurements = $Y_3 = 0.3 \qquad \sigma^2_{33} = 0.0$

$Y_7 = 0.2 \qquad \sigma^2_{77} = 0.0$

(d)

$$\{\mu_i\}'_u = \{0.5\ 0.5\ 0.5\ 0.5\ 0.5\ 0.5\ 0.5\ 0.5\ 0.5\ 0.5\ 0.5\ 0.5\}$$

$$\{\sigma^2_{ii}\}'_u = \{1.0\ 1.0\ 1.0\ 1.0\ 1.0\ 1.0\ 1.0\ 1.0\ 1.0\ 1.0\ 1.0\ 1.0\}$$

(e)

$$\{\mu_i\}'_c = \{0.40\ 0.35\ 0.30\ 0.28\ 0.25\ 0.23\ 0.20\ 0.28\ 0.35\ 0.43\ 0.50\ 0.50\}$$

$$\{\sigma^2_{ii}\}'_c = \{0.75\ 0.44\ 0.00\ 0.38\ 0.50\ 0.38\ 0.00\ 0.44\ 0.75\ 0.94\ 1.00\ 1.00\}$$

(f)

$$\{\mu_i\}'_k = \{0.28\ 0.29\ 0.30\ 0.28\ 0.25\ 0.23\ 0.20\ 0.21\ 0.23\ 0.24\ 0.25\ 0.25\}$$

$$\{\sigma^2_{ii}\}'_k = \{0.88\ 0.47\ 0.00\ 0.38\ 0.50\ 0.38\ 0.00\ 0.47\ 0.88\ 1.22\ 1.50\ 1.50\}$$

Figure 1. **Sample calculations of Bayesian updating procedure for a 12-node, one-dimensional flow field.**

At any unmeasured point, x_i, the value of Y_i is represented by a normal distribution with mean, μ_i, and uncertainty, σ_{ii}^2. After making measurements of Y, the uncertainty will be reduced significantly (perhaps to zero) at the measurement points, and it will be reduced somewhat in the vicinity of the measurement points due to the influence of the autocorrelation properties of Y. At any measured point, x_k, the value of μ_k in the vector of means becomes the measured value, Y_k, and in the absence of measurement error, $\sigma_{kk}^2 = 0$. The concepts presented in Figure 1 for a one-dimensional system apply equally well in two or three dimensions, except that the vector of means will be much longer and the autocovariance matrix more complex. Figures 1c through 1f will be discussed below in connection with Step 5.

It is important at this point to recognize the difference between the variance, σ, of the parameter Y, which represents the variability of the process, and the residual variance, σ_{ii}^2, after the estimation of Y_i, which represents the uncertainty in the value of the parameter at a given point. Variability is a property of the medium; uncertainty is felt by the analyst. In this paper, we use the notation μ, σ, and λ to describe the process, and $\{\mu_i\}$ and $[\sigma_{ij}^2]$ to describe the estimates.

In this paper, we recommend a Bayesian approach, whereby it is necessary to make prior estimates of $\{\mu_i\}$ and $[\sigma_{ij}^2]$. If such estimates are made prior to the taking of any measurements, and if the hydrogeologic environment is assumed to be statistically homogeneous, then the concept of ergodicity can be invoked to suggest that, for all i, $\mu_i = \mu$, and $\sigma_{ii}^2 = \sigma^2$, where μ and σ are the moments of the spatial stochastic process. That is, the best estimate of Y_i at any point i is the mean of the stochastic process, and the uncertainty is equal to the variability, σ. Ergodic arguments also suggest that the off-diagonal terms of the covariance matrix of the estimates, $[\sigma_{ij}^2]$, will depend on the parameters of the covariance model for the stochastic process: λ from equation (2), c and a from equation (3), or b_1 and b_2 from equation (4). For example, if equation (2) is invoked:

$$\sigma_{ij}^2 = \sigma_{ii}^2 e^{-|x_i - x_j|/\lambda} \tag{5}$$

The off-diagonal terms in Figure 1b are based on a linear autocorrelation function.

It is thus necessary to make prior estimates of the parameters μ, σ, and λ (or μ, σ, c, and a; or μ, σ, b_1, or b_2). If no measurements are available, these estimates are subjective. They will be based on (1) the analyst's experience and common sense, (2) available measurements from other sites in similar geologic media, and (3) measurements of other parameters at the site that are related to hydraulic conductivity (such as grain-size analyses, fracture apertures, seismic profiles, etc.). If it is known that there are distinctly different geologic units within the boundaries of the flow domain, each of those units would be treated as a separate, statistically homogeneous unit.

If prior estimates of μ, σ, and λ are to be based on initial measurement programs for Y, one can use the methods outlined in Step 5 to obtain $\{\mu_i\}$ and $[\sigma_{ij}^2]$.

Step 3. Incorporate scale effects using variance-reduction concepts

Because we are using a continuous-medium approach, our prior estimates of μ, σ, and λ are defined for mathematical points in space. When measurements are taken into account, they will be representative of a certain volume in space. For instance, the so-called point measurements obtained with single-well slug tests are representative of a scale of 0.1 to 1 m, whereas multiple-well pump tests reflect a scale of 10 to 100 m. Eventually, when the hydraulic conductivity is used in a finite-element model, the block sizes in the model may be on the order of 100 to 10,000 m. The statistics of the measured values are not immediately compatible with one another or with the needs of the model due to the fact that they are averaged over different volumes of a stochastic field. It is therefore necessary to adapt the parameter statistics to the scale of interest.

This step is accomplished by straightforward spatial integration (cf. de Marsily, *5*). Vanmarcke (*6*) provides a series of formulae that evaluate the integrals for simple geometric configurations, such as squares and rectangles, that are pertinent for a finite-element grid. Journel and Huijbregts (*7*) also provide formulae and type curves for this purpose.

For example, in two dimensions, if μ_1, σ_1, and λ_1 represent the prior estimates at the smaller of two scales, then μ_2 and σ_2 at the larger scale are given by:

$$\mu_2 = \mu_1 \tag{6}$$

$$\sigma_2{}^2 = \left[\frac{\Pi(\lambda_1)^2}{A_2} \right] \sigma_1{}^2 \tag{7}$$

where A_2 is the representative area at the larger scale. Vanmarcke also provides formulas for λ_2, and he provides equations for one-dimensional and three-dimensional systems, as well as two-dimensional. If formulae are not available, it is always possible to perform the integration numerically.

Step 4. Perform prior unconditional uncertainty prediction on travel time

Having used the subjective prior values of μ, σ, and λ to estimate values for all the terms of $\{\mu_i\}$ and $[\sigma_{ij}{}^2]$ in Step 2, we recognize that there are an infinite number of equally likely realizations of the field of Y values. We will use a subset of these and Monte Carlo simulation to make an unconditional uncertainty prediction of travel time. It is unconditional in the sense that it is not conditioned on the locations of any measurements.

There are several substeps: (1) Using the $\{\mu_i\}$ and $[\sigma_{ij}{}^2]$ values, generate m equally likely unconditional realizations of the yield of Y_i values; (2) convert the Y_i values to K_i values; (3) use Monte Carlo simulation with the finite-element model and the realizations of $\{K_i\}$ to produce m equally likely, steady-state, saturated flow fields; (4) calculate the groundwater travel time for the fastest flow path for each flow net; (5) develop a probability density function and/or a cumulative density function for travel time; and (6) determine the probability, X, that the travel time is less than 1,000 years.

It is this prior probability, X, that will hopefully be reduced by the gathering of additional data on K. In fact, it can also be reduced by additional data on hydraulic head, h, if such measurements are integrated with the application of inverse calibration procedures. This possibility will be addressed as Step 8, following presentation of the more direct assessment of the worth of hydraulic conductivity data.

There are a variety of alternatives to Monte Carlo analysis in substep (3) above. One can use one of the analytic stochastic simulation procedures such as the spectral perturbation technique (Gelhar, *8*) or first-order, second moment analysis (Dettinger and Wilson, *9*). These methods are less computer intensive than Monte Carlo simulation, but they are also less general. A Monte Carlo approach using particle tracking may be the easiest route to a travel-time calculation (Smith and Schwartz, *10*).

Step 5. Use hydraulic conductivity measurements to update estimates of uncertainty

Given a prior estimate of the vector of means $\{\mu_i\}$, and the covariance matrix $[\sigma_{ij}^2]$, together with a set of log hydraulic conductivity measurements at specific locations in the flow field, it is possible to produce updated estimates $\{\mu_i\}'$ and $[\sigma_{ij}^2]'$. In particular, one obtains updated estimates of the reduced uncertainty, σ_{ii}^2, at every element, i, in the finite-element network.

There are two steps. In the first step, one uses the data to calculate updated estimates of the moments of the spatial stochastic process, μ', σ', and λ', given the prior estimates, μ, σ, and λ. These can be obtained from a straightforward application of the Bayes theorem (Benjamin and Cornell, *14*). This application requires the specification of a likelihood function for each parameter, but this does not pose a problem as long as these parameters can be assigned a prescribed probability density function. The unconditional posterior estimates, $\{\mu_i\}_u'$ and $[\sigma_{ij}^2]_u'$, can then be determined from μ', σ', and λ'. In particular, $(\mu_i)' = \mu'$, and $(\sigma_{ij}^2)_u' = \sigma'$, for all i. The values of μ' and σ' may be larger or smaller than μ and σ. In the traditional kriging approach, the parameters of the variogram can be updated, but it is usual to do so with classical statistics on the enlarged data set, rather than by using the Bayes theorem.

In the second step, the data are used in a conditional sense, taking into account measurement locations. The updating can be carried out using the concepts of the Bayesian updating procedures provided by Hachich and Vanmarcke (*11*) and Massmann and Freeze (*12,13*), or by the traditional kriging approach described by Marsily (*5*). The two approaches to this second step are equivalent. We will follow the Bayesian updating terminology.

Let us assume that measurements of hydraulic conductivity, K, are made at p of the n blocks in a discretized flow system. The p measurements of K are converted to p measures of Y which are used to estimate the n actual values of Y by linear regression:

$$\{A\} = [B]\ \{Y\} + \{e\} \tag{8}$$

where:

$\{A\}$ = p × 1 vector of observed values
$\{Y\}$ = n × 1 vector of actual values
$[B]$ = p × n matrix of regression coefficients
$\{e\}$ = p × 1 vector of zero mean measurement errors

The matrix of regression coefficients, [B], can be used to incorporate measurement biases, or to allow for the fact that hydraulic conductivities may be inferred from some alternative parameter, such as grain size value. In the simplest case, for direct measurements with no bias, the matrix [B] would contain only zeros and ones.

The conditional posterior estimates, $\{\mu_i\}_c'$ and $[\sigma_{ij}^2]_c'$, are given by

$$\{\mu_i\}_c' = \{\mu_i\}_u' + [\sigma_{ij}^2]_u' [B]^T ([B] [\sigma_{ij}^2]_u' [B]^T + [E])^{-1} \bullet (\{A\} - [B] \{\mu_i\}_u') \quad (9)$$

$$[\sigma_{ij}^2]_c' = [\sigma_{ij}^2]_u' - [\sigma_{ij}^2]_u' [B]^T \bullet ([B] [\sigma_{ij}^2]_u' [B]^T + [E])^{-1} [B] [\sigma_{ij}^2]_u' \quad (10)$$

where [E] is the covariance matrix of the measurement errors.

The values of $(\sigma_{ii}^2)_c'$ are always less than or equal to $(\sigma_{ii}^2)_u'$ because the estimation is now conditioned by the knowledge of Y at the measured points. In general, the values of $(\sigma_{ii}^2)_c'$ are also less than the prior (σ_{ii}^2), unless the measurements have led to an increase in the estimated posterior value of the process variability, σ', over the prior, σ.

It is important to note that the reduction in $[\sigma_{ij}^2]$ depends only on the number and location of the measurements and not on the measured values. It is this feature of the Bayesian updating scheme that allows us to calculate the potential worth of additional data prior to actually making the measurements.

Figure 1d shows the unconditional posterior estimates for the prior given in Figure 1b, due to two error-free measurements made at nodes 3 and 7 (Figure 1c) of the 12-node one-dimensional flow field shown in Figure 1a. It was assumed that μ and σ have a Gaussian distribution function; the parameters μ and σ were updated using the Bayes rule and likelihood functions for a Gaussian pdf.

Figure 1e gives the conditional posterior estimates determined from equations (9) and (10). It is these values that would be used in the posterior predictions in Step 6.

It is worth noting that the differences in the unconditional updating (the first step of the two-step procedure) between the Bayesian approach and the classical statistical approach lead to significant differences in the final conditional posterior estimates. Figure 1f was determined from a kriging package assuming an unknown mean and the intrinsic hypothesis while the Bayesian equations are based on a known mean and second-order stationarity. The differences lie entirely in the first step; the second step of the procedure is identical for the two procedures. Kulkarni (*15*) has proposed a Bayesian kriging approach, but in a somewhat different context.

Figure 2 provides a hypothetical example that shows a two-dimensional mapping of the updating of $\{\mu_i\}$ to $\{\mu_i\}_c'$ and $\{\sigma_{ii}^2\}$ to $\{\sigma_{ii}^2\}_c'$. A comparison of Figures 2b and 2d shows a decrease in the uncertainty associated with estimates of Y_i over most of the flow field due to the influence of three error-free measurements of Y.

The degree of uncertainty reduction gained by a given measurement program depends on the geostatistical properties of the aquifer. Figure 3 is a hypothetical example based on Massmann and Freeze (*16*), that shows the effectiveness of a

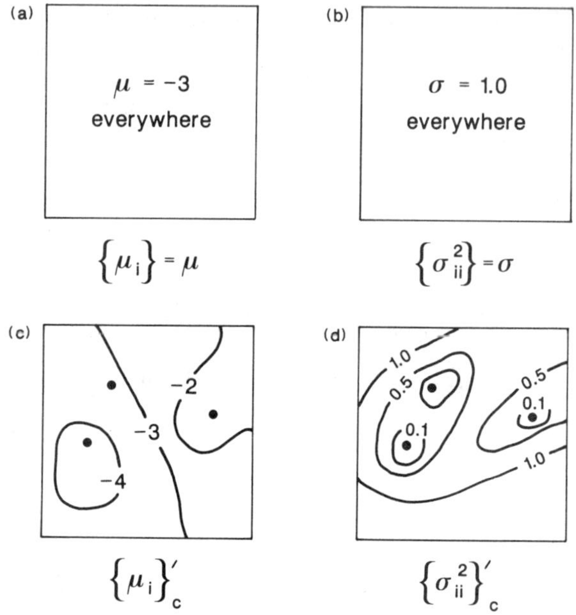

Figure 2. **(a) Prior mean log hydraulic conductivity.**
(b) Prior uncertainty in log hydraulic conductivity.
(c) Posterior mean log hydraulic conductivity.
(d) Posterior conditional uncertainty in log hydraulic conductivity.

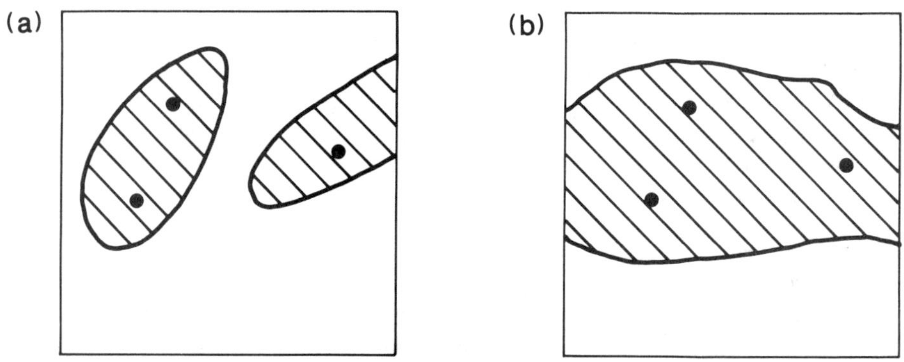

Figure 3. **(a) Region of 50% reduction in posterior conditional uncertainty in log hydraulic conductivity for media with short correlation length, λ.**
(b) Region of 50% reduction for media with longer correlation length.

three-point measurement program for two aquifers that are identical in every respect except for the correlation length, λ.

Step 6. Perform posterior conditional uncertainty prediction on travel time

The procedure outlined in Step 4 is now repeated with the updated $\{\mu_i\}_c'$ and $[\sigma_{ij}^2]_c'$, to provide a revised uncertainty prediction of the travel time, t′. An early example of this approach was published by Smith and Schwartz (*17*).

As was shown in Step 5, measurements of K_i (or Y_i) reduce the uncertainty in the input field. This reduction is transmitted through to the output field of hydraulic heads, by means of the Monte Carlo stochastic simulation. Figures 4a and 4b show a schematic contoured plot of the prior and posterior expected value for the estimates of hydraulic head in a regional aquifer due to the addition of three measurement points of K. These measurements would also be expected to reduce the uncertainty in predicted head values. In doing so, such measurements can lead to a reduction in X (Figure 4c).

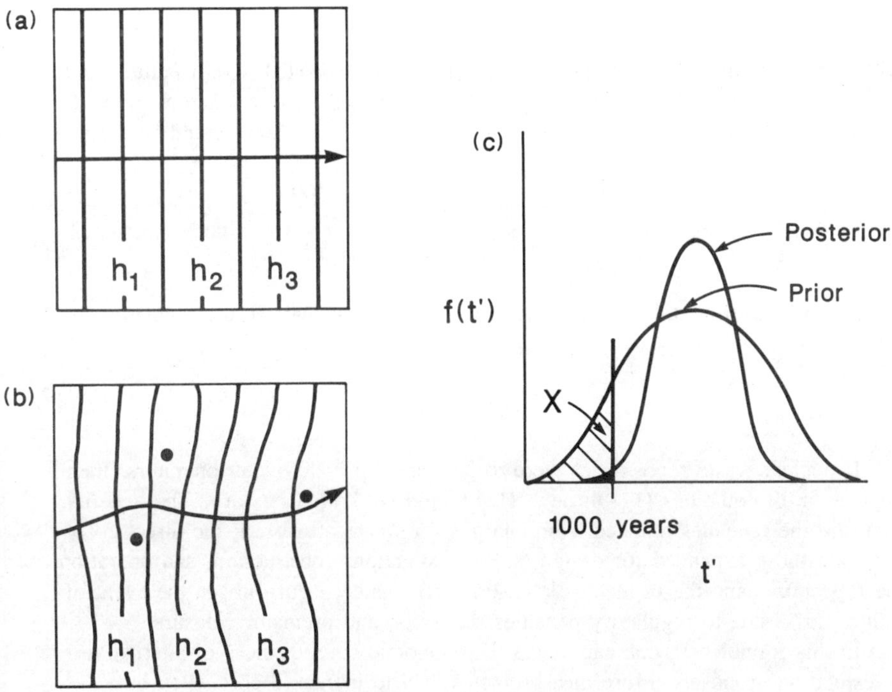

Figure 4. **(a) Unconditional simulation of hydraulic head field using prior log hydraulic conductivity distribution.**
(b) Conditional simulation of hydraulic head field using posterior log hydraulic conductivity distribution.
(c) Probability density function of pre-emplacement advective groundwater travel time for prior and posterior log hydraulic conductivity distributions.

Step 7. Use decision-theory framework to assess alternative courses of action and data worth

Thus far, we have addressed only the first question of the five posed at the beginning of this section. The remaining questions posed there are different from the first one in that they involve the assessment of alternatives. The investigation of such issues in a hydrogeologic context has recently begun (Massmann and Freeze, *13*).

Decision theory is suited to making decisions between alternative courses of action. These may involve alternative sites, alternative designs at a given site, or alternative data measurement networks in support of a given design. It is usual to define an objective function, Φ, as the discounted stream of benefits, B(t), costs, C(t), and risks, R(t):

$$\Phi = \sum_{t=0}^{T} \frac{1}{(1 + i)^t} [B(t) - C(t) - R(t)] \tag{11}$$

where i is the discount rate and T is the time horizon. The risks are defined as

$$R(t) = P_f(t) \quad C_f(t) \quad \gamma(C_f) \tag{12}$$

where P_f is the probability of failure, C_f is the cost associated with a failure, and γ is a normalized utility function which, for our purposes, can be taken as unity. For the travel time criteria, the probability of failure in year t, $P_f(t)$, would be defined as

$$\begin{aligned} P_f(t) &= Pr(t = t') && t \leq 1000 \\ P_f(t) &= 0 && t > 1000 \end{aligned} \tag{13}$$

where t′ is the travel time. The probability X, which appears in the regulatory criteria, is

$$X = \sum_{t=0}^{1000} P_f(t) \tag{14}$$

If nuclear waste repositories were to be run by private owner-operators, then application of equations (11) through (13) would be straightforward. The benefits, B(t), are the revenues realized from charges for waste processed; the direct costs, C(t), are those expended for design, site investigation, construction, and operation of the repository; and the probabilistic costs, $C_f(t)$, which occur only in the event of failure, arise due to regulatory penalties, lawsuits, and premature closure.

In this framework, one can assess the economic consequences of alternative site investigation proposals. More measurements lead to increased cost, C(t), but decreased risk, R(t). The reduction in risk results from a tightening of the probability density function for travel time as shown in Figure 4. Additional measurements should be curtailed when the additional costs are no longer balanced by the risk reduction that can be achieved.

A risk-cost-benefit economic framework is well suited to decision-making by private owner-operators with well-defined objective functions. It may be more difficult to apply in the public domain. For the repository case, the value of the probabilistic cost, $C_f(t)$, is problematical. Despite potential shortcomings, this approach, or variations on it, may represent the most rational way to assess alternatives in nuclear waste programs.

Step 8. Incorporate hydraulic head measurements

We now return to the issue of hydraulic head measurements. These too, when integrated with an inverse procedure, can lead to reduction in uncertainty in Y, and hence in travel time. They, too, have data worth, and like hydraulic conductivity, additional measurements should not be carried out when additional costs are no longer balanced by the resulting risk reduction.

The inverse method provides a structured approach for the construction of a hydrogeologic model, estimation of model parameters, and calculation of estimation errors associated with the parameter values. Central to the technique is the concept that hydraulic head measurements contain information on the spatial variation in hydraulic conductivity and that these data can be used to identify hydraulic conductivity variations with a smaller estimation error than an approach based only on the kriging of hydraulic conductivity measurements alone (Clifton and Neuman, *18*). An inverse simulation incorporates model calibration within the parameter estimation procedure, a step not included in the Bayesian updating of Step 5. Uncertainties in boundary conditions and boundary values can be included in the estimation procedure; however we limit our discussion here to hydraulic conductivity estimates, while recognizing that the more general problem is addressed in the same framework. Yeh (*19*) and Carrera and Neuman (*20*) provide recent reviews of inverse simulation in groundwater hydrology.

There are two ways to formulate the inverse; the parametric approach is based on direct estimation of hydraulic conductivity values in a finite number of zones, while the "geostatistical" approach relies on cokriging of hydraulic conductivity and hydraulic head measurements to obtain improved estimates of the spatial structure of hydraulic conductivity. Our discussion in this paper focusses on the parametric approach.

The parametric approach consists of first defining an interpretational model (an assumed boundary value problem with material properties assigned in a finite number of zones). Estimates of hydraulic conductivity values within each zone are then sought by minimizing the difference between observed and predicted values of hydraulic head, conditioned on prior measurements of hydraulic conductivity. This approach is formalized by minimizing an objective function of the form:

$$J = \{h - h^*\}^T [V_h]^{-1} \{h - h^*\} + \varepsilon \{Y - Y^*\}^T [V_y]^{-1} \{Y - Y^*\} \qquad (15)$$

where $\{Y\}$ is a vector of log conductivity values determined from the inverse, $\{Y^*\}$ is a vector of observed (and possibly estimated) values of log conductivity, $\{h^*\}$ is a vector of observed (and possibly estimated) values of hydraulic head, and $\{h\}$ is a vector of hydraulic head values calculated using the estimate $\{Y\}$. $[V_h]$ and $[V_y]$ are covariance matrices of pre-inverse estimation errors for hydraulic head and hydraulic conductivity. The parameter, ε, is a scaling factor that provides a relative weight to the hydraulic head and hydraulic conductivity data. This new notation simplifies equation (15) so that its structure is easily comprehensible. We note, however, that the terms involving Y in (15) have equivalence with terms already introduced in this paper using a more detailed notation. The vector, {Y*}, is equivalent to the vector of posterior conditional estimates of the mean log hydraulic conductivity, $\{\mu_i\}_c'$. The matrix $[V_y]$ is the posterior conditional estimate of the uncertainty in log hydraulic conductivity $[\sigma_{ij}^2]_c'$. The vector {Y*} is conditioned only on hydraulic conductivity measurements; the vector {Y} is conditioned on both hydraulic conductivity and hydraulic head measurements.

With an adequate data base, {Y*}, {h*}, $[V_y]$, and $[V_h]$ can be determined by kriging. The advantage here is that estimates of log conductivity and hydraulic head, and the estimation error, are provided for all regions of the flow domain, rather than just at measurement points. This step rests on the identification of the variogram for log conductivity, in that this allows inferences to be drawn on parameter values away from measurement points. In cases where data are insufficient to define a variogram, only values at measurement points are incorporated in equation (15), and $[V_h]$ and $[V_y]$ are diagonal matrices with values expressing the magnitude of measurement error.

Figures 5a and 5b show schematically how hydraulic head measurements incorporated in an inverse simulation might serve to increase the uncertainty reduction in hydraulic head estimates over that achievable with conductivity measurements alone. Figure 5c shows how this might impact on the pdf for travel time. We note that both Figures 4c and 5c indicate a reduction in X, with additional data. While this is usually the case, it is not always so; if additional data show that the posterior mean of the process, μ', is greater than the prior, μ, a decrease in travel time and an increase in X are possible.

Equations for calculating the estimation error for {Y}, and for the estimation error of hydraulic head values determined using the estimate {Y}, vary with the particular scheme used to solve equation (15). Note that, unlike the posterior hydraulic conductivity statistics estimated using Bayesian updating (Step 5), the statistics for log conductivity determined from an inverse simulation reflect the influence of assumptions about the nature of the flow system, location and type of boundary conditions, and zonation patterns adopted in the interpretational model. The development of criteria for selection among different interpretational models is currently an active research topic.

Potential Limitations of the Geostatistical Approach

In Table 1, we identify some of the assumptions and limitations associated with each of the steps in the proposed geostatistical methodology. We do so in the form of a set of questions that ought to be asked at each step in the sequence. In Table 1, we provide a brief comment on each issue, and we rank each issue as having lower or higher significance. The former have a less significant impact on the outcome of the geostatistical modeling process, and a lower research priority. The latter have a more significant impact on the outcome, and a higher research priority.

In the next section of the paper we provide a more detailed discussion of some of the issues we judge to be most significant.

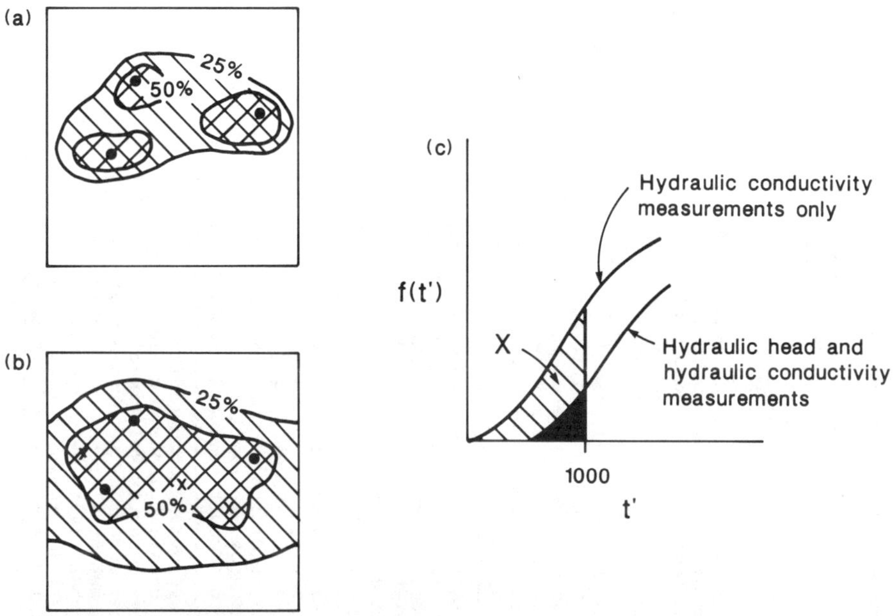

Figure 5. **(a) Region of 25% and 50% reduction in posterior conditional uncertainty in hydraulic head with measurements of hydraulic conductivity only.**
(b) Region of 25% and 50% reduction with measurements of hydraulic conductivity and hydraulic head.
(c) Effect on left tail of probability density function for travel time.

Issues of Special Significance

We have identified four issues of special significance from Table 1. These issues revolve around (1) the appropriateness of the Bayesian approach, (2) the application of inverse simulation, (3) identification of trends and zonations in hydraulic conductivity, and (4) problems associated with small-sample statistics. In this section, we discuss each in turn.

Appropriateness of the Bayesian approach

The siting, construction, and operation of a nuclear repository is a massive engineering undertaking. We agree with Benjamin and Cornell (*14*) that engineering decision-making is best carried out in a Bayesian framework. The primary difference between this framework and a classical statistical approach is the inclusion of subjective data in the analyses on which decisions are based. We hold that such data are included one way or another in all analyses (as a safety factor, for example), but the Bayesian framework incorporates it in a clear, objective, and scientific way. It is the only rational way to make use of the geologist's experience, based on prior studies of other sites.

Table 1. **Assumptions and limitations associated with geostatistical methodology.**

Step	Assumption or Limitation	Comments	Significance Lower	Significance Higher
Step 1				
Spatial Stochastic Process	Is hydraulic conductivity always lognormally distributed?	Hypothesis of lognormal distribution cannot be rejected for most field data. Such a distribution is expected for parameters that range over several orders of magnitude and cannot take on negative values. We suspect that this assumption does not have a very large influence on the final outcome of the analysis if X is not too small; for very small X, however, the influence of distribution shape on the tails may be larger; numerical tests should be carried out to test this hypothesis for small X.	✓	
	Is the hydraulic conductivity field usually stationary? What are the implications if it is not? Can trends and/or geological zonation be identified?	Much of the available data from porous-media has shown stationary behaviour, but there is reason to think that in fractured rocks, vertically decreasing trends exist. The identification of zonation is not usually a problem in layered sedimentary stratigraphy but may be a source of considerable uncertainty in fractured plutonic rock or in layered volcanic terrain. There may be problems in identifying trends or zonation under sparse data conditions. Once identified, however, methods are available to take these aspects into account geostatistically.		✓
	Is it adequate to use a univariate spatial stochastic process for K, or are uncertainties in porosity sufficiently large to require a bivariate process?	We believe that hydraulic conductivity is the parameter of greatest importance in uncertainty analysis for travel times. However, the implications of uncertainty in porosity and its potential cross-correlation with hydraulic conductivity deserves more attention. Effective porosity for sparsely fractured crystalline rock may be difficult to		

Table 1. **Assumptions and limitations associated with geostatistical methodology (continued).**

Step	Assumption or Limitation	Comments	Significance	
			Lower	Higher
		estimate. If the parameters are independent, the generation of conditional or unconditional fields can be made separately. If correlated and with identical pdf's, simple linear correlation can be used. If correlated with different pdf's, it is possible to generate the two fields simultaneously using rank correlation analysis (Iman et al, 22). There will be difficulty identifying correlation structure under sparse data conditions.	✓	
	What is the best auto-correlation structure to use? How should one decide?	With limited field data, it is difficult to reject any of the usual covariance functions. It is probably best to assume a form a priori based on experience gained in similar rock types where data may be more abundant. Exponential and spherical covariance functions have both been used successfully. The spherical model has the flexibility of two parameters, and the ability to include the nugget effect. If a nugget effect is added to the exponential model, both functions have enough flexibility to adjust experimental data. In any case, uncertainty in travel time is less sensitive to autocorrelation than to variance. It is also less sensitive to the form of the function than to its range.	✓	
Step 2				
Prior Distribution	Is a Bayesian framework appropriate for geostatistical studies associated with stochastic hydrogeological simulation?	This is an important philosophical issue with important operational implications. It is discussed in more detail in the text.		✓

Table 1. **Assumptions and limitations associated with geostatistical methodology (continued).**

Step	Assumption or Limitation	Comments	Significance Lower	Significance Higher
	How should the subjective prior be determined?	Knowledgeable hydrologists may differ over the nature of the covariance model and/or the values of the parameters. Methods for rationalizing subjective priors are available in other fields, and the needed technology transfer to bring these methods into use in hydrogeology is now underway (cf. Merkhofer and Runchal, 23). The form of the prior will have considerable influence on probabilities and decisions under conditions of sparse data availability.		✓
Step 3				
Scale effects	What are the pertinent representative volumes to use for the various hydraulic conductivity measurement methods in the variance-reduction formulae?	Too little research has been done on this question. Matheron (24) and Dagan (25) have addressed the problem of averaged permeability around a pumping well, but their results are too preliminary to be used directly. It is not clear whether results are particularly sensitive to this issue.		✓
	Can we identify embedded structures?	Under conditions of sparse data, it may be difficult to identify the existence of embedded structures. Once identified, however, the inclusion of nested structures is relatively straightforward (Journel and Huijbregts, 7). It is worth noting that in the integration performed to change the scale to that appropriate for a finite-element model, the small-scale structures will tend to disappear and only the large-scale ones will be significant. For any covariance function used, the lower scale structures will appear as a nugget effect. Sampling effort should therefore aim at identifying large-scale structures.	✓	

Table 1. **Assumptions and limitations associated with geostatistical methodology (continued).**

Step	Assumption or Limitation	Comments	Significance	
			Lower	Higher
Steps 4,6				
Stochastic Simulation	Are there theoretical issues associated with the conversion of normally-distributed Y_i values to lognormally distributed K_i values for use in the simulation model?	If one applies the transformation, $K_i=\exp(Y_i)$, it is well-known that the mean of the distribution of Y becomes the median of that of K. For this reason, many authors apply the correction, $K_i=\exp(Y_i+1/2\,\sigma_{ii})$. This leads to an unbiased estimate of $\bar{K}$, but then $Y_i=\ln(K_i)$ will be biased. We cannot have both quantities estimated simultaneously without bias. We note that, in 2D uniform flow, the average permeability is equal to the geometric mean (ie. the arithmetic average of Y_i, not K_i), therefore, using an unbiased estimate of Y_i seems the better alternative. This point deserves investigation by detailed stochastic simulations.	✓	
	Are the computer limitations on Monte Carlo methods so severe as to render the method impractical?	Monte Carlo methods appear to be within the capability of most large computer centers, especially if they have an array processor, for steady flow in 2D (and even in 3D, if the finite-element network is not too large). There are, as yet, severe limitations on 3D transient systems. Where geological stratigraphy and structure are complex, it may not be possible to design representative 2D sections for 3D flow problems. This is a particular problem in fractured plutonic rocks; perhaps less so in layered sedimentary and volcanic rocks.	✓	
	What are the limitations associated with spectral	There are limitations on the maximum variance that can be handled, and on the size and		

Table 1. **Assumptions and limitations associated with geostatistical methodology (continued).**

Step	Assumption or Limitation	Comments	Significance: Lower	Significance: Higher
	methods, and with first-order second moment methods?	complexity of the domain and its geological zonation. There are also limitations on the type of distributions and the form of the autocovariance function. It is not clear whether the necessary assumptions are likely to be satisfied at the waste sites under consideration. Certainly, these methods should be useful for initial assessments, and they do provide an alternative where Monte Carlo simulation is not feasible. They are much less expensive and computer intensive. The problems that may arise with sparse data are as yet uninvestigated. There is much ongoing research in this area.	✓	
	Are the values of X, determined with stochastic simulations, reliable?	The calculation of X (the probability of failing to meet the 1000-year travel-time criterion) is based on the tail of the simulated pdf for travel time, which is itself based on an input pdf for hydraulic conductivity that may not be particularly well-defined in its tails. This sensitivity to distributional tails deserves investigation.	✓	
Step 5				
Updating hydraulic conductivity	Are the Bayesian updating equations equivalent to the kriging equations?	The theories are almost identical. The intrinsic hypothesis that underlies kriging is slightly more general than the more classical stationarity assumption that underlies Bayesian updating. The primary difference lies in the Bayesian concept of a subjective prior. There is also a difference in the way that the parameters of		

Table 1. **Assumptions and limitations associated with geostatistical methodology (continued).**

Step	Assumption or Limitation	Comments	Significance	
			Lower	Higher
		the process are updated in the unconditional step. This leads to significant differences in posterior uncertainty; see the text.	✓	
	Should one use the two-step updating procedure (unconditional updating of μ, σ, λ; followed by conditional updating of (U_i) and $[\sigma_{ij}^2]$?	Classical updating procedures (both kriging and Bayesian) have not always included the initial unconditional updating of the parameters describing the variability of the process. We support the arguments of Kitanidis and Vomvoris (26) and Feinerman et al (27) that it is advisable to do so. Questions remain as to whether it should be done in a classical or Bayesian framework.		✓
Step 7 Decision theory	Is a risk-cost-benefit framework philosophically sound and politically acceptable for decisions involving nuclear waste?	There are philosophical arguments in the engineering literature between the safety-factor approach and the risk-balancing approach. We feel that the latter philosophy is best-suited to decision-making. The use of probabilistic criteria by regulatory agencies infers that they, too, accept a risk-based approach.	✓	
	What discount rate and time horizon should be used?	Decisions are very sensitive to these economic parameters, usually more so than to technical or hydrogeological parameters. Implications for nuclear repository siting and policy have been very little examined.		✓
	What value of X is going to prove acceptable to the regulatory agencies and the public?	Decision analyses will require specification by regulatory agencies of an acceptable value for X(the probability of not meeting the 1000-year travel time criterion), if uncertainty analysis in a risk-based framework is to be used to make decisions about alternative measurement programs and facility designs.	✓	

Table 1. **Assumptions and limitations associated with geostatistical methodology (continued).**

Step	Assumption or Limitation	Comments	Significance	
			Lower	Higher
	What probabilistic costs, $C_f(t)$, should be used in the risk-cost-benefit framework?	These costs, which must be borne in the event of a failure to meet the travel-time criterion, are very difficult to specify. Decisions may be sensitive to the values selected.		✓
	Are decisions based on probabilities determined from stochastic simulation robust?	In comparing alternatives, some alternatives may maximize the objective function over wide ranges of the decision space, others over only narrow ranges. When the probability of failure in the risk term comes from the tail of a stochastically-simulated pdf, a robust alternative may be desirable. Research is needed.	✓	
Step 8				
Incorporating hydraulic head	Will it be possible to calibrate regional-scale hydrogeologic models?	The advective models used in evaluating the 1,000-year travel time criterion are much simpler than those envisaged for full transport analyses, yet it is still not clear whether it will prove possible to calibrate them under the limited-data conditions expected. Experience in model calibration at this scale is limited.		✓
	Can a formalized inverse technique be adopted?	Inverse methods were developed primarily for surficial aquifer models. It is not yet known if these methods can be applied to far-field simulation models where the number of boreholes will be limited. It is not known whether there is likely to be sufficient numbers of hydraulic conductivity and hydraulic head measurements to estimate drift and variograms needed in the cokriging-geostatistical solution of the inverse problem. It may be that parametric approaches are more		

Table 1. **Assumptions and limitations associated with geostatistical methodology (continued).**

Step	Assumption or Limitation	Comments	Significance	
			Lower	Higher
		appropriate in cases with limited data. These issues are discussed in more detail in the text.		√
	What is the impact of measurement uncertainty in hydraulic head?	There is no difficulty in including measurement uncertainties in the geostatistical framework. However, many recent field results indicate that such uncertainties may be very large for hydraulic head measurements. The implications for model calibration and inverse simulation should be investigated.		√
	What is the role of subjective prior information in inverse simulation?	Prior information is used to constrain the parameter estimates. Solutions using subjective prior statistics should be compared with those based on classical statistics.	√	
	Should other variables be incorporated in the inverse framework?	Initial results suggest that other data sets, such as temperatures or geochemical patterns may be useful in constraining the estimates of hydrogeological parameters.	√	

The updating properties of the Bayesian approach, which can be carried out each time new data become available, are well suited to the iterative approach we espouse for the modeling/data-gathering sequence at a site. We feel that the first modeling efforts should precede or accompany initial site investigations. They can aid in the design of such investigations, and should not come on line only after such investigations are complete. Bayesian analysis is well suited to the evaluation of the worth of additional data, by incorporating into the decision process the cost of data collection and the value of the risk reduction it provides.

The potential paucity of data in nuclear waste repository investigations arises from cost considerations and from the desire not to alter the integrity of the geologic barriers by a large number of boreholes, which may in the future, if improperly sealed, lead to pathways for the migration of radionuclides. For this reason, indirect measurements in the form of geophysical surveys will be more readily available than direct data. Bayesian analysis is well suited to making best use of data with large error bounds as will be the case with the interpretation of geophysical data.

As a final positive argument, consider the time frame of the geologic disposal of nuclear waste. Implicit in our calculations is the assumption that a fractured porous medium, as we observe it today, will be unchanged over 1,000 or even 10,000 years. This basic assumption is questionable; in such a time frame, chemical precipitation or dissolution occurring naturally may alter the medium, or minor tectonic movements may slightly change the fracture apertures. It is therefore debatable whether a very precise characterization of the present situation is really representative of the future situation. A more generic ''prior'' estimation may be equally useful.

On the negative side, there is the subjectivity of the prior estimates, and of the form of the prior and the likelihood in the unconditional updating step. This subjectivity could easily be abused. Some scientists and engineers may feel that using subjective priors is unscientific or unprofessional. There is a question as to public and political acceptance of such techniques. The methodology is open to the charge that decision makers are trying to pretend that an unsatisfactory data base is satisfactory.

What is clear is that posterior probabilities are sensitive to whether classical statistics is used or a Bayesian approach is used. The ramifications of this decision—scientific, philosophical, and political—deserve investigation.

Applications of inverse simulation

Inverse simulation is a methodology for combining concepts of geostatistical analysis and model calibration. Most examples of inverse simulation reported in the literature are applied to two-dimensional, surficial aquifers. The data base commonly contains measurements from many tens of boreholes, and includes both hydraulic head and hydraulic conductivity estimates. Even in these instances, authors describe a variety of limitations in taking full advantage of the potential offered by inverse theory. Difficulties in the application of inverse simulation are related to (1) uncertainties in choosing an interpretational model for which parameter estimates are sought, (2) the number and location of measurement points, and (3) measurement errors in hydraulic head and hydraulic conductivity. These limitations are the same as those faced in more traditional trial-and-error calibrations, although their impact may not be as obvious as in a formalized inverse simulation.

Questions related to uniqueness and stability of the solution must be considered in every inverse simulation. If such problems occur, they may be due to selection of an inappropriate hydrologic model or to an attempt to identify parameters in a greater number of zones than can be justified on the basis of the available hydrologic data. Experience suggests that with a sparse data base, or boreholes irregularly distributed across the flow domain, it may only be possible to identify unique parameter values for simple models based on a small number of zones. The application of such models in probabilistic travel time calculations needs to be evaluated.

Noise in hydraulic head data originates from technical limitations in making accurate measurements of in situ pressures in deep boreholes, from errors introduced if long completion intervals are used in piezometers, and from perturbations in hydraulic head caused by heterogeneities in hydraulic conductivity on a scale smaller than that modeled in the inverse simulation. The greater the noise in the hydraulic head data, the smaller the reduction in estimation error that can be achieved with the inverse. Furthermore, with greater noise in the data, it may become more difficult to identify a unique set of model parameters. Underestimation of measurement error leads to misplaced confidence in model calculations. Studies are needed to provide realistic assessments of the accuracy of pressure measurements in deep piezometers.

Recent advances in equipment design and in multiport piezometers now permit numerous measurements of hydraulic head within single boreholes. It is not yet known to what extent detailed measurements of hydraulic head within single boreholes can play a role in reducing the total number of boreholes required for repository site assessment. Resolution of model parameters can potentially be improved if additional data sets, such as subsurface temperatures or geochemical data, are used as constraints in the inverse simulation.

A modeler introduces his personal judgments in an inverse simulation during the design of the interpretation model, in specifying upper and lower bounds on parameter values, and in assigning measurement error at data points. It is an important question to consider to what extent subjective prior information can be used in defining $\{\mu_i\}_c'$ and $[\sigma_{ij}^2]_c'$, while still obtaining meaningful results from the inverse approach. A related question concerns the relationship between the scale at which the heterogeneity in K is described using subjective prior information with Bayesian updating (Steps 1, 2, 3, and 5), and that which can be justified on the basis of hydrogeologic calibration (Step 8). Neither question has been resolved in the context of a nuclear waste site.

For near-field simulation at a repository site, with data from vertical shafts, horizontal adits, and their associated boreholes, there is likely to be a sufficient data base to apply inverse simulation techniques to aid in the resolution of hydraulic conductivity values. It is less clear at this time that for far-field simulation, with data from a limited number of widely spaced boreholes, that a formalized inverse method can be used successfully to build a predictive model of sufficient detail to provide meaningful travel time calculations, and to reduce uncertainties in model parameters.

Identification of trends and zonations of hydraulic conductivity

With the small number of boreholes expected, we anticipate difficulty in accurately locating the boundaries of hydrogeologic units and trends in hydraulic conductivities within hydrogeologic units. In layered sedimentary environments where layers are continuous and extensive, this may not be a serious problem, but in fractured igneous rock (both plutonic and volcanic) complex and unexpected discontinuous features may prove difficult to confirm, and even more difficult to characterize geostatistically. Such features may exert considerable influence on travel time. Current geostatistical methods do not sufficiently emphasize the uncertainty spawned by potential geologic misinterpretations.

One possible avenue of research would be to try to integrate the stochastic stratigraphy models used by petroleum geologists into our methodology. They use spatial stochastic process models to generate suites of equally likely stratigraphies in much the same way as hydrogeologists generate equally likely conductivity fields. Perhaps a marriage of these two approaches would be fruitful.

In fractured media, it is often assumed that hydraulic conductivity decreases with depth. In such cases, three possibilities are available: (1) zoning, (2) trend analysis, and (3) nonstationary geostatistics. The latter approach is described in de Marsily (*5*). All of them can, in principle, be carried out in a Bayesian framework where parameters are estimated a priori and updated a posteriori, but in all cases, the limited data base will limit the effectiveness of that approach.

Problems associated with small-sample statistics

Throughout Table 1 and in the discussions in this section, we have repeatedly drawn attention to the potential problems associated with the geostatistical methods (Bayesian or otherwise) when data networks are sparse and sample sizes small. In our opinion, this is the potential Achilles heel for the application of geostatistics at nuclear repository sites.

Among the steps and substeps where this problem may arise are the following:

1. Testing the hypothesis of a lognormal distribution for hydraulic conductivity.
2. Testing for stationarity of the hydraulic conductivity field, and/or identifying trends and zonal boundaries.
3. Determining the most appropriate form of the autocorrelation function and estimating the values of its parameters.
4. Identifying the existence of embedded structures.
5. Obtaining significant uncertainty reduction in the Bayesian updating step.
6. Application of the inverse procedures.

The stochastic approach is not a substitute for an inadequate data base. There is a need for much more research in the area of small-sample statistics.

As an example of the problem, consider that Journel and Huijbregts (*7*) suggest that a minimum of 30 pairs of points is necessary to estimate each value of the covariance function as a function of the lag and this would presumably be true for all formations at a site. This scale of data network is highly unlikely at a potential repository site.

In recent years, several efforts have been made to improve the robustness of the estimation of covariance functions or variograms (cf. Verly et al., *21*). One technique of interest in the Bayesian approach is the jackknife method, which gives an estimate of the "jackknife variance" of the value of a fitted parameter. It may be used as a measure of the uncertainty associated with the experimental determination of the parameter in the Bayesian updating equations.

Final Comment

We wish to reemphasize that we have limited ourselves in this paper to the consideration of advective transport, with hydraulic conductivity as the only uncertain hydrogeologic parameter. The framework we have set up is equally suited to transport that involves advection, dispersion, retardation, and decay. The additional parameters that arise, such as dispersivity, retardation factor, and the like, all exhibit spatial variability that leads to uncertainty, and they are all amenable to the geostatistical methodology we have described. This is true, both with respect to the conceptual strengths of the methodology and with respect to the potential operational weaknesses we have identified.

Literature Cited

1. Codell, R., *Draft generic position on groundwater travel time,* U.S. Nuclear Regulatory Commission, Washington, D.C. (1986).
2. Andrews, R. W., V. A. Kelley, J. A. McNeish, A. M. LaVenue, and J. E. Campbell, *Travel path/travel time uncertainties at salt sites proposed for high level waste repositories,* INTERA Technologies, Inc., TR-36, Austin, TX (1985).
3. Clifton, P. M., *Groundwater travel time uncertainty analysis,* Rockwell Hanford Operations, RHO-BW-ST-70-P, Richland, WA (1985).
4. Lin, Y. T., M. S. Tierney, and S. Sinnock, *Preliminary estimates of groundwater travel time and radionuclide transport at the Yucca Mountain repository site,* Sandia National Laboratories, SAND85-2701, Albuquerque, NM (1985).
5. Marsily, G. de, *Quantitative Hydrogeology,* Academic Press, New York, NY (1986).
6. Vanmarcke, E. H., *Random Fields: Analysis and Synthesis,* M.I.T. Press, Cambridge, MA (1983).
7. Journel, A. G. and C. Huijbregts, *Mining Geostatistics,* Academic Press, New York, NY (1978).
8. Gelhar, L., "Stochastic analysis of flow in heterogeneous porous media," in *Fundamentals of Transport Phenomena in Porous Media,* edited by J. Bear and Y. Corapcioglu, NATO ASI Series E, No.82, 657–717 (1984).
9. Dettinger, M. D. and J. L. Wilson, *Water Resources Res., 17,* 149–161 (1981).
10. Smith, L. and F. W. Schwartz, *Water Resources Res., 17,* 351–369 (1981).
11. Hachich, W. and E. H. Vanmarcke, *J. Geotech. Engin. Div. Proc. Am. Soc. Civil Engins., 109,* 373–385 (1983).
12. Massmann, J. and R. A. Freeze, *Water Resources Res., 23,* 351–367 (1987).
13. Massmann, J. and R. A. Freeze, *Water Resources Res.* (in press).
14. Benjamin, J. R. and C. A. Cornell, *Probability, Statistics and Decision for Civil Engineers,* McGraw-Hill, New York, NY (1970).
15. Kulkarni, R. B., "Bayesian kriging in geotechnical problems," in *Geostatistics for Natural Resources Characterization, Part 2,* G. Verly et al., eds., NATO ASI Series C, 182, Reidel Publishing, Dordrecht, The Netherlands, 775–786 (1984).
16. Massmann, J. and R. A. Freeze, *Water Resources Res., 23,* 368–380 (1987).
17. Smith, L. and F. W. Schwartz, *Water Resources Res., 17,* 1463–1479 (1981).

18. Clifton, P. M. and S. P. Neuman, *Water Resources Res.*, *18*, 1215–1234 (1982).
19. Yeh, W., *Water Resources Res.*, *22*, 95–108 (1986).
20. Carrera, J. and S. P. Neuman, *Water Resources Res.*, *22*, 199–210 (1986).
21. Verly, G., M. David, A. G. Journel, and A. Marechal, editors, *Geostatistics for Natural Resources Characterization*, Part 1, NATO ASI Series C, 182, Reidel Publishing, Dordrecht, The Netherlands, 1–140 (1984).
22. Iman, R. L., *Risk methodology for geological disposal of radioactive waste: A distribution-free approach to include rank correlation among input variables for simulation studies*, Sandia National Laboratories, Report SAND80-0157, Albuquerque, NM (1980).
23. Merkhofer, L. and A. Runchal, "Probability encoding: Quantifying judgemental uncertainty over hydrologic parameters for basalt," DOE/AECL Symposium on Geostatistical Sensitivity and Uncertainty Methods for Ground-Water Flow and Radionuclide Transport Modeling, this volume (1987).
24. Matheron, G., *Eléments pour une Théorie de Milieux Poreux*, Masson, Paris (1967).
25. Dagan, G., *Water Resources Res.*, *18*, 813–833 (1982).
26. Kitanidis, P. K. and E. G. Vomvoris, *Water Resources Res.*, *19*, 677–690 (1983).
27. Feinerman, E., G. Dagan, and E. Bresler, *Water Resources Res.*, *22*, 935–942 (1986).

Uncertainty in Forecasting Breakthrough of Fluid Transported Through Fractures

ROLAND N. HORNE
Department of Petroleum Engineering
Stanford University

Abstract

Tracer experiments in geothermal reservoirs emphasize the very great variability in rates of fluid movement through fractured rocks. This variability extends from the 10-meter to the kilometer-length scale. Thus tracer returns have been observed at some locations within hours at distances of up to 1 kilometer from the injection point, while other much nearer locations in the same formation do not observe the tracer until much later. In addition, transport rates have sometimes been extremely fast (up to 100 m/hr) even over such distances.

This paper discusses the conclusions reached after compiling the results of a large number of tracer tests in several different fractured reservoirs. It is evident in some cases that large-scale geological features, such as faults, are responsible for the variations in tracer return time. In other cases, there is no clear physical description that explains the differences. These results suggest that there will be no a priori way of forecasting transport rates in fractured systems without performing a tracer test.

Introduction

The movement of fluids through permeable rocks is of primary interest whether the fluid is groundwater, petroleum, geothermal steam, or a radioactive or chemical contaminant. The geothermal industry is involved in the transport of fluid through fractured volcanic rocks, and the experience accumulated over the past 30 years in the production and reinjection of steam and hot water from geothermal reservoirs has implications of broad interest in other fields.

That the porosity and permeability of geothermal reservoir rocks exist mainly in the fractures has been known for many years. Measurements of the permeabilities of cores recovered from geothermal wells are extremely low (of order 10^{-17} m^2) and are not characteristic of permeabilities inferred from well tests (which can be as large as 10^{-13} m^2). Porosities of cores are similarly low. Thus it has been concluded that, although fluid may be stored within the rock matrix, it is mobile primarily through the fractures only (at least within the time scale of interest to geothermal developers).

In spite of this knowledge, it was originally assumed that the degree of heterogeneity represented by the fractures was such that on the scale of an entire reservoir (which might cover tens of square kilometers), the flow would be as if through a porous medium. Calculations of heat and fluid recovery were traditionally made on this basis in the 1960s and early 1970s.

The recognition of the importance of the fractures coincided with the evolution of interest in reinjection. Until 1972, the waste hot water from geothermal power developments was disposed of to surface waters or ponds. Because the water was in very large quantities (thousands of tonnes per hour) and contained toxic substances (Axtmann [*1*]), surface disposal presented environmental difficulties. In addition, the loss of fluid from the reservoirs resulted in substantial drawdown of pressures, and there was concern that fluid reserves would be depleted far in advance of the recovery of the usable heat still contained in the rock. For these two reasons, reinjection was suggested as a means to prolong the useful life of the resource as well as a means to avoid the release of contaminants into the environment.

In early work involved in the design of a reinjection scheme for the Ahuachapán geothermal field in El Salvador, two significant new insights into geothermal development were obtained. A tracer test, using tritium, demonstrated that flow in the reservoir was highly heterogeneous (Einarsson et al. [*2*]). Tracer was recovered in a producing well 400 m distant from the injection well within 48 hours, while two similar wells did not receive any tracer until several months later (one well was 500 m distant from the injector, and the other was 1000 m). Based on these observations, Bodvarsson [*3*] made calculations on the "safe distance" between injectors and producers such that there would be no premature breakthrough of unheated injection water back into the production wells. Bodvarsson [*3*] used a model of flow in a fracture to conclude that a "safe distance" could be anywhere from 700 m to 4500 m, depending on the degree of fracturing in that particular direction. Thus it had been determined that (1) fractures caused heterogeneities in flow over scales as large as 1 kilometer, and (2) tracer tests were very useful in determining where these heterogeneities lie.

The work described by Einarsson et al. [*2*] and Bodvarsson [*3*] involved only three wells at Ahuachapán, and because large-scale reinjection into geothermal reservoirs only became widespread in the early 1980s, the full significance of their observations as to the fractured nature of geothermal reservoirs did not become evident until several years later. Due to the channeling of reinjected water through relatively small volumes of rock, unheated water was being returned to production wells in several different geothermal fields. The resulting loss of productivity became a serious concern to geothermal developers, who began to take pains to overcome the problem and avoid it in subsequent developments (Horne [*4*]).

Since the late 1970s, a large number of tracer tests have been reported for both existing and newly developed geothermal reservoirs. Collectively, these tracer tests confirm the original observations of Einarsson et al. [*2*] and have shown that fractures in the volcanic rocks (and in some cases sedimentary rocks) can be major conduits for flow and are different within reservoirs as well as between one reservoir and another. It has been seen that water can flow for distances as far as 1 kilometer at speeds of up to 100 m/hr. This has created significant uncertainty in the process of field development and has given birth to major research efforts in several countries. This paper summarizes the implications of the geothermal experience and the research into forecasting that has resulted from it.

The first part of the paper describes the effects of individual fractures crossing a wellbore, while the second illustrates the effects of fieldwide heterogeneity on tracer returns. The consequences of the effects of the preferential flows are discussed in the third section.

Fractures Intersecting Wells

The spacing between fractures within a geothermal reservoir is highly variable. However, because very large amounts of flow are required to develop a geothermal well economically, usually attention only focuses on the largest fractures. Large fractures intersecting the wellbore are commonly also known as "feed zones." When it comes to assessing feed zones, all (commercial) geothermal wells have at least one, and most have several more, over their productive length (which can vary from 300 m to 2000 m). Thus in terms of typical separations between major productive fractures, it is not an unreasonable generalization to say that they lie hundreds of meters apart, at least in the vertical sense.

It is another feature of active geothermal reservoirs (at least those of the liquid-dominated type) that the fluids are in upward motion due to natural convection over the heat source. As a consequence of this, the pressure gradient in the reservoir as a function of depth is almost always greater than hydrostatic pressure (typically by about 10%). This gives rise to an imbalance of pressures when the reservoir is penetrated by an open wellbore, which obviously can support a hydrostatic pressure gradient only if the fluid in the wellbore is stationary. It is therefore very common for geothermal wells to flow internally, usually upwards, from one feed zone to another, even when the well is closed at the surface.

These internal flows are a considerable nuisance when trying to make measurements of reservoir pressure and temperature, because they indicate that the well pressure is not the same as the reservoir pressure (except if there is only one feed zone, and even then only at that one location). The temperature of the fluid in the well is also not the same as that in the formation, because over the flowing section of the well the temperature will be constant (over the interval between the inflow and outflow points) and equal to the temperature of the inflowing water from only one feed zone (see Figure 1). At best, the reservoir pressure and temperature can be determined only at a single depth, namely that of the principal feed zone.

In spite of the fact that fluids produce into the wellbore only at isolated points, many standard well-testing interpretation techniques can be and are commonly applied to geothermal well tests. Many of these methods are based on a model of pressure transients in an isotropic, homogeneous porous medium. Although it is also common to see pressure transient responses characteristic of fractures (Ramey and Gringarten [5]), and sometimes double porosity behavior (Deruyck et al. [6]), at least half of the geothermal well tests show uniform porous medium, radial flow

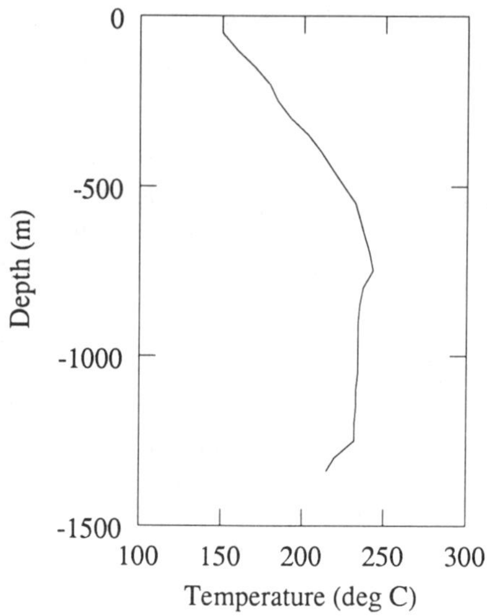

Figure 1. **Temperature profile for well M-9, Cerro Prieto, Mexico.**

response. This may lure us into the belief that fractures are so multiply connected out in the reservoir that pressure is transmitted more or less radially beyond a certain distance. This belief is further strengthened by observations from interference tests from one well to another, in which standard radial flow solutions are validly applied even more commonly.

That major heterogeneities are not evident in pressure transient tests is due to the diffusive nature of pressure transmission through permeable rocks. Sageev and Horne [*7*] demonstrated that a large impermeable body will affect the pressure interference at a well on the opposite side of the obstruction by only an imperceptible amount, even if the body has a radius ten times its distance from the wells. Thus, although pressure transient tests can be interpreted correctly to provide estimates of the permeability averaged over a large volume of the reservoir, the fracture permeability can be missed in a single interference test, even though the fractures may extend very large distances. It is only in the analysis of multiple interference tests that large-scale fracture zones become evident. Even though the tests themselves may show classical radial flow responses, the permeabilities between well pairs can be radically different.

Two examples of this have been published. The first was for the Beowawe geothermal reservoir in Nevada by Epperson [*8*], in which strong interference was observed along the Malpais fault but lesser effect perpendicular to it. The second example (Hoang et al. [*9*]) shows this more clearly for the Heber geothermal field in Southern California, in which a map of permeabilities inferred from well tests clearly shows high values along two fault zones. This second example is of additional interest since the Heber geothermal reservoir lies in a sequence of sandstones and shales, unlike most geothermal reservoirs which are in volcanic rocks. Until development began, it was assumed that the Imperial Valley reservoirs would behave much more uniformly than their counterparts in volcanic zones. More recently it has become evident that fractures and fracture zones play an important part in the performance of these sedimentary reservoirs too.

The paradoxical behavior of flow systems that seem relatively uniform in pressure transient tests but which experience rapid and major breakthrough of injected material through fractures is caused by the fundamental difference between two mechanisms. Well tests monitor the transmission of pressure pulses through the reservoir, whereas fluid breakthrough arises due to physical transport of the fluid mass. Pressure transients are a diffusive phenomenon and obey the diffusion equation (slightly compressible pressure transmission equation). Fluid flow, on the other hand, is a convective phenomenon and obeys the convection equation. Thus we expect measurements of the mass flow to expose much more detailed effects of the transport process. This expectation has been borne out in the tracer test observations described in the next section.

Fractures Between Wells

The pressure transients measured in a well test are transmitted through all of the connected fluid in the reservoir, whether the fluid is moving or stationary. A tracer test, on the other hand, monitors only that portion of the fluid that is actually moving between the point at which the tracer is injected and the points at which it is being recovered. Fluid that is immobile, or which is moving towards a location that is not being monitored, is invisible to the tracer test. However from the point of view of reinjection design, it is most useful to determine where the injected fluid is going with reference to the existing production wells. The portion of the traced fluid that does not return to the production wells is of lesser significance because it holds little concern of premature thermal breakthrough. For this reason, well-to-well tracer tests have been much more common than the injection-backflow kind of tracer test used in groundwater applications to estimate regional flows.

Discussed here are some typical responses of geothermal tracer tests in fractured geothermal reservoirs, based on results from El Salvador (Aumento et al. [*10*], Cuellar et al. [*11*], Einarsson et al. [2]); Japan (Horne [*12*], Inoue and Shimada [*13*], Ito et al. [*14*] and [*15*]); Iceland (Gudmundsson et al. [*16*]); the Philippines (Dobbie and Menzies [*17*], Sarit [*18*], PNOC [*19*]); and New Zealand (McCabe et al. [*20*]). Extensive tracer testing has also been carried out in the vapor-dominated geothermal system at the Geysers, California (Gulati et al. [*21*]), although published details are not as complete as those from other fields.

In general, the results of these tracer tests emphasize the strongly heterogeneous flow paths created by fractures—usually coincident with faults. Figure 2 from McCabe et al. [*20*] shows this particularly clearly for the set of tracer tests performed at Wairakei geothermal field in New Zealand. Figure 3 shows a tracer response characteristic of those at Wairakei (and most other geothermal fields). There is a single, sharply defined peak, suggesting only a single major flow path. Notice also the very rapid arrival time. Table 1 summarizes the responses of these various tests with respect to first tracer arrival, peak tracer arrival, peak tracer concentration, fraction of tracer recovered, and horizontal and vertical separation

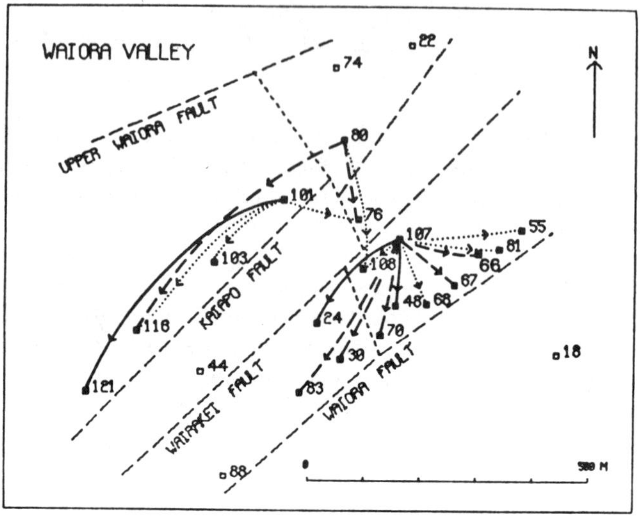

Figure 2. **Schematic representation of distributions to be sampled and how an individual particle moves through the continuum. Reprinted with permission (*19*).**

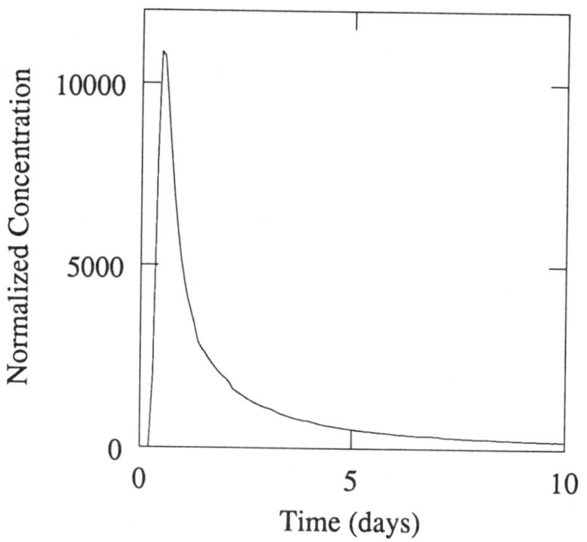

Figure 3. **Comparison of breakthrough curves generated using the discrete fracture model and the continuum model for four different cases of one-dimensional flow. Reprinted with permission (*19*).**

Table 1. **Summary of Wairekei tracer results.**

	First Arrival	Peak Arrival	Peak Concentration	Fractional Recovery	Horizontal Distance	Vertical Distance
WK80→76	4.0	8.7	88	0.0024	142	+119
80→108	5.5	10.0	16	0.0006	229	−33
80→116	3.3	7.6	230	0.0040	499	+97
WK101→76	2.5	7.0	10	0.0005	139	+114
101→103	2.0	5.0	30	0.0009	168	+140
101→116	2.5	7.5	23	0.0005	350	+92
101→121	1.2	2.5	10,500	0.0580	489	+585
WK107→24	0.2	0.4	10,000	0.0373	209	+389
107→30	4.5	9.0	55	0.0028	238	+236
107→48	0.3	0.7	2360	0.0133	117	+617
107→55	5.5	15.7	29	0.0018	216	+290
107→67	2.2	15.3	46	0.0032	126	+248
107→68	4.0	15.0	39	0.0007	124	+214
107→70	4.0	9.5	43	0.0025	174	+182
107→81	4.8	9.5	21	0.0009	178	+270
107→83	4.5	11.0	53	0.0034	326	+167
107→108	10.0	23.0	17	0.0001	84	−7

between injection and production points. Lovekin and Horne [22] used these results to optimize the hypothetical reinjection scheme using these wells. It was found that the optimum combination of injectors and producers selected was almost the same based on any of the tracer return parameters. On the other hand, using horizontal separation alone (without including tracer test results) gave an entirely different design that failed to avoid combinations of wells in which breakthrough of injected water had been rapid. This simple observation emphasizes our appreciation of the fact that the reservoir is not areally homogeneous.

Interestingly, Lovekin and Horne [22] found that using the vertical separation between injection and production points gave rise to the same optimum selection of wells as did the use of the tracer test results. Table 2 shows why this was so; all of the tracer test parameters characteristic of strong and rapid tracer breakthrough (early arrival, high concentration, large fractional recovery, and early peak arrival) are well correlated with each other and with the vertical distance between wells. The correlations are negative for arrival times; this would imply that arrival time is small if the distance is large, which is counterintuitive. However, since the correlation coefficients are not large, this means only that the first arrival and peak arrival are largely independent of vertical distance. On the other hand, there is a much stronger correlation between the fraction recovered and the vertical distance, suggesting that the fluid is generally moving downward, presumably due to negative buoyancy of the heavier, cooler water.

Notice in Table 2 that the horizontal separation is uncorrelated with any other parameter, i.e., the tracer return results are completely independent of areal separa-

Table 2. **Correlation matrix for Wairakei tracer results.**

	First Arrival	Peak Arrival	Peak Concentration	Fraction Recovered	Horizontal Distance	Vertical Distance
First Arrival	1.0000	0.8533	−0.5272	−0.5156	−0.2336	−0.6546
Peak Arrival		1.0000	−0.5753	−0.5573	−0.3448	−0.5746
Peak Concentration			1.0000	0.9760	0.3624	0.6686
Fraction Recovery				1.0000	0.4519	0.7117
Horizontal Distance					1.0000	0.1397
Vertical Distance						1.0000

tion of the wells. This result is even more significant than the correlation table shows, because the table includes only wells for which tracers actually broke through in measurable quantities. Thus there are several other monitored wells within similar distances for which no tracer was recovered. Thus the tracer returns are even less dependent on separation than is evident in the table.

Even though the tracer returns are independent of distance, they are not independent of location. Figure 2 shows that there is a distinct correlation between the fault locations and the strong returns. The seemingly paradoxical behavior along the Kaiapo fault in which the largest return (indicated by the solid arrow line) is over the largest distance is explained by the fact that well 121 is the deepest. McCabe et al. [*20*] explain the high recoveries in wells 24 and 48 (both of which are deep) as a flow down the Wairakei fault and back up the Waiora fault, postulating that the faults intersect at depth.

Another aspect of the fracture flow was evident in a series of tracer tests conducted in the Tongonan geothermal field in the Philippines (Figure 4). In June 1981, I^{131} injected into well 4R1 was recovered at wells 404, 401, and 108. A total of 16.29% of the injected tracer was recovered (compared to a maximum of only 6% in any of the Wairakei tests described above), with 11.45% being recovered in well 404 alone; 2.84% was recovered in well 401, and about 2% in well 108 (PNOC [*19*]). The arrival times indicate a minimum tracer speed of 57 m/hr for well 404, 30 m/hr for well 401, and 22 m/hr for well 108. The tracer responses at Tongonan differed from those at Wairakei in that each return was characterized by two or more peaks, whereas all but one of the Wairakei returns showed only one peak.

Tracer tests at another well at Tongonan, well 2R2, emphasize this two-path effect. Unlike the Wairakei tests, in which the tracer was injected at the well head, the Tongonan tests were performed by releasing the tracer downhole using a wireline sampling bottle. In March 1981, I^{131} was released in well 2R2 at the depth of its upper feed zone (400 m). In June 1981, a second test was conducted in which tracer was released into the lower feed zone at 1300 m depth. In the March test (upper feed zone), positive returns were measured at well 213 with a peak arrival at 19 hours and a recovery of 0.34%. In the June test (lower feed zone), much more of the tracer was recovered (1.68%) at well 213, but the first concentration peak did not arrive until 4.4 days. These results emphasize the individual behavior of single fractures intersecting the wellbore and demonstrate that the flow paths are not neces-

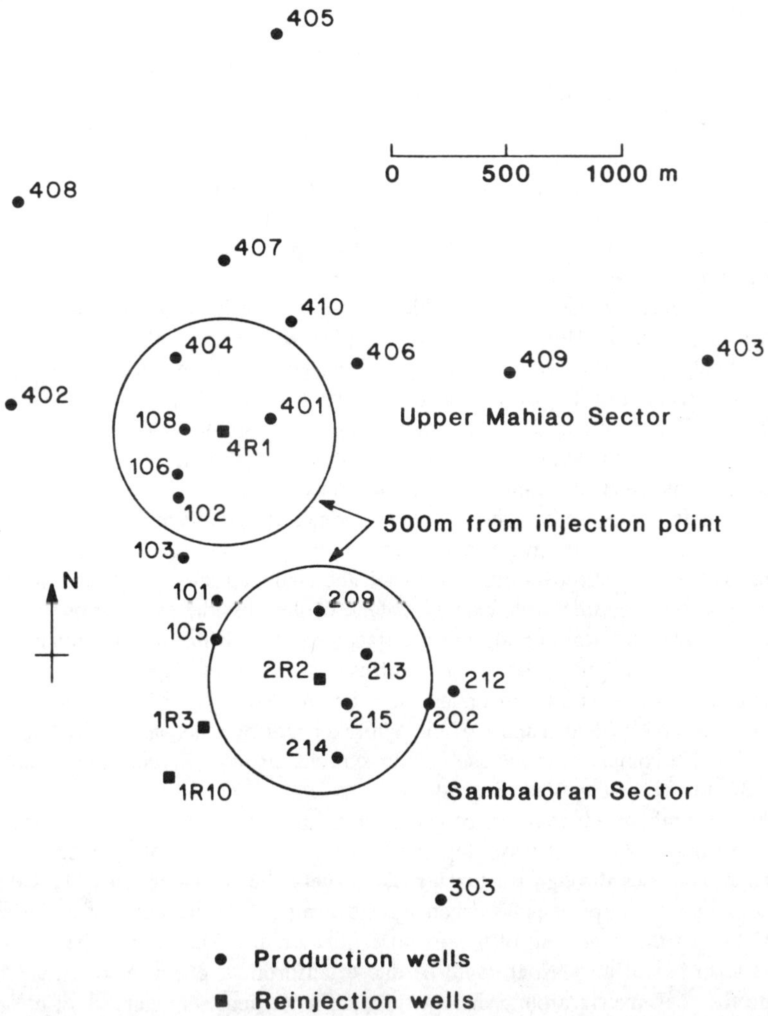

Figure 4. **Well location map of Tongonan geothermal field, the Philippines.**

sarily connected out in the reservoir (although the surface locations of wells 2R2 and 213 are only about 200 m apart). The injection rate into well 2R2 was constant at 200 tonnes/hr (Sarit [*18*]).

There are other examples of results similar to those described in these two geothermal fields. Many of these are available in the open literature, although many are still confined to proprietary company files and reports. The selection of these two examples is not intended to be comprehensive, but simply to point out some of the salient aspects of the problem.

Implications of Fracture Flow

Based on the observations described in the two previous sections, we are able to formulate a conceptual model of the way in which injected fluid (and perhaps naturally occurring fluid, too) flows through fractured geothermal reservoirs. The fractures clearly provide the major conduits for flow, both at the scale of a single wellbore as well as that of the entire reservoir. Fluids move along planar fault zones and are frequently constrained to flow in straight paths from one side of the reservoir to the other. Thus we can postulate that the tracer responses should be characteristic of linear flow in a planar fracture.

Several different models are available to describe the linear flow of a dissolved substance in a fracture. Horne and Rodriguez [*23*] demonstrated that Taylor dispersion was likely to be the dominant dispersive mechanism in the fracture itself and derived an extension of Taylor's dispersion coefficient (Taylor [*24*]) to include the linear planar flow configuration. This model was then used to model the tracer return profiles from the Wairakei tests by Fossum and Horne [*25*]. Although the profiles could be matched, matching was possible only by considering two separate paths with their corresponding responses superimposed upon each other. This was an unsatisfying result since it did not correspond to the basic concept. The inclusion of a second path was required in order to match the characteristic long "tail" in the return profile (see Figure 3, for example). One of the possible explanations for this delay in arrival of the trailing edge of the tracer spike is diffusion (Neretnieks [*26*]). Jensen and Horne [*27*] used the "matrix diffusion" model of Neretnieks et al. [*28*] to provide a response with a longer tail than the pure convection-dispersion model, and they were able to match most of the Wairakei returns with only a single path. Figures 5 and 6 compare the matches of Fossum and Horne [*25*] and Jensen and Horne [*27*] to one of the Wairakei tests.

Thus the high speed transport of geothermal water containing tracers appears to follow the same model as that of Neretnieks et al. [*28*], which was derived for transport of radionuclides through fractured rocks at very much slower rates. This demonstrates that one of the principal mechanisms governing the transport of the tracer is physical loss of the tracer out of the fracture flow stream. The precise mechanism of this loss term is still unclear. It could be due to adsorption, chemical reaction, diffusion into the rock matrix around the fracture, diffusion into stationary fluid in other fractures, degradation of the tracer, or ion exchange. All of these mechanisms are capable of giving rise to similar terms in the differential equations governing the model. Using a laboratory core with a fracture created along its axis, Johns [*29*] found that more tracer was retained in the core than flowed through. Research to pinpoint the site and mechanism of the holdup is ongoing.

Summary

Geothermal reservoirs are by their very nature highly permeable formations of rock. In reservoirs of commercial interest, fluids are produced in very large quantities and must flow at very high rates. It is not surprising, therefore, that almost all commercial reservoirs have major flow conduits within them. Based on drilling

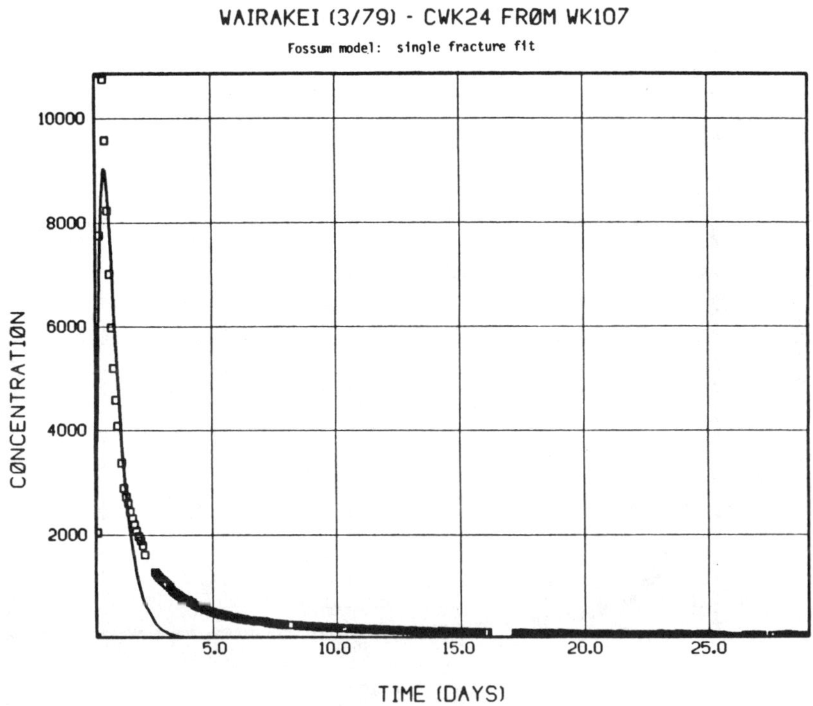

Figure 5. **Single path match of convection-dispersion model to WK24 tracer response, from Fossum and Horne [25].**

experience and on the results of tracer tests such as those described here, it is clear that faults and fractures provide the most permeable paths through the reservoir. That these paths provide short-circuit routes for injected fluids to pass quickly through the reservoir is of great commercial significance.

Many of the tracer returns have been easily correlated with faults previously recognized from geological mapping and from drilling records. There are counter-examples, however, in which the faults have been inferred from the tracer tests themselves. Remembering the example of the two kinds of response in well 2R2 in Tongonan, it is evident that the forecasting of injected water breakthrough would require the knowledge of the position of all major faults in three dimensions throughout the reservoir. Because this is usually only partially realizable, there seems no alternative but to carry out a series of tracer tests to determine these connections.

Acknowledgment

The writing of this paper has been funded by the U.S. Department of Energy under contract number DE-AT03-80SF11459 to the Stanford Geothermal Program.

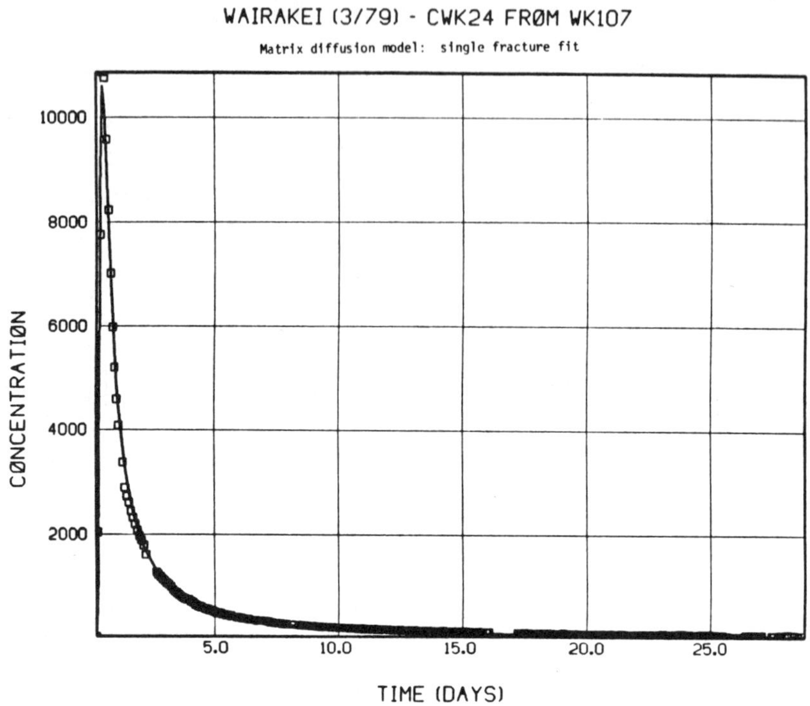

Figure 6. **Single path match of matrix-diffusion model to WK24 tracer response, from Jensen and Horne [27].**

Literature Cited

1. Axtmann, R. C., "Environmental Impact of a Geothermal Power Plant," *Science, 187,* 795–803 (19).
2. Einarsson, S. S., Vides, R. A., and Cuellar, G., "Disposal of Geothermal Waste Water by Reinjection," *Proceedings, 2nd U.N. Symposium on Geothermal Energy,* San Francisco, CA, 1349–1354 (1975).
3. Bodvarsson, G., "Thermal Problems in the Siting of Reinjection Wells," *Geothermics, 1,* 63–66 (1971).
4. Horne, R. N., "Reservoir Engineering Aspects of Reinjection," *Geothermics, 14,* 449–458 (1985).
5. Ramey, H. J., Jr., and Gringarten, A. C., "Effect of High-Volume Vertical Fractures on Geothermal Steam Well Behavior," *Proceedings, Second United Nations Symposium on the Development and Use of Geothermal Resources,* San Francisco, CA, May 20–29, 1759–1762 (1975).
6. Deruyck, B. G., Bourdet, D. P., DaPrat, G., and Ramey, H. J., Jr., "Interpretation of Interference Tests in Reservoirs with Double Porosity Behavior—Theory and Field Experiments," paper SPE 11025, *Proceedings, SPE Annual Fall Conference and Exhibition,* New Orleans, LA, September (1982).

7. Sageev, A., and Horne, R. N., "Interference Testing: Detecting an Impermeable or Compressible Sub-Region," *SPE Formation Evaluation,* June (1988).
8. Epperson, I. J., "1981 Interference Well Testing, Beowawe, Nevada," *Transactions Geothermal, Resources Council, 7,* 413–416 (1983).
9. Hoang, V. T., James, E. D., and Epperson, I. J., "Heber Geothermal Field: Reservoir and Production Characteristics," SPE 16339, *Proceedings 1987 SPE California Regional Meeting,* Ventura, CA, April 8–10, 151–158 (1987).
10. Aumento, F., Liguori, P. E., Choussy, M., Santana, A., Campos, T., and Escobar, D., "The Geothermal System of the Ahuachapán Field," paper B5 presented at the International Conference on Geothermal Energy, Florence, Italy (May 1982).
11. Cuellar, G., Choussy, M., and Escobar, D., "Extraction-Reinjection at Ahuachapán Geothermal Field, El Salvador," in *Geothermal Systems: Principles and Case Studies* (edited by L. Ryback, and L.J.P. Muffler), Wiley, New York (1981).
12. Horne, R. N., "Geothermal Reinjection Experience in Japan," *J. Pet. Tech. 34,* 495–503 (1982).
13. Inoue, K., and Shimada, K., "Reinjection Experiences in the Otake and Hatchobaru Geothermal Fields, Japan," *Proceedings 7th New Zealand Geothermal Workshop,* Auckland, New Zealand, 69–74 (1985).
14. Ito J., Kubota, Y., and Kurosawa, M., "On the Geothermal Water Flow of the Onuma Geothermal Reservoir," *Chinetsu* (Geothermal Energy), *14,* 15–20 (1977) (in Japanese).
15. Ito, J., Kubota, Y., and Kurosawa, M., "Tracer Tests of the Geothermal Hot Water at Onuma Geothermal Field," *Japan Geothermal Energy Association Journal, 15,* 87–93 (1978) (in Japanese).
16. Gudmundsson, J. S., Hauksson, T., Thorhallsson, S., Albertson, A., and Thorolfsson, G., "Injection and Tracer Testing in Svartsengi Field, Iceland," *Proceedings, 6th New Zealand Geothermal Workshop,* Auckland, New Zealand, November 7–9, 175–180 (1984).
17. Dobbie, T. P, and Menzies, A. J., "Geothermal Waste Water Reinjection Trials, Tongonan Geothermal Field, Republic of the Philippines," *Proceedings, 1979 New Zealand Geothermal Workshop,* Auckland, New Zealand, 134–140 (1979).
18. Sarit, A. D., "Well 2R2 Reinjection Test Results, Tongonan Geothermal Field, Republic of the Philippines," Project Report Geothermal 81.19, Geothermal Institute, University of Auckland, New Zealand (1981).
19. PNOC, "Reinjection Tests at Tongonan Geothermal Site," Philippines National Oil Company Energy Development Corporation Report, Manila, Philippines (1981).
20. McCabe, W. J., Barry, B. J., and Manning, M. R., "Radioactive Tracers in Geothermal Underground Water Flow Studies," *Geothermics, 12,* 83–110 (1983).
21. Gulati, M. S., Lipman, S. C., and Strobel, C. J., "Tritium Tracer Survey at the Geysers," Geothermal Resources Council, *Transactions, 2,* 237–239 (1978).
22. Lovekin, J., and Horne, R. N., "Optimization of Injection Scheduling in Geothermal Fields," *Transactions, Geothermal Resources Council, 11* (1987).
23. Horne, R. N., and Rodriguez, F., "Dispersion in Tracer Flow in Fractured Geothermal Systems," *Geoph. Res. Lett., 10,* 289–292 (1983).
24. Taylor, G. I., "Dispersion of Soluble Matter in Solvent Flowing Slowly Through a Tube," *Proceedings of the Royal Society, London, 219,* 186–203 (1953).
25. Fossum, M. P., and Horne, R. N., "Interpretation of Tracer Return Profiles at Wairakei Geothermal Field using Fracture Analysis," *Transactions, Geothermal Resources Council, 6,* 261–264 (1978).
26. Neretnieks, I., "Diffusion in the Rock Matrix: An Important Factor in Radionuclide Retardation," *J. Geoph. Res., 85,* 4379–4397 (1980).
27. Jensen, C. L., and Horne, R. N., "Matrix Diffusion and its Effect on the Modeling of Tracer Returns from the Fractured Geothermal Reservoir at Wairakei, New Zealand,"

Proceedings, 9th Stanford Geothermal Reservoir Engineering Workshop, Stanford, CA (December 1983).

28. Neretnieks, I., Eriksen, T., and Tahtinen, P., "Tracer Movement in a Single Fissure in Granitic Rock: Some Experimental Results and Their Interpretation," *Water Resources Research, 18,* 849–858 (1982).
29. Johns, R. A., "Experimental Investigation of the Flow of Tracers in Fractured Cores," Report SGP-TR-113, Stanford Geothermal Program, Stanford, CA (June 1987).

Validation of Stochastic Flow and Transport Models for Unsaturated Soils: Field Study and Preliminary Results

THOMAS J. NICHOLSON
Waste Management Branch
U.S. Nuclear Regulatory Commission
Washington, DC

PETER J. WIERENGA
New Mexico State University
Las Cruces, New Mexico

GLENDON W. GEE and
ELIZABETH A. JACOBSON
Pacific Northwest Laboratory
Richland, Washington

DONALD J. POLMANN
DENNIS B. McLAUGHLIN
and LYNN W. GELHAR
Massachusetts Institute of Technology
Cambridge, Massachusetts

Abstract

The U.S. Nuclear Regulatory Commission (NRC) is funding a field study to collect a data base specially designed to validate a stochastic unsaturated flow and transport theory. This theory, described in Mantoglou and Gelhar (*1*), (*2*), and (*3*), was developed earlier at the Massachusetts Institute of Technology (MIT) under funding by the NRC. The stochastic approach is an efficient methodology for characterizing and assessing spatial variability of the hydraulic properties of soils in the unsaturated zone. The important hydraulic properties are the saturated hydraulic conductivity, the specific water content, the soil moisture content, and moisture tension characteristics. They are derived from both laboratory studies using core samples and field measurements using permeameters, suction lysimeters, and tensiometers. At the New Mexico State University (NMSU) field experimental station, an observational trench 4.8 m × 26.5 m and 6 m deep has been excavated. Instrumentation located at various points from the trench face monitored the advancing wetting front and coincident tracers resulting from constant rate infiltration from an irrigation drip system adjacent to the trench. The measured hydraulic properties were analyzed, and directional semivariograms were developed to investigate the statistical characteristics of the hydraulic conductivity data set. The results indicated that the principal direction of anisotropy corresponds to the horizontal and vertical directions of flow. Measured soil moisture contents and tensions due to the advancing wetting showed qualitative agreement between the stochastic simulations and the observed field data, particularly for the horizontal tension distributions. Full validation has yet to be achieved. Further field experiments are being designed and will be conducted to examine transient flow and transport conditions.

Introduction

The NRC has funded a comprehensive field study to validate the stochastic theory developed by investigators at MIT under an earlier NRC funded study. This stochastic theory, developed to describe large-scale transient flow in unsaturated systems [Mantoglou and Gelhar (*1*), (*2*), and (*3*)], and its accompanying "mean flow and transport" models are presently being tested at a field site near Las Cruces, New Mexico. The need to develop the stochastic theory and associated numerical codes is derived from licensing requirements and guidance set forth in the low-level radioactive waste (LLW) siting and performance criteria of Chapter 10 of the Code of Federal Regulations, Part 61 (10 CFR Part 61), which states that an LLW site shall be capable of being fully characterized and modeled. Further, the variability and sensitivity of the data used in site performance assessments should be determined.

The stochastic theory and approach offer an efficient way of characterizing the spatial variability of the hydraulic and transport properties. The present field study is concerned primarily with "mean flow" characterization. Preliminary determination of transport properties derived from conservative tracer experiments is ongoing, while transport property analysis using a nonconservative tracer may be utilized in the future. The field study is designed to measure the variability of important hydraulic parameters in undisturbed soils located in the unsaturated zone. The important hydraulic parameters estimated from the field data are (1) the logarithm of the saturated hydraulic conductivity, $\ln K_s$; (2) the specific water capacity, C; and (3) the slope of the logarithm hydraulic conductivity versus water potential relationship, α.

This paper is organized to provide an overview of the regulatory research needs, validation strategy, stochastic theory, field plan and implementation, the field data including its assessed variability, and the stochastic and deterministic modeling activities to simulate the field site and experimental conditions. Much of the information reported is taken directly from an earlier NRC Technical Contractor Report [NUREG/CR-4622, Wierenga et al. (*4*)] entitled the same as this overview summary paper, and from unpublished Pacific Northwest Laboratory (PNL) research progress reports.

Regulatory research needs

The NRC has been directed by Congress in the Low-Level Radioactive Waste Policy Act of 1980, and the subsequent Amendments Act of 1985, to develop regulatory guidance for assisting the individual states in siting and assessing future LLW disposal facilities. Previously, the NRC issued technical siting and performance requirements in 10 CFR Part 61 for use in LLW sites. These technical criteria encourage LLW disposal in the unsaturated zone (above the regional water table) and also require that the site be capable of being modeled. Previous numerical simulation applications by the NRC staff have been for simplistic two-dimensional conceptual models in spatially homogeneous hydrogeologic units using existing deterministic ground-water codes (e.g., FEMWATER and UNSAT2). Research to examine more realistic conceptual models that would describe three-dimensional flow and solute transport in spatially variable hydrogeologic systems is needed. Also, there exists the need to examine uncertainty and sensitivity aspects of the performance assessment codes, especially the input data and its subsequent numerical simulation results. Therefore, a research study was begun in 1984 at MIT to develop a stochastic theory and the accompanying numerical methods to predict large-scale unsaturated flow and transport within three-dimensional regions characterized by spatially variable hydrogeologic properties. The NRC-sponsored research work performed at MIT under the direction of Dr. Lynn Gelhar has provided both the theory and the models to successfully simulate lateral flow and tension isopleths in heterogeneous unsaturated soils [Mantoglou and Gelhar (*1,2,3*)].

The application of this developed stochastic flow theory and the derived models to actual laboratory and field data is an important aspect of the NRC-sponsored confirmatory research. Therefore, a validation study was funded to develop a specially designed laboratory and field data base for examining and testing the stochastic flow and transport models. The experimental study reported in this paper was developed by the PNL and NMSU investigators, in cooperation with NRC research staff and MIT researchers.

Objectives

The objective of the confirmatory research field-based study is to collect a detailed data base for validating stochastic flow and transport models for unsaturated soils. The theoretical stochastic method; the approach for selecting an adequate data base upon which stochastic models can be tested; and the details of the field program, including the plot design and initial test configurations of the field trench, are all integral parts of the validation process. The data base developed during this study has been useful for testing the detailed stochastic transport theory developed by Gelhar and his associates [Gelhar (*5*), Mantoglou and Gelhar (*6*), Yeh et al. (*7,8,9*)]. The field study has produced a detailed characterization of the hydraulic properties of a heterogeneous layered soil of alluvial origin. The transport of a water-coincident contaminate has also been studied.

Anticipated benefits

It is becoming apparent that conventional deterministic methods and their required input parameters do not adequately account for the variable nature of subsurface flow. For this reason, stochastic models are currently being tested to better characterize subsurface variability of flow and transport.

By using the data sets obtained from the field study it will be possible to test other stochastic models as they become available. Further understanding of these systems utilizing the stochastic models can enable both the licensee and the NRC staff to better quantify the uncertainties associated with subsurface flow systems. The evaluation of such models will assist the NRC staff in assessing the potential for unsaturated flow and transport at nuclear waste facilities [high-level waste (HLW), LLW, and uranium recovery], and in reviewing other stochastic models that may be used in site characterization and performance assessment portions of waste site operator (licensee) reports to the NRC. This evaluation also includes assessment of the adequacy of the site characterization data base, model-selection assumptions, input, and expected uncertainty (e.g., error band ranges) of results.

The stochastic modeling strategy, developed in the MIT research study, has the advantage that it is fully predictive, that is, the input parameters required for the model can be determined without using observations of variables that are being predicted (e.g., tension, moisture content, solute concentration). Thus, the approach avoids the pitfalls of curve-fitting methods and consequently has the prospect of providing predictions at time and space scales which are much larger than are feasible in direct observations for site characterization. Another important feature of this stochastic modeling approach is its ability to predict not only the mean behavior but also a measure of the reliability of the predictions expressed in terms of variance of the predicted quantities.

Validation Strategy

As part of this overall framework, the NMSU field experiments were devised to test the predictive capabilities of such models. The validation process used in the study consisted of the following steps:

1. Collect data on the spatial variability of soil hydraulic properties at the field site.
2. Carry out statistical analyses of these data to determine the parameters for the stochastic model.
3. Develop stochastic model predictions using numerical simulations of the mean flow and solute transport equations. From these mean solutions, the variance of the predicted quantities, moisture tension for example, is also predicted.
4. Observe the distribution of moisture and solute concentration during the field experiment.
5. Compare the observed and predicted moisture and solute distributions in terms of the following validation criteria:
    a. a qualitative comparison of the bulk characteristics; for example, the degree of horizontal spreading of moisture as a result of tension-dependent anisotropy and hysteresis
    b. a comparison of the predicted mean distributions of moisture and solute concentration with the corresponding observations at several times during the experiment, and
    c. a comparison of the variations of the observations around the predicted mean solution with the predicted variances of moisture content.

This last point is a very important element of the stochastic theory because it provides an objective, quantitative acceptance criterion that can be applied to assess the significance of the differences between the model prediction and the actual observations. The overall process of model validation is illustrated schematically in Figure 1. Note that because the stochastic modeling approach is truly predictive, it can be used to make preliminary predictions as was done in the design of the large-scale field experiment.

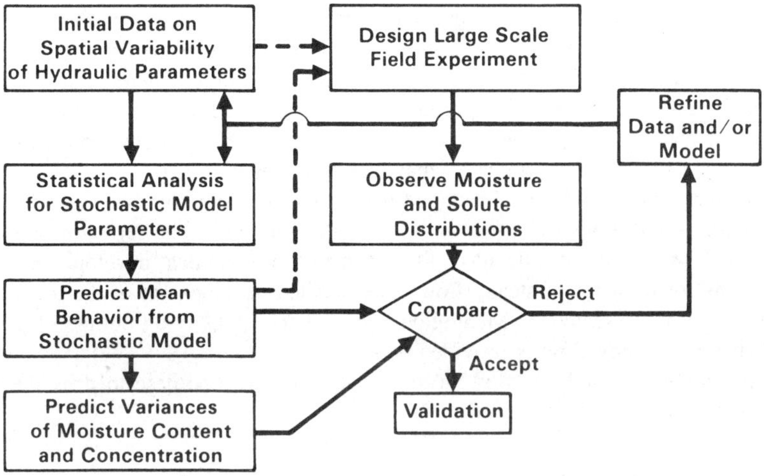

Figure 1. **Stochastic model validation process.**

Stochastic Theory

A stochastic approach to the analysis and prediction of flow and solute transport in unsaturated soils is motivated by the complicated natural heterogeneity of soil hydraulic properties that are always observed in natural earth materials. The determination of the degree of heterogeneity at the field site (e.g., unsaturated hydraulic conductivity as a function of capillary tension) is a principal objective of the validation effort.

It is assumed that the unsaturated hydraulic conductivity of the soil is represented by

$$\ln K = \ln K_s - \alpha \psi \tag{1}$$

where K is the unsaturated hydraulic conductivity, K_s is the saturated conductivity, α is the soil parameter, and ψ is the capillary tension. Within the stochastic framework, the hydraulic parameters such as $\ln K_s$ and α are represented as three-dimensional, statistically anisotropic, homogeneous, random fields. Therefore, parameters such as the log saturated hydraulic conductivity will have a covariance function of the form.

$$R_{ff}(\vec{\xi}) = E[f(\vec{x} + \vec{\xi})\, f(\vec{x})] \tag{2}$$

where $f = \ln K_s - E(\ln K_s)$ is the perturbation of $\ln K_s$ around the mean, $\vec{\xi}$ is the separation vector, and $\vec{x}$ is the position vector. A simple covariance model that represents a three-dimensional, statistically anisotropic medium is

$$R_{ff}(\vec{\xi}) = \sigma_f^2 \exp[-(\xi_1^2/\lambda_1^2 + \xi_2^2/\lambda_2^2 + \xi_3^2/\lambda_3^2)^{1/2}] \tag{3}$$

where σ_f^2 is the variance of ln K_s, and the λ_is are correlation scales in the three directions. The quantities σ_f^2 and the three λ_is are the statistical parameters that now characterize the variability of, in this case, saturated hydraulic conductivity. The same statistical parameters are required for each of the hydraulic parameters being considered; that is, in Equation (1), α is also characterized by a covariance function parameterized in a fashion similar to that shown in Equation (3).

The stochastic theory was developed using these random parameters in the classical flow and solute transport equations for unsaturated soils. The series of papers by Yeh et al. (*7,8,9*) present a detailed discussion of the theoretical development and applications for the case of steady flow. The detailed development of the theory for unsteady flow is presented by Mantoglou and Gelhar (*1*). Only an outline of the approach for unsteady flow is provided here.

It is widely accepted that flow in unsaturated soils can be described by the following partial differential equation:

$$C(\psi) \frac{\partial \psi}{\partial t} = \frac{\partial}{\partial x_i} K(\psi) \frac{\partial(\psi + z)}{\partial x_i} \tag{4}$$

where $C(\psi)$ is the specific moisture capacity, t is time, x_i are the spatial coordinates, and z is the vertical position. The parameters K and C in Equation (4) are now taken to be three-dimensional, statistically homogeneous random fields, and the resulting stochastic differential equation is solved by the perturbation approach, which uses spectral representation techniques. This leads to a solution for the perturbation in the capillary tension as a function of the input statistical properties of the hydraulic parameters K and C. From this solution, the expected value of Equation (4), the effective mean flow equation, is found, and the variance of tension is evaluated in terms of the statistical parameters that characterize the input variability. In the case of the mean flow, the result is an effective large-scale hydraulic conductivity and an effective specific moisture capacity. Both of these mean flow characteristics are dependent on the statistical properties of the variations of hydraulic conductivity and specific moisture capacity.

Figure 2 illustrates the kind of effects that are predicted in the mean flow equation. The figure shows that the effective hydraulic conductivity exhibits tension-dependent anisotropy, as well as hydraulic hysteresis. These results illustrate new phenomena that have been discovered by the stochastic approach.

This same kind of stochastic analytical approach is applied to the solute transport equation using the random velocity field that is found from the solution of the flow equation. The large-scale mean transport equation is then solved, and the result is an enhanced mixing effect characterized by a macrodispersion tensor. Mantoglou and Gelhar (*6*) also developed results for this solute transport case, although only for steady-state flow. One of their results for the longitudinal macrodispersivity in a stratified soil with vertical mean infiltration is

$$A_{11} = \frac{(\sigma_f^2 + \sigma_a^2 H^2)\,\lambda_1}{\gamma^2\,(1 + A\,\alpha_T)} \tag{5}$$

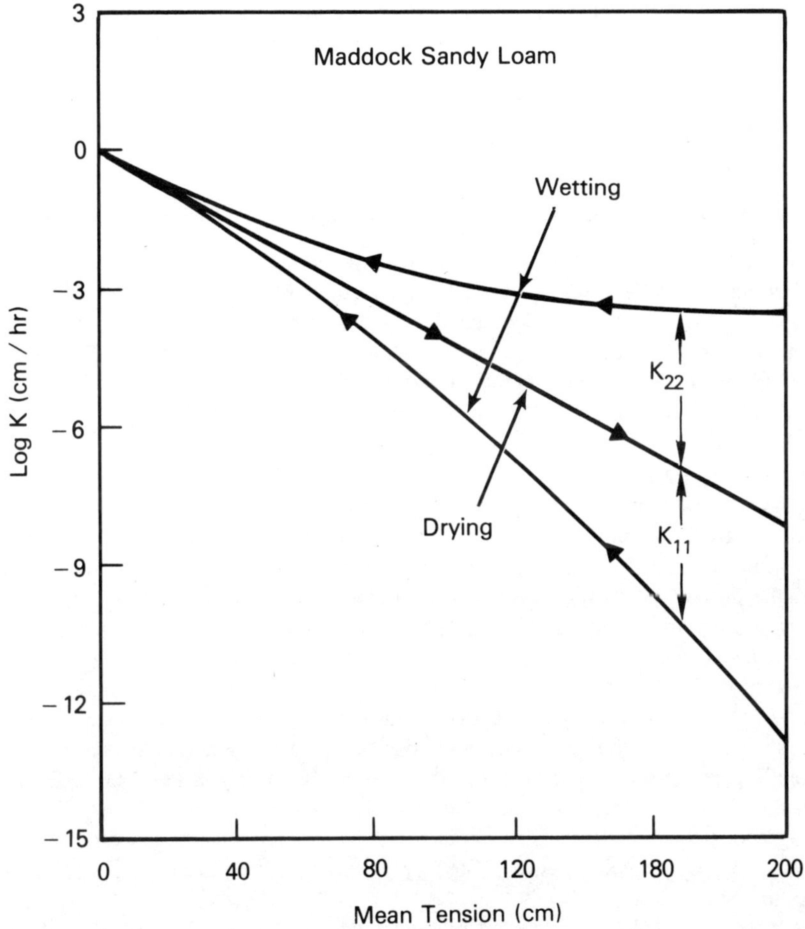

Figure 2. **Effective hydraulic conductivities of Maddock sandy loam for unsteady flow, showing predicted hysteresis ($\bar{K}_{11}$, is the vertical conductivity, and $\bar{K}_{22}$ the horizontal conductivity; for $\partial H/\partial t = \pm 0.01$ cm/s).**

where

σ_a^2 = var (α)

H = mean tension

λ_1 = vertical correlation scale

γ = a flow normalization factor

A = E(α)

α_T = local transverse dispersivity.

The primary goal of the proposed field experiments is to evaluate whether results such as those in Figure 2 and Equation (5) are valid for field conditions. To evaluate predictive results of this type at a given field site, the statistical parameters in covariance expressions such as Equation (3) (i.e., the variance and correlation scales) must be estimated from data on the spatial variability of the pertinent parameters.

Methods for Evaluating Spatial Variability of Model Parameters

To make effective use of the potential predictive capabilities of stochastic models for flow and solute transport, the randomly fluctuating behavior of controlling hydraulic properties must first be statistically characterized. Assuming that fluctuating hydraulic properties can be represented as statistically homogeneous random fields, it is anticipated that means and spatial covariance functions will provide sufficient statistical characterization of the variability associated with heterogeneity.

Semivariogram approach

The stochastic analytical approach assumes that the hydraulic properties, ln K_s, α, and C, are three-dimensional, statistically anisotropic, and homogeneous random fields. The statistical parameters required to characterize the spatial variability of the hydraulic properties are the means, variances, and correlation scales (λ_i, i=1,2,3,) assuming a covariance model like that presented in Equation (3). The field experiment is designed so that the spatial variability of the hydraulic properties will be characterized for a vertical plane (i.e., two-dimensional). Further assumptions will have to be made to obtain the statistical parameters in the third direction.

The first step in the data analysis is to investigate the assumption that the hydraulic properties are statistically homogeneous. If a trend in the data is discovered, the trend can be removed and the modified data treated as a statistically homogeneous random field. The covariance and cross-covariance structure of the hydraulic properties will be investigated by calculating directional semivariograms for the properties.

Let $Z(\vec{x})$ and $W(\vec{x})$ represent random functions for two of the hydraulic properties. The directional experimental semivariograms for the function $Z(\vec{x})$ can be calculated from

$$\gamma(\vec{\xi}) = 1/2 \, E\,[Z(\vec{x} + \vec{\xi}) - Z(\vec{x})]^2 \tag{6}$$

where $\vec{\xi}$ is the displacement vector [Journel and Huijbregts (*10*); David (*11*)]. The directional experimental cross-semivariograms for functions $Z(\vec{x})$ and $W(\vec{x})$ can be calculated from the data by using

$$\gamma_{zw}(\vec{\xi}) = 1/2 \, E\,[Z(\vec{x} + \vec{\xi}) - Z(\vec{x})]\,[W(\vec{x} + \vec{\xi}) - W(\vec{x})] \tag{7}$$

[Meyers (*12*)].

Several theoretical semivariograms can be fitted to the experimental results to determine the range and correlation lengths in several directions [Depner (*13*)]. This information is then used to indicate the principal directions and the corresponding integral scales. If the principal directions are different from the assumed axes in the model, a decision will be made as to which integral scales are more appropriate to use in the model.

Field Plan and Implementation

The purpose of the field studies is to provide a data base that can be used to validate existing and yet-to-be developed flow and transport models for the movement of water and chemicals through unsaturated soil and geologic profiles at large scales. Initial experiments for this study have been conducted in a lysimeter facility at the NMSU College Ranch, which is 40 km northeast of Las Cruces, New Mexico. Preliminary laboratory and large-column studies of solute transport in uniform soils have been completed [Wierenga et al. (*4*)].

Field site

The field site is on a basin slope of Mount Summerford, which is at the north end of the Doña Ana Mountains. The Doña Ana Mountains form a domal uplift complex of the younger rhyolitic and the older andesitic volcanics intruded by monzonite [Gile et al. (*14*)]. The sediments on the mid-piedmont have young soils (6500 to 4000 years B.C.) of the Doña Ana and Onite series (coarse-loamy Typic Haplargids). Sediments on the lower-piedmont slopes are older soils (mid–late Pleistocene) dominated by the Berino series (fine-loamy Typic Haplargids). The geologic features, geomorphic surfaces, soil series, and vegetation types found in the area around the field test area are typical of those of many areas in southern New Mexico and are similar to arid and semiarid areas of the southwestern United States.

The climate in the region is characterized by an abundance of sunshine, low relative humidity, and an average Class A pan evaporation of 239 cm/yr. Average annual precipitation is 23 cm with 52% of the rainfall occurring between July 1 and September 30. The average monthly maximum air temperature is highest in June at 36 °C and lowest in January at 13 °C.

Field experiment setup

A trench 26.5 m long, 4.8 m wide, and 6.0 m deep was dug in undisturbed soil. Water containing a conservative tracer was applied at a controlled rate of 1.76 cm/day on the surface of one side of the trench. The movement of water below the soil surface was monitored with neutron probes and tensiometers and also by visual observation of the water movement on the trench wall. Soil solution samples were taken to determine the movement of the tracer below the surface of the soil. Because of the extensive layering of the soil, it was expected that water and solute movement would occur horizontally, as well as vertically downward. The extent of the horizontal movement versus downward movement was qualitatively compared with predictions based on the stochastic theory. More rigorous comparisons are planned. During construction of the trench, the physical properties of the soil at the site were determined in sufficient detail to allow estimation of the statistical parameters needed in the stochastic models and comparative deterministic model.

Figure 3 presents a top view of the trench, showing a 4-m by 9-m area adjacent to the trench on which water was applied. The walls of the trench were excavated vertically and have been reinforced to prevent cave-in. The trench and adjacent irrigated area have been covered to prevent rainfall and surface runoff recharge and to minimize evaporation.

Irrigation system

The drip irrigation system similar to the one shown in Figure 4 consists of 60 irrigation lines placed parallel to the trench face 0.15 m apart. Each irrigation line contains 13 drippers 0.30 m apart. The irrigation scheme was designed to irrigate at a rate of 0.02 m/day in four equal periods. Based on a designed surface flux of 2 cm/day, which is equivalent to 720 l/day (190 gal/day), the system ran four times a day for a period of 17 minutes. The applied water was metered onto the plot to allow for water balance calculations in combination with neutron probe moisture readings.

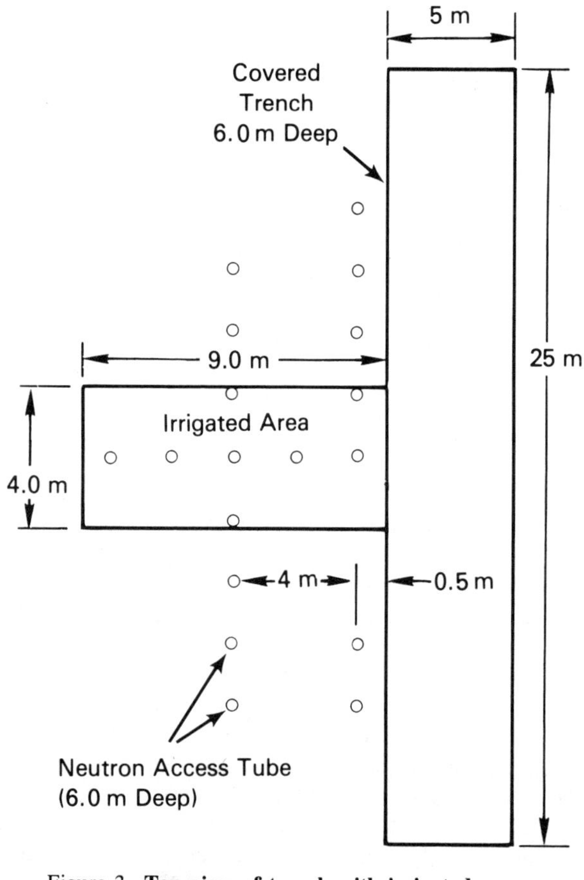

Figure 3. **Top view of trench with irrigated area.**

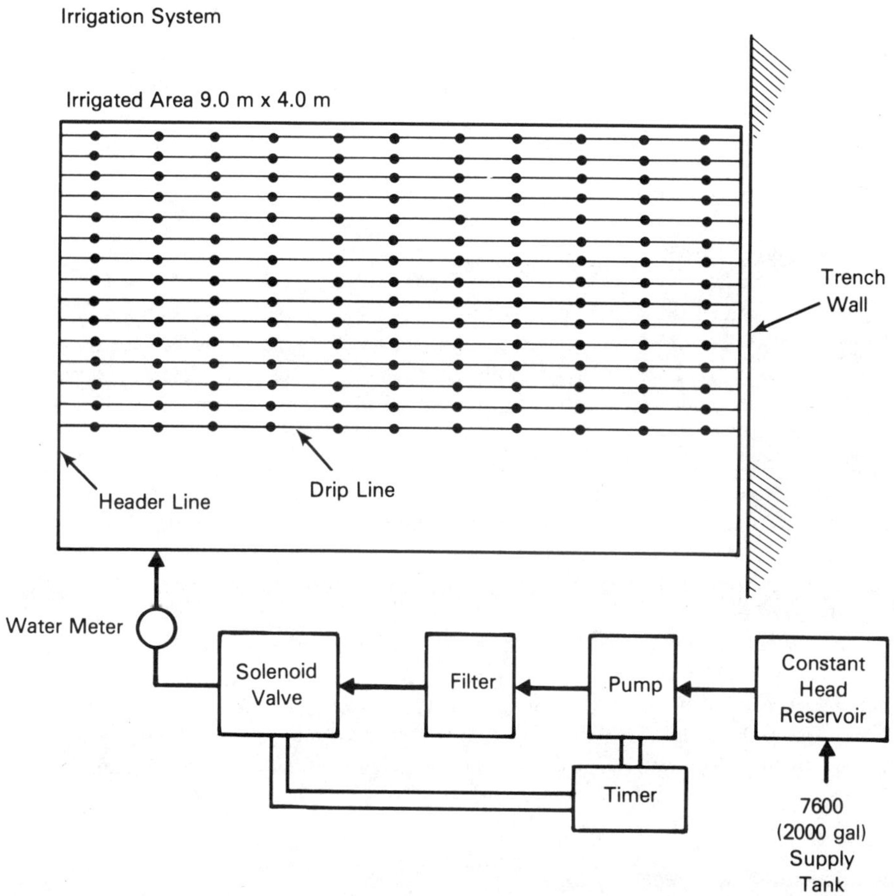

Figure 4. **Irrigation system layout for the trench experiment.**

Laboratory and field uniformity tests of the drip irrigation system prior to the experiment indicated extremely small nonuniformities and a random distribution to the nonuniformities. The calculated effective flux rate was 1.76 cm/day or 635 l/day for the 75 days of application.

Water content distribution

The water content in the soil was determined with neutron probes. Neutron access tubes were installed vertically in two rows parallel to the trench face (Figure 3). The first row is 0.5 m from the trench wall and the second row is 4.5 m from the trench wall. There is a third row of neutron access tubes positioned perpendicular to the trench and along the center of the irrigated area (Figure 3). These tubes are also spaced at 2-m intervals. The total number of access tubes is 18. All access tubes extend to a depth of approximately 6 m.

The neutron probes were calibrated in situ by taking simultaneous measurements of water content of the soil. These measurements were obtained by gravimetric sampling during and after installation of the access tubes and by taking neutron count rates in the access tubes. The bulk density of the soil was obtained from core samples taken from each of the soil horizons during excavation of the trench.

Soil sampling scheme

During excavation of the trench, core samples were taken of each soil horizon along the length of the trench at 0.5-m intervals. A total of 50 core samples were taken per soil horizon. Nine distinct soil horizons were sampled, resulting in 450 core samples. In situ saturated hydraulic conductivity values were determined with the borehole method adjacent to each site where core samples were taken. The cores were taken to the laboratory where the soil water retention curve, the saturated hydraulic conductivity, and the bulk density were determined for each core. For a few cores from each horizon, the unsaturated hydraulic conductivity will be determined using the one-step outflow method of Kool et al. (*15*). When the soil cores were taken, additional soil was collected and analyzed for particle size distribution. At three locations in the trench, core samples were taken at intervals of 20 cm down to a depth of 6 m.

A total of 90 cores were taken at these three locations to determine the vertical distribution in soil water-retention data and bulk density. The soil water-retention data have been fitted to the van Genuchten water-retention model [van Genuchten (*16*)]. Using the saturated hydraulic conductivity and the water-retention curves, values for the unsaturated conductivity can be calculated with this same model.

Solute sampling scheme

During the initial 10 days of the irrigation period, a conservative tracer, tritium, was added to the water applied to the surface of the soil adjacent to the trench (see Figure 3). The tritium movement through the soil was determined by collecting soil solution samples through suction tubes installed at various depths below the soil surface. The suction tubes, consisting of 1-cm-outside diameter (OD) by 8-cm-long porous ceramic cups attached to 1-cm-OD plastic pipe were installed through the trench wall (Figure 5). The suction tubes and tensiometers were installed in a grid with 0.5-m horizontal spacing between suction tubes and tensiometers. The tensiometers and suction tubes were inserted 0.5 m into the wall of the trench. During the course of the experiment, additional suction tubes or tensiometers were installed where necessary. The tensiometers have a septum stopper at the top [Marthaler et al. (*17*)] for use with a handheld pressure transducer (Soil Measurement Systems, Las Cruces, New Mexico). Breakthrough curves were obtained from the suction tube samples.

To date, two discrete breakthrough plots have been obtained from the top two sampler levels. These indicate a remarkably consistent and uniform plume that spreads (diffuses) with depth.

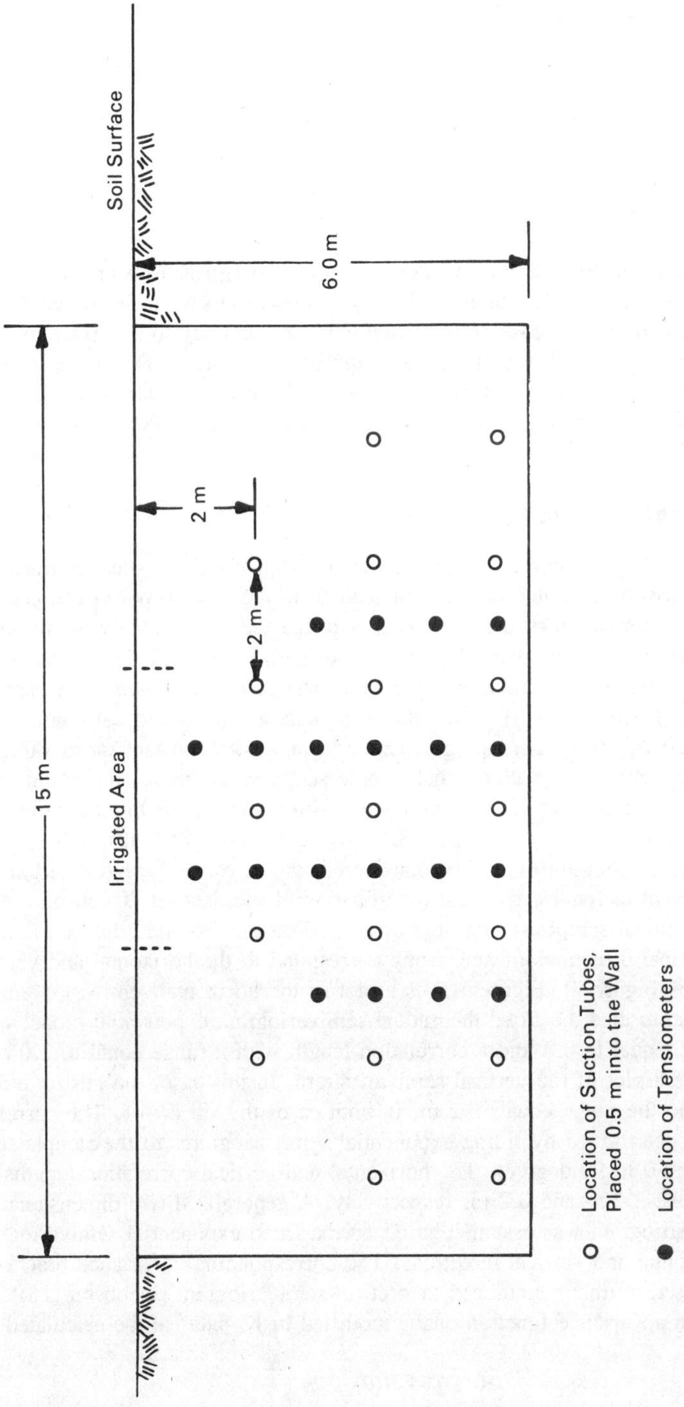

Figure 5. Side view of trench with location of tensiometers and suction lysimeters.

Field Data and Its Assessed Variability

At the Las Cruces field site the following data have been collected:

1. Field-saturated hydraulic conductivity (K_s) measurements using Guelph permeameters (measurements taken at 607 different locations in the trench face during the excavation using nine horizontal and three vertical transects; the data measurements along each horizon collected at 0.5-m intervals)
2. Particle size distribution
3. Bulk density
4. Soil moisture retention curve determinations (performed on the core samples taken at the same locations as the Guelph permeameter measurements).

The in situ $K_s$values range from a minimum of 9 cm/day to a maximum exceeding 50,000 cm/day (which was assigned as the limiting value). This suggests a wide range (i.e., more than three orders of magnitude) in the field hydraulic conductivity values. For the geostatistical analyses, ln K_s values were evaluated as discussed in the next section.

Geostatistical Analysis

Directional semivariograms were used to investigate the statistical characteristics of the ln K_s two-dimensional data set obtained from the Guelph permeameters. The statistical characterization of the ln K_s data required the estimation of the mean and covariance function for the data. A trend in the ln K_s data was indicated by the lack of distinct sills for the directional sample semivariograms calculated from the data (see Figure 6) [Jacobson (*18*)]. Thus, the ln K_s data are not statistically homogeneous. The "best fit" trend consisting of small regions with constant mean values was identified by calculating the directional sample semivariograms for the modified data set and looking for distinct sills of similar magnitude. These modified data can be considered to be statistically homogeneous. Although more detailed trends (e.g., a trend for each soil horizon) could be considered, the "best fit" trend determined here was thought to require the least modification of the data set [Jacobson (*18*)].

The directional sample semivariograms based on the modified data set indicated that the principal directions of anisotropy correspond to the horizontal and vertical directions (see Figure 7). Figure 7(a) is a plot of the horizontal semivariogram. The dashed line represents the fitted theoretical semivariogram exponential model with a nugget effect, equal to 0.4 and a correlation length with a range equal to 2.0 m. Figure 7(b) is a plot of the vertical semivariogram. In this case, the nugget effect equals 0.4 and the range equals 0.2 m. In both cases the sill is 1.1. The correlation lengths were determined by fitting exponential semivariograms to the sample semivariograms for 0 to 90 degrees. The horizontal and vertical correlation lengths were calculated to be 2.0 m and 0.2 m, respectively. A generalized two-dimensional exponential semivariogram was assumed based on the fitted exponential semivariograms for the horizontal and vertical directions. The corresponding covariance function was calculated based on this generalized theoretical semivariogram [Jacobson (*18*)].

Thus, the covariance function of the modified ln K_s data can be calculated from

$$R(\vec{\xi}) = R(0) - \gamma(\vec{\xi}) \tag{8}$$

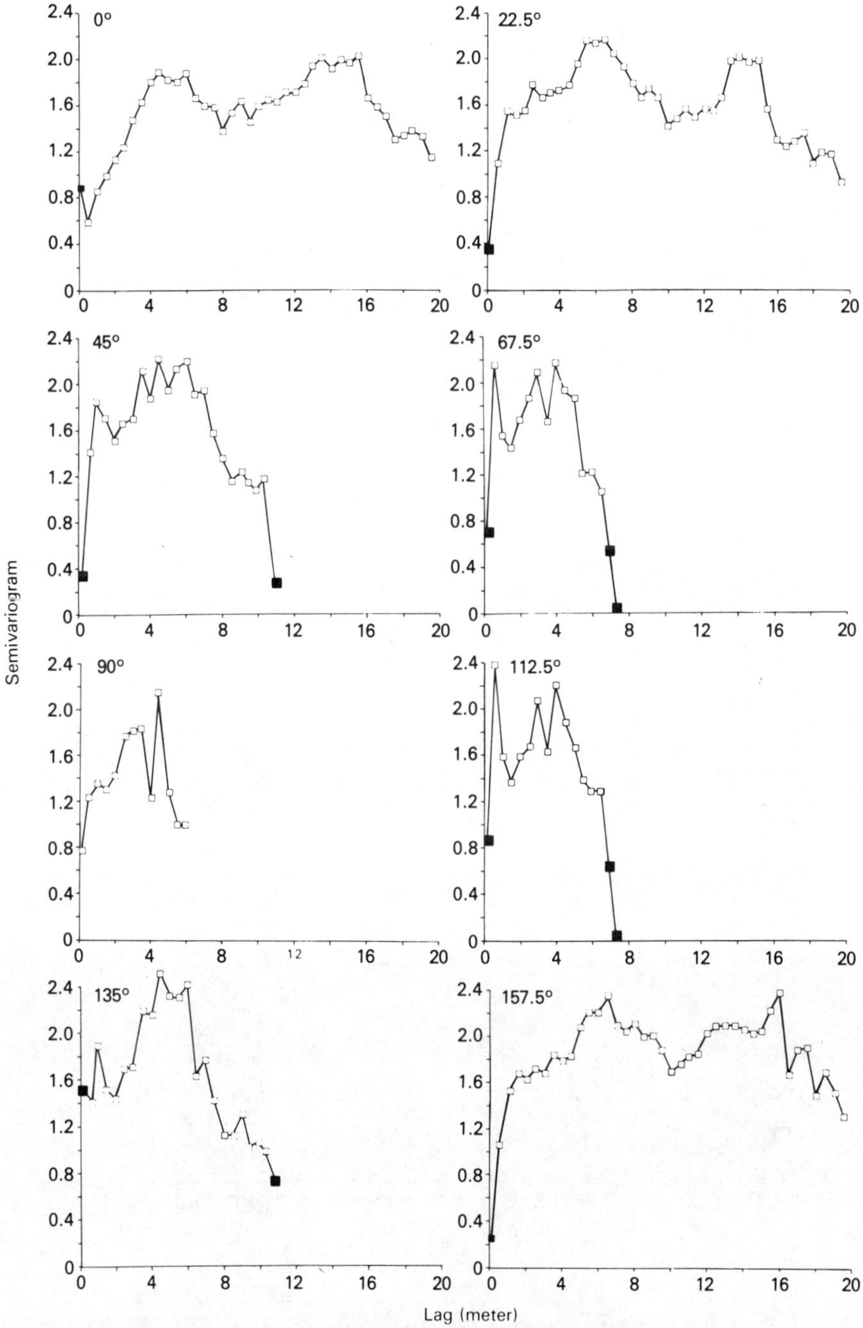

Figure 6. **Directional sample semivariograms based on two-dimensional ln K_s data. The lag is the separation distance between data pairs.**

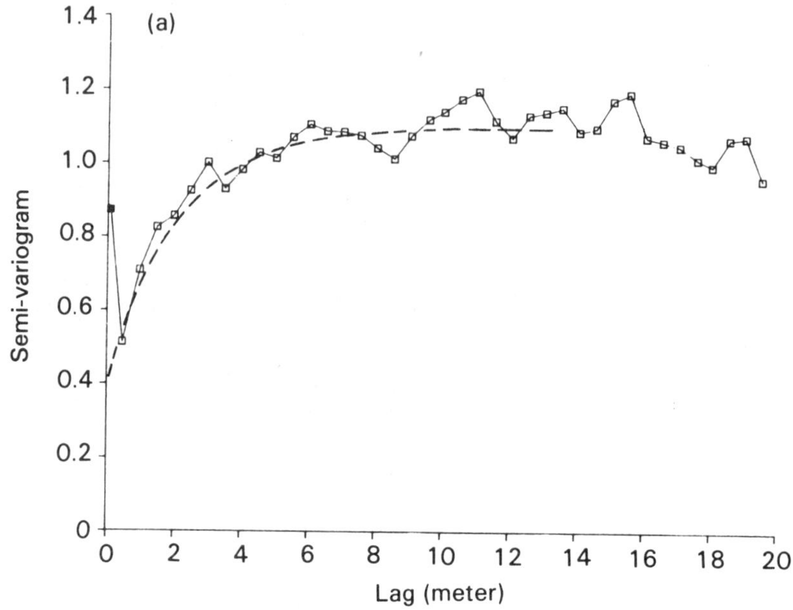

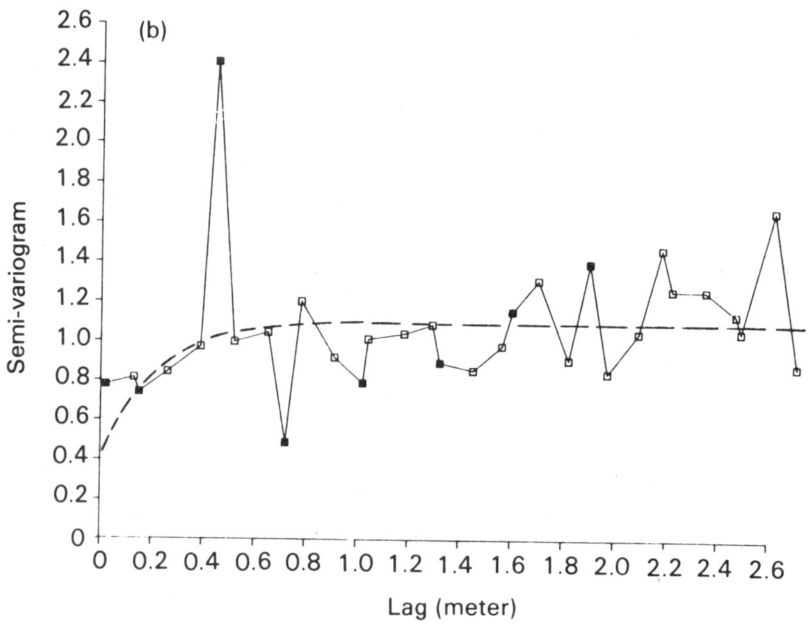

Figure 7. **Comparisons of theoretical exponential semivariograms to sample semivariograms for directions of 0 and 90 degrees shown in parts (a) and (b), respectively (modified data set).**

From Figure 7, R(0) = 1.1 is equal to the sill of the semivariogram (i.e., sill is defined as the true variance and is found when the semivariogram reaches a constant value) and

$$\gamma(\vec{\xi}) = 0.4 + 0.7\left(1 - \exp\left[-\left(\left(\frac{x}{2.0}\right)^2 + \left(\frac{z}{0.2}\right)^2\right)^{1/2}\right]\right) \tag{9}$$

where 0.4 is the nugget value (i.e., the nugget value represents a variance that may be caused by measurement errors and/or spatial variations in the data on a scale smaller than the sampling interval).

Combining R(0) and $\gamma(\vec{\xi})$ using equation (8) yields

$$R(\vec{\xi}) = 1.1 - \left\{0.4 + 0.7\left(1 - \exp\left[-\left(\left(\frac{x}{2.0}\right)^2 + \left(\frac{z}{0.2}\right)^2\right)^{1/2}\right]\right)\right\}$$

or

$$R(\vec{\xi}) = 0.7\exp\left[-\left(\left(\frac{x}{2.0}\right)^2 + \left(\frac{z}{0.2}\right)^2\right)^{1/2}\right] \tag{10}$$

A total of 607 ln K_s values were available for this statistical analysis based upon Guelph permeameter data. This large number of data points allowed determination of the correlation length for the horizontal direction due to the spatial resolution of the data in this direction. However, the vertical spatial resolution was not as detailed. Thus, the vertical correlation length is not as clearly defined.

Wetting Front Advance Analysis

A detailed water budget analysis was conducted using daily total flowmeter readings and daily neutron meter moisture contents, and taking into account the previous spatial distribution of moisture content. Thus far, an excellent agreement has been shown between the applied volume of water and the water measured directly in the subsurface for the first 20 days of the experiment.

The wetting front observed on the trench face was recorded (a grid with 0.5 m-square elements was superimposed on the trench face) for the 72-day experiment (Figure 8). Figure 8 shows the locations of the wetting front based on noticeable soil color differences. The wetting front reached the 6-m depth (trench floor) at approximately day 55. Simultaneously, the wetting front advance and moisture distributions were interpreted from daily tensiometer and neutron probe data. Figure 9 shows the progress of the wetting front using tensiometer data. The advance of the wetting front, in both the parallel and perpendicular directions to the trench face, was also plotted using neutron probe data. Further analysis of specific tension isopleths associated with the wetting front will be developed and compared to the deterministic and stochastic mean flow model simulations.

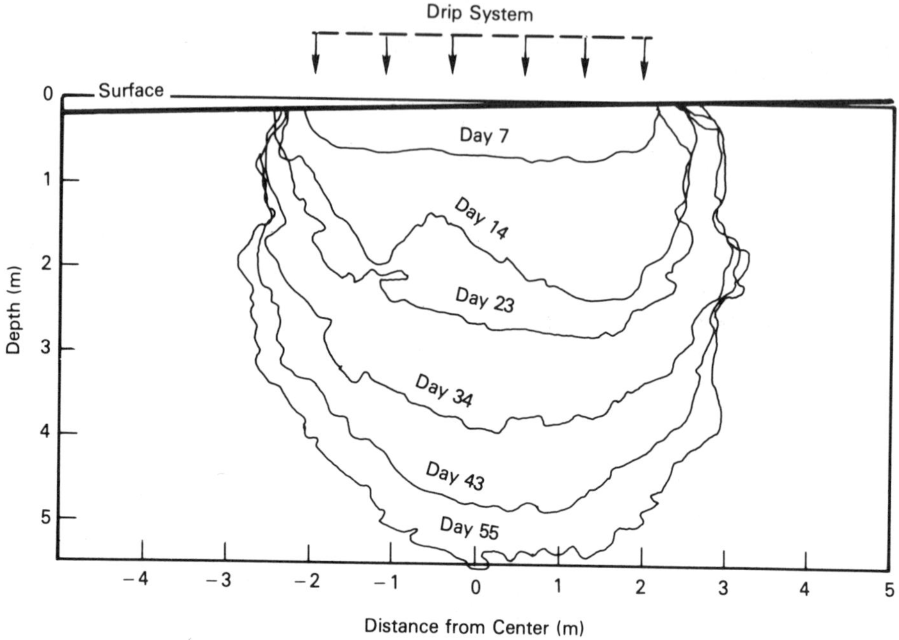

Figure 8. **Plot of visually observed advancing wetting front. Dashed line indicates location of irrigated surface above the trench face.**

Stochastic Modeling Activities

A "mean flow" stochastic model was utilized by the MIT investigators to simulate a physical system analogous to the field experiment. An important aspect to the modeling was to determine the anticipated lateral spreading of the applied moisture. In order to compare the simulated wetting front to the observed field results, tensiometer data were used since the simulations provide soil tension results. Stochastic model simulation results for 5- and 10-day periods were performed. Plots of the 500- and 100-cm tension isopleths are presented for comparison to the actual field tension data (see Figure 10). Uncertainty in the initial pressure conditions resulted in use of an approximate initialization value of 1,000-cm tension uniformly distributed. This corresponds to a relatively dry, low-conductivity initial condition. Figure 10 illustrates the simulated results for a two-dimensional vertical plane parallel to the trench face. Note that the left edge of Figure 10 represents a line of symmetry in the flow domain that corresponds to the centerline of the wetted strip.

Figure 11 is a sample plot of the MIT deterministic simulations. This figure depicts a physical domain identical to that of Figure 10. Note that initial conditions are equavilent and that the deterministic analysis does not reveal the lateral spreading evident in the stochastic simulation results. Also the deterministic mode simulates vertical propagation more rapidly.

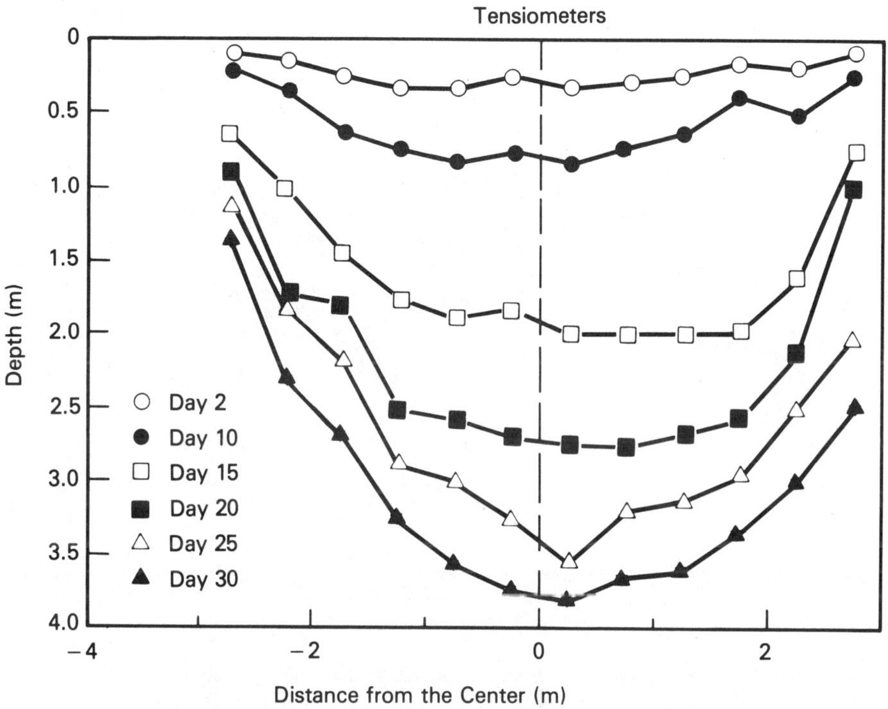

Figure 9. **Plot of advancing wetting front using tension data from tensiometers 0.5 m into and parallel to the trench face.**

A qualitative comparison has been made between the deterministic and stochastic "mean flow" model simulations and the measured tensiometer data from the field experiment. This assessment is based on a preliminary review of the point measurements of field tension taken at times of 5 and 10 days from tensiometers located 0.5 m into the trench face (on a sampling grid 0.5 m square in the vertical plane). These field data were compared with the contoured simulation results of Figures 10 and 11; the 500-cm tension contour represents a relatively dry value within the measurement range of the tensiometer instrumentation. A general review of these point data and the two modeling solutions indicates that the stochastic results compare more favorably to the field measurements. These comparisons suggest that the stochastic model is better able to simulate conditions of lateral and vertical spreading, such as those observed in the field experiment, relative to the deterministic model. In particular, compared to the deterministic method, the stochastic model exhibits a stronger horizontal component of spreading and milder spatial gradients of tension in the horizontal direction compared to the vertical direction. In addition, vertical propagation of the wetting front is slower in the stochastic analysis, more closely matching field results than does the deterministic case. These bulk characteristics, especially concerning the drier regions of the system, suggest that the

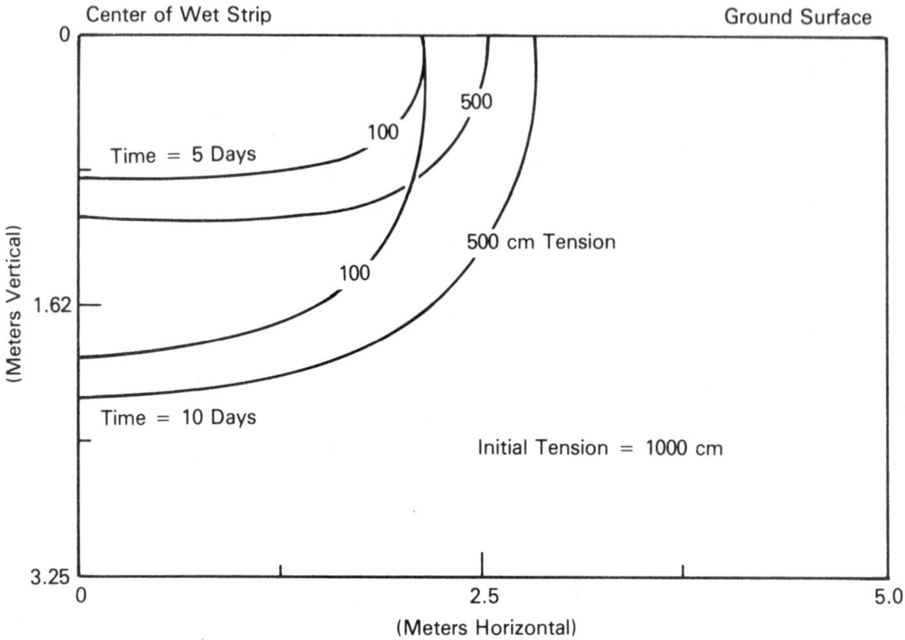

Figure 10. **Numerical simulation of wetting front propagation by solving effective mean equations using stochastic theory. Plot of contours of equal tension.**

stochastic model is able to represent the observed field phenomena in a more complete and accurate manner. More detailed quantitative comparisons with the field data are not justified because of the preliminary nature of model input parameters (e.g., conductivity and moisture retention functions; initial conditions of tension and moisture content) used in the simulations.

Summary

The outlined field study, the associated numerical analyses of field data, and the simulated flow studies illustrate a relatively new approach to validating ground-water flow models. An experiment that was specifically designed to provide data from the monitoring of temporal and spatial distributions of moisture contents, soil tensions, and solute concentrations was performed to examine and test the stochastic theory. A team of field experimentalists and modelers met on numerous occasions to determine the experimental design and analytical methods to be used. A great deal has been accomplished to date through analyses of the detailed field data, observed wetting front advances, tritium plume breakthroughs, and numerical simulations.

A large amount of collected data has yet to be analyzed, and further simulations are anticipated. The computer simulations required extremely long processing times, in some cases approaching actual field experiment periods. More efficient numerical

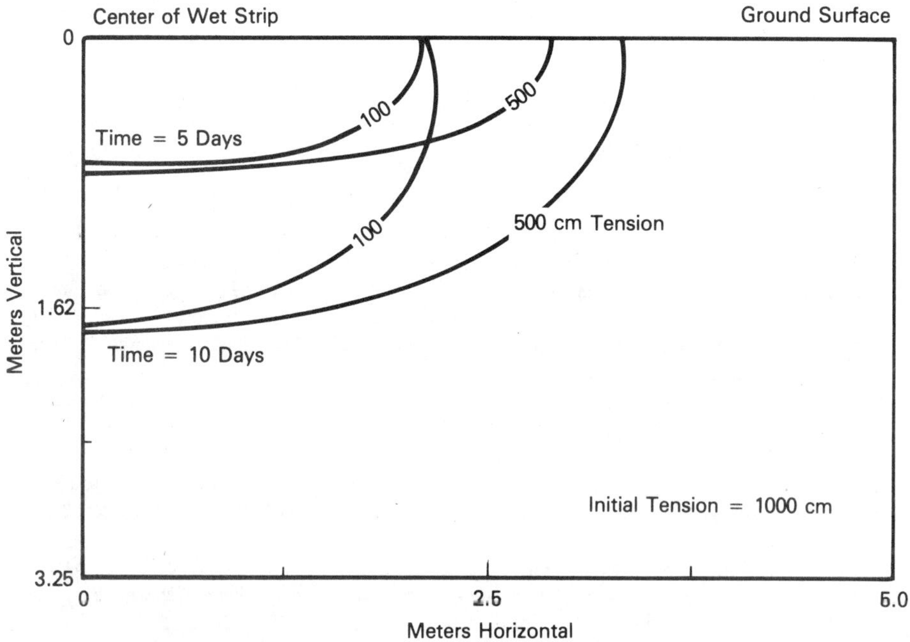

Figure 11. **Numerical simulation of wetting front propagation by solving deterministic van Genuchten equations (variances = 0). Plot of contours of equal tension.**

techniques and modeling schemes need to be investigated if the entire 75-day experiment duration is to be simulated. The work to date is still preliminary, with full validation of the flow and transport theory yet to be achieved.

A better analysis of in situ and laboratory soil moisture retention curves for the entire range of very wet to very dry conditions is needed. In addition, the ambient field tensions need to be determined for the very dry conditions associated with the desert environment in which the experiment has been conducted. Additional future field experiments are being planned to utilize both transient, finite pulse wetting events and subsequent soil moisture and solute redistributions as opposed to the steady infiltration experiment reported here.

Literature Cited

1. Mantoglou, Aristotelis, and Lynn W. Gelhar, "Stochastic Modeling of Large-Scale Transient Unsaturated Flow Systems," *Water Resources Research,* Vol. 23, No. 1, pp. 37–46 (1987).
2. Mantoglou, Aristotelis, and Lynn W. Gelhar, "Capillary Tension Head Variance, Mean Soil Moisture Content, and Effective Specific Soil Moisture Capacity of Transient Unsaturated Flow in Stratified Soils," *Water Resources Research*, Vol. 23, No. 1, pp. 47–56 (1987a).

3. Mantoglou, Aristotelis, and Lynn W. Gelhar, "Effective Hydraulic Conductivities of Transient Unsaturated Flow in Stratified Soils," *Water Resources Research*, Vol. 23, No. 1, pp. 57–68 (1987b).
4. Wierenga, P. J., L. W. Gelhar, C. S. Simmons, G. W. Gee, and T. J. Nicholson, *Validation of Stochastic Flow and Transport Models for Unsaturated Soils: A Comprehensive Field Study*, NUREG/CR-4622, U.S. Nuclear Regulatory Commission, Washington, DC (1986).
5. Gelhar, L. W., "Stochastic Analysis of Flow in Heterogeneous Porous Media," in *Fundamentals of Transport Phenomena in Porous Media*, J. Bear and M. Corapcioglu, eds., Martinus Nijhoff Publ., Dordrecht, Netherlands, pp. 675–717 (1984).
6. Mantoglou, A., and L. W. Gelhar, "*Large Scale Models of Transient Unsaturated Flow and Contaminant Transport Using Stochastic Methods*," Technical Report No. 299, Ralph M. Parsons Laboratory, Department of Civil Engineering, Massachusetts Institute of Technology, Cambridge, Massachusetts (January 1985).
7. Yeh, T.-C. Jim, L. W. Gelhar, and A. L. Gutjahr, "Stochastic Analysis of Unsaturated Flow in Heterogeneous Soils 1. Statistically Isotropic Media," *Water Resour. Res.*, Vol. 21, No. 4, pp. 447–456 (1985a).
8. Yeh, T.-C. Jim, L. W. Gelhar, and A. L. Gutjahr, "Stochastic Analysis of Unsaturated Flow in Heterogeneous Soils 2. Statistically Anisotropic Media," *Water Resour. Res.*, Vol. 21, No. 4, pp. 457–464 (1985b).
9. Yeh, T.-C. Jim, L. W. Gelhar, and A. L. Gutjahr, "Stochastic Analysis of Unsaturated Flow in Heterogeneous Soils 3. Observations and Applications," *Water Resour. Res.*, Vol. 21, No. 4, pp. 465–471 (1985c).
10. Journel, A. G., and Ch. J. Huijbregts, *Mining Geostatistics*, Academic Press, New York (1978).
11. David, M., *Geostatistical Ore Reserve Estimation*, Elsevier Scientific Publishing Company, New York (1977).
12. Meyers, D. E., *Joint Estimation of Random Functions: The Matrix Form: Res. Rep.*, Department of Mathematics, University of Arizona, Tucson, Arizona (1981).
13. Depner, J. S., "Estimation of the Three-Dimensional Anisotropic Spatial Covariance of Log Permeability Using Single-Hole and Cross-Hole Packer Test Data from Fractured Granites," Masters Thesis for Department of Hydrology and Water Resources, University of Arizona, Tucson, Arizona (1985).
14. Gile, L. H., J. H. Hawley, and R. B. Grossman, *Distribution and Genesis of Soils and Geomorphic Surfaces in a Desert Region of Southwestern New Mexico, Guidebook*, 156 pp., Soil Sciences Society of America and New Mexico State University, Las Cruces (1970).
15. Kool, J. B., J. C. Parker, and M. Th. van Genuchten, "Determining Soil Hydraulic Properties from One-Step Outflow Experiments by Parameter Estimation," *Soil Sci. Soc. Amer. J.*, Vol. 49, pp. 1348–1354 (1985).
16. van Genuchten, M. Th., *Calculating the Unsaturated Hydraulic Conductivity with a New Closed-Form Analytical Model*, Report 78-WR-08, Department of Engineering, Princeton University, 63 pp. (1980).
17. Marthaler, H. P., W. Vogelsanger, F. Richard, and P. Wierenga, "A Pressure Transducer for Field Tensiometers," *Soil Sci. Soc. Amer. J.*, Vol. 47, pp. 624–627 (1983).
18. Jacobson, E. A., *Statistical Characterization of Spatial Variability of Hydraulic Properties*, PNL Progress Report to NRC (FIN B2887) (August 1987).

Parameter Sensitivity and Importance for Radionuclide Transport in Double-Porosity Systems*

DAVID TOMASKO
Earth Sciences Division
Sandia National Laboratory
Albuquerque, New Mexico**

M. REEVES, V. A. KELLEY, and J. F. PICKENS
INTERA Technologies, Inc.
Austin, Texas

Abstract

A type-curve methodology using four dimensionless parameters is developed for modeling radionuclide transport through a fractured, double-porosity system. A dimensionless velocity and a dimensionless dispersion coefficient are used to characterize transport through the fractures, while a dimensionless solute storage enhancement coefficient and a dimensionless time constant for matrix diffusion are used to characterize the coupling of the matrix to the fractures.

Using the dimensionless parameters, parameter-variation curves are developed to exhibit the general response of a performance measure, nuclide travel time, with respect to the entire range of parameter variability. Parameter sensitivities are then calculated as the normalized derivatives of the parameter-variation curves for the dimensionless set of base-case values. The parameter sensitivities are then recast to find the response of the performance measure in terms of a natural set of dimensional parameters that characterize the double-porosity transport model. As a final step, importance coefficients are defined to ascertain the relative ranking of the natural set of base-case parameters.

The method is demonstrated by presenting the results obtained by modeling the transport of an infinitely long-lived radionuclide through the Culebra Dolomite Member of the Rustler Formation, the most transmissive zone located above the Waste Isolation Pilot Plant (WIPP) in southeastern New Mexico. Parameter ranges and base-case values depict the uncertain properties of the Culebra, while double-porosity transport simulations with the SWIFT II computer model predict the corresponding range in travel time, the performance measure adopted for this study.

* This work was supported by the U.S. Department of Energy (DOE) under contract DE-AC04-76DP00789.

**Currently with Energy and Environmental Systems Division, Argonne National Laboratory, Argonne, Illinois.

The computed parameter sensitivities and importance provide valuable guidance to the site characterization and repository performance assessment programs at the WIPP site, and underscore the importance of hydraulic and conservative tracer tests for reducing performance measure uncertainties.

Introduction

Uncertainties in the predicted breakthrough times at a given location for a radioactive solute transported through a double-porosity system are anticipated because of uncertainties in the physical parameters of the system. The extent of the effect of individual parameter uncertainties on a given breakthrough calculation is a function of both the sensitivity of the parameter, which can be defined using a normalized derivative, and the parameter's range.

For the purposes of this study, we will investigate the sensitivity of double-porosity transport calculations to underlying physical parameters, and we will rank the parameters according to their importance, which we define as the absolute value of the product of the parameter's sensitivity and range. This ranking can then be used to optimally reduce the uncertainty in the performance measure of the study, breakthrough time, by applying field and laboratory work to reduce the relative range of the most important parameters.

We will first describe the methodology used to calculate parameter sensitivity and importance in a double-porosity system, and will then present the results of breakthrough calculations performed for the transport of a sorbing solute through the Culebra Dolomite Member of the Rustler Formation which is located above the Waste Isolation Pilot Plant (WIPP) in southeastern New Mexico (Figure 1).

Double-Porosity Transport

Streltsova-Adams (1978) has characterized a double-porosity system as one in which the primary medium, composed of low-permeability material, provides a high degree of storativity, and the secondary medium provides a high-permeability path of low storativity.

Transport of an infinitely long-lived radioactive sorbing solute through a fracture can be characterized by the equation:

$$K \phi \frac{\partial C}{\partial t} = -u \frac{\partial C}{\partial X} + D \frac{\partial^2 C}{\partial X^2} - \Gamma/\rho \tag{1}$$

and transport in the surrounding matrix is given by the relation:

$$K'\phi' \frac{\partial C'}{\partial t} = D'\nabla^2 C' \tag{2}$$

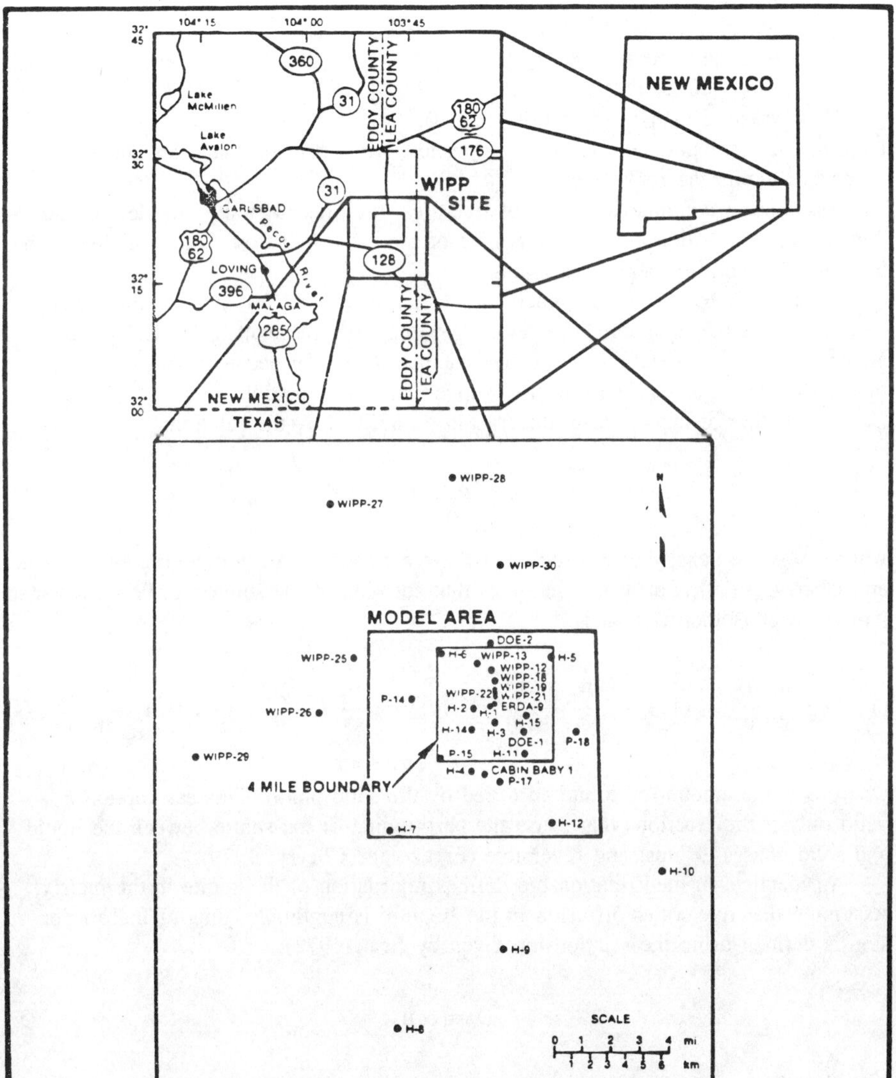

Figure 1. **Site location for the Waste Isolation Pilot Plant (WIPP) showing Culebra borehole locations for regional hydrologic characterization.**

In Equations 1 and 2,

C = solute concentration (M/M)
X = longitudinal distance within the fractures (L)
t = time (t)
K = dimensionless retardation
ϕ = porosity (L^3/L^3)
u = Darcy velocity (L^3/L^2t)
D = dispersion coefficient (L^2/t)
ρ = fluid density (M/L^3)
Γ = fracture-matrix interaction (M/L^3t).

Unprimed quantities denote fracture properties, while primed quantities denote properties within the rock matrix.

Assuming a continuous system of fractures, the present study considers a one-dimensional simulation along a streamline of the calibrated flow field as indicated by Equation 1; lateral dispersion is ignored.

Equation 2, by use of the general Laplacian operator (∇^2), permits the matrix to be treated as either a sphere or a slab. For simplicity, the matrix discussion will assume the case of a slab, while actual calculations performed with the SWIFT II model (Reeves et al., 1986a,b,c) use a spherical representation.

Retardation of the solute in the fracture is defined by the relation:

$$K = 1 + \rho_r K_d (1 - \phi)/\phi \tag{3}$$

where ρ_r is the density of the rock and K_d is a standard partition coefficient (Freeze and Cherry, 1979). Equation 3 assumes that sorption of the solute can be represented with a linear isotherm. That is,

$$(1 - \phi) \frac{\rho_r}{\phi} \frac{\partial S}{\partial t} = (1 - \phi) \frac{\rho_r}{\phi} \frac{\partial S \partial C}{\partial C \partial t} = (1 - \phi) \frac{\rho_r}{\phi} K_d \frac{\partial C}{\partial t} = (K - 1) \frac{\partial C}{\partial t} \tag{4}$$

where S is the amount of solute adsorbed by the solid phase. This assumption is valid only if the reactions that cause the partitioning of the solute beween the liquid and solid phases are fast and reversible (Freeze and Cherry, 1979).

An identical primed relationship defines retardation of the solute in the matrix. Assuming that free-water diffusion in the fracture is negligible, longitudinal dispersion is defined using the relationship given by Bear (1972):

$$D = \alpha u \tag{5}$$

where α is defined as the fracture dispersivity (L).

Transport in the matrix is assumed to be governed by diffusion only (Equation 2). The coefficient of diffusion used is defined as:

$$D' = \theta' \phi' d' \tag{6}$$

where

θ' = matrix tortuosity (L^2/L^2)

d' = free-water diffusivity (L^2/t).

The interaction between the fracture and matrix is obtained by a simple mass balance and is given by the following relationship:

$$\frac{\Gamma}{\rho} \cong (1 - \phi)\, \gamma\, D' \left. \frac{\partial C'}{L' \partial X'} \right|_{X' = 0} \tag{7}$$

where

X' = distance measured from the fracture-matrix interface into the matrix (L)

L' = matrix-block length equivalent to one-half the distance between adjacent fractures (L).

Quantity γ indicates either spherical ($\gamma = 3$) or slab ($\gamma = 1$) geometry. Here, the value of γ is not essential to the type-curve concept and, for simplicity, $\gamma = 1$ is chosen.

For a fracture, ϕ is much less than one. Therefore, Equation 7 can be approximated as:

$$\frac{\Gamma}{\rho} \cong D' \left. \frac{\partial C'}{L' \partial X'} \right|_{X' = 0} \tag{8}$$

Initial and boundary conditions for Equations 1 and 2 are given by:

$$C(X, t = 0) = 0. \tag{9}$$

$$C(X = 0, t) = C_0 \tag{10}$$

$$C'(X, X', t = 0.) = 0. \tag{11}$$

$$C'(X, X' = 0., t) = C(X, t) \tag{12}$$

$$\frac{\partial C'}{\partial X'}(X, X' = L', t) = 0. \tag{13}$$

Equations 1 and 2 are numerically solved for initial and boundary conditions 9 through 13 using the SWIFT II computer code (Reeves et al., 1986a,b,c).

Dimensionless Variables and Parameters

Several dimensionless variables can be introduced for solute transport in a fractured system. These variables can be based on either physical properties or quantities related to performance measures specified in Federal regulations (U.S. Environmental Protection Agency [1985] and U.S. Nuclear Regulatory Commission [1983]). The following three dimensionless variables will simplify later sensitivity and importance calculations:

χ = X/X_c where X_c = 3280 m, the average distance traveled by a solute in the Culebra Dolomite Member of the Rustler Formation from a point above the centroid of the waste-emplacement area within the WIPP site to the accessible environment along a streamline of the calibrated flow field (Figure 2)

τ = t/t_c where t_c = 1000 years

and

C = C/C_0 where C_0 is the time-independent solute concentration at X = 0. After transformation to Laplace space, transport Equations 1 and 2 and interac-

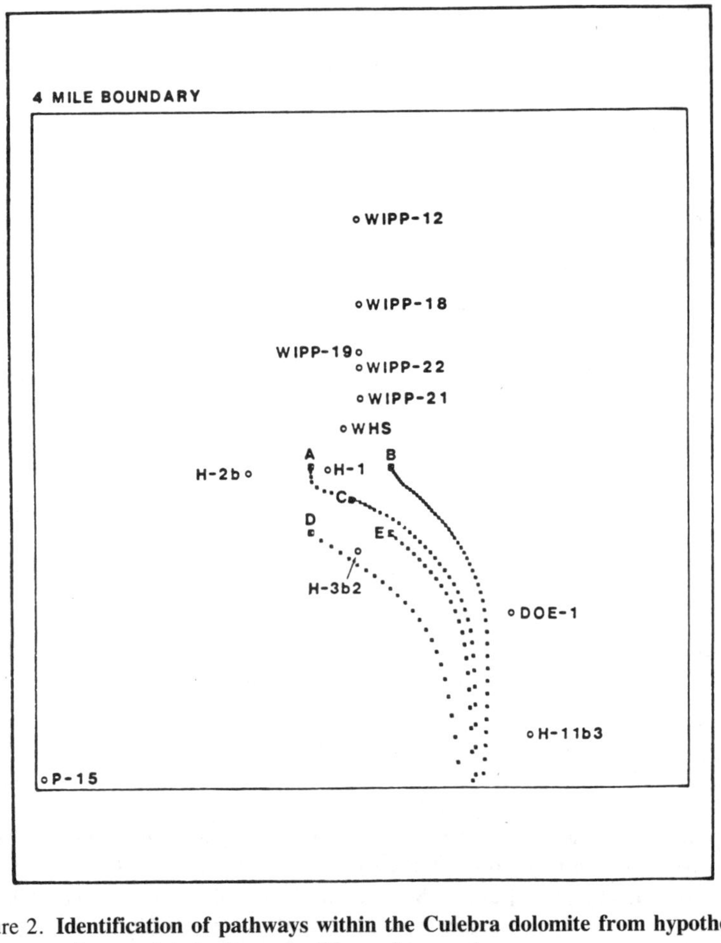

Figure 2. **Identification of pathways within the Culebra dolomite from hypothetical release points to the accessible environment.**

tion Equation 8 combine to yield the following equation in dimensionless form:

$$\eta \frac{\partial^2 \hat{C}}{\partial \chi^2} - \nu \frac{\partial \hat{C}}{\partial \chi} = \hat{C} \left[s + (\varkappa' - 1) \sqrt{s/\tau_c'} \quad \tanh \sqrt{s\tau_c'} \right] \tag{14}$$

where

s = Laplace-transform variable.

$\hat{C}$ = Laplace-space concentration.

Equation 14 is not directly used for the calculations. As discussed earlier, all calculations were performed with the SWIFT II code (Reeves et al., 1986a,b,c). Equation 14, however, can be used to identify four additional dimensionless parameters of interest.

Two of these dimensionless parameters characterize transport within the fractures. The first is a dimensionless velocity given by the following expression:

$$\nu = (u\ t_c)/(K\ \phi\ X_c). \tag{15}$$

The other is a dimensionless dispersivity which appears in the dimensionless dispersion coefficient for the fracture:

$$\eta = \beta\ \nu. \tag{16}$$

where

β = dimensionless dispersivity given by:

$$\beta = \alpha/X_c. \tag{17}$$

The remaining two dimensionless parameters characterize the coupling of the matrix to the fractures. The first, termed the dimensionless storage enhancement, is defined by the relation:

$$\varkappa' = (K\phi + K'\ \phi')/(K\phi). \tag{18}$$

The other fracture-matrix coupling parameter is defined by the equation:

$$\tau_c' = t_c'/t_c \tag{19}$$

where

$$t_c' = (L'^2 K')/(D'\ \theta'). \tag{20}$$

Physically, τ_c' represents the time required for the solute to penetrate from the fracture-matrix interface to the full length of the matrix block, and for the concentration there to reach approximately 68% of its interfacial value. In the remainder of this article, τ_c' will be identified as the dimensionless diffusion time.

Equation 14 indicates that there are two distinct modes of single-porosity transport that bound double-porosity behavior. If the matrix does not participate in the transport process ($\phi' = 0$), $\varkappa' = 1$ and the right-hand side of the equation reduces to $\hat{C}s$. This term indicates single-porosity transport within the fracture only. For this case, $\tau_c' = \infty$. That is, an infinite time is required for matrix diffusion. On the other hand, if the dimensionless time (τ varies as s^{-1}) becomes sufficiently large compared to the matrix-diffusion time (τ_c'),

$$\lim_{s \to 0} \tanh \sqrt{s\tau_c'} = \sqrt{s\tau_c'} \tag{21}$$

and the right-hand side of Equation 14 becomes $\varkappa' \hat{C}s$. This term represents single-porosity transport through the combined fracture-plus-matrix porosity. In this case, $\tau_c' = 0$, and the solute is instantaneously equilibrated between the matrix and the fracture. Figure 3 illustrates the limiting behavior of the double-porosity formulation.

Type Curves

Figures 3 through 6 present type curves of dimensionless breakthroughs at the accessible environment boundary ($\chi = 1$) as functions of dimensionless time, t/t_c. These four figures typify the general appearance and behavior of such curves relative to the four dimensionless parameters: diffusion time, τ_c'; storage enhancement, $\varkappa'$; velocity, ν; and dispersivity, β. Note, subscript zero refers to a predefined base-case

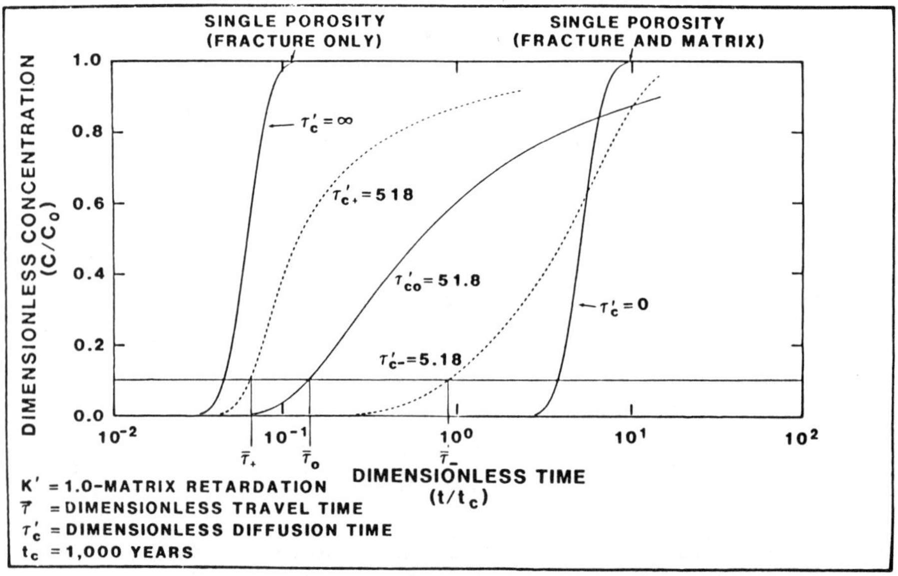

Figure 3. **Type curves demonstrating the variation of breakthrough with dimensionless diffusion time.**

value for the calculation, subscripts + and − refer to calculations shifted from the base-case value in the plus and minus directions, respectively, and − superscripts refer to dimensionless travel times for breakthrough concentrations of 10%. The curves are useful for demonstrating the qualitative sensitivities to variations in the dimensionless parameters and are not intended to quantify the effects for any real flow system.

As discussed previously, Figure 3 shows the two-single porosity curves forming an approximate envelope about the double-porosity behavior. Single-porosity fracture-plus-matrix changes to double-porosity and then to single-porosity fracture-only behavior as τ_c' increases from zero to infinity.

A horizontal line is shown intersecting each curve of Figure 3 at the 10% concentration level. For the purposes of this study, the times that correspond with the 10% concentration levels are defined as the breakthrough times for the performance measure, and will be used in calculating parameter sensitivity and importance. Additional calculations reported by Reeves et al. (1987) indicate that the conclusions presented in this document are valid for breakthrough times based on relative concentrations which vary from 1 to 50%.

Each point of interception in Figure 3 of the performance-measure curve with a double-porosity curve identifies a unique dimensionless travel time associated with a predefined performance measure value. Quantities $\bar{\tau}_+$, $\bar{\tau}_-$, and $\bar{\tau}_0$ show three such dimsionless travel times which correspond to dimensionless diffusion times of 518, 5.18, and 51.8, respectively. These values will be used later to generate parameter-variation curves, which will then provide individual parameter sensitivities.

Figure 4 illustrates the effect of storage enhancement within the matrix ($\varkappa'$). As anticipated, storage enhancement retards the fracture-plus-matrix single-porosity breakthrough by a factor of $\varkappa'$ relative to the fracture-only single-porosity breakthrough. Increasing the storage in the matrix increases the opportunity for matrix diffusion, resulting in a general delay in solute breakthrough.

Figure 5 shows the effect of dimensionless contaminant velocity in the fracture, ν, on the breakthrough calculation. Decreasing ν leads to a simple translation of the single-porosity envelope to later times. By virtue of a general increase in solute travel time, the opportunity for diffusion increases, producing a general lowering of the solute concentration.

Figure 6 shows the effect of increasing the dimensionless dispersivity, β. Because dispersivity can be conceptualized as being related to the variation in the solute velocity distribution within the fracture network along the travel path, material moving with a greater-than-average velocity breaks through earlier than material moving at the average velocity, and material moving with a less-than-average velocity breaks through later than material moving at the average velocity. Therefore, when β is increased, the range of velocities about the average is increased, and the range of travel times is appropriately smeared towards both earlier and later arrival times.

Parameter-Variation Curves

Plots of the relative travel time $\bar{\tau}/\bar{\tau}_0$ obtained from the interception points of the type curves and a predefined dimensionless breakthrough concentration versus the

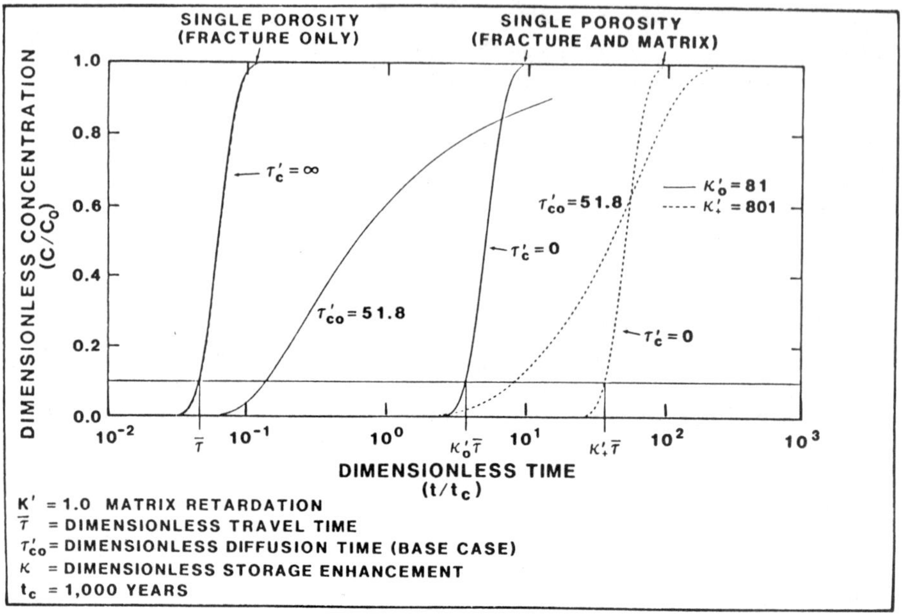

Figure 4. **Type curves demonstrating the variation of breakthrough with dimensionless storage enhancement.**

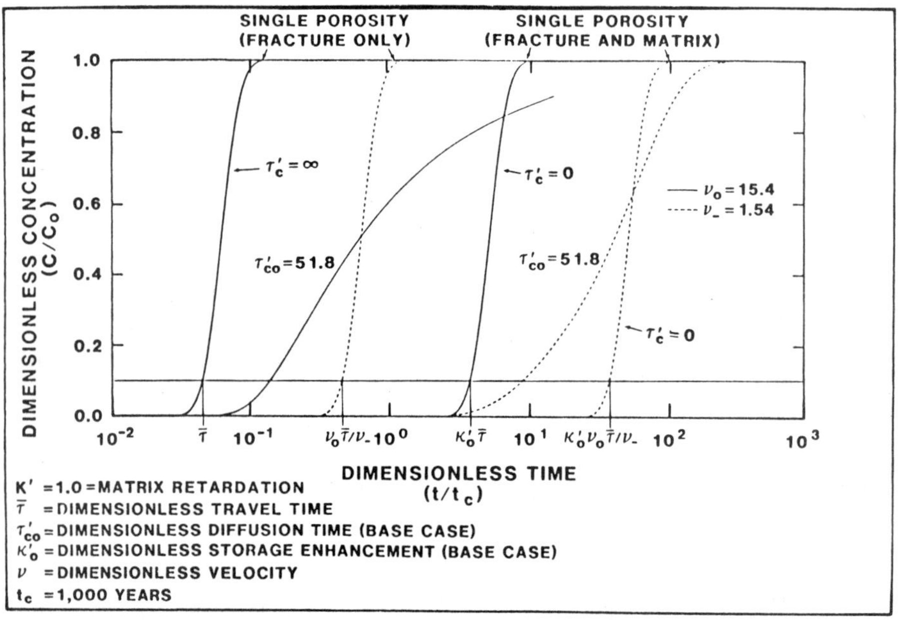

Figure 5. **Type curves demonstrating the variation of breakthrough with dimensionless velocity.**

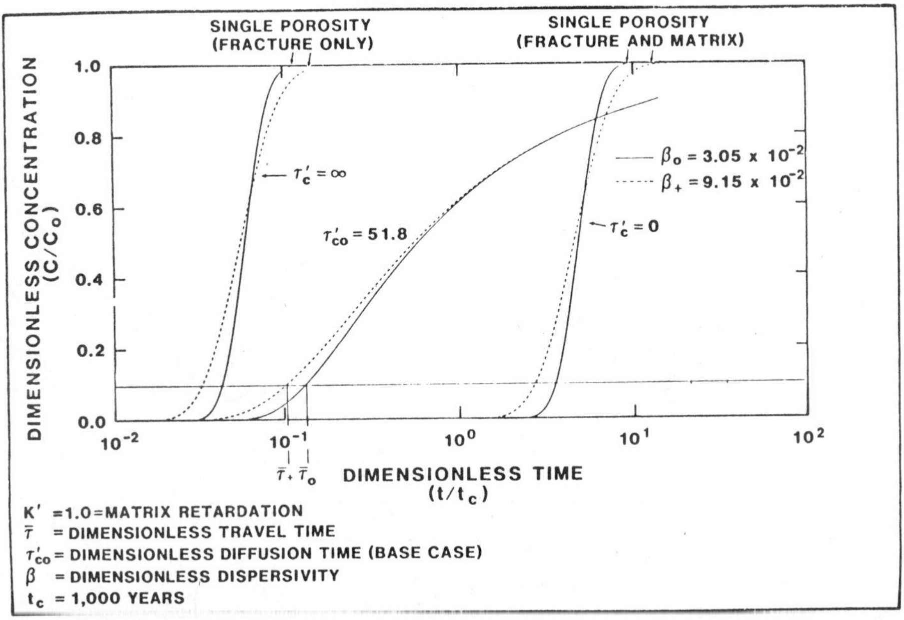

Figure 6. **Type curves demonstrating the variation of breakthrough with dimensionless dispersivity.**

full relative range of a given dimensionless variable (τ'_c/τ'_{c0}, ν_0/ν, $\varkappa'/\varkappa'_0$, or β/β_0) provide parameter-variation curves. A typical parameter-variation curve demonstrating relative travel time as a function of dimensionless diffusion time (τ_c'/τ_{c0}') is shown in Figure 7. The points plotted correspond to the data shown in Figure 3.

Figure 7 is presented for illustrative purposes. In actual practice, additional equally spaced points are determined near the base-case value (1,1) to facilitate taking an accurate numerical derivative for sensitivity calculations at the base-case point.

Parameter-variation curves show the response surface along one of four axes in dimensionless-parameter space. They can potentially reveal highly sensitive regions of nonlinearity that can be distant, in parameter-space, from the base-case location. Such curves complement determinations of differential sensitivities and parameter-importance coefficients, which depend on single-point, base-case evaluations and which might otherwise give no indication of highly nonlinear regions.

Parameter Sensitivity and Importance

Dimensionless-parameter sensitivity is represented by a 4 × 1 sensitivity vector, ψ. Normalized derivatives obtained from parameter-variation curves define each element of the vector:

$$\psi_i = \frac{\xi_i}{\bar{\tau}}\bigg|_0 \frac{\partial \bar{\tau}}{\partial \xi_i}\bigg|_0 \tag{22}$$

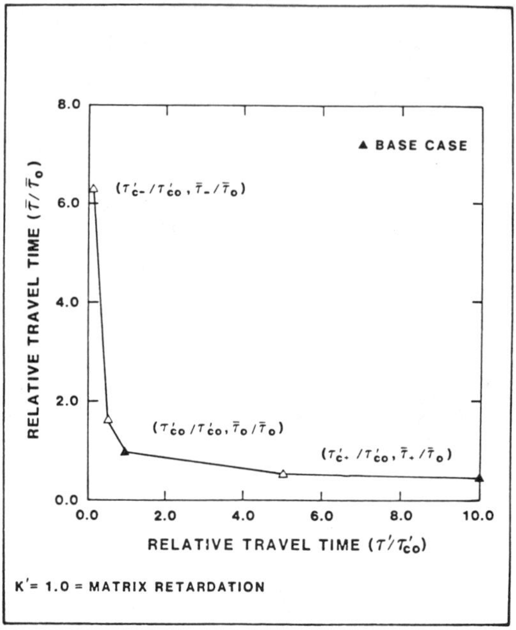

Figure 7. **Parameter-variation curve demonstrating travel time as a function of dimensionless diffusion time.**

where

$\bar{\tau}$ = dimensionless-travel time

$\xi_i = \tau'_c, \varkappa', \nu$, and β for i = 1 to 4,

respectively.

The normalized derivatives defined by Equation 22 are calculated using a two-point central difference about the base-case location, as indicated by the subscript 0. If the response is steep, additional calculations are performed at decreasing separations from the base-case value until a two-digit accuracy is achieved between successive calculations.

Sensitivity with respect to natural (dimensioned) parameters is represented by a 7 × 1 sensitivity vector S. Again, normalized derivatives about the base-case value define each element of the vector:

$$S_i = \frac{P_i}{\bar{\tau}}\bigg|_0 \frac{\partial \bar{\tau}}{\partial P_i}\bigg|_0 \qquad (23)$$

where

$P_i = D', \theta', L', \phi', \phi, \alpha$, and u for i = 1 to 7, respectively.

The sensitivity vector, S, is obtained from the dimensionless sensitivity vector, ψ, by means of a 7 × 4 transformation matrix, A:

$$S = A \psi \tag{24}$$

Normalized derivatives about the base case, again, define each element of A:

$$A_{ij} = \left.\frac{P_i}{\xi_j}\right|_0 \left.\frac{\partial \xi_j}{\partial P_i}\right|_0 \tag{25}$$

Parameter importance for the dimensional variables is represented by a 7 × 1 importance vector, I. This vector is obtained from the sensitivity vector S as follows:

$$I = | U S | \tag{26}$$

where the vertical bars denote absolute values. U is a diagonal matrix based on a relative-uncertainty matrix (relative range):

$$U = u_i \delta_{ij} \tag{27}$$

Here,

$$u_i = R_i/P_{i0} \tag{28}$$

where

$$R_i = \text{range of uncertainty for parameter i}$$

and

$$P_{i0} = \text{base-case value for parameter i.}$$

As indicated by Equation 26, a highly sensitive, but certain parameter ($u_i = 0$) has a zero importance. Contrastingly, a parameter with a small sensitivity ($S_i \cong 0$) and a relatively large uncertainty can be very important.

In order to reduce the uncertainty in a calculated performance measure, additional field and laboratory work can be performed to better define the base-case values of the physical parameters and reduce their range. Logically, reductions in the performance measure uncertainty should be accomplished by reducing the uncertainties in the parameters according to their rank in importance, i.e., starting with the most important and progressing to the least. In actual practice, the uncertainty-reduction process is constrained by both physical limitations and budgets. However, once the parameter importance ranking has been established, the analyst can optimally design future studies by taking all factors into account.

Sensitivity and Importance Calculations for the WIPP Site

The Waste Isolation Pilot Plant, located in southeastern New Mexico (Figure 1), is currently being evaluated to further quantify its performance in providing permanent storage for transuranic (TRU) wastes. As part of this quantification, sensitivity and importance calculations were performed by Reeves et al. (1987) for transport of an infinitely long-lived radioactive solute through the Culebra dolomite, the most transmissive member of the Rustler Formation (Mercer, 1983).

The Culebra dolomite is a dolostone predominantly composed of medium silt-sized dolomite crystals. The crystals are uniformly sorted and tightly packed. Detrital components such as quartz, mica, clay minerals, and gypsum have been identified in core samples (Core Laboratories, 1986). Vugs and pits, possibly bedding-plane related, range in size from less than 1 mm to as large as 10 cm. Of special significance to the present study's conceptualization is the following observation: outcrops, cores, shaft mappings, hydraulic tests (Beauheim, 1986, 1987a,b), and tracer-test interpretations (Kelley and Pickens, 1986) indicate that a significant portion of the Culebra above the WIPP site is fractured.

For this study, a high-transmissivity fracture-controlled path is assumed to provide the means for transport of long-lived radionuclides through the Culebra to the accessible environment. The solute is injected at a constant rate under a natural-head gradient into a position in the Culebra that coincides with the upward projection of the centroid of the WIPP facility. The solute is then transported one-dimensionally along a flow path to the accessible environment (Figure 2, path C). Flow paths for this study were obtained from calculations performed by Haug et al. (1987) using a two-dimensional regional conceptualization of the Culebra. Both matrix diffusion and sorption retard the movement of the solute along the flow path. Properties along the flow path were assumed to be homogeneous and the Darcy flux in the fracture was assumed to be the average flux observed along the flow path of the two-dimensional simulation.

Table 1 presents the base-case values and ranges for the nine parameters used to characterize solute transport in the Culebra. Parameter ranges and base-case values depict the uncertain properties of the Culebra, while simulations with the SWIFT II computer code exhibit the corresponding ranges in travel time, defined as the time for the dimensionless solute concentration at the accessible environment boundary to reach 10% of its initial injection value. In general, the rationale used to select the base-case values and ranges for the parameters is based on tracer-test interpretations (Kelley and Pickens, 1986), regional hydraulic-test simulations (Haug et al., 1987), laboratory measurements (Core Laboratories, 1986), and literature reviews (Reeves et al., 1987). The values shown in Table 1, thus, represent best estimates for the base-case values and associated uncertainties for the Culebra dolomite. A complete discussion of the selection rationale is given by Reeves et al. (1987). Dimensionless transformations of the parameter base-case values and ranges are given in Tables 2 through 4.

Figure 8 shows base-case breakthrough curves for each of four values of matrix retardation ($K' = 1.0$, 150, 1490, and 1.49×10^5) that correspond to matrix-partition coefficients, K_d', of 0., 10, 100, and 10,000 ml/g. Due to limited surface area for reaction within the fractures, all calculations assumed a fracture retardation

Table 1. **Summary of the range and base-case values adopted for each double-porosity transport parameter.**

Parameter	Symbol	Base Case	Range	Units	Relative Range[c]
Free-water diffusivity	D'	2×10^{-5}	$4 \times 10^{-6} - 9 \times 10^{-5}$	cm^2/s	4.30
Matrix tortuosity	θ'	0.15	0.05 – 0.5		3.00
Matrix-block length	L'	1.2	0.25 – 7.0	m	5.63
Matrix porosity	ϕ'	0.16	0.07 – 0.30		1.44
Fracture porosity	ϕ	2×10^{-3}	$2 \times 10^{-4} - 2 \times 10^{-2}$		9.90
Longitudinal dispersivity	α	100	50 – 300	m	2.50
Fracture flux	u	3.2×10^{-9}	$2.2 \times 10^{-9} - 4.7 \times 10^{-9}$	m/s	0.78
Matrix distribution coefficient	k_d'	(a)	(a)	ml/g	(d)
Fracture distribution coefficient	k_d	(b)	(b)	ml/g	(d)

(a) This study uses four fixed values of the matrix distribution coefficient: $k_d' = 0, 10, 10^2$, and 10^4 ml/g.

(b) This study neglects sorption in the fracture (see text) and sets $k_d = 0$.

(c) Magnitude of range divided by base-case value, a dimensionless quantity.

(d) Not applicable.

Table 2. **Evaluation of base-case values of the dimensionless transport parameters in terms of matrix retardation K′.**

Parameter	Equation[a]	Base Case
Diffusion time	$\tau_c' = L'^2K'/\theta D't_c$	$0.152K'$
Storage enhancement	$\varkappa' = 1 + K'\phi'/K\phi$	$1 + 80.0K'$
Dimensionless velocity	$\nu = (u/K\phi)(t_c/x_c)$	15.4
Dimensionless dispersivity	$\beta = \alpha/x_c$	0.0305

(a) Characteristic dimensions are $t_c = 1{,}000$ y and $x_c = 3{,}280$ m.

Table 3. **Evaluation of ranges of the dimensionless transport parameters as a function of matrix retardation K′.**

Natural Parameter				Dimensionless Parameter Range[a]			
	Symbol	**Range**	**Unit**	$\tau_c' = L'^2K'/\theta'D't_c$	$\varkappa' = 1 + K'\phi'/K\phi$	$\nu = (u/K\phi)(t_c/x_c)$	$\beta = \alpha/x_c$
Free-water diffusivity	D'	$7 \times 10^{-6} - 9 \times 10^{-5}$	cm^2/s	$(0.0338 - 0.435)K'$			
Matrix tortuosity	θ'	0.05 – 0.5		$(0.0457 - 0.457)K'$			
Matrix-block length	L'	0.25 – 7.0	m	$(0.00661 - 5.18)K'$			
Matrix porosity	ϕ'	0.07 – 0.30			$(1 + 35K') - (1 + 150K')$		
Fracture porosity	ϕ	$2 \times 10^{-4} - 2 \times 10^{-2}$			$(1 + 8\,K') - (1 + 800K')$	1.54 – 154	
Fracture dispersivity	α	50 – 300	m				0.0152 – 0.0915
Fracture flux	u	$2.2 \times 10^{-9} - 4.7 \times 10^{-9}$	m/s			10.6 – 22.6	
Maximum Range				$(0.00661 - 5.18)K'$	$(1 + 8K') - (1 + 800K')$	1.54 – 154	0.0152 – 0.0915

(a) Characteristic dimensions are $t_c = 1{,}000$ y and $x_c = 3{,}280$ m.

Table 4. **Summary of range and base-case values computed for the dimensionless parameters of each fixed value of the retardation.**

Retardation			Dimensionless Parameter(a)			
K′			τ_c'	$\varkappa'$	ν	β
1.00	Base		0.152	81.0	15.4	3.05×10^{-2}
	Range	Lower	6.61×10^{-3}	9.00	1.54	1.52×10^{-2}
		Upper	5.18	801	154	9.15×10^{-2}
1.50×10^{2}	Base		22.9	1.20×10^{4}	(b)	(b)
	Range	Lower	0.994	1.20×10^{3}	(b)	(b)
		Upper	7.79×10^{2}	1.20×10^{5}	(b)	(b)
1.49×10^{3}	Base		2.27×10^{2}	1.20×10^{5}	(b)	(b)
	Range	Lower	9.88	1.20×10^{4}	(b)	(b)
		Upper	7.74×10^{3}	1.20×10^{6}	(b)	(b)
1.49×10^{5}	Base		2.27×10^{4}	1.19×10^{7}	(b)	(b)
	Range	Lower	9.87×10^{2}	1.19×10^{6}	(b)	(b)
		Upper	7.74×10^{5}	1.19×10^{8}	(b)	(b)

(a) Definition of parameters:
τ_c' = diffusion time,
$\varkappa'$ = storage enhancement,
ν = dimensionless velocity,
β = dimensionless dispersion.
(b) Value does not vary from that for K′ = 1.00.

value (K) equal to 1.0 (Reeves et al., 1987). As indicated in Figure 8, the effect of matrix retardation is very important, and the time required for breakthrough ($C/C_0 = 0.1$) is directly proportional to K′. This type of behavior does not generally apply to double-porosity solute transport. The behavior seen here is a direct consequence of the double-porosity parameters used to describe transport in the Culebra (Reeves et al., 1987). While the effects of matrix retardation are clearly significant, calculations performed by Reeves et al. (1987) indicate that parameter sensitivity and importance are, for all practical purposes, independent of the value used for K′. That is, for any given value of matrix retardation, the sensitivities and importance of the other parameters are nearly invariant.

Figure 9 shows type curves that demonstrate the variation of breakthrough with dimensionless diffusion time, τ_c', for a K′ value of 150. These curves characterize solute breakthrough for the full range of values for dimensionless diffusion time. All other parameters are set to their base-case values (Table 3). The transport behavior

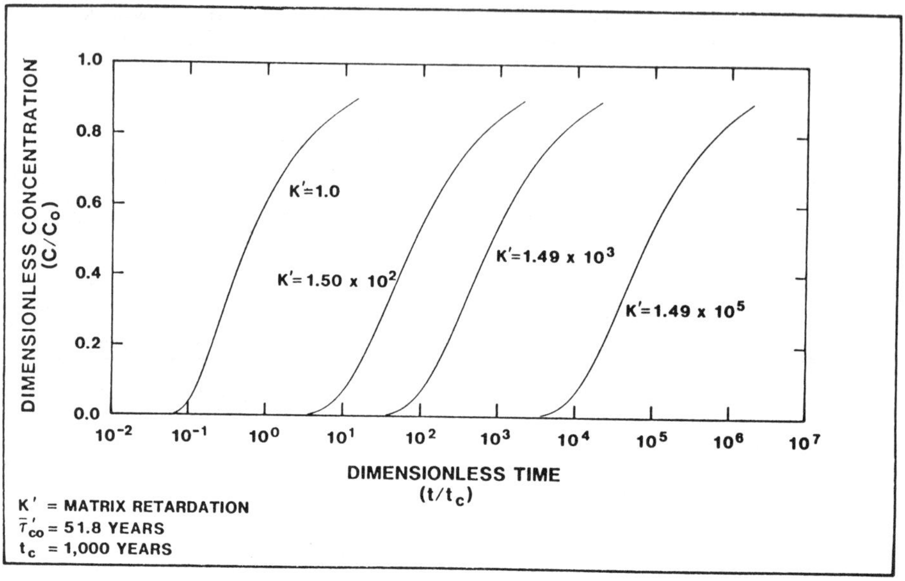

Figure 8. **Base-case breakthrough curves for fixed values of matrix retardation.**

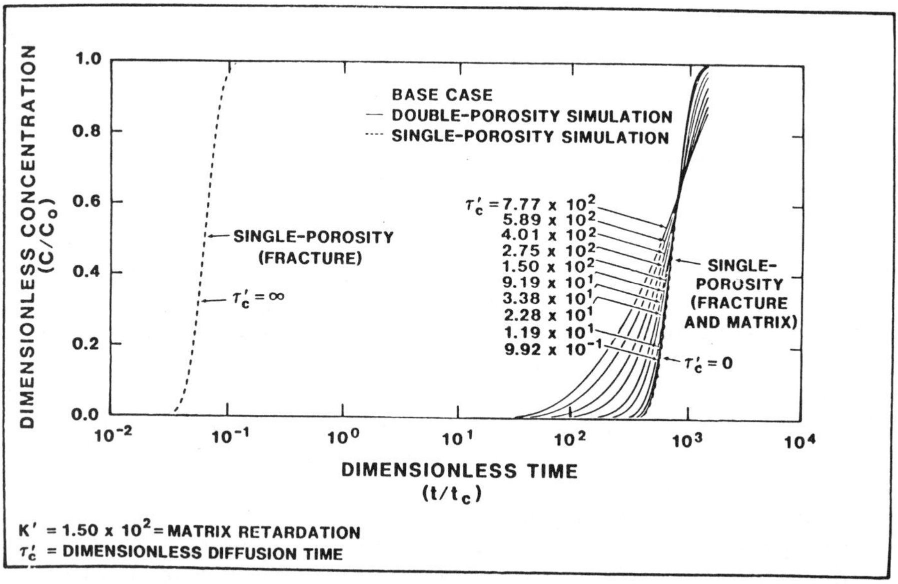

Figure 9. **Type curves demonstrating the variation of breakthrough with dimensionless diffusion time for the Culebra dolomite.**

for the base-case value of τ_c' (dotted line) is very similar to that calculated for the single-porosity fracture-plus-matrix limit. This finding appears to explain the single-porosity behavior noted earlier for the linear variation observed in matrix retardation (Figure 8).

Figure 10 recasts the type curves shown in Figure 9 into the form of a parameter-variation curve at a breakthrough time corresponding to $C/C_0 = 0.1$. Data points are shown for matrix retardation of 150 and 1.0. Differences between the two sets of points are negligible, as are parameter-variation curves produced by matrix retardations of 1490 and 1.49×10^5 as mentioned above.

Some nonlinearity is indicated in Figure 10, with the base-case value lying in the most rapidly varying portion of the curve. Overall response near the base-case value is, however, sufficiently linear to make use of the differential approach for approximating the sensitivity of the parameter. As seen in Figure 10, a sensitivity value computed about the base-case represents a nearly maximum value for the prescribed range of diffusion time. Results indicate that increasing τ_c' by a factor of 15 decreases the travel time to approximately one-half of the base-case value, independent of K′.

Figures 11 through 15 give type and parameter-variation curves for the following dimensionless parameters: storage enhancement, $\varkappa'$; velocity, ν; and dispersivity, β. As used in this discussion, storage refers to both solid-phase (sorbed) and fluid-phase concentrations. For a fracture, this storage is proportional to the product of fracture

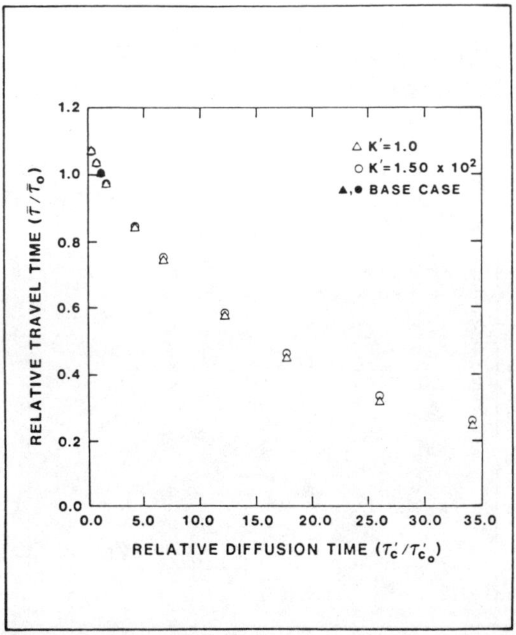

Figure 10. **Parameter-variation curves showing travel time as a function of relative diffusion time for the Culebra dolomite.**

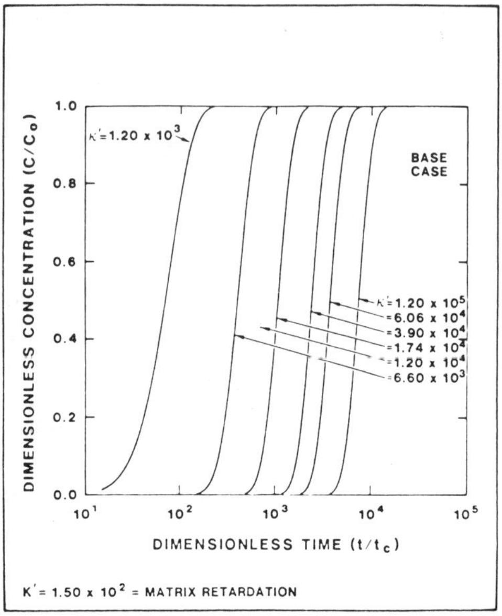

Figure 11. **Type curves demonstrating the variation of breakthrough with dimensionless storage enhancement for the Culebra dolomite.**

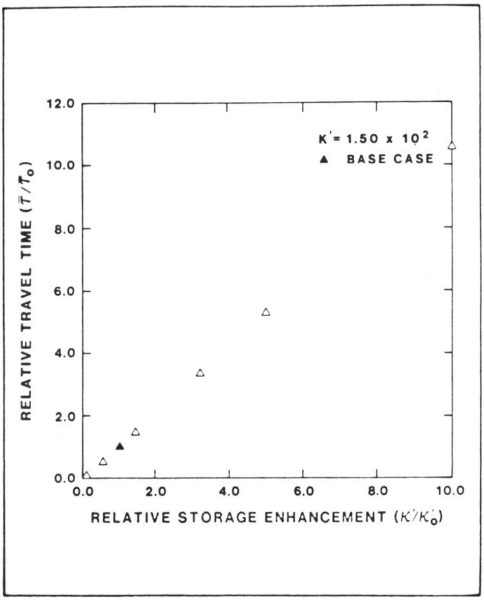

Figure 12. **Parameter-variation curve showing travel time as a function of relative storage enhancement for the Culebra dolomite.**

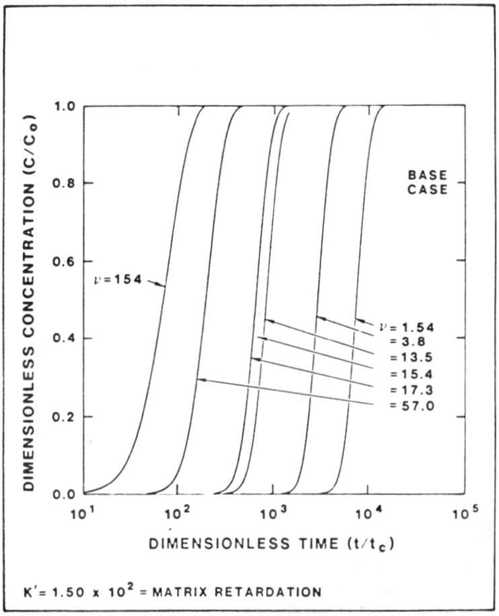

Figure 13. **Type curves demonstrating the variation of breakthrough with dimensionless velocity for the Culebra dolomite.**

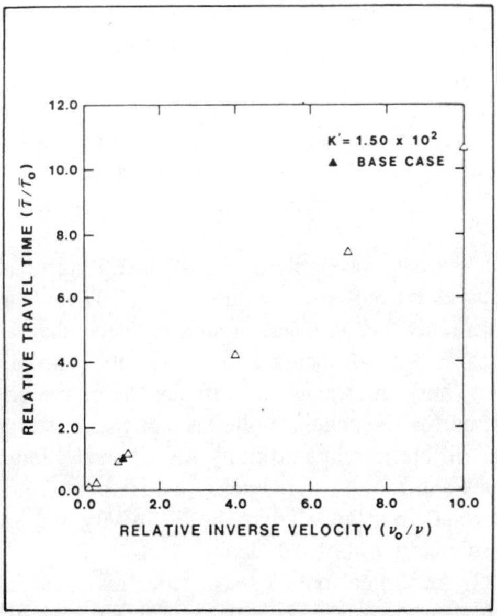

Figure 14. **Parameter-variation curve showing travel time as a function of relative inverse velocity for the Culebra dolomite.**

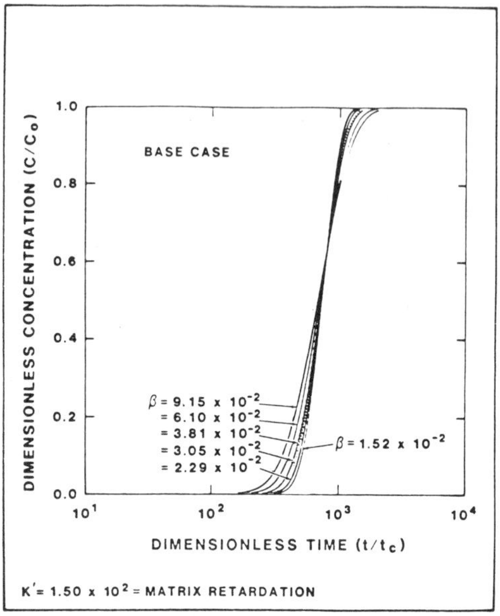

Figure 15. **Type curves demonstrating the variation of breakthrough with dimensionless dispersivity for the Culebra dolomite.**

retardation, K, and porosity. Dimensionless storage enhancement, $\varkappa'$, represents the factor by which the rock matrix increases the total storage relative to its value for the fracture. Dimensionless velocity and dispersivity represent simple transformations of their natural-parameter counterparts. For each of the dimensionless parameters, parameter-variation curves vary insignificantly with matrix retardation, and only curves for $K' = 150$ are shown. In all cases, the parameter-variation curves are seen to be nearly linear, thus allowing use of the differential approach for calculating parameter sensitivity.

Sensitivities about the base-case values (Equation 22) are obtained from the parameter-variation curves by numerical differentiation. Table 5 lists the values calculated for the Culebra dolomite. A negative sign indicates that travel time (performance measure) decreases with an increase in the variable, while a positive sign indicates the opposite. Thus, an increase in diffusion time, velocity, or dispersivity decreases the travel time for the solute, while an increase in storage enhancement increases it. The large differences in sensitivity for $\varkappa'$ and τ_c' indicate that for the chosen base-case values and ranges, the Culebra matrix primarily functions as a storage unit, and that its kinetic attributes, given by diffusivity and tortuosity, do not, in general, interfere significantly with this function.

Using Equation 23, the dimensionless sensitivities given in Table 5 are transformed back into their natural-space counterparts (Table 6). Multiplying the absolute value of each sensitivity by its appropriate relative range then yields parameter-importance coefficients. For added clarity, Table 6 ranks parameter importance with

Table 5. **Parameter sensitivity for the dimensionless parameters.**

Parameter	Sensitivity[a]
Diffusion time	−0.064
Storage enhancement	1.1
Dimensionless velocity	−1.1
Dimensionless dispersivity	−0.17

(a) Results are valid for retardations K′ = 1, 1.50×10^2, 1.49×10^3, and 1.49×10^5.

Table 6. **Parameter sensitivities and parameter-importance coefficients for the natural parameters.**

Parameter	Symbol	Sensitivity[a]	Importance Coefficient[a]	Importance Ranking
Free-water diffusivity	D′	0.064	0.28	4
Matrix tortuosity	θ'	0.064	0.19	5
Matrix-block length	L′	−0.13	0.72	3
Matrix porosity	ϕ'	1.1	1.5	1
Fracture porosity	ϕ	(b)	(c)	6,7
Fracture dispersivity	α	−0.17	0.025	6,7
Fracture flux	u	1.1	0.83	2

(a) Results are valid for retardations K′ = 1, 1.50×10^2, 1.49×10^3, and 1.49×10^5.

(b) Less than 0.0125.

(c) Less than 0.124.

rank 1 representing the parameter having the most importance and, hence, the most need for accurate characterization for regional transport calculations. Matrix porosity ranks as the most important parameter, and is followed by Darcy flux in the fracture, and matrix-block length. Fracture porosity and longitudinal dispersivity rank as the least important parameters of all. The two matrix kinetic parameters, diffusivity and tortuosity, attain intermediate to low importance for the Culebra dolomite. Rankings are the same for all values of matrix retardation used in the calculations.

With respect to regional double-porosity solute transport, the Culebra, though apparently fractured, appears to possess excellent retardation properties for the assumed system conceptualization, range of properties, and flow field. The rock matrix serves strictly a storage function; kinetic effects in the matrix are minimal. In fact, this study shows a nearly ideal behavior of the matrix on a regional scale under the conditions of a low-pressure natural-gradient breach scenario. It indicates that under the assumed conditions, transport times through the Culebra dolomite from a point located above the WIPP facility to the accessible environment can be approximated using a single-porosity approach with a porosity equal to the combined porosity of the fracture and matrix.

Summary and Conclusions

In this paper we presented the development of a methodology for calculating the sensitivity and importance of parameters controlling the breakthrough time at the accessible environment of a solute in a fractured, double-porosity system. The calculations were facilitated by the use of four dimensionless parameters, and the use of both type and parameter-variation curves.

The methodology was then demonstrated for transport of a sorbing solute through a fractured portion of the Culebra Dolomite Member of the Rustler Formation at the WIPP site. The results of these calculations indicate that the most important parameters for a natural-gradient breach scenario and the assumed base-case values, ranges, and flow field are matrix retardation, matrix porosity, Darcy flux in the fractures, and matrix-block length. Longitudinal dispersion in the fracture had a significantly lower importance as did the matrix kinetic properties of diffusivity and tortuosity.

By ranking the importance of the transport parameters for the Culebra dolomite simulation, we can optimally design future field and laboratory work to reduce the uncertainty in transport travel-time calculations. Although the results presented in this paper are highly relevant to programs at the WIPP site, they are very site and scenario dependent. Parameter sensitivities and importance for other sites, or other scenarios, require their own specific set of calculations and analyses.

Literature Cited

1. Bear, J., 1972. *Dynamics of Fluids in Porous Media.* American Elsevier Publishing Company, New York, New York, 764 p.
2. Beauheim, R. L., 1986. *Hydraulic-Test Interpretations for Well DOE-2 at the Waste Isolation Pilot Plant (WIPP) Site.* Sandia National Laboratories, SAND86-1364, 84 p.

3. Beauheim, R. L., 1987a. *Analysis of Pumping Tests of the Culebra Dolomite Conducted at the H-3 Hydropad at the Waste Isolation Pilot Plant (WIPP) Site.* Sandia National Laboratories, SAND86-2311.
4. Beauheim, R. L., 1987b. *Interpretations of Single-Well Hydraulic Tests Conducted at and Near the Waste Isolation Pilot Plant (WIPP) Site, 1983–1987.* Sandia National Laboratories, SAND87-0039 (in preparation).
5. Core Laboratories, Inc., 1986. ''Special Core Analysis Study for INTERA Technologies, WIPP Site,'' File Number: SCAL 203-850073. Core Laboratories Inc., Aurora, Colorado.
6. Freeze, R. A., and J. A. Cherry, 1979. *Groundwater,* Prentice-Hall, Inc., Englewood Cliffs, New Jersey.
7. Haug, A., V. A. Kelley, A. M. LaVenue, and J. F. Pickens, 1987. *Modeling of Ground-Water Flow in the Culebra Dolomite at the Waste Isolation Pilot Plant (WIPP) Site: Interim Report.* Sandia National Laboratories, Contractor Report SAND86-7167.
8. Kelley, V. A., and J. F. Pickens, 1986. *Interpretation of the Convergent Flow Tracer Tests Conducted in the Culebra Dolomite at the H-3 and H-4 Hydropads at the Waste Isolation Pilot Plant (WIPP) Site.* Sandia National Laboratories, Contractor Report SAND86-7161.
9. Mercer, J. W., 1983. *Geohydrology of the Proposed Waste Isolation Pilot Plant Site, Los Medaños Area, Southeastern New Mexico.* U.S. Geological Survey, Water-Resources Investigation Report 83-4016, 113 p.
10. Reeves, M., D. S. Ward, N. D. Johns, and R. M. Cranwell, 1986a. *Theory and Implementation for SWIFT II, the Sandia Waste-Isolation Flow and Transport Model, Release 4.84.* Sandia National Laboratories, NUREG/CR-3328 and SAND83-1159, 189 p.
11. Reeves, M., D. S. Ward, N. D. Johns, and R. M. Cranwell, 1986b. *Data Input Guide for SWIFT II, the Sandia Waste-Isolation Flow and Transport Model for Fractured Media, Release 4.84.* Sandia National Laboratories, NUREG/CR-3162 and SAND83-0242, 144 p.
12. Reeves, M., D. S. Ward, P. A. Davis, and E. J. Bonano, 1986c. *SWIFT II Self-Teaching Curriculum: Illustrative Problems for the Sandia Waste-Isolation Flow and Transport Model for Fractured Media.* Sandia National Laboratories, NUREG/CR-3925 and SAND84-1586, 96 p.
13. Reeves, M., V. A. Kelley, and J. F. Pickens, 1987. *Regional Double-Porosity Solute Transport in the Culebra Dolomite: An Analysis of Parameter Sensitivity and Importance at the Waste Isolation Pilot Plant (WIPP) Site.* Sandia National Laboratories, Contractor Report SAND87-7105 (in preparation).
14. Streltsova-Adams, T. D., 1978. ''Well Hydraulics in Heterogeneous Aquifer Formations.'' *Advances in Hydroscience,* V. T. Chow, ed., Vol. 11, pp. 357–423.
15. U.S. Environmental Protection Agency, 1985. ''Environmental Standards for the Management and Disposal of Spent Nuclear Fuel, High-Level and Transuranic Radioactive Wastes (Final Rule),'' 40 CFR Part 191, *Federal Register,* Vol. 50, No. 182, pp. 38066–38089, September 19, 1985.
16. U.S. Nuclear Regulatory Commission, 1983. ''Disposal of High-Level Radioactive Wastes in Geologic Repositories: Licensing Procedures,'' 10 CFR Part 60, *Federal Register,* June 30, 1983.

The Sensitivity Index as a Screening Tool in the Analysis of Ground-Water Contaminant Transport

ROBERT A. FJELD and BEN L. SILL
Clemson University
Clemson, South Carolina

BRIAN B. LOONEY
E. I. du Pont de Nemours and Company
Savannah River Laboratory
Aiken, South Carolina

Abstract

A method is presented that incorporates features of Tomovic's sensitivity index to facilitate the time-dependent sensitivity analysis of ground-water contaminant transport in a radioactive risk assessment application. The basic approach involves specification of nominal, minimum, and maximum values of the transport parameters, calculation of contaminant concentrations corresponding to the parameter values, and, finally, calculation of the sensitivity index as a function of time. A key aspect of the method is definition of an appropriate sensitivity index. In the application presented here, an index is defined that emphasizes deviations from nominal concentrations that are significant relative to the maximum nominal concentration and, at the leading edge of the contaminant breakthrough curve, deemphasizes the uncertainties, which are large on a relative basis but small on an absolute basis.

The method is illustrated using a simple one-dimensional ground-water advection-dispersion model and is applied to a radioactive waste disposal site. An uncertainty analysis of the same site is performed by Monte Carlo simulation for comparison. The sensitivity index is found to be a practical, easily implemented screening tool for identifying the critical parameters, determining the time periods over which parameter uncertainty is important, and obtaining an upper bound estimate of the magnitude of the uncertainty. Precise estimates of the uncertainty can subsequently be obtained by more elegant techniques. Although the application presented here involves low-level radioactive waste, the method and general conclusions regarding the use of a sensitivity index in ground-water transport modeling might be useful in evaluating high-level or hazardous waste disposal problems as well.

Introduction

The Savannah River Laboratory recently completed an effort to assess the impact of closure options for radioactive, hazardous, and mixed waste sites at the Savannah River Plant (SRP). A key element of this effort is the estimation of health risks due to various pathways of contaminant transport over the time period from 100 to 1000 years following closure of a site. The risk assessments are performed with a multipathway risk assessment code (*1*) that includes a variety of atmospheric, surface water, and ground-water pathways. For a given transport pathway, the code returns a nominal value of risk for a specified set of nominal values of the input parameters. A separate effort is being conducted to evaluate model sensitivity and uncertainty.

The approach selected for performing the sensitivity/uncertainty analysis is influenced by some general characteristics of the problem. One is that the potential number of sites at which the technique could be applied is quite large. This, coupled with the fact that the time period for the analysis ranges over a 1000-year time span, tends to mitigate against computationally intensive approaches. Another is that the uncertainty in some of the parameters is quite large, in some instances varying over four orders of magnitude, and that few site-specific data are available for quantifying the uncertainty. This effectively eliminates first order error propagation as an alternative. Finally, the solution to the contaminant transport model is essentially a step breakthrough curve, modified by radioactive decay. Although the relative uncertainty is extremely large at the leading and trailing edges of the step, the uncertainty during those periods is of little practical concern because of the extremely low concentrations.

This paper is concerned with a sensitivity/uncertainty analysis technique that can serve as a screening tool to identify the parameters having the greatest impact on uncertainty, to determine bounds on the uncertainty of the model, and to determine the time periods during which the sensitivity is of practical concern. The initial screening could then be followed by one of the more rigorous techniques that are available (*2–4*). The objectives of the paper are (1) to present the technique, (2) to illustrate the technique using a ground-water transport problem, and (3) to compare the technique to a Monte Carlo simulation. Based on the comparison, the sensitivity index was found to be a practical tool to identify the critical parameters in a model, as well as the time periods of importance.

Sensitivity Analysis Technique

The approach adopted for the study is largely intuitive, although it incorporates two features of "Tomovic" sensitivity analysis (*5–7*). One is the use of nominal, maximum, and minimum values of parameters rather than parameter distributions. This feature is appropriate given the general sparsity of data for quantifying parameter uncertainty. The other is the use of a relative parameter, the sensitivity index, to quantify uncertainty in calculated contaminant concentrations.

The procedure is as follows: First, nominal, minimum, and maximum values of the input parameters are specified; and the contaminant concentration is determined for the nominal values. The minimum and maximum values of a given parameter are then substituted for its nominal value, keeping the remaining values constant. This yields bounds on concentration, i.e., C_t^{pmin} corresponding to the minimum value of parameter p, and C_t^{pmax} corresponding to the maximum value of parameter p. All of these concentrations vary over time. The value of C_t^{pmin} may be greater or less than C_t^{pmax}, depending on the functional dependence of concentration on p.

The specific definition selected for the sensitivity index is largely intuitive and is dictated by the objectives. Here, the objectives are not only to identify the critical parameters, but also to quantify bounds. The following definition is proposed:

$$S_t = \frac{C_t - C_t^{nom}}{C_{max}^{nom}} \tag{1}$$

where S_t is the sensitivity index, C_t^{nom} is the concentration based upon nominal values of the parameters, C_t is either C_t^{pmin} or C_t^{pmax}, and C_{max}^{nom} is the maximum instantaneous value of concentration based upon nominal values of the parameters (i.e., the nominal concentration at the "maximum" time).

The sensitivity index defined by Equation (1) is essentially the deviation of the bounding concentration from the nominal, scaled to the maximum nominal concentration. The use of the maximum nominal concentration for scaling is appropriate in this exercise because it minimizes the importance of deviations from the nominal concentration that are large on a fractional basis if those concentrations are $<<C_{max}^{nom}$. The sensitivity index is close to zero when the difference between C_t and C_t^{nom} is small compared to the maximum nominal concentration. The index is negative when C_t is less than the nominal concentration and is positive when it is greater. As illustrated in a subsequent section, this definition permits identification of the parameters having the largest impact on sensitivity and quantification of the bounds on predictions. For some models, the combined effect of uncertainty in all parameters can be determined by simultaneously substituting various combinations of maximum and minimum parameter values for nominal values. However, this must be done with caution because C_t^{pmin} (or C_t^{pmax}) may be less than, equal to, or greater than C_t^{nom}, depending on the relationship between C_t and p at a given time.

Ground-Water Transport Model

Generic waste site

The ground-water pathway incorporated into the PATHRAE risk assessment code was developed for the generic waste site depicted in Figure 1. The contaminant leaches from the site at a rate $\lambda_L I_o$, where I_o is undecayed inventory, and $\lambda_L(y^{-1})$ is fractional leach rate. Water percolating through the unsaturated zone at a volumetric flow rate $q_c(m^3/y)$ carries the contaminant vertically to the unsaturated zone. There it displaces and assumes the same velocity as the ground-water. The contaminant then undergoes advection at the mean ground-water velocity and dispersion as it migrates toward a point of use at a well or an outcrop.

Nominal, minimum, and maximum values of select parameters for a generic site are presented in Table 1. The nominal values in the table, although of the same order of magnitude as typical values for SRP waste sites, are selected for convenience in illustrat-

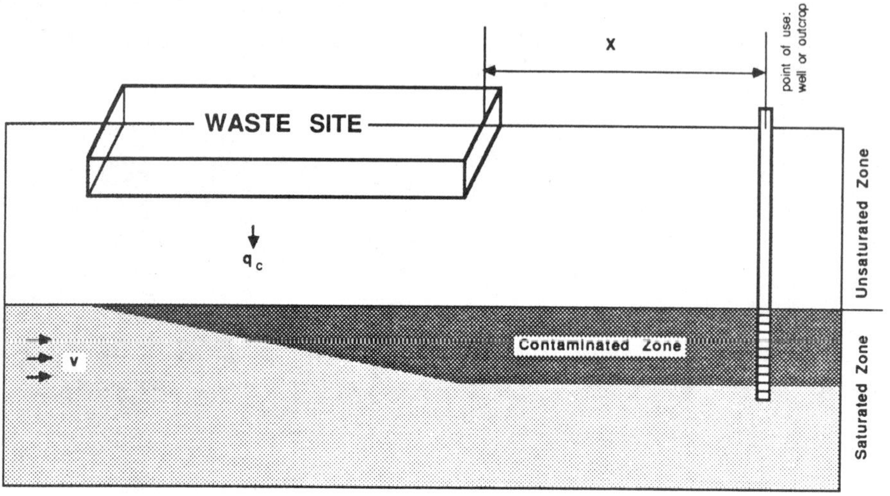

Figure 1. **Generic waste site.**

ing the sensitivity analysis. Minimum and maximum values are taken as a factor of 10 less than or greater than the nominal values for all parameters except the leach rate, for which a factor of 100 is used. These uncertainties are probably high for most parameters and for most sites. The transport time in the unsaturated zone is not varied in this example because the half-life is taken to be long compared to transport times. For a decaying species, uncertainty in the transport time in the unsaturated zone can obviously have a large impact on concentration uncertainty.

Transport equation

The transport model for the ground-water pathway is an adaptation and extension of a solution by Burkholder and Rosinger (*8*) to the one-dimensional advection-dispersion

Table 1. **Nominal, minimum, and maximum values of parameters used in generic sensitivity analysis.**

Parameter	Symbol	Units	Nominal	Minimum	Maximum
Retarded velocity	v/R	m/y	10	1	100
Longitudinal dispersivity	α_ℓ	m	10	1	100
Leach rate	λ_L	y^{-1}	0.01	0.0001	1
Inventory/flow rate	I_o/q_c	s/m^3	1	0.1	10
Transport time in unsaturated zone	t_u	y	10	10	10

equation for a point, decaying band source. The extensions incorporated into the model include plug flow transport through the unsaturated zone, the summation of point sources to approximate a line source, modification of the inventory to account for continuous addition of material to the site over its lifetime, and a multiplicative approximation to account for transverse dispersion. For the purposes of the present study, the only extension of the basic Burkholder and Rosinger solution is plug flow transport through the unsaturated zone.

The one-dimensional transport equation can be written as

$$\frac{\partial C}{\partial t} = -\frac{v}{R}\frac{\partial C}{\partial t} + \alpha_\ell \frac{v}{R}\frac{\partial^2 C}{\partial t^2} - \lambda C \tag{2}$$

where t is time, v is ground-water velocity, α_ℓ is longitudinal dispersivity, R is retardation factor, C is concentration, λ is decay constant, and the x-direction is aligned with the ground-water velocity vector. The decaying band boundary conditions are

$$C(x,0) = 0$$

$$\frac{\partial C}{\partial t}(\infty, t) = 0$$

$$C(0,t) = \begin{cases} \dfrac{\lambda_L I_o}{q_c}\exp(-\lambda t) & 0 \le t \le T \\ 0 & t > T \end{cases} \tag{3}$$

where $T = 1/\lambda_L$. The solution is

$$C(x,t) = \begin{cases} \dfrac{\lambda_L I_o}{2q_c}\exp(-\lambda t_d) g(t) & 0 \le t \le T \\ \dfrac{\lambda_L I_o}{2q_c}\exp(-\lambda t_d)[g(t) - g(t-1/\lambda_\ell)] & t > T \end{cases} \tag{4}$$

where

$$g(t') = \text{erfc}\left[\frac{Rx - vt'}{2\sqrt{\alpha_\ell v R t'}}\right] + \exp\left[\frac{x}{\alpha_\ell}\right] \text{erfc}\left[\frac{Rx + vt'}{2\sqrt{\alpha_\ell v R t'}}\right].$$

The time, t, is travel time in the saturated zone and the decay time, t_d, is given by

$$t_d = t_u + t, \tag{5}$$

where t_u is travel time in the unsaturated zone. It can be shown that Equation 4 is equivalent to solution C-13 in van Genuchten and Alves (*9*) for the special case of a contaminant whose decay rate in the waste site is equal to its degradation rate in the aquifer.

Results

The general behavior of the contaminant pulse is illustrated in Figure 2, where the nominal concentration is given as a function of time for half-lives of infinity and 10 years. The waste is introduced into the site at t = 0. The concentration front arrives at a receptor 100 m downgradient approximately 20 years later after 10-year travel times in both the unsaturated and saturated zones. The pulse width is approximately 100 years (corresponding to the fractional leach rate of 0.01 y^{-1}).

The sensitivity of concentration to v/R is shown in various ways in Figure 3. The calculations are based upon nominal values of all parameters except v/R, for which nominal values are replaced by minimum and maximum values. C_t^{nom}, C_t^{pmin}, and C_t^{pmax} are presented in Figure 3a, as a function of time. Although this plot clearly illustrates the effect of uncertainty in v/R on predictions of concentration, it would be difficult to compare the relative sensitivity due to uncertainty in this parameter to that due to another. This problem can be overcome by using an index such as

$$\text{index} = \begin{cases} (1 - C_t^{pmin})/C_t^{pmax} & (\text{for } C_t^{pmin} < C_t^{pmax}) \\ (1 - C_t^{pmax})/C_t^{pmin} & (\text{for } C_t^{pmin} > C_t^{pmax}) \end{cases} \tag{6}$$

as suggested by Hoffman and Gardner (*5*). A value of 1 indicates high sensitivity. The Hoffman and Gardner sensitivity index for v/R is given in Figure 3b. The sensitivity is large as the front approaches, decreases as the pulse passes, and increases afterwards. The high sensitivity prior to the following passage of the pulse is of little practical significance because the concentrations are extremely low. Also, little can be concluded about the actual magnitude of the concentration limits. This can be partially remedied by defining an index as the ratio of C_t^{pmin} or C_t^{pmax} to C_t^{pnom} as shown in Figure 3c. The fractional difference between nominal and bounding concentrations is clearly demonstrated. However, here again, the large fractional differences that occur as the contaminant pulse approaches and leaves the point of use are not of practical importance in a risk assessment application because the actual concentrations are so small compared to the maximum concentrations as to be of no consequence.

The problem of high, but inconsequential, sensitivity at relatively low concentrations is overcome by the definition given in Equation 1. The sensitivity index as a function of time due to uncertainty in v/R, α_ℓ, λ_L, and I_o/q_c is presented in Figure 4. The plots are drawn with identical scales to facilitate comparisons. For the parameter values specified in Table 1, some general observations can be made concerning the instantaneous sensitivity. First, there is an early period of low sensitivity as the contaminant front migrates to the point of use. This is followed by a period of approximately 10 to 100 years during which the sensitivity index is significant as the contaminant pulse passes. Then, following passage of the pulse, the sensitivity index becomes small because the concentration approaches zero. Second, the sensitivity index clearly indicates that, taken

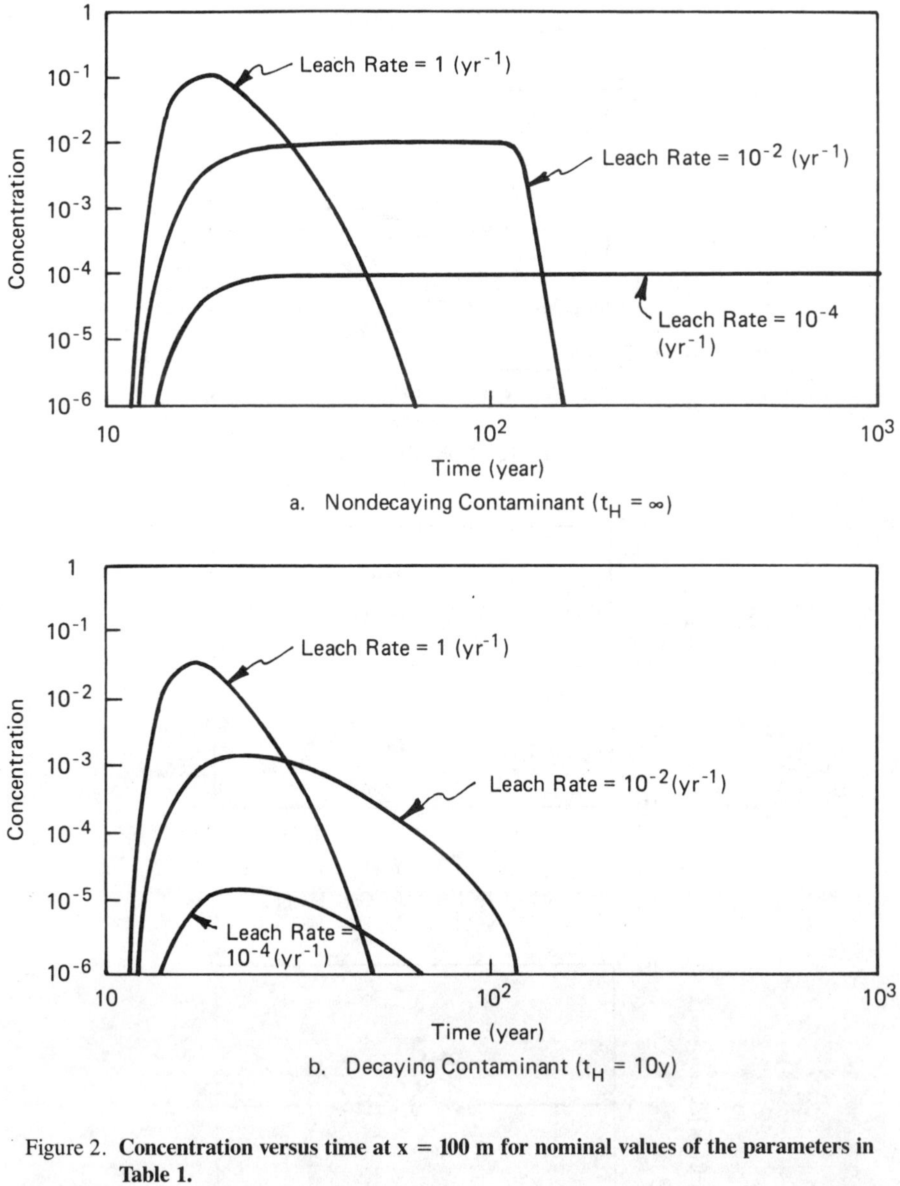

Figure 2. **Concentration versus time at x = 100 m for nominal values of the parameters in Table 1.**

individually, the parameters causing the greatest uncertainty in concentration are λ_L and I_o/q_c.

The combined sensitivity due to simultaneous uncertainty in all of the parameters is shown in Figure 5. Calculations are based on the minimum value of α_ℓ and maximum values of I_o/q_c and v/R, and curves are given for all three values of λ_L (nominal, minimum, and maximum). For $\lambda_L = 1$, the sensitivity index is extremely high for a very

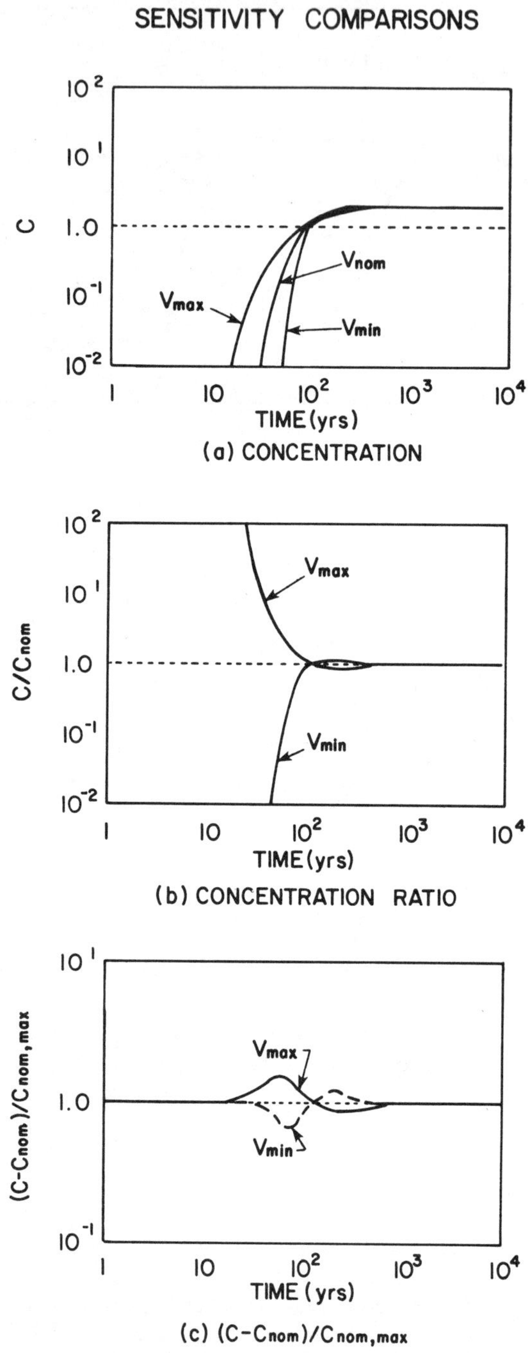

Figure 3. **Three methods of expressing the sensitivity of concentration to uncertainty in v/R.**

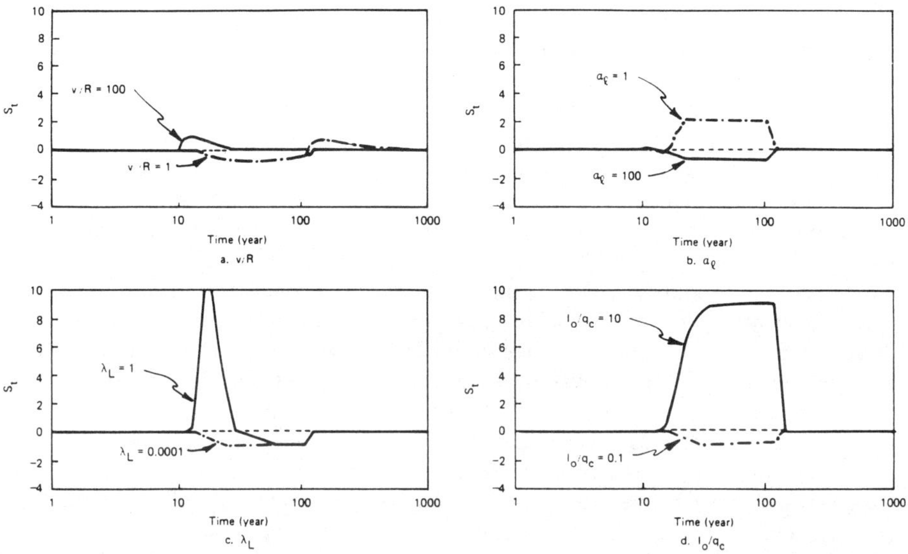

Figure 4. **Sensitivity of concentration to (a) v/R, (b) α_ℓ, (c) λ_L, and (d) I_0/q_c as expressed by the sensitivity index (Equation 1).**

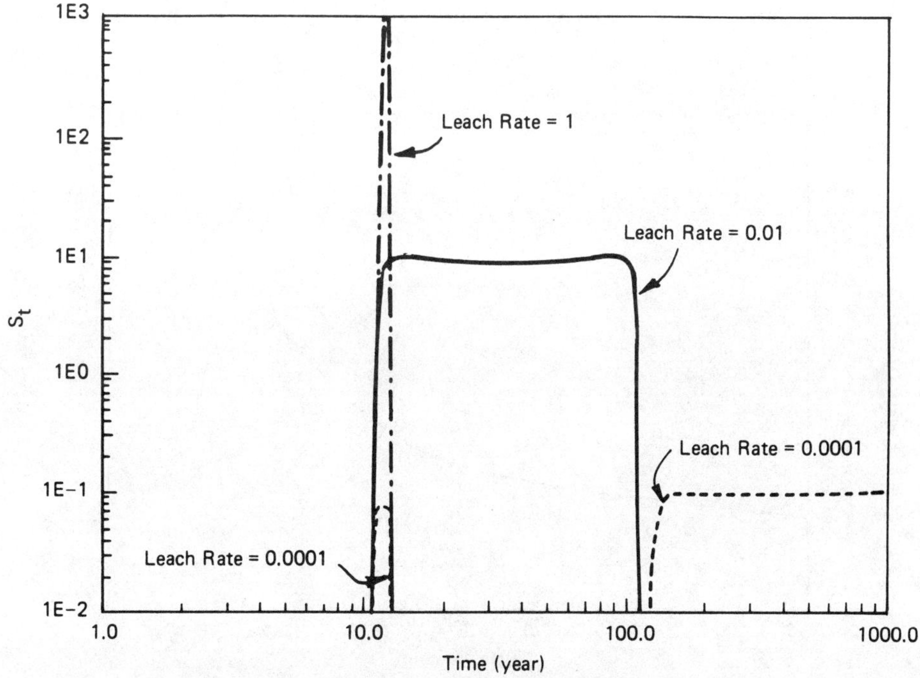

Figure 5. **Sensitivity of concentration due to the combined uncertainty in v/R, α_ℓ, λ_L, and I_o/q_c as expressed by the sensitivity index (Equation 1).**

short period of time. This would result from the unlikely error of simultaneously underestimating λ_L by a factor of 100, underestimating v/R and I_o/q_c by a factor of 10, and overestimating α_ℓ by a factor of 10. The principal value of the sensitivity index in this cumulative calculation is to identify, in a relatively simple way, the time at which the model is most sensitive to the parameters. It can also be used to make an upper bound estimate of the magnitude of the uncertainty. In this instance, the upper bound would be 1000 times the maximum nominal concentration. However, this upper bound estimate only applies to a very short time period. The upper bound during other time periods is much smaller.

Monte Carlo simulations were performed to provide a comparison against which results of the sensitivity index could be judged. The Monte Carlo calculations require parameter distributions. These were constructed as follows. The geometric means were set equal to the respective nominal parameter values in Table 1, and the minima and maxima specified in Table 1 were set at three geometric standard deviations from the mean. The minima and maxima specified for v/R, I_o/q_c, and λ_L were absolute bounds and the distributions were truncated at these values. This was not done for α_ℓ because there are data points that lie outside these values (*10*). An example of a cumulative probability plot of the concentration distribution resulting from the combined uncertainty in all of the parameters is given in Figure 6. The scale on the left side of the graph is the number of standard deviations below (negative values) or above (positive values) the median. The corresponding cumulative probability is given on the right side. This figure represents the cumulative probability for a receptor located 100 m downgradient at 25 years. The median concentration is approximately 5×10^{-3}. This is slightly less than

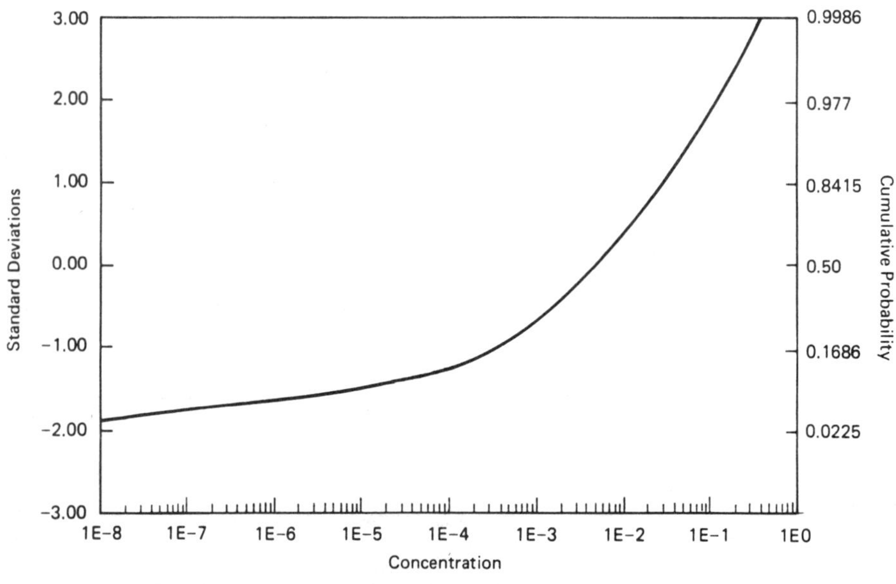

Figure 6. **An example of a cumulative probability plot resulting from the propagation of parameter distributions through Equation 4 using the Monte Carlo method.**

the nominal concentration at this time, which is approximately 8×10^{-3}. The 99.85% concentration, i.e., that concentration corresponding to a cumulative probability of 99.85%, is approximately 3×10^{-1}.

Plots of the 99.85% concentration for individual parameters are presented in Figure 7. The nominal concentration is included for comparison. Note that uncertainties in v/R and α_ℓ serve to increase the potential width of the pulse, but do not affect the maximum concentration. Uncertainty in λ_L and I_o/q_c can result in a factor of ten uncertainty in the maximum concentration at the 99.85% level. In addition, α_ℓ can lead to an extension of the pulse width to very large times. The combined uncertainty for these factors is presented in Figure 8. Note that the magnitude of the 99.85% concentration is similar in the combined uncertainty plot and the individual plots. Although the Monte Carlo calculations yield a much more precise quantification of the uncertainty than the sensitivity index, the general conclusions regarding the importance of λ_L and I_o/q_c and the time at which the effects are most critical are supported by the Monte Carlo simulation.

Summary

A screening technique has been presented to facilitate the time-dependent sensitivity/uncertainty analysis of ground-water contaminant transport in a radioactive risk assessment application. The technique involves calculation of a sensitivity index using

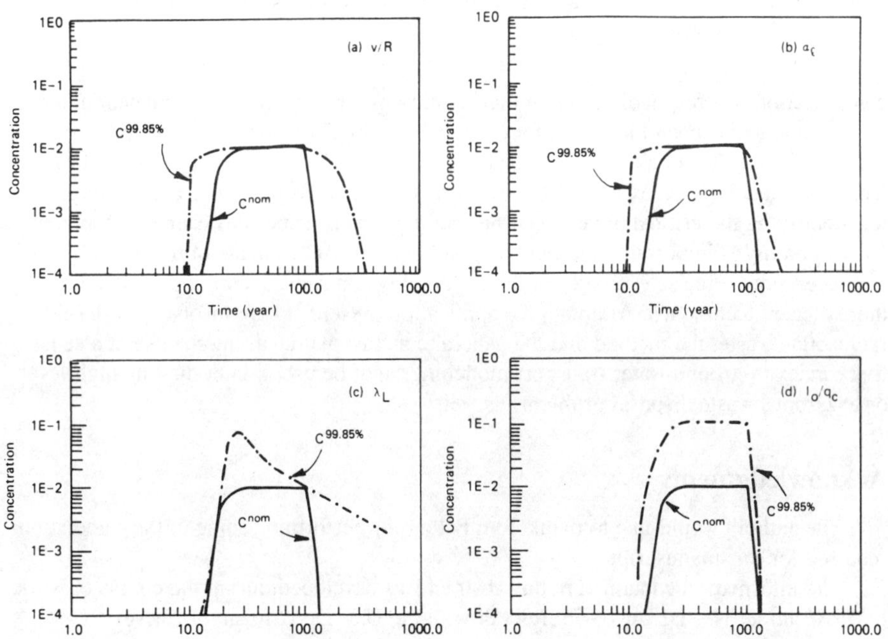

Figure 7. **C^{nom} and $C^{99.85\%}$ as a function time due to uncertainty in (a) v/R, (b) α_ℓ, (c) λ_L, and (d) I_o/q_c.**

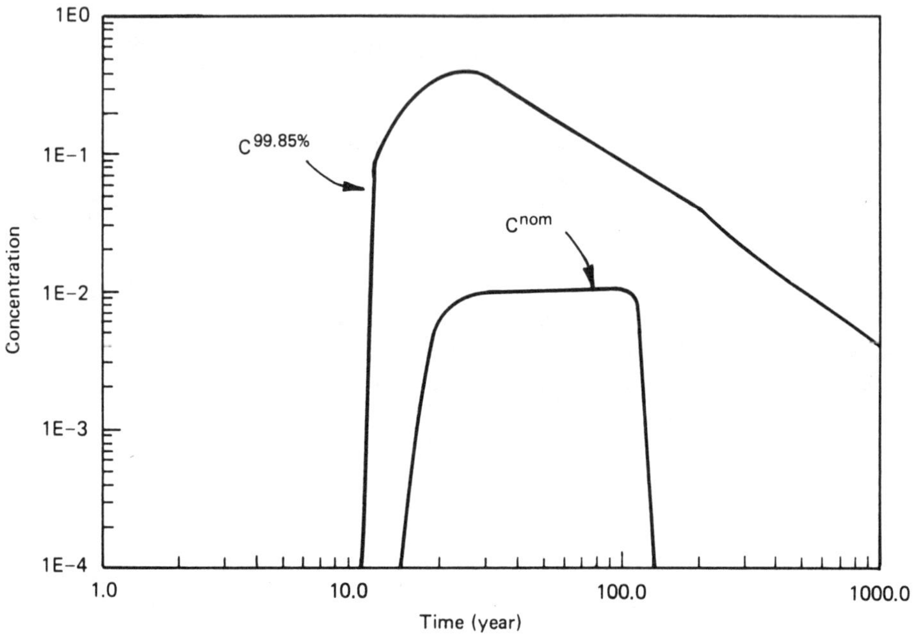

Figure 8. **C^{nom} and $C^{99.85\%}$ as a function time due to the combined uncertainty in v/R, α_ℓ, λ_L, and I_o/q_c.**

concentrations corresponding to nominal, minimum, and maximum parameter values. The technique has been illustrated for a one-dimensional ground-water transport problem and the results compared to a Monte Carlo uncertainty analysis of the same problem. The sensitivity index is found to be a practical, easily implemented screening tool for identifying the critical parameters, determining the time periods over which parameter uncertainty is important, and obtaining an upper bound estimate of the magnitude of the uncertainty. Precise estimates of the uncertainty can subsequently be obtained by more elegant techniques. Although the application presented here involves low-level radioactive waste, the method and the general conclusions regarding the use of a sensitivity index in ground-water transport modeling might be useful in evaluating high-level or hazardous waste disposal problems as well.

Acknowledgments

The authors would like to thank Tom Pollog for performing some of the calculations required for this manuscript.

The information contained in this abstract was developed during the course of work under Contract No. DE-AC09-76SR00001 with the U.S. Department of Energy.

Literature Cited

1. Merrell, G. B., V. C. Rogers, K. K. Nielson, and M. W. Grant, *The PATHRAE performance assessment code,* Rogers and Associates Engineering Corporation Report RAE-8469/3, Salt Lake City, UT (1985).
2. INTERA Environmental Consultants, Inc., *A proposed approach to uncertainty analysis,* ONWI-488, prepared for Office of Nuclear Waste Isolation, Battelle Memorial Institute, Columbus, OH (1982).
3. Gilbert, R. O., *Statistics for environmental studies,* PNL-SA-12523, Battelle Memorial Institute, Richland, WA (1984).
4. Oblow, E. M., F. G. Pin, and R. Q. Wright, *Nucl. Sci. Eng., 94,* 46 (1986).
5. Hoffman, F. O. and R. H. Gardner, "Evaluation of uncertainties in radiological assessment models," in *Radiological Assessment: A Textbook on Environmental Dose Analysis,* Till, J. E. and H. R. Meyer (ed.), U.S. Nuclear Regulatory Commission, Washington, DC (1983).
6. Tomovic, R., *Sensitivity Analysis of Dynamic Systems,* McGraw-Hill Book Company, New York, NY (1963).
7. Tomovic, R. and M. Vukobratovic, *General Sensitivity Theory,* American Elsevier Publishing Company, Inc., New York, NY (1970).
8. Burkholder, H. C. and E.L.J. Rosinger, *Nucl. Tech., 49,* 150 (1980).
9. van Genuchten, M. Th. and W. J. Alves, *Analytical solutions of the one-dimensional convective-dispersive solute transport equation,* U.S. Department of Agriculture Technical Bulletin 1661 (1982).
10. Gelhar, L. W., A. Mantoglou, C. Welty, and K. R. Rehfeldt, *A review of field scale physical solute transport processes in saturated and unsaturated media,* EPRI-4190, Electric Power Research Institute, Palo Alto, CA (1975).

Effect of Variable Fracture Permeability/Matrix Permeability Ratios on Three-Dimensional Fractured Rock Hydraulic Conductivity

TODD C. RASMUSSEN, T.C. JIM YEH and DANIEL D. EVANS
Department of Hydrology and Water Resources
University of Arizona, Tucson, Arizona

Abstract

A sensitivity analysis is implemented using the boundary integral method to determine the relative contribution to total liquid flow through a rock block both by flow through the rock matrix and by flow through discrete fractures embedded within the rock matrix. Total liquid flow through a rock block with embedded fractures of arbitrary density, orientation, permeability, shape, and areal extent is simulated by discretizing the fracture-matrix boundary and the exterior surface of the rock sample using boundary elements. Steady liquid flow through fractures is computed by creating nodes along the outer rim of each fracture and then routing between nodes within the fracture plane using a two-dimensional boundary integral formulation. Flow between fractures is calculated by assigning boundary nodes along fracture intersections. Liquid flow through the rock matrix is calculated by employing a three-dimensional boundary integral formulation for nodes on fracture boundaries and on the exterior surfaces of the rock block. All nodes within the network are linked using the three-dimensional boundary integral formulation. Constant matrix permeability, k_m, and fracture permeability, k_f, within individual fractures are assumed. This assumption provides the ability to account for steady liquid flow through fractures and the rock matrix when complex, random shapes of matrix blocks between fractures are present.

Simulations are performed to investigate the effects on the global vertical hydraulic conductivity, K_z, of (1) variable k_f/k_m ratios due to variable water saturation or to geochemical effects such as dissolution or chemical precipitation; (2) fracture orientation with respect to a vertical gradient vector, J_z; and (3) fracture length relative to the flow domain. Results for both two- and three-dimensional flow domains indicate that when individual fractures fully bisect the flow domain and are aligned parallel, perpendicular, or diagonal to the flow direction, the global vertical hydraulic conductivity, K_z, can be related to k_f and k_m using the arithmetic, harmonic, or geometric means, respectively, with account taken for the widths of the two media. The influence of fractures which do not fully divide the matrix is substantially reduced over fractures which bisect the flow domain. Application of the technique is motivated by fracture sets observed at the Apache Leap field site in tuff near Tucson, Arizona. Extensive near-vertical fractures at the site are complemented by a minor set of randomly oriented fractures.

Introduction

Liquid flow and solute transport through porous media can be substantially affected by the presence of both high- and low-permeability zones. The permeability contrasts can result from the presence of fractures, which either provide high-velocity pathways due to enlarged openings or form barriers due to chemical precipitation or air-filled voids under unsaturated conditions. Due to the concern that fractures may provide high-velocity conduits for wastes disposed in fractured porous media to reach the accessible environment, a methodology must be developed for characterizing the movement of water and solutes in fractured rock.

As part of a program to evaluate characterization procedures for unsaturated flow and transport in fractured rock, liquid flow experiments are being performed at the Apache Leap Tuff Site located in slightly welded tuff near Tucson, Arizona. Figure 1 presents fractures intersecting nine boreholes in slightly welded tuff at this site. Figure 2 is a plot of the normals of fracture surfaces. Two fracture sets are apparent in the figures; a predominately vertical north-south trending fracture set with strikes of approximately 345° (±40°) and dips of approximately 75° (±15°), and a second, random fracture set with no preferential strike or dip. Due to the possible influence of the observed fracture sets on fluid flow at variable water saturation, a fracture-matrix flow characterization methodology is deemed important to the overall characterization of flow at this site.

A number of approximation techniques have been developed for the purpose of modeling liquid flow through both a porous rock matrix and fractures embedded within the matrix. Long et al. (*1*) assumed that a representative elementary volume (REV) can be defined for the purpose of estimating the hydraulic properties of a fractured medium. A potential shortcoming of using the REV concept is that the size of the REV in fractured media may be either difficult to determine or larger than the scale of field testing.

Gelhar and Axeness (*2*) used stochastic theory to estimate the effective hydraulic conductivity tensor for conditions of saturated flow in a statistically anisotropic medium with arbitrary orientation of stratification. The tensor is shown to be of second rank and symmetric. Parameters for the stochastic model include the mean, variance, and covariance function of the natural logarithm of the saturated hydraulic conductivity, as well as the orientation of the stratification. Stochastic theory has also been used by Yeh et al. (*3,4,5*) to describe steady unsaturated flow in a heterogeneous medium. They demonstrate that the effective hydraulic conductivity of a statistically anisotropic medium has tensorial properties, and also that the anisotropy of the hydraulic conductivity is dependent upon the moisture content of the medium. This results from the variation in pore-size distributions within the soil medium.

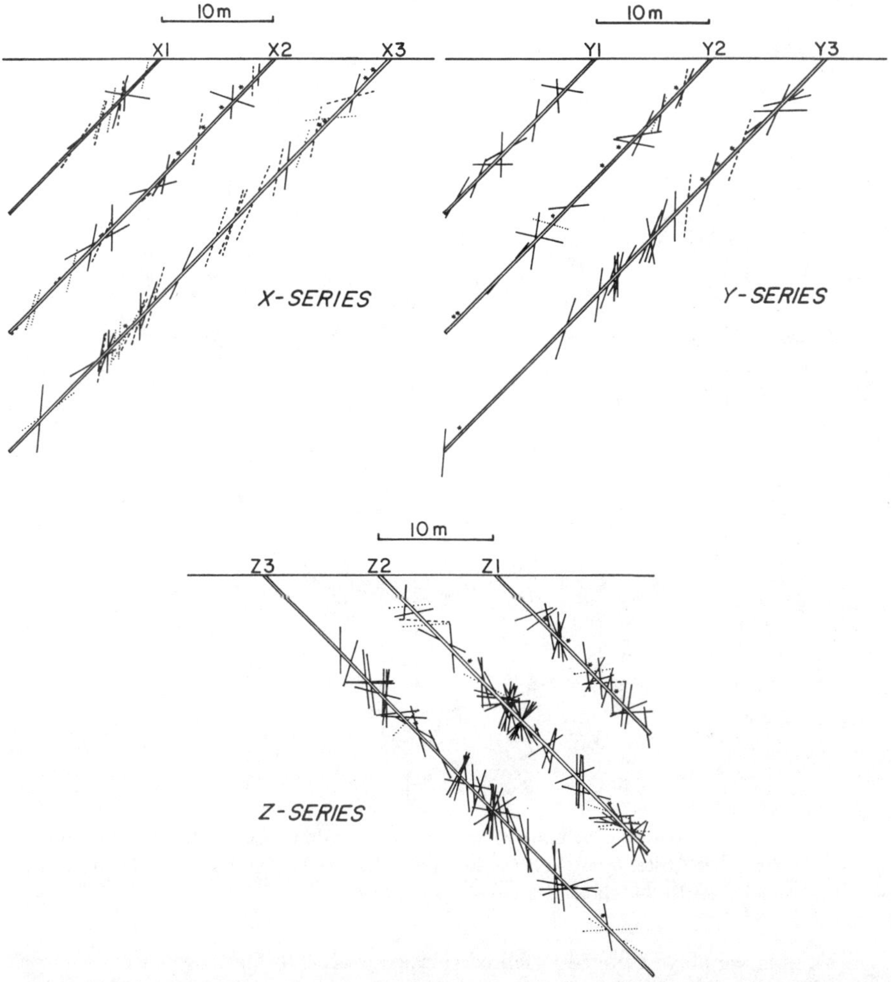

Figure 1. **Observed fracture traces in the borehole plane in slightly welded tuff at the Apache Leap Site near Tucson, Arizona.**

Extensions of these results to fractured media are possible if hydraulic properties of fractures are available. Coakley et al. (*6*), in a simulation of fracture void space geometries using a two-point correlation function of void spaces and contact areas obtained from metal injection data, showed that the variability of equivalent permeability and anisotropy of equivalent permeability decrease as the fracture size is increased. Another technique used to examine coupled fracture and matrix flow under saturated conditions by Bibby (*7*), Huyakorn et al. (*8*), and Moench (*9*) is the dual-porosity formulation which decouples the two flow components and solves for flow in each medium separately and then recouples the two solutions.

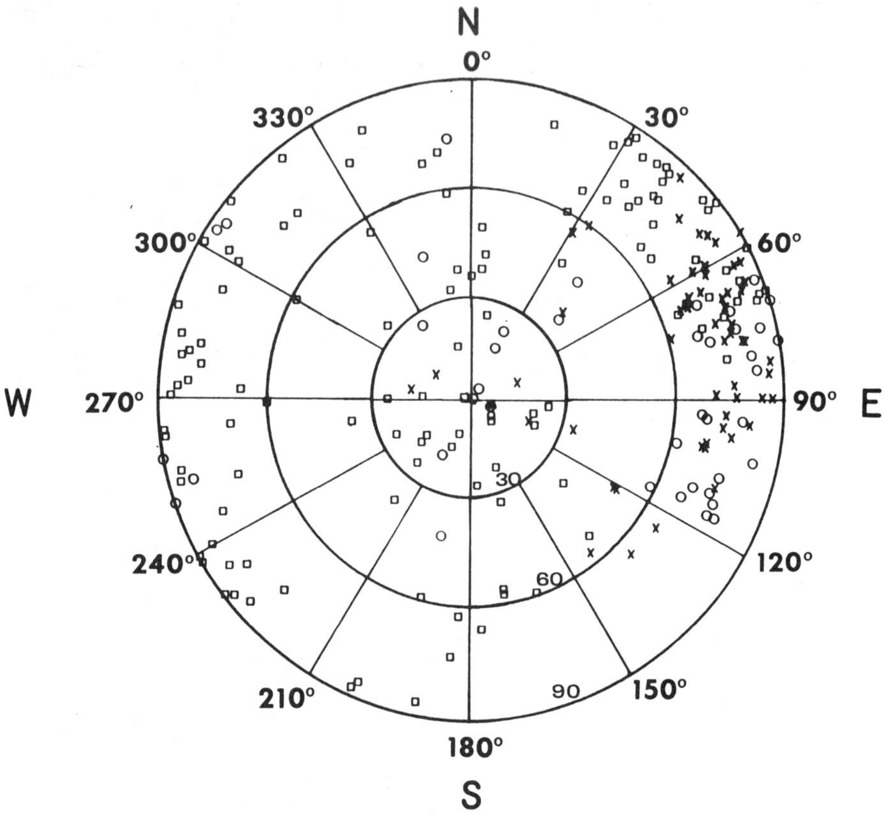

Figure 2. **Lower stereonet projection of fracture orientations observed at the Apache Leap Tuff Site, from X-series boreholes (crosses), Y-series boreholes (circles), and Z-series boreholes (squares).**

Flow in variably saturated fractured rocks was recently investigated by Wang and Narasimhan (*10*). Based on a measured fracture aperture distribution, they derived a theoretical relationship between fracture hydraulic conductivity and pressure head which considers the influence of asperities. With this theoretical relationship they used a numerical model to simulate flow in regularly spaced fractures of infinite areal extent bounded by matrix blocks. The model provides estimates of changes in saturation levels, in permeabilities, in darcian flow velocities in fractures and at fracture-matrix interfaces, and in effective fracture-matrix flow areas. One possible shortcoming of the model is that in the geologic formation of interest the fractures are randomly oriented with limited areal extent, rather than of infinite areal extent with orthogonal orientations of uniform density. Complex matrix shapes must be incorporated in order to fully understand flow in variably saturated fractured rocks.

Of interest to researchers is the determination of whether the stochastic and continuum models adequately describe flow through a discrete fracture network. If these models can be shown to adequately describe such flow, then existing porous medium models can be used. To this end, experiments have been performed using formulations which explicitly incorporate fracture flow.

It is demonstrated in this paper that the boundary integral method can be used to discretize and simultaneously solve steady liquid flow through randomly oriented, discrete fractures embedded within a porous medium. Experience with modeling two-dimensional flow within three-dimensional networks of discrete fractures by Huang and Evans (*11*) and by Rasmussen (*12*) has resulted in the capability to discretize random networks of fractures embedded within an impermeable matrix. Fracture flow is characterized by defining physical and hydraulic parameters for each fracture. The physical properties of fractures include the orientation, areal extent, location of fracture center, shape, and permeability. Because properties of only a few fractures can be identified in field-scale problems, statistical techniques must be employed to characterize sets of fractures. For each fracture parameter, probability distribution functions (pdf's) are defined using distributions and coefficients which best approximate observed characteristics of the fracture network at the site of interest.

The hydraulic properties of interest for steady liquid flow within a fractured rock mass are the matrix permeability, k_m, and the fracture permeability, k_f, which can vary as a function of geochemical processes, such as dissolution and precipitation, or as the result of variable saturation levels. While Kilbury et al. (*13*) have provided physical evidence to indicate that k_f can also vary from point to point within a single fracture, that case is not considered here. To investigate the effect of variable k_f, a sensitivity analysis is performed which evaluates changes in global vertical hydraulic conductivity, K_z, due to a global unit head gradient, J_z, as a function of fracture orientation, length, and ratio between k_f and k_m. The k_f/k_m ratio varies widely, with very large numbers for systems in which fracture flow dominates, with values of unity for fractures with no appreciable hydraulic influence, and with values near zero for fractures which are filled with chemical precipitates or are unsaturated.

Boundary Integral Method

The boundary integral method (BIM) has been widely applied within the field of fluid hydraulics and subsurface flow modeling (Lafe et al., *14*), although normally only for two-dimensional applications. A few three-dimensional applications have been reported (Liggett and Liu, *15*; Brebbia et al., *16*; Huyakorn and Pinder, *17*), as well as applications to flow through fractured rock (Shapiro and Andersson, *18*; Elsworth, *19*,*20*). Such studies have examined flow through homogeneous media, or through layered media with homogeneous properties within each layer. Recent advances have also provided the ability to examine two-dimensional flow through heterogeneous media (Lafe and Cheng, *21*). To illustrate the BIM method, an application is presented for solving the Laplace equation in two and three dimensions, respectively:

$$\nabla^2 u(x) = 0 \qquad \text{in } \Omega \tag{1a}$$

and

$$\nabla^2 u(x) = 0 \qquad \text{in R} \tag{1b}$$

where

u(x) spatially variable fluid head, m;
x position vector, m;
Ω two-dimensional spatial domain; and
R three-dimensional spatial domain.

Assigned head and flux boundary conditions are, respectively (Figure 3):

$$u(x) = \bar{u}(x) \qquad \text{on } \Gamma \text{ or } \Omega \tag{2a}$$

or

$$q(x) = \bar{q}(x) \qquad \text{on } \Gamma \text{ or } \Omega \tag{2b}$$

where

$\bar{u}(x)$ assigned head boundary condition, m;
$q(x)$ spatially variable fluid flux vector, m/s;
$\bar{q}(x)$ assigned flux boundary condition, m/s;
Γ one-dimensional spatial domain.

The fluid flux can be related to the fluid head using the tensorial form of Darcy's law in two and three dimensions, respectively:

$$q(x) = -T \nabla u(x) \tag{3a}$$

and

$$q(x) = -K \nabla u(x) \tag{3b}$$

where

T transmissivity tensor, m^2/s; and
K hydraulic conductivity tensor, m/s.

An approximate solution for fluid head and flux which minimizes the error between the true and estimated head and flux over the flow domain is obtained by solving the weighted residual statement (Brebbia et al., *16*) in both two and three dimensions, respectively:

$$\int [u^*(x,x^*) \; \nabla^2 u(x)] \; d\Omega = 0 \tag{4a}$$

and

$$\int [u^*(x,x^*) \; \nabla^2 u(x)] \; dR = 0 \tag{4b}$$

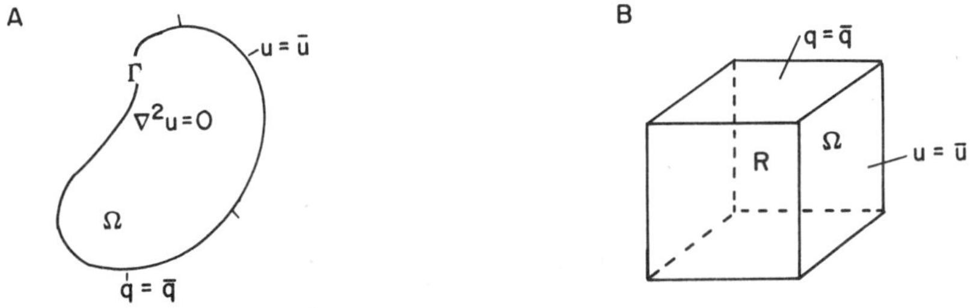

Figure 3. **Flow domain and boundary conditions for two-dimensional (A) and three-dimensional (B) media. Symbols are defined in text.**

where

$u^*(x,x^*)$ weighted residual function, m;

x^* position vector of weighted residual function.

The weighted residual function is dependent on position within the flow domain and on the physical and fluid properties within the flow domain. In order to transform Equation 4 into an expression which can be evaluated, Green's second identity is used in two and three dimensions, respectively:

$$\int (f\,\nabla^2 g - g\,\nabla^2 f)\,d\Omega = \int (f\,\delta g/\delta n - g\,\delta f/\delta n)\,d\Gamma \tag{5a}$$

and

$$\int (f\,\nabla^2 g - g\,\nabla^2 f)\,dR = \int (f\,\delta g/\delta n - g\,\delta f/\delta n)\,d\Omega \tag{5b}$$

where

f,g arbitrary functions; and

n direction normal to boundary.

Assigning $u(x)$ to g and $u^*(x,x^*)$ to f results in:

$$\int (u^*(x,x^*)\,\nabla^2 u(x) - u(x)\,\nabla^2 u^*(x,x^*))\,d\Omega = \tag{6a}$$

$$\int (u^*(x,x^*)\,\delta u(x)/\delta n - u(x)\,\delta u^*(x,x^*)/\delta n)\,d\Gamma$$

and

$$\int (u^*(x,x^*)\,\nabla^2 u(x) - u(x)\,\nabla^2 u^*(x,x^*))\,dR = \tag{6b}$$

$$\int (u^*(x,x^*)\,\delta u(x)/\delta n - u(x)\,\delta u^*(x,x^*)/\delta n)\,d\Omega$$

The first half of the integral on the left-hand side of Equation 6 is equal to zero (from Equation 4), while the remaining half is calculated by noting that (Brebbia et al., *16*):

$$\nabla^2 u^*(x,x^*) = -\,2\pi\,\Delta(x,x^*) \tag{7a}$$

and

$$\nabla^2 u^*(x,x^*) = -4\pi\, \Delta(x,x^*) \tag{7b}$$

in two and three dimensions, respectively, where $\Delta(x,x^*)$ is the Dirac delta, with properties:

$$\Delta(x,x^*) = \begin{cases} 0 & x^* \neq x \\ \infty & x^* = x \end{cases} \tag{8}$$

Substituting Equation 7 into Equation 6 and noting that:

$$\int u(x)\, \Delta(x,x^*)\, d\Omega = u(x^*) \tag{9a}$$

and

$$\int u(x)\, \Delta(x,x^*)\, dR = u(x^*) \tag{9b}$$

yields:

$$c(x^*)\, u(x^*) + \int u(x)\, q^*(x,x^*)\, d\Gamma = \int q(x)\, u^*(x,x^*)\, d\Gamma \tag{10a}$$

and

$$c(x^*)\, u(x^*) + \int u(x)\, q^*(x,x^*)\, d\Omega = \int q(x)\, u^*(x,x^*)\, d\Omega \tag{10b}$$

where $q(x) = \delta u(x)/\delta n$ and $q^*(x,x^*) = \delta u^*(x,x^*)/\delta n$. For positions internal to the flow domain, $c(x^*)$ equals 2π in two dimensions and 4π in three dimensions. For positions along a one-dimensional boundary (Brebbia et al., *16*; Elsworth, *19*):

$$c(x^*) = 2\pi - \alpha \tag{11}$$

where α is the exterior angle at location x.

By discretizing the boundaries of the flow domain, a relationship between flow and head can be used which replaces the integral terms of Equation 10 with numerical summations for nodes along the boundaries (Figure 4):

$$c(x^*)\, u(x^*) + \Sigma\, u(x)\, q^*(x,x^*) = \Sigma\, q(x)\, u^*(x,x^*) \tag{12a}$$

and

$$c(x^*)\, u(x^*) + \Sigma\, u(x)\, q^*(x,x^*) = \Sigma\, q(x)\, u^*(x,x^*) \tag{12b}$$

Equation 12 can be written more concisely as (Brebbia, *22*):

$$H\, u = G\, q \tag{13}$$

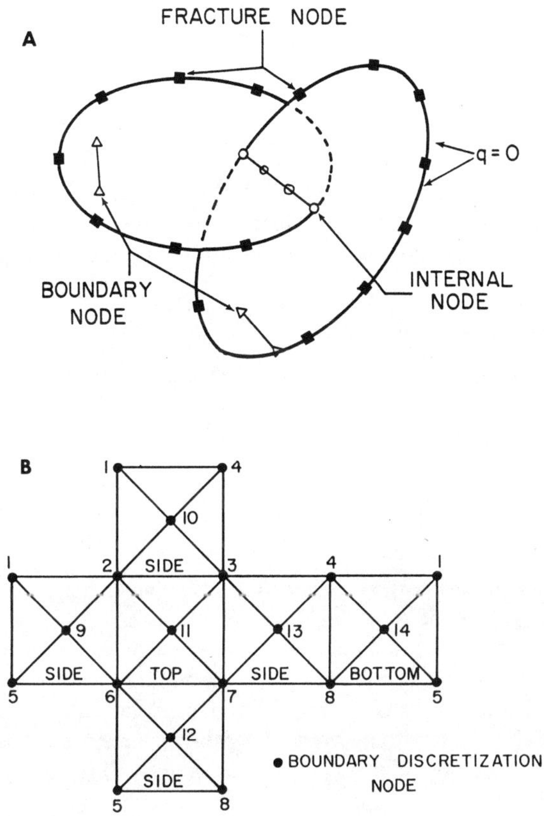

Figure 4. **Boundary discretization schemes for two-dimensional planar fractures (A) and three-dimensional matrix (B).**

where

u head on boundary, located at discrete boundary positions, m;

q flux normal to boundary, located at discrete boundary positions, m/s; and

G,H boundary integral coefficient matrices, with

$$G = \begin{bmatrix} g_{11} & g_{12} & g_{13} & \cdots & g_{1m} \\ g_{21} & g_{22} & g_{23} & \cdots & g_{2m} \\ g_{31} & g_{32} & g_{33} & \cdots & g_{3m} \\ \cdot & \cdot & \cdot & & \cdot \\ g_{m1} & g_{m2} & g_{m3} & \cdots & g_{mm} \end{bmatrix} \tag{14a}$$

and

$$H = \begin{bmatrix} h_{11} & h_{12} & h_{13} & \cdots & h_{1m} \\ h_{21} & h_{22} & h_{23} & \cdots & h_{2m} \\ h_{31} & h_{32} & h_{33} & \cdots & h_{3m} \\ \cdot & \cdot & \cdot & & \cdot \\ h_{m1} & h_{m2} & h_{m3} & \cdots & h_{mm} \end{bmatrix} \tag{14b}$$

where m is the number of discretization intervals along the boundary.

For constant head and flux along discrete boundary segments, the elements of *G* and *H* can be found for two- and three-dimensional flow domains using Equation 12:

$$g_{ij} = \int u^*(x,x^*)\, d\Gamma \tag{15a}$$

$$h_{ij} = \int q^*(x,x^*)\, d\Gamma \tag{15b}$$

and

$$g_{ij} = \int u^*(x,x^*)\, d\Omega \tag{15c}$$

$$h_{ij} = \int q^*(x,x^*)\, d\Omega \tag{15d}$$

where i and j are indices corresponding to the position in the *G* and *H* matrices.

Linear interpolation of head and flux along elements will more accurately account for variations along the boundary than constant interpolation functions. Higher order interpolation will also improve the accuracy of the procedure, at the expense of numerical convenience. For a linear head and flux variation along an element, the head and flux at any point along the element are determined for two- and three-dimensional flow domains using, respectively (Figure 5):

$$u(e) = [u_1\,(1 - e) + u_2\,(1 + e)]\,/\,2 \tag{16a}$$

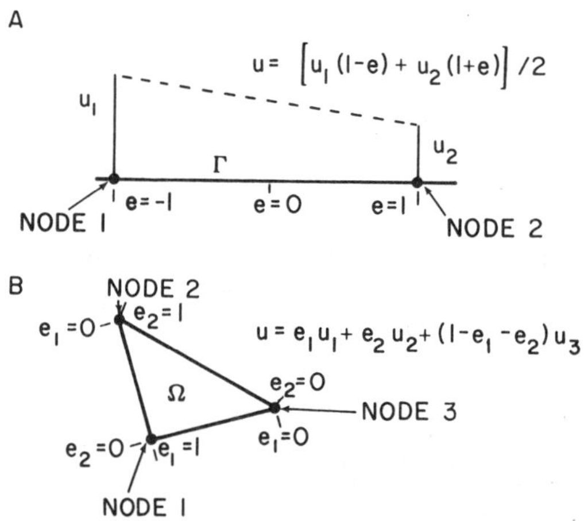

Figure 5. **Linear interpolation functions for one- (A) and two- (B) dimensional boundaries.**

$$q(e) = [q_1 (1 - e) + q_2 (1 + e)] / 2 \tag{16b}$$

and

$$u(e_1,e_2) = u_1 e_1 + u_2 e_2 + u_3 (1 - e_1 - e_2) \tag{16c}$$

$$q(e_1,e_2) = q_1 e_1 + q_2 e_2 + q_3 (1 - e_1 - e_2) \tag{16d}$$

where

- u,q head and flux at any position along a boundary line or surface element;
- e linear interpolation coordinate $(-1 < e < 1)$ along boundary line element;
- e_1,e_2 linear interpolation coordinates $(0 < e_1 < 1; 0 < e_2 < 1)$ along boundary surface element;
- u_1,u_2 head at endpoints of boundary line element;
- q_1,q_2 flux at endpoints of boundary line element;
- u_1,u_2,u_3 head at corners of triangular boundary surface element; and
- q_1,q_2,q_3 flux at corners of triangular boundary surface element;

The only remaining requirement is to determine the weighted residual functions, $u^*(x,x^*)$ and $q^*(x,x^*)$. For two- and three-dimensional homogeneous flow domains with directionally constant transmissivity and hydraulic conductivity, the functions are taken to be the fundamental solutions, or, respectively:

$$u^*(x,x^*) = \ln(T^{1/2}/r) / 2\pi T \tag{17a}$$

$$q^*(x,x^*) = \delta r/\delta n / 2\pi r \tag{17b}$$

and

$$u^*(x,x^*) = - 1/4\pi K r \tag{17c}$$

$$q^*(x,x^*) = \delta (1/r)/\delta n / 4\pi \tag{17d}$$

where r is the scaler distance between x and x^*. Huyakorn and Pinder (*17*, p. 317) present analytic solutions to the integration of Equation 17 for two-dimensional flow, as required by Equation 15. For conditions of three-dimensional flow the integrations are performed numerically using Gaussian integration over triangular areas (Cowper, *23*).

For flow between two intersecting fractures, additional internal boundary elements are introduced to represent the line of intersection between the fracture planes (Figure 6). From mass balance considerations and equivalence of heads along the boundary, a larger matrix can be formed which is composed of submatrices of flow within each fracture plus the flow across the line of intersection. The pure-fracture flow matrix equation in each fracture is:

$$[G_1 \quad G_{1i}] \begin{Bmatrix} q_1 \\ q_i \end{Bmatrix} = [H_1 \quad H_{1i}] \begin{Bmatrix} u_1 \\ u_i \end{Bmatrix} \tag{18a}$$

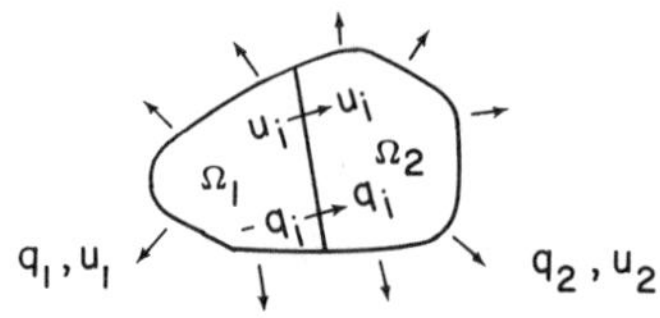

Figure 6. **Multiple flow domain geometry.**

and

$$[G_{2i} \quad G_2] \begin{Bmatrix} {}^{-}q_i \\ q_2 \end{Bmatrix} = [H_{2i} \quad H_2] \begin{Bmatrix} u_i \\ u_{2i} \end{Bmatrix} \tag{18b}$$

where

- G_1,H_1 boundary integral coefficient matrices between elements along exterior surfaces of fracture 1;
- G_2,H_2 boundary integral coefficient matrices between elements along exterior surfaces of fracture 2;
- G_{1i},H_{1i} boundary integral coefficient matrices between elements along exterior surfaces of fracture 1 and interfacial elements;
- G_{2i},H_{2i} boundary integral coefficient matrices between elements along exterior surfaces of fracture 2 and interfacial elements;
- u_1,q_1 flow and heads along exterior boundary of fracture 1;
- u_2,q_2 flow and heads along exterior boundary of fracture 2; and
- u_i,q_i flow and heads along interfacial boundary between fractures.

Combining Equations 18a and 18b yields:

$$\begin{bmatrix} G_1 & G_{1i} & O \\ O & {}^{-}G_{2i} & G_2 \end{bmatrix} \begin{Bmatrix} q_1 \\ q_i \\ q_1 \end{Bmatrix} = \begin{bmatrix} H_1 & H_{1i} & O \\ O & H_{2i} & H_2 \end{bmatrix} \begin{Bmatrix} u_1 \\ u_i \\ u_2 \end{Bmatrix} \tag{19}$$

Equation 19 is appropriate for two intersecting fractures. Networks of intersecting fractures require more interfacial elements, with a matrix structure composed of blocks of nonzero elements, alternating with blocks of zero elements corresponding to nodes which do not lie in the same fracture plane. Equation 19 is solved by selecting appropriate boundary conditions at all noninterfacial nodes (either prescribed head or flux) and by reducing the resulting global set of equations to the form:

$$A\,a = B\,b = b' \tag{20}$$

where

- A,B square matrices of known boundary integral coefficients;
- a vector of unknown boundary conditions;

b vector of known boundary conditions; and

b' vector of known coefficients and boundary conditions.

Equation 20 can be solved using Gaussian elimination or other direct solvers.

Once the unknown head and fluxes are determined, the head at any position internal to the flow domain can be calculated using:

$$u_i = \sum_j (q_j g_{ij} - u_j h_{ij}) \tag{21}$$

Fracture flow which incorporates flow into and out of the fracture from the matrix is governed by the Poisson equation:

$$\nabla^2 u_f + q_m = 0 \tag{22}$$

where q_m is the net flow into the fracture from the matrix through both walls of the fracture. Flow between the fracture and the matrix is spatially variable over the fracture. The boundary integral method can be employed to evaluate the magnitude of the flow and the head distribution within the fracture and surrounding matrix by reforming the weighted residual statement of Equation 4a as:

$$\int [u^*(x,x^*)\ \nabla^2 u(x)]\ d\Omega = \int [u^*(x,x^*)\ q_m(x)]\ d\Omega \tag{23}$$

The fracture flow matrix equation thus becomes:

$$F + H\,u = G\,q \tag{24}$$

where F is the contribution to fluid flow from the rock matrix into the fracture, integrated over all triangular fracture elements on the surface of the fracture (Figure 7). The value of F is calculated from:

$$F = \sum_i \left[\sum_j w_j\ (q_m\ u^*)_j\right] A_i \tag{25}$$

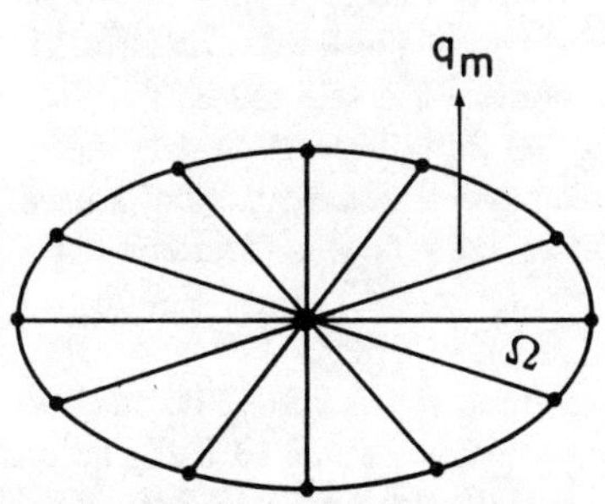

Figure 7. **Fracture surface discretization geometry showing net flux term, q_m, representing flow between the fracture and matrix.**

where

- i index over all fracture area elements;
- j counter over all numerical integration points on the fracture area element;
- w_j Gaussian integration weighting factor for integration points (from Cowper, *23*); and
- A_i area of fracture element.

Flow through the matrix is solved using the three-dimensional boundary integral equation:

$$G_m q_m = H_m u_m \tag{26}$$

where the subscripts denote matrix flow components calculated using three-dimensional formulation. Equations 24 and 26 are coupled using mass balances between interface nodes and the equivalence of heads for fracture and matrix nodes.

Two-Dimensional Porous Medium Application

Before proceeding with an analysis of flow through two-dimensional planar fractures embedded within a three-dimensional porous matrix, it is instructive to examine a simplified example of flow through a one-dimensional linear fracture embedded within a two-dimensional porous medium. In this example a square with unit length sides is constructed so that two opposing sides are assigned no-flow boundary conditions, the top surface is assigned a unit hydraulic head, and the bottom surface is assigned a hydraulic head of zero. The resulting global gradient is only in the vertical direction, and the magnitude of this global vertical gradient, J_z, is one. The matrix permeability, k_m, of the interior flow region is assigned a unit value and the matrix region is assigned a width, b_m, of 0.99. The center of a fracture with an arbitrary aperture, b_f, of 0.01 is placed in the center of the unit square, and length, rotation, and fracture permeability, k_f, parameters are assigned.

Fracture orientation is allowed to vary in such a manner that the fracture is either parallel, perpendicular, or diagonal to the direction of flow. Also, fracture length is allowed to vary from fully to partially dividing the flow domain. Finally, k_f/k_m is varied from 10^{-5} to 10^5. Table 1 summarizes the flow test parameters.

To discretize the flow domain, nodes are placed around the periphery of the unit square, as well as around the rim of the linear fracture. The location of nodes along the perimeter of the matrix domain are selected so that linear head and flow changes are reproduced. (Errors may be introduced if the true head and flow are varying in a nonlinear manner and a linear interpolation function is used.) From experience, it has been determined that nodes must be placed at corners and near where fractures and matrix boundaries are closest to each other. Figure 8 illustrates the flow domain geometry and nodal discretization schemes.

The number of nodes required to discretize the fracture surface is determined by comparing calculated flows with flows obtained using an analytic solution for a known problem. In this case, a flow regime with a fracture perpendicular to the direction of the gradient completely bisects the flow domain. The aspect ratio of the fracture is the proportion between the distance between nodes to the width of the

Table 1. **Fracture parameters for flow experiments.**

Simulation	1	2	3	4	5	6
Length	1.	1.4	1.	0.5	0.7	0.5
Rotation	0°	45°	90°	0°	45°	90°

All Simulations

Fracture center is located at center of flow domain.

Fracture aperture, b_f, is 0.01. Matrix width, b_m, is 0.99.

Fracture permeability/matrix permeability ratios of (10^{-5}, 10^{-3}, 10^{-1}, 1, 10^1, 10^3, 10^5) used for each simulation.

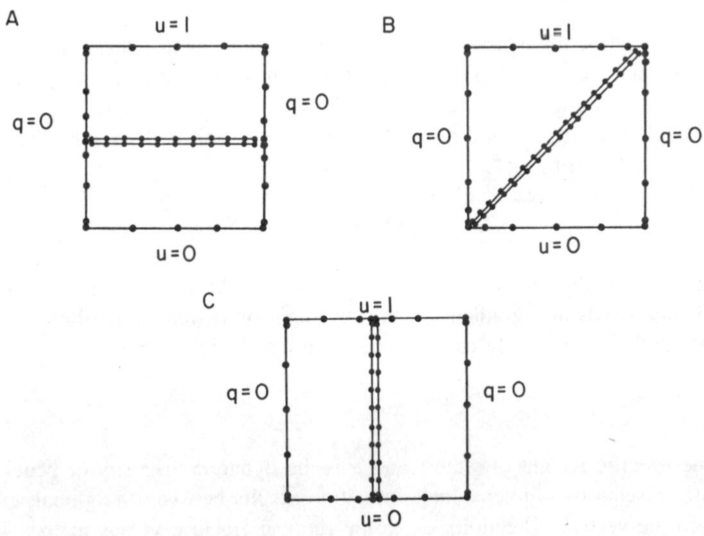

Figure 8. **Two-dimensional flow geometry showing no flow and constant head boundaries for fractures perpendicular (A), diagonal (B), and parallel (C) to direction of head gradient.**

flow domain (Figure 9). For an aspect ratio near one, good accuracy is expected, with decreasing accuracy as the ratio departs from unity. Table 2 presents simulation results for two-dimensional flow error analysis. It can be observed that the accuracy of the procedure decreases rapidly when the aspect ratio is greater than 10:1. An aspect ratio less than 10:1 is used in the simulation experiments which follow.

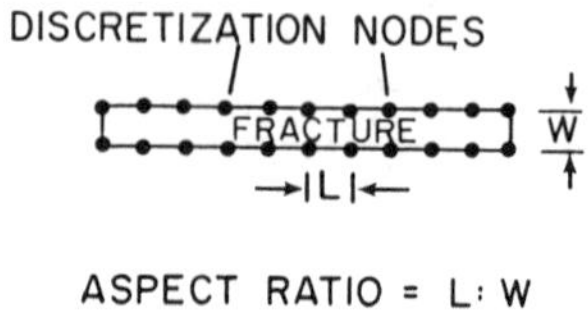

Figure 9. **Definition of aspect ratio as proportion of distance between nodes to width of fracture.**

Table 2. **Precision of boundary integral method as a function of aspect ratio, defined here as the ratio of the distance between nodes to the width of the flow domain.**

Aspect Ratio	1:1	5:1	10:1	20:1	25:1
2-D Error (percent)	0.000	0.003	0.20	5.9	25.4

Once the geometry, boundary conditions, and the number and location of boundary nodes have been established, simulations can be performed to determine the global vertical hydraulic conductivity, K_z, defined as the measured flux through the domain, Q, divided by the unit area, A, and also the unit hydraulic gradient imposed across the domain, J_z:

$$K_z = Q / (A\ J_z) \tag{27}$$

This measure of the hydraulic conductivity is consistent with the interpretation of a flow test in which heads and gradients within the interior region cannot be examined. The global hydraulic conductivity can also be related to k_z by noting:

$$K_z = k_z\ \gamma/\mu \tag{28}$$

where γ is the specific weight of water, and μ is the dynamic viscosity of water.

Figure 10 presents two-dimensional simulation results between the global permeability in the vertical direction, k_z, to the ratio of fracture versus matrix permeabilities, k_f/k_m. It can be concluded that for flow simulations with a fracture which fully bisects the flow domain (i.e., simulations 1, 2, and 3) that:

- K_z is directly related to k_m when:
  - The fracture fully divides the flow domain perpendicular to the direction of flow, and k_f is high; and
  - The fracture fully divides the flow domain parallel to the direction of flow, and the k_f is low.

  K_z can be calculated from the harmonic average of k_f and k_m when the fracture is perpendicular to the direction of flow. The harmonic average, k_h, is calculated using:

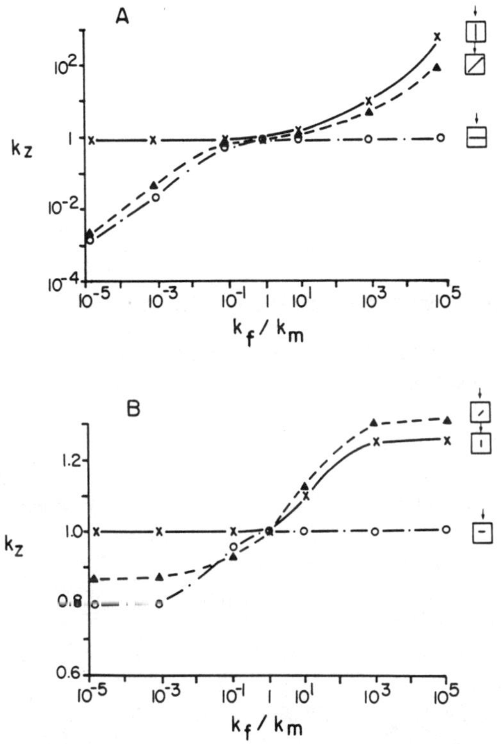

Figure 10. **Results of simulations performed using the two-dimensional flow geometry for fractures fully (A) and partially (B) dividing the flow domain.**

$$k_h = (b_f + b_m) / (b_f/k_f + b_m/k_m) \tag{29a}$$

- K_z is directly related to k_f when:
  - — The fracture fully divides the flow domain perpendicular to the direction of flow, and k_f is low; and
  - — The fracture fully divides the flow domain parallel to the direction of flow, and k_f is high.

  K_z can be calculated from the arithmetic average of k_f and k_m when the fracture is parallel to the direction of flow. The arithmetic average, k_a, is calculated using:

$$k_a = (b_f\, k_f + b_m\, k_m) / (b_f + b_m) \tag{29b}$$

- K_z is directly related to k_f when the fracture fully divides the flow domain diagonal to the direction of flow. K_z can be calculated from the geometric average of k_h and k_a when the fracture is diagonal to the direction of flow.

The geometric average, k_g, is calculated using (Bear, *24*):

$$k_g = 1 / (\cos^2 \alpha / k_h + \sin^2 \alpha / k_a) \tag{29c}$$

where α is fracture orientation ($\alpha = 0$ for a perpendicular fracture). For a fracture with a length one-half of the flow width (i.e., simulations 4, 5, and 6), the effects of fracture orientation are similar to those for a fully divided flow domain, but the effects are reduced.

Three-Dimensional Porous Medium Application

To evaluate the boundary integral procedure for its ability to estimate three-dimensional flow properties of a fractured medium, a simplified three-dimensional flow domain is investigated by defining a cube with unit length sides, four of which are no-flow boundaries and the top and bottom are assigned a constant head of one and zero, respectively (Figure 11). Again, the matrix permeability is assigned a value of unity and the center of a fracture is located at the center of the flow domain and the fracture is allowed to vary in length and orientation, as described in Table 1.

An analysis of the effect of nodal density on simulation error is performed in a manner similar to that performed for two-dimensional flow. The aspect ratio is again defined as the distance between nodes divided by the thickness of the flow domain. Results for three-dimensional aspect ratio error analyses are presented in Table 3. An aspect ratio less than 10:1 (corresponding to an error of less than 5 percent) is used in the simulation experiments described next.

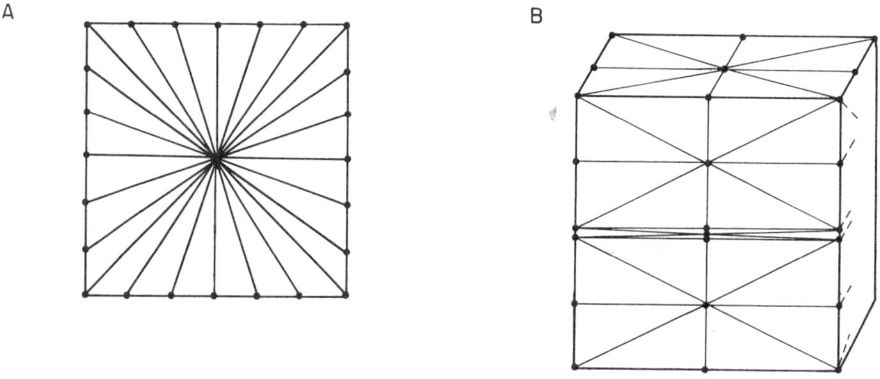

Figure 11. **Three-dimensional flow geometry showing fracture (A) and matrix (B) boundary surface discretization strategies.**

Results for experiments conducted using the same conditions (i.e., simulations 1 through 6; presented previously in Table 1) as for the two-dimensional experiments are presented in Figure 12. Table 4 compares simulation and analytic results using Equation 29. Note that the simulation results compare favorably except for the case

Table 3. **Precision of boundary integral method as a function of aspect ratio, defined here as the ratio of the distance between nodes to the width of the flow domain.**

Aspect Ratio	1:1	5:1	10:1	20:1	25:1
2-D Error (percent)	0.000	0.484	4.70	7.2	17.8

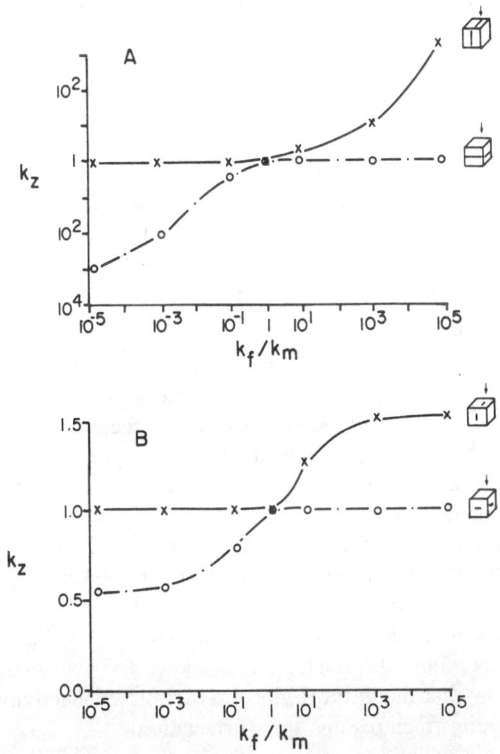

Figure 12. **Results of simulations performed using the three-dimensional flow geometry for fractures fully (A) and partially (B) dividing the flow domain.**

when the fractures are placed diagonally across the flow domain and a large fracture/matrix permeability ratio is used. In this case, the error is attributed to the positioning of one end of the fracture directly upon the upper flow surface, with no intervening matrix. By providing a direct connection between the upper flow boundary and the fracture, an estimated global permeability higher than the theoretical value is to be expected.

Table 4. **Analytic and simulation results for variable fracture/matrix permeability ratios using three-dimensional boundary integral method. Fully bisecting fractures.**

k_f/k_m	Perpendicular k_h	Perpendicular k_z	Parallel k_a	Parallel k_z	Diagonal k_g	Diagonal k_z
10^{-5}	0.001	0.0018	0.99	0.98	0.002	0.003
10^{-3}	0.091	0.023	0.99	0.99	0.17	0.07
10^{-1}	0.92	0.90	0.99	0.99	0.95	0.96
10^{-0}	1.	1.	1.	1.	1.	0.99
10^{+1}	1.01	1.02	1.09	1.20	1.05	1.07
10^{+3}	1.01	1.02	10.1	9.79	1.84	6.23
10^{+5}	1.01	1.02	1000.	1013.	2.02	12.40

Summary

Hydraulic flow properties of a combined fracture-matrix system are characterized using a proposed three-dimensional boundary integral procedure. The procedure accounts for flow through fractures embedded within a porous matrix. The only discretization required is along the boundaries of the fractures and the exterior surface of the matrix block being examined. An average unit vertical head gradient is induced across the block, and the resultant flow is used to estimate the global vertical hydraulic conductivity, K_z.

The effects of variations in fracture permeability on K_z are simulated by allowing the fracture permeability to vary from 10^{-5} to 10^5 relative to the matrix permeability. The variations in fracture permeability can result from geochemical processes such as dissolution and precipitation of minerals, or the result of variable water saturation levels within the fracture. It is shown for both two- and three-dimensional flow that the vertical global hydraulic conductivity is a function of the length of intervening fractures and their orientation.

When fractures are of infinite areal extent and the permeability/matrix permeability ratio, k_f/k_m, is large, K_z is directly related to k_f when all the fractures are oriented vertically, but is unaffected by the fractures if they are all oriented horizontally. For small k_f/k_m ratios, the roles of vertical and horizontal fractures reverse, with horizontal fractures controlling the magnitude of K_z, and vertical fractures having no effect.

Literature Cited

1. Long, J.C.S., J. S. Remer, C. R. Wilson, and P. A. Witherspoon, "Porous Media Equivalents for Networks of Discontinuous Fractures," *Water Resour. Res.*, *18*(3):645–658 (1982).

2. Gelhar, L. W. and C. L. Axeness, "Three-Dimensional Stochastic Analysis of Macrodispersion in Aquifers," *Water Resour. Res., 19*(1):161–180 (1983).
3. Yeh, T.-C.J., L. W. Gelhar, and A. L. Gutjahr, "Stochastic Analysis of Unsaturated Flow in Heterogeneous Soils, 1, Statistically Isotropic Media," *Water Resour. Res., 21*(4):447–456 (1985).
4. Yeh, T.-C.J., L. W. Gelhar, and A. L. Gutjahr, "Stochastic Analysis of Unsaturated Flow in Heterogeneous Soils, 2, Statistically Anisotropic Media With Variable Alpha," *Water Resour. Res., 21*(4):457–464 (1985).
5. Yeh, T.-C.J., L. W. Gelhar, and A. L. Gutjahr, "Stochastic Analysis of Unsaturated Flow in Heterogeneous Soils, 3, Observations and Applications," *Water Resour Res., 21*(4):465–472 (1985).
6. Coakley, K., K. Muralidhar, J.C.S. Long, and L. R. Myer, "Equivalent Permeability of Statistically Simulated Single Fractures," *Conference on Geostatistical, Sensitivity, and Uncertainty Methods for Ground-Water Flow and Radionuclide Transport Modeling,* San Francisco, CA, September 15–17 (1987).
7. Bibby, R., "Mass Transport of Solutes in Dual-Porosity Media," *Water Resour. Res., 17*(4):1075–1081 (1981).
8. Huyakorn, P. S., B. H. Lester, and C. R. Faust, "Finite Element Techniques of Modeling Groundwater Flow in Fractured Aquifers," *Water Resour. Res., 19*(4):1019–1035 (1983).
9. Moench, A. F., "Double-Porosity Models for a Fissured Groundwater Reservoir With Fracture Skin," *Water Resour. Res., 20*(7):831–846 (1984).
10. Wang, J.S.Y. and T. N. Narasimhan, "Hydrologic Mechanisms Governing Fluid Flow in a Partially Saturated, Fractured, Porous Medium," *Water Resour. Res., 21*(12):1861–1874 (1985).
11. Huang, C. and D. D. Evans, *A 3-Dimensional Computer Model to Simulate Fluid Flow and Contaminant Transport Through a Rock Fracture System,* NUREG/CR-4042, U.S. Nuclear Regulatory Commission, 108 pp. (1985).
12. Rasmussen, T. C., "Computer Simulation of Steady Fluid Flow and Solute Transport Through Three-Dimensional Networks of Variably Saturated, Discrete Fractures," in Evans, D.D. and T.J. Nicholson (Eds.), *Flow and Transport Through Unsaturated Fractured Rock,* AGU Geophysical Monograph 42, pp. 107–114 (1987).
13. Kilbury, R. K., T. C. Rasmussen, D. D. Evans, and A. W. Warrick, "Water and Air Intake of Surface-Exposed Rock Fractures in Situ," *Water Resour. Res., 22*(10): 1431–1443 (1986).
14. Lafe, O. E., J. A. Liggett, and P.L-F. Liu, "BIEM Solutions to Combinations of Leaky, Layered, Confined, Unconfined, Nonisotropic Aquifers," *Water Resour. Res., 17*(5):1431–1444 (1981).
15. Liggett, J. A. and P.L-F. Liu, *The Boundary Integral Equation Method for Porous Media Flow,* George Allen and Unwin, 255 pp. (1983).
16. Brebbia, C. A., J.C.F. Telles, and L.C. Wrobel, *Boundary Element Techniques, Theory and Applications in Engineering,* Springer-Verlag, 464 pp. (1984).
17. Huyakorn, P. S. and G. F. Pinder, *Computational Methods in Subsurface Flow,* Academic Press, 473 pp. (1983).
18. Shapiro, A. M. and J. Andersson, "Steady State Fluid Response in Fractured Rock: A Boundary Element Solution for a Coupled, Discrete Fracture Continuum Model," *Water Resour. Res., 19*(4):959–969 (1983).
19. Elsworth, D., "A Model to Evaluate the Transient Hydraulic Response of Three-Dimensional Sparsely Fractured Rock Masses," *Water Resour. Res. 22*(13):1809–1819 (1986).

20. Elsworth, D., "A Boundary Element-Finite Element Procedure for Porous and Fractured Media Flow," *Water Resour. Res., 23*(4):551–560 (1987).
21. Lafe, O. E. and A.H.H-D. Cheng, "A Perturbation Boundary Element Code for Steady State Groundwater Flow in Heterogeneous Aquifers," *Water Resour. Res., 23*(6):1079–1084 (1987).
22. Brebbia, C. A., *The Boundary Element Method for Engineers,* Pentech Press, 189 pp. (1978).
23. Cowper, G. R., "Gaussian Quadrature Formulas for Triangles," *Int. J. Math. Engr., 7*(3):405–408 (1973).
24. Bear, J., *Hydraulics of Groundwater,* McGraw-Hill Book Co., 569 pp. (1979).

4 Statistical Methods for Modeling Flow and Transport in Fractured Media

Three-Dimensional Geometric Modelling of a Fracture Network

JEAN-PAUL CHILÈS
Bureau de Recherches Géologiques et Minières
Orléans, France

Abstract

Fluid-flow studies of fractured rocks require a three-dimensional modelling of the fracture network. Because the complexity of reality does not allow a complete description of the actual network, it is therefore necessary to use Monte-Carlo models. Various basic models can be used depending on the situation: random infinite fractures (planes), random finite fractures (polygons or discs), and locally self-similar fractal networks. Other models of easy simulation can be derived by mixing, clustering, and regionalizing the parameters. These models are reviewed. Emphasis is placed on the determination of the parameters by means of morphological, statistical, geostatistical, and fractal tools that can be used with one-dimensional data (boreholes) or two-dimensional data (outcrops and drift walls). The last section explains how to condition a three-dimensional simulation on the data.

Introduction

To study rock mechanics and subsurface flow in deep hard rocks such as granite, we must know the characteristics of the fracture field. Applications are more and more complex, and some of them involve a complete description of the fracture field (cf., e.g., Long, 1986). As fractures are not directly accessible, but accessible only at their intersections with boreholes, outcrops, or drifts, one of the first steps in studying a given site is to model the fracture field and to simulate it.

The modelling of a fracture field is complicated. Fracturing can be studied at various scales: large faults, fractures, and microfissures. At each scale, two distinct features can be studied:

1. The fracture network: Each element is considered as relatively simple (e.g., a planar fracture), and the study is concerned with the imbrication of the various elements.
2. The single fracture: Generally an open fracture is not a pair of parallel plates; there are voids and contact areas that depend on stress and have a dramatic influence on conductivity (cf. Feuga, 1986).

The single fracture is usually studied by means of geostatistical or fractal methods. See, e.g., Sayles and Thomas (1977) and Mandelbrot et al. (1984) for metal fractures, and Gentier (1986) and Coakley et al. (1987) for rock fractures.

This paper is focused on the fracture network. A three-dimensional (3-D) study domain, even limited, can correspond to several thousand or million fractures. Few of these fractures intersect the boreholes, outcrops, or drift walls where the network is observed, so deterministic modelling is not possible, except for major faults. Hence we consider only stochastic modelling. The first tools used to characterize a network were mainly statistical: distribution of orientation, trace length, spacing, and aperture. In the last years there has been an increasing use of geostatistical tools to study the regionalized behavior of the network. Now fractal methods are also being used. All these approaches enable a realistic modelling and simulation of reality that is usually complex. We examine all these aspects of modelling, with a final section devoted to the conditioning of a 3-D simulation by one-dimensional (1-D) or two-dimensional (2-D) observed fracture intersections.

Fractures are considered here as planar. This is of course a simplification, but one that does not usually modify the conclusions on connectivity or stability. According to the scale, they are considered either infinite or convex, so that fracture traces are either straight lines or segments.

Mathematical Morphology Approach

Mathematical morphology provides a means of fully characterizing a fracture network. As we consider here stochastic models, let us denote by A the (random) set of all the fractures (fracture walls and void spaces). If we neglect the aperture of the fractures, this set will be fully determined by the knowledge of the functional moment, Q(B), for any (nonrandom) set B, where Q(B) is defined by:

$$Q(B) = \text{Prob}\,\{B \cap A = \varnothing\} = \text{Prob}\,\{B \text{ does not intersect any fracture}\} \qquad (1)$$

This characterization is the exact equivalence of the spatial law for random functions. The random set A is usually assumed to be stationary. Q(B) is then unchanged by translation of B and can be estimated from the sole realization of the fracture network.

In practice, it is not possible to consider any set B. Conrad and Jacquin (1973) used this approach to study a 2-D network, with B being a segment, a disc, or a crown. The moments Q(B) were measured with a texture analyzer. This is a very powerful method that deserves to be rediscovered. Its implementation is not always obvious, and hence other tools, less powerful but simpler, are usually used.

Statistical Characterization

Reality is 3-D, but the observations are 1-D (boreholes) or 2-D (outcrops and walls). Therefore modelling is performed in two steps:

1. Characterization of the 1-D or 2-D data.
2. Link with a 3-D model.

In this section, we only consider the first step. It can be applied to all fractures, but it is generally advisable to subdivide the fractures into several directional sets and to study them separately. So the first task is the study of fracture orientation.

Orientation

The orientation of a fracture plane can be defined by its strike and dip, or by its pole, i.e., by the unit vector normal to the fracture. We will use the latter definition. As fracture orientation is an undirected directional variable, it is usually visualized on an equiareal lower (or upper) hemisphere projection diagram, e.g., a Schmidt diagram (Phillips, 1960).

When interpreting such a diagram, we have to take account of the geometric bias resulting from the orientation of the surveyed boreholes or outcrops. Fractures that are normal to the surveyed line or surface occur more frequently than oblique fractures. In order to reduce this kind of bias, sampling along various orientations is recommended. If this is not possible each recorded fracture will be weighted by the inverse of the a priori probability of observing this fracture in case of a uniform distribution of the orientation. If, for example, a borehole of length l is surveyed, the a priori probability for a fracture with a pole at θ angle with the borehole is proportional to $l|\cos\theta|$.

Even if possible biases are corrected, the pole distribution is seldom uniform. Indeed the orientation of fractures is related to tectonic history. The first tectonic episode has often created two conjugate sets of fractures, whereas the last episodes have reactivated previous fractures. This is the reason why a pole diagram usually displays several sets of orientation (Figure 1). It is generally recommended to study these sets separately, as individual sets are expected to be more homogeneous than combined sets (e.g., the trace length can differ from one set to the other). Several cluster analysis methods have been developed to automatically define the various sets (cf., e.g., Bailey, 1975; Schaeben, 1984). But structural geologists can often relate fractures to tectonic history and can help to physically ground the definition of sets.

The modelling of orientation within a cluster can be made by using a truncated Fisher distribution (cf., e.g., Lewis and Fisher, 1982), or simply with a bivariate Gaussian distribution in the plane tangent to the central orientation of the cluster.

In the following, we most often consider the characteristics of a set of fractures (i.e., fractures that correspond to a given class of orientation).

Trace length

If fractures cannot be considered as infinite at the scale of the study, the distribution of trace length is an important parameter that can be obtained from 2-D surveying (outcrops of drift walls). But the experimental trace length distribution is affected by two biases:

1. Censoring: When both ends of a trace are not observable, as is often the case in a drift with vertical fractures, the length recorded is a censored length;
2. Truncation: Traces that are shorter than some cutoff length are not recorded, either to save time or because it is difficult to discriminate between natural fractures and fractures due to mining when these are of very limited extent.

Both types of bias are well known, and methods for correcting the mean distribution or deriving the underlying distribution have been developed (Baecher and Lanney, 1978;

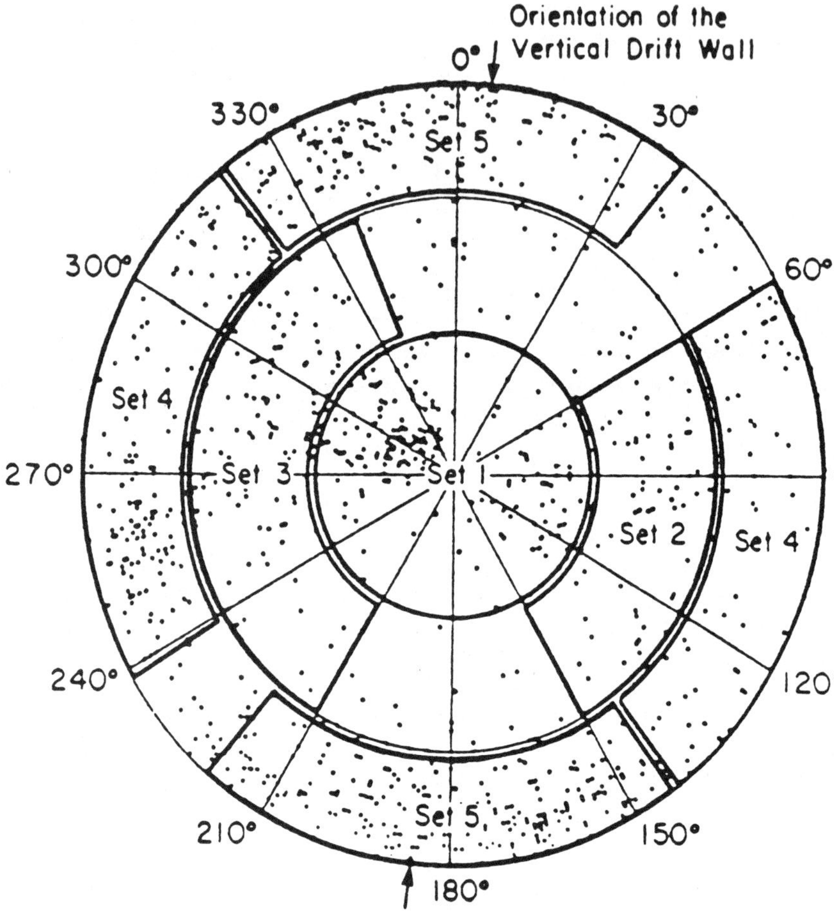

Figure 1. **Diagram of poles to fractures on a Schmidt equal-area stereo net (1178 recorded fracture traces on a 80-m × 2-m drift wall).**

Cruden, 1977; Laslett, 1982; Baecher, 1983; Pahl, 1981). Pahl's approach separates the traces into three classes according to the number of visible endpoints (0, 1, or 2). His distribution-free estimation method of the mean trace length can be extended to the estimation of the distribution itself (Massoud, 1987; cf. Figure 2). Usually this distribution is found to be approximately lognormal.

Spacing

Spacing is measured by the separation between successive intersections of fractures with a surveying line, either for an individual set or for all fractures. It is defined for infinite fractures as well as for finite ones. The inverse of the mean spacing is the average density of intersections along the surveying line. If the spacing distribution is exponential, it is summarized by its mean. Although many authors found exponentially

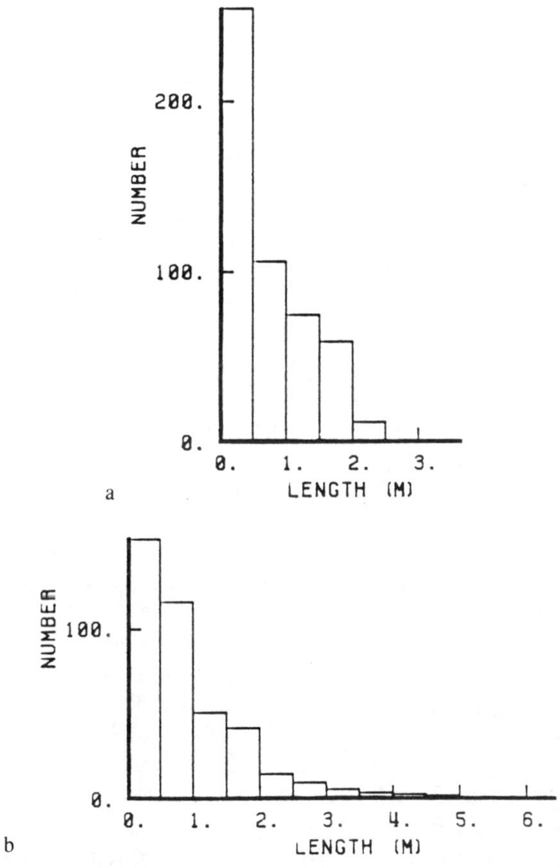

Figure 2. **Histogram of trace length of a set of subvertical fractures: (a) observed (and so censored and truncated); (b) reconstructed (i.e., after bias corrections). The vertical axis is the number of fractures in each class.**

distributed data sets, this is not always the case, as we will see further, and the spacing histogram must be computed at each study.

Note that in the 1-D case, if the successive spacings are independent, the knowledge of the spacing distribution is strictly equivalent to the knowledge of the functional moment Q(B) defined by (1) for any B, and characterizes the phenomenon.

Aperture

Aperture is an important parameter for flow calculations, but is difficult to handle, for two main reasons:

1. Aperture is not constant: There are void spaces and contact areas, and flow between two parallel plates separated by the mean aperture has nothing to do with the actual flow.

2. Because of stress release, the measured aperture has nothing to do with an in situ aperture.

Thus, the study of the apparent aperture is not satisfactory. Indirect methods are used, e.g., water-injection tests (Smith et al., 1987; Snow, 1970). In some cases, however, the measured aperture has a meaning. If the void spaces are filled with calcite or another material, the aperture is replaced by a thickness that is not affected by stress release. The thickness is often found to have a lognormal distribution and to be correlated with the size of the fracture (cf. Loiseau, 1987, for correlation coefficients around 0.90).

Basic Models

All models can be deterministic or stochastic. The first models were deterministic, as illustrated by the orthogonal model of Snow (1965) that is defined by three orthogonal directions of equidistant planes. Then came models with purely random parameters, such as the Poisson flat process (Priest and Hudson, 1976). As we know that fracture orientation is not purely random, we will adopt stochastic models with a distribution of orientation that is not necessarily uniform.

All the models described in this section share common assumptions:

1. All fractures are planar (assumption applicable to the whole paper).
2. All fracture locations are equally probable.
3. All fracture orientations are independent of fracture locations.

These basic models will be described succinctly. A comprehensive description from a rock mechanics point of view can be found in Dershowitz (1984).

Random planes

In some cases, the fractures are long enough compared to the studied domain, so that they can be considered as infinite. The simplest stochastic model is the Poisson flat model, defined by a single parameter, density λ, which is the average intersection surface of the network with a unit volume. If we want to take account of a nonuniform distribution of the fracture orientation, we add the probability density function $f(\omega)$ of the pole orientation.

If we consider planes whose poles lie within an elementary solid angle $d\omega$ centered on direction ω, their intersections with the line of direction ω that cuts across the origin of the coordinate system is a 1-D Poisson process with density $\lambda\ f(\omega)d\omega$. So a simulation of a 3-D Poisson flat process can be very simply achieved in two steps (Figure 3):

1. Simulation of a 1-D Poisson process of density λ;
2. For each point of the Poisson process, selection of a random orientation according to p.d.f. $f(\omega)$: The corresponding random plane is the plane orthogonal to this orientation so that its algebraic distance to the origin is equal to the abscissa of the point of the 1-D Poisson process.

The simplicity of this model is very gratifying with its ease of parameter fitting and availability of numerous properties of this kind of network through geometric probability theory (Kendall and Moran, 1963; Santalo, 1976). It is sometimes used in hydrological applications (e.g., Andersson et al., 1984). The counterpart of this simplicity is lack of realism in most cases. Misuse of this model produces a too systematic block network and an excessively connected fracture network.

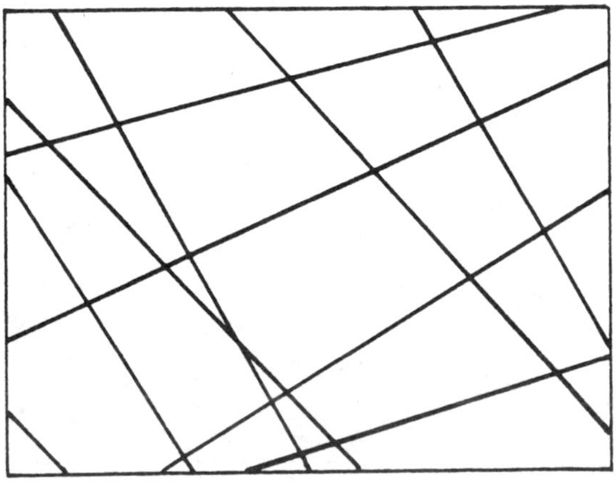

Figure 3. **Section of a 3-D model of random planes (Poisson flat process).**

Random coplanar polygons

Usually fractures are studied at a scale where they cannot be considered as infinite. One way to get finite fractures is to add a second step to the preceding model by subdividing each plane into a network of polygons, each of which will be randomly tossed as fracture or intact rock. Two methods can be applied (Figure 4):

1. Each Poissonian flat is subdivided into polygons by an independent Poisson line process (which can be simulated with the same method as the Poisson flat process); this model was developed by Veneziano (1978).
2. Polygons are defined by the intersection of Poissonian flats rather than by an independent process; this model was developed by Dershowitz (1984).

These models are simple: They have just one or two more parameters: persistence (the proportion of polygons selected as fractures), and for model 1, the Poisson line density.

The size of individual polygons is well known through geometric probability theory. It is related to the density of Poisson lines in model 1 and to that of Poisson planes in model 2. But the size of fractures also depends on persistence, which conditions the mean number of elementary polygons forming a fracture. As model 2 has one degree of freedom less than model 1, Dershowitz gives it more flexibility by adding planes with zero persistence: these planes provide additional intersections on the other planes, allowing reduction of the size of the elementary polygons.

In model 2, fracture intersections always occur at Poissonian flat intersections. So a fracture can end when encountering another fracture, which is not the case in model 1. These two models correspond to different phenomena. But they share the property of providing coplanar fractures, which can be understood as the result of weakness planes in rock.

Random non-coplanar polygons

Model 2 is based on a random subdivision of space into nonoverlapping polyhedra, i.e., three-dimensional tessellation, and on a random selection of polyhedron faces to be

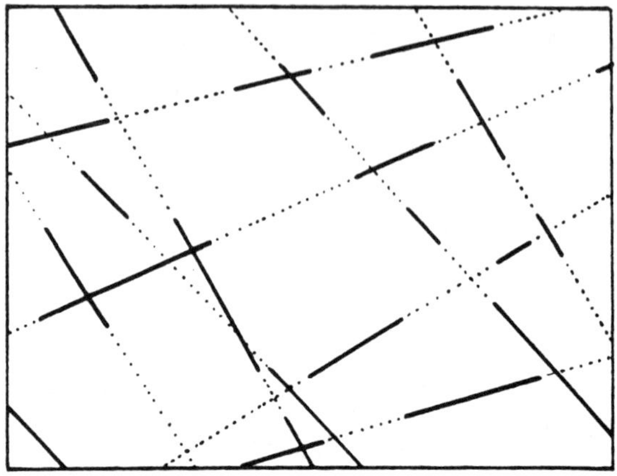

a

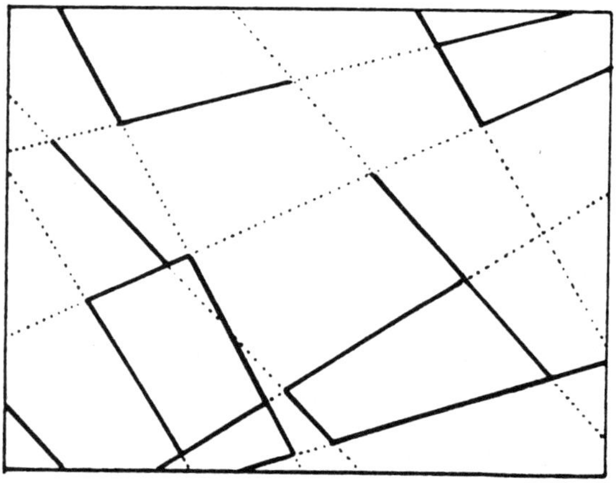

b

Figure 4. **Sections of 3-D models of random coplanar polygons, delimited within random planes: (a) model 1 (Veneziano): polygons delimited by independent processes; (b) model 2 (Dershowitz): polygons delimited by random planes.**

kept. It leads to coplanar fractures because of the definition of the polyhedra. If a model of non-coplanar fractures is needed, one only has to replace this particular tessellation by another tessellation model that does not produce coplanar polyhedron faces (Figure 5). The best known model is the three-dimensional Voronoi model: It is obtained from a 3-D Poisson process of seeds that grow at a constant rate up to formation of polyhedra. The Delaunay model is also based on a Poisson process, but the Poisson points now represent block vertices. In this model, all block faces are triangles; in the former they are hexagons. The Voronoi model with variable growth rate would result in variable-sided polygons. The geometric properties of these models are also well known (cf. Santalo, 1976; Serra, 1982).

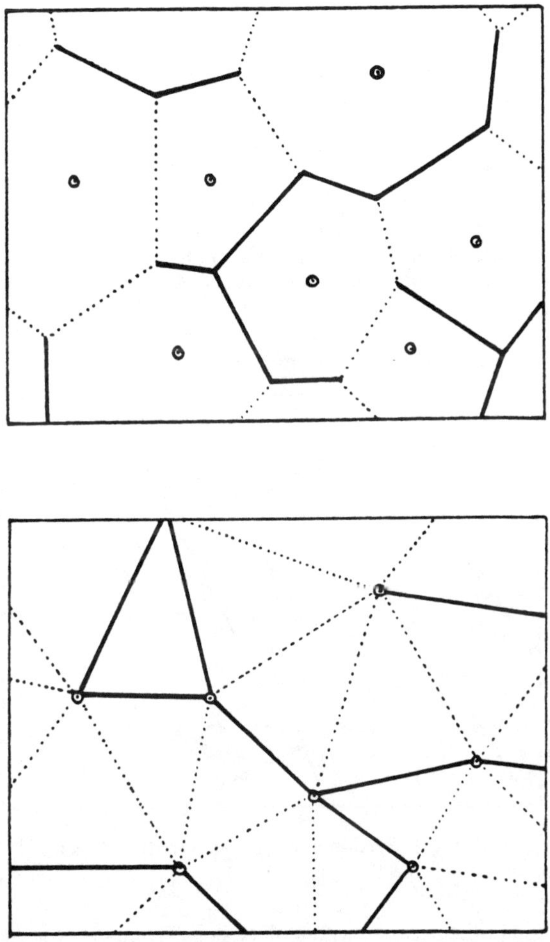

a

b

Figure 5. **Section of a 3-D model of random non-coplanar polygons: (a) Voronoi polyhedra (growth process from Poisson seeds); (b) Delaunay tessellation (block vertices from a Poisson process).**

These models produce networks with a uniformly distributed orientation. It is possible to account for a specific orientation distribution by adapting persistence to the orientation of the block face. These models are however scarcely used in 3-D, probably because they are based on a set of points that have no physical existence (at least for the Voronoi model). So the fracture parameters (what can be measured) are only indirectly connected to the model parameters. From this point of view, the random disc model remains closer to reality.

Random discs

This model (Figure 6) was introduced for rock mechanics applications by Baecher et al. (1977). It is a particular case of the Boolean model (see, e.g., Matheron, 1967; Serra, 1982). Each fracture is assumed to be a circular (or elliptic) disc, and each is defined by its central location, diameter, and orientation:

1. The disc centers constitute a 3-D Poisson process.
2. The diameter is considered as random with a given distribution.
3. As in the previous models, orientation is random.
4. All fractures are assumed to be independent of each other.

This model has also a limited number of parameters. They are directly related to the shapes of fractures.

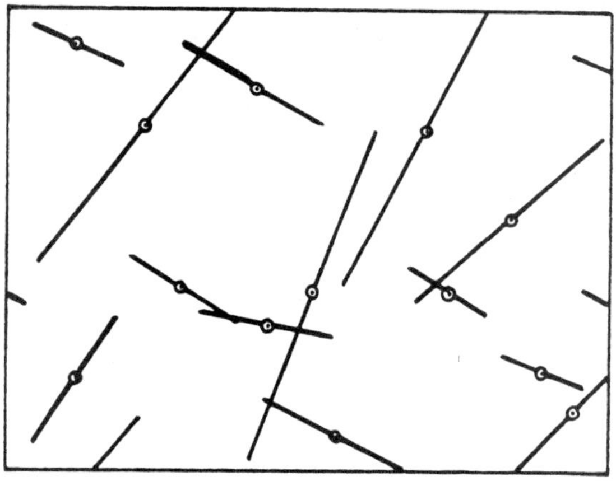

Figure 6. **Section of a 3-D random disc model.**

The intersection of the network with a plane is a 2-D model of segments centered on the points of a 2-D Poisson process. The parameters of the 3-D and 2-D models (Poisson process densities, disc diameter and trace length distributions) are related by simple stereological relationships (see Warburton, 1980). The 3-D model parameters are hence easily deduced from 2-D measurements.

In this model, blocks are formed only when the fracture density or disc size is sufficiently high to cause intersections between triplets of discs rather than pairs. Blocks are not shaped as convex polyhedra. So this model seems suitable for rock masses that do not exhibit distinct rock blocks.

Fractal Approach

The fractal theory of Mandelbrot (1983) is now widely used in the earth sciences. Applications to fracture network modelling are, however, not numerous in spite of the kinship of the terms "fracture" and "fractal." Fractal models use a single parameter, the fractal dimension.

Fractal dimension—box dimension

The exact definition of fractal dimension, or Hausdorff dimension, does not provide a simple means to measure it in a fracture network. But several alternative definitions exist that yield the same value as the Hausdorff dimension in the case of self-similar fractals. For fracture networks, the best tool is the box-counting dimension, or simply box dimension, initially called similarity dimension, as defined by Mandelbrot (1985).

To define this box dimension, suppose that we are studying a 2-D pattern (a network of traces in our case). We can cover the study domain with a lattice made of square boxes of side r, and count the number N(r) of boxes intersecting the pattern (Figure 7). If the pattern is a rectangle of area S and r is small enough, we have approximately $N(r) = S/r^2$. Likewise, if the pattern is reduced to a segment of length L parallel to one side of the boxes, we have $N(r) = L/r$. In both cases, the Euclidean dimension (2 or 1), is D so that N(r) has the form $N(r) = A/r^D$, or equivalently:

$$\log N(r) = \alpha - D \log r$$

So D is the absolute value of the log-log plot slope of N versus r. This relation is generalized to more complex patterns, where D can be a noninteger value. D is the box dimension.

Let us mention some applications to fracture network: Barton and Larsen (1985) found a dimension between 1.12 and 1.16 for three tuff outcrops; Jacquin and Adler (1987) gave a dimension of 1.8 to an outcrop in a quarry; Thomas (1987), with another method, concluded the fractal behavior of several calcareous or granitic outcrops to have a dimension slightly above 1.

Scale dependence

In practice, one must be very careful when analyzing a log-log plot. The mechanical use of the method always provides a graph, even if the phenomenon is not fractal. Of course in this case the graph is not necessarily linear, but it may look approximately lin-

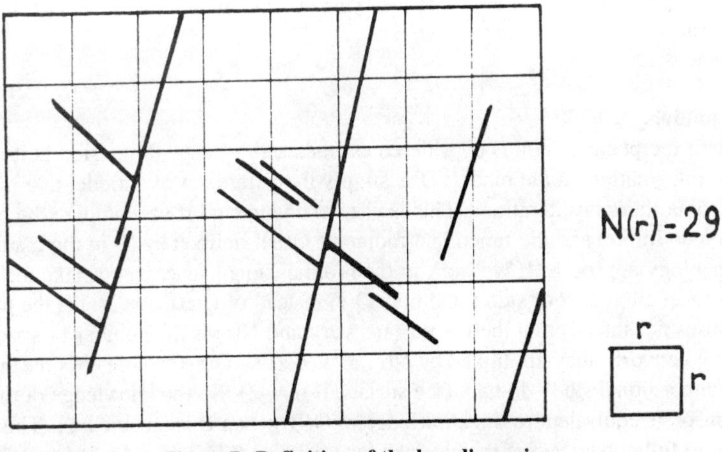

Figure 7. **Definition of the box dimension.**

ear at the scale of the study. Figure 8 displays the log-log plot for a survey on a 10-m by 10-m area in a granite massif. The absolute slope, i.e., the box dimension, varies from 1 for small r values to 2 for large r values. Indeed for small values (around 1 cm), the linear shape of traces prevails, whereas for large values (several meters) any r × r box intersects the network, leading to a box dimension of 2 at this scale. Similar conclusions would be drawn on the log-log plot of any of the 3-D basic models of the preceding section (with a box dimension varying from 2 to 3).

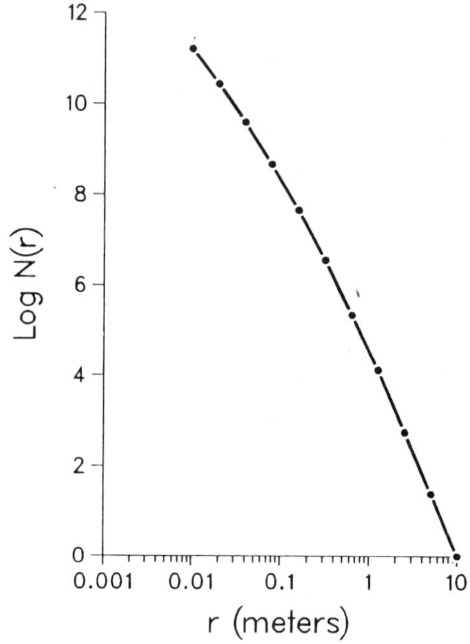

Figure 8. **Log-log plot of N(r) for a 2-D surveying on 10 m by 10 m. All traces are taken into account.**

Fractal models

Even if the phenomenon is considered as fractal, the fractal dimension D itself yields no information on the model: D is simply the parameter of a model that must be chosen from other considerations. This is easy to understand if we compare the box dimension definition and the functional moment, Q(B), defined by (1) in the mathematical morphology approach: If we work in the n-dimensional space, let us take for B the n-dimensional cubic box of side r and note Q(r) instead of Q(B). As Q(r) is the proportion of boxes not intersecting the fracture network and N(r) is the number of boxes intersecting the network, they are linked by $N(r) = V\,r^n\,(1 - Q(r))$, where V is the volume of the n-dimensional study domain (the surface if $n = 2$). So the knowledge of the fractal dimension is equivalent to the knowledge of Q(B) when B is a cubic box. This is not sufficient to fully characterize the network, as we know that it would require Q(B) for any set B (cf. the study of the same network by Conrad and Jacquin, 1973, and by Jacquin and Adler, 1987).

An exception is the 1-D case (the fracture network is reduced to a network of points) when successive spacings can be considered as independent. This is why the few attempts to simulate a fracture network with a fractal model start from a 1-D model: Thomas (1987) simulates infinite planar fractures from a fractal simulation of intersections with a sampling line. Chilès (1987) simulates finite fractures of fixed orientation and random length in a 2-D case; the method takes account of a scale-dependent box dimension (Figure 9). These methods are not yet operational for realistic 3-D simulation and need to be perfected.

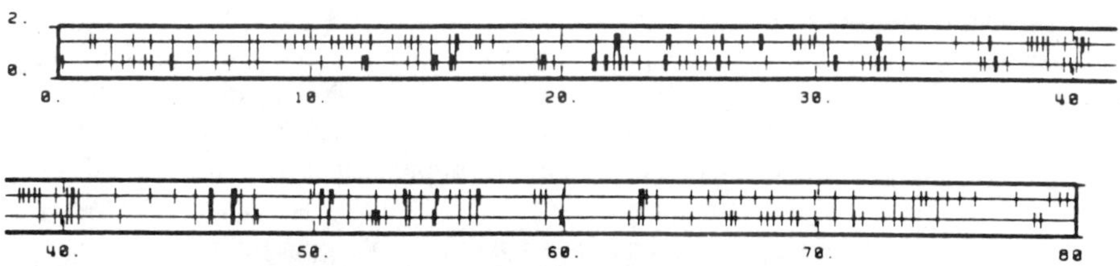

Figure 9. **Simulation of a set of subvertical fracture traces on an 80-m by 2-m drift wall with a model with scale-dependent box dimension. Only the intersections with two lines are simulated. The distance between these two lines corresponds to the mean trace length. The figured trace lengths are conventional.**

Geostatistical Approach

The above basic models provide networks of independent fractures or at least nearly independent fractures (in the models of random coplanar polygons, independence is not complete, as all the polygons of a common plane have the same orientation). These models are a good starting point for the study of actual fields, because fractures are discontinuities, the location of which has a random character. But many phenomena, although random, are also partially structured and require a geostatistical approach (Matheron, 1965). So the assumptions of independence and of constant density that are the basis of these models must be questioned, and if necessary weakened. Among the first writers on the subject were Miller (1979), who found a range of around 5 to 10 m on the variogram of spacings obtained from line surveying, and La Pointe (1980). (As in many publications now, the term "variogram" is used instead of "semi-variogram," for the sake of simplicity, and also for theoretical considerations; see Chilès, 1979.)

Spacing

The intersection of any of the basic models with a line provides a Poisson process of points; successive spacings are independent. Moreover they follow an exponential distribution, so that their relative variance equals 1. Instead of simply computing the mean spacing, it is worth verifying if the actual data do satisfy these properties. This can be achieved with the variogram referenced to the fracture sequence number introduced by Miller (1979): The spacing values are not assigned to points on the sampling line, but to successive integer abscissae, with the first observed spacing assigned to the abscissa $x = 1$, the second one to $x = 2$, and so on. The h distance of variogram $\gamma(h)$ is then measured in number of fractures. In a Poisson process this variogram is a pure nugget effect with sill equal to m^2 (m is the mean spacing).

Figure 10 presents the results obtained by Massoud (1987) for a set of subvertical fractures in the Fanay granitic massif (France). The variogram is actually flat, but its sill is twice as high. The histogram is obviously not exponential. Observation of fracture traces along drift walls shows that this is due to clustering. Fractures are grouped in clusters causing short spacing between fractures from the same cluster and long spacing when the two fractures belong to two distinct clusters. The same phenomenon has already been observed by Snow (1970).

Fracture trace density

Another property of the Poisson process is that the number of points in disjoint segments (or volumes in space) are independent variables with a Poisson distribution. So if we subdivide a surveying line into equal segments of length l, and for distances $h = l$, $2l$, $3l, \ldots$, we compute the variogram of the number of intersections per segment, this variogram will also exhibit a pure nugget effect. If a 2-D survey is available, the same property holds for the variogram of the number of trace centers per rectangle (this is always true for the random disc model, and if the rectangle is large enough for the other models). Figure 11 illustrates such a variogram obtained in Fanay by Massoud (1987) from a survey of traces on 2-m-high drift walls. We can clearly see:

- a nugget effect, corresponding to the random part of the fracture location;
- a first range of several meters, which may correspond to the extension of fracture clusters;
- longer range components, which can only be accounted for by a regionalized variation of the fracture density.

Fanay data were subdivided into five sets. One of them displays purely random behavior, but the other four provide structured variograms. Loiseau (1987) found similar results for fractures observed in gneiss.

Fracture parameters

Other fracture parameters can also be studied by geostatistics. For the pole orientation of fracture sets, Miller (1979) and Massoud (1987) found variograms with a short-range component (several meters). This can be understood as the fact that fractures belonging to a cluster tend to have the same orientation, whereas cluster orientations are less correlated. (In the case of Fanay data, this is substantiated by observation.)

More General Models

In order to take account of the complexity of reality, several simple generalizations of the basic models can be used.

Mixed models

In their 2-D study, Conrad and Jacquin (1973) observed long fractures and short

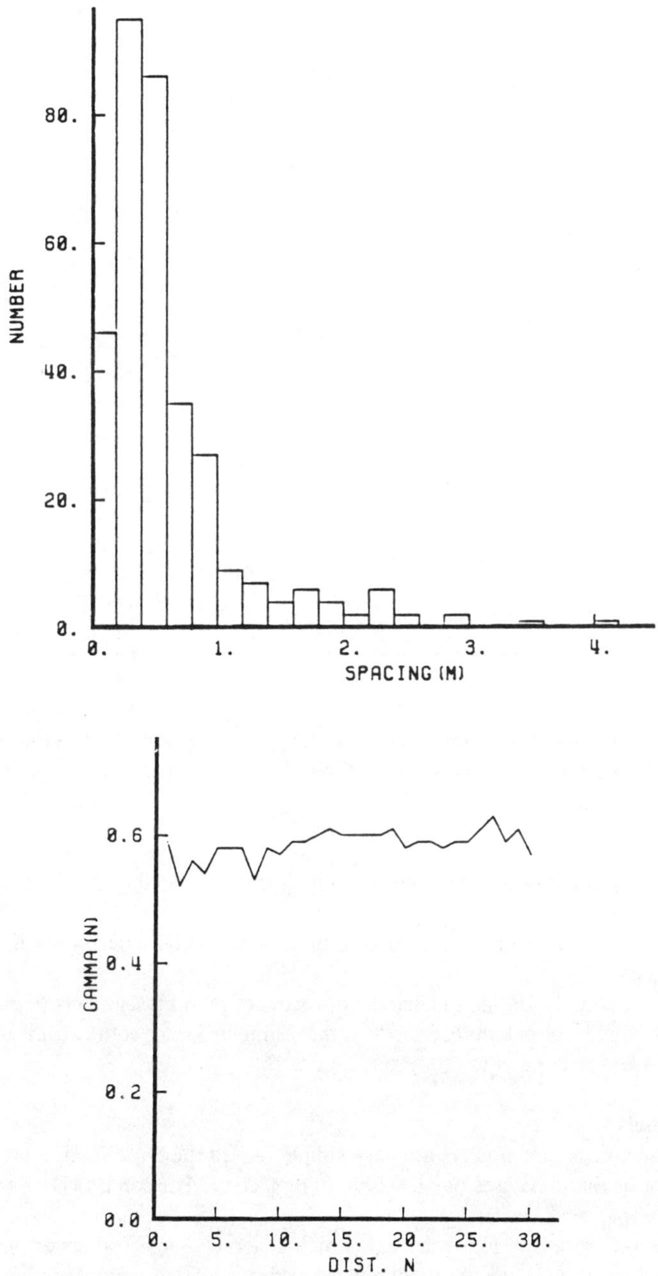

Figure 10. **Histogram and variogram of spacing for a set of subvertical fractures (Fanay data). The variogram is referenced to fracture sequence number. Note the incompatibility with a Poisson process assumption.**

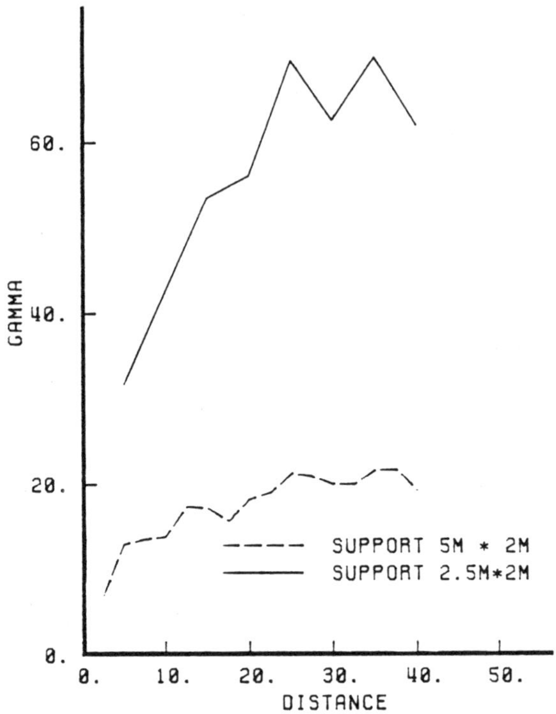

Figure 11. **Variogram of the number of trace centers per rectangle for two rectangle sizes (Fanay data, horizontal fracture set).**

ones that usually ended when encountering a long fracture. So they used a hierarchical model:

- a primary network of infinite Poisson lines, which subdivides the plane into polygons;
- within each polygon, an independent process of random segments (the 2-D version of the 3-D random disc model); the segments are of course truncated if they encounter the edge of the polygon (Figure 12).

Cluster models

Models with fracture clusters are very simple. As all the basic models are based on Poisson point or flat processes one has only to replace the Poisson process by a procedure in two steps:

- a Poisson process of lower density, which generates seeds (points or planes);
- around each seed, generation of a cluster of independent points or planes (Figure 13).

This kind of model has two new parameters: the mean number of elements of a cluster (the distribution of this number is difficult to determine, and so a Poisson distribution is usually assumed); and the dispersion of the elements around the seed (for similar rea-

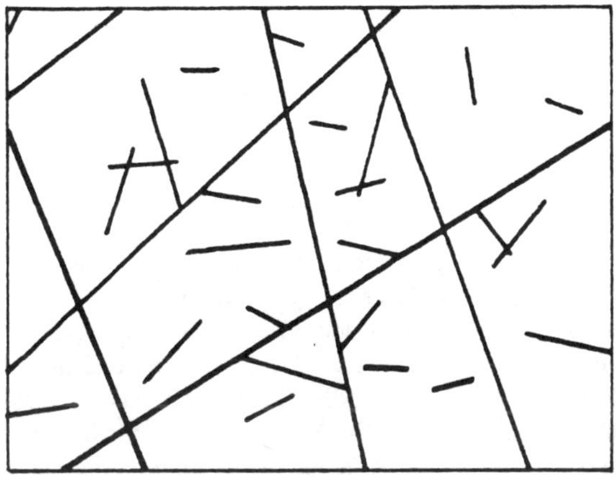

Figure 12. **2-D hierarchical model of Conrad and Jacquin (1973): (a) Poisson line process; (b) independent random segment processes within each polygon.**

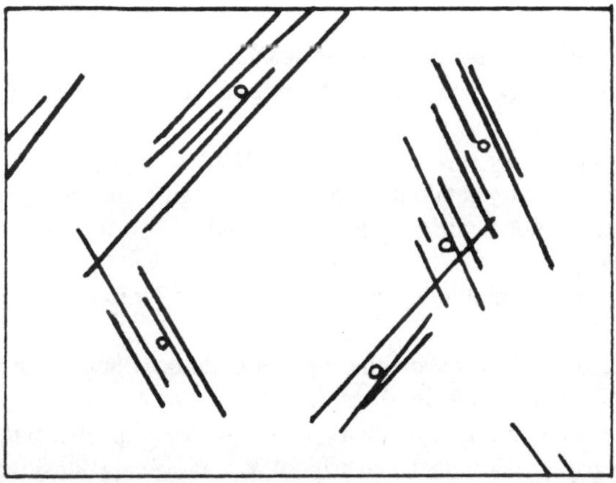

Figure 13. **2-D section of a hierarchical cluster model: (a) Poisson seed process; (b) independent disc clusters around each seed.**

sons, a Gaussian distribution is usually assumed). The density of seeds is taken as the ratio of the density of elements by the mean number of elements per cluster.

Regionalization of density

If the fracture density is not constant, there are two ways to introduce a regionalization of the model, depending on the type of model:

1. For models based on Poisson points (random disc model, Voronoi), the density λ of disc centers can be replaced by a stationary random function $\lambda(x)$ with variogram $\gamma(h)$.

2. For models based on Poisson flats (random coplanar polygons), we can similarly replace persistence p by a stationary random function (the randomization of the density of planes would bring less flexibility).

In this case, a parameter (e.g., the fracture density) is replaced by a mean value plus a variogram.

Parameter fitting

These models are easy to simulate: Regionalization can be achieved by a classical geostatistical simulation of the parameter (cf., e.g., Mantoglou, 1987; Matheron, 1973); mixed models and cluster models simply add one step more. The real problem is fitting the various additional parameters. In some cases, it is not more difficult than for the other parameters: For their 2-D mixed model, Conrad and Jacquin (1973) use the functional moment Q(B); for a model of Poissonian twin planes, Matheron (1971) gives the spacing distribution of the twin planes from the distribution of the spacing of twin points that can be measured along a borehole. But usually the parameter fitting is more indirect. It can be carried out without major problem when the parameters of the model remain close to observable quantities. From this point of view, models of random planes or random discs offer much flexibility as illustrated in the following.

Example of disc clusters with regionalized density

This is a rather complex model, as it generalizes the random disc model in two ways. Its parameters are the following:

1. The cluster centers or seeds locally form a Poisson process of density $\lambda(x)$; this density is a stationary random function with mean λ and variogram $\gamma(h)$.
2. Each cluster includes n discs; n is random and follows a Poisson distribution with mean θ.
3. The disc centers spread independently around the seeds according to an isotropic Gaussian distribution.
4. All fractures of a cluster show the same orientation, but the cluster orientations are independent and have the same distribution.
5. All disc diameters are independent and follow the same distribution.

Orientation distribution, diameter distribution, and average density of disc centers $\lambda\theta$ are determined for each studied fracture set as in the basic random disc model. Apart from complications due to a 3-D modelling from 1-D or 2-D data, there are two difficulties in obtaining the other parameters:

1. Several clusters may overlap, and it is not necessarily clear to which seed a fracture belongs; so splitting of the overall density $\lambda\theta$ into λ and θ is not obvious.
2. The number of disc centers within a volume of size V centered at point x is not equal to the local density $\lambda(x)$ multiplied by θV, but to a realization of a random variable with mathematical expectation $\lambda(x)\theta V$; so the variogram of $\lambda(x)$ is not directly accessible.

But when all parameters are fixed, it is possible to define the theoretical value of the variograms of fracture spacing and trace center number. This enables one to fit the remaining parameters by a trial-and-error method.

The theoretical spacing variogram (or simply the spacing relative variance) is studied by a 1-D Monte-Carlo technique. This requires as a preliminary the calculation of the distribution of 1-D clusters from the distribution of 3-D clusters (not all fractures of a cluster intersect a straight line). It is easy to demonstrate that the simultaneous probability of 1-D clusters of m points that are part of a 3-D cluster of n discs (n ≥ m) is

$$\lambda_{mn} = \lambda \binom{n}{m} p_n \int\int P(y, z)^m [1 - P(y, z)]^{n-m} \, dy \, dz \ (0 < m \leq n) \tag{2}$$

We consider here constant cluster density λ. The three coordinates x, y, z are explicit. The surveying line is the Ox axis. $p_n = e^{-\theta} \theta^n/n!$ is the marginal distribution of n. P(y, z) is the probability for a disc of a seed located at point (x, y, z) to intersect the Ox axis, and can be obtained from simple considerations of geometrical probability (to be published). The marginal density of 1-D clusters of m points is then

$$\lambda'_m = \sum_{n \geq m} \lambda_{mn}$$

Figure 14 presents the relative variance of fracture spacing as a function of θ (mean number of discs of the 3-D clusters): this enables a good evaluation of θ if the dispersion of a cluster is not too high, which is the normal use of this kind of model. If density λ is not constant, the experimental relative variance must be computed locally.

The derivation of the theoretical variogram of trace center number, slightly long, is explained in Massoud (1987) and Hestir et al. (1986). Figure 15 presents this variogram

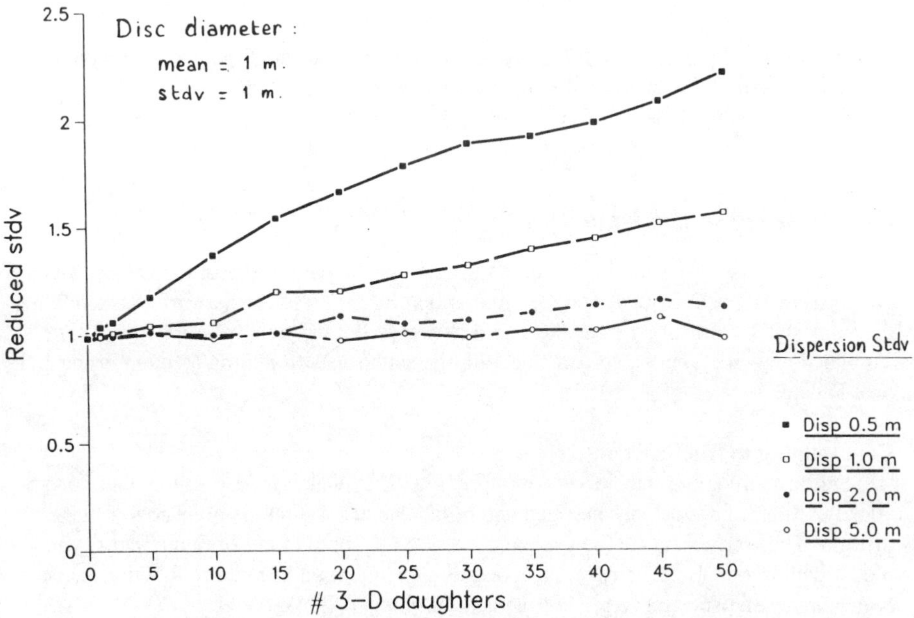

Figure 14. **Theoretical relative variance of fracture spacing for a model of disc clusters, as a function of the average number of discs (or daughters) per 3-D cluster, for various dispersions of the disc centers. The disc diameters are lognormal with mean = 1 m and standard deviation = 1 m.**

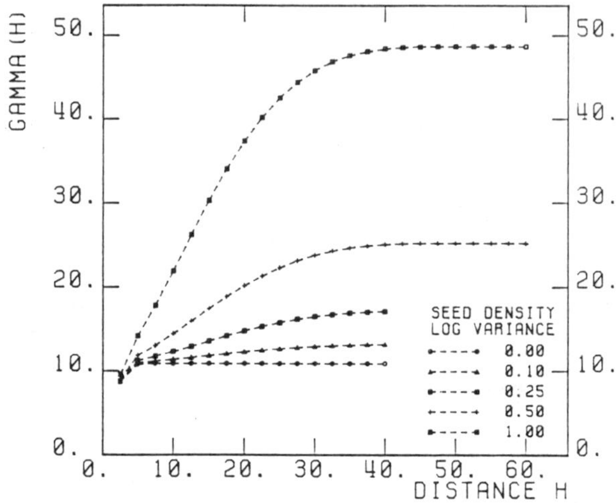

Figure 15. **Theoretical variogram of trace center number for a regionalized model of disc clusters, for various variances of the seed density. The seed density is assumed to be lognormal; its logarithm is assumed to have a cubic variogram with a range of 50 m. The average number of discs per 3-D clusters is 20, and the seed density is 0.05 per m³. The disc diameters are lognormal with mean = 1 m and standard deviation = 1 m.**

for several sets of parameters. If θ is known, a good evaluation of the dispersion of the disc centers and of the variogram $\gamma(h)$ of density $\lambda(x)$ can be made.

The detailed application of this methodology to actual data collected in the Fanay granitic site is described in Massoud (1987) and Hestir et al. (1987).

Conditioning to the Data

The tools we have examined enable fitting of all the parameters of the chosen 3-D model from 1-D or 2-D data, and the simulation of a 3-D fracture network that will have the same statistical properties as the actual network. But it would be still better if the simulated network would also coincide with the actual network at the location of the 1-D or 2-D data.

Conditioning to fracture density

Seismic tomography methods are improving. They can at least give the relative variations of the fracture density between two boreholes and define the precision of the method. These new data can be used to condition the simulation of fracture density in a model with regionalized density. The conditional simulation of random functions is well known in geostatistics (cf. e.g., Journel and Huijbregts, 1978).

Conditioning to observed fractures

Conditioning to observed fractures can be easily made for models directly based on Poisson processes, because all points, lines, or flats of a Poisson process are independent of each other. So the general outline of a conditioning is the following (Figure 16):

1. Produce a nonconditional simulation.
2. Reject all the fractures that intersect the surveyed lines or surfaces, and retain all those that do not intersect.
3. Add the actual fractures that have been surveyed.

The simplest case is the Poisson flat model, because fractures are infinite. It has been used by Andersson et al. (1984). For a random disc model, step 3 brings a slight complication, as the extension of the fracture must be determined from an intersection or a trace; in other words, we need the joint distribution of disc diameter and disc center location conditioned by observation (point or trace length). This results from geometric probability calculations (to be published).

With a disc cluster model, the same approach can be followed. But as independence is now at the cluster level, the rejection defined in step 2 must be applied at this level: We must reject all the fractures of the cluster as soon as one of them intersects the surveyed lines or surfaces. Conversely, in step 3, we must add to the surveyed fractures a simulation of the other fractures belonging to the same clusters (Figure 17).

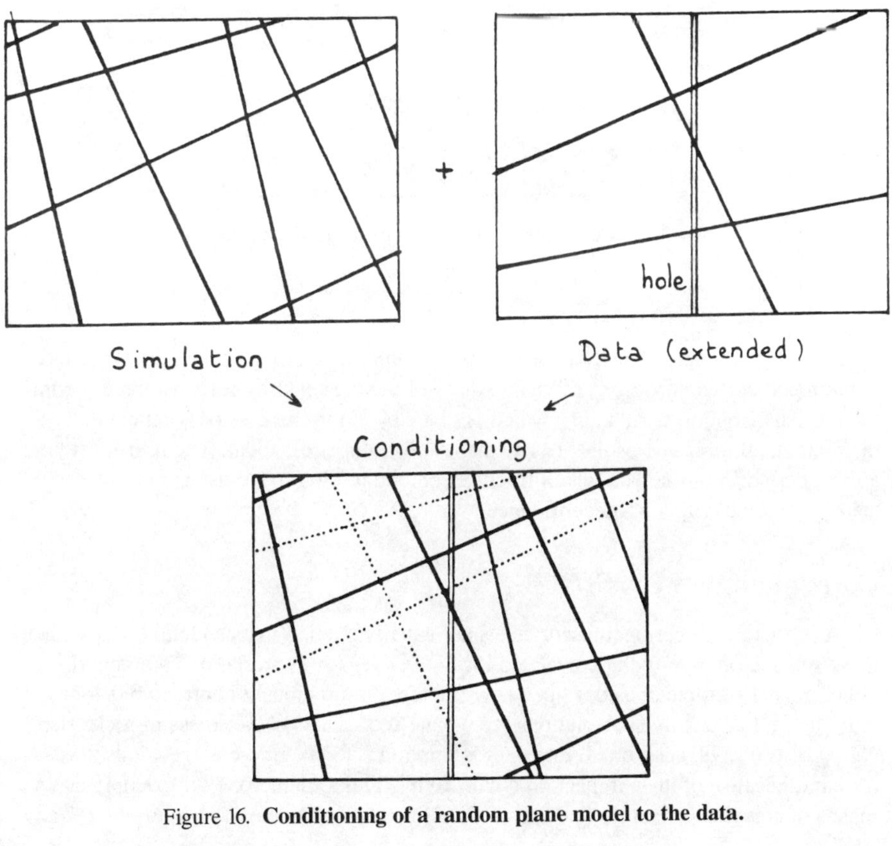

Figure 16. **Conditioning of a random plane model to the data.**

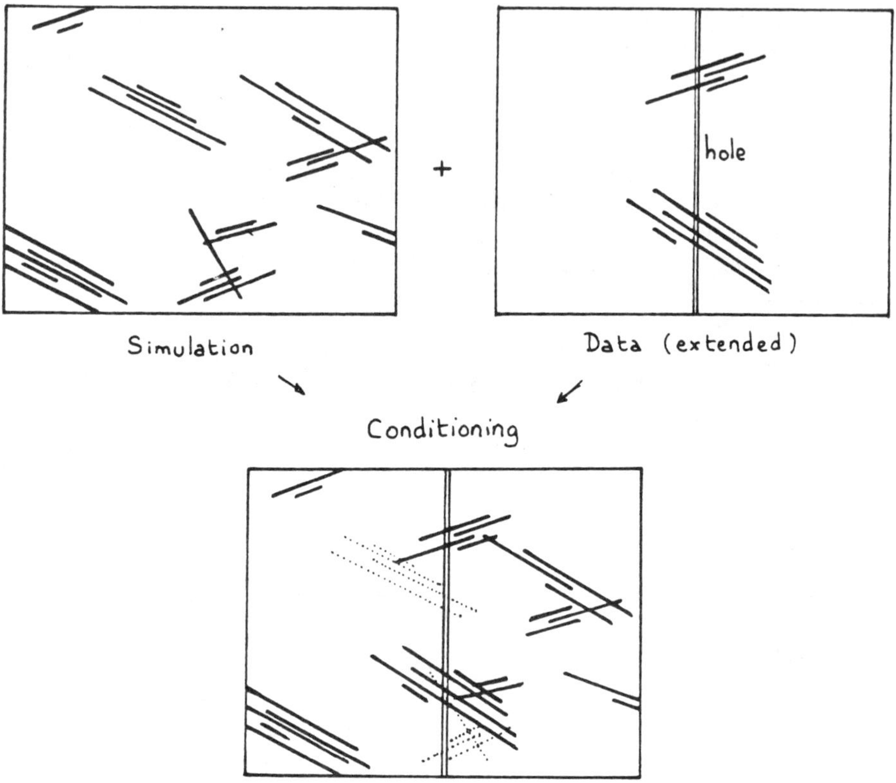

Figure 17. **Conditioning of a disc cluster model to the data.**

To achieve this, we need the distribution of the total number n of discs of a cluster conditioned by the number m of visible discs of the cluster; this derives directly from the joint distribution of m and n, which is given by (2) in the case of borehole data. So the theoretical tools are available. The main constraint is practical: The user must be able to define from the data which fractures belong to the same cluster; but this can usually be done with a good confidence.

Correlation Between Sets

The various directional fracture sets are usually studied independently. But in fact these sets are often found at correlated locations. Negative correlations (competition effect: recent tectonic episodes open new fractures in previously nonfractured zones only) as well as positive correlations (conjugate sets, zones of weakness in rock; cf., e.g., Loiseau, 1987) can be observed. It is important to take these correlations into account, because of their impact on connectivity. This can be done for example by means of correlated fracture densities.

Conclusion

Stochastic methods are now operational and can realistically model and simulate the geometry of actual fracture networks. A set of more or less sophisticated models is available. The parameters of the 3-D models can be defined from 1-D and 2-D surveying. Simulation reproduces the stochastic properties of the model and can be conditioned to the actual data.

These techniques can still be improved. For example, it should be possible to take into account the chronology of brittle tectonic episodes that generated fractures, to model variations in the shape of the network in the vicinity of major faults or dykes (cf. Delaney et al., 1986), and to develop the relation with tomography as this technique improves. This is very important for the feasibility study of potential waste disposal sites.

But the major problem is probably not really the generation of the fracture field, but the huge number of fractures that can be generated in a medium-size 3-D domain and the impossibility of taking all these fractures into account for flow calculations even with powerful computers. This problem already exists if we consider each fracture as a pair of parallel plates, and moreover if we want to take into account the morphology of the individual fractures. Two approaches can be followed. First we could try to define an equivalent continuous medium that would reproduce the flow behavior of the fractures of the first structural order; but the existence of this medium and of a representative elementary volume remains to be demonstrated (cf. Long, 1986). The other approach is the development of channel models. When fractures are subjected to high normal stress, flow becomes concentrated in channels; this apparent complication can in fact bring about some simplifications. If we can simulate the channels (with a 2-D model similar to our 3-D model; the problem is the determination of its parameters), we can replace the network of plane elements by a network of linear elements. Many other questions remain open if we consider mechanical and/or thermal phenomena. Advances in stochastic modelling of network geometry require an interdisciplinary approach.

Literature Cited

1. Andersson, J., A. M. Shapiro, and J. Bear, 1984. "A stochastic model of a fractured rock conditioned by measured information," *Water Resources Research*, Vol. 20, No. 1, pp. 79–88.
2. Baecher, G. B., 1983. "Statistical analysis of rock mass fracturing," *Mathematical Geology*, Vol. 15, No. 2, pp. 329–348.
3. Baecher, G. B., and N. A. Lanney, 1978. "Trace length biases in joint surveys," *Proceedings of the 19th U.S. Symposium on Rock Mechanics*, A.I.M.E., pp. 56–65.
4. Baecher, G. B., N. A. Lanney, and H. H. Einstein, 1977. "Statistical description of rock properties and sampling," *Proceedings of the 18th U.S. Symposium on Rock Mechanics*, A.I.M.E., pp. 5C1-1–5C1-8.
5. Bailey, A. I., 1975. "A method of analyzing polymodal distributions in orientation data," *Mathematical Geology*, Vol. 7, No. 4, pp. 285–293.
6. Barton, C. C., and E. Larsen, 1985. "Fractal geometry of two-dimensional fracture networks at Yucca Mountain, southwestern Nevada," *Proceedings of the International Symposium on Fundamentals of Rock Joints*, Bjorkliden, pp. 77–84.
7. Chilès, J. P., 1979. "Le variogramme généralisé," Internal note N-612, Centre de Géostatistique, Fontainebleau.

8. Chilès, J. P., 1987. "Fractal and geostatistical methods for modeling of a fracture network," MGUS 87 Conference, Redwood City, California. (To be published in *Mathematical Geology*, August, 1988.)
9. Coakley, K., K. Muralidhar, and J.C.S. Long, 1987. "Statistical simulation of single fracture void space geometry and calculation of fluid flow," DOE/AECL Conference on Geostatistical Sensitivity and Uncertainty Methods for Ground-water Flow and Radionuclide Transport Modeling, San Francisco.
10. Conrad, F., and C. Jacquin, 1973. "Représentation d'un réseau bidimensionnel de fractures par un modèle probabiliste: Application au calcul des grandeurs géométriques des blocs matriciels," *Revue de l'I.F.P.*, Vol. XXVIII, No. 6, Nov.–Dec., pp. 843–890.
11. Cruden, D. M., 1977. "Describing the size of discontinuities," *Int. J. Mechanics, Mineral Science and Geomechanics Abstracts*, Vol. 14, pp. 133–137.
12. Delaney, P. T., D. D. Pollard, J. I. Ziony, and E. H. McKee, 1986. "Field relations between dykes and joints: Emplacement processes and paleostress analysis," *Journal of Geophysical Research*, Vol. 91, No. B5, pp. 4920–4938.
13. Dershowitz, W. S., 1984. "Rock joint systems," Ph.D. dissertation, Massachusetts Institute of Technology, Cambridge, Massachusetts, 764 p.
14. Feuga, B., 1986. "Hydrogeologic modelling of fractured rock," *Jornadas sobre modelos matematicos aplicables al almacenamiento de residuos radioactivos*, Escuela Superior de Engenieros de Minas de Madrid.
15. Gentier, S., 1986. "Morphologie et comportement hydromécanique d'une fracture naturelle dans un granite sous contrainte normale," Thesis, Université d'Orléans, 2 vol., 637 p.
16. Hestir, K., D. Billaux, J.C.S. Long, and J. P. Chilès, 1987. "Statistical modeling of a three-dimensional fracture network," DOE/AECL Conference on Geostatistical Sensitivity and Uncertainty Methods for Ground-Water Flow and Radionuclide Transport Modeling, San Francisco.
17. Hestir, K., J. P. Chilès, J. Long, and D. Billaux, 1986. "Three-dimensional modeling of fractures in rocks; from data to a regionalized parent-daughter model," American Geophysical Union meeting, San Francisco, California. (AGU Geophysical Monograph volume on unsaturated fractured rock, 1987.)
18. Jacquin C. G., and P. M. Adler, 1987. "Structure of porous networks and fluid flow in rock masses," Workshop on Forced Fluid Through Strong Fractured Rock Masses, Garchy, France.
19. Journel, A. G., and C. J. Huijbregts, 1978. *Mining geostatistics*, Academic Press, London, 600 p.
20. Kendall, M. G., and P. A. Moran, 1963. *Geometrical probability*, Hafner, New York.
21. La Pointe, P. R., 1980. "Analysis of the spatial variation in rock mass properties through geostatistics," *Proceedings of the 21st U.S. Symposium on Rock Mechanics*, Rolla, Missouri, pp. 570–580.
22. Laslett, G. M., 1982. "Censoring and edge effect in areal and line transect sampling of rock joint traces," *Mathematical Geology*, Vol. 14, No. 2, pp. 125–140.
23. Lewis, T., and N. I. Fisher, 1982. "Graphical methods for investigating the fit of a Fisher distribution to spherical data," *Geophys. J. R. Astr. Soc.*, Vol. 69, pp. 1–13.
24. Loiseau, P., 1987. "Etude structurale et géostatistique des gneiss de la région du Cézallier: modélisation tridimensionnelle de réseaux de fractures; application à l'écoulement des fluides," Thesis, Université d'Orléans, 200 p.
25. Long, J.C.S., 1986. "Modeling of fluid flow and transport in fracture networks," *Jornadas sobre modelos matematicos aplicables al almacenamiento de residuos radioactivos*, Escuela Superior de Engenieros de Minas de Madrid.
26. Mandelbrot, B. B., 1983. *The fractal geometry of nature* (2nd ed.), Freeman, New York, 468 p.
27. Mandelbrot, B. B., 1985. "Self-affine fractals and fractal dimension," *Physica Scripta*, Vol. 32, pp. 257–260.

28. Mandelbrot, B. B., D. Passoja, and A. Paullay, 1984. "Fractal character of fracture surfaces of metals," *Nature*, Vol. 308, pp. 721–722.
29. Mantoglou, A., 1987. "Digital simulation of multivariate two- and three-dimensional stochastic processes with a spectral turning bands method," *Mathematical Geology*, Vol. 19, No. 2, pp. 129–149.
30. Massoud, H., 1987. "Modélisation de la petite fracturation par les techniques de la géostatistique," Thesis, E.N.S. des Mines de Paris, 189 p.
31. Matheron G., 1965. *Les variables régionalisées et leur estimation*, Masson, Paris, 306 p.
32. Matheron, G., 1967. *Eléments pour une théorie des milieux poreux*, Masson, Paris.
33. Matheron, G., 1971. "Les polyèdres poissoniens isotropes," *Actes du 3e Colloque Européen* sur la Fragmentation, Cannes, 5–8 Oct., II-9, pp. 509–534.
34. Matheron, G., 1973. "The intrinsic random functions and their applications," *Advances in Applied Probability*, No. 5, pp. 439–468.
35. Miller, S. M., 1979. "Geostatistical analysis for evaluating spatial dependence of fracture set characteristics," *Proceedings of the 16th APCOM Symposium, SME-AIME*, New York, pp. 537–545.
36. Pahl, P. J., 1981. "Estimating the mean length of discontinuity traces," *Int. J. Rock Mechanics, Mineral Science and Geomechanics Abstracts*, Vol. 18, pp. 221–228.
37. Phillips, F. C., 1960. *The use of stereographic projection in structural geology* (2nd edition), Edward Arnold, London, 86 p.
38. Priest, S. D., and J. Hudson, 1976. "Discontinuity spacings in rock," *Int. J. Rock Mechanics, Mineral Science and Geomechanics Abstracts*, Vol. 13, pp. 135–148.
39. Santalo, L. A., 1976. *Integral geometry and geometric probability*, Addison-Wesley, Reading, Massachusetts.
40. Sayles, R. S., and T. R. Thomas, 1977. "The spatial representation of surface roughness by means of the structure function: A practical alternative to correlation," *Wear*, Vol. 42, pp. 263–276.
41. Schaeben, H., 1984. "A new cluster algorithm for orientation data," *Mathematical Geology*, Vol. 16, No. 2, pp. 139–153.
42. Serra, J., 1982. *Image analysis and mathematical morphology*, Academic Press, London, 628 p.
43. Smith, L., C. Mase, and F. Schwartz, 1987. "An evaluation of techniques for estimating fracture aperture," DOE/AECL Conference on Geostatistical Sensitivity and Uncertainty Methods for Ground-Water Flow and Radionuclide Transport Modeling, San Francisco.
44. Snow, D. T., 1965. "A parallel plate model of fractured permeable media," Ph.D. dissertation, University of California, Berkeley, 331 p.
45. Snow, D. T., 1970. "The frequency and apertures of fractures in rock," *Int. J. Rock Mechanics and Mineral Science*, Vol. 7, pp. 23–40.
46. Thomas, A., 1987. "Structure fractale de l'architecture de champs de fractures en milieu rocheux," *Comptes Rendus de l'Académie des Sciences de Paris*, t. 304, Série II, No. 4, pp. 181–186.
47. Veneziano, D., 1978. "Probabilistic model of joints in rock," Unpublished manuscript, Massachusetts Institute of Technology, Cambridge, Massachusetts.
48. Warburton, P. M., 1980. A stereological interpretation of joint trace data. *Int. J. Rock Mechanics, Mineral Science and Geomechanics Abstracts*, Vol. 17, pp. 181–190.

Some Recent Observations of Channeling in Fractured Rocks—Its Potential Impact on Radionuclide Migration

IVARS NERETNIEKS, HARALD ABELIN, and LARS BIRGERSSON
Department of Chemical Engineering
Royal Institute of Technology
Stockholm, Sweden

Abstract

During the last few years several observations in the field have been made to indicate that water flow is very unevenly distributed in fractured crystalline rocks. In a field experiment in the Stripa international research project it was noted that in several individual fractures that were studied in some detail, most of the naturally flowing water emerged in a few spots.

In an experimental drift 700 m^2 of the walls and ceiling were covered with more than 375 plastic sheets. Water flowrates were measured in all sheets. The results showed large areas that were practically dry (about 70%), and in the areas where water flows the flowrate varied considerably between different spots. Tracer tests performed in both the above experiments also showed some clear channeling effects.

Observations over more than 5 km in two long tunnels, the Bolmen and Kymmen tunnels, show that the water emerges in distinct spots. In the tunnels the spots are clearly seen because iron hydroxides precipitate from the water. The spots are typically less than 1 m long along a fracture or fracture intersection. Typically there are one or a few such spots over an area of 100 m^2. A detailed mapping of the Kymmen tunnel has given quantitative information on the number and flowrates of channels.

In the site of the Repository for Reactor Waste, Slutföutförvar för Reaktoravfall (SFR), that was recently excavated, a detailed mapping of all wet spots shows that there is on the average about one wet spot every 85 m^2. The observed wet spots account for most of the water inflow. There is additional diffuse inflow of water, and 18 spots over an area of 14,000 m^2 carry 50% of the water.

These and similar observations have led us to investigate the potential impact of pure channel flow on radionuclide transport and retardation. A simple model was made based on the assumption that the channels are independent over a certain distance. The transport properties in such a bundle of channels have been shown to be different in some important respects from the properties of a highly fractured medium. Model calculations show that the few fast channels with little sorption area may carry sorbing radionuclides at a rate that may not let the nuclides decay as well as if the rock were more, and more evenly, fractured.

Background and Introduction

During the last decade there has been an increasing interest in the hydrology of deep-lying rocks because of their potential use for siting repositories for high-level radioactive waste (HLW). In Sweden most of the rocks studied are granites and gneisses. The Swedish Nuclear Fuel Company, Svensk Kärnbränsleförsörjning (SKB), has studied the deep-lying rocks since late 1976 and has gathered a wealth of information from more than 20 downhole measurements in holes down to more than 800 m depth. In addition a mass of information has been gathered by observations on exposed rock surfaces. The observations show that the rock is fractured on different scales. Large shear zones at distances of several kilometers surround large rock blocks. The large blocks of rock have smaller fracture zones surrounding smaller rock blocks. The latter contain a large number of small fractures or agglomerates of fractures. Typically the small fractures are at spacings of tens of cm to a meter or more. Many of the fractures are closed and carry little or no water. The rock matrix has a very low hydraulic conductivity compared to the fractures. Most of the water that moves in the rock will move in fractures. The matrix is porous, however, and dissolved species in the water may intrude the matrix by molecular diffusion much faster than the species would be carried by flow in the matrix.

Assessment of radionuclide transport from repositories

The water moving in the fractures in the rock may transport dissolved species such as radionuclides from a repository for radioactive waste. The assessment of how much and at what rate the nuclides are carried by the moving water is usually based on calculations of the flowrate of the water in the rock, its velocity, its pathways, and on the retardation of the radionuclides by physical and chemical interactions with the rock.

In two recent safety analyses for HLW repositories, SKBF (1983) and Nagra (1985), the water flowrate calculations were based on conventional hydrologic models applying Darcy's law to a medium that was assumed to be well described by a porous medium equivalent. The results from such calculations are flowrates and pathways in all parts of the rock included in the model.

The flowrates and flow paths are used in the transport models. These models describe the movement of the radionuclides with the water. The radionuclides do not, however, move with the velocity of the water in general. Their velocity will deviate from the average velocity because of variations in the water velocity and the mixing of different portions of the water. Some portion of the nuclides may move faster and some slower than the average velocity because of these processes, which are summarized in the term "dispersion." Most of the nuclides are expected to be cationic or neutral in the waters present in the crystalline rocks investigated. The nuclides will adsorb and/or ion-exchange on the negatively charged surfaces of the rock minerals. These processes may considerably retard the radionuclides, which in some instances are expected to move many orders of magnitude slower than the water. For a given flowrate the retardation will be larger for a nuclide if there are more exposed surfaces with which the nuclides interact. One of the crucial questions thus is how much fracture surface the flowing water encounters. In addition to interaction with the fracture surface, the radionuclides may diffuse into the rock matrix and sorb onto the inner surfaces of the rock matrix. The inner surfaces are much larger (many orders of magnitude) than the fracture surfaces in contact with the flowing water. If the inner surfaces are accessed then the retardation may increase considerably.

Some Recent Field Observations and Experiments

Overview

Within the international Stripa project two tracer experiments have been performed. One experiment, the 2D experiment, was made in three different fractures with a migration distance of 5 to 10 m. The flow and tracer transport in the fracture was very uneven. In the 3D tracer experiment 11 tracers were injected above a drift that was covered with 375 plastic sheets over an area of 700 m^2. Water and tracer flowrates were measured in the sheets. In the SFR-repository, a detailed mapping of all visible leaks in the ceilings was made. The flowrate from each spot was estimated or measured. Detailed fracture mapping and characterization of the leaks were done. In the 4500-m-long drilled Kymmen tunnel the same type of observations were made. Another tunnel at Bolmen was visually inspected, but no detailed mapping of the leaks was made. The observations clearly show that the water emerges in narrow spots in fractures or fracture intersections. By far, the largest part of a fracture carries very little water.

A large number of downhole hydraulic measurements have been performed in the Swedish program (SKBF, 1983). These measurements also show that most of the fractures are closed and that a very small number of intercepted channels would carry most of the flow (Neretnieks, 1987). Laboratory-scale measurements on natural fractures taken from the Stripa experimental mine also show evidence of channeling (Neretnieks et al., 1982). In the descriptions of the sites in the following sections, the flowrate distribution and other characteristics are quantified where possible.

Stripa

The Stripa experimental mine is located in central Sweden at a depth of 350 m in water-saturated granitic rock.

2D experiment. In the Stripa 2D experiment (Abelin, 1986), water flow observed in three fractures intersecting drifts was noted to flow very unevenly in those fractures. Figure 1 shows the distribution of water flow. The figures adjacent to the hollow arrows indicate the water flowrate in an approximately 0.7-m-long section of the fracture. It can be seen that more than 90% of the water flows in 5 of the 27 observation sections and that there often are dry sections between the sections carrying water.

Nonsorbing tracers were injected into the fracture, and the concentration of the tracer in the collected water was measured. Sorbing tracers were also injected, but these were retarded so much that they were not found in the water. The fracture was

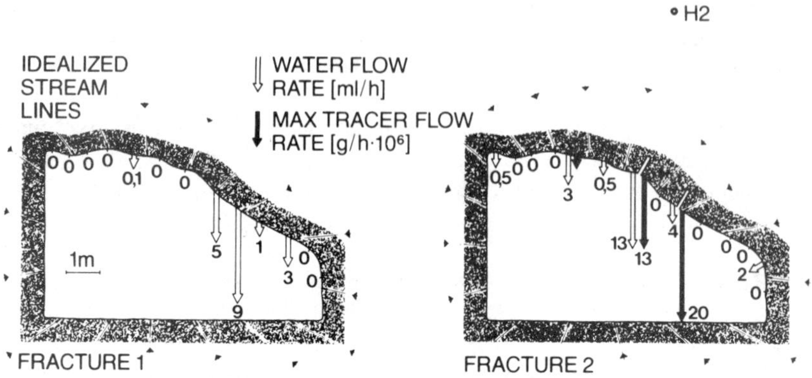

Figure 1. **Experimental drift intersecting the plane of the fracture. The arrows indicate the location of water flowrate.**

excavated and the concentration of the tracers on the surface of the fracture was measured. Figure 2 shows the distribution of strontium on the surface. The observations show that the water and tracers move in a few paths and that most of the fracture is nearly dry.

3D experiment. In the recently finished 3D experiment in Stripa (Abelin et al., 1987), a 75-m-long drift was excavated at 360 m below the ground level. Two cross arms, each 12.5 m long, were also excavated. The drift was covered with more than 375 individual plastic sheets. Figure 3 shows a view inside the drift with the plastic sheets.

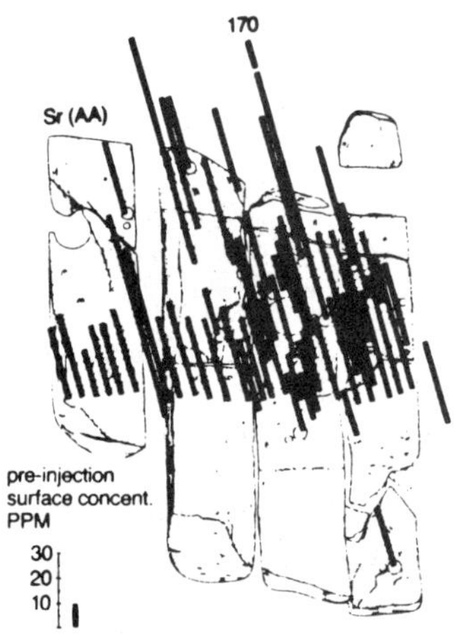

Figure 2. **Surface concentration of strontium on the excavated fracture. The length of a core in the figure is about 0.6 m.**

Figure 3. **View into 3D drift with plastic sheet arrangement.**

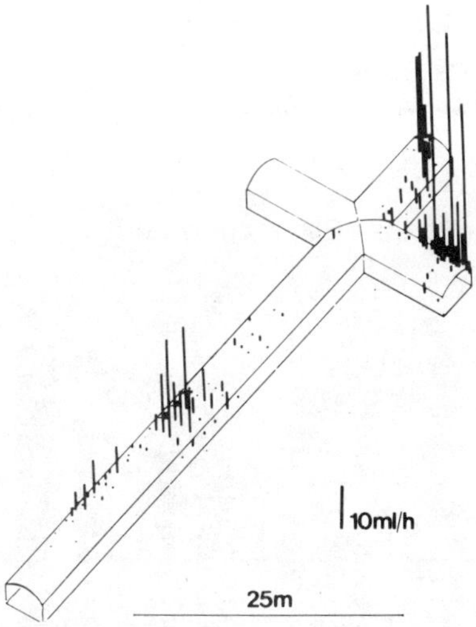

Figure 4. **Flowrate distribution over 375 sheets in the drift. The length of the bars is proportional to the flowrate in a sheet.**

The natural water flowrate was monitored for nearly 3 years. Figure 4 shows the flowrates to each individual sheet. It is seen that the flowrate varies considerably between different areas and between different sheets in the different areas. On this scale at least there is a strong concentration of flow in some areas, whereas other areas are dry or have very small flowrates. Table 1 shows the flowrate distribution among the 375 sheets in the drift proper according to flowrate ranges. Figure 5

Table 1. **The distribution of flowrates in the Stripa 3D drift among the 375 sheets.**

Flowrate/Sheet (ml/h)		Number of Sheets	Flowrate Total	Cumulative	
				% Area	% Flowrate
<0.1		266	0	100.0	100.0
0.1–	<0.2	13	1.3	29.1	100.0
0.2–	<0.4	7	1.6	25.6	99.8
0.4–	<0.8	10	5.6	23.7	99.6
0.8–	<1.6	20	22.6	21.1	98.7
1.6–	<3.2	19	46.1	15.7	95.2
3.2–	<6.4	14	63.0	10.7	88.1
6.4–	<12.8	12	108.5	6.9	78.4
12.8–	<25.6	7	124.0	3.7	61.6
25.6–	<51.2	6	215.0	1.9	42.5
> = 51.2		1	60.0	0.3	9.3
Total		375	647.7		

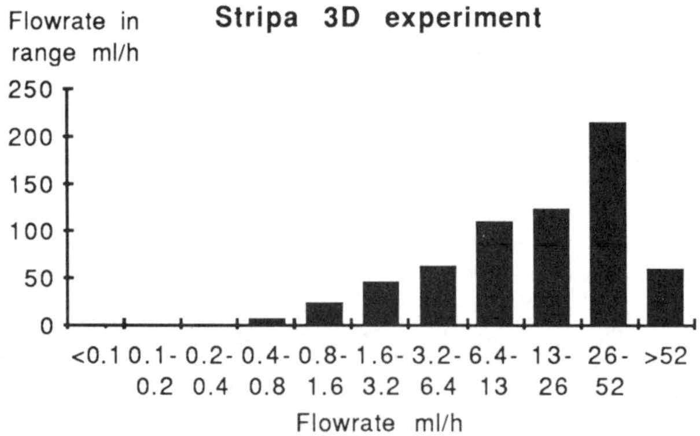

Figure 5. **Sum of flowrates in the different flowrate ranges in the Stripa 3D drift.**

shows a histogram of water flowrates in the different ranges. It can be seen in Table 1 and Figure 5 that 50% of the flow takes place in about 3% of all the covered area.

After the sheets had been mounted and water collected for some time, three vertical holes were drilled 70 m upward. The holes were sealed off except in nine sections where nonsorbing tracers were injected in two long zones. The injection was kept up for more than 18 months and the monitoring for more than 2 years. Injection points are shown in Figure 6. Analysis of the tracers in the waters collected in the sheets indicated none found in the wet right-hand arm or in the bottom of the drift. The six tracers that reached the drift during the experiment emerged in the central part of the drift. This is illustrated in Figures 7 and 8 for the tracer Eosin B.

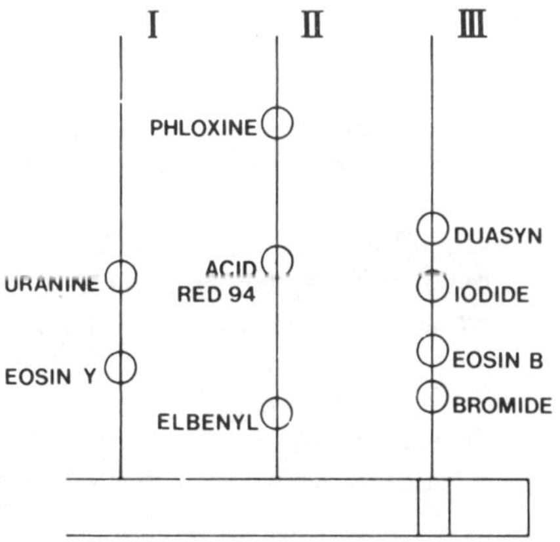

Figure 6. **Injection points for the nine tracers.**

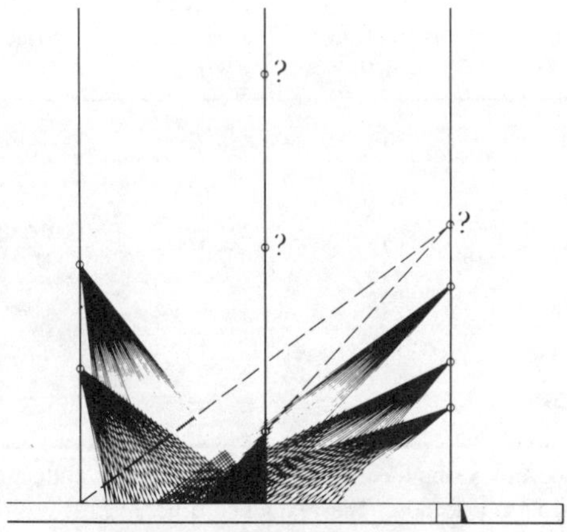

Figure 7. **Overview of the locations at which the tracers arrived.**

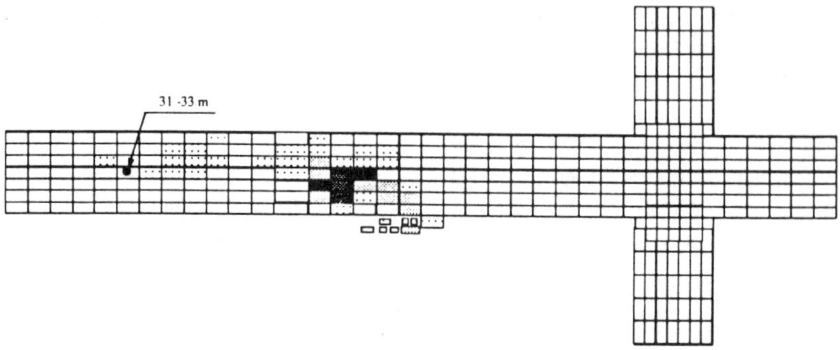

Figure 8. **Sheets at which the tracer Eosin B arrived. Darker spots indicate higher concentration.**

The measured flowrates and residence times of the tracers allow the flow porosity to be determined. This is shown on Table 2 and in Figure 9, where the porosity is plotted as a function of distance.

The hydraulic conductivity of the rock adjacent to the drift is evaluated based on the assumption that the rock mass may be described as a homogeneous porous medium and that Darcy's law is valid. We need to introduce some simplifying assumptions because the head boundary conditions are not known exactly due to the presence of drifts and tunnels of the old mine. For a homogeneous porous medium the simplifying assumptions would introduce only minor errors and these are generally accepted in most other cases when the properties of fractured rocks are studied.

Table 2. **Flow porosity determined from flowrate and tracer residence time.**

Tracer	Residence Time (hours)	Flowrate in Section (ml/h)	Length of Section (m)	Distance to Injection Point (m)	Porosity (10^5)*
Eosin B	3000	176	22	19	3.4
Uranine	5500	153	23	32	2.0
Elbenyl	2000	178	10	10	15.5
Eosin Y	6000	198	27	18	6.9
Iodide	7000	200	31	29	2.6
Bromide	2500*	200	31	13	4.5

*The bromide residence times obtained from the fits using the collection time are in error because of interference from iodide. The residence time 2500 h was estimated by visually observing the breakthrough curves for the early times when the iodide had not yet reached the collection sheets in any appreciable concentration. The value is only approximate.

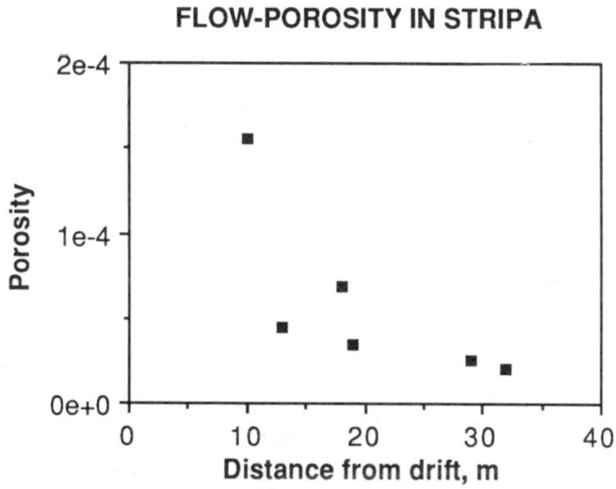

Figure 9. **Flow porosity as a function of distance to the injection point.**

The water flowrate monitoring in the sheets gives information on the upper part of the experimental drift. This drift and a 140-m-long access drift were sealed off with tight walls, and the flowrate and the humidity of incoming and leaving air was measured. These ventilation tests gave information on the water flowrate to the uncovered part of the ventilation drift and to the whole access drift.

Using the above flowrate data and heads data obtained by measuring the hydraulic heads in the open sections in the injection holes, the hydraulic conductivities of the different parts of drifts were determined. The results are summarized in Table 3 below.

The flowrates vary considerably over the length of the drift, as do the hydraulic conductivities. The data in Table 3 have been averaged over more than 100 m of drift. The variations within the experimental drift are shown in Figure 4. It is clearly seen that within a 100-m section the hydraulic properties of the rock are far from constant. The difference between the access drift and the experimental drift is a factor of 4. This also indicates that the 100-m scale may not be sufficient to obtain reasonably constant average values.

Table 3. **Hydraulic conductivity of the experimental and the access drift.**

Location	**Conductivity × 10^{11} m/s**
Experimental drift	
Ceiling + upper sides	0.44–0.84
Floor + lower sides	2.0 –3.8
Average	1.2 –2.3
Access drift	0.51–0.96

SFR

The final repository for the reactor waste in Sweden, the SFR, is being built at the Forsmark site in mideastern Sweden 50 m under the bottom of the Baltic Sea. Bolvede and Christiansson (1987) have made a detailed survey of the water flow in the rock in the SFR drifts and tunnels.

Observations at the SFR. It was found that under the prevailing hydraulic gradient the ceilings of the 4 main repository tunnels have an inflow of 30 l/min. This flowrate emerges from 164 different spots; 41 of the spots are at fracture intersections or small holes (points). Figure 10 shows the tunnel system in SFR and the spots where flow was observed.

Table 4 and Figure 11 summarize the results. Note that the flowrates in the SFR are in l/min and are thus several orders of magnitude larger than in Stripa. The wet spots were of six types:

- Horizontal or near-vertical fracture with flow mainly from points
- Horizontal or near-vertical fracture with flow from all the fractures
- Near-vertical fracture with flow mainly from the lower end of the fracture
- Intersecting fractures with flow mainly from the point of intersection
- Diffuse flow
- Flow from a point or a small hollow where clay has been washed away.

The wetted fracture widths observed vary typically between 1 and 3 m, although a few larger widths were also found. The total area of observation in these tunnels was 14,000 m^2.

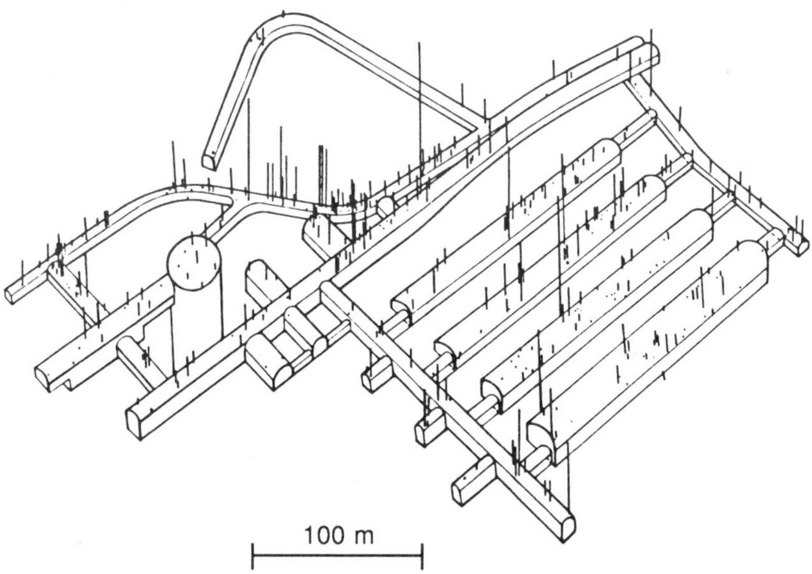

Figure 10. **Overview of the SFR site. The length of the bars is proportional to the flowrate.**

Table 4. **Distribution of flowrates in the four tunnels and the silo of SFR.**

Flowrate/Channel (l/min)	Number of Channels	
	Total	(Points, types 4 & 6)
<0.1	67	21
0.1– <0.2	38	10
0.2– <0.4	41	3
0.4– <0.8	12	2
0.8– <1.6	4	4
> = 1.6	2	1

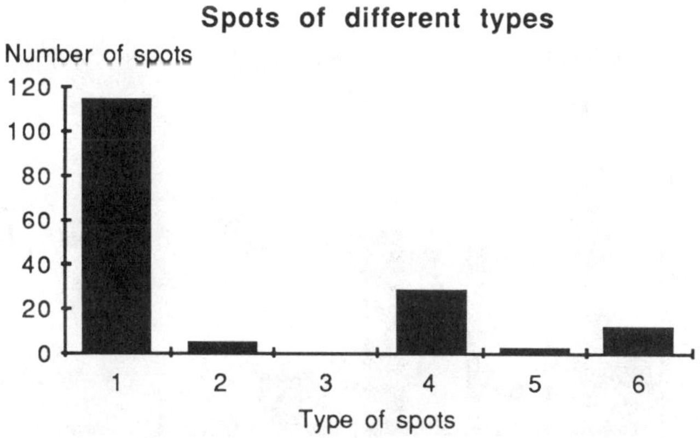

Figure 11. **Histogram of the different types of spots in the four main tunnels of SFR.**

Flowrate distribution. Table 5 shows that nearly 50% of the flowrate can be accounted for in 18 spots over an area of 14,000 m^2. Of these spots 7 belong to the type 4 and 6, i.e., "point" spots, as shown in Figure 12.

On the average there is only one channel per 85 m^2. Of these a few carry most of the water, and 50% of the flow takes place in less than 10% of the channels. It was noted that many of the channels with the highest flowrates were very narrow. The water emerged essentially in a small point in the rock. The wetted surface of the channels per volume of the rock is very small in this case if only those channels carrying most of the water are accounted for. Assuming that the tunnels behave like one long tunnel uninfluenced by the presence of other excavations and that flow is radial, the hydraulic conductivity can be approximately assessed from the radial flow equation. The hydraulic conductivity is on the order of 8×10^{-9} m/s.

Table 5. **Summary of the flowrates in different flowrate ranges in the four main tunnels of the SFR for all types of spots.**

Flowrate Range, l/min	Number of Spots	Flowrate, l/min	Fraction of Spots	Fraction of Flow	Flow per Spot, l/min
<0.1	67	2.53	0.41	0.08	0.038
0.1– <0.2	38	3.85	0.23	0.13	0.101
0.2– <0.4	41	9.30	0.25	0.31	0.227
0.4– <0.8	12	6.30	0.07	0.21	0.525
0.8– <1.6	4	4.50	0.02	0.15	1.125
> = 1.6	2	4.00	0.01	0.13	2.000
Total	164	30.48	1.00	1.00	

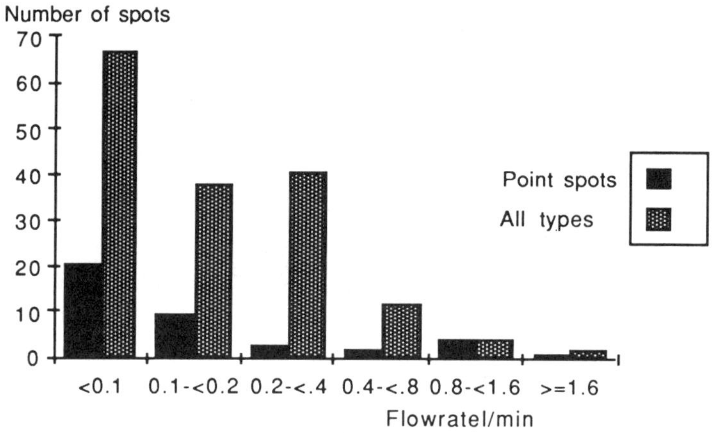

Figure 12. **Histogram of number of spots carrying different ranges of flow. All types, as well as type 4 and 6 spots, are shown.**

Kymmen and Bolmen

A detailed survey of the Kymmen tunnel has recently been made by Palmqvist (1987). The observations are summarized in the following.

Observations. The Kymmen tunnel is a full-face drilled tunnel 4.6 m in diameter. The tunnel will transport water to a hydroelectric power plant. The tunnel has a length of 4500 m and traverses four major rock types: gneiss-granite (Gg), 13% of length; gneiss-granite with slabs of amphibolite (GgA), 40%; fine-grained quartz-porphyry called leptite (L), 40%; and amphibolite (A), 7%.

The rock is fractured and contains 63 charted fracture zones. Water is observed to flow in localized spots in fractures and in many fracture zones where the frequency of water-conducting spots is larger. Many of the fractures have fracture-coating materials. In fracture zones rock alteration minerals such as clays have been observed; 11 of the 63 fracture zones have been injected with concrete to decrease the water leakage. The tunnel has an overburden that varies between 50 m and several 100 m.

The tunnel has been mapped visually, and these observations are supported by 38 photographs. All observations are indicated in maps of scale 1:200 and characterized with a simple coding system in addition to a description of the directions and dips of the fractures. Fracture-filling and coating materials are also noted. The coding system is described in Table 6.

The information on the size of the rock blocks may be used to determine the maximum contact area between the water moving in the fracture zone and the rock. Sorbing nuclides are retarded proportionally to the wetted surface, and this information is needed to assess radionuclide migration rates.

The "typical" water flowrate figures shown in Table 7 are approximate. More accurate figures may be obtained. In the observations in Kymmen it was not possible to perform the measurements because of the very tight time schedule. The tunnel was coming on line shortly after the drilling was finished.

Flowrate distribution. Tables 8 and 9 show the flowrate distribution.

Channel characteristics. The different leaky spots have been characterized. The clear majority of the leaks larger than V2, 368 leaks, have very small extent; they are called point leaks. Only 29 leaks have an extent larger than 0.2 m. In addition there is a large number of V1 leaks and some diffuse wet areas, but these account for only a small fraction of the total flowrate. Table 10 shows the numbers and lengths of the point and "curtain" leaks.

Hydraulic conductivity. Table 11 gives the estimated hydraulic conductivity and spot density. The average hydraulic conductivity is of the same magnitude as that in the SFR. The density of wet spots is one spot per 163 m^2. In the SFR it was one per 85 m^2. This difference is not necessarily significant, because the mapping has

Table 6. **Fracture zone characterization.**

K1	Parallel fracturing with fracture spacing > 10 cm
K2	Parallel fracturing with fracture spacing < 10 cm
K3	Rock blocks with a size typically > 20 cm
K4	Rock blocks with a size typically < 20 cm
K5	Entirely crushed zones of rock

Table 7. **Water flowrate characterization.**

		Flowrate Range (l/min)	"Typical" (l/min)
V1	Infrequent local drops	<0.01	0.005
V2	Frequent drops—sipping water	0.01–0.2	0.04
V3	Clearly flowing water	0.2 –1.5	0.5
V4	Strongly flowing water	1.5 –6.5	3
V5	Very strong flow	> 6.5	10

Table 8. **Number and character of spots in Kymmen divided according to rock type.**

Rock	Total Length of Tunnel (m)	V2	V3	V4	V5	Total Flowrate (l/min)*
Gg	640	11/1.7	5/0.8	—	—	1.9
GgA	1891	68/3.6	34/1.8	3/0.2	—	29
L	1896	163/8.6	94/5.0	3/0.2	4/0.2	103
A	69	2/2.9	—	—	—	0.08

*"Typical" values from Table 7 on water flowrate characterization have been used.

Table 9. **Number of spots with water flow larger than V1 (outside/within fracture zones).**

Rock	Total Length of Tunnel/of Zone (m)	V2	V3	V4	V5	Flowrate Outside/in Zone (l/min)*
Gg	640/4	11/0	5/0	—	—	1.9/0
GgA	1891/132	54/13	26/9	1/2	—	18/11
L	1896/519	60/103	47/47	1/2	0/4	29/74
A	69/17	0/2	—	—	—	0/0.08

Table 10. **Number of point leaks and "curtain" leaks.**

	Number of Point Leaks	Number of Curtain Leaks	Total Length, m	Minimum Length, m	Maximum Length, m
V2					
in zones	106	12	9.0	0.5	4
outside zones	119	6	8.7	0.2	4
V3					
in zones	55	0	—	—	—
outside zones	68	11	14.7	0.2	6
V4					
in zones	4	0	—	—	—
outside zones	2	0	—	—	—
V5					
in zones	4	0	—	—	—
outside zones	0	0	—	—	—

Table 11. **Estimated hydraulic conductivity (K_p) and spot density in the tunnel.**

	K_p (m/s)	Spots/1000 m²			
		V2	V3	V4	V5
In zones	$66 \cdot 10^{-9}$	17.6	8.2	0.59	0.59
Outside zones	$4.6 \cdot 10^{-9}$	2.26	1.43	0.036	0
Average	$11 \cdot 10^{-9}$	3.74	2.06	0.061	0.061

been done by different teams and the flowrate ranges in the compilations are different.

The fracture zones make up 672 m of the total length of 4496 m; 207 m of the crushed zone length has been injected with concrete. The crushed zones carry 85 l/min, and the section outside the crushed zones carries 49 l/min.

Radionuclide Transport Models

Overview

It is usually implicitly assumed that the fractured rock even at depth is sufficiently fractured to allow an averaging to be performed of the properties in a volume of rock such that it is meaningful to assign an average value of the property to a "point" in the rock. The properties of interest for flow are hydraulic conductivity and porosity. The size of the volume over which the averaging is performed is called a "representative elementary volume" (REV). It is implicitly assumed that the REV is considerably smaller than the rock volume studied.

To calculate the flowrate in the rock volume, Darcy's law is used together with the appropriate boundary conditions to calculate the flowrate and flow directions in "all" points in the rock mass of interest. Known variations of properties in space, such as the presence of fracture zones and other large-scale features, are readily included. In principle, only the amount of available information and the size of the computer limit the amount of detail that can be included in the model calculations. Having thus obtained a detailed map of the flowrates, flow paths, and if needed the actual flow velocities, this information is then used directly in the transport models.

Advection-dispersion models

One of the simplest transport models is based on the assumption that the dissolved species is carried by the water with the average velocity of the water and that the variation of the species transport around the average velocity is a random process similar to molecular diffusion. The concentration of a species in the water is then described by

$$\frac{\partial c}{\partial t} + u \frac{\partial c}{\partial z} = D_L \frac{\partial^2 c}{\partial z^2} \tag{1}$$

The third term accounts for dispersion, and it is implicitly assumed that the dispersion coefficient, D_L, is constant in time and in space. Equation (1) can be used to calculate how a radionuclide that does not interact with the solid material will move with the water. Decay is not accounted for in Equation (1) but is easily added when needed.

The advection-dispersion model has been shown to be reasonably applicable in many laboratory investigations using unconsolidated beds of sand or other particles. It has been used to model transport in sedimentary rocks with some success. It is reasonable to assume that it is locally applicable to transport in individual open fractures or even in fractures that are filled with a porous matter such as clay or fine-grained crystalline matter.

Sorption on surfaces. When the dissolved species sorbs on the surfaces of the solid material it will be retarded in relation to the water velocity. The retardation is proportional to the sorption capacity per surface area and the exposed surface area. Equation (1) may be made to include this effect. For a linear instantaneous equilibrium the equation is modified to become

$$R_a \frac{\partial c}{\partial t} + u \frac{\partial c}{\partial z} = D_L \frac{\partial^2 c}{\partial z^2} \tag{2}$$

where R_a is the retardation factor due to surface sorption.

$$R_a = 1 + K_a \cdot a \tag{3}$$

K_a is the surface sorption coefficient (concentration of species per m^2 surface in equilibrium with a water with a given concentration per m^3), and a is the surface-to-volume ratio (m^2 of wetted surface per m^3 of water). If the water velocity is u the nuclide velocity will be

$$u_N = u/R_a = u/(1 + K_a \bullet a) \tag{4}$$

We would like to draw the reader's attention to the following aspect of how the nuclide velocity is influenced by sorption. Equation (4) seemingly says that the nuclide velocity is proportional to the water velocity. For a given rock with a known flowrate per m^2, the Darcy velocity U_o (which has been obtained from the hydraulic conductivity and the known hydraulic heads), the velocity "u" in the fractures is inversely proportional to the fracture aperture, assuming that the number of fractures is constant.

$$u = U_o \bullet S/\delta \tag{5}$$

where δ is the fracture aperture and S is the average fracture spacing. The wetted area of a fracture per *volume of water* "a" is

$$a = 2W \bullet L/(W \bullet L \bullet \delta) = 2/\delta \tag{6}$$

W is the width and L is the length of the fracture. Inserting Equations (5) and (6) into (4) gives

$$u_N = \frac{U_oS}{\delta(1 + K_a \frac{2}{\delta})} = \frac{U_oS}{2\,K_a} = \frac{U_o}{K_a a'} \tag{7}$$

$$\text{for } K_a 2/\delta >> 1$$

where a′ is the wetted fracture area per *volume of rock*, a′ = 2/S. For the sorbing nuclides the term $2 \bullet K_a/\delta$ usually is >> 1, and then the nuclide velocity is independent of the water velocity in the rock for a given flux (Darcy velocity). The nuclide velocity is then determined primarily by the flux, the surface sorption coefficient, and the wetted fracture surface in the rock. It may be noted that the flow porosity of the rock does not influence the velocity of sorbing nuclides for a given Darcy velocity and fracture spacing under these circumstances.

The flux may be assessed as discussed above. K_a is determined by laboratory measurements but a′, the wetted surface, must be assessed by field observations on actually flowing fractures. This is usually not done. Only in a few instances have such determinations been done (SKBF, 1983).

Sorption in the bulk of the rock. The rock matrix is porous. The nuclides could conceivably penetrate the rock between the fractures if diffusion were fast and if the time were long enough. If the matrix is fully penetrated not only the wetted fracture surfaces would be available for sorption, but also all the inner surfaces of the rock matrix. The retardation coefficient then could be written

$$R_d = 1 + K \bullet (1 - \varepsilon_f)/\varepsilon_f \tag{8}$$

K is the volumetric sorption coefficient and the nuclide velocity becomes

$$u_N = u/R_d \approx u \bullet \varepsilon_f/K = U_o/K \tag{9}$$

For $\varepsilon_f << 1$, which is true for fractured crystalline rock, $R_d \approx K/\varepsilon_f$, and the right-hand side of the approximation of Equation (9) is permissible. K is measured in the laboratory under such conditions that it is ensured that the rock samples have been thoroughly penetrated by the sorbing nuclides. Again it is seen that the flow porosity does not influence the velocity of the sorbing species to an appreciable extent for a given Darcy velocity.

Matrix diffusion. In the two previous cases two extreme assumptions were used. In the case of surface sorption it was assumed that only the fracture surface in contact with the flowing water was available for sorption. In the case of bulk sorption, it was assumed that all the rock between two or more fractures was ''immediately'' equilibrated with the water in the fractures. In practice it may be expected that the rock nearest to the fracture surface will be penetrated by the dissolved diffusing species. The penetration depth will increase with increasing contact time (Neretnieks, 1980). It was shown that the rock volume, which may be expected to be accessed by diffusion during contact times of hundreds or more years, may have a considerably larger retardation effect than surface sorption alone would allow. On the other hand, the penetration depths for strongly sorbing nuclides, such as plutonium and neptunium, are not expected to be more than a few tens of centimeters even for contact times of hundreds of thousands of years. Most of the rock matrix will thus not be penetrated and will not help in retarding the nuclides. The amount of rock accessed will depend on the sorption and diffusion properties of the matrix and the nuclides and also on the size of the surface area in contact with the mobile water from which the nuclides may diffuse into the matrix. It was shown by Neretnieks (1980) that also in this case the amount of wetted surface per volume of rock is one of the primary variables, whereas the flow porosity as such in practice will have a negligible influence on the nuclide velocity for most sorbing nuclides.

For the ''matrix diffusion'' case when only a limited part of the rock participates in the sorption, the amount of rock that is reached by the nuclides will increase proportionally to the square root of time in many cases of interest. When there is matrix diffusion there will be no constant retardation factor. The retardation factor, if at all meaningful to define, will seem to increase with time. A special case with flow in channels with matrix diffusion into the channel walls is discussed later in connection with some sample calculations.

Channeling model

A ''channeling model'' (Neretnieks, 1983) is used in the sample calculations. The model states in essence that if the channel characteristics are known as well as the frequency of channels, which carry a certain flowrate, the transport in every channel is calculated individually and the sum of the effluents is calculated by adding the effluent concentration of all channels. In this case where the channels have been grouped in flowrate ranges, the effluent from every group is calculated and multiplied by the number of channels in every group.

$$c = \Sigma f_i \cdot c_i \tag{10}$$

f_i is the fraction of flow in group "i" of the channels, and c_i is the effluent concentration from a channel in that group.

Figure 13 illustrates a rock mass with independent channels and how the different channels may carry a nuclide with different residence times in the different channels. The fast channels will allow little decay whereas the slow ones will allow decay. The concentration at the effluent from a channel will depend on the retardation and possible decay of the nuclides during the passage through the channel. The main retardation effect is due to sorption within the rock matrix. Hydrodynamic dispersion is neglected in this case because "dispersion" due to the flowrate distribution will be considerably larger than that due to hydrodynamic dispersion.

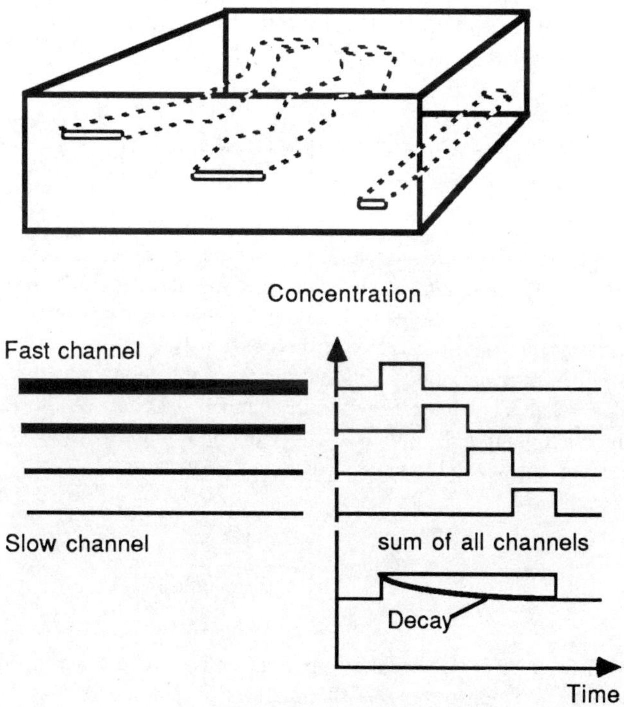

Figure 13. **Visual channel model.**

The effluent concentration from a single channel has been modeled (Neretnieks, 1980). The model describes the following case: Water with the nuclide flows with a constant velocity u_f in a thin fracture with aperture δ. The concentration in the water in the fracture c_f will change in time and space as described by the advection-dispersion equation. There is in this case, however, the added influence of the porous matrix. The nuclide will diffuse into the micropores of the matrix and sorb on the inner surfaces of the matrix. This is given by Equation (11):

$$\frac{\partial c_f}{\partial t} + u_f \frac{\partial c_f}{\partial z} = \frac{2\,D_e}{\delta} \frac{\partial c_p}{\partial x}\Big|_{x=0} - \lambda c_f \tag{11}$$

The diffusion into (and out of) the rock matrix is given by Equation (12):

$$\frac{\partial c_p}{\partial t} = D_a \frac{\partial^2 c_p}{\partial x^2} - \lambda c_p \tag{12}$$

The initial and boundary conditions state that initially there are no nuclides in the water in the fracture and also no nuclides in the pore waters of the rock matrix. The boundary conditions state that at time zero the nuclide concentration is suddenly raised to c_o at the inlet of the fracture. If the inlet concentration decreases by radioactive decay, the solution to the above equations becomes

$$c = c_o e^{-\lambda t} \operatorname{erfc}\left(\frac{D_e t_w}{\delta (D_a (t - t_w))^{1/2}}\right) \tag{13}$$

In the form Equation (13) is written, the water residence time, t_w, enters explicitly. An analysis of the equation shows, however, that for times much longer than the water residence time, the water residence time and the fracture aperture do not influence the results by the same reason as in the previously discussed cases with surface sorption and full bulk penetration. This is seen from the following: the water residence time t_w = channel volume/flowrate in channel = Length•Width•Aperture/Flowrate. Thus in Equation 13 above t_w/δ = Length•Width/Flowrate = LW_{fr}/Q = Wetted fracture surface/2•Flowrate. Equation (13) becomes

$$c = c_o e^{-\lambda t} \operatorname{erfc}\left(\frac{D_e L W_{fr}}{Q (D_a (t - t_w))^{1/2}}\right) \tag{14}$$

The wetted fracture surface is the important entity because it is from the surface in contact with the flowing water that the nuclides diffuse into and sorb in the rock matrix. For sorbing nuclides and for contact times longer than the water residence time, the water residence time as such has little influence. This helps considerably in the subsequent analysis because it is not well known. The most important entities to assess are

- Water flowrate in a channel, Q
- Wetted surface of a channel, A_{fr}
- Diffusivity in the rock matrix, D_e
- Sorption coefficient in the matrix, K.

The latter two entities have been measured extensively and are taken from SKBF (1983).

Some Example Calculations

In this section some example calculations are made to illustrate the effects of channeling and of the limited sorption surface on the migration of a sorbing nuclide. Pu-240 has been chosen for the example. The channeling characteristics of the SFR are used, and it is assumed that each channel has a width of 1 m. The influence of path length, porosity, and average flowrate is calculated. Furthermore the other extreme case of flow, i.e., very frequent mixing, is also investigated.

Migration of Pu-240 in a rock with the same properties as at the SFR

Equation (13) is the basis for calculating the concentration in each channel. The mixed effluent is calculated by Equation (10). The distribution of flowrates among channels is taken from Table 5. The total flux Q in the rock after the repository has been closed and as part of the regional flow system is taken to be 1 and 10 $l/m^2/y$. The volumetric sorption coefficient and the effective diffusivity are taken from SKBF (1983) to be 13,500 and 5×10^{-14} m^2/s, respectively. The results of the calculations are given in Table 12 and Figure 14.

The case of frequent mixing is simulated by using Equation (13) but averaging all the flow in the rock over the channels as if the channels had equal flowrates. This has the same effect as if the waters in the channels mix frequently and have access to all the surfaces of all the channels.

Table 12. **Maximum concentrations c/c_0 of Pu-240 at different distances z and different flowrates Q.**

Distance z (m)	50	100	200	400	800	1600
Channeling case						
Q $l/m^2/y$						
1	0.21	0.11	0.05	0.018	0.003	1.3×10^{-4}
10	0.67	0.55	0.40	0.25	0.14	0.07
Mixing case						
Q $l/m^2/y$						
1	0.08	0.011	1.6×10^{-4}	10^{-7}	$<10^{-9}$	$<10^{-9}$
10	0.64	0.48	0.29	0.12	0.02	8×10^{-4}

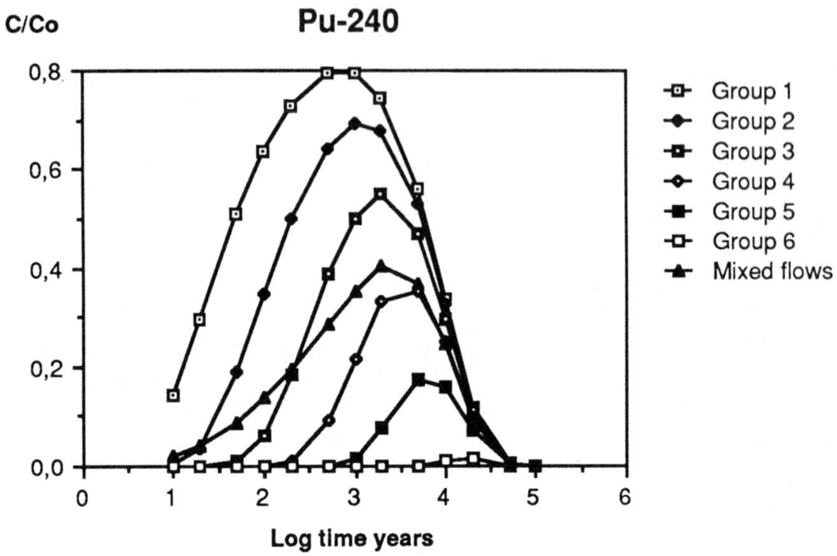

Figure 14. **Effluent curves from the six groups of channels and the mixed effluent (channeling) for Pu-240 migration with a flowrate of Q = 1 l/m²/y at 50-m distance. Group 1 refers to the largest flowrate range in Table 5, group 2 to the next largest, etc.**

Discussion and Conclusions

The field observations available at present indicate that water flows very unevenly in crystalline rocks. The flow seems to take place in channels with widths of a few meters at most, and often much smaller widths. Quite frequently the larger flowrates seem to occur in tubelike channels at fracture intersections. The observations have been performed mostly on rock surfaces in tunnels, and it cannot be ruled out and is even probable that the presence of tunnels disturbs the rock stresses in such a way that the channeling changes. Observations in deep drillholes give similar qualitative information. There are consistently only a few channels that carry a large fraction of the water flowrate. From the point of view of radionuclide transport and retardation, it is the fast narrow channels that carry the nuclides quickly and perhaps, in some circumstances, do not give the nuclides time to decay significantly.

In the calculated example there is a negligible difference whether or not there is mixing in the channels for short migration distances. For the longer distances there is a difference of several orders of magnitude. There is a considerable difference between the cases of low- and high-average water flowrates. The influence of the wetted area is also seen directly from the structure of the argument in Equation (13). The argument contains the group WL/Q. A decrease of W by a factor of 10 has the same effect as an increase in Q by the same factor or a decrease of L. The diffusivity or the sorption coefficient has to change by a factor of 100 to have the same effect as a change of a factor of 10 in the other entities.

The results were obtained for a water residence time determined for a flow porosity of 10^{-4}. A change of a factor of ten up or down is not noticeable in the results. This is because the concentration maxima occur at 2,000 to 10,000 years, whereas the water residence time is several orders of magnitude smaller.

The size of the wetted surface used in the example where 164 channels 1-m wide exist in a cross-sectional area of 14,000 m^2 is equivalent to a rock with parallel fractures open along all its length every 85 m. This is not in contradiction to the observations made in deep holes in crystalline rock if only those fractures are counted where most of the flow takes place.

The calculations are based on the assumption that the flow path structure and geometry of the channels in the rock are similar to those observed at the tunnel walls. There are indications that some channels may close due to the changes in rock stress induced by the presence of the tunnels. This must be further investigated.

Very few observations on channeling are available. The methods of interpretation are in their infancy, but because of the large potential impact of the sorption surface it is concluded that channeling should be investigated further both theoretically and by field observations.

Acknowledgments

The support of the SKB Company is gratefully acknowledged. The mapping of the tunnels at Forsmark by VIAK Company, the mapping of the tunnel at Kymmen by Bergab Company, and the discussions with the geologists responsible for the work have been a necessary base for this paper. Their enthusiastic cooperation is gratefully acknowledged.

Literature Cited

Abelin, H., Birgersson, L., Gidlund, J., Moreno, L., and Neretnieks, I., 1987. *Final Report on the Stripa 3D Experiment Part 3,* Stripa Project Report 87-21, NEA/OECD, Dept. Chem. Eng., Royal Institute of Technology, S 100 44 Stockholm, Sweden.

Abelin, H., 1986. "Migration in a Natural Fracture: An In Situ Experiment in a Natural Fracture," Ph.D. Thesis, Dept. Chem. Eng., Royal Institute of Technology, S 100 44 Stockholm, Sweden.

Bolvede, P., and Christiansson, R., 1987. *SKB Forsmarksarbetena SFR, Vattenförande Sprickor inom Lagerområdet* (Water-bearing fractures in the repository area), VIAK Company, Stockholm, January.

Nagra Project Gewähr, 1985. *Endlager für Hochaktive Abfälle: Das System der Sicherheitsbarriären,* Nagra, Baden, Switzerland, January.

Neretnieks, I., 1980. "Diffusion in the Rock Matrix: An Important Factor in Radionuclide Retardation?" *J. Geophys. Res., 85,* 4379–4397.

Neretnieks, I., Eriksen, T., and Tähtinen, P., 1982. "Tracer Movement in a Single Fissure in Granitic Rock: Some Experimental Results and Their Interpretation," *Water Resources Res., 18,* 849–858.

Neretnieks, I., 1983. "A Note on Fracture Flow Mechanisms in the Ground," *Water Resources Res., 19,* 364–370.

Neretnieks, I., 1987. ''Channeling Effects in Flow and Transport in Fractured Rocks—Some Recent Observations and Models,'' Paper presented at the GEOVAL Conference in Stockholm, April, 1987 (Proceedings in print 1987).

Palmqvist, K., 1987. *Kymmens kraftverk. Fullortsborrad tunnel. Hydrogeologisk kartering och utvärdering,* Report in preparation, Bergab AB Company, Göteborg, Sweden.

SKBF, 1983. *Final Storage of Spent Nuclear Fuel,* KBS-3, Report by Swedish Nuclear Fuel Supply Company, Stockholm, Sweden, May.

Transport of Dissolved Substances in Fissured Granite

LUTZ LIEDTKE and ARNO PAHL
Federal Institute for Geosciences and Natural Resources
Hannover, Federal Republic of Germany

Abstract

Several research projects have been under way in the Grimsel rock laboratory (Switzerland) since 1983. In the Fracture System Flow Test project, the development of methods and equipment for studying flow rate in fissured rock is being continued. Ten cored boreholes up to 135 m long were drilled for this purpose. The configuration of these holes, serving as injection boreholes and observation holes in specific rock sections, was designed according to the geological structures in the rock around the tunnels of the rock laboratory.

Since 1986, tracers have been added to the injected water so that detailed measurements on injected water volumes, pressure heads, and flow rates could be made. Nonlinear flow rates are used for calculating the size of the fissures, the storage capacity, and the permeability tensor under non-steady-state conditions.

The calculations were done with a finite-element program. With this program fissures can be modeled using tubular, discoid, or any other spatial element. Variation of the individual parameters provides an indication of the quality of the in situ experiments and the numerical models. The methods and results of the experiments will be described.

Introduction

A joint project between the Federal Institute for Geosciences and Natural Resources (BGR, Hannover), the National Cooperative for the Storage of Radioactive Waste (NAGRA, Baden), and the Research Centre for Environmental Sciences (GSF, Braunschweig) [*1*,*2*] has been under way since 1983. Whereas some projects have already been completed, others are planned to run up to the end of 1990. The projects are designed to develop and to test methods for studying the hydrogeological, rock mechanical, and geophysical properties of crystalline rock formations. In the Fracture System Flow Test the BGR is developing methods and equipment for studying flow rates in fissured rock.

Gallery Construction and Boreholes

The underground site for this test is 400 to 500 m below the rock surface, located in the central Aare granite of the Aare Massif in the Swiss Alps close to the Grimsel Pass. Crucial for the selection of this location was a water-bearing joint zone encountered by one of the numerous exploration boreholes sunk in 1980. The test site at 1730 m above sea level was constructed in 1983/84 by blasting.

Water flowed from the fissures in the lowest section of the test site. The fissures, faults, and cleavage planes were mapped in detail on a special map of the floors, walls, and roofs. In the process it was clearly shown that the intersection of the rock by the fissures is limited. Only a few fissures were observed to extend a significant distance through the rock. This structural analysis was the basis for deciding the location of the first boreholes. The encountered fractures and fracture systems, their spatial distribution, and the injection tests in the first boreholes were used to define the exact location and direction of the subsequent boreholes. The configuration of the existing ten boreholes, shown in Figure 1, was designed in this way. The boreholes are generally 43 to 68 m long with a diameter of 85 mm. The central borehole is approximately 136 m in length [*4,5,6*].

The boreholes are inclined downwards to facilitate subsequent tests in the water-saturated zone.

Geohydraulic Zones

On the basis of the hydraulic injection tests and the geological data, five zones can be differentiated (Figure 2).

1. The northern zone shows moderately fissured and partially hydrothermally metamorphosed granite.

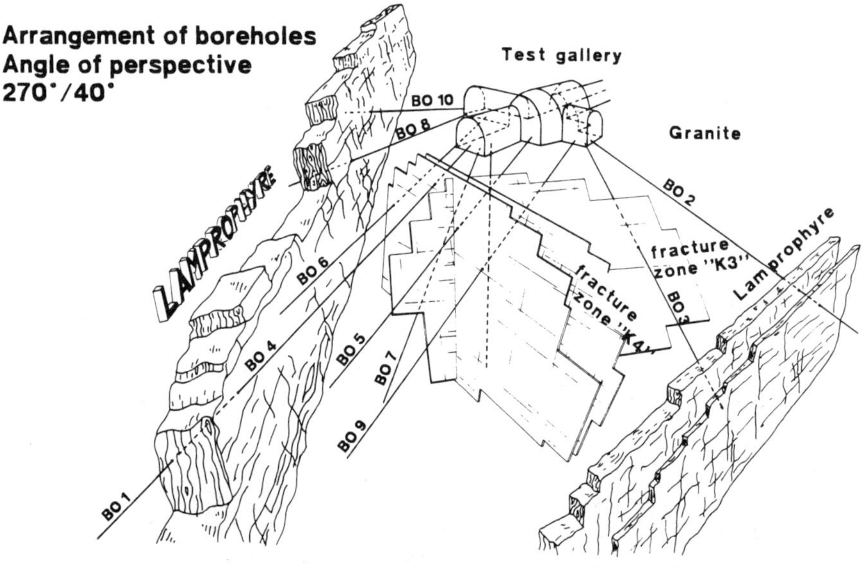

Figure 1. **Arrangement of boreholes. Angle of perspective 270°/40°.**

2. South of zone 1 is a thick lamprophyre zone, subdivided by xenoliths of varying thickness and showing a strong thermal influence. Petrographically this lamprophyre is a kersantite, the most frequently occurring lamprophyre type in the central Aare Massif.
3. Most of the test configuration is in the central zone. The exposed granite is

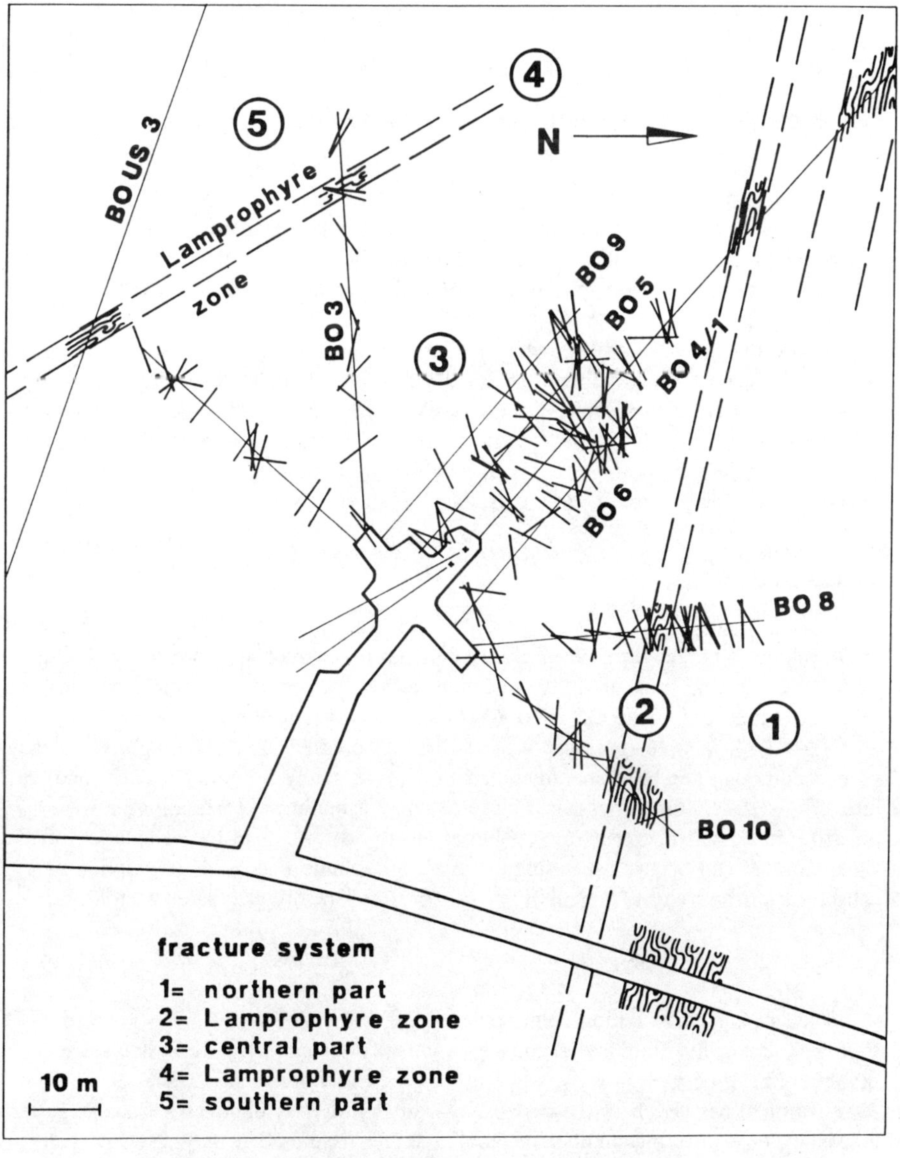

Figure 2. **Plan of the gallery area of the experiment.**

relatively highly fissured. About 20 to 30 m to the north and northwest of the cavern the zone is highly jointed to shattered.

4. This zone is jointed to the south by a biotite slate belt, only a few meters thick. The biotite is extremely slaty, shows very fine banding, is partially folded, and in parts is scaly in character.
5. The southern zone consists of a very hard, solid granite intersected by a few hydraulically significant joint sets.

The tests described herein were performed in zone 3. Interconnection with the other zones could not be proved to date.

Description of Equipment and Its Positioning

In principle, it is assumed that hydrogeologically critical zones can be sealed with pneumatic packers. Probes equipped with three sensors measuring pressure, temperature, and electrical conductivity of the water are installed in the packers. The measurement data are collected and stored centrally in a computer outside the borehole. The water is injected into a borehole with concurrent measuring in the other boreholes [*3*].

Before the main tracer tests were performed the transmissivity had to be calculated. The injected water volume for boreholes 5 and 9 is shown in Figures 3 through 5 as a typical example. On the basis of these results the injection element consisting of two 2-m-long packers and a probe was installed at a depth of 29 m in borehole 9. The probes and packers in neighboring boreholes 5 and 4 were positioned as close as possible to the hydraulically active fissures.

Transmissivity

Transmissivity T is the integral of the rock transmissivity along a 1-m section of the borehole, whereas pore permeability and fissure permeability are not differentiated. To determine transmissivity, injection tests were performed with increasing and decreasing pressure heads (35 m to 80 m) and at two temperatures (50 to 55 °C and 25 to 30 °C). The water pressure was gradually increased after reaching quasi steady-state conditions. The test time amounted to approximately 3 hours for each temperature. The results, summarized in Figure 5, show a nonlinear correlation between pressure heads and injected water volume. Up to 50% error in the calculation of the water volume can occur, depending on injection volume under the assumption of constant transmissivity. Transmissivity is more realistically calculated as follows:

$$T = 10^{-(4.24 + 0.0045 \times H)}(\mathrm{m/s})$$

The pressure distribution in the neighboring boreholes 4 and 5 is shown in Figure 6. Basically, four intersecting hydraulically effective fissure systems are involved in this test (Figure 7), whereby the rock is more or less fractured in the intersection zones. Each of the disks shown in Figure 7 symbolizes a fracture system consisting of a differing number of parallel single fissures that may also intersect each other.

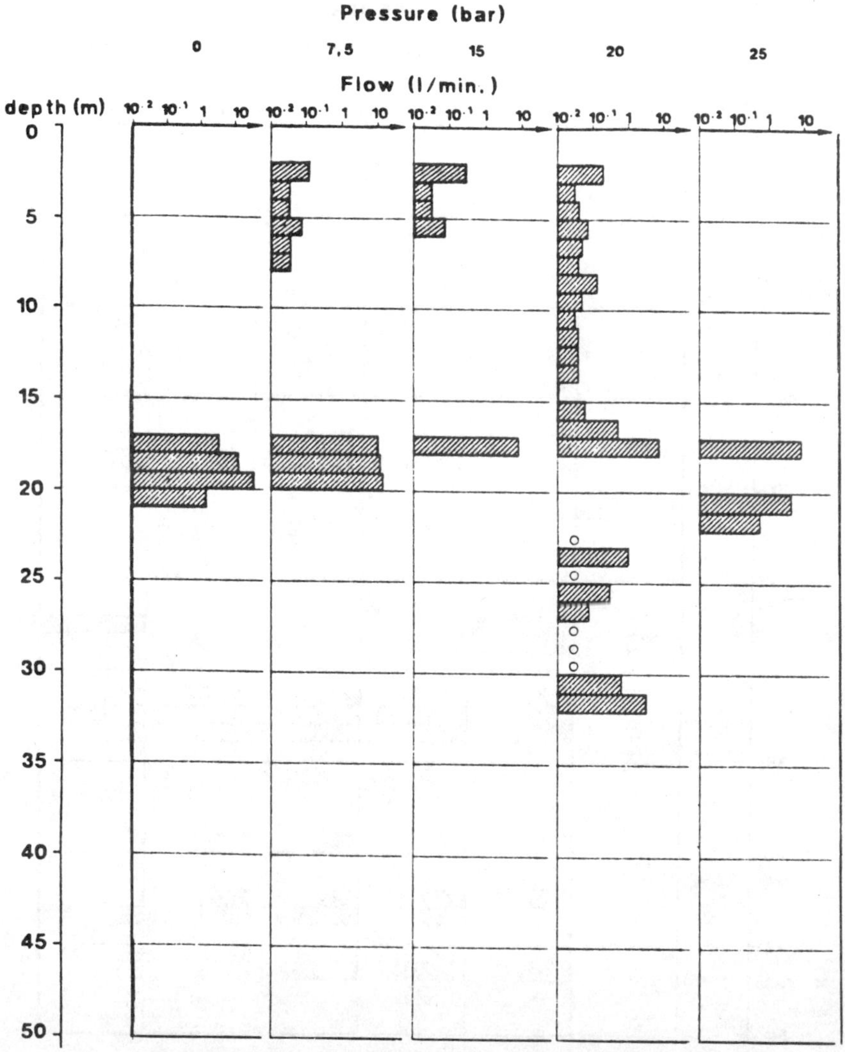

Figure 3. **The amount of water injected in borehole BO 5.**

Tracer Tests

Tracer tests with brine (100 g NaCl/m³ H_2O) were performed to calculate the water flow rates in the represented fracture system. The injection point in borehole 9 at depth 29 m (Figures 1 and 2) and the water volume of 15 l/min remained constant. During the tests all boreholes with one exception were sealed with pneumatic and mechanical packers so that most of the injected brine could flow only from one borehole. Because time lags occurred due to flow times and dispersion effects, fresh water and brine were injected alternately. The following three tests were performed:

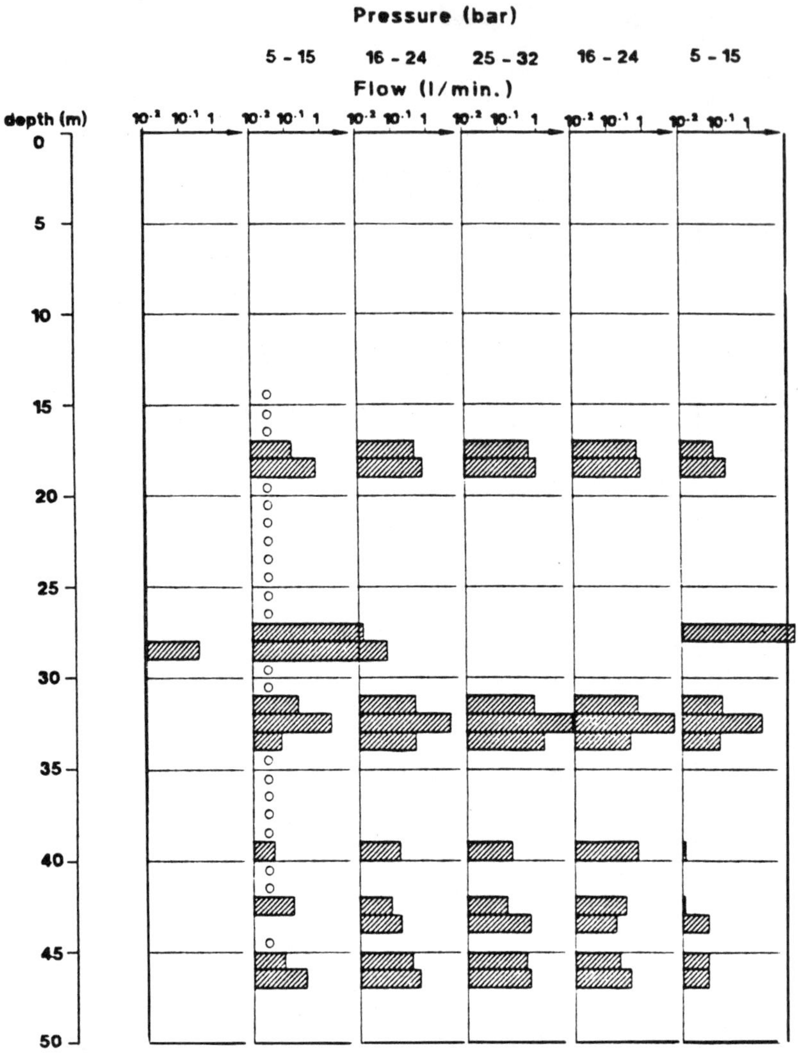

Figure 4. **The amount of water injected in BO 9.**

I. Brine was injected over a 3-hour period in borehole 9, and the rock was flushed subsequently for 22 hours with fresh water. The brine concentration of the discharged water was measured both at the borehole mouth, Figure 8, and at neighboring borehole 5 by means of numerous probes. As an example, the measurement in a borehole is shown in Figure 9 for the first 200 min. Penetration times lasted at the borehole mouth 35 min. and in the borehole 20 min. The water, therefore, has a duration of approximately 20 min over an estimated 9-m flow path through fissure intersections. The

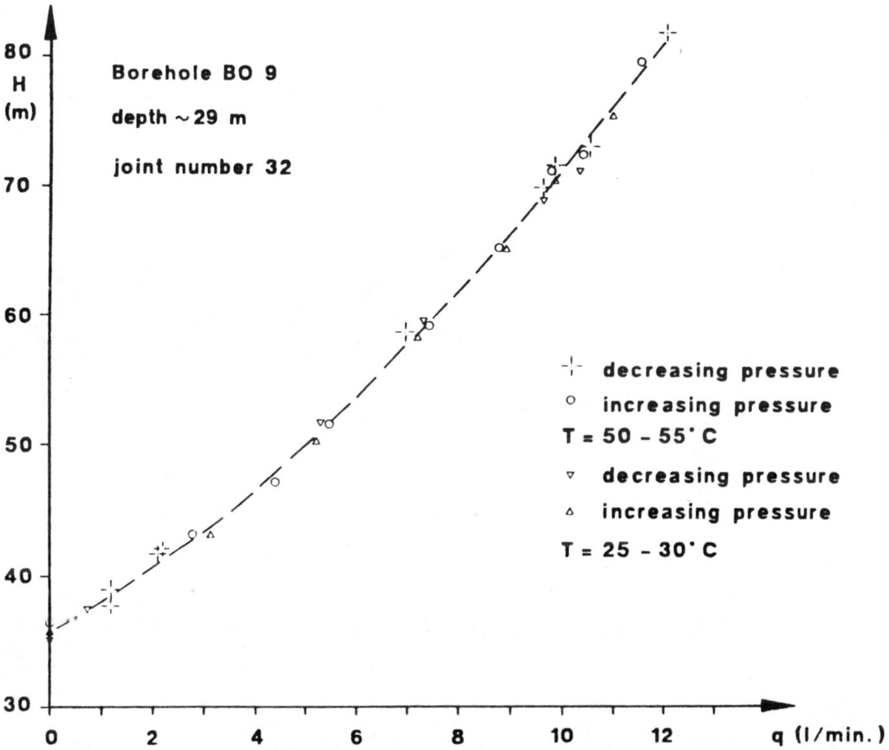

Figure 5. **Injection pressure as a function of the amount of water.**

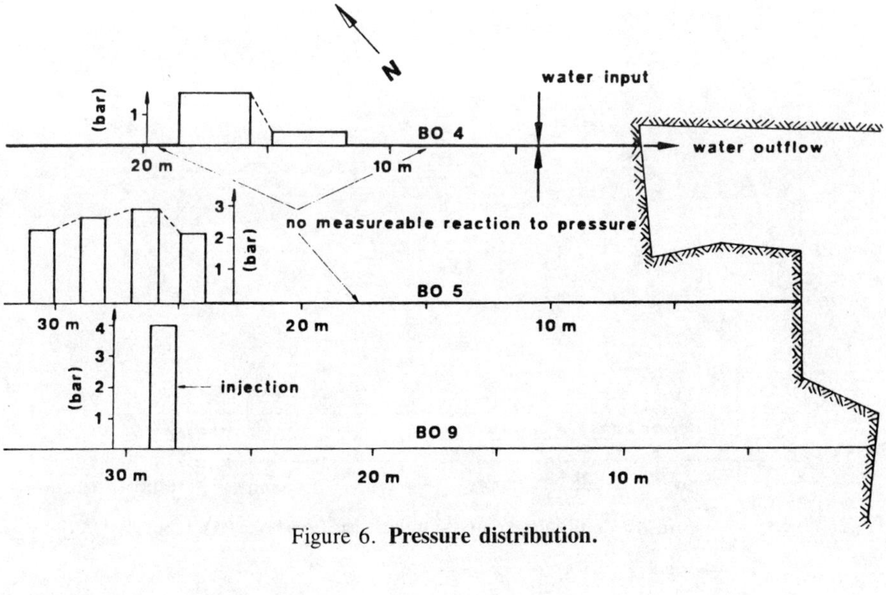

Figure 6. **Pressure distribution.**

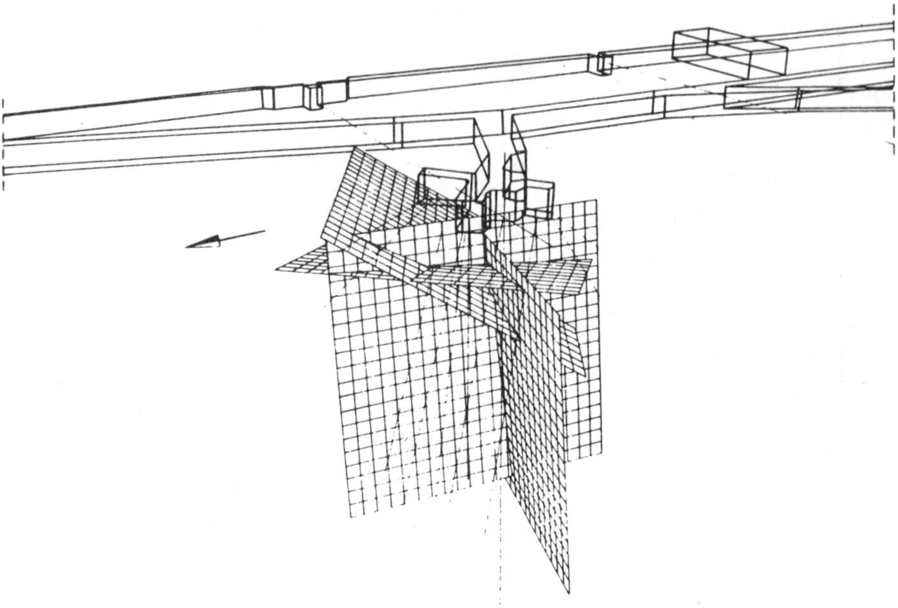

Figure 7. **Representation of the joints involved in the tracer experiments.**

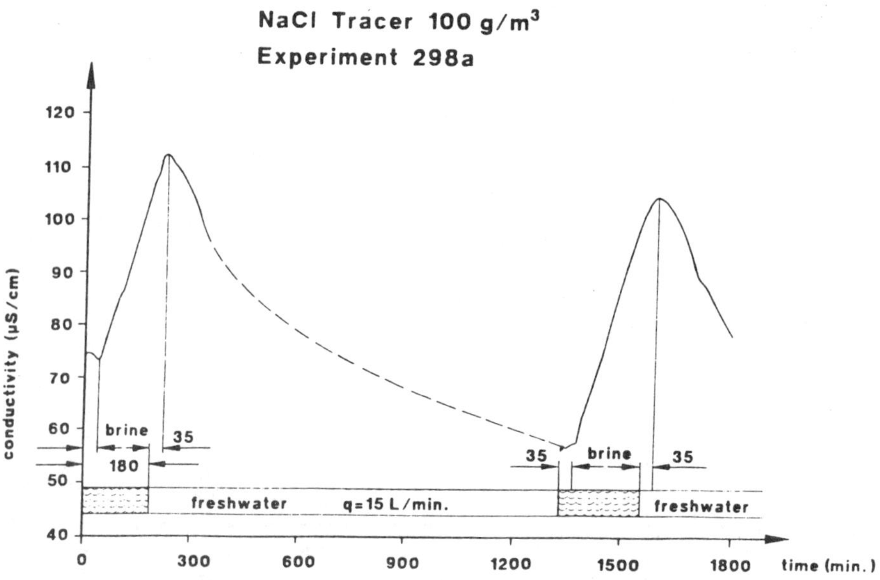

Figure 8. **Conductivity at the mouth of borehole BO 5.**

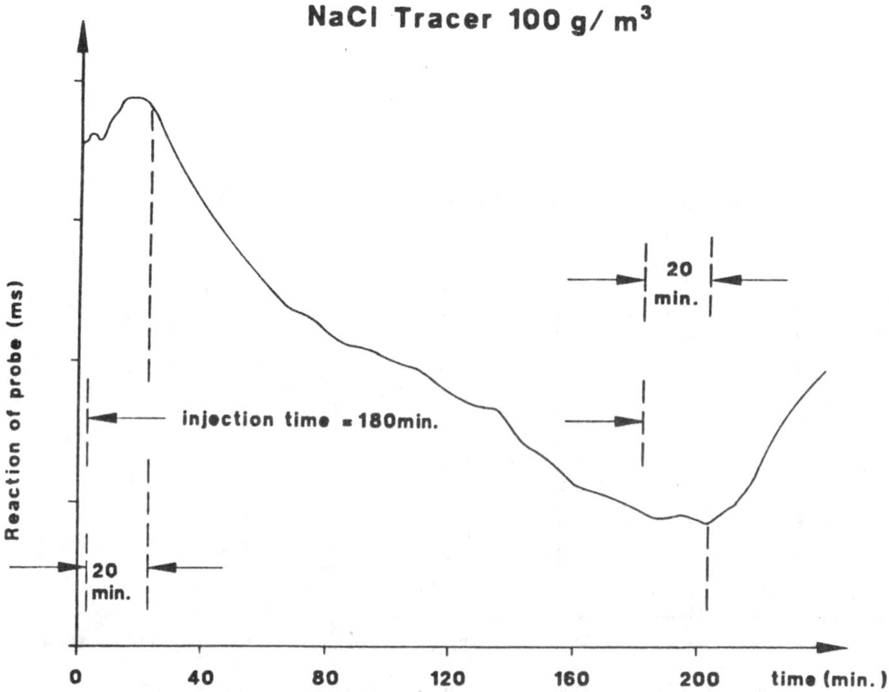

Figure 9. **Experiment 298a: reaction of probe 39, depth = 19 m.**

discharge volumes, pressure heads, and injection volumes remain constant during the test. Steady-state flow behavior can therefore be assumed.

II. The same holds for the following test. The injection point, the injected water volume, and the pressure head and measuring points were not altered. Brine with electrical resistances of 180 μS/cm and 14 μS/cm was injected alternately for 1-hour periods, Figure 10. The penetration times for this rapid change in salt concentration could only be calculated by changes in the flank of the curve. These changes at the borehole mouth vary between 25 and 26 min.

III. In the third test, neighboring borehole 5 was sealed with a number of packers. The injected brine could exit only through a fissure at 4 m depth in borehole 4. The section below 4 m was sealed by packers as in all other boreholes. The injection period with brine amounted to 5 hours and with fresh water to 19 hours (Figure 11). The cycle was repeated three times with subsequent flushing of the rock for a number of days with fresh water. In this way the electrical conductivity of the outflowing water was lower than the natural conductivity of the ground water (70 to 90 μS/cm). All three cycles are shown in Figure 11, and an exaggerated cycle is shown in Figure 12. The penetration time of approximately 80 min is averaged and can be regarded as representative. The fluctuations amount to $\pm$20 min.

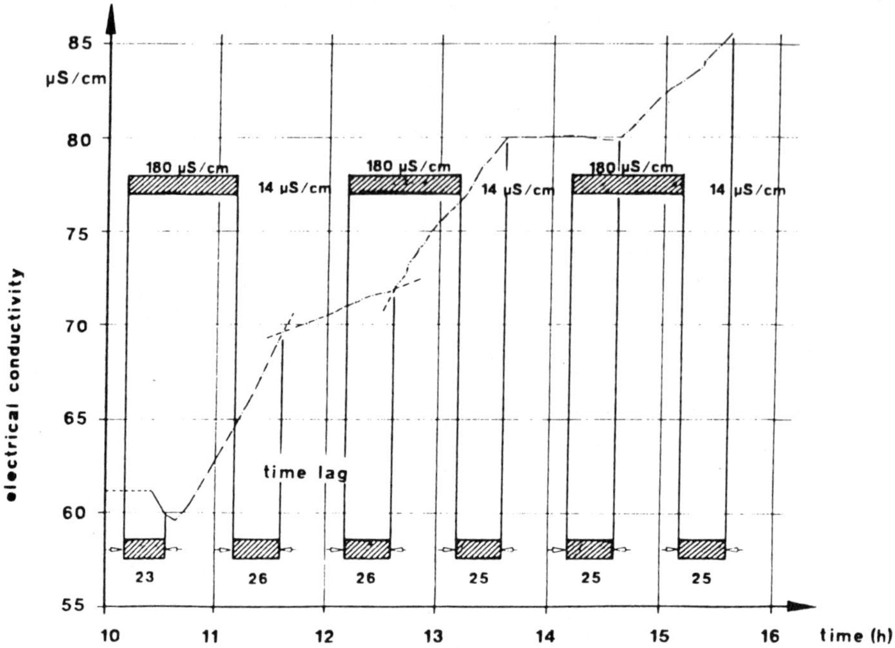

Figure 10. **Water flow: q = 15 l/min, salt 100 g/m³ = 180 μS/cm from BO 9 (29 m depth) to BO 5 (0 m depth).**

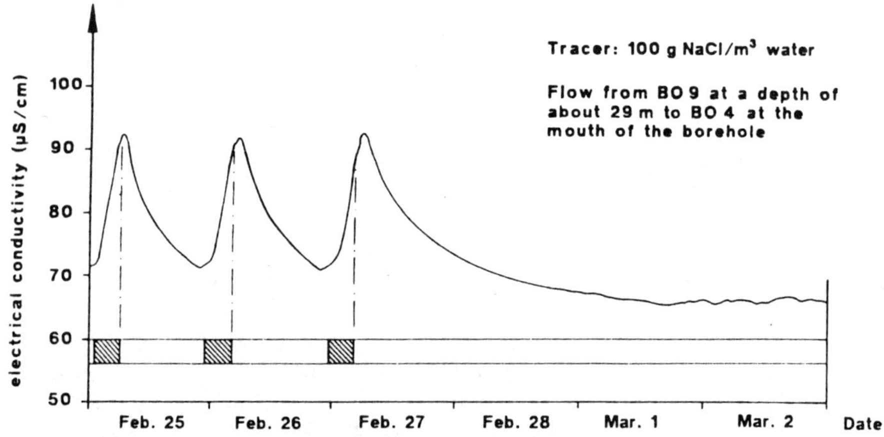

Figure 11. **Electrical conductivity as a function of time (3 cycles).**

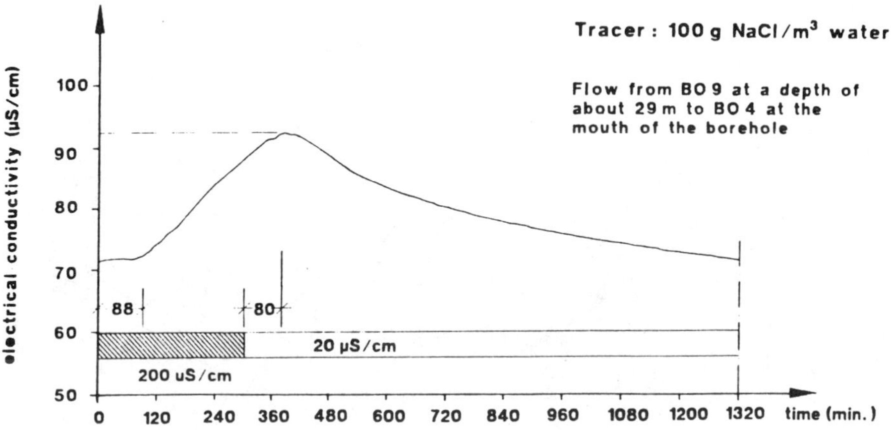

Figure 12. **Electrical conductivity as a function of time (1 cycle).**

Calculating Fissure Width

Using the finite element method (FEM), a hydraulic calculation was performed using the measured penetration times, water volumes, conductor pipe levels, and geometrical boundary conditions. The fissure model is simplified to two intersecting fissures on a plane along the line of intersection. The disk has variable thickness and consists of 132 isoparametric 2-D elements. The injection and discharge points of the water are defined as inflow and outflow. The FE grid extends beyond the water inflow and outflow points by 12 to 16 m, thereby taking the boundary conditions into consideration. A water-saturated condition is assumed. Comparing the measured times and the integrated time from the length/flow rate and elements together with transmissivity variations produces the calculated fissure thickness of 2.2 mm. The conductor pipe levels of approximately 80 m at the injection point were kept constant for the variations. The calculated conductor pipe levels are shown in Figure 13 from two perspectives. The rapid decrease of pressure heads in the fissures around the borehole correlates well with the measured values.

The flow rate distribution was calculated at four integration points in each element, but for purposes of clarity only one per element is shown on Figure 14. The calculated and average measured flow rate distribution along the shortest flow paths from borehole 9 to borehole 4 are shown on Figure 15.

The calculated fissure thickness $d = 2.2$ mm and transmissivity $k_f = 2 \cdot 10^{-3}$ m/s must be regarded as integrals. The thickness and hence transmissivity vary significantly within the fissure system because of, among other things, leaching and solid matter deposition. Furthermore, the fissure system is also to be regarded conditionally as a geometrically irregularly curved structure. Intersections within fissures of same direction, all treated as a fissure set (Figure 7), are conceivable.

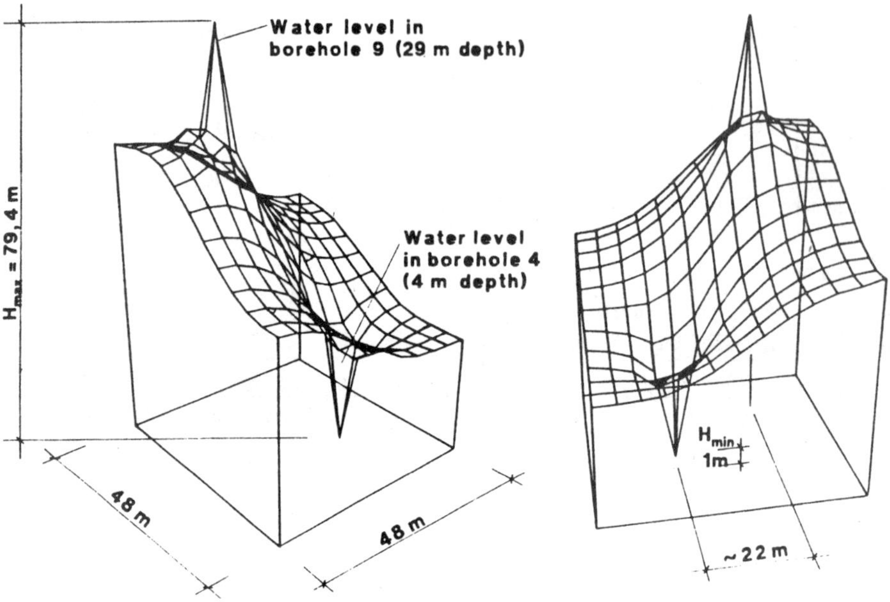

Figure 13. **Potentiometric surface in a joint: q = 15 l/min, $k_f = 2 \cdot 10^{-3}$ m/s, S = 10^{-8}, joint width 2.2 mm.**

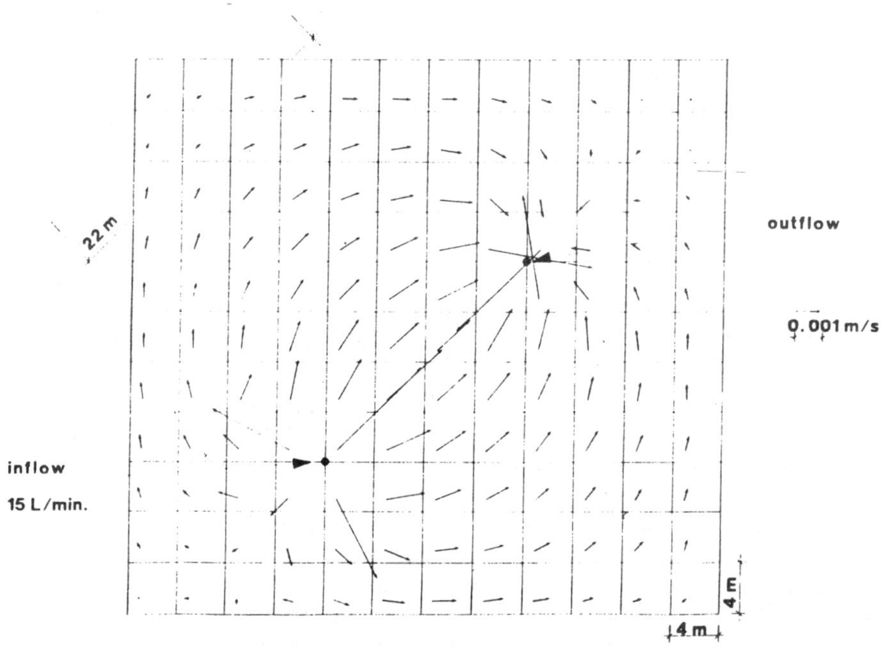

Figure 14. **Distribution of flow velocities: joint width = 2.2 mm, $k_f = 2 \cdot 10^{-3}$ m/s.**

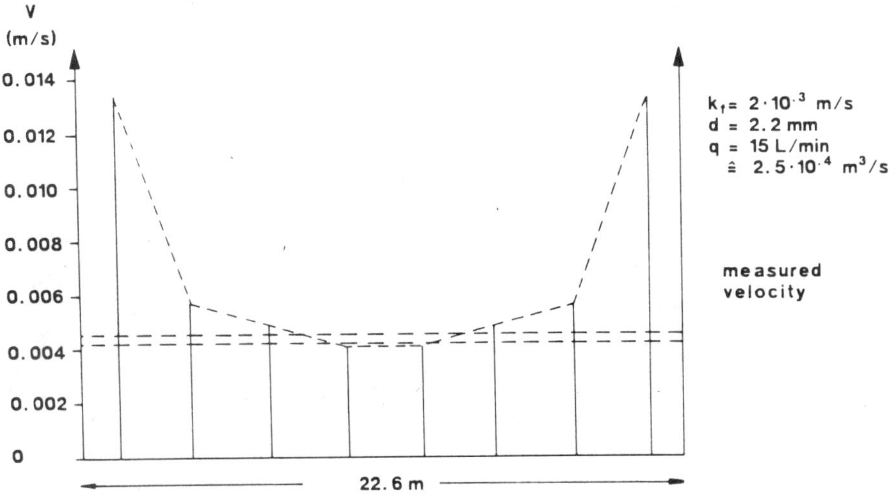

Figure 15. **VE 300: Distribution of flow rates along the shortest flow paths.**

Prospects

The Fracture System Flow Test research project will be continued for 3 years. In addition to further development of equipment and the geological mapping of fissures, slickensides, and fissure depositions or leaching, further development of the test procedure and the program for calculating flow processes using the FEM is planned. The emphasis is on the calculation of transport times and dispersion characteristics of dissolved matter in fissured rock.

Literature Cited

1. NAGRA, BGR, GSF, *Grimsel Test Site. Overview and Test Programs*, Technical Report 86-46 (1985).
2. Brewitz, W. and Pahl, A., "German Participation in the Grimsel Underground Rock Laboratory in Switzerland," *International Symposium on the Siting, Design and Construction of Underground Repositories for Radioactive Wastes*, Hannover, Federal Republic of Germany (March 1986).
3. Liedtke, L. and Pahl, A., "Water Injection Test and Finite Element Calculations of Water Percolation Through Fissured Granite," *Proceedings of a Joint CEC-NEA Workshop*, Brüssels, Belgium (May 1984).
4. Pahl, A., Bräuer, V., Heusermann, St., Kilger, B. and Liedtke, L., "Results of Engineering Geological Research in Granite," *5th Int. Congress of the IAEG*, Buenos Aires, Argentina (October 1986).
5. Liedtke, L. and Pahl, A., *Untersuchung der Permeabilität von geklüftetem Granit im Felslabor Grimsel, Schweiz,* Deutsche geologische Gesellschaft, Königstein, Federal Republic of Germany (1985).
6. Pahl, A., Bräuer, V., Heusermann, St., Kilger, B. and Liedkte, L., "Results of Engineering Geological Research in Granite," *Bulletin of the International Association of Engineering Geology,* No. 34, Paris (1986).

Application of Stochastic Methods for the Simulation of Solute Transport in Discrete and Continuum Models of Fractured Rock Systems

LESLIE SMITH
Department of Geological Sciences
University of British Columbia
Vancouver, British Columbia

FRANK SCHWARTZ
Department of Geology
University of Alberta
Edmonton, Alberta

CHARLES MASE
Department of Earth Sciences
University of Waterloo
Waterloo, Ontario

Abstract

This paper addresses two issues currently limiting the application of solute transport models in fractured rock systems. Existing continuum models, developed originally for porous media, invariably assume dispersion is an isotropic process, although dispersion is known to be an anisotropic process in fractured rock. A continuum approach is described that can account for anisotropic dispersion in a network of finite-length fractures. The technique is based on particle-tracking methods. Statistics describing particle transport are collected from a discrete fracture model. This discrete model is a small-scale prototype of the much larger system of interest. These statistics are used to control the pattern of transport in the larger continuum and enable us to mimic in the larger system the particle transport that is observed in the smaller system. The second issue addressed concerns the magnitude of potential errors in the interpretation of hydraulic and tracer tests to estimate fracture aperture. Discrete network models require data on fracture aperture to transform a geometric model of the network into a hydraulic model. Numerical experiments are described that suggest that large differences in aperture estimates based on hydraulic and tracer tests are possible because of differing sensitivity to the effects of network geometry and spatial variability in aperture. Difficulties arise from the application of an axisymmetric flow model to characterize the velocity field within the fracture isolated for testing.

Introduction

Performance assessments for the siting and design of underground nuclear waste repositories require that hydrogeologic models be used to estimate rates at which radionuclides may migrate from a repository to the accessible biosphere. As rocks with low permeability are targeted as candidate sites, it is probable that fractures will play an important, if not dominant, role in providing pathways for radionuclide migration. There are two modeling approaches to the simulation of fluid flow and solute transport in fractured media. The continuum approach treats a fractured rock mass as if it were equivalent to a porous medium, while network models are based upon an explicit representation of the fractures.

If fracture densities are much greater than those that mark the onset of percolation, then a continuum approach can be adopted. In this case, parameters must be formulated to characterize advective, dispersive, and chemical processes occurring within representative elemental volumes (REV) of the rock mass. At sites underlain by sparsely fractured rock, or those where a significant proportion of the fractures are sealed by mineral precipitates, considerable transport distance may be required before solutes encounter a representative sample of fluid velocities. Until solutes have migrated over that distance, continuum models are not applicable. It is also conceivable that several scales of fracturing occur within the boundaries of the flow domain, with a relatively few, areally extensive fractures exerting a disproportionately large influence on patterns of groundwater flow. These situations can be identified as sub-REV problems and in these instances, discrete network models are appropriate. A characteristic of these environments is the potential for solute transport to be significantly influenced by individual fractures and their interconnections with other fractures forming the network.

Stochastic concepts underlie the development of discrete network models. Probability distributions are used to define fracture locations, dimensions, and orientations. Techniques are in place to identify the statistical structure of a fracture network from both borehole data and mapping in adits. Two examples from the recent literature are given by Rouleau and Gale (*1*,*2*) and Long and Billaux (*3*). The locations of known fractures can be included using a conditional simulation technique [Andersson and Thunvik (*4*)]. From these geometric parameters, and given a probability distribution defining fracture aperture, the hydraulic and transport properties of the network can be calculated. Model output reflects uncertainty introduced into the analysis because of limited knowledge of the geometry of the fracture network [see, for example, Smith and Schwartz (*5*)].

Several key issues currently limit the practical application of solute transport models in fractured media. In the continuum approach, the model must be formulated in a way that takes account of the influence of fractures on patterns of spreading. Results of two-dimensional simulations show that dispersion in a fractured rock mass is an anisotropic process. That is, spreading patterns in fractured media depend upon the orientation of the hydraulic gradient relative to the orientation of the fracture sets forming the network [Schwartz et al. (*6*)]. Existing codes, developed originally for porous media, invariably assume dispersion is an isotropic process, with dispersivity values that are independent of the orientation of the hydraulic gradient.

Discrete network models, by the very nature of their formulation, can properly account for the effects of fracturing on dispersion. However, the application of discrete models leads to a different set of problems. These models are based on the assumption that it is possible to measure fracture aperture in field settings. In addition, there are limits on the size of a domain that can be modeled as a finite number of fractures can be included in the solution scheme.

This paper focuses on two issues. First, we describe a continuum model to simulate dispersion within a fractured rock mass, taking account of the anisotropic nature of the transport process. This continuum model is designed for far-field calculations. Second, we examine the reliability of several techniques that are used to obtain in situ estimates of fracture aperture from hydraulic and tracer tests. For the purpose of this paper, we consider only advection and dispersion, and neglect the impact of diffusion and chemical processes in the examples that are presented. Dual porosity formulations, which allow for fluid flow in matrix blocks between the fractures, are also not considered in our discussion.

A Continuum Model for Dispersion in Fractured Rock

In this section we outline a continuum approach that accounts for the influence of fracturing on dispersion. The modeling concept proposed is statistical; it involves the application of particle-tracking techniques to simulate transport within a continuum. As the particle-tracking method was originally conceived [Ahlstrom et al. (*7*); Smith and Schwartz (*8*)], solute transport is simulated by adding reference particles to a velocity field and moving them individually in simple deterministic and probabilistic steps. These two components of motion represent advection and dispersion, respectively. Concentrations determined from the spatial distribution of particles are also a solution to the well-known advection-dispersion equation. An important feature of the particle-tracking method is that it is possible to modify the character of particle motion to represent the more complex patterns of dispersion that occur in fractured rocks.

The question of how to describe dispersion during solute transport through a fractured rock is addressed by observing particles moving in a discrete fracture model. This network is a small subdomain from a much larger continuum. By characterizing particle motion in the subdomain, we have a way of incorporating the influence of fracture geometry in the continuum model. This numerical approach avoids the necessity of developing an anisotropic dispersion tensor for transport in a connected network of discontinuous fractures. The components of such a tensor, and whether or not its coefficients could be calculated in a practical sense, have yet to be determined. The modeling technique proceeds in two steps. First, a transport simulation is carried out for the discrete subdomain model. Second, statistics on particle motion are transferred to the larger scale, continuum model. Predictions of solute travel times and species concentrations are then developed with the continuum model.

The discrete subdomain model

Step one begins with the generation of a realization of a fracture network within a subdomain of the region of flow. Probability distributions describing the geometry of the fracture network are assumed to be identical to those of the larger continuum. This assumption of statistical homogeneity is the key element linking the two models. A modified version of a discrete network model described by Schwartz et al. (*6*) is used here. This model has been extended to deal with more general boundary conditions and includes an updated sparse-matrix solver for flow, which makes it possible to work with larger fracture networks. Our results to date are based on a network constructed from two sets of orthogonal fractures.

The subdomain for the discrete network has a square shape. Constant values of hydraulic head are assigned to all nodes formed where fractures intersect the side boundaries. Actual values of hydraulic head assigned to these nodes are linearly interpolated from values specified at the four corners. These boundary conditions are similar to those used by Long et al. (*9*). By assigning appropriate values of hydraulic head to the corners of the subdomain model, it is possible to generate flow with a specific orientation relative to the attitude of the fracture sets. These values of hydraulic head are calculated within the code in such a way that the magnitude of the gradient remains the same in the chosen direction of flow. Typically, seven different orientations of flow are examined in a realization of the fracture network. This scheme enables us to collect statistics at 15° intervals over a 90° sector of flow (Figure 1). For consistency with later notation (see Figure 1), the statistics are presented as though the orientation of the gradient is fixed and the network is being rotated.

Statistics on the reference particles are collected by tracking the particle swarm as it moves through the discrete network. A separate set of the statistics is collected each time the orientation of the mean hydraulic gradient is changed relative to the

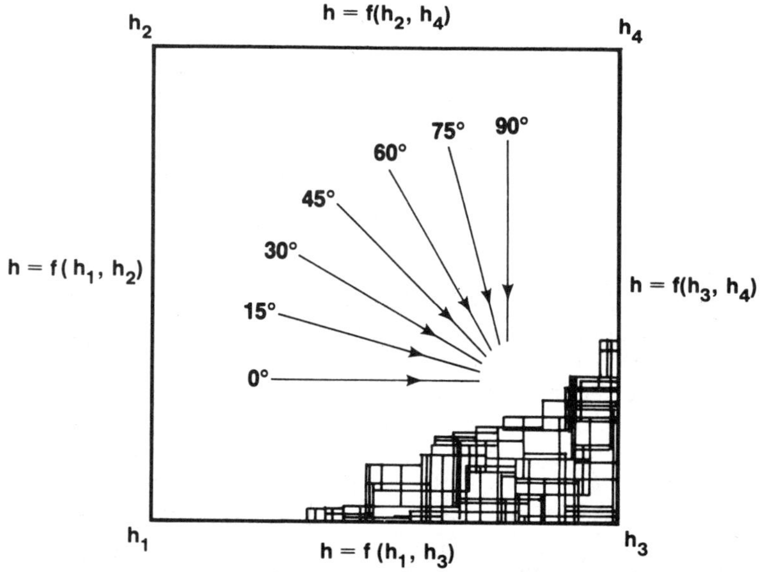

Figure 1. **Example of a discrete fracture network within a square flow domain. Boundary conditions as shown make it possible to orient flow at various angles to the network. Adapted from (*19*).**

two fracture sets (Figure 1). In a typical simulation, 500 reference particles are added as an instantaneous pulse at a single upstream node. The following statistics are collected to characterize particle motion: (1) frequency distributions on lengths between fracture intersections for the two sets (r_a and r_b); (2) distributions characterizing the frequency of particles choosing one of four possible directions of motion for each of the seven different orientations of the network relative to the gradient (d_{ij}:i = 1,4; j = 1,7); and (3) frequency distributions on the negative logarithm of the fluid velocity as a function of the particle direction (i) and the orientation of the hydraulic gradient (j). These velocity distributions are characterized by a mean pv_{ij} and standard deviation s_{ij}. Velocity values are normalized by dividing by the hydraulic gradient that is imposed on the subdomain. The direction of motion (d_{ij}) is labeled in Figure 2.

The subdomain must be large enough so that a sufficient number of fracture intersections are present in the subdomain model to obtain representative statistics on particle motion. Effectively, the subdomain must approximate an REV for solute transport. This assumption can be crudely examined by identifying expected symmetries in the particle statistics.

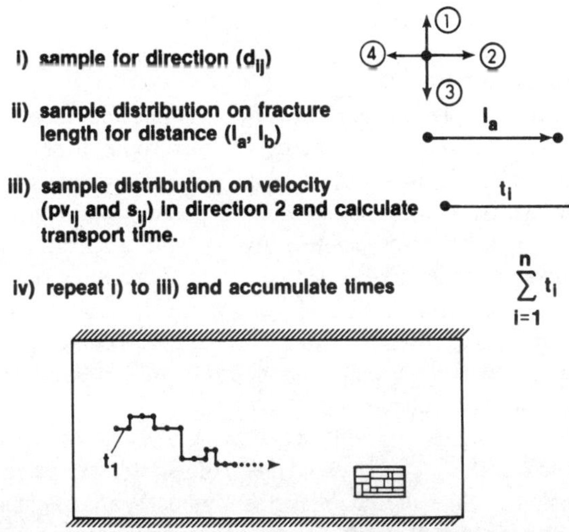

Figure 2. **Schematic representation of distributions to be sampled and how an individual particle moves through the continuum. Reprinted with permission (*19*).**

The continuum model

To simulate mass transport in the continuum, it is necessary to first determine how the magnitude and orientation of the hydraulic gradient vary through the region of flow. These data are obtained by solving the steady state groundwater flow equation, assuming the domain is a homogeneous (possibly anisotropic) porous medium. The local hydraulic gradient is calculated from the resulting flownet.

The steps involved in mimicking particle motion in the continuum are summarized in Figure 2. The direction that a particle P_k, located at point (x,y), will take during part of a time step Δt is determined by sampling the frequency distributions on particle direction (d_{ij}), given the orientation of the local hydraulic gradient in the vicinity of (x,y). Recall that the subdomain statistics were calculated for only seven discrete orientations (0°, 15°,..., 90°). To refine the particle-tracking technique, we use the orientation (α) of the gradient at (x,y) to develop a unique frequency distribution d_i. This latter distribution is formed by linear interpolation from the appropriate bounding values of d_{ij}. The actual choice of direction (i.e., 1, 2, 3, or 4) is determined by randomly sampling d_i.

Having determined a direction for transport, the appropriate distributions for length (r_a or r_b) are sampled to estimate how far the particle moves in the selected direction. Finally, by sampling the velocity distribution (pv_{ij} and s_{ij}) and then scaling the velocity by the magnitude of the local hydraulic gradient, we can determine the time required to reach the next hypothetical fracture intersection. The combination of velocity and travel distance provides the travel time for P_k over the hypothetical segment of the network. These steps of moving the particle and accumulating values of travel time over each fracture segment are repeated until the particle completes the time step Δt and another particle is transported. One time step is completed when all particles are moved.

Testing the procedure

By working with a simple fracture network, it is possible to compare plume shapes calculated using de Jong's (*10*) theory of anisotropic dispersion with those simulated using our modeling approach. de Jong's theory provides a straightforward way to calculate components of the anisotropic dispersion coefficient D_{ij} for a regular network of infinite length fractures, subject to flow with a specified hydraulic gradient. The test networks consist of two sets of parallel fractures oriented at 90° with respect to each other. Fractures in both sets are spaced at intervals of 1.0 m and have a hydraulic conductivity of $10^{-3.2}$ m/s. This simple geometry also permits us to calculate the necessary statistics for the continuum model, without resorting to a simulation using the discrete submodel.

Components of the dispersivity tensor (δ_{xx}, δ_{xy}, δ_{yy}) and the principal values (δ_{11}, δ_{22}) are compared in Table 1. The lines labeled "de Jong" are calculated from his theoretical relationships, while the lines labeled "model" are calculated using summary statistics on particle distributions observed in a continuum simulation. Four cases are presented for different orientations of the fracture set relative to a fixed hydraulic gradient. Correspondence is excellent in all four cases.

A second procedure for testing the code involves examining how well breakthrough curves from the continuum model fit those obtained from a discrete fracture model of the same size. This discrete model is the same one used to collect the moving particle statistics. In effect, the discrete fracture model provides data not unlike a field or laboratory experiment.

A total of four one-dimensional and two two-dimensional sets of results are compared. The continuum trials are carried out using data set D-l (Table 2). The flow regime for the one-dimensional case is generated with $\alpha = 0°$ and θ_a aligned at 0°, 15°, 30°, and 45°, respectively, relative to the x-axis. The flow domain for the two-dimensional simulation is shown in Figure 4(a).

Table 1. **Comparison of calculated dispersivities using de Jong's theory and our model. In all cases, the hydraulic gradient is parallel to the x-axis (α = 0°).**

Case		Dispersivity (m) δ_{xx}	δ_{xy}	δ_{yy}	δ_{11}	δ_{22}
$\theta_a = 45°$	de Jong	0.0	0.0	0.35	0.35	0.0
$\theta_b = -45°$	Model	0.0	0.03	0.35	0.35	0.0
$\theta_a = 30°$	de Jong	0.10	0.17	0.29	0.39	0.0
$\theta_b = -60°$	Model	0.10	0.16	0.28	0.38	0.0
$\theta_a = 15°$	de Jong	0.44	0.26	0.15	0.59	0.0
$\theta_b = -75°$	Model	0.42	0.25	0.15	0.57	0.0
$\theta_a = 0°$	de Jong[1]	0.0	0.0	0.0	0.0	0.0
$\theta_b = -90°$	Model	0.01	0.0	0.0	0.01	0.0

[1]Singularity in de Jong's method.

Table 2. **Summary of moving particle statistics obtained for the discrete subdomain. Data set D1.**

Distribution		θ_a 0°	15°	30°	45°	60°	75°	90°
	j	1	2	3	4	5	6	7
d_j	i = 1	3250	900	250	72	28	6	1
	2	25600	23700	21000	18200	12460	7322	3250
	3	3250	7322	12460	18200	21000	23700	25600
	4	1	6	26	72	250	900	3250
pv_{ij}	i = 1	3.640	3.675	3.705	3.740	3.740	3.740	3.740
	2	3.210	3.235	3.275	3.345	3.445	3.560	3.640
	3	3.640	3.560	3.445	3.345	3.275	3.235	3.210
	4	3.740	3.740	3.740	3.740	3.705	3.675	3.640
s_{ij}	i = 1	.345	.350	.360	.370	.370	.370	.370
	2	.130	.135	.150	.160	.205	.255	.345
	3	.345	.255	.205	.160	.150	.135	.130
	4	.370	.370	.370	.370	.360	.350	.345

In general, the breakthrough curves simulated with the continuum model match data from the discrete model reasonably well [Figures 3 and 4(b), 4(c)]. Unlike nearly all other existing continuum models, our approach appears to be able to simulate the anisotropic behavior associated with mass transport in the fracture networks.

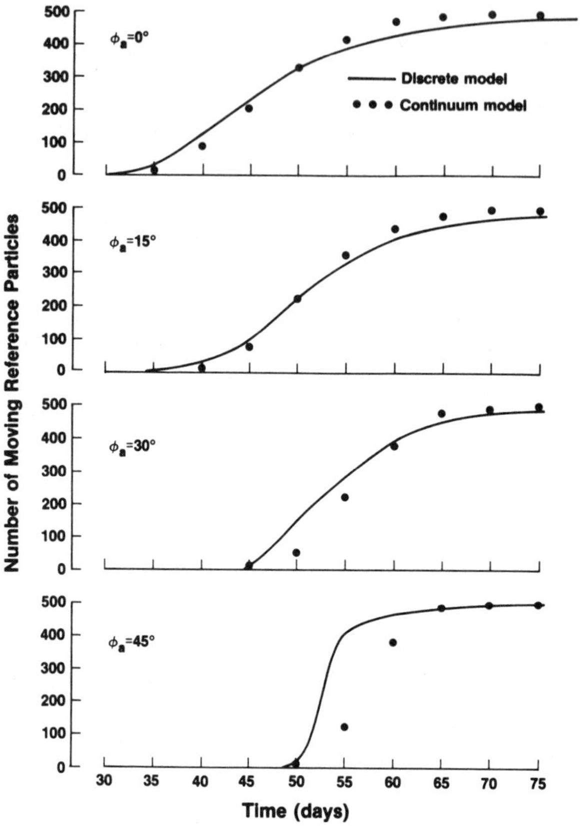

Figure 3. **Comparison of breakthrough curves generated using the discrete fracture model and the continuum model for four different cases of one-dimensional flow. Reprinted with permission (*19*).**

There is obviously not the close agreement that one usually obtains between numerical and analytical models. The differences we observe occur because the continuum model based on an "averaged" set of frequency distributions is not exactly comparable to the discrete network, which is characterized by relatively significant local variability.

Additional work is required to develop this continuum approach as a practical method. There is an initial requirement for a more general discrete-network model, allowing for nonorthogonal fractures with a random component to fracture orientation within each set. In principle, the modeling concept can be extended to three dimensions, although computational aspects have yet to be considered.

Estimation of Fracture Aperture

A key parameter in the continuum model described in the previous section, and in any discrete fracture model, is fracture aperture. Information is required on frac-

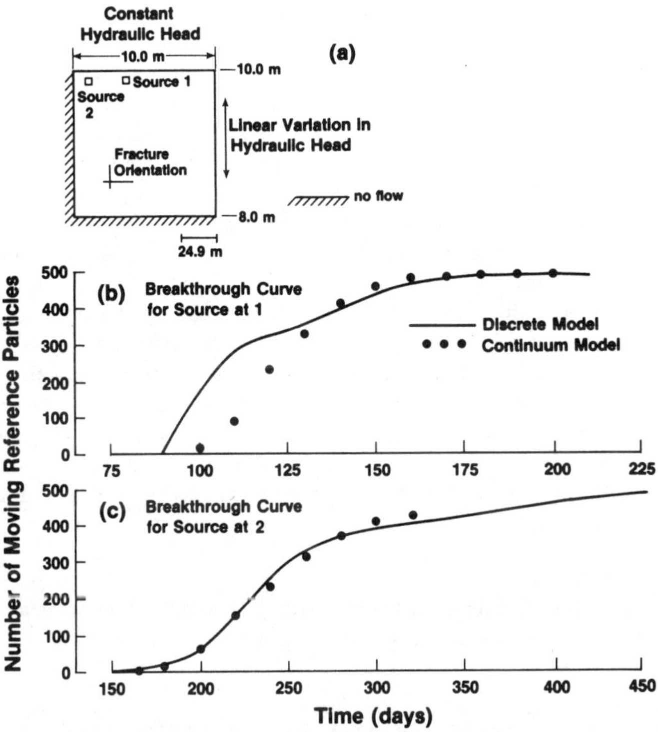

Figure 4. **(a) Sketch of the flow domain, illustrating the dimensions, fracture orientations, and boundary conditions on fluid flow. Particles are released at sources 1 and 2. (b) and (c) compare the breakthrough curves for the discrete and continuum models with sources at positions 1 and 2, respectively. Reprinted with permission (*19*).**

ture aperture to transform a geometric model of the fracture network into a hydraulic model. For these modeling approaches to be viable, it is essential that methodologies be available to characterize the hydraulic properties of individual fractures. Mechanical measurements of aperture in the disturbed region around a borehole are of questionable reliability. Current practice relies on hydraulic and/or tracer tests to obtain in situ estimates of fracture aperture. Results of both steady state withdrawal tests and injection-withdrawal tests are reported in the recent literature.

To estimate fracture aperture, a model must be adopted that describes the flow field within the plane of the fracture. Two key assumptions are adopted in practice. First, the cubic flow law is used to express the volume flow rate within the fracture as a function of the hydraulic gradient. Second, it is assumed that the variation in the hydraulic gradient reflects an axisymmetric flow, so that the hydraulic gradient is inversely proportional to the radial distance from the withdrawal well. See Hodgkinson and Lever (*11*) and Novakowski et al. (*12*) for a review of the mathematical formulation of the equations used to calculate fracture aperture.

An aperture estimate derived from hydraulic head data is frequently referred to as a "hydraulic aperture." If the estimate is derived from the results of a tracer test, then the estimate is labeled a "tracer aperture." Hydraulic calculations are based on differences in hydraulic head at two boreholes for given pumping rates. Calculations based on tracer tests reflect an integration of the fluid velocities along the pathline followed by the tracer in moving from the injection to the withdrawal point.

Radially symmetric models are appropriate in the idealized case of laminar flow in a smooth, planar fracture of large areal extent. However, the potential exists for deviations from this simple model because of distortion of the flow field by the effects of fractures of finite dimension, by intersections with other fractures acting as line sources or sinks of fluid, and by rough fracture walls. In the following paragraphs, we describe a numerical experiment that examines the magnitude of potential errors in the interpretation of hydraulic and tracer tests to obtain estimates of fracture aperture. Differences in estimates of aperture that are based on hydraulic and travel time data are highlighted. In the examples to be presented, it is assumed two boreholes penetrate the fracture identified for field testing. The experiment focuses on factors distorting the radial flow field. The validity of the cubic law is not examined; we view this as an important but separate issue.

Withdrawal test

Figure 5 illustrates a fracture network positioned within an impermeable rock block. This section represents a vertical slice through the center of the rock mass (at

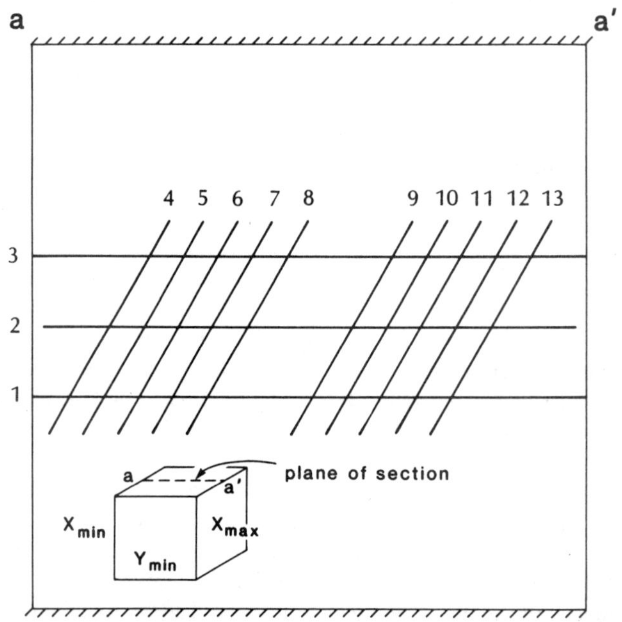

Figure 5. **Discrete fracture network used in example simulating a withdrawal test. Domain is 16m × 16m × 16m.**

Y = 8 m). The network consists of three horizontal fractures, connected by 10 cross fractures that intersect the horizontal fractures at an angle of 60°. Tests are simulated assuming the injection and withdrawal boreholes are open only to fracture plane 2. The bounding edges of this fracture are impermeable. The withdrawal well is located at the center of fracture plane 2 (with Q = 0.18 l/min). Fracture planes 1 and 3 extend to the external boundaries and are assigned equal values of hydraulic head on the X face. Fluid moves in from these constant head boundaries to meet the pumping demand of the withdrawal well. The Y face of the rock block is impermeable. Although highly idealized, this network contains features characteristic of real networks that can be expected to perturb an axisymmetric flow field. Solutions for the hydraulic head distribution and solute travel time through this three-dimensional fracture network are obtained using a model described by Smith et al. (*13*, *14*).

Field evidence suggests that, even for "open" fractures, there is considerable variation in the aperture of a planar fracture due to the combined effects of undulating fracture surfaces, mineral precipitates, and closure under in situ stresses [e.g., Abelin et al. (*15*); Brown and Scholz (*16*)]. This aperture variation leads naturally to concepts of "channelling" within a fracture network. To model rough-walled, open fractures, aperture values can be represented as a spatial stochastic process. Results for a single realization are described here. For the horizontal fractures and cross fracture 12, a lognormal distribution is specified with a mean of −3.30 and a standard deviation of 0.25 (base 10 logarithms for aperture in meters). An isotropic correlation length of 2 m is assumed. Aperture values for the other nine cross fractures are generated from LN[−5.50, 0.25]. Cross fracture 12 forms the dominant link between fracture plane 2 and the other fractures within the network. Values of the standard deviation and correlation length are selected here only to illustrate concepts, and do not represent a particular geologic environment. Similar calculations for fluid flow in a single roughwalled fracture are presented by Brown (*17*).

Figure 6a is a contour plot of the aperture variation within fracture plane 2. Note that there is a crescent-shaped channel of higher aperture located in the central region of the fracture plane.

Figure 6b is a plot of the steady state hydraulic head field within fracture plane 2. The values shown have been normalized, with the hydraulic head at the withdrawal well assigned a value of zero, and the maximum value of hydraulic head within the fracture plane assigned a value of one. The plot demonstrates that the general pattern of fluid flow is determined by the geometry of the fracture network and the connections to the constant head boundaries, with the aperture variation acting to perturb the magnitude of the local hydraulic gradient. Due to the influence of the geometry of the network, the flow field is decidedly nonsymmetric.

A plot of hydraulic aperture is shown in Figure 6c. Here, contours represent aperture values that would be calculated as a function of the location of an injection borehole. The plot is constructed by specifying a number of regularly spaced injection locations, noting the hydraulic head at each location, and using a tracking particle to trace the path and travel time from each release point to the withdrawal well. From these data, both hydraulic and tracer apertures can be calculated. Two features can be noted. First, as the location of the injection borehole is varied, the value of the hydraulic aperture does not reflect the spatial variation in aperture within the

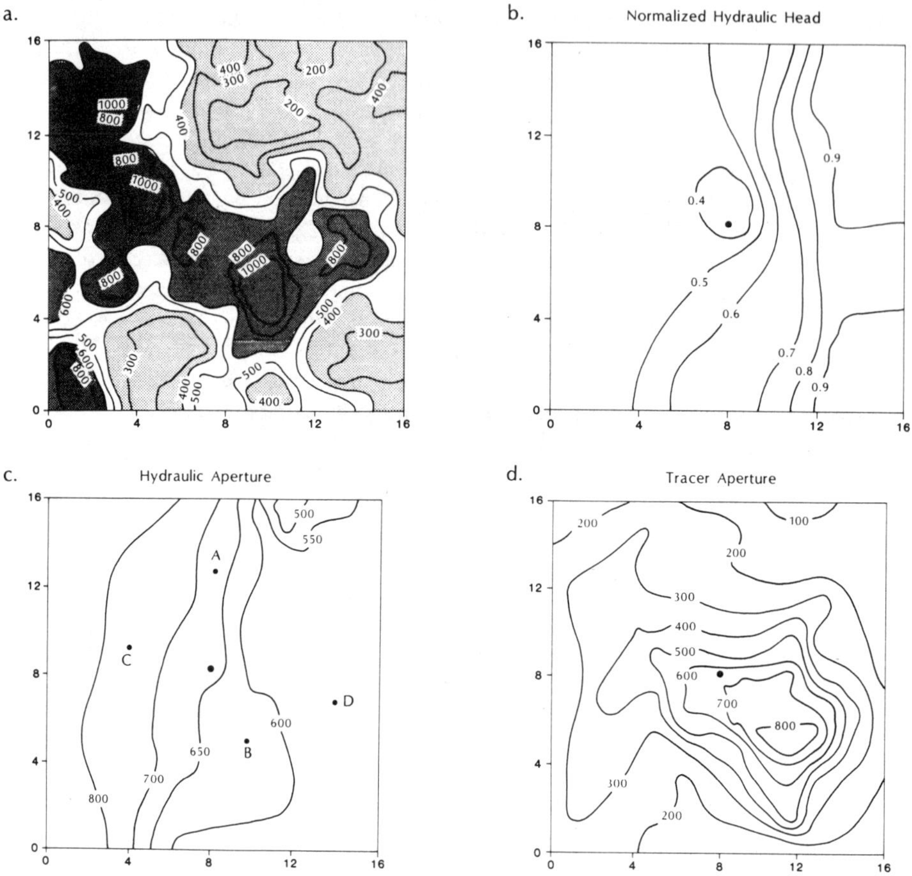

Figure 6. **Results of aperture calculations for fracture plane 2. (a) True aperture variation. (b) Normalized hydraulic head field due to fluid withdrawal from well at center of the fracture plane. (c) and (d) are estimates of aperture from hydraulic and tracer tests, respectively. See text for explanation of contour values. Aperture values in microns. Dimensions of the fracture are expressed in meters. Adapted from Smith *(18).***

fracture plane. Second, the magnitude of the hydraulic aperture is influenced primarily by aperture values in the close vicinity of the withdrawal well, this being the region in which the greatest decline in hydraulic head occurs. Other examples, not shown here, suggest the hydraulic aperture is only mildly sensitive to the distortion of the radial flow pattern due to network effects.

Values of the tracer aperture are plotted in Figure 6d. The tracer aperture is sensitive to the location of fracture intersections that provide a source of fluid to meet the pumping demand (a network effect), and to the relative positioning of the lower and higher aperture zones within the fracture plane. Estimates are a smoothed version of true values.

Figures 6c and 6d can be compared to identify relationships between aperture estimates based on hydraulic and tracer data. Consider the position of injection boreholes at points labeled A, B, C, and D in Figure 6c. Point A is located within a region of the fracture plane with a lower aperture than the mean value. Here the estimate of the hydraulic aperture would exceed the tracer aperture by approximately 400 μ. The hydraulic aperture does not record the lower aperture region in the vicinity of the injection borehole because this model reflects conditions only in the immediate vicinity of the withdrawal well. In general, the hydraulic aperture will exceed the tracer aperture if the withdrawal well is located in a zone of the fracture plane that is more permeable than the injection borehole. The travel time of a tracer released at point A records the presence of the lower aperture zone, leading to the lower value for the tracer aperture. At point B, estimates of the tracer aperture would exceed the hydraulic aperture by approximately 200 μ because the injection point is located in a zone with a higher aperture than that of the withdrawal well, and it is positioned in the main "channel" that supplies fluid to the withdrawal well. At points C and D, the hydraulic aperture exceeds the value estimated from a tracer test. For point D, the tracer aperture underestimates the true value because limited circulation within the dead-end segments of the fracture plane leads to large values for travel time, while the hydraulic aperture does not see the effects of an injection point located within a dead-end segment of the fracture.

Large differences between hydraulic and tracer apertures are possible because of the differing sensitivities to the effects of the network geometry and spatial variations in aperture. Additional examples, including an evaluation of aperture estimates based on fluid volume balance relationships, are presented by Smith et al. (*18*).

Injection-withdrawal test

Example two is a simulation of an injection-withdrawal test. The fracture network is shown in the inset diagram of Figure 7. This section represents a vertical slice through the center of the rock mass (at Y = 8 m). Except for the smaller number of cross fractures and their location, the geometry of the network and the boundary conditions on the rock block are the same as in the previous example. It is assumed that the central, horizontal fracture has been isolated at the well bores. To focus on network geometry effects, all fractures are assigned a uniform aperture of 500 μ. Figure 7 shows a plot of normalized hydraulic head when $Q_1 = Q_2$. All five cross fractures are positioned between the injection and withdrawal wells. With $Q_1 = Q_2$, sufficient fluid is supplied to the fracture plane to meet the pumping demand. However, it is possible that some fraction of the injected fluid (carrying tracer) will exchange with the "far-field" fracture network.

The estimated hydraulic aperture is 520 μ. By assuming the first arrival of tracer at the withdrawal well reflects migration along a straight line path connecting the two wells; the aperture calculated from tracer data is 860 μ. The difference in these values is due to the exchange of fluid mass at the fracture intersections, which causes reduced fluid velocities and longer transit times across the region of fracture plane 2 cut by the cross fractures, relative to that in a confined axisymmetric flow model. In this example, approximately 14% of the injected fluid exits along the constant head boundary on the left-hand side of the rock mass, while an equal volume of fluid

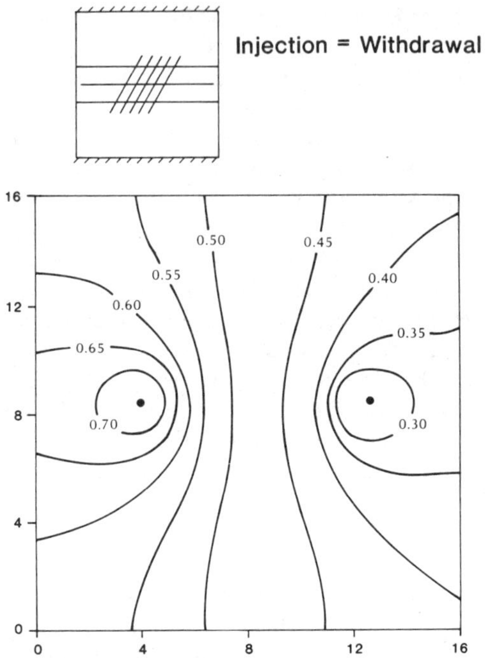

Figure 7. **Normalized hydraulic head field for the injection-withdrawal simulation. Position of wells indicated by solid circles. Upper inset plot illustrates network geometry.**

enters the fracture network and moves to the withdrawal well from the right-hand side. The greater the "leakage" into or from the cross fractures, the larger the error in the estimate of the aperture from travel time calculations. The hydraulic aperture is not sensitive to this perturbation in the radial flow pattern. When the cross fractures are assigned an aperture of 5 μ, both hydraulic and tracer tests yield values close to the true value of 500 μ. Fluid exchange with other fractures in the network is most likely to be recognized by less than 100% recovery of the tracer at the withdrawal well. Unfortunately, providing unique interpretations to explain that loss in terms of a network geometry is probably not a realistic goal at this time.

Summary

A continuum approach has been developed that can represent anisotropic dispersion in a network of fractures. The technique requires that statistics describing the character of particle transport be collected from a discrete network model of a fracture system. This discrete model is in practice a small-scale prototype of a much larger system of interest. These statistics are used to control the pattern of transport in the larger continuum and enable us to mimic in the larger system the particle transport that is observed in the smaller system.

Testing of the model has verified the accuracy of the modeling concept. We are able to duplicate patterns of anisotropic dispersion predicted by de Jong's theoretical model for transport in equidistant fractures of infinite length. Comparisons of the two procedures were conducted for a variety of different orientations of this simple network. The model was also reasonably successful in simulating the pattern of dispersion in the discrete network from which the transport statistics were collected. We have initiated experiments to examine in detail the factors that control plume shapes and anisotropic dispersion in fractured media.

Two examples were presented to demonstrate the potential for errors in estimates of fracture aperture using conventional testing and interpretation. Large differences between hydraulic and tracer apertures are possible because of differing sensitivities to the effects of network geometry and spatial variations in aperture. If aperture estimates are to be derived using results of hydraulic and tracer tests, it appears that an approach is needed to provide for additional constraints on the model adopted to describe the flow system. Such techniques conceivably require multiple borehole responses, integration with geophysical techniques that sense the migration of a tracer within the plane of a fracture, and more sophisticated interpretational models than those based on a simple axisymmetric assumption.

Acknowledgments

The work reported here has been funded by Atomic Energy of Canada Limited and the Natural Sciences and Engineering Research Council of Canada.

Literature Cited

1. Rouleau, A., and J. E. Gale, "Statistical characterization of the fracture system in the Stripa granite, Sweden," *Int. J. Rock Mech.*, *22* (1985).
2. Rouleau, A., and J. E. Gale, "Stochastic discrete fracture simulation of groundwater flow into an underground excavation in granite," *Int. J. Rock Mech.*, *24* (1987).
3. Long, J.C.S., and D. M. Billaux, "From field data to fracture network modeling: An example incorporating spatial structure," *Water Resour. Res.*, *23*(7) (1987).
4. Andersson, J., and R. Thunvik, "Predicting mass transport in discrete fracture networks with the aid of geometrical field data," *Water Resour. Res.*, *22* (1986).
5. Smith, L., and F. W. Schwartz, "An analysis of the influence of fracture geometry on mass transport in fractured media," *Water Resour. Res.*, *20*(9) (1984).
6. Schwartz, F. W., L. Smith, and A. S. Crowe, "A stochastic analysis of macroscopic dispersion in fractured media," *Water Resour. Res.*, *19* (1983).
7. Ahlstrom, S. W., H. P. Foote, R. C. Arnett, C. R. Cole, and R. J. Serne, *Multicomponent mass transport model: Theory and numerical implementation (Discrete-Parcel-Random Walk Version)*, PNL-2127, Pacific Northwest Laboratory, Richland, WA (1977).
8. Smith, L., and F. W. Schwartz, "Mass transport 1. A stochastic analysis of macroscopic dispersion," *Water Resour. Res.*, *16* (1980).
9. Long, J.C.S., J. S. Remer, C. R. Wilson, and P. A. Witherspoon, "Porous media equivalents for networks of discontinuous fractures," *Water Resour. Res.*, *18* (1983).
10. de Jong, G. de Josselin, *Dispersion of a point injection in an anisotropic porous medium*, New Mexico Tech (1972).

11. Hodgkinson, D. P., and D. A. Lever, ''Interpretation of a field experiment on the transport of sorbed and nonsorbed tracers through a fracture in crystalline rock,'' *Radioactive Waste Management and the Nuclear Fuel Cycle, 4* (1983).
12. Novakowski, K. S., Evans, G. V., Lever, D. A., and K. Raven, ''A field example of measuring hydrodynamic dispersion in a single fracture,'' *Water Resour. Res., 21* (1985).
13. Smith, L., Mase, C. W., and F. W. Schwartz, ''A model for transport in networks of planar fractures,'' Inter. Assoc. Hydraulic Res., Sympos. on the Stochastic Approach to Subsurface Flow, Paris, France, June (1985).
14. Smith, L., Mase, C. W., and F. W. Schwartz, ''A numerical model for transport in networks of planar fractures,'' Inter. Assoc. Hydrogeologists, Sympos. on Hydrogeology of Rocks of Low Permeability, Tucson, Arizona (1985).
15. Abelin, H., Neretnieks, I., Tunbrant, S., and L. Moreno, *Final Report on the Migration in a Single Fracture*, Stripa Project Rept. 85-03, Nuclear Fuel Safety Project, Stockholm, Sweden, May (1985).
16. Brown, S. R., and C. H. Scholz, ''Broad bandwidth study of the topography of natural rock surfaces,'' *J. Geophys. Res.*, *90*, 12575–12582 (1985).
17. Brown, S. R., ''Fluid flow through rock joints: The effects of surface roughness,'' *J. Geophys. Res.*, *92*(82), 1337–1347 (1987).
18. Smith, L., Mase, C. W., and F. W. Schwartz, ''Estimation of fracture aperture using hydraulic and tracer tests,'' Proc., Rock Mechanics, 28th U.S. Symposium, Tucson, Arizona, Balkema Pub., 453–463 (1987).
19. Schwartz, F. W., and L. Smith, ''A Continuum Approach for Modeling Mass Transport in Fractured Media,'' *Water Resour. Res., 24,* 8 (1988).

Equivalent Permeability of Statistically Simulated Single Fractures

KEVIN COAKLEY, K. MURALIDHAR,
JANE C. S. LONG, and LARRY R. MYER
Earth Sciences Division
Lawrence Berkeley Laboratory
University of California
Berkeley, California

Abstract

We have developed a statistical model for simulating void spaces and contact areas in a single fracture. We have identified the two-point correlation function as a connectivity statistic for the void spaces. Given an observed pattern of fracture void spaces and contact areas in the plane of the fracture (from a metal injection experiment), we simulate other statistically similar void space geometries (with variable or constant aperture distributions), which have the same two-point correlation function as the observed void space geometry. We use finite-difference procedures to compute equivalent permeabilities of the real and simulated fractures. We solve for fluid flow in these complex geometries in two ways: first by assuming Darcy's law and then by solving the three-dimensional Navier-Stokes equations for laminar flow. When assuming Darcy's law, we equate local hydraulic conductivity to aperture squared. While solving the Navier-Stokes equations, we use a coordinate transformation technique. In our present work, we assign small aperture values to contact areas in order to circumvent computer time problems. Both flow models account for complex boundary shapes. We show that the equivalent permeabilities computed with the Navier-Stokes model are always less than the equivalent permeabilities computed with the Darcy's law model. However, solution with the Navier-Stokes model is much more difficult than with the simpler and quicker Darcy's law model. Using the Darcy model, we study the variability of permeability and scale effects. We compare the equivalent permeability of simulated fractures to an observed fracture. We also present a lower bound for the variability of equivalent permeability, using Darcy's law. The bound applies to fractures with contact area and void spaces with constant apertures and incorporates knowledge of contact area percentage, scale, and spatial correlation. We then extend the bound to the case where voids have variable aperture distributions.

Introduction

To understand flow through a single fracture, we have modeled void space and contact area geometry and then solved the equations of flow within the defined space. Rather than assuming a simplified or idealized void space geometry, we simulate void space geometries that are statistically similar to an observed void space geometry as observed in the fracture. We solve for flow and determine equivalent permeability with the commonly used Darcy's law, as well as the Navier-Stokes equations. We compare equivalent permeabilities computed in each flow model to determine the relative advantage of solving the fundamental, but computationally intensive, Navier-Stokes equations rather than solving the computationally fast, but empirical, Darcy's law. We also study how fracture permeability variability and anisotropy depend on scale, using Darcy's law. We also present a lower bound for the variability of equivalent permeability, using Darcy's law, based on the Power Average Model (Journel et al., 1986). The bound is for fractures with contact area and void spaces with constant apertures. The bound incorporates knowledge of contact area, scale, and correlation. We then extend the bound for variable void space aperture distributions using a critical path analysis (Charlaix et al., 1987; Katz and Thompson, 1986).

Geometry

Previous models for single fracture void space geometry range from the simple parallel plate (Witherspoon et al., 1980) to the more complicated equivalent resistor network (Tsang, 1984) and models based on fractals (Brown, 1987; Nolte et al., 1987), which also give complex geometries.

We have developed a statistical model of fracture void space geometry based on an observed pattern of voids in the plane of a fracture void space from a metal injection experiment (Pyrak-Nolte et al., 1987). We compute the two-point correlation function of the void space and simulate statistically similar void spaces that have the same two-point correlation function. The two-point correlation function (Berryman and Blair, 1986), $S_2(\vec{r})$gives the probability that two points, separated by $\vec{r}$, both lie in the void space. In our model, we assume an isotropic two-point correlation function. The two-point correlation function and the correlogram (Journel and Huijbregts, 1978) give equivalent information. As a consistency check, we compare histograms of contact area island sizes for the observed and simulated geometries. Given information about the aperture distribution, we could also simulate variable aperture void space geometries. Before describing the details of the simulation, we first describe our data.

Data

Our data come from a metal injection experiment (Pyrak-Nolte et al., 1987) performed on a cylindrical core sample of quartz monzonite from the Stripa mine in Sweden. The core measured 11.6 cm in diameter and 18.4 cm in height. The sample contained a single natural fracture orthogonal to the long axis of the core. Hot molten Wood's metal was injected into the voids of the fracture, which was under a

10 MPa stress. This corresponds to an in situ stress at 300 m below the surface. After the metal solidified, the fracture was taken apart and both the upper and lower fracture surfaces were photographed. Because the metal stuck to either the upper or lower fracture surface, we superimposed two images to get our composite void space geometry. We used a Zeiss Image Analysis System to digitize the images of bright metal on darker rock. The fracture was discretized into pixels with side lengths of 0.21 mm.

We classified each pixel as a void space or contact area depending on whether the pixel's grey level intensity was above or below a chosen threshold. We chose a threshold that gave a digital image with roughly the same shapes and contact area percentage as the visual images of metal on rock. However, we have not studied how to select the best threshold in this work. Sources of error include nonuniform illumination of the fracture; rotation, tilt, and translation of the two fracture surfaces relative to each other before superimposition; subjective choice of threshold; and classification errors due to void spaces having shapes different from our square pixels. We have not studied these errors in this work.

We took the inner section of the fracture, 128 pixels by 128 pixels, as our data set. The observed void space geometry, shown in Figure 1, has 18.3% contact area.

Simulation

We simulate void spaces with constant or variable aperture distributions by clipping a correlated Gaussian process defined on a 128 by 128 pixel geometry. Our method is similar to other indicator simulation techniques (Journel and Isaaks, 1984).

In our simulation, we first assign an independent Gaussian random variable to each pixel. We then convolve this "white noise" process with a radially symmetric filter that has exponentially decaying weights. Hence, a moving average filter assigns the $(i,j)^{th}$ pixel a new Gaussian random variable, which is the weighted average of all the independent Gaussians previously assigned to pixels within a circular neighborhood of the $(i,j)^{th}$ pixel. We then clip our correlated Gaussian process; pixels with random variables above a chosen cutoff are void and the remaining are contact. We summarize as follows:

1. Assign each pixel (i,j) (i=1, 128; j=1, 128) an independent Gaussian X_{ij}. Each Gaussian has mean zero and variance one.
2. Convolve X with a filter H to get correlated Gaussian process Y.

$$Y_{ij} = \sum_{k=-D}^{D} \sum_{l=-D}^{D} X_{i-k\ j-l}\, H_{kl} \tag{1}$$

where

$$H_{kl} = \begin{cases} C \exp(-\beta\sqrt{k^2 + l^2}) & \text{for } \sqrt{k^2 + l^2} \le D \\ 0 & \text{otherwise} \end{cases} \tag{2}$$

The normalization factor, C, is chosen so that each of the correlated Gaussians, Y_{ij} (i = 1, 128; j = 1, 128), has mean zero and variance one.

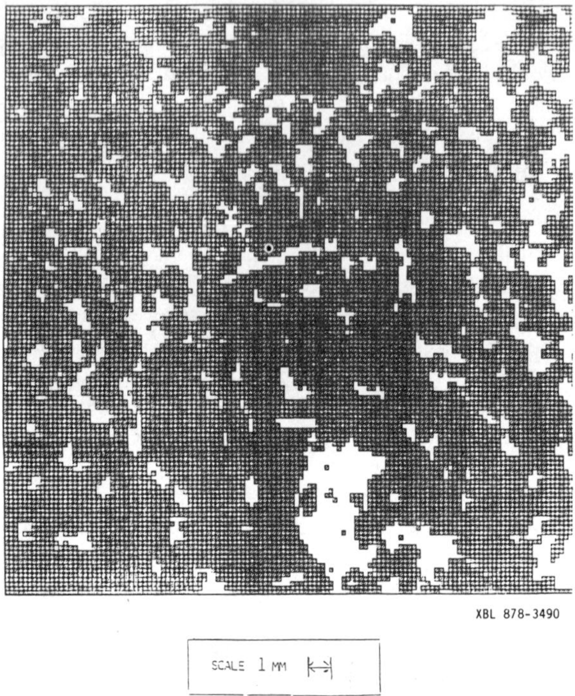

Figure 1. **Observed void space geometry, 128 by 128 pixels. (1 pixel length is 0.21 mm. Contact area = 18.3%. Voids are black pixels. Data are from metal injection experiment [Pyrak-Nolte et al., 1987].)**

3. Clip

$$I_{ij} = \begin{cases} 1 & Y_{ij} \geq Y_0 \text{ void} \\ 0 & Y_{ij} < Y_0 \text{ contact} \end{cases} \tag{3}$$

In our simulation, the radius of our circular neighborhood, D, is 7 pixel lengths. The exponential damping factor, β, which determines how strongly to weight Gaussians assigned to neighboring pixels is 0.75 (inverse pixel length). We arrived at these values by trial-and-error comparison of the two-point correlation functions of the observed and simulated void space geometries. The value of 0.75 for β gave better results than did values of 1.0 or 0.5 for β. However, we have not carefully studied what range about 0.75 would give satisfactory results. Different β values led to different conclusions in this work. We found that for $\beta = 0.75$, a larger circular neighborhood with $D = 10$ gave very similar results. However, when we used a smaller circular neighborhood with $D = 3$, we got noticeably worse results.

Because the metal injection experiment (Pyrak-Nolte et al., 1987) reveals only connected voids, the two-point correlation function of the observed void spaces may differ from the two-point correlation function of all the voids. However, when simulating void space geometries with the same contact area as the observed void space geometry, 18.3%, only a small percentage of isolated voids, i.e., voids with no immediate north, south, east, or west neighbors, results—perhaps 0.1%. Hence, in our model the statistics of the observed voids and the statistics of all the voids differ insignificantly. Nonetheless, we remove isolated void spaces in our simulation using the Hoshen-Kopelman (1976) algorithm. The algorithm identifies all the distinct clusters of void space pixels using a method that relabels interconnected clusters.

We simulated three void space geometries statistically similar to the observed void space geometry (see Figs. 2, 3, 4). By adjusting the clipping level, all contact area percentages agreed. We also compared the two-point correlation functions (see Fig. 5) and contact area island size histograms (see Fig. 6).

To get a variable aperture distribution, we transformed the random variables assigned to the void pixels. In our example, we chose the exponential transform so as to get a correlated lognormal void space aperture distribution. We chose this transformation for convenience and as a first step. The next step will be to obtain data on the void space aperture histogram and spatial correlation function. These will dictate which transformation to use. However, casts of single fractures made by Gale (1987) do suggest a lognormal distribution for void space apertures.

We assign apertures to the $(i,j)^{th}$ pixel as follows:

$$b_{ij} = \begin{cases} \exp(\mu + \sigma Y_{ij}) & \text{when } I_{ij} = 1 \text{ void} \\ 0 & \text{when } I_{ij} = 0 \text{ contact} \end{cases} \quad (4)$$

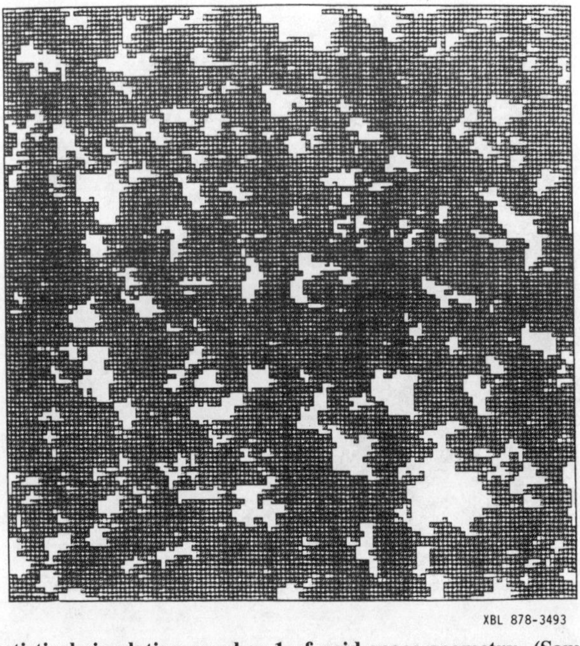

Figure 2. **Statistical simulation number 1 of void space geometry. (Same scale as in Figure 1. Contact area = 18.3%.)**

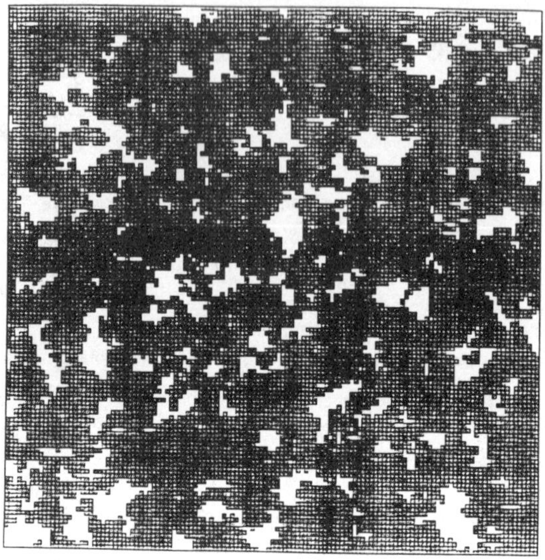

Figure 3. **Statistical simulation number 2 of void space geometry. (Same scale as in Figure 1. Contact area = 18.3%.)**

Figure 4. **Statistical simulation number 3 of void space geometry. (Same scale as in Figure 1. Contact area = 18.3%.)**

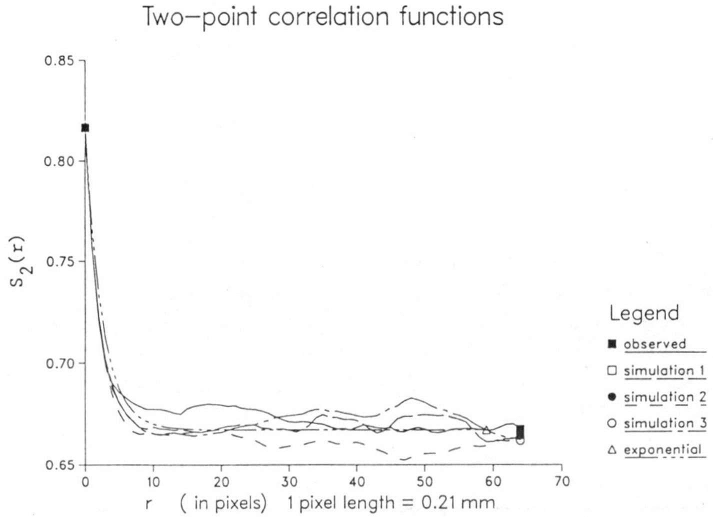

Figure 5. **Two-point correlation functions of observed (solid line) and three simulated void space geometries. (We also plot exponential fit function $p^2 + p(1 - p)e^{-\lambda r}$ where $\lambda = 0.4$, $p = 0.817$.)**

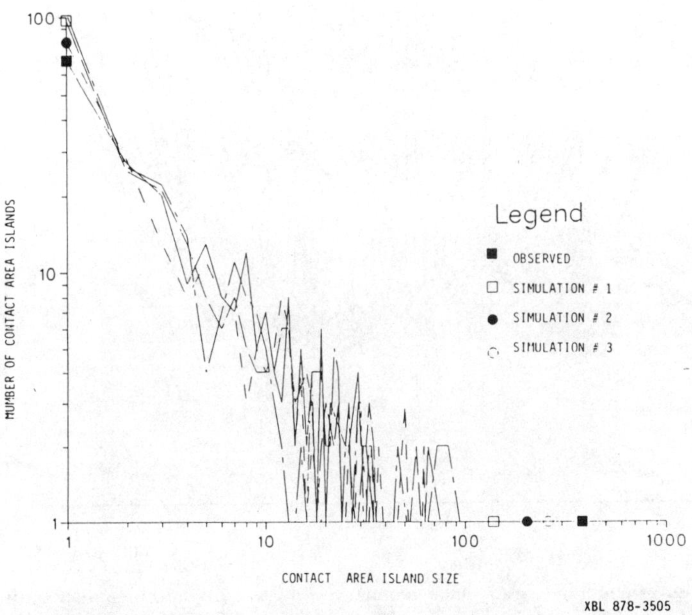

Figure 6. **Contact area island size histograms for observed and simulated void space geometries.**

In our simulation, we take $\mu = 0.1$ and $\sigma = 0.25$ (see Fig. 7). In Figure 7, contact areas are elevated and large void space apertures are depressed. This convention allows a better perspective of the geometry. Contact area is 17.7%. Unlike for the simulated void space geometries with constant apertures, we did not remove isolated voids.

Flow

Previous efforts to improve understanding of flow through a single fracture have usually emphasized better models of void space geometry but not improving upon the commonly used Darcy's law flow model (Tsang, 1984). Brown (1987) used the Reynolds equation to study flow in a fracture, but the Reynolds equation does not contain any viscous terms. In our work, we try to model both the details of void space geometry and the details of flow. We solve for flow using Navier-Stokes equa-

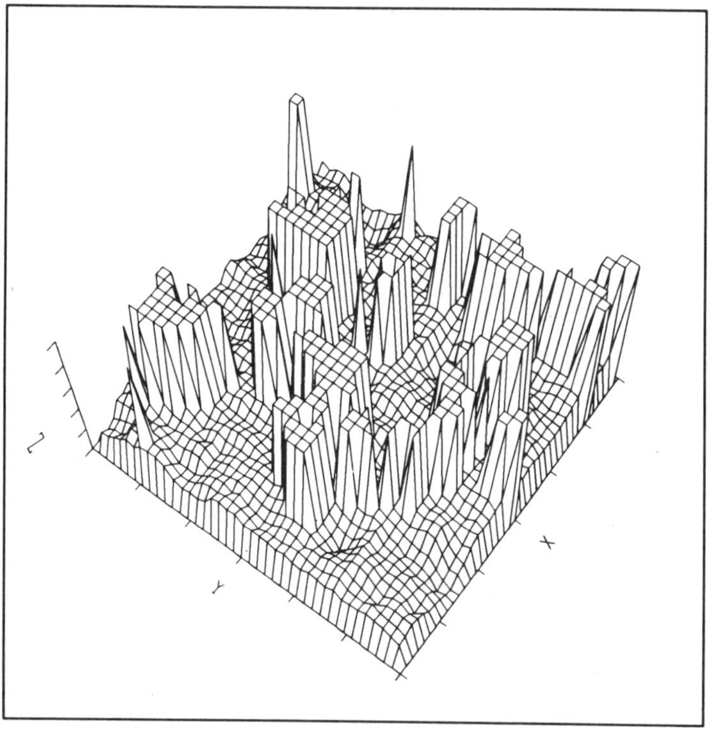

XBL 878-3498

Figure 7. **Plot of simulated void space geometry with variable aperture distribution. (Data set is not tapered near boundaries. Fracture geometry is 41 by 41 pixels; length = 6. Contact areas appear as plateaus. Void space apertures range from 0.64 to 1.0.)**

tions rather than just Darcy's law. This is an improvement because the Navier-Stokes equations are based on Newton's second law for fluids (Schlichting, 1979), whereas Darcy's law is empirical in nature.

We have developed a finite difference scheme to solve the Navier-Stokes equations. We note that recently, Rothman (1988) introduced a cellular-automaton approach to solve the Navier-Stokes equations in two dimensions for complex geometries that should be compared to our technique.

We compare equivalent permeabilities computed in the Navier-Stokes flow model and the Darcy's law flow model for a variety of complex geometries.

Governing equations

The Navier-Stokes equations in coordinate free form are given as follows:

$$\underset{\sim}{u} \cdot \underset{\sim}{\nabla}\, \underset{\sim}{u} = -\underset{\sim}{\nabla}\, p + \frac{1}{Re} \nabla^2 \underset{\sim}{u}\ , \tag{5}$$

$$\underset{\sim}{\nabla} \cdot \underset{\sim}{u} = 0, \tag{6}$$

where

Re = Reynolds number = $\dfrac{\sqrt{\Delta p}}{\rho} \dfrac{h}{\nu}$

Δp = pressure difference driving flow

ρ = density of fluid

ν = kinematic viscosity of fluid

h = characteristic dimension such as the maximum aperture

In fracture hydrology applications, the Reynolds number is expected to fall below 20.

To solve the Navier-Stokes equations, we replace the continuity equation (Eq. (6)) by Poisson's equation of pressure

$$\nabla^2 p = \frac{1}{Re} \nabla^2 D - \underset{\sim}{\nabla} \cdot (\underset{\sim}{u} \cdot \underset{\sim}{\nabla}\, \underset{\sim}{u}) \tag{7}$$

where

$$D = \underset{\sim}{\nabla} \cdot \underset{\sim}{u}$$

We then map complex flow domains into simple ones with a coordinate transformation (Thompson, 1982; Muralidhar and Long, 1987). We solve the Navier-Stokes equations in the transformed domain using a finite difference method. We invert the matrix arising from the numerical scheme by the Gauss-Seidel method. Because the size of our finite-differences mesh influences results, we refine the mesh until equivalent permeability stabilizes. Computation in three dimensions is very expensive. If we can simplify the Navier-Stokes equations, we may arrive at equations that are practical to solve.

We can simplify if:

1. Nonlinear acceleration terms, given by $\underset{\sim}{u} \bullet \underset{\sim}{\nabla} \underset{\sim}{u}$ in Equation (5), are generally small.
2. Tortuosity due to variable aperture is small compared to tortuosity of flow going around contact areas, that is, if the component of velocity normal to the fracture plane is small compared to the in-plane velocities.
3. The fluid velocity quickly attains a parabolic shape (with an adjustable mean) across the aperture. The length over which this happens is called the development length and is less than Re/20 for laminar flow (Schlichting, 1979). Schlichting shows how an initially uniform velocity profile becomes parabolic for a parallel plate geometry. For a plate separation h, the length over which this happens, L, is $L/h \approx Re/20$. In our work, the characteristic length h is the maximum aperture, which is always 1.

These observations suggest a reduction of the full equations governing flow based on the following assumptions:

1. Inertial terms, $\underset{\sim}{u} \bullet \underset{\sim}{\nabla} \underset{\sim}{u}$, are identically zero.
2. The component of velocity normal to the fracture plane (defined by x and y coordinates) is zero.
3. The velocities u and v in the fracture plane can be expressed as,

$$\begin{aligned} u(x,y,z) &= 6\bar{u}(x,y)\, z\, (h-z)/h^2 \\ v(x,y,z) &= 6\bar{v}(x,y)\, z\, (h-z)/h^2 \end{aligned} \tag{8}$$

where h is the local aperture.

Assumptions (1) and (2) lead to a fracture permeability always larger than the full Navier-Stokes permeability because inertial terms and additional tortuosity, which would lead to additional pressure drop, are ignored. Our fracture geometries have length 6 and maximum aperture of 1. Hence the development length, for Re = 5, is 0.25, which is much less than the fracture length of 6. Our geometries are not as simple as the parallel plate geometry and the validity of assumption (3) should be checked; however for some geometries more complex than the parallel plate geometry, Muralidhar and Long (1987) verified the parabolic velocity assumption.

Integrating the Navier-Stokes equations across the fracture from 0 to h, we obtain the Reduced Navier-Stokes equations:

$$\begin{aligned} \frac{1}{Re}\left(u_{xx} + u_{yy} - \frac{12u}{h^2} \right) &= hP_x \\ \frac{1}{Re}\left(v_{xx} + v_{yy} - \frac{12v}{h^2} \right) &= hP_y \end{aligned} \tag{9}$$

$$\frac{\partial}{\partial x}(hP_x)+\frac{\partial}{\partial y}(hP_y) = \frac{1}{Re}\left(\frac{24}{h^3}\right)(uh_x + vh_y)$$

where u and v are now local fluxes.

We recover Darcy's law by further neglecting the in-plane tortuosity given by terms u_{xx}, u_{yy}, v_{xx}, v_{yy}. The pressure equation for darcian flow is then:

$$\frac{\partial}{\partial x}(h^3 P_x) + \frac{\partial}{\partial y}(h^3 P_y) = 0 \tag{10}$$

We call this a Darcy's law flow model because flow is linearly related to pressure drop. This equation is also known as the Reynolds Lubrication equation. Our Darcy's law model is a local cubic law (Witherspoon et al., 1980) model for flow.

We compare the Navier-Stokes (Eqs. (5) and (6)), Reduced Navier-Stokes (Eq. (9)), and Darcy's law (Eq. (10)) flow models by computing the equivalent permeability of some moderately complex test geometries. We then compare the Navier-Stokes and Reduced Navier-Stokes flow models by computing equivalent permeability for a more complex geometry. For the more complex geometry, we did not solve the Navier-Stokes equations because the solution required too much computer time.

We expect that the different flow models will give different equivalent permeabilities. For example, Muskat (1946) shows that for flow through sands, a Darcy's law model applies only when Re is equal to or less than 1. This may also be true in our geometry but it is important to note that our definition of Re differs from Muskat's. Muskat used the average grain diameter of the sands to define a characteristic length, whereas we used the maximum aperture to define a characteristic length. Thus for the same velocity and channel size, Muskat's Re is most likely smaller than ours. In this work, we solve for flow with Re equal to 5. So, we must examine our results to see if Darcy's law applies at Re = 5.0.

To solve these flow equations, (9) or (10), we must specify boundary conditions. For consistency, we impose the same boundary conditions for all the geometries we study.

In the Navier-Stokes computations, we set $\underset{\sim}{u} = 0$ on impermeable boundaries including the surface of the contact areas and the bounding surfaces of the fracture that determine the aperture. We specify pressure drop across the fracture and determine local pressure gradients from the momentum equations. On inflow boundaries, $\underset{\sim}{u} = (u_m, 0, 0)$ where u_m is the mean flux through the fracture (determined iteratively). For the outflow boundary, we have the gradient condition, $\underset{\sim}{\nabla}\, \underset{\sim}{u} \cdot \underset{\sim}{n} = 0$, where $\underset{\sim}{n}$ is the normal vector for the outflow boundary.

In the Darcy computations, we specify pressure drop across the fracture and forbid flow to pass through the fracture side boundaries, i.e., $\underset{\sim}{u} = (u_x, 0, 0)$ on the sides. In all flow computations, we apply the pressure drop in either the east-west or north-south direction.

We compute relative equivalent permeabilities in all cases. For a particular geometry, we first normalize each aperture by dividing it by the maximum aperture. Hence, the normalized apertures range from zero to one. After applying a dimensionless unit pressure drop across the fracture, which has dimensionless length six, we compute the flow, Q, through the fracture. We then compute the flow, Q_{MAX} through the fracture with all apertures set to one. We normalize the equivalent permeability.

$$K_{eq} = \frac{Q}{Q_{MAX}}$$

Thus, normalized equivalent permeability ranges from zero to one.

To demonstrate our method, we solved for flow using the Navier-Stokes equations for a geometry consisting of four constrictions (Fig. 8). In Figure 8, large apertures are elevated above the smaller apertures. After generating the grid (Fig. 9), we solved for the velocity field (Fig. 10).

We then compared the different flow models, Navier-Stokes, Reduced Navier-Stokes, and Darcy, by solving for flow and equivalent permeability for the same four-constriction geometry. We varied the pinch, i.e., minimum aperture, and found that the flow models disagree more as minimum aperture decreases (see Fig. 11), and tortuosity in the vertical plane becomes more important.

We also solved for flow using the Reduced Navier-Stokes equations and Darcy's law for a tapered version of the variable void space geometry that we simulated (Fig. 7). We did not solve the Navier-Stokes equations because of time considerations.

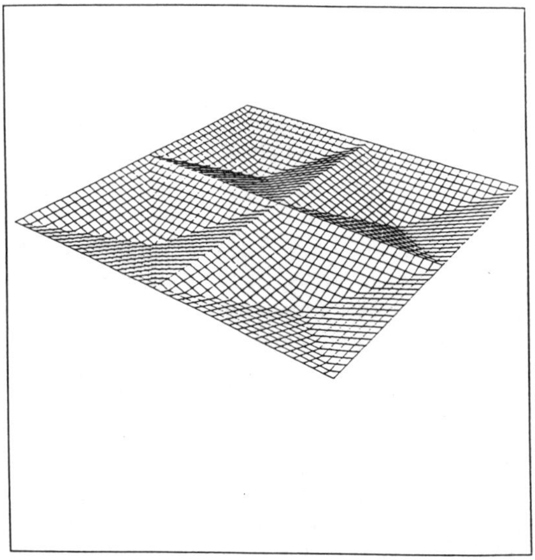

XBL 878-3499

Figure 8. **Three-dimensional view of aperture distribution of geometry with four constrictions. (Larger apertures appear elevated.)**

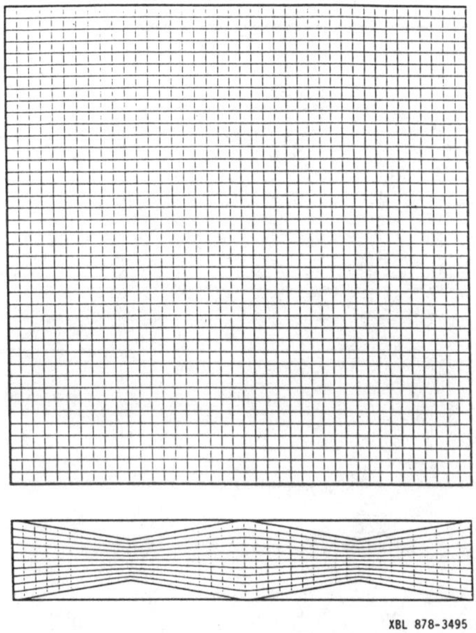

Figure 9. **Grid generation for four-constriction geometry.**

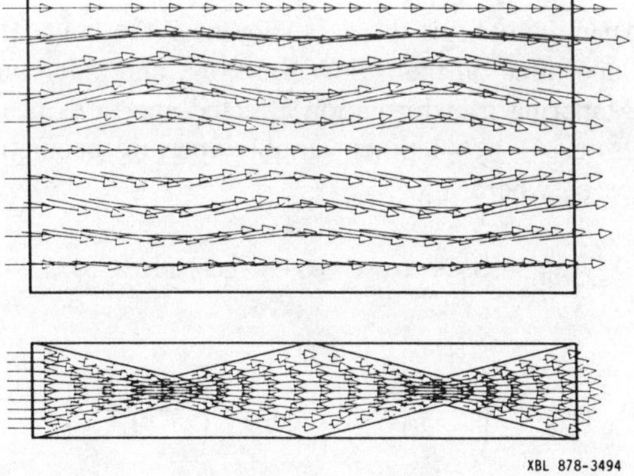

Figure 10. **Velocity vector plot of fluid flow through geometry with four constrictions (Figure 8). (Minimum opening = 0.28. Maximum opening = 1.0. Fracture length = 6. Reynolds number = 5. Navier-Stokes equations are solved.)**

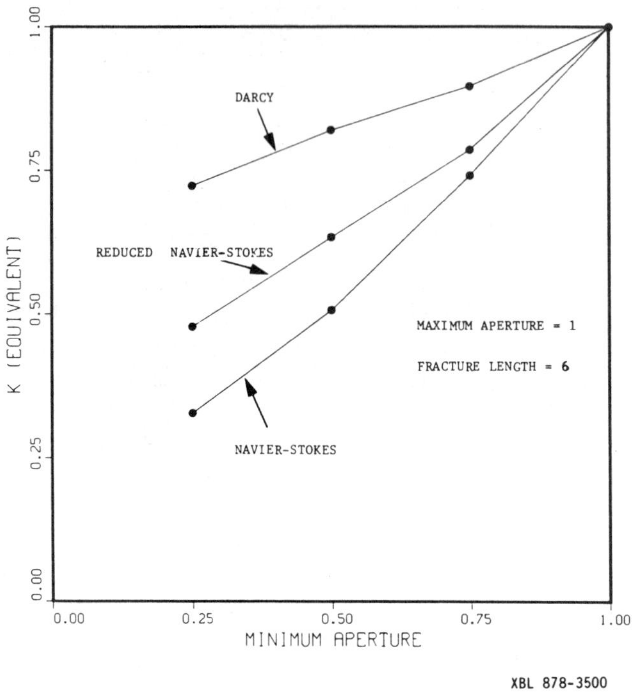

Figure 11. **Plot of equivalent permeability as a function of minimum aperture of geometry with four constrictions. (Comparison of three flow models—Navier-Stokes, Reduced Navier-Stokes, and Darcy's law.)**

Without tapering, we could not get a numerically stable solution of the Reduced Navier-Stokes equations for the geometry shown in Figure 7. This failure may be due to the difficulty of satisfying our boundary condition, $\underset{\sim}{\nabla}\ \underset{\sim}{u} \cdot \underset{\sim}{n} = 0$, for a complex geometry. Tapering simplifies the geometry near the boundary.

After tapering, apertures on the inflow, outflow, and side boundaries were equal to one exactly. The tapering transformation affected apertures near the boundaries most. After tapering our 41 by 41 pixel variable aperture geometry, the $(i,j)^{th}$ pixel's aperture, b_{ij} becomes b_{ij}^T where

$$b_{ij}^T = b_{ij} + (\alpha_i + \alpha_j - \alpha_i\alpha_j)(1 - b_{ij});$$

with

$$\alpha_i = \left(\frac{i - 21}{20}\right)^6, \quad \alpha_j = \left(\frac{j - 21}{20}\right)^6$$

for

$$i = 1, 41 \text{ and } j = 1, 41.$$

On the boundaries, either i = 1, i = 41, j = 1, or j = 41. Hence, on the boundaries, $(\alpha_i + \alpha_j - \alpha_i \alpha_j)$ equals one and tapered apertures equal one. But for apertures closer to the middle of the fracture, $(\alpha_i + \alpha_j - \alpha_i \alpha_j)$ is very small and the tapered aperture is almost the same as the original aperture. Figure 12 shows the tapered geometry. For better viewing, contact areas appear as plateaus and large apertures are depressed.

For the Darcy's law flow computation, we set contact area pixel apertures to zero. When solving the Reduced Navier-Stokes equations, we could not directly assign to contact area apertures zero values. Instead, we gradually reduced contact area apertures from a large value, 0.64, to a very small value, 0.01, and extrapolated to find the equivalent permeability for zero-valued contact area apertures. We did this to ensure a numerically stable solution.

Surprisingly, we found that the equivalent permeabilities determined by the two flow models, Darcy's law and the Reduced Navier-Stokes equations, were close, 0.4883 and 0.4803, respectively. We note that Darcy's law allowed larger velocities to develop inside the fracture (Fig. 13) than did the Reduced Navier-Stokes equations (Fig. 14). Fluid velocities and equivalent permeability should be greater when using the Darcy's law model than when using the Reduced Navier-Stokes equations because Darcy's law ignores tortuosity which the Reduced Navier-Stokes accounts for. The full Navier-Stokes solution is expected to give permeabilities less than 0.4803 because of additional pressure drop due to additional tortuosity and inertial terms.

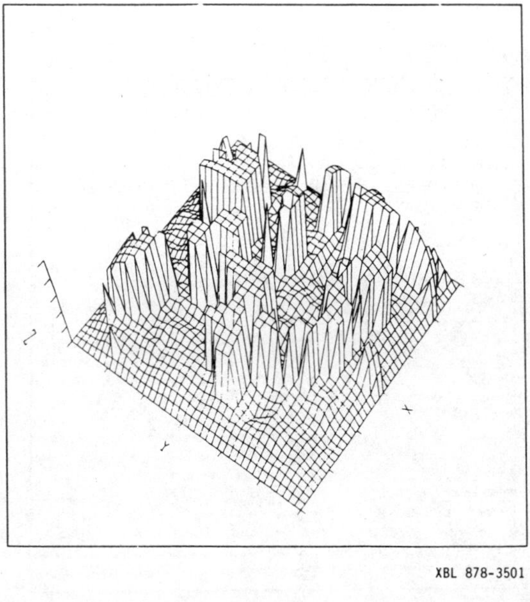

Figure 12. **Plot of simulated void space geometry with variable aperture distribution. (Data set is tapered near boundaries. Fracture geometry is 41 by 41 pixels; length = 6. Contact areas appear as plateaus. Void space apertures range from 0.64 to 1.0.)**

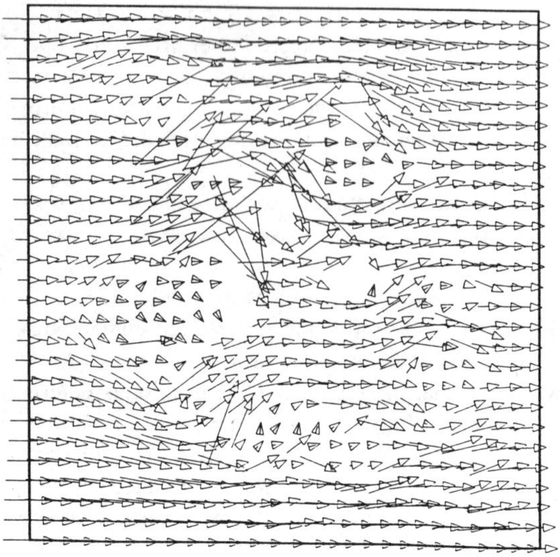

Figure 13. **Velocity vector plot of fluid flow in simulated variable aperture void space geometry (Figure 12) using Darcy's law. (K_{eq} = 0.4883.)**

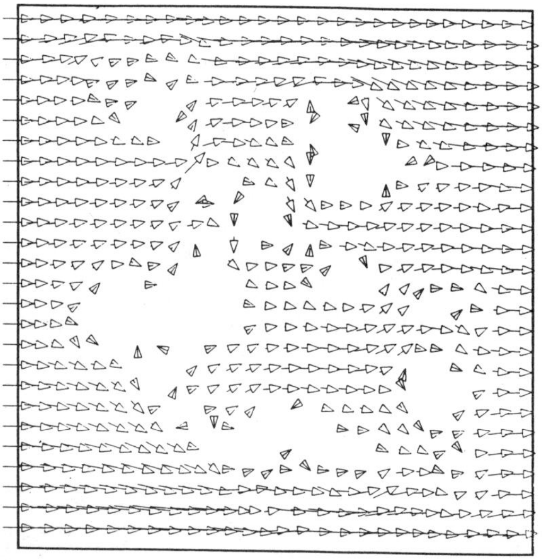

Figure 14. **Velocity vector plot for simulated variable aperture void space geometry (Figure 12) using reduced Navier-Stokes equations. (K_{eq} = 0.4803.)**

Statistics of Equivalent Permeability

Consistency check

Earlier, we showed that the two-point correlation functions and the contact area histograms for the observed and simulated void space geometries matched well (Figs. 5 and 6). As a further consistency check, we assigned to voids unit-valued apertures and to contact areas zero-valued apertures and solved for flow, using Darcy's law, for the simulated and observed void space geometries.

For each geometry, we computed two equivalent permeabilities; one corresponded to an east-west pressure drop and the other corresponded to a north-south pressure drop. We found that computed equivalent permeability, K_{eq}, depended on the size of the finite-difference equation mesh. In Figure 15, we plot K_{eq} against mesh size for one of the simulated geometries. Based on this plot, we conclude that a mesh size of 255 by 255 determines K_{eq} to a precision of a few percent.

Figure 16 shows computed K_{eq}'s for the simulated and observed void space geometries at meshes of size 128 by 128 and 255 by 255. In Table 1, we list K_{eq}'s computed with a 255 by 255 mesh. Based on Figure 16 and Table 1, we conclude that our simulation model gives geometries consistent with the observed one; the K_{eq}'s are consistent.

Analytic bounds for equivalent permeability

Upper and lower bounds, using a Darcy's law model, for equivalent permeability (Beran, 1968) tend not to be useful for conductances that have broad distributions (Berryman and Blair, 1986; Madden, 1976). Madden (1976) describes how some of these bounds, based on energy dissipation minimization principles, lead to the arithmetic mean and harmonic mean of the conductance distribution as upper and lower bounds. Berryman and Blair (1986) discuss how the two-point correlation function bounds permeability for systems with two-point correlation functions isotropic in three dimensions. The bound is

$$K \le \frac{2}{3}\left(\int_0^\infty dr\ r\ [S_2(r) - p^2]/(1 - p)^2 \right)$$

where K is permeability, $S_2(r)$ is the two-point correlation function, and p is porosity. However, because our system is not isotropic in three dimensions, we cannot apply this result to our problem. Therefore, numerical simulations offer a better route to study the variability of equivalent permeability.

Variability and scale

Average equivalent permeability and the variability of equivalent permeability depend on scale (Long et al., 1982). Desbarats (1986) studied the equivalent permeability of sandstone-shale distributions using a Darcy flow model. He discretized areas into pixels, or volumes into cubes, and assigned unit conductances or very small conductances to the pixels or cubes using a geostatistical correlation structure. He showed that as the side length of the area or volume of his sample increased relative to the correlation length, equivalent permeability stabilized. Further, Desbarats found

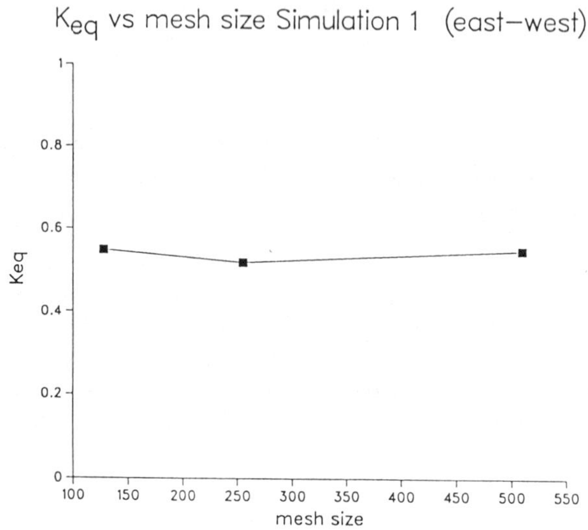

Figure 15. **Plot of equivalent permeability against finite-difference mesh size for simulation 1 geometry (Figure 2). (We apply an east-west pressure drop. Darcy's law is used. Voids have unit conductances and contact areas have zero-valued conductances.)**

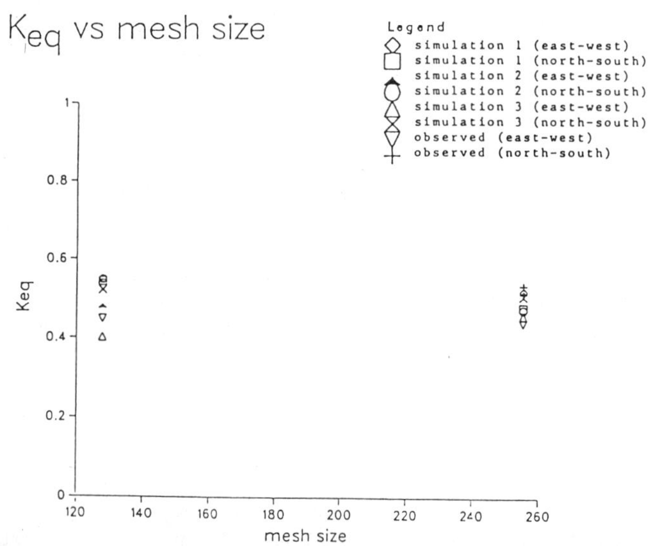

Figure 16. **Equivalent permeability of observed (Figure 1) and simulated (Figures 2, 3, 4) void space geometries against size of mesh used in finite-difference solution of Darcy's law. (Voids have unit-valued conductances and contact areas have zero-valued conductances.)**

Table 1. **K_{eq} for simulated and observed geometries with 255 by 255 mesh.**

	K_{eq} (East/West)	K_{eq} (North/South)
Observed	0.4337	0.5338
Simulation 1	0.5188	0.4771
Simulation 2	0.5223	0.4709
Simulation 3	0.4561	0.5054

that the coarser he discretized areas of volumes, the more he underestimated equivalent permeability.

We study how the variability of equivalent permeability depends on scale for our simulated fracture void space geometries. In general, there are two sources of variability of K_{eq}. First, for a pattern with a fixed contact area percentage, the spatial arrangement of the contact area pixels determines equivalent permeability. Hence, variability of the empirical correlation structure of simulated patterns leads to variability of K_{eq}. Second, average K_{eq} depends on contact area percentage. Hence, variability of contact area percentage is also a source of variability of K_{eq}. We study both sources of variability in this work.

To study the first source of variability, we compute K_{eq} for patterns with the same contact area percentage. We study the variability of equivalent permeability for 10 by 10 pixel, 30 by 30 pixel, and 64 by 64 pixel patterns. Each of these patterns is a subset of a large pattern that we simulated with our clipped correlated Gaussian process model.

To get our patterns, we first simulated large 200 by 200 pixel void space geometries using our clipped correlated Gaussian method. From each large 200 by 200 pixel geometry, we selected nonoverlapping subareas, with 20% contact area, that covered areas of 10 by 10 pixels and 30 by 30 pixels.

To get geometries of size 64 by 64 pixels, we partitioned the 128 by 128 pixel simulated geometry shown in Figure 2 into 4 quadrants. Each quadrant had contact area nearly, but not exactly, equal to 20%. Because we are studying variability at fixed contact area percentage, we adjusted the computed K_{eq}'s after computing a power average (Journel et al., 1986) trend (Fig. 17). We computed the residuals about the power average trend. We added the residuals to the trend, evaluated at 20% contact area, to get the adjusted equivalent permeabilities.

For each geometry, we computed K_{eq} for both an east-west and a north-south pressure drop. We define the magnitude of the difference between the north-south and east-west K_{eq}'s as our anisotropy statistic. The simulated geometries, on average, have isotropic correlation structure because we used a radially symmetric moving average filter in our simulation. Hence, we expect anisotropy to diminish as size increases. For small-scale samples, there might be significant anisotropy.

In Table 2, we list average K_{eq} ($<K_{eq}>$), the standard deviation of K_{eq} ($\sigma_{K_{eq}}$), and the average anisotropy (ΔK_{eq}) at each scale. In Figures 18 and 19, we plot K_{eq} and anisotropy of K_{eq} at three scales. As anticipated, our study shows the variability of K_{eq} and the anisotropy of K_{eq} diminish as fracture size increases.

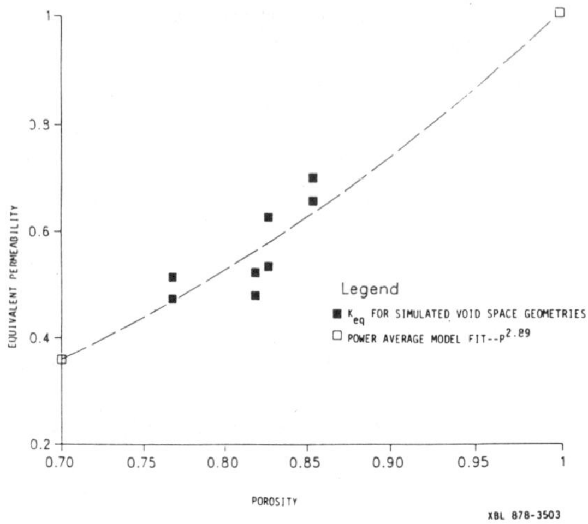

Figure 17. **Power average model fit to equivalent permeabilities of 64 by 64 pixel geometries. (Each geometry has contact area close to 20%. Trend is $K_{eq} \approx p^{2.89}$. Darcy's law is used. Voids have unit conductances and contact area conductances are zero.)**

Lower bound for variability of equivalent permeability

In the earlier section, we studied the variability of equivalent permeability for simulated geometries with 20% contact area. For a fixed clipping level, our simulation model gives patterns with contact area percentage which fluctuate about some average. In this section, we focus on how fluctuations in contact area percentage propagate through as fluctuations in K_{eq}. We will present lower bounds for the variability of equivalent permeability by studying theories that predict equivalent permeability as a function of contact area percentage. The bounds are lower bounds because we are not incorporating the variability of K_{eq} due to the many ways to arrange a fixed percentage of contact area pixels within a pattern.

Our first lower bound for variability of K_{eq} is based on the Power Average Model (Journel et al., 1986). We first consider voids that all have constant aperture and then extend our results to the variable aperture case using a critical path analysis (Charlaix et al., 1987; Katz and Thompson, 1986).

Table 2. **Variability of K_{eq} and anisotropy of K_{eq} at 20% contact area.**

Scale	$<K_{eq}>$	$\sigma_{K_{eq}}$	$<\Delta K_{eq}>$	No. of Realizations
10 by 10	0.468	0.181	0.200	6
30 by 30	0.492	0.088	0.074	7
64 by 64	0.524	0.049	0.052	4

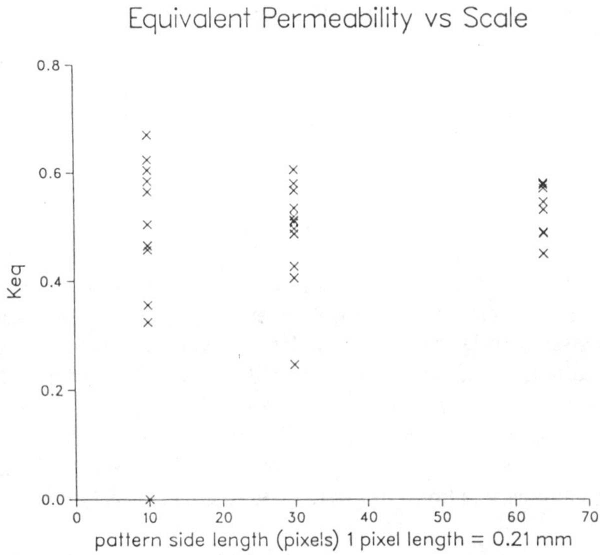

Figure 18. **Plot of equivalent permeability of N by N pixel simulated geometries. (N = 10, 30, 64. Darcy's law is used. For each geometry, two K_{eq}'s are computed. We apply an east-west pressure drop as well as a north-south pressure drop.)**

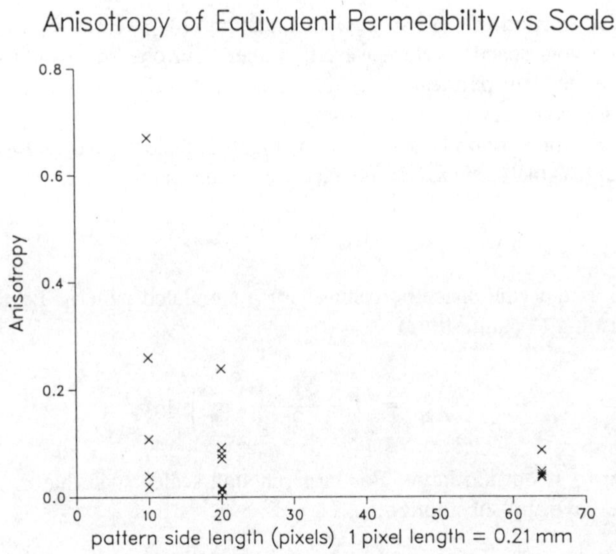

Figure 19. **Plot of anisotropy of equivalent permeability for N by N pixel simulated geometries. (N = 10, 30, 64. Contact area = 20%. Anisotropy is the magnitude of the difference between the east-west K_{eq} and the north-south K_{eq}. Darcy's law is used.)**

For local hydraulic conductivities that have a binary distribution, the Power Average Model predicts

$$< K_{eq} >^{w} = p\, K_{hi}^{w} + (1 - p)K_{low}^{w} \tag{11}$$

where

$$-1 \leq w \leq 1.$$

The local hydraulic conductivity equals K_{hi} or K_{low} with respective probabilities p and 1 − p. The parameters in the model depend on scale and correlation of the conductances. Deutsch (1988) found, by simulation, that w depends on scale (although it stabilizes for sufficiently large scales) and void correlation structure.

We assign zero conductances to contact areas. We assume the Power Average Model is valid. Because Deutsch fit a Power Average Model with $w > 0$ for geometries with correlation structure similar to ours, we expect $w > 0$. Hence, we assume $w > 0$. With these assumptions, we derive from Equation (11) that

$$< K_{eq} >^{w} = p \tag{12}$$

or

$$< K_{eq} > = p^{1/w}$$

where p is the fraction of void spaces. In Equation (11), we set K_{hi} to one and K_{low} to zero. If the void space aperture were b, rather than one, we would multiply $p^{1/w}$ by b^2 to get equivalent permeability. But we consider unit void space apertures and unit void space conductances in this analysis.

To account for variability about the average equivalent permeability $p^{1/w}$, we include a residual that depends on contact area, scale, and correlation structure.

$$K_{eq} = p^{1/w} + \varepsilon \tag{13}$$

We define $\hat{p}$ as the void space percentage for a simulated pattern. By the propagation of errors formula (Young, 1962),

$$\sigma_{K_{eq}}{}^{2} \approx \left(\frac{\partial < K_{eq} >}{\partial p}\, \sigma_{\hat{p}} \right)^{2} + \sigma_{\varepsilon}{}^{2}$$

provided that $\sigma_{\hat{p}}$ is not too large. For large enough scales such that $\sigma_{\hat{p}}$ is small and w is stable, we have, by substitution,

$$\sigma_{K_{eq}}{}^{2} \approx \left(\frac{1}{w}\, p^{\frac{1}{w} - 1} \right)^{2} \sigma_{\hat{p}}{}^{2} + \sigma_{\varepsilon}{}^{2} \tag{14}$$

We now evaluate the fluctuation of $\hat{p}$ for our simulated geometries and compute a lower bound for $\sigma_{K_{eq}}$. The two-point correlation function of our simulated void space geometries is well described by an exponential model (Fig. 5).

$$S_2(\vec{r}) = p^2 + p(1 - p)e^{-\lambda|\vec{r}|}, \tag{15}$$

where $\lambda = 0.4$ and $p = 0.817$. A simulated geometry will have a void space percentage, $\hat{p}$ which fluctuates about p. Assuming this exponential form for $S_2(\vec{r})$, we derive an asymptotic expression for $\sigma_{\hat{p}}$.

$$\sigma_{\hat{p}} \approx \frac{\sqrt{p(1 - p)}}{N} \sqrt{1 + 2\pi e^{-\lambda}\left(\left(\frac{1}{\lambda}\right) + \frac{1}{\lambda^2}\right)} \tag{16}$$

where the fracture is N by N pixels. Thus, an approximate lower bound for $\sigma_{K_{eq}}$ is

$$\sigma_{K_{eq}} \geq \left(\frac{1}{w} p^{\frac{1}{w} - 1}\right) \sqrt{1 + 2\pi e^{-\lambda}\left(\frac{1}{\lambda^2} + \frac{1}{\lambda}\right)} \frac{\sqrt{p(1 - p)}}{N} \tag{17}$$

We normalize $\sigma_{K_{eq}}$ by the expected value of K_{eq} giving a coefficient of variation for our simulated patterns.

$$\frac{\sigma_{K_{eq}}}{< K_{eq} >} \geq \left(\frac{1}{wp}\right) \sqrt{1 + 2\pi e^{-\lambda}\left(\frac{1}{\lambda^2} + \frac{1}{\lambda}\right)} \frac{\sqrt{p(1 - p)}}{N} \tag{18}$$

Our result says that the variability of K_{eq} for our simulated patterns diminishes as the pattern size increases. We have incorporated the effects of scale, contact area percentage, and correlation in our analysis.

Equation (17) suggests the following result. For large enough fractures so that Equation (16) is valid, $\sigma_{K_{eq}}$ (the variability of K_{eq} for our ensemble of statistically similar void space geometries) will exceed σ_{min} unless fracture area exceeds N_0^2 where

$$N_0 = \frac{1}{\sigma_{min}} \left(\frac{1}{w} p^{\frac{1}{w} - 1}\right) \sqrt{1 + 2\pi e^{-\lambda}\left(\frac{1}{\lambda^2} + \frac{1}{\lambda}\right)} \sqrt{p(1 - p)}\,. \tag{19}$$

Note that σ_{min} is arbitrary. If one requires small variability among the geometries taken from the statistical ensemble, then the area of the patterns, N^2, must exceed a large critical area N_0^2.

Substituting the values of the λ and p which give a good fit to our observed void space two-point correlation function, we now estimate the variability of K_{eq} for simulated patterns of size N by N pixels.

$$\sigma_{K_{eq}} \geq \frac{4.8}{N} \tag{20}$$

We now estimate total variability of K_{eq} for simulated patterns, at the 64 by 64 pixel scale, by incorporating the variability due to variations in the spatial arrangement of contact areas within a pattern and the variability due to variations in the contact area percentage. From Table 2, the variability of K_{eq} due to spatial arrangement variability is 0.049. From Equation (20), the variability of K_{eq} due to fluctuations of contact area is 0.075. Hence, assuming independence of the two sources of variability, we estimate total variability as:

$$\sigma_{K_{eq}} \approx \sqrt{(0.075)^2 + (0.049)^2} = 0.09$$

At the 64 by 64 pixel scale, both sources of variability of K_{eq} are comparable.

Lower bound for the variable aperture case

Before, we estimated the variability of K_{eq} for simulated void space geometries where void spaces had constant apertures. Here, we extend our methods to consider variable aperture void space distributions. We now consider absolute equivalent permeability rather than relative equivalent permeability.

Suppose that the voids have apertures ranging from b_{min} to b_{max} and contact areas again have zero apertures. Then the lower bound for $\sigma_{K_{eq}}$ falls in the range:

$$\left(\frac{1}{w}\, p^{\frac{1}{w}-1}\right)\sqrt{1 + 2\pi e^{-\lambda}\left(\frac{1}{\lambda^2} + \frac{1}{\lambda}\right)}\;\frac{\sqrt{p(1-p)}}{N}\left(b_{min}^2,\ b_{max}^2\right). \tag{21}$$

This range could be very broad.

For broad variable aperture distributions, the smallest conductances may control the flow. Following the ideas of (Charlaix et al., 1987; Katz and Thompson, 1986; Kirkpatrick, 1979; Ambegakor et al., 1971), we identify a cutoff conductance, b_c^2, which controls the flow. We set all conductances less than b_c^2 to zero, and all above b_c^2 to b_c^2. With the Power Average Model, we form the functional

$$K_{eq} = b_c^2\,(1 - F(b_c))^{\frac{1}{w}} \tag{22}$$

where $1 - F(b_c)$ is the fraction of conductances above b_c^2. We then maximize the functional to estimate K_{eq}. Because all conductances above b_c^2 are set to b_c^2, the functional underestimates equivalent permeability.

The functional takes its maximum value at b^* defined by

$$b^* = \frac{2w(1 - F(b^*))}{f(b^*)} \tag{23}$$

where

$$f(b^*) = \left.\frac{dF}{db}\right|_{b=b^*}$$

Hence, our estimate for equivalent permeability is:

$$\hat{K}_{eq} = b^{*2}\,(1 - F(b^*))^{\frac{1}{w}} \tag{24}$$

Because the correlogram of a clipped Gaussian process is nearly the same for all clipping levels that leave between 10% and 90% contact area (Switzer, 1977), we can compute a bound for the variability of equivalent permeability. Because the correlogram is nearly the same, the two-point correlation function has nearly the same form. Hence, we replace p with $1 - F(b^*)$ in Equations (15) to (18) to get:

$$\sigma_{K_{eq}} \geq b^{*2}\left|\frac{1}{w}\,(1 - F(b^*))^{\frac{1}{w}-1}\right|\sqrt{1 + 2\pi e^{-\lambda}\left(\frac{1}{\lambda^2} + \frac{1}{\lambda}\right)}\;\frac{\sqrt{F(b^*)(1 - F(b^*))}}{N} \tag{25}$$

However, if the fraction of conductors that control the flow is too small, the Power Average Model may not apply.

For disordered systems, consisting of conducting and nonconducting material, Percolation Theory (Kirkpatrick, 1973; Stauffer, 1985) predicts conductivity when the conductors just barely connect. For a system with a fraction p of conducting material and a fraction $1 - p$ of nonconducting material, Percolation Theory predicts that

$$K = K_0\,(p - p_c)^t$$

where p_c is the percolation threshold and t depends only on dimension (according to universality arguments) and K_0 is a constant approximately equal to one. In two dimensions, $t \approx 1.3$. When square pixels are independently assigned as conducting or nonconducting, $p_c \approx 0.59$ (Orbach, 1986). For correlated assignment schemes, as in our problem, p_c is shifted.

We estimate K_{eq} and $\sigma_{K_{eq}}$ using a critical conductance approach and Percolation Theory. We again follow ideas developed by Ambegakor et al., 1971; Charlaix et al., 1987; and Katz and Thompson, 1986. Our Percolation Theory/critical path analysis functional for equivalent permeability is

$$K_{eq} = K_0\,b^2\,(F(b_{crit}) - F(b))^t \tag{26}$$

Maximizing the functional for equivalent permeability, we get

$$\hat{K}_{eq} = K_0\,b^{*2}\,(F(b_{crit}) - F(b^*))^t \tag{27}$$

and

$$\frac{\sigma_{K_{eq}}}{< K_{eq} >} \geq \frac{t}{N} \sqrt{\frac{F(b^*)(1 - F(b^*))}{(F(b_{crit}) - F(b^*))^2} \left\{ 1 + 2\pi e^{-\lambda} \left(\frac{1}{\lambda^2} + \frac{1}{\lambda} \right) \right\}} \tag{28}$$

where

$$b^* = \frac{2(F(b_{crit}) - F(b^*))}{tf(b^*)} \tag{29}$$

To illustrate our technique, consider a fracture with the following aperture distribution

$$F(b) = (1 - p) + p\, b^k$$

$$0 \leq b \leq 1 \tag{30}$$

$$k > 0$$

where again F(b) is the fraction of aperture less than or equal to b. With this distribution, we have a contact area fraction equal to $1 - p$. Critical path analysis for the Power Average Model gives

$$\hat{K}_{eq} = b^{*2}\, p^{*1/w} \tag{31}$$

$$\frac{\sigma_{K_{eq}}}{< \hat{K}_{eq} >} \geq \frac{1}{wp^*} \sqrt{1 + 2\pi e^{-\lambda} \left(\frac{1}{\lambda^2} + \frac{1}{\lambda} \right) \frac{\sqrt{p^*(1 - p^*)}}{N}} \tag{32}$$

where

$$b^* = \left[\frac{2w}{k + 2w} \right]^{1/k}, \tag{33}$$

$$p^* = \left[\frac{k}{k + 2w} \right] p, \tag{34}$$

$$p^* = 1 - F(b^*)$$

Critical path analysis for the Percolation Theory Model gives

$$\hat{K}_{eq} = K_0\, b^{*2}\, (p^* - p_{crit})^t \tag{35}$$

$$\frac{\sigma_{K_{eq}}}{< K_{eq} >} \geq \frac{t}{N} \sqrt{\frac{p^*(1 - p^*)}{(p^* - p_{crit})^2} \left\{ 1 + 2\pi e^{-\lambda} \left[\frac{1}{\lambda^2} + \frac{1}{\lambda} \right] \right\}} \tag{36}$$

where

$$b^* = \left[\left(\frac{2}{2 + tk} \right) \left(1 - \frac{p_{crit}}{p} \right) \right]^{1/k}, \tag{37}$$

$$p^* = p_{crit} + \left(\frac{tk}{2 + tk} \right) (p - p_{crit}), \tag{38}$$

$$p_{crit} = 1 - F(b_{crit}).$$

By taking a critical path analysis, we have estimated a lower bound for the variability of equivalent permeability of our simulated void space geometries. We accounted for the effects of scale, spatial correlation, and void space aperture distribution.

Summary

This work shows the importance of specifying fracture void space geometry accurately and identifying the relevant physics to select a practical flow model. Darcy's law always gives the largest equivalent permeability, and the Navier-Stokes equations give the smallest equivalent permeability. The Reduced Navier-Stokes equations give an intermediate value. For tracer studies, this means that Darcy's law overestimates velocity. We can rescale the equivalent permeability computed using Darcy's law to get the equivalent permeability computed using the Navier-Stokes equations as long as flow remains linearly related to gradient of pressure.

The three flow models gave significantly different equivalent permeabilities for the four-constriction geometry (Figs. 8 and 11). For our more complex variable aperture geometry (Fig. 12), Darcy's law and the Reduced Navier-Stokes equations gave nearly the same equivalent permeabilities. The full Navier-Stokes solution was not obtained for this case.

We have succeeded in simulating fracture void space geometries statistically similar to an observed fracture void space geometry; both the two-point correlation function (Fig. 5) and contact area size histograms match well (Fig. 6). Further, when we assign all the voids a constant aperture, the average equivalent permeabilities of the simulated and observed geometries using Darcy's law agree reasonably well (Table 1). We identified the two sources of variability of equivalent permeability for simulated void space geometries where voids had constant apertures. We studied both the variability due to the many ways to arrange contact areas in a simulated pattern (Table 2) as well as the variability due to fluctuations of contact area percentage (Eq. (17)). We found that both sources of variability diminished as we increased the size of the simulated geometry.

Because real fracture void spaces will have variable aperture distributions, we need more sophisticated procedures than assigning each void a constant aperture and then computing flow. A critical path analysis combined with the Power Average or Percolation Model offer such promise to estimate the variability of equivalent permeability analytically.

Literature Cited

1. Ambegakor, V., B. I. Halperin, and J. S. Langer, "Hopping Conductivity in Disordered Systems," *Phys. Rev. B.*, Vol. 4, p. 2612 (1971).
2. Beran, M. J., *Statistical Continuum Theories*, Interscience, New York (1968).
3. Charlaix, E., E. Guyon, and S. Roux, "Permeability of a Random Array of Fractures of Widely Varying Apertures," *Transport in Porous Media*, Vol. 2, No. 1 (February 1987).
4. Berryman, J. G. and S. C. Blair, "Use of a Digital Image Analysis to Estimate Fluid Permeability of Porous Materials. Application of Two-Point Correlation Functions," *J. Appl. Phys.*, Vol. 60, pp. 1930–1938 (1986).
5. Brown, S., "Flow Through Rock Joints: The Effect of Surface Roughness," *J. Geophys. Res.*, Vol. 92, pp. 1337–1347 (1987).
6. Desbarats, A. J., "Stochastic Modeling of Flow in Sand-Shale Sequences," Ph.D. Thesis, Stanford University (1986).
7. Deutsch, C., "Calculating Effective Absolute Permeability in Sandstone/Shale Sequences," submitted to *J. of Pet. Tech.* (1988).
8. Gale, J. E., "Comparison of Coupled Fracture Deformation and Fluid Flow Models with Direct Measurements of Fracture Pore Structure and Stress Flow Properties," *28th U.S. Symposium on Rock Mechanics*, I. W. Farmer, J.J.K. Daemen, C. S. Desai, C. E. Glass, and S. P. Neuman, Eds., pp. 1213–1222 (1987).
9. Hoshen, J. and R. Kopelman, "Percolation and Cluster Distribution, Part I," *Phys. Rev. B.*, *8*, p. 3438 (1976).
10. Journel, A. G., C. D. Deutsch, and A. J. Desbarats, *Power Averaging for Block Effective Permeability,* SPE 15128, Society of Petroleum Engineers (1986).
11. Journel, A. G. and C. Huijbregts, *Mining Geostatistics*, Academic Press, London, 600 pp. (1978).
12. Journel, A. G. and E. H. Isaaks, "Conditional Indicator Simulation: Application to a Saskatchewan Uranium Deposit," *Math Geol.*, Vol. 16, No. 7, pp. 685–718 (1984).
13. Katz, A. J. and A. H. Thompson, "A Quantitative Prediction of Permeability in Porous Rock," *Phys. Rev. B*, Vol. 34, No. 11, pp. 8179–8181 (1986).
14. Kirkpatrick, S., *Ill-Condensed Matter*, R. Balian, R. Maynard, and G. T. Toulouse, Eds., North-Holland, Amsterdam (1979).
15. Kirkpatrick, S., "Percolation and Conduction," *Reviews of Modern Physics*, Vol. 45, No. 4, pp. 574–588 (1973).
16. Long, J.C.S., J. S. Remer, and P. S. Witherspoon, "Porous Media Equivalents for Networks of Discontinuous Fractures," *Water Resources Research*, Vol. 18, No. 3, pp. 645–658 (1982).
17. Madden, T. R., "Random Networks and Mixing Laws," *Geophysics*, Vol. 41, No. 6a, pp. 1104–1125 (1976).
18. Muralidhar, K. and J.C.S. Long, *A Scheme for Calculating Flow in Fractures Using Numerical Grid Generation in Three-Dimensional Domains of Complex Shapes*, LBL-24453, Lawrence Berkeley Laboratory, Berkeley, California (1987).
19. Muskat, M., *The Flow of Homogeneous Fluids Through Porous Media*, J. W. Edwards, Inc., Ann Arbor, Michigan (1946).

20. Nolte, D. D., L. J. Pyrak-Nolte, and N.G.W. Cook, ''The Fractal Geometry of the Flow Paths in Natural Fractures in Rock and the Approach to Percolation,'' accepted for publication in *Special Issues: Fractals in Geophysics* in PAGEOPH (1987).
21. Orbach, R., ''Dynamics of Fractal Networks,'' *Science*, Vol. 231, pp. 814–819 (1986).
22. Pyrak-Nolte, L. J., L. R. Myer, N.G.W. Cook, and P. A. Witherspoon, ''Hydraulic and Mechanical Properties of Natural Fractures in Low Permeability Rock,'' *Proceedings of Sixth International Conference on Rock Mechanics* (Montreal, Canada), A. A. Balkema, Rotterdam, Boston (1987).
23. Rothman, D., ''Cellular-Automaton Fluids: A Model for Flow in Porous Media,'' *Geophysics,* Vol. 53, No. 4, pp. 509–518 (1988).
24. Schlichting, H., *Boundary-Layer Theory*, McGraw-Hill, New York (1979).
25. Stauffer, D., *Introduction of Percolation Theory*, Taylor and Francis, Philadelphia and New York (1985).
26. Switzer, P., ''Estimation of Spatial Distributions from Point Sources with Application to Air Pollution Measurement,'' *International Statistical Institute*, Vol. 47, pp. 121–147 (1977).
27. Thompson, J. F., Ed., *Numerical Grid Generation*, Elsevier Science Publishing Company, New York (1982).
28. Tsang, Y. W., ''The Effect of Tortuosity on Fluid Flow through a Single Fracture,'' *Water Resour. Res.*, Vol. 20, pp. 1209–1215 (1984).
29. Witherspoon, P. A., Y. S. Wang, and J. E. Gale, ''Validity of Cubic Law for Fluid Flow in a Deformable Rock Fracture,'' *Water Resour. Res.*, Vol. 16, pp. 1016–1024 (1980).
30. Young, H. D., *Statistical Treatment of Experimental Data*, McGraw-Hill, New York, 171 pp. (1962).

5 Geostatistical Methods

MUCK—A Novel Approach to Co-Kriging

ISOBEL CLARK
Geostokos Ltd.
London, England

KAREN L. BASINGER and
WILLIAM V. HARPER
Battelle Office of Nuclear Waste Isolation
Columbus, Ohio

Abstract

Multivariate geostatistical estimation—"co-kriging"—has been used in practice for well over a decade. The technique as developed by Matheron has been documented in many case studies.

There are disadvantages to the standard approach, however, not the least of which is that the variables which are measured must be sited at the same geographical location. For example, it is necessary that if an exploration project is considering several minerals, they will tend to measure all of them on the same samples.

There are cases in which this type of information is not available. Indeed, the motivation for using a multivariate approach may be to specifically improve estimation in areas where few variables have been measured.

A variation in co-kriging will be presented which does not place this restriction on the modeling and estimation processes. A case study will be given which includes the added complication of strong trend. Hence the acronym for this new method—Multivariate Universal Co-Kriging.

Introduction

The term "co-kriging" appears to have been coined by Matheron himself in his original work, *The Theory of Regionalized Variables and Its Applications* (1971). A review of this original approach is useful before starting to generalize.

It is common in geological sampling to measure more than one variable at each sample location. For mining purposes, for example, many mineral grades might be measured, plus perhaps the width of the lithological unit. In hydrology, one might possibly measure porosity and permeability, hydraulic head, grain size, and so on. Some of these variables might be correlated. Some of them may not have been measured at all locations. If the value of a variable at an unsampled location is to be estimated, the spatial structure (semivariogram) for that variable must be established. The sample values may then be used to produce a "kriging" estimator of the variable at that location.

Suppose, however, that this variable has been fairly sparsely sampled; it may be more expensive to obtain values, or perhaps just very difficult. Suppose, in addition, that this variable is significantly correlated with one or more of the other variables that have been measured. Matheron studied the question as to whether these correlated variables could be used to estimate the variable of particular interest. He developed the concept of a "cross semivariogram" between two variables and, from this structure, a set of kriging equations, which he dubbed "co-kriging." The details of this theory are thoroughly documented by Matheron (1971) and in numerous papers, notably those by Guarascio (1976) and Myers (1982, 1983, 1984), so that only those points that are relevant to the present discussion need be discussed here.

When Matheron developed co-kriging, he supposed that there would be (effectively) a single sampling campaign. Some or all of the variables of interest would be measured at each location. The "cross semivariogram" was defined as follows:

$$\gamma_{gf}(h) = 0.5\ E[(g_i - g_j)\ (f_i - f_j)]$$

where g_i, g_j are the measured values of variable g at locations i and j

where f_i, f_j are the measured values of variable f at locations i and j

and h is the distance between the two samples.

Under the usual intrinsic hypothesis and the assumption of no significant drift, the estimation of the value g_o at an unsampled location gives rise to a set of co-kriging equations. The solution of these equations produces a minimum variance unbiased estimator in the form of a linear combination of the neighboring values of both variables. This set of equations can be generalized with reasonable ease to the case where a significant drift is present (Myers, 1982).

Although co-kriging does not seem to be in routine use in geological applications, well-documented case studies illustrate the use of the technique in various situations (cf. Carr and McCallister, 1985; and Unal and Haycocks, 1986).

Some problems associated with co-kriging

The theory of co-kriging seems to be well founded. However, it has been noted by various authors (Carr et al., 1985) that its application in practice gives rise to some real difficulties. These are mostly encountered in the calculation and modelling of the cross semivariogram graph. Myers (1982) suggests that difficulties in semivariogram modelling can be reduced by using a pseudo-cross semivariogram comprising the *sum* of the values squared (rather than just the difference). This graph is modelled and then "adjusted" by subtracting the single-variable semivariogram models and dividing by two.

Myers (1982) states that "it is simpler to utilize cross variograms than to use cross covariances, but this is not always possible." He formulates his approach to co-kriging, therefore, almost completely with cross covariance functions and generalizes to the semivariogram function. These calculations require stationarity of both mean and variance rather than the less restrictive intrinsic hypothesis. He also states that the cross covariance function and the single-variable covariance functions must obey the Cauchy-Schwartz inequality in the same way that a normal covariance must be less than the product of the two standard deviations. At first sight, this seems sensible, but it is difficult to show mathematically other than at the origin of the graph. Myers also asserts that the Cauchy-Schwartz inequality should hold between the cross semivariogram and the univariate semivariograms. There is no justification for this statement.

In fact, given the formulation of the traditional cross semivariogram function, there does not seem to be any practical reason why it should be nonnegative. If the two variables under consideration are inversely correlated at a significant level, then the cross semivariogram function would have a stable form but would show negative values for $\gamma(h)$. This means that we have to model a ''semi-variance'' graph that generally decreases as the distance increases. This could lead to some very strange kriging weightings for distant samples. Negative kriging variance might also be encountered. If the variables are effectively uncorrelated, the graph will oscillate around zero—a very unsatisfactory behavior from the point of view of building a stable set of kriging equations.

The problems arising in the usage of the cross semivariogram graph might be due to the original formulation of the function, rather than to instabilities in the modelling or the sample configuration. The choice of the traditional form follows naturally when the derivation is made first for the covariance function and only then generalized to the semivariogram graph. It can easily be shown that the above semivariogram definition is obtained by subtracting the cross covariance function from the normal statistical covariance at distance zero. The authors propose the reverse procedure, deriving the cross semivariogram function from the concept of the single-variable semivariogram function and the intrinsic hypothesis.

An Alternative Formulation

The semivariogram for, e.g. variable g, is usually defined as

$$\gamma_g(h) = 0.5 \, E[(g_i - g_j)^2]$$

using the same notation as before. This definition is justifiable on many grounds—most of them mathematical. Intuitively, this approach may be justified as follows: the ultimate goal is to produce an estimate for an unsampled location that will be as close as possible to the actual value; the difference between the estimate and the actual value is the ''estimation error''; the best estimator will be one with the smallest errors—in mathematical terms, mean zero and minimum variance. In short, a kriging system seeks to minimize the expected value of the squared errors, i.e., the squared difference between the known sampled values and the unknown unsampled value. Therefore, it seems only sensible to build a continuity or ''predictability'' model to describe the squared difference between values at different locations.

For variable f the equivalent form is

$$\gamma_f(h) = 0.5 \, E[(f_i - f_j)^2]$$

It is suggested that the cross semivariogram be defined as

$$\gamma_{gf}(h) = 0.5\ E[(g_i - f_j)^2]$$

This formulation follows the intuitive logic described for the ordinary semivariograms. The only real difference between the single-variable semivariogram and the two-variable cross semivariogram is that the values at different sample locations will be values of different variables.

A possible difficulty in the practical application of either type of cross semivariogram will occur if the two variables are of widely differing scales. This problem is analogous to that encountered in calculating classical covariance between such variables, where precision may be lost due to the numerical values being of different orders of magnitude. A suggested solution is to "standardize" the individual variables to give each an average of zero and a sill on the univariate semivariogram of unity. The situation is further complicated in the presence of trend and is suggested as a topic for intensive study at a later date.

The formulation of the cross semivariogram function shown above has some advantages over the classical Matheron method. First, and perhaps most important, it cannot become negative. If the samples are highly negatively correlated, the graph will rise precipitately rather than fall below zero. If the variables are uncorrelated, the differences will show as a constant "nugget effect"—a flat horizontal line on the graph. This function, then, may be modelled in the same way as a normal single-variable semivariogram with no need to introduce intermediate graphs.

It does not appear easy to establish a relationship between this type of cross semivariogram and the cross covariance function. The mathematics is not particularly complex, but it involves the knowledge of a stationary mean value for each of the two variables. If a covariance exists, i.e., if the variables are stationary over the study area, then of course the mean will be stationary. However, the construction of the cross semivariogram function does not require that the mean be stationary and may, therefore, be more geologically valid.

Our formulation of the cross semivariogram has been expressed as though there were no significant drift or trend over the study area. This is not really a restriction. If there is a significant trend, then it must be removed in some way before *any* of the semivariograms can be calculated. In that case, then all semivariograms are calculated as the difference between the residual values at each sample location.

The major advantage

It might appear that the reformulation of the cross semivariogram function is justified on mathematical grounds alone. The removal of the extra step in the modelling is an unexpected but welcome bonus. However, it must be said that neither of these features is the main motivation for the development of a new approach to cokriging.

To explain the reasons for the current development, the motivation for the original development of co-kriging must be examined. The major application for co-kriging over the last two decades or so has been to estimate one particular variable that has been undersampled compared to the others. For example, it is relatively easy to measure base-metal content in a massive sulfide deposit. It is, however, costly and time consuming to measure the specific gravity of every drill-core section. On the other hand, it is reasonably easy to establish a correlation between metal value and specific gravity. Co-kriging is one way in which the density might be estimated more accurately without recourse to vastly increased sampling costs. The other major application (Myers, 1982) appears to be in estimating linear combinations of various parameters at one time.

There is a fundamental difference between these requirements and those of the program that initiated the present research. The study concerned the potentiometric levels within various aquifers in a particular study area with a view to ultimately predicting fluid flow and travel path parameters. Because of the nature of the sampling available, those boreholes having reliable observations of the potentiometric pressure in one aquifer tended to be at different locations from the boreholes used to sample another aquifer. The sample data, then, could not be used to calculate a semivariogram such as that described in the classical geostatistical literature because these require multiple measurements at a significant proportion of the sample locations.

Our reformulation of the cross semivariogram enables *all* potential pairs of sample values to be used, including values at the same location should they be available. This feature is invaluable in helping to establish the nugget effect, which must be present in a cross semivariogram. Both a more stable shape for the graph (because it is based on more pairs) and a more efficient estimate for all parameters required for the model-building stage are obtained. Standard semivariogram models may be used without problems, and they have been found to suit all of the cross semivariograms encountered so far.

Co-kriging

If the proposed form of the semivariogram is acceptable and the underlying intrinsic hypothesis that ''difference in value'' is a function only of distance between sample locations, some kind of optimal estimator can be developed. Only the linear estimator with or without significant trend will be considered in this paper, i.e., those methods known as ''ordinary kriging and universal kriging.'' These techniques are based on the assumption that a linear combination of the sample values provides a sufficient estimator for an unsampled location, and that a good measure of the accuracy of this estimator is the variance of the error of estimation thus produced. The kriging equations are a direct result of minimizing this estimation variance under the constraint that the estimator be unbiased.

Carrying out such a derivation on the basis of the new cross semivariogram model presents little difficulty other than that of algebraic complexity. The resulting set of kriging equations are identical to those traditionally used for co-kriging. Myers (1982) formulated the co-kriging system as a set of matrices that is invaluable for clarifying the situation when all measurements are made at all locations. It becomes a little unwieldy when there are a large number of ''missing'' or undersampled variables, because rows and columns must be deleted from each matrix. In this situation, where very few of the locations can be assumed to have more than one variable measured, the matrix approach is of limited use. The equations have, therefore, been expressed in an algebraic notation and then reformulated into matrices for the purposes of computer programming.

A brief appendix is attached that provides the algebraic expressions for the co-kriging equations when estimating the value of one variable at an unsampled location. A short discussion is given here.

In formulating the layout for the co-kriging equations, the following information is required:

1. The location of the sample (geographical coordinates).
2. The variable measured at this location.
3. The value observed for this variable.

It is obvious from the above that if more than one variable *has* been measured at a single location, it should be treated as more than one sample in the data set. The bulk of the co-kriging equations then look identical to an ordinary kriging system:

1. The left-hand matrix will contain semivariogram values for all possible combinations of sample values.
2. The right-hand "vector" will contain the semivariogram values between each sample and the unsampled location being estimated.
3. The left-hand "vector" will contain the weighting factors allocated to each sample to produce the optimal estimation of the unknown value.

There are two major differences between this and the usual kriging system of equations. First, one must carefully keep track of which variable is measured on each sample so that semivariogram values from the correct model will be calculated, whether single-variable or cross semivariogram. This presents few problems. Second, one must apply the usual constraints to ensure that the estimator remains unbiased. In the ordinary kriging system, the weighting factors are constrained to sum to unity, thus ensuring unbiasedness in the absence of trend. This adds one equation to the system and one "unknown" to the vector of weighting factors. In universal kriging, a constraint is added for each mathematical term in the expression of the trend, so that the estimator remains unbiased. If the trend contains six terms, six equations will be added to the kriging system and six "unknowns" to the final solution.

In co-kriging, similar constraints must be used to ensure that the estimator is unbiased. Considering the case without trend, previous authors (op. cit.) have suggested the following conditions:

1. Identify which samples have been measured for the same variable as that being estimated. The sum of the weighting factors for these samples must be unity.
2. For *each* other variable, identify the samples for which this variable has been measured. The sum of the weighting factors for these samples must be zero. This process is repeated for each variable encountered in this sample set.

These conditions are extremely restrictive. There is no reason, for example, why the individual variables should all have factors summing to zero. If the study area does exhibit stationarity and the mean value for each variable could be estimated, the constraints could follow any number of different patterns. However, this question is outside the current scope of this paper. In addition, adoption of the above constraints leads to some great simplifications when computer programming.

In short, then, to ensure that the co-kriging estimator is unbiased, there must be as many constraints as there are observed variables in the sample set, or at least in the set being used for estimation. In the case where significant trend exists—universal co-kriging—constraints must be added for each term in the (polynomial) trend that has been identified for each variable. For example, if there are three variables with no trend on the first, three terms in the trend for the second, and six terms for the third, then $1 + 3 + 6 = 10$ equations must be added to the co-kriging system. That is, ten extra rows and columns to the left-hand matrix, and ten items to each of the right- and left-hand vectors.

Considering ordinary kriging without drift as a special case of universal kriging with one term (constant) in the trend, the process of co-kriging might be summarized as follows. A set of equations is developed that is remarkably similar to those of single-variable kriging, except that the constraints on the weighting factors require the addition of enough equations to cover all of the terms in all of the trends for all of the variables being used in the estimation. Otherwise, the only noticeable difference is that there are more sample values available in making each estimate.

Multivariate co-kriging

In his 1982 paper, Myers pointed out that the estimation of one (possibly undersampled) variable was not really the prime purpose of co-kriging. When expressed in matrix form, it is clearly seen that some or all of the variables may be estimated at the same time without a significant increase in computing time or cost. This is simply because the "left-hand matrix" of intersample semivariogram values will remain the same no matter which variable is actually being estimated at the unsampled location. That is, values can be estimated for all of the variables at a single location and the matrix on the left-hand side would be the same.

The two vectors involved in the co-kriging system will change, of course. The vector on the right-hand side contains the semivariograms between each sample and the value at the unsampled location. This will obviously change if a different variable is being estimated. The variable under estimation will also affect the constraints on the weighting factors. If estimated values are required for several variables at the unsampled location, there will be as many right-hand-side vectors as there are variables. In direct consequence, there will be as many solution vectors as there are variables. However, since the left-hand side remains the same, the solution for all of the vectors may be carried out simultaneously. The primary implication of this is that many variables may be estimated for the cost of one.

Multivariate universal co-kriging

Co-kriging may be used to estimate an inadequately sampled single variable, or it may be used to estimate several variables simultaneously. In the latter case, the estimation of each variable is improved, in that more information may be used, and the cost is scarcely greater than estimating any one of the variables singly. To avoid confusion as to the purpose of this estimation process, it is proposed that the term "multivariate" be used to emphasize that more than one variable is being estimated. Because the technique is also applicable in the presence of significant trend, the traditional term "universal" should be included; thus the term "multivariate universal co-kriging" or MUCK, is derived.

In summary, then, a new definition of the semivariogram graph is proposed which meshes intuitively with the concept of the original semivariogram in use for single variables. The inclusion of this new form results in a system of equations remarkably similar to that obtained using the old form. This system of equations is used to obtain "optimal" estimates for all of the variables at each unsampled location simultaneously. This approach is referred to as multivariate universal co-kriging (MUCK).

A Case Study—The Pennsylvanian and Wolfcamp Aquifers

The area of study is a region in the northern portion of Texas in the Panhandle region (Figure 1). The sample data are potentiometric values from deep brine aquifers called the Pennsylvanian and the Wolfcamp, which in some areas are almost 8,000 feet below the ground surface. The variable considered in this study is the potentiometric or piezometric level, expressed as the elevation (in feet) above sea level of water in a well at that location.

Deaf Smith County in northern Texas is a possible site for a future high-level nuclear waste repository. The Pennsylvanian and Wolfcamp aquifers underlie the salt bed, which would be the host for such a repository (cf. Bair et al., 1985). The

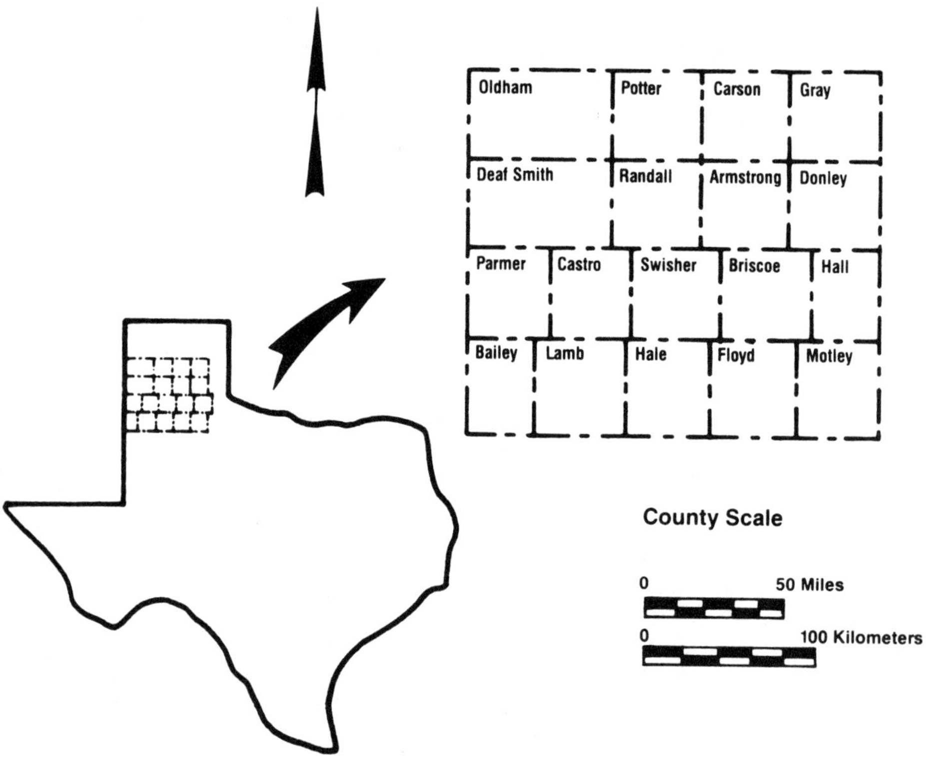

Figure 1. **Location map for the geostatistical analysis.**

analysis of these aquifers is a vital component in the characterization of this site, because a breach of the repository could cause hazardous material to be released into the aquifers. One of the requisite parameters in determining flow rates and directions is the potentiometric level, because fluids tend to flow from areas of higher potentiometric values to those that are lower.

The data used in this report have been collected and culled by the Office of Nuclear Waste Isolation's contractor, Stone & Webster Engineering Corporation (SWEC). Their refinement of the original sample data included deleting depressured, local grossly overpressured, and underpressured data so that the resulting data base more accurately represented the original flow system at steady-state conditions before oil and gas production (Bair et al., 1985). A total of 109 data values were available for the Pennsylvanian aquifer, whereas the Wolfcamp retained 85 in the final data set. It is worth noting, at this point, that the sample locations for the two aquifers are quite different.

The first step in any geostatistical analysis is to calculate experimental semivariograms and fit (if possible) a valid model to the graph. In the multivariate case, univariate semivariograms must be constructed for each aquifer and a cross semivariogram between the two sets of potentiometric data.

Both aquifers exhibit very strong trends, as would be expected with this type of variable. For both the Pennsylvanian and the Wolfcamp aquifers, a linear trend was sufficient to describe the study area. Experimental semivariograms are calculated on the residuals from these trends. Figures 2 and 3 show the calculated semivariogram

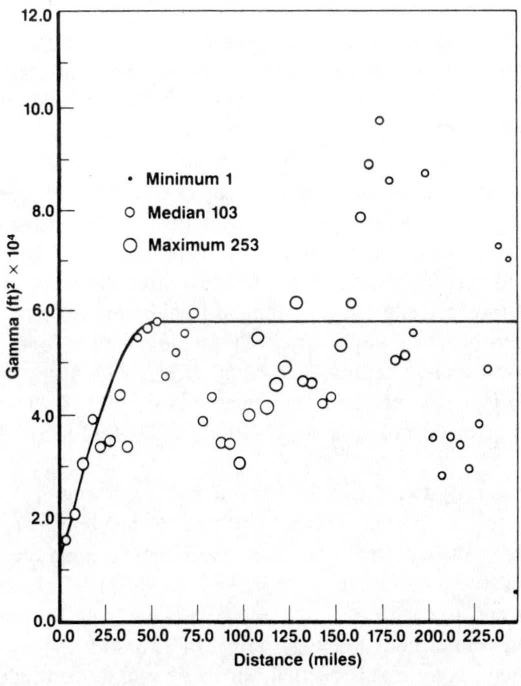

Figure 2. **Semivariogram model for Pennsylvanian data.**

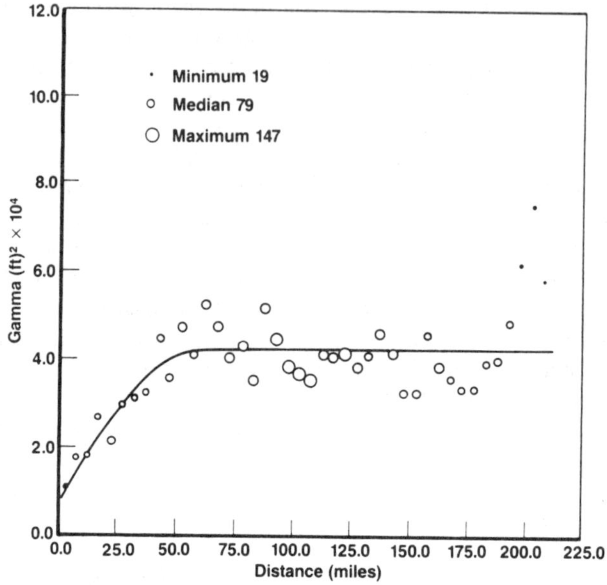

Figure 3. **Semivariogram model for Wolfcamp data.**

graphs for the residual data in the two aquifers and the models fitted to them. Comparing the model with the data in each case, it can be seen that the model tends to be a little higher than one would intuitively expect. This is a function of the trend surface fitting, which tends to result in a downward bias in the observed variation between sample values. The "true" sill is determined by a cross-validation process using universal kriging (UK) on the original data values. This process consists of removing a data value from the sample set, estimating the value at that location from the surrounding samples, and comparing the estimate with the "true" value. This procedure is repeated for every data value. In ideal circumstances, the resulting statistics should average zero and have a standard deviation of unity.

The models fitted to these graphs are both simple spherical ones. The Pennsylvanian residuals show a range of influence of 50 miles, with a nugget effect of 12,000 ft^2 and a sill (for the spherical component) of 46,000 ft^2. The Wolfcamp has a slightly longer range at 60 miles, a nugget effect of 11,000 ft^2, and a sill of 34,000 ft^2.

Because there are only two variables considered in this study, one cross semivariogram needs to be constructed. Figure 4 shows the resulting experimental semivariogram when taking the difference between the Pennsylvanian and the Wolfcamp data. It should be remembered that the data being used in this calculation are the residuals from a linear trend in each case. A model has been fitted to the graph. It is spherical with a range of 35 miles, a nugget effect of 15,000 ft^2 and a sill of 28,000 ft^2.

A cross-validation exercise was carried out using the above models and the multivariate universal co-kriging (MUCK) system. In this study, an average of zero and

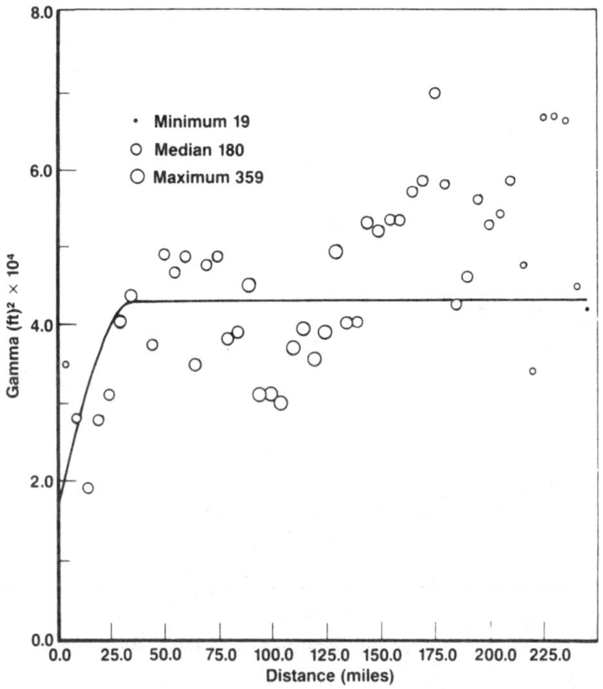

Figure 4. **Cross variogram for Pennsylvanian and Wolfcamp data.**

a standard deviation of 1.13 were obtained. These figures were considered acceptable for the purposes of this study.

The second stage of the analysis is to produce estimates of unsampled locations for each variable. To illustrate this process, a square grid of points has been estimated right across the study area. This grid can be used to produce maps for the potentiometric surface of each aquifer. Because a kriging technique is being used, a standard error is obtained for each grid point, enabling the construction of maps of the "reliability" of the estimation of the potentiometric level.

Figure 5 shows the MUCK-generated map for estimating the potentiometric head values in the Pennsylvanian aquifer. This estimation is carried out using both Pennsylvanian and Wolfcamp samples, and this is emphasized by the posting of the sample locations, showing intersections in the Pennsylvanian as solid circles and those in the Wolfcamp as stars. For comparison, Figure 6 shows the map estimated by universal kriging using only sample values for the Pennsylvanian aquifer. Visual comparison of the two maps shows local variation in many of the contours, particularly in those areas where the additional sampling is included by the MUCK estimation. This is particularly noticeable in the northeastern Panhandle and the central portion of the study area. The contours change very little in those areas where the Wolfcamp samples merely augment the existing Pennsylvanian sampling.

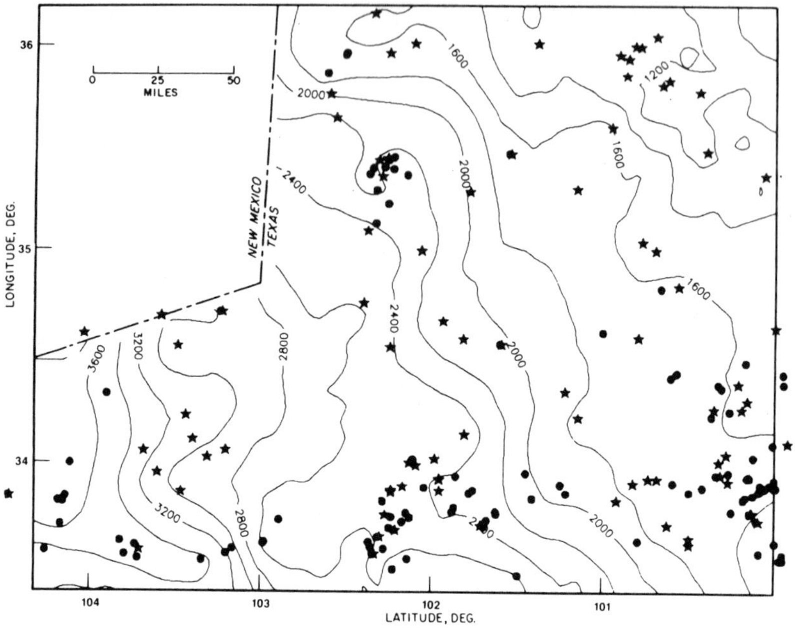

Figure 5. **Co-kriging estimates for potentiometric level—Pennsylvanian aquifer.**

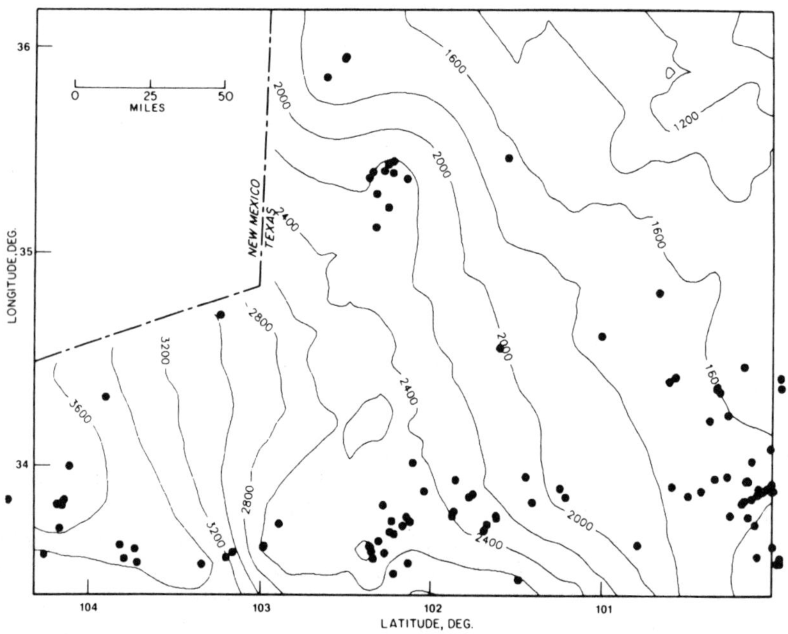

Figure 6. **Kriging estimates for potentiometric level—Pennsylvanian aquifer.**

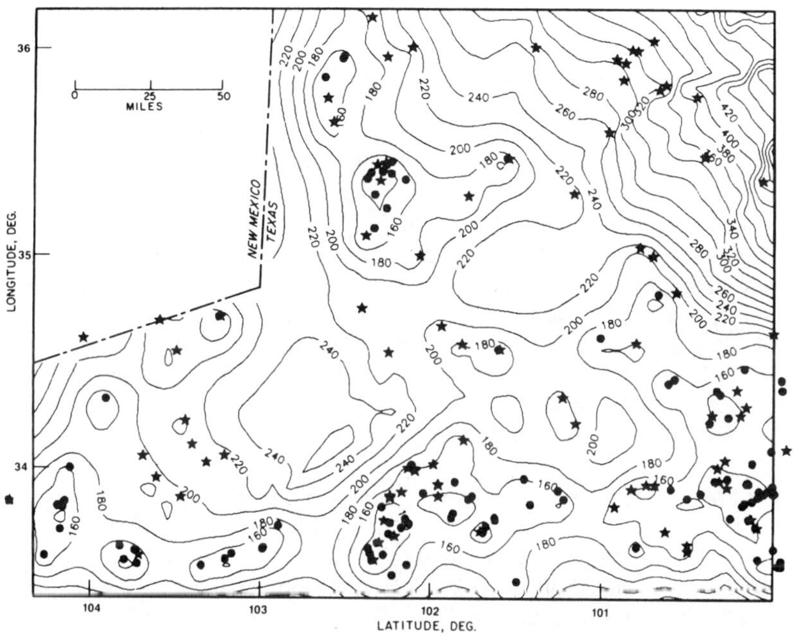

Figure 7. **Co-kriging standard errors for potentiometric level—Pennsylvanian aquifer.**

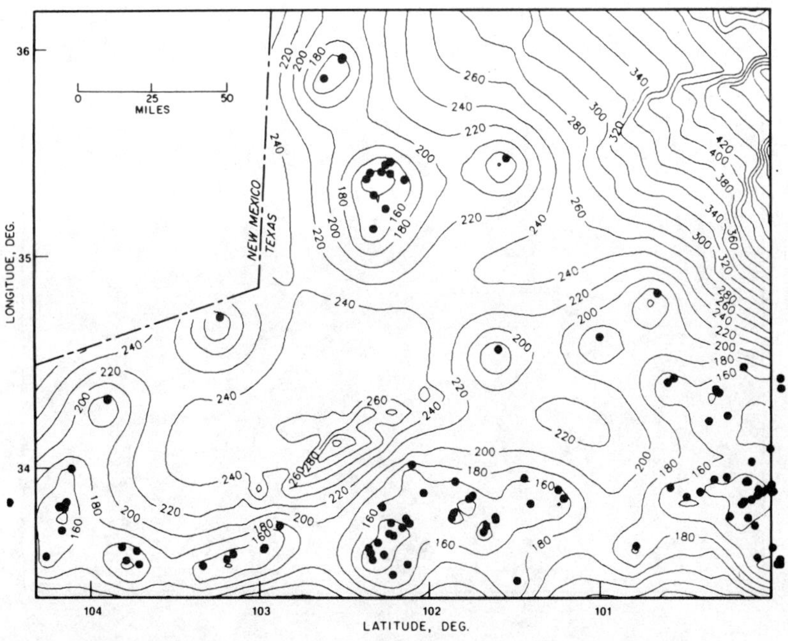

Figure 8. **Kriging standard errors for potentiometric level—Pennsylvanian aquifer.**

Figures 7 and 8 show the corresponding "standard error" maps for Figures 5 and 6, respectively. Visual comparison of this pair of maps reinforces the interpretation above. The inclusion of the Wolfcamp samples significantly reduces the standard errors in those areas where Pennsylvanian samples are scarce.

The MUCK system simultaneously produces estimates and standard errors for the Wolfcamp aquifer while mapping the Pennsylvanian. Figures 9 and 11 show the estimated and standard error maps respectively. A separate universal kriging exercise must be undertaken to produce Figures 10 and 12 for the Wolfcamp alone.

Visual inspection of each of these pairs of maps reveals that the incorporation of additional sampling, albeit for a different variable, can change the estimation of the potentiometric surface. However, this does not provide a quantitative measure of what, if any, advantage has been gained by using MUCK rather than the more conventional UK. In theory, MUCK should provide smaller standard errors than UK because more information is used in the production of the estimates, both in terms of numbers of data and in the additional modelling required for the cross semivariogram. If the multivariate approach has significantly improved the standard errors, the result will be a tighter confidence banding around the estimates of the surfaces.

The two methods were compared as follows. The standard error at each of the grid points was squared to provide the estimation variance. For each grid point, then, there is a univariate universal kriging estimation variance (UKEV) found by UK and a multivariate estimation variance found by MUCK. The UKEV is divided

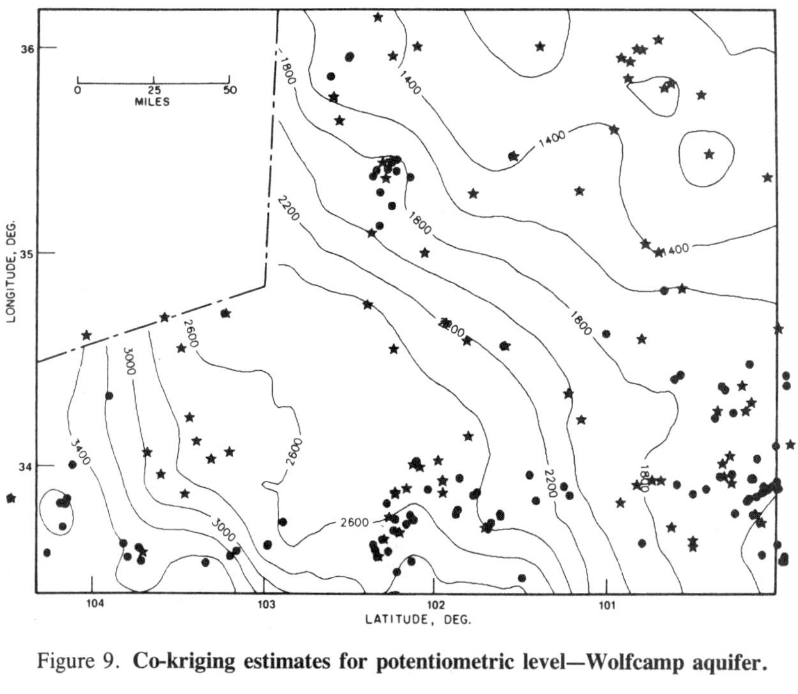

Figure 9. **Co-kriging estimates for potentiometric level—Wolfcamp aquifer.**

by the MUCK estimation variance to produce a variance ratio between the two methods. A variance ratio of exactly 1 indicates a location where the multivariate approach adds no extra reliability. Any value greater than 1 indicates a location where MUCK has improved the estimate.

Figure 13 shows a contour map of the variance ratio between UK and MUCK for the Pennsylvanian data. Both Pennsylvanian and Wolfcamp sample locations are shown on the map (circles and stars, respectively) to emphasize the multivariate nature of the MUCK procedure. It can be seen that the highest ratios are in locations where the Wolfcamp samples fill gaps left in the Pennsylvanian sampling. This map illustrates the reduction in estimation variance achieved by including sample data obtained from a different, but significantly correlated, variable. The variance ratios range from 1.0 to over 1.8, with an average over the study area of 1.15. A simplistic interpretation of this figure would be that the average increase in "efficiency" of the estimates is around 15%. Perhaps more accurately, we should take the square root of this figure and expect a general decrease in "standard error" of around 7.5%.

Figure 14 shows the corresponding variance ratio map for the Wolfcamp aquifer. Again, it can be seen that "improvement" is most noticeable at those locations where Pennsylvanian samples exist in gaps in the Wolfcampian sampling pattern. Here, the average variance ratio is 1.09, resulting in approximately 4% general improvement in the final confidence levels.

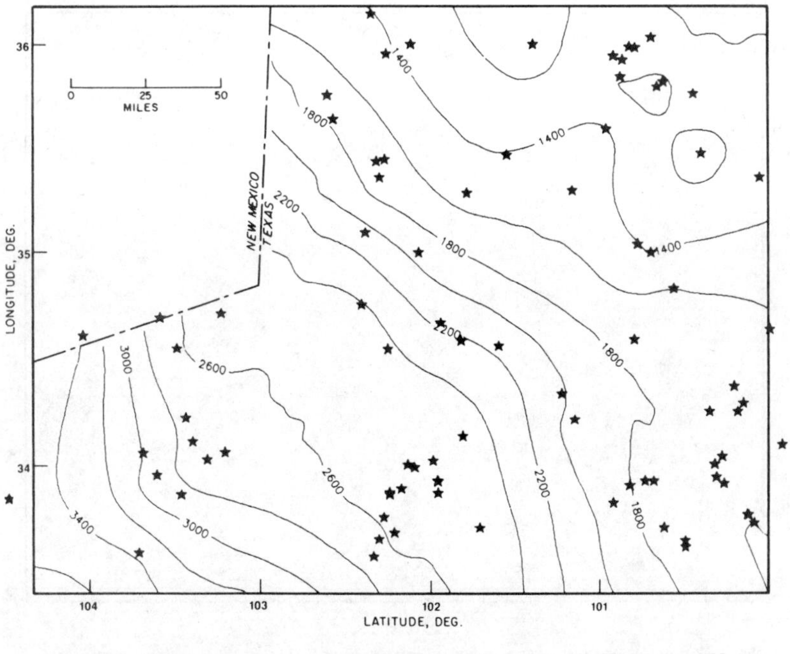

Figure 10. **Kriging estimates for potentiometric level—Wolfcamp aquifer.**

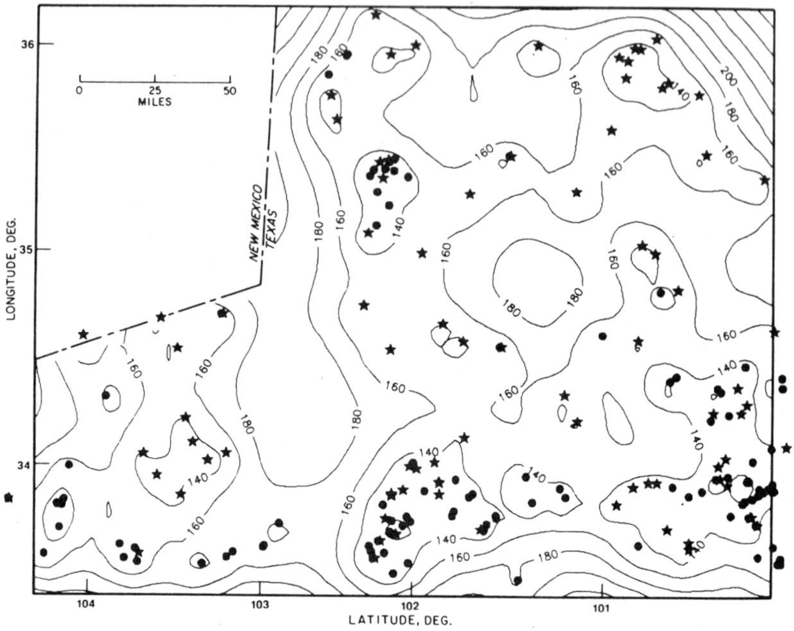

Figure 11. **Co-kriging standard errors for potentiometric level—Wolfcamp aquifer.**

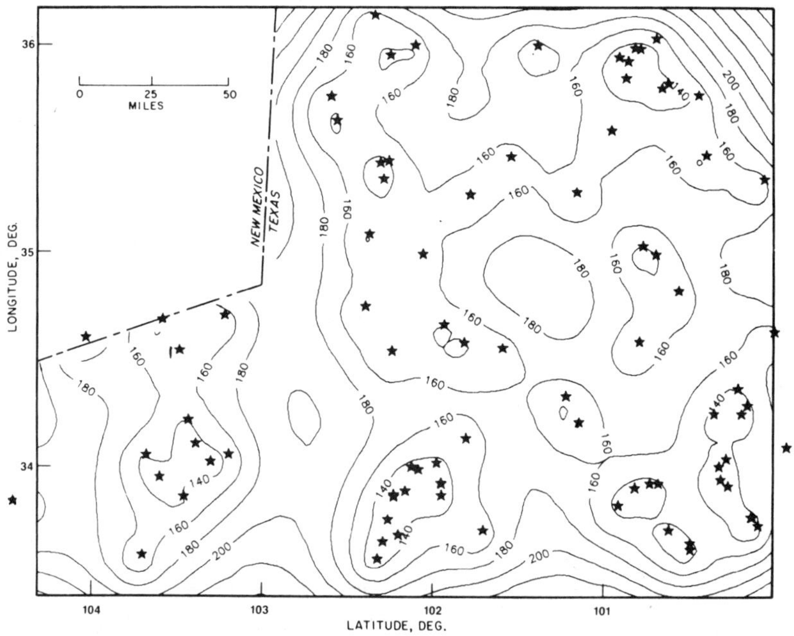

Figure 12. **Kriging standard errors for potentiometric level—Wolfcamp aquifer.**

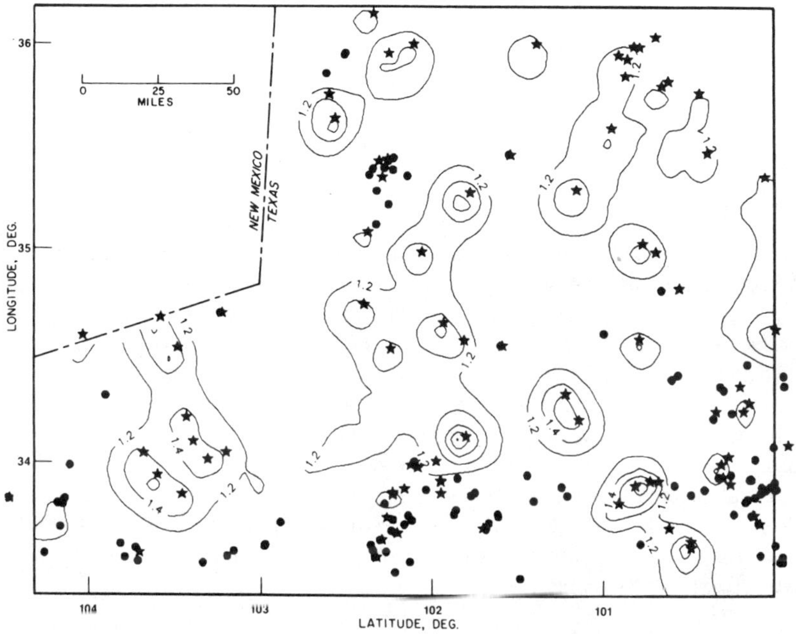

Figure 13. **Ratio between kriging and co-kriging variances—Pennsylvanian aquifer.**

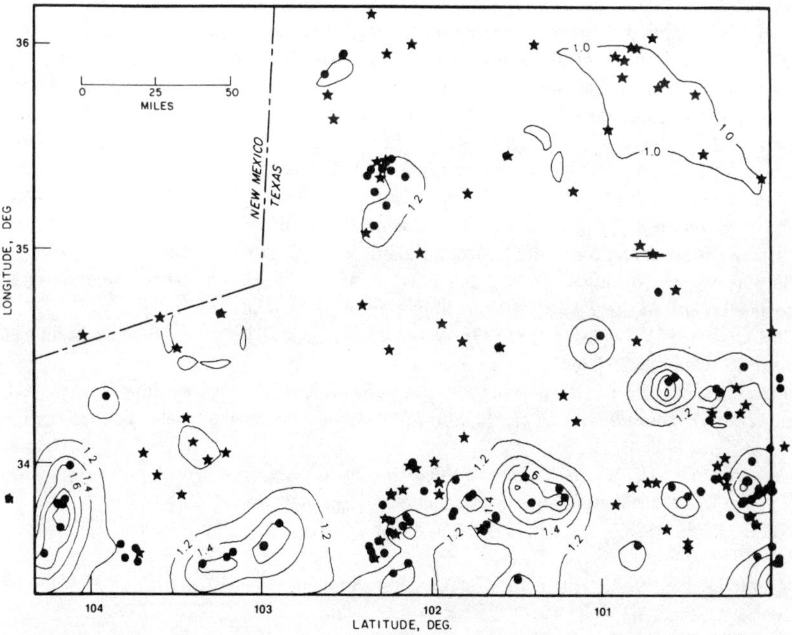

Figure 14. **Ratio between kriging and co-kriging variances—Wolfcamp aquifer.**

This case study shows that the estimation of a spatial variable can be enhanced using samples from another correlated variable sampled at different locations. It should be remembered that any reduction in uncertainty at this stage of estimation of the hydrogeologic variables will carry through to any later sensitivity and/or simulation analyses.

Summary

It has been suggested that estimation of spatially distributed variables might be improved if cross correlations between variables could be included in the estimation process. Traditional co-kriging has been considered and found to be inapplicable to the problem under study. A new formulation of the cross semivariogram is proposed that not only is applicable to this particular study, but also avoids many of the problems associated with previous co-kriging methods.

The new cross semivariogram model is used to produce a minimum variance unbiased estimation technique that has been dubbed multivariate universal co-kriging, partly to distinguish it from traditional co-kriging. The case study for the Texas Panhandle illustrates that MUCK will improve estimation to some extent throughout the study area but will have significant impact on those areas that have been undersampled for a particular variable.

Literature Cited

1. Bair, E. S., O'Donnell, T. P., and Picking, L. W., 1985. *Hydrogeologic Investigations Based on Drill-Stem Test Data: Palo Duro Basin Area, Texas and New Mexico*, BMI/ONWI-566, February, prepared by Stone & Webster Engineering Corporation for Office of Nuclear Waste Isolation, Battelle Memorial Institute, Columbus, Ohio.
2. Carr, J. R., and McCallister, P. G., 1985. "An Application of Co-Kriging for Estimation of Tripartite Earthquake Response Spectra," *Mathematical Geology*, *17*, 5, 527–545.
3. Carr, J. R., Myers, D. E., and Glass, C. E., 1985. "Co-Kriging—A Computer Program," *Computers and Geosciences*, *11*, 2, 111–127.
4. Harper, W. V., and Furr, J. M., 1986. *Geostatistical Analysis of Potentiometric Data in the Wolfcamp Aquifer of the Palo Duro Basin, Texas*, BMI/ONWI-587, April, Office of Nuclear Waste Isolation, Battelle Memorial Institute, Columbus, Ohio.
5. Harper, W. V., Basinger, K. V., and Furr, J. M., 1986. *Geostatistical Analysis of Potentiometric Data in the Pennsylvanian Aquifer of the Palo Duro Basin, Texas*, BMI/ONWI-680, January, 1988, Office of Nuclear Waste Isolation, Battelle Memorial Institute, Columbus, Ohio.
6. Guarascio, M., 1976. "Improving the Uranium Deposit Estimations (the Novazza Case)," in *Advanced Geostatistics in the Mining Industry*, M. Guarascio et al. (Eds.), D. Reidel, 351–367.
7. Matheron, G., 1971. *The Theory of Regionalized Variables and Its Applications*, Cahier No. 5, Centre de Morphologie Mathematique de Fontainebleau, 211 pp.
8. Myers, D. E., 1982. "Matrix Formulation of Co-Kriging," *Mathematical Geology*, *14*, 3, 249–257.
9. Myers, D. E., 1983. "Estimation of Linear Combinations and Co-Kriging," *Mathematical Geology*, *15*, 5, 633–637.
10. Myers, D. E., 1984. "Co-Kriging—New Developments," in *Geostatistics for Natural Resources Characterization*, G. Verly et al. (Eds.), D. Reidel, 295–305.

11. Unal, A., and Haycocks, C., 1986. "Geostatistical Coal Resource Characterization Using Intervariable Correlations," in *Proc. 19th APCOM Symposium,* Pennsylvania State University, April, SME-AIME (publ.), 231–239.

Mathematical Appendix

Ordinary kriging (OK)

To estimate the value, T, at an unsampled location A, we form a linear estimator:

$$T^* = \sum w_i g_i$$

where g_i is the observed value for sample S_i, w_i is the weighting factor given to sample i, and T* denotes an estimate of T.

The error incurred in the estimation may be expressed as (T − T*). It can be shown that (in the absence of trend):

$$E[T - T^*] = 0 \text{ if } \sum w_i = 1$$

The variance of the estimation errors then becomes

$$\sigma_\varepsilon^2 = E[(T - T^*)^2]$$

which, with a little algebraic manipulation, reduces to

$$\sigma_\varepsilon^2 = 2 \sum w_i \, \gamma(S_i,A) - \sum \sum w_i w_j \, \gamma(S_i,S_j) - \gamma(A,A)$$

where $\gamma(A, S_i)$ is the semivariogram value between the unsampled location and the ith sample, and so on.

Minimization of this variance with respect to the weighting factor, w_i, and subject to the unbiasedness constraint results in a system of equations as follows:

$$\sum w_j \, \gamma(S_i,S_j) + \lambda = \gamma(S_i,A) \text{ for } i = 1,2,3 \ldots n$$

$$\sum w_j = 1$$

This is the ordinary kriging system. The use of these weighting factors also results in a simplification of the estimation variance:

$$\sigma_{ok}^2 = \sum w_i \, \gamma(S_i,A) + \delta - \gamma(A,A)$$

Universal kriging (UK)

The universal kriging system is similar in form, with extra Lagrangian multipliers (λ) added for each term in the trend. For example, if a linear trend existed in a two-dimensional plane this would be expressed as

$$m(x,y) = b_o + b_1 x + b_2 y$$

Knowledge of the b coefficients is not necessary for the kriging stage, only the form of the trend. The constraints for unbiasedness then become

$$\sum w_j = 1$$

$$\sum w_j x_j = x_A$$

$$\sum w_j y_j = y_A$$

and $$\sum w_j \gamma(S_i, S_j) + \lambda_o + \lambda_1 x_i + \lambda_2 y_i = \gamma(S_i, A) \text{ for } i=1,2,3 \ldots n$$

The universal kriging variance simplifies to

$$\sigma_{uk}^2 = \sum w_i \gamma(S_i,A) + \lambda_o + \lambda_1 x_A + \lambda_2 y_A - \gamma(A,A)$$

Multivariate universal co-kriging (MUCK)

We illustrate this approach with a three-variable example. Suppose we wish to estimate the value of variable g at location A. We have n observations on g available in the immediate neighborhood. We also have m observations on another variable, f, and t observations on a third variable, r. We extend the above notation so that (for example) S_{fi} denotes the ith sample with a measured value of variable f and A_g the value of variable g at location A.

Our estimator now becomes

$$T^* = \sum w_i g_i + \sum v_j f_j + \sum u_k r_k$$

One set of constraints resulting in an unbiased estimator (in the absence of trend) is

$$\sum w_i = 1 \qquad \sum v_j = 0 \qquad \sum u_k = 0$$

Of course, if we were estimating the value of variable f at this location, the $\{w_i\}$ would sum to zero and the $\{v_j\}$ would sum to 1, etc.

The general form of the estimation variance becomes

$$\begin{aligned}\sigma_\varepsilon^2 = {} & 2 \sum w_i \gamma(S_{gi},A_g) + 2 \sum v_j \gamma(S_{fj},A_g) + 2 \sum u_k \gamma(S_{rk},A_g) \\ & - \sum \sum w_i w_p \gamma(S_{gi},S_{gp}) - 2 \sum \sum w_i v_q \gamma(S_{gi},S_{fq}) \\ & - \sum \sum v_j v_q \gamma(S_{fj},S_{fq}) - 2 \sum \sum w_i u_\ell \gamma(S_{gi},S_{r\ell}) \\ & - \sum \sum u_k u_\ell \gamma(S_{rk},S_{r\ell}) - 2 \sum \sum v_j u_\ell \gamma(S_{fj},S_{r\ell}) - \gamma(A_g,A_g)\end{aligned}$$

Minimization of this variance subject to the unbiasedness constraints gives

$$\sum w_i\, \gamma(S_{gi},S_{gp}) + \sum v_j\, \gamma(S_{fj},S_{gp}) + \sum u_k\, \gamma(S_{rk},S_{gp}) + \lambda_g = \gamma(S_{gp},A_g)$$

$$\sum w_i\, \gamma(S_{gi},S_{fq}) + \sum v_j\, \gamma(S_{fj},S_{fq}) + \sum u_k\, \gamma(S_{rk},S_{fq}) + \lambda_f = \gamma(S_{fq},A_g)$$

$$\sum w_i\, \gamma(S_{gi},S_{r\ell}) + \sum v_j\, \gamma(S_{fj},S_{r\ell}) + \sum u_k\, \gamma(S_{rk},S_{r\ell}) + \lambda_r = \gamma(S_{r\ell},A_g)$$

$$\sum w_i = 1$$

$$\sum v_j = 0$$

$$\sum u_k = 0$$

and a simplified co-kriging variance of

$$\sigma^2_{ock} = \sum w_i\, \gamma(S_{gi},A_g) + \sum v_j\, \gamma(S_{fj},A_g) + \sum u_k\, \gamma(S_{rk},A_g) + \lambda_g - \gamma(A_g,A_g)$$

With this formulation it is fairly simple to generalize to the universal co-kriging form, including the trend (if any) for each variable as extra constraints. We find that this formulation also highlights terms that remain constant, regardless of which variable is being estimated, and enables us to proceed to the multivariate case with ease.

For matrix formulation of these equations, see Myers (1982).

Some Techniques for Stochastic Modeling of Three-Dimensional Fracture Networks

KEVIN HESTIR and JANE LONG
Earth Sciences Division
Lawrence Berkeley Laboratory
Berkeley, California

JEAN-PAUL CHILÈS
Bureau de Recherches
Géologiques et Minières (BRGM)
Orléans, France

DANIEL BILLAUX
Earth Sciences Division
Lawrence Berkeley Laboratory
Berkeley, California
on leave from BRGM

Abstract

This paper describes some new techniques for stochastic modeling of three-dimensional fracture networks. We use geostatistical simulation methods to reproduce features of the spatial structure of the rock such as the variation of fracture density and fracture orientation in space. For an example of the method we use mapped fracture data from the Fanay-Augères mine, in Limousin, France. Two different sections of a drift wall, S1 and S2, were mapped. The S1 section is wet, and the S2 section is dry. For each case, the fractures are divided into five different sets and each set is modeled separately. The fractures in each set are represented as discs placed randomly in space. The diameter of each disc is chosen independently from a fixed probability distribution determined from the trace length distribution. For the location of discs a point process called the parent-daughter process is used. This process gives a clumping or swarming of fractures not found in the usual Poisson model. The orientation of the discs is characterized as a fluctuation about the mean orientation for the set. This fluctuation has a spatial structure that is simulated with geostatistics. Geostatistical simulations of the two fracture systems are under way. The connectivities of the simulations will be assessed to see if there is any correlation with the fact that the S1 section of drift is wet and the S2 section is dry.

Introduction

Over the past few years, attempts have been made to model the hydrology of fractured rock by collecting data on the individual fractures and synthesizing a model of the hydraulically active fracture network. Numerical models that have been developed for this purpose include our own (Long et al., 1982; Long and Witherspoon, 1985) as well as those of Robinson (1984), Dershowitz (1984), Rouleau (1984), Cacas et al. (1987), and others. These models have in some cases been applied to the analysis of field data, for example Rouleau (1984), Cacas et al. (1987), and Long and Billaux (1987). In Long and Billaux, we explored some techniques for using geostatistics to model some of the spatial structures commonly observed in fracture systems. In particular, we modeled the spatial variation of fracture density in a two-dimensional case.

In this paper, we extend the techniques to the case of three-dimensional fractures where we can also simulate the variation of fracture density, and fracture orientation in space. Data on the conductivity of individual fractures are extremely difficult to obtain, and we did not have such data. Therefore we do not attempt to stochastically generate this parameter. As data on the conductivity of individual fractures are so rare, perhaps the best way to assign this parameter in the model is to use a backfitting method such as that employed by Cacas et al. (1987). We do not pursue this procedure in this work. Instead, we concentrate on defining the network geometry in space and examining its connectivity.

One of the major problems in stochastic fracture modeling is the inability to obtain appropriate data directly. An ideal situation for fracture network simulation would be to have three-dimensional data for some large piece of the network. These data would give the geometry and conductivity of each fracture, including any channeling that may occur within the fractures. With this hypothetical information we could generate simulations that would approximately match the three-dimensional flow system. That is, if we knew all the geometry, we could model the flow.

This sort of ideal fracture data is not available. Lack of three-dimensional data makes the problem we face in constructing a simulation much harder. Typically the data we have from fractured rock provide a one-dimensional view through a three-dimensional system, i.e., a borehole. At best, we may have a two-dimensional sample such as a drift wall mapping.

With "perfect" two-dimensional data we would perhaps have a solvable problem. In this case we would still need to determine the three-dimensional shape of the fractures and the nature of flow through those fractures. Unfortunately, even interpreting correct statistics from two-dimensional data is difficult due to biases in sampling techniques.

The lack of three-dimensional data compounded with the less than "perfect" nature of two-dimensional data creates the requirement that we make some reasonable assumptions about the three-dimensional structure of the fracture network. These assumptions are used to build our conceptual model for the fracture mesh. The simulation is then based on the conceptual model.

The conceptual model we use here represents fractures as thin discs in three-dimensional space. The location of a disc is defined as the location of its center. The placement of fracture centers is determined by a stochastic process called the parent-daughter point process which is described below. Diameters of discs are assumed to be independent random variables. Orientation of fractures is taken to be a stationary random field. We will base the parameters of this model on the analysis of drift wall mapping data collected at the uranium mine of Fanay-Augères, in Limousin, France.

This paper gives the details of the conceptual model and the construction of these simulations. In a future paper, we will present the actual numerical simulations of fracture networks and an examination of the connectivity of these simulations.

Our presentation is divided into the following parts:

1. A description of the data.
2. Description of a conceptual model for the system.
3. Use of the data to determine the parameters required by the conceptual model.
4. Description of the techniques for numerical simulation of the fracture systems.
5. Conclusions.

Fanay-Augères Data

Fanay-Augères is a uranium mine owned by Cogema Co. and located in Limousin, France in the granite massif of Saint-Sylvestre. Since 1980 this mine has been used as a test facility to develop methods and tools for investigating mass and heat transfer in granitic rocks (Barbreau et al., 1985; Lassagne, 1983).

We focus on the data collected in a long section of a drift, about 3 m in diameter at the 320-m level (Figure 1). In this drift fractures on the east wall have

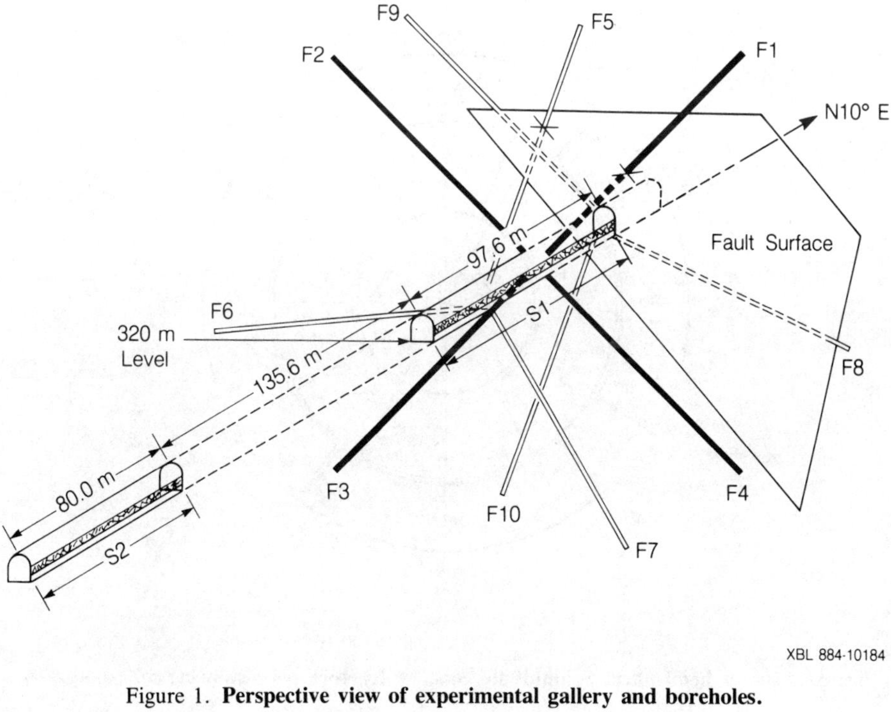

Figure 1. **Perspective view of experimental gallery and boreholes.**

been mapped over two sections, S1 and S2, totaling 180 m in length. S1 is 100 m in length and S2 is 80 m. We construct two simulations, one using data from S1 and one using data from S2. For any fracture trace which intersected a 2-m-high rectangle, the visible trace length, number and location of visible endpoints, orientation, and morphology were recorded. For 80% of the mapped fractures both endpoints were visible.

Ten boreholes 50 m long were drilled in three radial patterns (Figure 1). Oriented core was obtained from these holes and the fractures logged. More than 220 steady-state permeability tests were performed in these holes between packers spaced at various distances. In summary, these data provided information about the location, size, and characteristics of about 7000 fractures.

A usual step in fracture analysis is to divide fractures into sets and model each set separately. We have followed this course as explained in Long and Billaux (1987). Gros (1982) gives an analysis of the fracturing at Fanay-Augères that identifies seven major tectonic episodes from which he defined five major sets of fractures by their orientation. However, 60% of the fractures do not fall into any of the tectonic classes. This classification is clearly not adequate for our purposes. Therefore, we have broadened the definition of the sets so that 98% of the fractures mapped in S1 (1059 out of 1076) and 95% of the fractures mapped in S2 (1115 out of 1177) are included. The resulting classification, based on orientation, is illustrated with a lower hemisphere equal area stereonet in Figure 2. Figure 3 shows the fracture traces from S1 and S2.

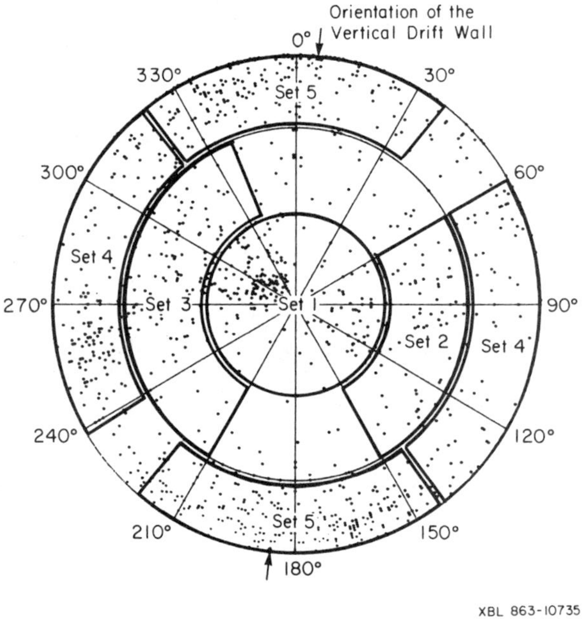

Figure 2. **Lower hemisphere Schmidt stereonet of fracture poles showing set boundaries.**

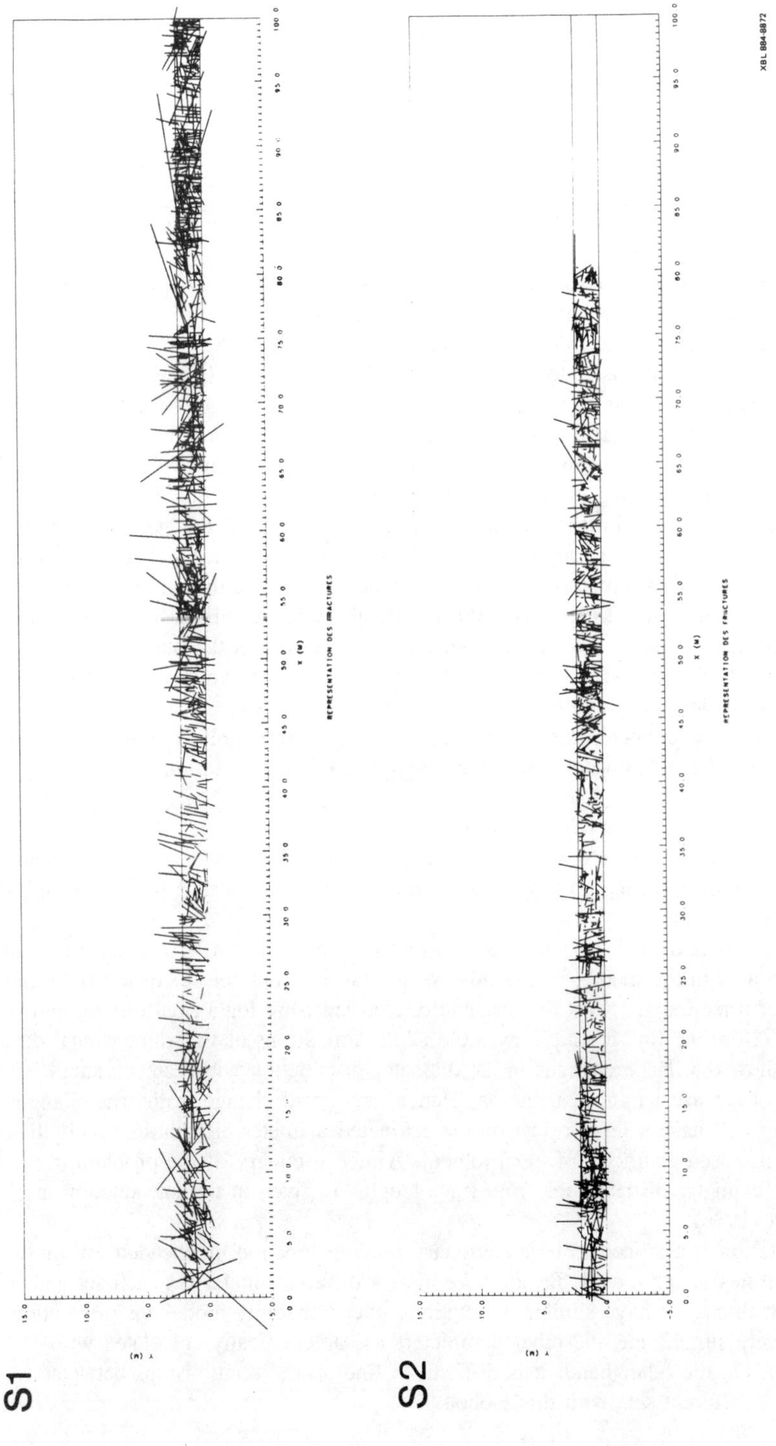

Figure 3. Traces of fractures from sections S1 and S2.

In summary, there are two data sets, one for S1 and one for S2. Each data set consists of the location and truncation of trace endpoints, and for each trace, an orientation and set classification.

Description of the Conceptual Model

The choice of a stochastic conceptual model for a three-dimensional fracture system involves specifying:

1. Individual characteristics of the fractures such as shape, size, orientation, aperture, or transmissivity.
2. The arrangement of the fractures in space, defined by the type of point process their locations follow and the way they truncate each other.

When dealing with three-dimensional fracture systems, many of these choices must be made somewhat arbitrarily because we can only take one-dimensional or two-dimensional samples from the three-dimensional system. However, once these choices are made, we can derive some of the statistical relationships between the parameters of the three-dimensional model we have chosen and the statistics that can be measured in the field. Except in the simple Poisson case, these statistical relationships do not give us a direct estimate of the parameters of the three-dimensional model, but rather enable us to derive the statistical properties on the plane from a guess at these parameters. So by trial and error we can match the observed statistics. We typically find that there are many different choices for parameters of the three-dimensional model that give the same two-dimensional statistics. We will call each of these choices a *parameter solution.* Hence, in statistical terminology, our problem is ill-posed, i.e., the data do not determine a unique parameter solution.

Individual characteristics of the fractures

We model fractures as circular discs in space. The disc assumption is reasonable in the absence of evidence on the actual shape of fractures, because it is the simplest possible assumption.

The distribution of disc diameters is assumed to be lognormal, independent of all other quantities in the model. Using this assumption it is possible to fit the field distribution of trace lengths with the distribution derived from lognormal distribution of diameters. Here we find a simple example of the limitations of two-dimensional data. One can show that large changes in the diameter distribution will result in small changes in the trace length distribution. Hence, any small change in the trace length distribution will have a large effect on the estimated diameter distribution. This illustrates the ill-posed character of our problem. A nice summary of the problem of estimating diameter distributions from trace lengths is given in the introduction in O'Sullivan (1986).

Orientation is assumed to be a stationary random function independent of all other quantities in the model. Because we have a different model for each set and fractures within a set have similar orientations, the probability model for orientation is a relatively simple one and other parameters are automatically correlated with orientation. On the other hand, it is difficult to find spatial relationships between fractures of different sets with this model.

Point processes and placement of fractures

Our approach to choosing a point process for locating fracture centers is to examine semivariograms of the three-dimensional point processes projected onto the two-dimensional plane. We will call these variograms the *trace density semivariograms.* We choose the theoretical three-dimensional point process that has a trace density semivariogram closest to the trace density semivariogram from the data. This analysis is discussed below.

A point process is defined as a random phenomenon, each realization of which is a set of various points dispersed in space, with no infinite accumulation in any bounded subset. The two-dimensional sample we work with is a realization of fracture traces as shown in Figure 3 for S1 and S2. Placing points at the centers of traces creates the two-dimensional point process which is the projection of the original three-dimensional process we wish to determine.

For any point process **N**, we define the random variable **N**(B) as the number of points inside a set B. To examine the structure of point processes in two dimensions we use the following definitions. Let $\vec{x}$ denote a point in two-space, $\mathbf{R}^2$. $B_{\vec{x}}$ will denote a small subset of $\mathbf{R}^2$, for example a square, centered on the point $\vec{x}$. **N** will denote a point process which is a two-dimensional projection of a three-dimensional point process. We define the semivariogram, $\gamma(\vec{x}, \vec{y})$, of $\mathbf{N}(B_{\vec{x}})$ via

$$\gamma(\vec{x}, \vec{y}) = \frac{1}{2} E((\mathbf{N}(B_{\vec{x}}) - N(B_{\vec{y}}))^2).$$

For the point processes we consider, $\gamma(\vec{x}, \vec{y})$ is a function of only the difference, $\vec{x} - \vec{y}$. This is the standard property of second-order stationarity. For a discussion of the geostatistical terminology concerning semivariograms see, for example, Journel and Huijbregts (1978). This terminology is used freely in the following.

Below, several candidate point processes and their trace density semivariograms are described. For a general overview of point processes the interested reader should consult Ripley (1981) or Stoyan et al. (1987).

The Poisson process

The simplest and most widely used point process for placement of fractures is the Poisson process. The projection of a Poisson process into two-dimensional space is again a Poisson process. If a point process **N** is Poisson with density λ, then **N**(B) is a Poisson random variable with rate λA where A = area (B). This implies that $E(\mathbf{N}(B_{\vec{x}})) = \mathrm{Var}(\mathbf{N}(B_{\vec{x}}))$. Also for a Poisson process, $\mathbf{N}(B_1)$ and $\mathbf{B}(B_2)$ are independent random variables when B_1 and B_2 are disjoint. This implies the semivariogram of $\mathbf{N}(B_{\vec{x}})$ is a pure nugget when $|\vec{x} - \vec{y}|$ is large enough so that $B_{\vec{x}}$ and $B_{\vec{y}}$ are disjoint.

The regionalized Poisson process

This is a Poisson process with a density varying in space. The density is a stationary random function, which, if Gaussian, can be characterized by its semivariogram and constant mean. In this case the semivariogram of $\mathbf{N}(B_{\vec{x}})$ now has a structure, but locally, when the random density has a small variance, $\mathbf{N}(B_{\vec{x}})$ is approximately equal to its mean as in the pure Poisson process. Such point processes are also known as doubly stochastic Poisson processes. Doubly stochastic processes are very general and theoretically include the parent-daughter processes. However, in our case the density can have a large variance and is best characterized using the processes described below.

The parent-daughter process

Starting from a Poisson process, a cloud of points (or daughters) is placed around each point (called a parent or seed) of the Poisson process. The number of points in each cloud is a Poisson random variable, each point being placed in relation to the seed independently of the other points according to a specified probability distribution. This process enables us to model the fact that fractures often occur in swarms. In this case, the semivariogram of $\mathbf{N}(B_{\vec{x}})$ still shows a pure nugget, but the variance of $\mathbf{N}(B_{\vec{x}})$ is greater than its mean.

The regionalized parent-daughter process

This is a parent-daughter process generated by parents from a regionalized Poisson process. For this process, the semivariogram of $\mathbf{N}(B_{\vec{x}})$ has a structure, and locally the variance of $\mathbf{N}(B_{\vec{x}})$ is greater than its mean.

Choice of a point process

The choice of the point process for our model is based on the trace density semivariograms derived from the data. Figure 4 shows the trace density semivariograms from S1. In all cases we take the support, B, to be a 2-m × 5-m rectangle.

The experimental semivariograms of **N** show two different behaviors, depending on the set. One is a true nugget effect, and the other is a semivariogram with steps, which is indicative of nested structures. The step semivariograms are the most frequent, and these are incompatible with a simple Poisson point process. For all cases, the ratio of the variance of **N** to its mean is generally between 4 and 5. For a simple Poisson process the ratio should be one. The regionalized Poisson process could represent the fracture centers but the underlying random rate would have to fluctuate in a way that would be difficult to describe.

In a relatively simple way, the regionalized parent-daughter process can explain both the structured aspect of all of the semivariograms for **N**, and the high ratio of the variance of **N** to its mean. The nested aspect of the semivariograms will then be due to the presence of the clouds (or swarms) of fractures, the shorter range in the semivariogram being related to the size of these clouds, and the longer range being indicative of the structure of the parents. This model was chosen.

Parameter Inference: Fitting the Model

As mentioned in the introduction, the biggest problem in fitting the model to the data is using two-dimensional information to draw inferences about three-dimensional quantities. Statistical tools developed for this purpose are described in this section.

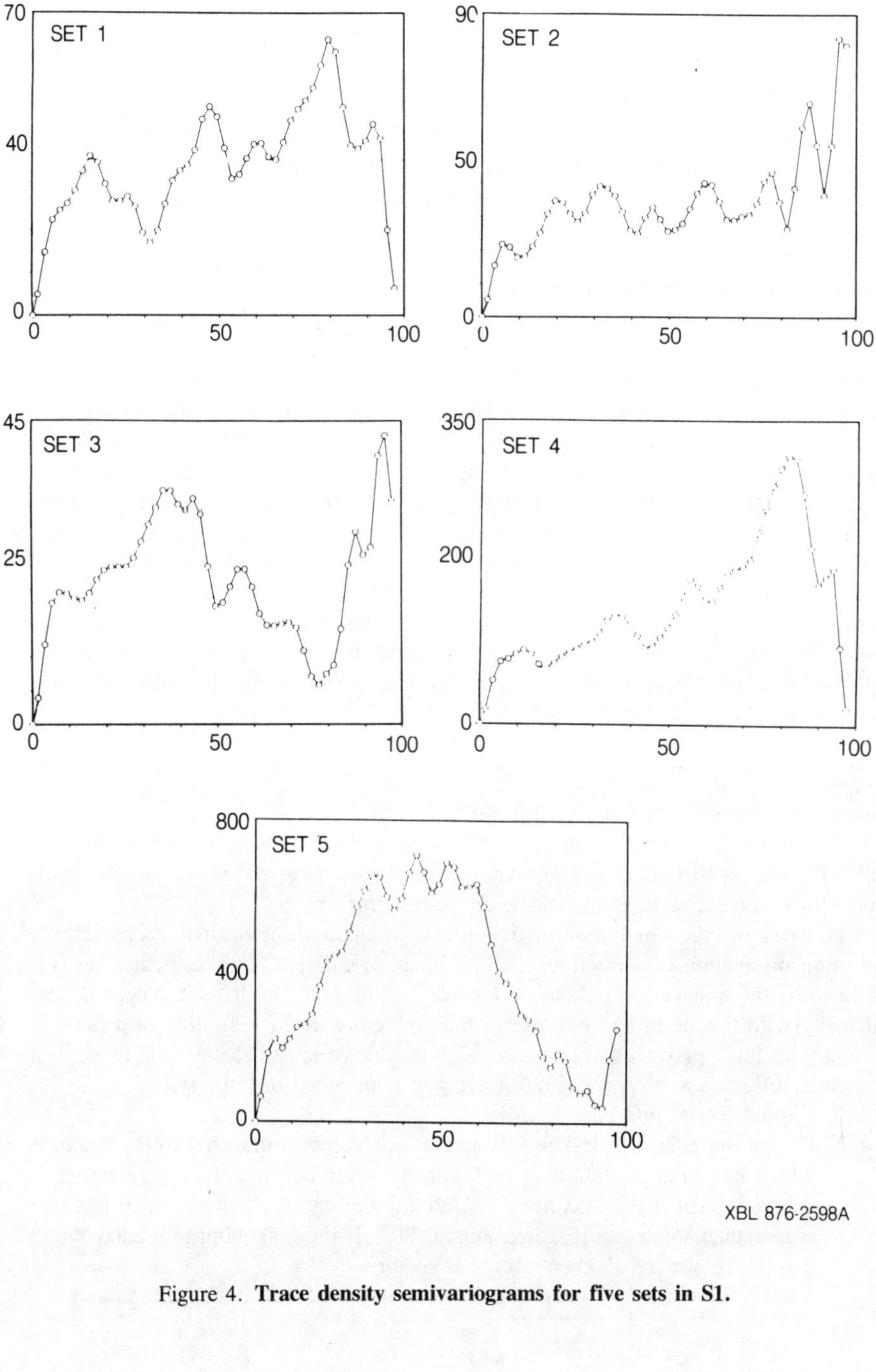

Figure 4. **Trace density semivariograms for five sets in S1.**

The modeling is done set by set so we will have each of the parameter values for each set in S1 and S2. The quantities we need to find are

- Parameters that describe the stochastic process of fracture orientation.
- Average and standard deviation of the lognormal diameter distribution.
- Parameters that describe the parent-daughter process of fracture locations in space.

When we have specified all of these quantities we have determined a parameter solution and fitted the model to the data. In this section we discuss the determination of each of these model parameters. We begin by describing the censoring of data from small traces to compensate for measuring biases.

Censoring of small trace length data

We wish to compare our simulation of the S1 section of the drift with the S2 section. Accordingly, we want to eliminate any differences in the way data were recorded in the sections. Recall Figure 3 is a plot of all the fracture traces recorded in S1 and S2. The traces in these two cases look similar. However, one difference between S1 and S2 is a higher fraction of smaller fractures appearing in S2. Because these data were collected by several groups of workers we suspect that this difference is the result of biases in the recording of data from small traces. Also, because fractures occur down to microscopic scales one must define a cutoff length below which no fracture traces are considered. When we censor all data from traces with lengths below 0.3 m from each of S1 and S2 we find similar histograms for the remaining trace lengths. Hence, this censoring seems to equalize these measurement biases. We therefore use only these censored data realizing that we have an underestimation of the density of traces in our sampling plane. After censoring there is a total of 1008 fractures for all 5 sets in S1 and 769 fractures for all 5 sets in S2.

Parameters for orientation distribution

We model the orientation as a random field in three-dimensional space. Specifically, the idea is to find an average orientation for each set and model the deviation from the average as a random field with spatial structure.

To carry out the orientation modeling and simulation one must first characterize the mean orientation of a fracture set. The mean orientation for a particular set is taken to be the unit vector parallel to the sum of the poles in that set. There is some difficulty with this definition because of the ambiguity in the definition of a pole; if $\vec{x}$ is a pole for a particular fracture so is $-\vec{x}$. So we must calculate the average in a slightly different way. A meaningful average is obtained in three steps:

1. Choose a trial reference vector.
2. Define the pole of a fracture to be the unit vector orthogonal to the fracture which has an angle less than 90° with the reference vector. The reference vector is kept if the maximum angle between any two poles defined this way is less than some cutoff, for example, 90°. If the maximum is greater than the cutoff, we try another reference vector.
3. Once we have found a reference vector and used it to define the poles we

sum those poles and normalize the sum. This is the pole of the mean orientation.

To continue the modeling and simulation of orientation we next have to describe the deviations of the poles in a fracture set from the mean orientation of that set. To do this the mean orientation $\vec{p}$ (Tables 1a and 1b) is calculated for each set and the frame of reference is rotated so that $\vec{p}$ is on the z axis. The rotation matrix we use appears in Mardia (1972) and corresponds to a rotation first about the y-axis then about the z-axis. A scatter plot of the x and y coordinates of the rotated poles represents the distribution of the orientations about the mean. This is shown for the S1 data in Figure 5.

Finally, to simulate the orientation, we simulate values of x and y. A brief description of the procedure for geostatistical simulation is given in the Appendix. As described in the Appendix, the first step in simulation is to transform x and y into Gaussian variables. Such a transformation is called anamorphosis. After anamorphosis, x and y are essentially uncorrelated as shown in Tables 2a and 2b. Hence, for purposes of simulation x and y can be taken to be independent.

The next step in the simulation of x and y is the characterization of their semivariograms. Example semivariograms are shown for S1 in Figure 6. One can see that the nugget effect in all of them is fairly large. Some of this nugget effect is due to measurement errors, which were evaluated by Massoud and Chilès (1985). The rest of the nugget is simulated by a white noise. The structured part is simulated on a regular grid following techniques outlined in the Appendix. The relevant parameters for S1 and S2 are given in Tables 1a and 1b. The structured parts of x and y for a given fracture are then found by choosing the values at the simulated grid point closest to the fracture center.

Finally, the orientation of a given fracture is taken as the mean for the set plus a component of white noise plus the structured component from the geostatistical simulation.

Determination of the disc diameter distribution

The trace length data are used to find the diameter distribution. As we have mentioned before the determination of the diameter distribution from trace lengths is an ill-posed problem. One reason for this is that the sampling area is finite so that many traces are truncated by the boundaries of this area. Hence, information about the upper tails of the trace length distribution and diameter distribution is not available. A second reason is that the process of finding the diameter distribution from the trace length distribution is sensitive to any biases or noise in the data. This has the effect that a small change in the trace length distribution causes a large change in the resulting diameter distribution.

A functional relationship between the distribution of fracture trace lengths and the distribution of fracture diameters not accounting for truncation effects was developed by Warburton (1980), and is given by

$$f(1) = 1/\overline{D} \int_{1}^{\infty} \frac{g(D)}{\sqrt{D^2 - 1^2}} \, dD .$$

Table 1a. **Mean orientations and parameters of reduced orientation variograms for S1.**

Set	Mean orientation		x variogram**			y variogram**		
	θ* °	ϕ °	nugget	sill	range m	nugget	sill	range m
1	289	9.6	1.0	1.5	40	0.75	1.3	10
2	274	44.2	1.0	1.25	40	1.0	1.0	-
3	79	46.2	1.0	1.0	-	1.0	1.0	-
4	351	87.1	0.75	1.12	50	0.75	1.5	50
5	263	88.0	0.8	1.0	50	0.8	1.25	40

*These are standard spherical coordinates with positive x-axis pointing north and positive y-axis pointing west.

Table 1b. **Mean orientations and parameters of reduced orientation variograms for S2.**

Set	Mean orientation		x variogram**			y variogram**		
	θ* °	ϕ °	nugget	sill	range m	nugget	sill	range m
1	231	4.5	.75	1.2	13	1.0	1.0	-
2	250	45.7	.75	1.25	40	0.9	1.4	18
3	66	44.9	1.0	1.9	50	0.9	0.9	-
4	1	88.8	0.95	1.20	30	1.10	1.10	-
5	276	81.8	0.8	0.8	-	1.10	1.50	50

*These are standard spherical coordinates with positive x-axis pointing north and positive y-axis pointing west.

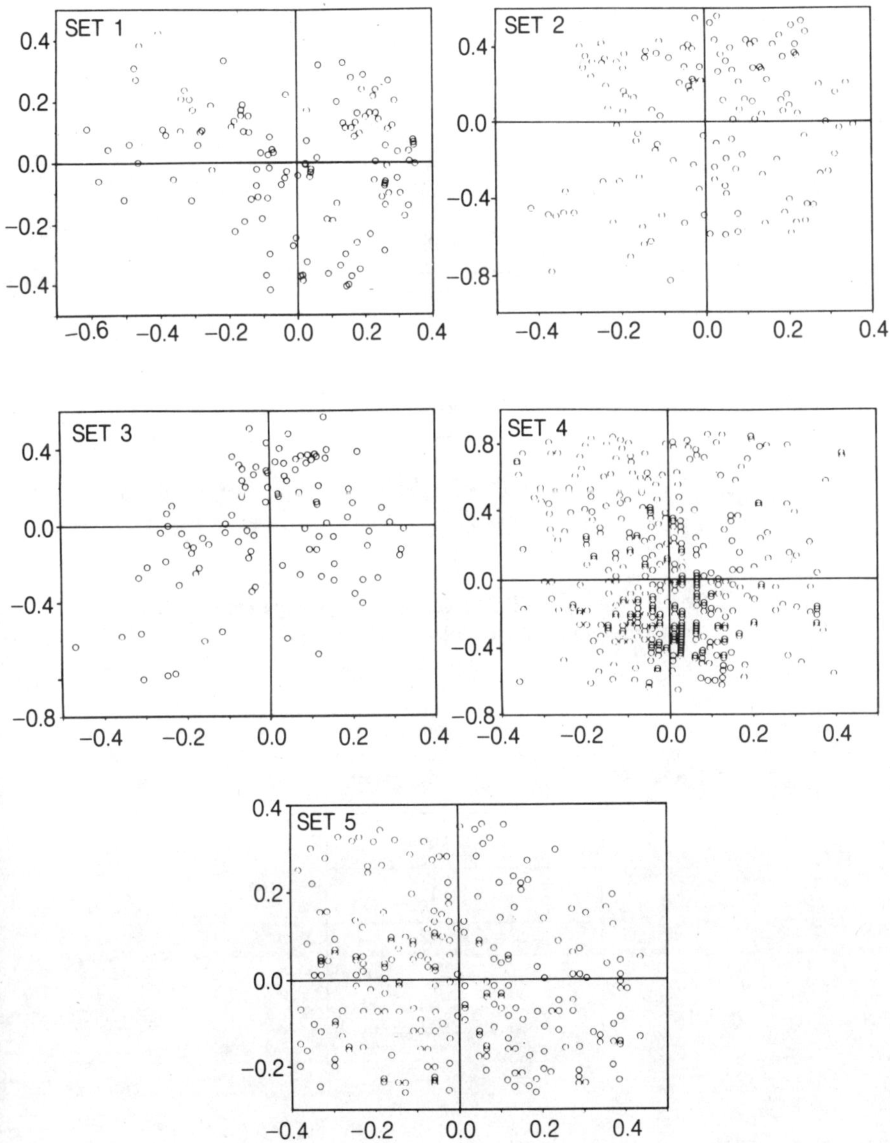

XBL 876-2601A

Figure 5. **Scatter plot of rotated orientations before anamorphosis for five sets in S1.**

Table 2a. **Correlation coefficients for x and y components of the orientation data after anamorphosis, S1.**

Set	Correlation coefficient
1	-.136
2	0.092
3	0.283
4	-0.141
5	-0.128

Table 2b. **Correlation coefficients for x and y components of the orientation data after anamorphosis, S2.**

Set	Correlation coefficient
1	0.194
2	-0.068
3	-0.126
4	0.196
5	-0.038

where

D = diameter of a given fracture,

$g(D)$ = pdf (probability density function) of fracture diameters,

$f(l)$ = pdf of fracture traces,

$$\overline{D} = \text{mean of } D \equiv \int_0^\infty D\, g(D)\, dD.$$

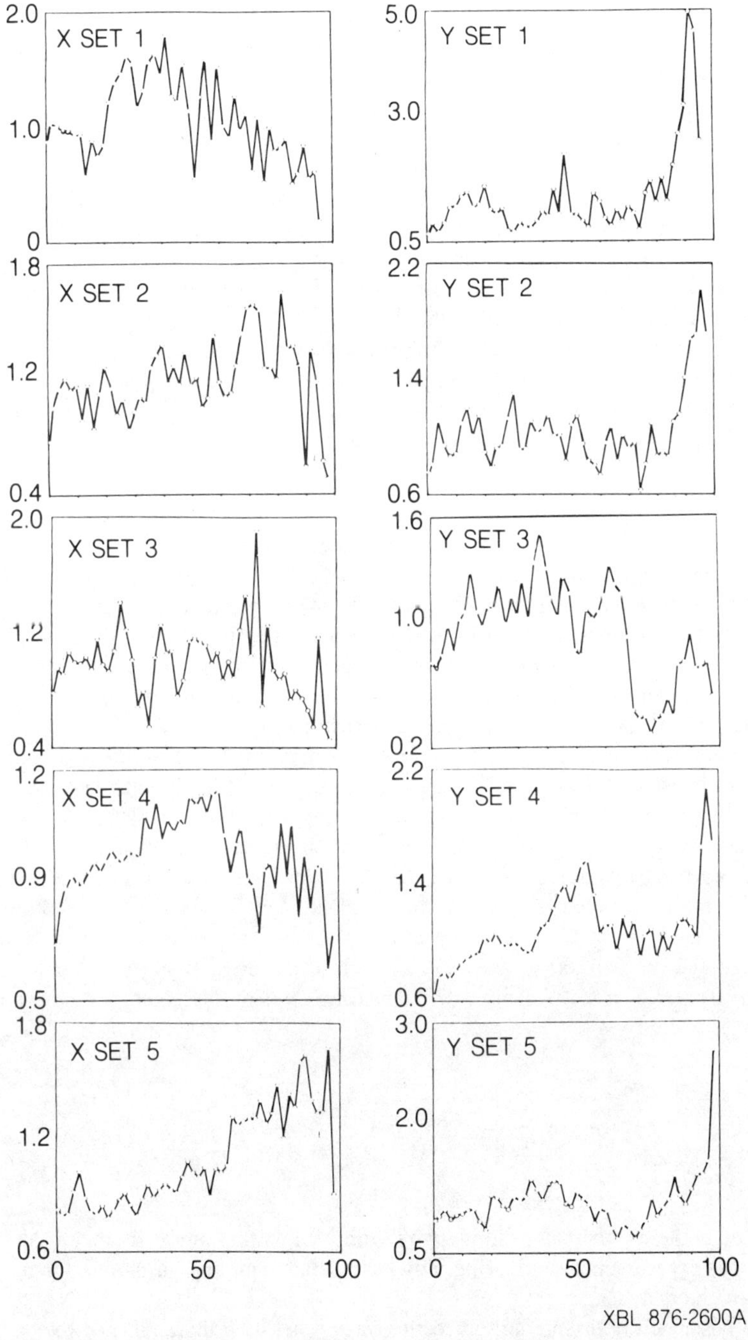

Figure 6. **Semivariograms of orientation after anamorphosis for five sets in S1.**

So, given a particular probability density function for fracture diameters, g(D), we can evaluate the integral to obtain the density of trace lengths f(l). One can then compare the computed density of trace lengths to that obtained from the field data. By trial and error, one can find a reasonable, but not necessarily unique distribution for fracture diameters. This process ignores truncation effects.

To include the effects of truncation, we use a Monte Carlo procedure that simulates traces in the drift. First, we assume a lognormal distribution for fracture diameters with a given standard deviation and mean. Next a number of discs with this diameter distribution are placed in space with orientations chosen at random from measured orientations in the drift. We calculate the intersections of these fractures with a rectangle representing the drift wall. The resulting traces are analyzed to get a trace length distribution including truncation percentages and separate distributions for traces where one or two endpoints appear. These distributions are compared to those from the data. This process is repeated for a range of values of mean and standard deviation of diameters until a good match is found between actual trace lengths and simulated trace lengths.

Calculation of the trace density semivariogram from the data

To complete the model fitting we need to find the parameters of the parent-daughter point process which describes the location of fracture centers. We have said that these parameters are found by matching plane density semivariograms. Figure 4 shows the trace density semivariograms calculated from the S1 data. A similar computation was done for S2. We describe how these semivariograms were calculated.

The drift mappings are in rectangles measuring 100 m × 2 m and 80 m × 2 m respectively (Figure 3). Length is in the x direction and width in the y direction. Hence, 0 m ≤ x ≤ 100 m or 80 m and 0 m ≤ y ≤ 2 m. To calculate the experimental trace density semivariogram we count the number of traces in a window that measures 5 m × 2 m and is centered at the point $\vec{x}$, then multiply this number by an "orientation correction factor" that adjusts for the fact that the fractures are not perpendicular to the sampling plane. Let $E_{\vec{x}}$ denote the resulting number of traces adjusted for orientation in the window centered at $\vec{x}$. We slide the window centered on $\vec{x}$ = (x,y), across the length of the drift, fixing y = 1 and letting x range from 2.5 m to 97.5 m (or 77.5 m for S2) using increments of size 1 m. γ(h) is estimated via

$$\gamma(h) \approx \frac{1}{2}\frac{1}{n_h}\left\{\sum_{\{(\vec{z},\vec{w}):|\vec{z}-\vec{w}|=h\}} (E_{\vec{z}}, - E_{\vec{w}})^2\right\}$$

where n_h is the number of window pairs with centers that are a distance h apart. Typically, experimental semivariograms have small sampling variations for h less than half the drift length.

To describe the orientation correction factor, let θ be the angle between a fracture pole and the sampling plane measured in degrees. We always take $0° \le \theta < 90°$. If a fracture with angle θ is placed randomly in space the probability that the fracture intersects the sampling plane is proportional to $\cos\theta$. Hestir et al. (1987) use this result to show that the orientation correction factor for a particular window is

$$\frac{1}{\overline{\cos(\theta)}},$$

where $\overline{\cos(\theta)}$ is the average of the $\cos(\theta)$'s for all fractures in the window.

Theoretical trace density semivariogram for the parent-daughter model

We now describe how to find the parameters for the parent-daughter model. These parameters define the distributions of the density of parents and of the number of daughters per parent, and the dispersion of the daughters around the parents. In the Appendix, we derive the theoretical two-dimensional trace density semivariogram results from a given parameterization of the three-dimensional parent-daughter model. This enables us to establish the relationship between the parameters of the parent-daughter model and the experimental semivariogram. These formulas were originally worked out in Deverly (1986) for a model of nugget deposits. In our derivation for discs, the trace density semivariogram is calculated by conditioning on a fixed realization of parents (i.e., assume there is a fixed realization of parents) and then removing the conditioning. The end results along with some notation are shown below.

Let $\mathbf{x}$ be a point in three-dimensional space, R^3 and $\Lambda_p(\mathbf{x})$ be the stationary random process for the parent rate. We assume that Λ_p is a Gaussian process with covariance structure defined by a spherical semivariogram with a given range and sill. Also let

$\overline{\lambda}_p$ = $E(\Lambda_p(\mathbf{x}))$, the average of Λ_p.

$C_{\Lambda_p}(\mathbf{h})$ = $E(\Lambda_p(\mathbf{x} + \mathbf{h})\Lambda_p(\mathbf{x}))$, this is the non-centered covariance of Λ_p.

λ_D = rate of daughters, i.e., the number of daughters is taken to be Poisson with rate λ_D.

$F(\mathbf{x})$ = probability distribution function for the location of a daughter relative to the parent. In our model we always assume the displacements $(x,y,z) = \mathbf{x}$, are independent, identically distributed, mean 0, Gaussian random variables.

$B, \hat{B}$ be windows in the sampling plane.

$Q(B)$ = the number of traces in B.

$\Psi(B,\mathbf{x})$ = probability that a daughter from a parent at $\mathbf{x}$ intersects B.

The parameters that specify the model are

- The average, range, and sill of the parent rate Λ_p,
- The rate, λ_D, of the Poisson number of daughters around each parent, and
- The standard deviation of the Gaussian dispersion of daughters about the parents.

In terms of these parameters we find, as shown in the Appendix:

$$E(Q(B)) = \overline{\Lambda}_p\lambda_D \int_{R^3} \Psi(B,\mathbf{x})d\mathbf{x} = \overline{\Lambda}_p\lambda_D \,\text{area}(B)\overline{D} \qquad (1)$$

and

$$E(Q(B)Q(\hat{B})) = \lambda_D{}^2 \int_{R^3}\int_{R^3} \Psi(B,\mathbf{x})\,\Psi(\hat{B},\mathbf{y})C_{\Lambda_p}(\mathbf{x}-\mathbf{y})d\mathbf{x}d\mathbf{y} \tag{2}$$

$$+\ \overline{\Lambda}_p\lambda_D{}^2 \int_{R^3} \Psi(B,\mathbf{x})\Psi(\hat{B},\mathbf{x})d\mathbf{x}$$

$$+\ \overline{\Lambda}_p\lambda_D \int_{R^3} \Psi(B\cap\hat{B},\mathbf{x})d\mathbf{x}\ .$$

Numerical modeling: Generation of the fracture mesh

Once the parameter solution is available, the following steps are taken to create a realization of the fracture system.

1. A simulation of Λ_p is made in a three-dimensional region. We assume that Λ_p is a Gaussian random field with average large enough in comparison to its standard deviation so that negative values for rate occur with low probability. Negative values are set at zero. The covariance structure is specified by a spherical semivariogram having a range and sill given in the parameter solution.
2. Parents are simulated on small subregions, $\Delta V_{\mathbf{x}}$, centered at points $\mathbf{x}$, with constant volume $\Delta\mathbf{x}$. To simulate the parents in $\Delta V_{\mathbf{x}}$, we first find the value of $\Lambda_p(\mathbf{x})$ (if $\Lambda_p(\mathbf{x}) < 0$ we set $\Lambda_p(\mathbf{x}) = 0$).
3. Next, we pick the number of parents in ΔV_x according to a Poisson process of rate $\Lambda_p(\mathbf{x})\Delta\mathbf{x}$.
4. Each parent in ΔV_x is placed at random in ΔV_x according to a uniform distribution.
5. For each parent we pick a number of daughters according to a Poisson distribution with rate λ_D.
6. Next, the random locations of fractures (daughters) are determined. This is done by choosing displacements from the parent in the x, y, and z directions independently according to a Gaussian distribution with mean 0 and variance specified by the parameter solution.
7. The orientation is determined in three steps. Start with the mean for the set. Add the structured part from the orientation simulation described above. Finally, add a random component as determined by the nugget of the orientation semivariogram.
8. The fracture diameter is chosen from its global distribution.

To check the simulations, we can take a sample of generated fractures and calculate the sample average for diameter and orientation to make sure they fall within reasonable limits. We can also record the number of fractures generated in the generation volume V, and check to see that this falls in a reasonable range. To do this note that

$$E(\mathbf{F}(V)) = \overline{\Lambda}_p \lambda_D \mathrm{volume}(V)$$

and

$$\mathrm{Var}(\mathbf{F}(V)) = \overline{\Lambda}_p \lambda_D \mathrm{volume}(V) + \lambda_D^2\, \overline{\Lambda}_p \int_{R^3} \Psi^2(V,\mathbf{x}),d\mathbf{x}$$

$$+ \lambda_D^2 \int_{R^3}\int_{R^3} \left[C_{\Lambda_p}(\mathbf{x}-\mathbf{y}) - \overline{\Lambda}_p^2\right] \Psi(V,\mathbf{x})\, \Psi(V,\mathbf{y})\, d\mathbf{x}d\mathbf{y}\ .$$

where $\mathbf{F}(V)$ denotes the number of fractures occurring in volume V. Equations (1) and (2) can be used to verify these expressions with $\Psi(V,\mathbf{x})$ replacing $\Psi(B,\mathbf{x})$.

Connectivity Study

We can also make a study of the connectivity in our mesh simulations. Observations at Fanay show that the section of the drift called S1 is wet and the section labeled S2 is dry. We expect that our connectivity studies will reflect this difference.

A measure of connectivity is defined as follows. The first step is to place an artificial fracture somewhere in the center of our simulation; this is our starting point which we call the level 0 fracture. Level 1 fractures are defined to be all fractures that intersect the level 0 fracture. In general, level k fractures are defined to be all fractures that intersect a level k-1 fracture and have not been previously assigned to a level. The recorded number of fractures in each level is our measure of connectivity. We call these the level sizes.

One way to view this measure of connectivity is as a simple summary of the number of paths through which water can flow if injected into the level 0 fracture. Some work needs to be done to examine growth and the variability of the level sizes in the simulations.

A practical limitation of this idea is that the number of fractures that can be examined to find level sizes is limited by computing facilities. Also, of course, this is a method that is suited to examination of simulations and not to "real life" fracture meshes. Its relevance is therefore dependent on the relevance of the simulation.

Conclusion

We have described some techniques for the simulation of a fracture network based on data from a real fracture system. The simulation reproduces the spatial variability of fracture density and orientation. To accomplish this, we have developed a stochastic model that incorporates this spatial variation. We have also described how the model can be fit to the data and have introduced a measure of connectivity for fracture mesh simulations.

Acknowledgment

This work is the product of cooperation between Lawrence Berkeley Laboratory (LBL) and the French Bureau de Recherches Géologiques et Minières (BRGM). Support was provided to LBL by the U.S. Department of Energy under Contract No. DE-AC03-76SF00098 and to BRGM by the Agènce Française pour les Economies d'Energie. The data used here were taken in the Fanay-Augères mine owned by Cogema under contract with the French Commissariat à l'Energie Atomique. The authors are indebted to Paul Witherspoon of LBL and Bernard Feuga of BRGM.

Literature Cited

1. Alfaro, M., 1980. "The Random Coin Method: Solution to the Problem of Simulation of a Random Function in the Plane," *Mathematical Geology, 12*.
2. Barbreau, A., B. Come, S. Derlich, E. Durand, G. de Marsily, P. Peaudecerf, and G. Vouille, 1985. "Experiments Performed on Granite in the Underground Research Laboratory at Fanay-Augères, France," *Proc. Internat. Symp. on Coupled Processes Affecting the Performance of a Nuclear Waste Repository,* Lawrence Berkeley Laboratory, Berkeley, CA, Sept.
3. Cacas, M. C., E. Ledoux, G. de Marsily, and B. Tillie, 1987. *Etude de l'Effet d'Echelle en Milieu Fissuré,"* Report of Ecole Nationale Superieure des Mines de Paris.
4. Dershowitz, W., 1984. "Rock Joint Systems," Ph.D. dissertation, Massachusetts Institute of Technology, Cambridge, MA.
5. Deverly, F., 1984. "Echantillonnage et Géostatistique," thèse, Ecole Nationale Superieure des Mines de Paris.
6. Gros, Y., 1982. *Etude Structurale de la Mine de Fanay-Augères—Géométrie, Cinématique et Chronologie des Fractures,* BRGM report.
7. Hestir, K., Jean-Paul Chilès, Jane Long, and Daniel Billaux, 1987. "Three Dimensional Modeling of Fractures in Rock: From Data to a Regionalized Parent Daughter Model," *AGU Monograph on Flow in Unsaturated Rock.*
8. Journel, A. G. and Ch. J. Huijbregts, 1978. *Mining Geostatistics,* Academic Press Inc., London.
9. Lassagne, D., 1983. *Essai de Caractérisation du Milieu Fracturé en Massif Granitique,* BRGM report 83 SGN 576 GEC, 45060 Orléans, France.
10. Long, J.C.S. and D. M. Billaux, 1987. "From Field Data to Fracture Network Modeling—An Example Incorporating Spatial Structure," *Water Resources Research, 23*(7) pp. 1201–1216.
11. Long, J.C.S. and P. A. Witherspoon, 1985. "The Relationship of Degree of Interconnection to Permeability in Fracture Networks," *Journal of Geophysical Research, 90*(B4), pp. 3087–3098.
12. Long J.C.S., J. S. Remer, C. R. Wilson, and P. A. Witherspoon, 1982. "Porous Media Equivalents for Networks of Discontinuous Fractures," *Water Resources Research, 18*(3), pp. 645–658.
13. Mardia, K. V., 1972. *Statistics of Directional Data,* Academic Press Inc., London.
14. Massoud, H. and J. P. Chilès, 1985. *Apprôche Stochastique de l'Erreur Liée à la Position et à l'Orientation d'une Fracture,* BRGM technical note.
15. O'Sullivan, F., 1986. A Statistical Perspective on Ill-posed Inverse Problems, *Statistical Science,* Vol. 1, No. 4, pp. 502–527.
16. Ripley, B. D., 1981. *Spatial Statistics,* John Wiley and Sons, New York.

17. Robinson, P., 1984. "Connectivity, Flow and Transport in Network Models of Fracture Media," Ph.D. dissertation, St. Catherine's College, Oxford.
18. Rouleau, A., 1984. "Statistical Characterization and Numerical Simulation," Ph.D. dissertation, Univ. of Waterloo, Waterloo, Ontario.
19. Stoyan, D., W. S. Kendall, and J. Mecke, 1987. *Stochastic Geometry and Its Applications,* John Wiley and Sons, New York.
20. Warburton, P. A., 1980. "A Stereological Interpretation of Joint Trace Data," *Int. Jour. Rock Mech. Min. Sci. and Geomech. Abstr.*, Vol. 17, pp. 181–190, Pergamon Press Ltd., Oxford.

APPENDIX

Geostatistical Simulation Procedure

To describe the process of geostatistical simulation we first define a few terms. Let $\mathbf{X}(\bar{t})$ be a random field in n-dimensional space, $n = 1, 2, 3$. We have $\gamma(\bar{h}) = 1/2\ E(\mathbf{X}(\bar{t} + \bar{h}) - \mathbf{X}(\bar{t}))^2$. It is assumed that $\gamma(\bar{h}) = \gamma(|\bar{h}|) = \gamma(h)$, where $h = |\bar{h}|$, that is, γ is isotropic. Generalizations to anisotropic γ can be made but are not discussed here.

Now assume $\mathbf{X}$ is observed at points $\vec{t}_1, \vec{t}_2, \ldots, \vec{t}_n$. We can estimate $\gamma(h)$ with

$$\gamma(h) \approx \frac{1}{2}\frac{1}{n_h}\left\{ \sum_{\{(\vec{t}_i,\vec{t}_j)\,:\,|\vec{t}_i-\vec{t}_j| = h\}} (\mathbf{X}(\vec{t}_i) - \mathbf{X}(\vec{t}_j))^2 \right\}$$

where n_h is the number of points $\vec{t}_1, \vec{t}_2, \ldots, \vec{t}_n$ that are distance h apart (within a certain tolerance). Also, we can see if $\mathbf{X}$ is a Gaussian process by examining the histogram of the observed values $\mathbf{X}(\vec{t}_1), \mathbf{X}(\vec{t}_2), \ldots, \mathbf{X}(\vec{t}_n)$.

If $\mathbf{X}$ is a Gaussian process with semivariogram γ, then the simulation of $\mathbf{X}$ is a straightforward procedure using either the turning bands method (Journel and Huijbregts, 1978) or the random spheres method (Alfaro, 1980). We use the random sphere method in our work.

If $\mathbf{X}$ is not a Gaussian process we first transform it into one. This is done by using a transformation f, called an anamorphosis (or graphical transformation). Specifically, f is a one to one function constructed so that the histogram of $f(\mathbf{X}(\vec{t}_1)), f(\mathbf{X}(\vec{t}_2)), \ldots, f(\mathbf{X}(\vec{t}_n))$ has a Gaussian shape. It is then assumed that $f(\mathbf{X})$ is a Gaussian process. Details of the construction of f can be found in Journel and Huijbregts (1978, pp. 478).

Once the anamorphosis f, is found the next step is to simulate the transformed process $f(\mathbf{X})$. This is done as described above by estimating the semivariogram of $f(\mathbf{X})$ using

$f(X(\vec{t}_1))$, $f(X(\vec{t}_2))$, . . . , $f(X(\vec{t}_n))$ and employing the random spheres method. Finally, the resulting simulation of f(**X**) is transformed by f^{-1} yielding the simulation of **X**.

Calculation of the Theoretical Trace Density Semivariogram for the Parent-daughter Model

Below we use notation given in the section on parameter inference. Divide space into small disjoint sets ΔV_k centered at $\mathbf{x}_k$ with volume $\Delta \mathbf{x}_k$. Let P_k denote the number of parents occurring in ΔV_k. Let $Q_k(B)$ denote the number of daughters appearing in B from parents occurring in ΔV_k. To condition on Λ_p we suppose Λ_p is a fixed function on $\mathbf{R}^3$, $\Lambda_p : \mathbf{R}^3 \to \mathbf{R}^+$, ($\mathbf{R}^+$ = the set of non-negative real numbers). E_{Λ_p} will denote expectation given a fixed Λ_p. Also note that when Λ_p is fixed, P_k is a Poisson random variable with rate $\Lambda_p(\mathbf{x}_k)\Delta\mathbf{x}_k$.

We do a second conditioning by assuming that P_k is fixed for each k, in other words, the ensemble of parent points is fixed. $E_{\Lambda_p,\Omega}$ will denote expectation given a fixed Λ_p and fixed ensemble, Ω. Now, for fixed Λ_p and Ω the following statements hold:

1) $Q_k(B)$ is a Poisson random variable with rate $P_k\lambda_D\Psi(B,\mathbf{x}_k)$, hence $E_{\Lambda_p,\Omega}(\ Q_k(B)\ ) = P_k\lambda_D\Psi(B,\mathbf{x}_k)$.

2) For $k \neq j$, $Q_k(B)$ and $Q_j(B)$ are independent.

3) $Q_k(B_1)$ and $Q_j(B_2)$ are independent for any k and j when B_1 and B_2 are disjoint.

4) $E_{\Lambda_p}(\mathbf{X}) = E_{\Lambda_p}(\ E_{\Lambda_p,\Omega}(\ \mathbf{X}\ )\ )$ and $E(\ \mathbf{X}\ ) = E(\ E_{\Lambda_p}(\ \mathbf{X}\ ))$.

Finally, notice that

$$Q(B) = \sum_k Q_k(B)$$

$$Q(\hat{B}) = \sum_k Q_k(\hat{B})$$

and

$$Q(B)Q(\hat{B}) = \sum_{k \neq j} Q_k(B)Q_j(\hat{B}) + \sum_k Q_k(B)Q_k(\hat{B}) \ .$$

Now, conditioning on Λ_p and Ω yields

$$E_{\Lambda_p,\Omega}(Q(B)) = \sum_k P_k \lambda_D \Psi(B,\mathbf{x}_k) \quad .$$

For calculation of $E_{\Lambda_p,\Omega}(\ Q(B)Q(\hat{B})\ )$ first recall the fact that if N_1, N_2 and N_3 are independent Poisson random variables with rates λ_1, λ_2, and λ_3 then

$$E(\ (N_1 + N_2)(N_2 + N_3)\ ) = (\lambda_1 + \lambda_2\ )(\ \lambda_2 + \lambda_3\ ) + \lambda_2 \quad .$$

This implies that

$$E_{\Lambda_p,\Omega}(\ Q_k(B)Q_k(\hat{B})\ ) = P_k^2 \lambda_D^2 \Psi(B,\mathbf{x}_k)\Psi(\hat{B},\mathbf{x}_k) + P_k \lambda_D \Psi(B \cap \hat{B},\mathbf{x}_k)$$

which gives

$$E_{\Lambda_p,\Omega}(Q(B)Q(\hat{B})) = E_{\Lambda_p,\Omega}(\sum_{k \neq j} Q_k(B)Q_j(\hat{B})\ ) + E_{\Lambda_p,\Omega}(\ \sum_k Q_k(B)Q_k(\hat{B})\ )$$

$$= \sum_{k \neq j} \lambda_D^2 P_k P_j \Psi(B,\mathbf{x}_k)\Psi(B,\mathbf{x}_j)$$

$$+ \sum_k P_k^2 \lambda_D^2 \Psi(B,\mathbf{x}_k)\Psi(\hat{B},\mathbf{x}_k) + \sum_k P_k \lambda_D \Psi(B \cap \hat{B},\mathbf{x}_k) \quad .$$

Next we remove the conditioning on the ensemble Ω,

$$E_{\Lambda_p}(\ (Q(B)\ ) = E_{\Lambda_p}(\ (E_{\Lambda_p,\Omega}(\ (Q(B)\ )\ ) = \sum_k \Lambda_p(\mathbf{x}_k)\Delta \mathbf{x}_k \lambda_D \Psi(B,\mathbf{x}_k)$$

and

$$E_{\Lambda_p}(\ Q(B)Q(\hat{B})\ ) = E_{\Lambda_p}(\ E_{\Lambda_p,\Omega}(\ Q(B)Q(\hat{B})\ )\ )$$

$$= \sum_{k \neq j} \lambda_D^2 \Lambda_p(\mathbf{x}_k)\Lambda_p(\mathbf{x}_j)\Psi(B,\mathbf{x}_k)\Psi(B,\mathbf{x}_j)\Delta \mathbf{x}_k \Delta \mathbf{x}_j$$

$$+ \sum_k (\ (\Lambda_p(\mathbf{x}_k)\Delta \mathbf{x}_k)^2 + \Lambda_p(\mathbf{x}_k)\Delta \mathbf{x}_k\ )\lambda_D^2 \Psi(B,\mathbf{x}_k)\Psi(\hat{B},\mathbf{x}_k)$$

$$+ \sum_k \Lambda_p(\mathbf{x}_k)\Delta \mathbf{x}_k \lambda_D \Psi(B \cap \hat{B},\mathbf{x}_k) \quad .$$

Lastly, we remove the conditioning on Λ_p.

$$E(\ Q(B)\ ) = E(\ E_{\Lambda_p}(\ Q(B)\ )\ ) = \sum_k \overline{\Lambda}_p \Delta \mathbf{x}_k \lambda_D \Psi(B,\mathbf{x}_k)$$

and

$$E(\ Q(B)Q(\hat{B})\ ) = E(\ E_{\Lambda_p}(\ Q(B)Q(\hat{B})\ )\ ) =$$

$$= \sum_{k \neq j} \lambda_D^2 \Psi(B,\mathbf{x}_k)\Psi(\hat{B},\mathbf{x}_j) C_{\Lambda_p}(\ \mathbf{x}_j - \mathbf{x}_k\ )\Delta\mathbf{x}_k \Delta\mathbf{x}_j$$

$$+ \sum_k \lambda_D \overline{\Lambda}_p \Psi(B,\mathbf{x}_k)\Psi(\hat{B},\mathbf{x}_k)$$

$$+ \sum_k (\Delta\mathbf{x}_k)^2(\text{ bounded term })$$

$$+ \sum_k \overline{\Lambda}_p \Delta\mathbf{x}_k \lambda_D \Psi(B \cap \hat{B},\mathbf{x}_k)\ .$$

Now, letting $\Delta\mathbf{x}_k \to 0$ gives the result

$$E(\ Q(B)\ ) = \overline{\Lambda}_p \lambda_D \int_{R^3} \Psi(B,\mathbf{x})d\mathbf{x} = \overline{\Lambda}_p \lambda_D \text{area}(B)\overline{D} \tag{1}$$

and

$$E(\ Q(B)Q(\hat{B})\ ) = \lambda_D^2 \int_{R^3}\int_{R^3} \Psi(B,\mathbf{x})\ \Psi(\hat{B},\mathbf{y}) C_{\Lambda_p}(\mathbf{x} - \mathbf{y}) d\mathbf{x}d\mathbf{y} \tag{2}$$

$$+ \overline{\Lambda}_p \lambda_D^2 \int_{R^3} \Psi(B,\mathbf{x})\Psi(\hat{B},\mathbf{x})d\mathbf{x}$$

$$+ \overline{\Lambda}_p \lambda_D \int_{R^3} \Psi(B \cap \hat{B},\mathbf{x})d\mathbf{x}\ .$$

Interdistance Statistic

Using conditioning methods as those employed above one can derive some interdistance statistics. One result in this direction shows that for S equal to the sum of interdistances between all possible pairs of trace centers in a particular window

$$E(S) = \frac{1}{2}\overline{\Lambda}_p \lambda_D^2 \int_{R^3} \Psi^2(B,\mathbf{x})h(\mathbf{x})d\mathbf{x} + \frac{1}{2}\lambda_D^2 \int_{R^3}\int_{R^3} C(\mathbf{x} - \hat{\mathbf{x}})H(\mathbf{x},\hat{\mathbf{x}})\Psi(B,\mathbf{x})\Psi(B,\hat{\mathbf{x}})d\mathbf{x}d\hat{\mathbf{x}}\ .$$

where

$$\mathbf{x} = (x,y,z)$$

$$\hat{\mathbf{x}} = (\hat{x},\hat{y},\hat{z})\ ,$$

the sampling window is defined in the x-direction by $a < x < b$, $\sigma =$ standard deviation of the displacement of a daughter from a parent,

$$h(\mathbf{x}) = h(x) = \frac{1}{c_x^2} \int_{a-x}^{b-x} \int_{a-x}^{b-x} |u - w| \frac{1}{2\pi\sigma^2} e^{\frac{-1}{2\sigma^2}(u+w)} dudw$$

$$c_x = (2\pi\sigma^2)^{-½} \int_{a-x}^{b-x} e^{\frac{-1}{2\sigma^2}u^2} du \ .$$

and

$$H(\mathbf{x}, \hat{\mathbf{x}}) = H(x,\hat{x})$$

$$= \frac{1}{c_x c_{\hat{x}}} \int_{a-x}^{b-x} \int_{a-\hat{x}}^{b-\hat{x}} \frac{1}{2\pi\sigma^2} |u + x - w - \hat{x}| e^{\frac{-1}{2\sigma^2}(u^2 + w^2)} dudw \ .$$

A formula for Var(S) can also be developed. Although not implemented yet, these formulas can be used to eliminate some combinations of parameter values that give matching trace density semivariograms.

Nonparametric Approach for Localized Stochastic Model of Rock Joint Systems

DAE S. YOUNG and STEVEN F. HOERGER
Department of Mining Engineering
Michigan Technological University
Houghton, Michigan

Abstract

Geostatistics was extended to regionalize random unit vectors and applied to develop a localized stochastic model of joint systems in geological formations. The characteristic parameters of rock joints regionalized in the model are the vectorial variable of orientation and the scalar variables of joint spacing, maximum length, average length, and surface roughness. Indicator kriging (IK) was incorporated with the local block model development to estimate the probability distributions of those rock joint parameters.

The pole vector of a joint plane projected on the upper reference hemisphere was regionalized directly and its spatial variability was measured by the difference vector. The probability distribution of pole vectors was plotted on the tangent plane and a two-dimensional cutoff system was developed to define its indicator variables.

Actual IK for pole vectors was carried out by Mononodal IK (or Proportional IK) to economize the IK process and to overcome a potential problem of kriging systems due to a large number of cutoff points compared to the number of available data for kriging. The application of IK to other scalar parameters is a straightforward process estimating their local probability distributions within each block. Consequently, a complete set of joint parameters was modeled and their probability distributions were estimated for each local block.

The IK procedure used in this joint modeling was verified through a case study made on an open pit mine, where rock joint systems were surveyed completely on the pit walls. An IK estimate of the local statistical distribution based on only one quarter of the survey data was compared with the actual sample histogram of survey data within the same block. A positive comparison was reached and this confirmed the accuracy of both the IK joint model for the open pit mine and the IK procedure employed.

The localized stochastic model of rock joint systems in geological formations has a broad range of applications as presented in this paper.

Introduction

Discontinuities in geological formations have a major influence on various geotechnical analyses, such as analyses for structural stabilities of rock exposed at a slope or in an underground excavation, geohydrological modeling of fluid flow through fractured rock formations, or oil and gas reservoir modeling. This has been recognized in numerous research works by Goodman et al. (*1*), Piteau (*2*), Louis (*3*), Warren and Root (*4*), Evans and Ubani (*5*), etc. Most of their work was done based on deterministic models of joints, in which the joint geometries and physical properties were simplified and represented as a single, regular, planar discontinuity with a constant characteristic property. In reality, however, many characteristics of rock joints are best described in probabilistic terms, and there are intrinsic advantages in geotechnical approaches that directly employ the relevant statistical distributions. Recently, the spatial variability of joint parameters was studied in depth by introducing vectorial variograms and kriging techniques [Young and Hoerger (*6*) and Young (*7*)], and the local variability of joint parameters was emphasized in the geotechnical analysis. The impact of local variations on the structural stability was demonstrated through a case study on a pit slope stability where the local behavior of slopes predicted from the local cell block model was significantly different from the global slope stability [Hoerger and Young (*8*)].

Considering the aforementioned studies, an ultimate model of rock joint systems is a localized stochastic model for geotechnical analyses that will lead to an analytical solution in local probabilistic terms. In this paper, the local stochastic model of rock joint systems observed in a porphyry copper mine was developed by using the indicator kriging algorithm (IK) [Journel (*9*), Lemmer (*10*)]. A complete set of joint parameters was fully described in the form of statistical distributions for each cell block in the model. Rock joints were characterized by such parameters as orientation, spacing, maximum length, average length, and surface roughness. The capability and reliability of joint models from IK were tested positively when the IK estimate of joint parameter distributions within local blocks was compared with those of actual joint survey data in the same blocks. The details of this model development are presented in this paper.

Joint Survey Data

The joint systems in rock formations at an open pit porphyry copper mine were completely surveyed on pit walls by the cell mapping technique [Call et al. (*11*), Cicchini et al. (*12*)]. A total of 939 cells of 15 m x 15 m square were surveyed throughout the pit as shown in Figure 1. Joint parameters surveyed in the field and reduced for geotechnical analyses are briefly described below.

Orientation. The orientation was surveyed in terms of true dips and dip directions, and it was measured along the longest trace of the joint group by following the standard method [Call et al. (*11*) or ISRM (*13*)]. Representative joint groups in a cell were identified in the field by the field geologist (or sampler). The pole vector of a joint plane was calculated from these measurements to intercept the upper reference hemisphere. The pole was regionalized directly in this geostatistical modeling and

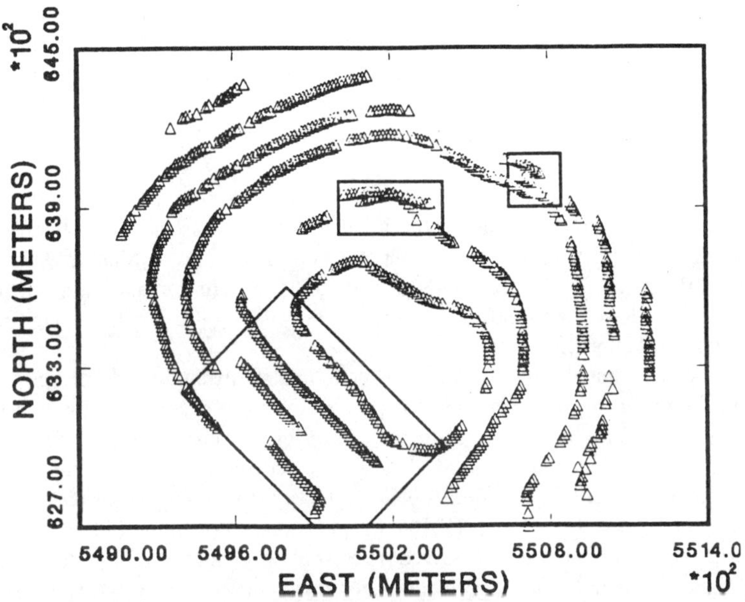

Figure 1. **Location map of 939 cells surveyed in the pit.**

projected on Grossman's tangent plane for its statistical distribution [Grossman (*14*), Young (*7*)].

Maximum Length. The maximum length of the joint trace was measured along the longest trace of the joint group that was observed within a cell window. No statistical modeling or data reduction was made on this parameter.

Spacing. The spacing parameter was measured in the field by the perpendicular distance between joints in a group identified in a cell. It is an average of the field survey data and is surveyed for each group separately.

Surface Roughness. Roughness was measured in angular terms along the true dip of the longest trace observed in a joint group identified in a cell. It is an apparent maximum angular variation of the dip from its overall dip and measures the relative irregularity of a joint surface, which could be a significant portion of the effective mechanical strength of a joint.

Average Length. The average length of joints was estimated from a statistical modeling of joint traces exposed on pit walls. A simple and usual statistical model of joints was employed for this purpose, where joints were assumed to be flat discs, parallel to each other, and centered on random points whose volume density has a Poisson distribution. By knowing (or assuming) the trace length distribution from the field survey data, the mean length of uncensored and untruncated trace length can be calculated from the censored and truncated field sample data [Pahl (*15*), Priest and Hudson (*16*), Warburton (*17*, *18*)]. For this open pit model, an elliptic disc shape and negative exponential distribution of traces were used in the statistical model to calculate the average length. A negative exponential distribution of trace lengths was proved from the field data [Pryor (*19*)] and also used in numerous joint data published by Baecher et al. (*21*), Call et al. (*11*), Barton (*20*), etc.

Joint Families

An extensive cluster analysis on the joint orientation was made to identify the joint families in the pit by using the computer code FRACTAN [Shanley and Mahtab (*22*)], which applies Poisson's distribution test. Three major families were identified from FRACTAN and the eastward family in Figure 2 was selected for further geostatistical modeling in this paper.

The selected joint family contains 780 cells and its mean attitude of poles was in the angular coordinate of: co-latitude (or plunge) of 15.29°. and co-longitude (or trend) of 91.64°. This is equivalent to the dip of 74.71° and the dip direction of 91.64° for the mean attitude of joint planes projected on a stereonet. The spatial distribution of 780 cells in this family was uniform throughout the pit [compare Figure 3 with Figure 9 in Young (*7*)] and no particular domain could be observed. Therefore, the whole pit was considered as a single domain for this joint model development. Statistics of joint parameters for this selected family are summarized in Table 1.

Nonparametric Approach: IK

When poles are projected on the upper reference hemisphere, they are random unit vectors, scattered around the mean attitude of the joint family. Very little is

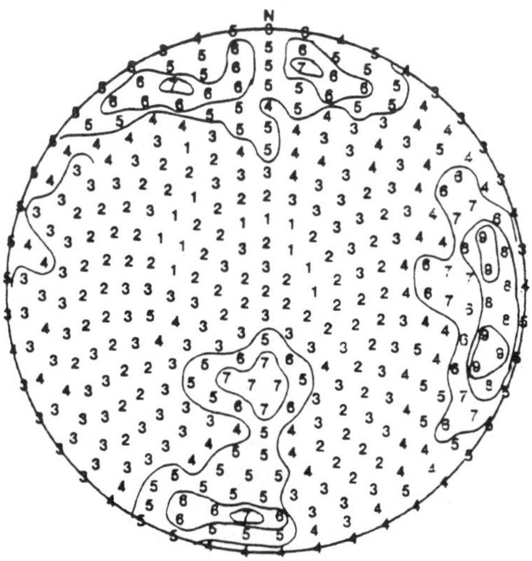

Figure 2. **Contours of poles projected on a stereonet.**

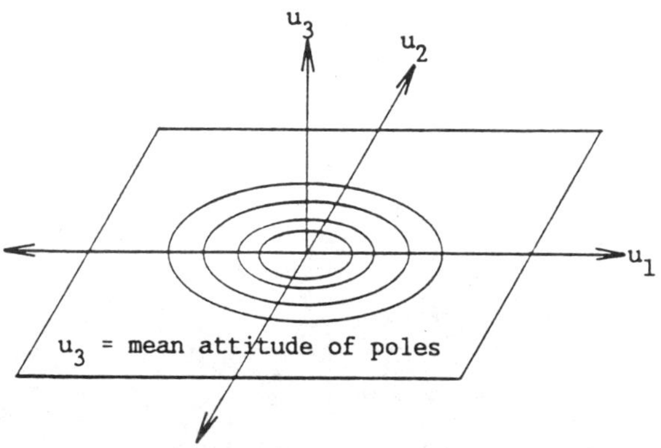

Figure 3. **Concentric ellipses of probability regions in Grossman's tangent plane.**

known about the statistical distribution of poles on the sphere, and it requires highly complicated mathematics to define its probability density functions in three-dimensional space [Arnold (*23*), Bingham (*24*), Fisher (*25*), Watson (*33*)]. Therefore, a nonparametric approach is highly desirable for the statistics of pole vectors. In addition, naturally occurring rock joint pole vectors are spatially correlated and the spatial variability has to be incorporated with the statistics or modeling of joints in rock formations.

Considering the above aspects, IK is a better choice for developing a localized and stochastic model of joints, which may be the ultimate model of joints in geological formations for various geotechnical analyses, because the localized and probabilistic analysis is achievable by applying the current deterministic geotechnical methods on this type of model [Young and Hoerger (*34*)]. In the following sections, the geostatistical operations of IK for the development of a local stochastic joint model are presented, which consist of both vectorial and scalar variables (or joint parameters).

Pole vectors

Recently a simple binormal distribution of joint orientations (or poles) was studied by Grossman (*14*). When the pole was projected on the tangent plane at the mean attitude of a joint family, a bivariate normal distribution was accepted from Kolmogorov's goodness-of-fit tests. In this bivariate normal distribution, the dispersion of poles around the mean attitude (u_3) is measured by the tangent of the angular difference, and concentric ellipses of probability regions are drawn on Grossman's

Table 1. **Statistics of the joint set selected.**

Parameters	Spacing	Roughness	Average Length	Maximum Length	Mean Attitude of Poles
Mean	0.845 m	3.17	0.545 m	2.47 m	co-lat. = 15.29°
Variance	0.582	7.34	0.235	2.76	co-long. = 91.64°

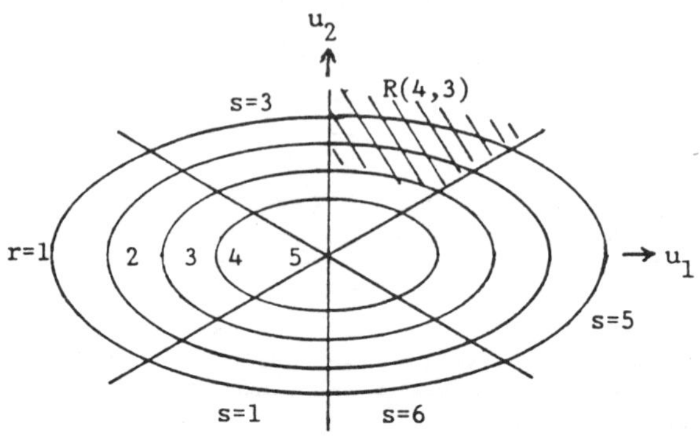

Figure 4. **(6-sector × 5-region) quadrilateral cutoff system used in IK for pole vectors.**

tangent plane as shown in Figure 3. Based on the concentric probability regions, a two-dimensional cutoff system was developed for pole vectors and applied to the joint modeling in this paper.

Cutoff Systems and Indicator Variables. Regardless of the exact shape of the statistical distribution in the tangent plane (or spherical surface), two-dimensional class intervals are required to construct a discrete histogram of pole vectors. This means that a two-dimensional cutoff system is needed to define the indicator variable of poles. Considering the elliptic probability regions in Grossman's tangent plane, it is natural to lay out a quadrilateral cutoff area, which results from slicing these concentric ellipses in radial directions (see Figure 4).

Grossman's bivariate normal distribution may not be general for every joint family, but it provides an useful guideline for developing a two-dimensional cutoff system for the pole vector. As a nonparametric approach, IK does not require one to know the exact statistical distribution of pole vectors. The important aspects in the two-dimensional cutoff system are the sequential order of numbering cutoff areas and maintaining the consistent order for both indicator variables and IK operations in this modeling.

The sequential order of defining indicator variables of pole vectors with the two-dimensional cutoff system was followed in this way: the sector number, s, was counted clockwise and commenced from the minor axis (u_2) of the probability contour ellipse, and the region number, r, was increased toward the center of the concentric ellipse and started from the outmost region (see Figure 4). Then, the cumulative cutoff area, R(s,r) = Δ below the r-region in the s-sector was integrated toward the coordinate origin and cumulated sector by sector.

Consequently, the indicator variable of a pole vector, **z**(x), at a location, x, in a stationary domain of geological formations can be defined as

$$i(x;\ s,r) = 1 \quad \text{if } \mathbf{z}(x) \subset R(s,r) = \Delta$$

$$i(x;\ s,r) = 0 \quad \text{otherwise} \tag{1}$$

This is very similar to the indicator variable of scalar variables [Journel (*9*, *26*)] and IK is repeatable for poles, because the indicator variable itself is a unitless indicator number and does not carry a sense of direction in vectors. Further details on indicator variables of pole vectors are referred to in Young (*27*).

In this modeling, six 60°-angular sectors and five regions in Grossman's tangent plane were used to describe the pole vector distribution and define the indicator variable. Thus, there are 30 quadrilateral cutoff areas and 30 corresponding indicator variables for the IK operation. IK will estimate the relative frequencies above these cutoff areas. The numbers and sizes of cutoff areas should be arranged to cover a significant portion of the tangent plane and to represent the specific dispersion of pole vectors around the mean attitude (u_3) of the joint family under study. The 30 cutoff areas used in this modeling were specified in Table 2, including the orientation of coordinate axes on the tangent plane.

Table 2. **(6 × 5) quadrilateral cutoff system: radii of elliptic regions and corresponding angular difference from the mean attitude of poles (u_3).**

Region No.	**1**	**2**	**3**	**4**	**5**	**Orientation**	
Probability (%)	77.6	54.6	28.7	7.4	0.0	Co-lat.	Co-long.
Radius in Units							
Major (u_1)	0.612	0.445	0.291	0.139	0.0	3.97°	1.08°
Minor (u_2)	0.406	0.296	0.193	0.093	0.0	73.60°	257.44°
Angular Diff. (°)							
Major	31.47	24.00	16.20	7.93	0.0		
Minor	22.12	16.49	10.92	5.29	0.0		

Major and minor standard deviation: $\sigma_M = 0.354$, $\sigma_m = 0.234$.

Variogram and Indicator Variogram. The spatial variability of pole vectors was measured by the vector difference, which is identical to the angular difference as poles are projected on the upper unit reference hemisphere. Then, the variography of pole vectors can be defined in a similar way as applied to scalar variables in the usual geostatistics [Young (*7*), Journel and Huijbregts (*28*)]:

$$2\gamma\,(x,h) = E\{||\vec{z}\,(x) - \vec{z}\,(x+h)||^2\} \tag{2}$$

The geostatistics or variogram structural analysis of the usual scalar variables can be repeated for the pole vector, because the variography turns out in scalar terms again.

The experimental variogram for pole vectors in the selected joint family was modeled by an isotropic spherical variogram, whose parameters are a sill of 0.205 and 250 m in the range value (see Table 3). Therefore, the pole vector is regionalized and geostatistics is applicable to its interpretations and modeling.

Table 3. **Spherical model parameters of variograms.**

Joint Parameters	Orientation	Spacing	Roughness	Average Length	Maximum Length
Sill Value	0.145	0.53	6.87	0.23	2.65
Nugget Value	0.06	0.275	4.00	0.135	1.60
Range (m)	250	250	250	250	250

By following the identical definition for an indicator variogram [$\gamma_I(h)$] for scalar variables such as the usual ore grades, $\gamma_I(h)$ of a pole vector can be computed based on the indicator variable in Equation (1) and its structural analysis can be repeated in a way that is similar to scalar variables. Due to 30 quadrilateral cutoff areas in the tangent plane and the corresponding 30 indicator variables, a set of 30 indicator variograms should be found according to the original IK [Journel (*9,26*)]. In reality, however, it may be impractical, if not impossible, to fit models to so many experimental $\gamma_I(h)$'s based on a limited number of pole vectors.

This was simplified in this joint modeling by using only a single representative variogram for all cutoffs as is done in Mononodal IK [Lemmer (*10,29*)] and Median IK [Journel (*30*)] in place of modeling indicator variograms and performing IK operations for each cutoff. Instead of computing and modeling indicator variograms for each cutoff area separately, all indicator variograms were calculated and multiplied by a cutoff-dependent constant to have a sill of 0.25, which is the potential maximum sill of an indicator variogram. The sill of 0.25 of this indicator variogram was obtained by multiplying a proportional factor to every experimental indicator variogram. The multiplier was

$$0.25/\{F(\Delta) \cdot [1-F(\Delta)]\}$$

where $F(\Delta)$ is a cumulative frequency in the cutoff area Δ.

The variograms that are indicator variograms normalized to a sill of 0.25 are called raised indicator variograms and are denoted by $\gamma_{RI}(h)$ in this paper. The raised indicator variograms for all cutoffs can be plotted together using the same coordinate axes in order to fit a single variogram model which is representative of the variogram structure of each cutoff's indicator variable.

Raised indicator variograms were calculated for all 30 cutoffs and a single model was fitted as given in Figure 5. The model fitted on these $\gamma_{RI}(h)$'s was a spherical model with a range, a, of 250 m:

$$\begin{aligned} \gamma_{RI}(h) &= 0.16 + 0.09\,[(3/2)(h/a) - (1/2)(h/a)^3], \text{ if } h \le a \\ \gamma_{RI}(h) &= 0.25, \text{ if } h > a \end{aligned} \tag{3}$$

Scalar variables

Other joint parameters including spacing, surface roughness, and average and maximum lengths are in scalar terms, and the usual geostatistical operation was followed for their variograms. For each parameter, a single variogram model was fitted

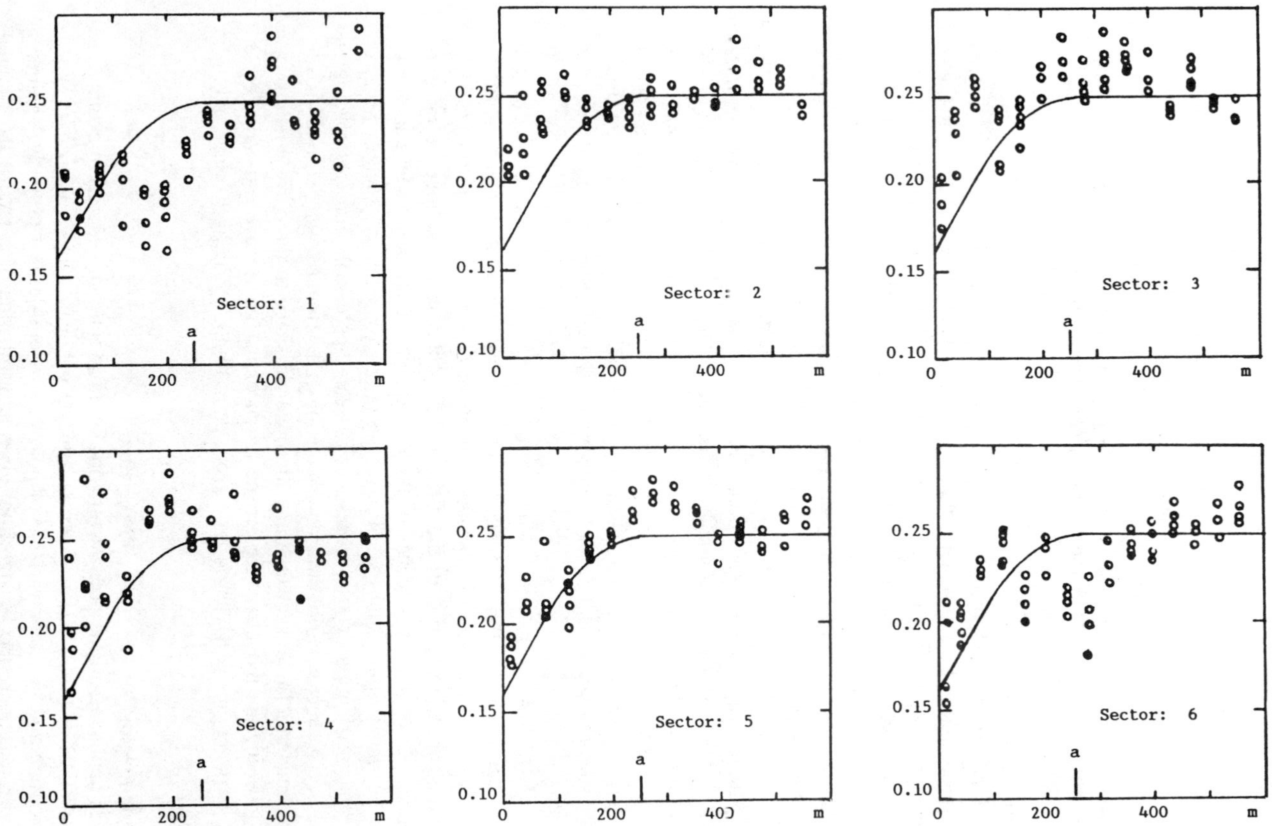

Figure 5. **Raised indicator variograms for 6 sectors for pole vectors (abscissa = distance, h and a = 250 m of range).**

that was representative of raised indicator variograms. Spherical variograms were modeled with an unique range value of 250 m for these scalar parameters, and variogram model parameters are given in Table 3.

As seen in the orientation variograms, nugget values are high in these variograms. However, the unique range of 250 m indicates that the variograms identified a unique spatial correlation in this natural joint family and the joint by itself is a random outcome from random phenomena observed in the natural geological formations. Thus, the joint parameters, which are artifacts used to describe the joint property conveniently, should not alter the characteristic structures of joint variograms.

Indicator variogram

As done for the pole vector in the previous section, raised indicator variograms also were computed for the scalar parameters. As expected from their variograms, isotropic spherical models with an unique range of 250 m were fitted well as shown in Figure 6. These spherical variogram parameters are summarized in Table 4.

The raised indicator variogram approach has many advantages over the conventional approach of one indicator variogram per cutoff, particularly for estimation of two-dimensional (bivariate) probability distributions. In general, the input data size available for geotechnical analysis is not large enough to compute the many meaningful variograms required for conventional IK (in this case 30 indicator variograms). The raised indicator variogram approach does not require fitting a model to any single experimental indicator variogram; instead, the modeling process uses experimental variograms from all cutoffs as inputs, giving the modeller a much larger set of information to use in determining adequacy of model fit. Additionally, the raised indicator variogram approach leads to many IK advantages that are discussed in the next sections.

Outside of this paper, a few points were checked to ensure the validity of geostatistical operations for this IK modeling of joint systems. Histograms of joint parameters were studied and the proper number of cutoff points were selected for IK to ensure adequate resolution of the estimated distributions throughout the range of data values. Also, the usual shapes of these histograms as published by numerous authors for other joint systems were observed in this joint family, such as negative exponential distribution of joint traces and Poisson-type distributions for spacing and roughness. The vector variogram for orientation (Eq. 2) and scalar variograms for other parameters (Table 3), which are all similar in shape to the corresponding raised indicator variogram models, were proved by checking the average estimation error and the variance of the deviation ratio from point-to-point kriging [Young and Hoerger (*6*), Young (*7*)].

Indicator kriging

IK is applicable to the pole vector (or any unit vector in general) to estimate its local statistical distributions, as given in a recent publication by Young (*27*). When the variogram and indicator variable of pole vectors are defined properly, as done in the previous sections, the IK algorithm gives an estimation of ϕ (A,Δ), the cumulative frequency of pole vectors [**z**(x)] within a local area A in a stationary domain of geological formations whose projections on the tangent plane are located within the

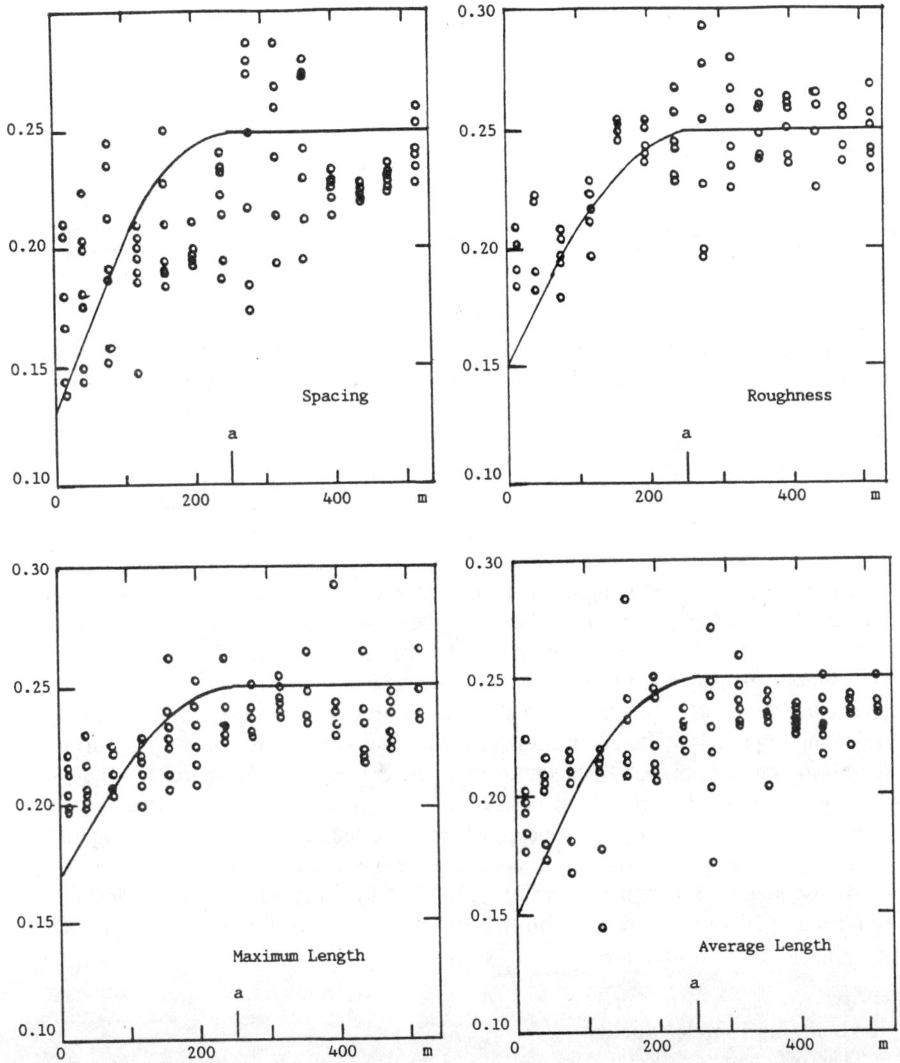

Figure 6. **Raised indicator variograms for scalar joint parameters (range, a = 250 m, and abscissa = distance, h).**

cumulative cutoff area, Δ, below the cutoff point, r-region in the s-sector [R(s,r) = Δ]. $\phi(A,\Delta)$ is enough to characterize the local probability distribution of pole vectors and develop a local stochastic model of joints. The IK algorithm was implemented using an ordinary kriging (OK) system to estimate $\phi(A,\Delta)$ in this paper. It is important to remember here that the sequential order of numbering r-regions and s-sectors, and cumulating R(s,r) = Δ in the two-dimensional cutoff system, should be consistent for both indicator variables and corresponding IK operations.

Table 4. **Spherical model parameters for raised indicator variograms.**

Joint Parameters	Poles	Spacing	Roughness	Average Length	Maximum Length
Sill Value	0.25	0.25	0.25	0.25	0.25
Nugget Value	0.16	0.13	0.15	0.15	0.17
Range (m)	250	250	250	250	250

The actual IK process for estimating pole vector probability distributions was analogous to using Mononodal IK (or proportional IK) for estimating scalar probability distributions [Lemmer (*10*,*29*)]. The key feature of these krigings is that a single variogram is used for the kriging of all indicator variables. Such an approach is possible when all indicator variograms demonstrate the same shape and differ only by a cutoff dependent multiplicative constant (proportionality between indicator variograms). The multiplier has no effect on the weighting factors that solve the OK system, so regardless of what multipliers were used in creating the raised indicator variograms, from which a single raised indicator variogram model was distilled, the one set of weighting factors from the OK system is applicable to the optimum estimation of every indicator variable defined by the cutoff system. IK for scalar parameters was also performed with one raised indicator variogram and one kriging system for all cutoffs. In addition to significantly reducing computer time, the single variogram approach to IK greatly reduces problems of order relations by ensuring that all incremental probabilities sum to one and that (in the absence of negative weighting factors) all incremental probabilities are in the zero-to-one interval [Lemmer (*10*,*29*)]. In this work and subsequent related IK estimations [Young and Hoerger (*34*)], no order relations problems were observed.

Consequently, a complete stochastic model of this joint family can be developed by repeating IK to estimate probability distributions for every joint parameter and every local block in the model.

IK results and model quality

The efficiency of IK on this joint modeling was tested by comparing IK estimates with the actual samples in three comparison areas (blocks A, B, and C in Figure 1). Area A has a size of 400 m × 200 m and was located where enough samples are available in a small area to generate local sample statistics. It was digitized with 32 regularly spaced points (8×4). Then, IK was done at each of these 32 points by using only one-quarter of the samples in this joint family. The average of 32 IK estimates was compared with the actual distribution of three-quarters of the samples retained from the kriging. In this way the locality of IK estimates, as well as the IK procedure itself, was tested simply by comparing two cumulative histograms of each parameter as shown in Figure 7. A very good agreement was observed for each joint parameter and a positive conclusion was reached for the IK capability in this local stochastic modeling of the joint systems. For the comparison

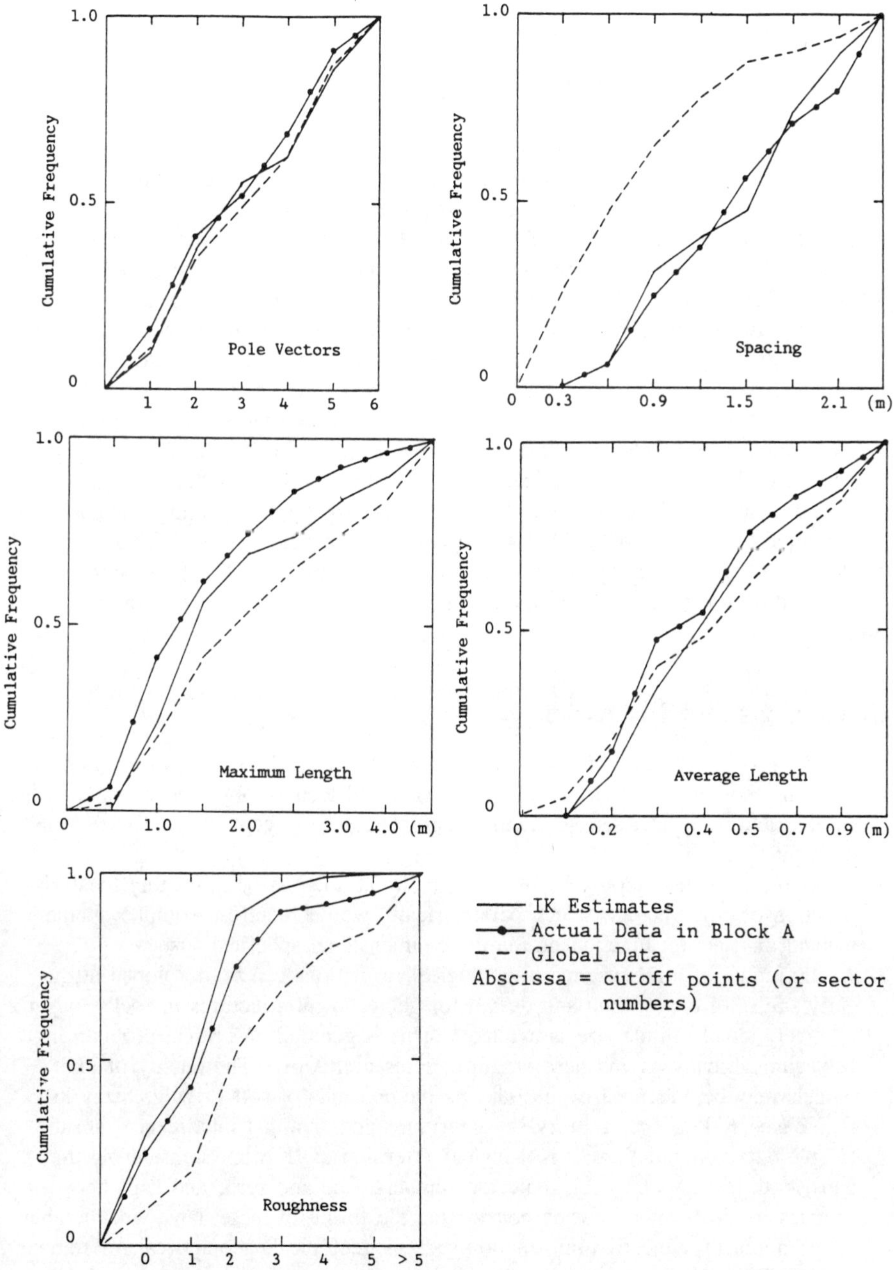

Figure 7. **Cumulative histograms for joint parameters in comparison block A.**

block B, which has an area of 200 m × 200 m, similar tests and conclusions were repeated as done in block A, but they were excluded from this paper.

Another test for the unbiasedness and stationarity conditions was made in the large area C, which has an area of 566 m × 850 m, and where 96 points were placed at regular intervals (8×12). Similarly, IK was repeated at each of the 96 points within area C based on one-quarter of the joint samples. Then, the average of the IK estimates was compared with the actual sample statistics obtained from the rest of the three-quarters of samples that were not used in IK. The graphical comparisons are given in Figure 8, which confirms positive test results.

Based on these tests, it is concluded that IK with the raised indicator variogram is a practical technique to develop a localized stochastic model of joint systems in geological formations. To build a complete stochastic model of joint systems, the IK employed here should be repeated for every family identified in the geological formation. In this open pit case, there are three joint families and IK should be repeated three times. A few points were checked without being presented in this paper when one-quarter of the joint samples were selected for IK. They were selected systematically by taking one after every three samples while following the list of the computer data file. Checking was accomplished throughout the comparisons between the selected one-quarter of samples and the whole samples, by such comparisons as experimental histograms and statistics, contour maps of pole vectors on stereonets, location maps for uniform spatial distribution over the pit, etc. The tests confirmed that the selection process did not destroy any characteristic structures in the joint family.

Conclusions and Discussion

1. IK is a proven technique for rock joint modeling, a local stochastic modeling of joints in geological formations. The precision and locality observed in the statistical distribution of characteristic model parameters are acceptable as seen in this open pit case.
2. As a nonparametric approach, IK offers a simple method of estimating local distribution of pole vectors, which parametrically would require a complex mathematical distribution function on the three-dimensional spherical surface.
3. An IK model of joint systems in geological formations can be developed at the early stages of exploration and design for geotechnical structures in rocks, when a relatively small sample size is available. This is generally a critical problem in geotechnical analyses and deserves further research work. The quality of joint models may be enhanced by introducing the concepts of soft kriging, fuzzy kriging, Bayesian kriging, or fuzzy set algorithms and artificial intelligence into the sample data manipulation. It is a logical extension of IK when considering the gaps, both the gaps between indicator numbers, one and zero, and gaps between samples in the field; and when considering the filling of these gaps with the abundant amount of subjective information such as geological, geophysical, or remote sensing data.
4. A local stochastic model that characterizes the geological formation in probabilistic terms may be an ultimate model for geotechnical analyses. A realistic solution

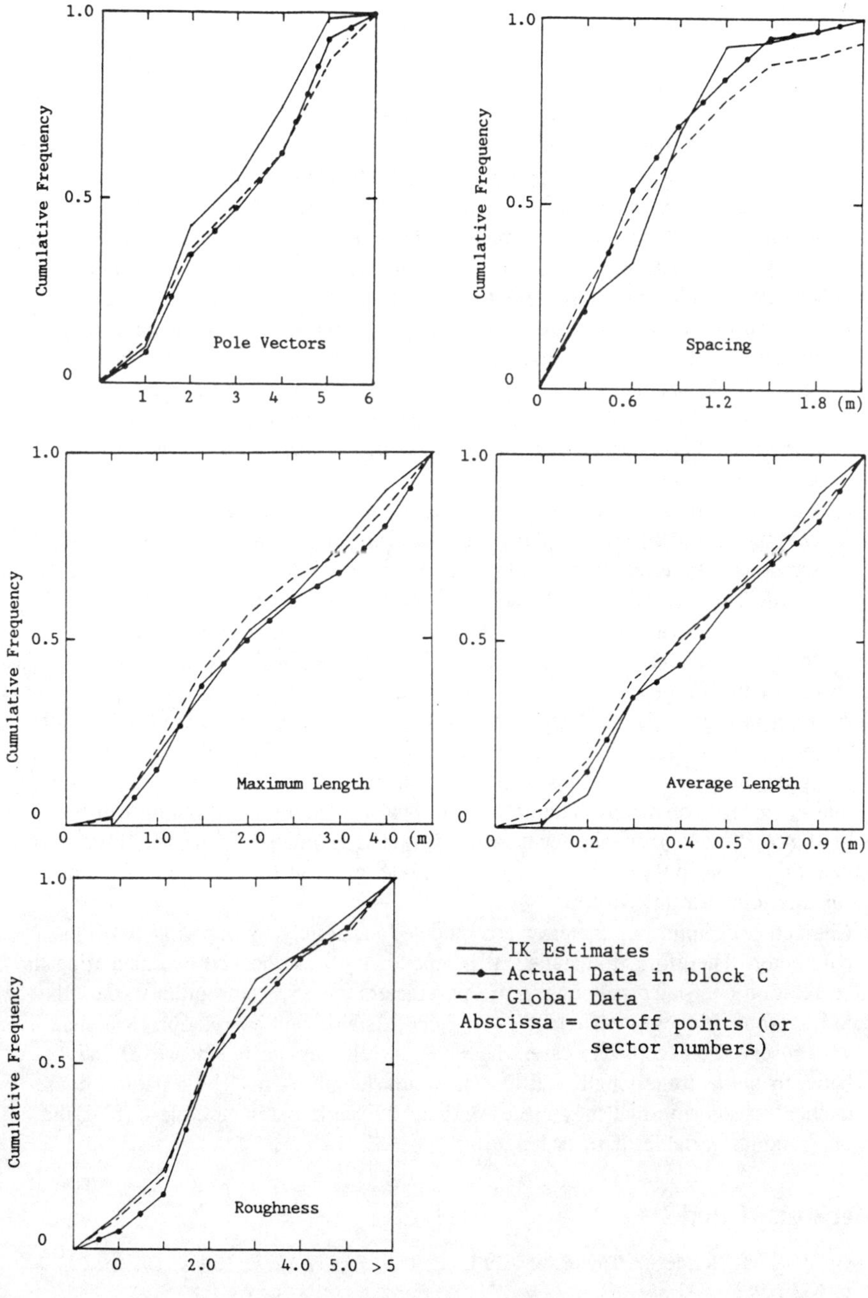

Figure 8. **Cumulative histograms for joint parameters in comparison block C.**

in probabilistic terms (or risk assessments in geotechnical analyses) may be obtainable for geotechnical problems by applying the current deterministic engineering directly onto this stochastic model [for example see Young (*31*), Young and Hoerger (*34*)].

5. An IK model of joint systems in rock masses has a broad range of applications in various geotechnical fields, including:
   a. Geotechnical structural stability analysis—a kinematic stability analysis on surface and subsurface structures excavated in fractured rocks can be localized and presented in probabilistic terms from the current deterministic engineering coupled with the stochastic simulation technique. Also, the geotechnical structures can be optimized at an early design stage, optimum in their sizes and shapes as well as their relative locations to other peripheral structures.
   b. Fluid flow through fractures—the quality of input models for geohydrological modeling and reservoir modeling could be improved through IK. The equivalent cell-block concept for multiphase flow through fractures could be improved with a realistic IK estimate and its deterministic flow model could be turned out as a probabilistic model when it is applied to a stochastic input model of fractured media.
   c. Cavability analysis—the fragment size distribution of caved materials from the caving method of mining plays a critical role in designing a mining system and selecting mining equipment. But it is not predictable yet until the mine is excavated. This problem may be solved from the IK joint model, since the fragment size distribution may be computable from the knowledge of natural fracture distributions in rock masses. It is also applicable to the precalculation of blasting fragmentation.
6. As seen in Figure 7, the permanence of the distribution of average trace lengths is good and its shape is close to the negative exponential distribution, which was accepted in other publications mentioned before. It is also a one-parameter distribution (inverse of mean). Therefore, the local probability distribution can be predicted directly from the local mean, which is estimated from OK. But this possibility was not pursued in this open pit modeling, and IK was applied uniquely for this joint parameter, too.
7. Characteristic joint parameters were modeled separately by applying IK to each parameter. Therefore, the statistical independency was checked by calculating the correlation coefficients among entire parameters [SPSSX subroutine was used (*32*)]. None of these coefficiencies revealed a significant correlation (less than 0.1 was observed), except one case where the correlation coefficient was 0.887 between the average length and the maximum length. It may be a natural consequence considering that they are essentially the same parameters describing the one random variable of trace length.

Literature Cited

1. Goodman, R. E., R. L. Taylor, and T. L. Brekke, *J. Soil Mech. Found. Div. ASCE, 94,* p. 637 (1968).
2. Piteau, D. R., "Geological factors significant to the stability of slopes cut in rock,"

presented at So. Afr. Inst. Min. Metall., Symp., Planning Open Pit Mines, Johannesburg (1970).

3. Louis, C., *A study of groundwater flow in jointed rock and its influence on the stability of rock masses,* Rock Mechanics Research Report No. 10, Imperial College of Sci. and Tech., London (1969).
4. Warren, J. E. and P. J. Root, *Soc. Pet. Eng. Jour.*, p. 525 (Sept. 1963).
5. Evans, R. D. and E. A. Ubani, ''An equivalent cell block porosity and permeability concept for modeling multiphase flow in naturally fractured, lenticular reservoirs,'' presented at SPE/DOE Low Permeability Reservoirs Symp., Denver (1987).
6. Young, D. S. and S. F. Hoerger, ''Spatial variability of rock fractures and modeling by geostatistics,'' presented at Int. Assoc. Eng. Geol. Symp., Beijing (1987).
7. Young, D. S., *Math. Geol., 19* (6), 467 (1987).
8. Hoerger, S. F. and D. S. Young, ''Predicting local rock mass behavior using geostatistics,'' presented at the 28th U.S. Rock Mech. Symp., Tucson (1987).
9. Journel, A. G., *Math. Geol., 15* (3), 445 (1983).
10. Lemmer, I. C., ''The Mononodal cutoff,'' Ph.D. Thesis, Stanford Univ., Stanford (1984).
11. Call, R. D., J. P. Savely, and D. E. Nicholas, ''Estimation of joint set characteristics from surface mapping data,'' presented at the 17th U.S. Rock Mech. Symp., Salt Lake City (1976).
12. Cicchini, P. E., at Call and Nicholas, Inc., Tucson, Ariz., personal communication (1987).
13. ISRM Commission on Standardization of Laboratory and Field Tests, Suggested Methods for the Quantitative Description of Discontinuities in Rock Masses, *Int. J. Rock Mech. Min. Sci. and Geomech. Abstr., 15*, p. 139 (1978).
14. Grossman, N. F., ''The bivariate normal distribution on the tangent plane at the mean attitude,'' in *Proc. of Int'l. Symp. Fundamentals of Rock Joints*, Bjorkliden, Sweden (1985).
15. Pahl, P. J., *Int. J. Rock Mech. Min. Sci. and Geomech. Abstr., 18* (3), 221 (1981).
16. Priest, S. D. and J. A. Hudson, *Int. J. Rock Mech. Min. Sci. and Geomech. Abstr., 18* (3), 183 (1981).
17. Warburton, P. M., *Int. J. Rock Mech. Min. Sci. and Geomech. Abstr., 17* (3), 181 (1980).
18. Warburton, P. M., *Int. J. Rock Mech. Min. Sci. and Geomech. Abstr., 17* (4), 305 (1980).
19. Pryor, Paul, at Call and Nicholas, Inc., Tucson, Ariz., personal communication (1987).
20. Barton, C. M., *A geotechnical analysis of rock structure and fabric in the CSA Mine, Cobar, New South Wales,* Applied Geomechanics Tech. Paper No. 24, CSIRO, Australia (1977).
21. Baecher, G. B., N. A. Lanney, and H. H. Einstein, ''Statistical description of rock properties and sampling,'' in *Proc. 18th U.S. Rock Mech. Symp.*, Golden (1977).
22. Shanley, R. J. and M. A. Mahtab, *FRACTAN: A computer code for analysis of clusters defined on the unit hemisphere,* U.S.B.M. IC 8671 (1975).
23. Arnold, K. J., ''On spherical probability distribution,'' Ph.D. Thesis, M.I.T., Boston (1941).
24. Bingham, C., ''Distributions on a sphere and on the projective plane,'' Ph.D. Thesis, Yale Univ., New Haven (1964).
25. Fisher, R. A., *Proc. Royal Soc. of London, A217,* p. 295 (1953).
26. Journel, A. G., ''The indicator approach to estimation of spatial distributions,'' presented at the 17th APCOM, Denver (1982).

27. Young, D. S., *Math. Geol., 19* (6), 481 (1987).
28. Journel, A. G. and Ch. J. Huijbregts, *Mining Geostatistics,* p. 600, Academic Press, London (1978).
29. Lemmer, I. C., *Math. Geol., 18* (7), 589 (Part I) and 605 (Part II) (1986).
30. Journel, A. G., "The place of non-parametric geostatistics," in *Geostatistics for Natural Resources Characterization,* G. Verly et al. (ed.), Reidel Publ. Co., Dordrecht (1983).
31. Young, D. S., *Int. J. Min. Geol. Eng., 3* (4), 215 (1985).
32. SPSSX Subroutine—Pearson Correl. Coeff. in *SPSSX Introductory Statistics Guide,* by M. J. Norusis, p. 92, McGraw-Hill, New York (1983).
33. Watson, G. S., *Statistics on Spheres,* John Wiley, New York (1983).
34. Young, D. S. and S. F. Hoerger, "Geostatistics applications to rock mechanics," in *Proc. 29th U.S. Rock Mech. Symp.,* Minneapolis (1988).

Evaluating the Effect of Sampling and Spatial Correlation on Ground-Water Travel Time Uncertainty Coupling Geostatistical, Stochastic, and First Order, Second Moment Methods

ROBERT W. ANDREWS, A. MARSH LaVENUE, and JERRY A. McNEISH
INTERA Technologies, Inc.
Austin, Texas

Abstract

Ground-water travel time predictions at potential high-level waste (HLW) repositories are subject to a degree of uncertainty due to the scale of averaging incorporated in conceptual models of the ground-water flow regime as well as the lack of data on the spatial variability of the hydrogeologic parameters. Both of these sources of uncertainty are related to the spatial correlation of these parameters, in particular the permeability. The present study describes the effect of limited observations of a spatially correlated permeability field on the predicted ground-water travel time uncertainty. Varying permeability correlation lengths have been used to investigate the importance of this geostatistical property on the tails of the travel time distribution.

This study uses both geostatistical and differential analysis (first order, second moment) techniques. Following the generation of a spatially correlated permeability field which is considered "reality," semivariogram analyses are performed upon small random subsets of the generated field to determine the geostatistical properties of the field represented by the observations. Kriging is then employed to generate a kriged permeability field and the corresponding standard deviation of the estimated field conditioned by the limited observations. Using both the "real" and "kriged" fields, the ground-water flow regime is simulated and ground-water travel paths and travel times are determined for various starting points. These results are used to define the ground-water travel time uncertainty due to path variability. The variance of the ground-water travel time along particular paths due to the variance of the permeability field estimated using kriging is then calculated using the first order, second moment method. The uncertainties in predicted travel time due to path and parameter uncertainties are then combined into a single distribution.

The results from this study indicate that it is possible to determine the uncertainty of the predicted ground-water travel time with a limited number of observations. However, the uncertainty is greatly affected by the correlation within the field of interest. For example, the number of observations and the difficulty to determine the geostatistical properties of a given field increase as the correlation length decreases. In addition, changes in the correlation length result in similar changes in the travel time variance. Increasing the correlation length (assuming the variance of the parameter remains the same) results in an increase in the uncertainty of the travel time, while decreasing the correlation length reduces the uncertainty of the travel time.

Introduction

One of the principal sources of uncertainty in predictions of the postclosure performance of potential high-level waste (HLW) sites is the spatial variability of hydrogeologic parameters in the vicinity of the site. While hydrogeologic parameters such as permeability and effective porosity have some effect on the predicted EPA cumulative release rate and the NRC engineered barrier system release rate, they have a significant effect on the predicted ground-water travel time from the edge of the disturbed zone to the accessible environment. The purpose of the present paper is to extend earlier work we and others have performed addressing the uncertainty in the predicted ground-water travel time, by incorporating both the uncertainty in the geostatistical properties of the permeability field (in particular the correlation scale) as well as the number of observations used to characterize the geostatistical properties of this field.

Two basic methods may be utilized to quantify the uncertainty in predicted ground-water travel times. These methods may be classified as the differential approach (or first order, second moment method) and the statistical sampling approach. The differential approach predicts the mean and variance of the travel time based on the mean and variance of the system parameters (e.g., permeability, porosity, etc.) and the sensitivity of the travel time to these system parameters. The sensitivity derivatives can be readily generated utilizing direct sensitivity methods such as the adjoint technique. The statistical sampling approach produces a probability distribution function (PDF) of the travel time from multiple simulations (or realizations) of the system behavior.

Both the differential and statistical sampling uncertainty analyses techniques have been applied to evaluate the NRC ground-water travel time uncertainty at the Deaf Smith County bedded salt site in the Texas Panhandle. The statistical sampling results are presented in Andrews et al. (*1,2*), while the differential analysis results are summarized in LaVenue and Andrews (*3*). To date, the studies evaluating the ground-water travel time uncertainty at the Deaf Smith County site have focused on vertical cross-sectional models, given that the relatively large downward hydraulic gradients cause all ground-water travel paths to cross the varied lithologies beneath the host salt (Lower San Andres [LSA] Unit 4). However, we have also noted in these studies that the most significant parameter affecting the "likely" travel path to the accessible environment and, as a result, the travel time uncertainty along the likely travel paths, is the horizontal permeability distribution in the carbonate interbed beneath the host salt (the LSA Unit 4 interbed).

The two-dimensional vertical cross-sectional models have necessitated the treatment of the LSA Unit 4 interbed as a one-dimensional feature. Even given this limitation, it was possible to discern that the permeability variability (in particular the correlation distance) of this interbed had a significant effect on the predicted travel time uncertainty. As the travel time uncertainty controls the likelihood that the site is able to meet the NRC travel time criteria, it is therefore important to evaluate the effect of the horizontal permeability correlation on the expected travel time uncertainty within the LSA Unit 4 interbed. However, given the fact that we are limited in our ability to precisely quantify the permeability correlation length, we have also examined the effect a relatively small number of sampling points (e.g., boreholes) have on the predicted travel time uncertainty for differing correlation lengths. The ultimate aim of this study is to determine the best means of quantifying the spatial correlation length, including the number and location of sampling points.

Approach

The approach taken in the present study to evaluate the effect of spatial correlation on ground-water travel time uncertainty consists of the following steps:

1. Generate realizations of a stationary random permeability field for different correlation lengths. In this study, we used one realization for each correlation length. While we realize that additional realizations would be useful to verify the robustness of the selected field, this was unfortunately not possible.
2. Simulate the ground-water flow regime through the generated fields (considered reality).
3. Determine the ground-water travel paths and the travel times for various starting points in the "real" fields.

In the present study, we have investigated random permeability fields with correlation lengths (λ) to e^{-1} correlation of 650 and 2000 m for the exponential semivariograms employed. These distances correspond to values expected for a dolomite such as the LSA Unit 4. While we have also used a correlation length of 1000 m, those results did not add additional insight to the study and therefore are not reported herein. For both random fields we have assumed that the permeability distribution is lognormally distributed with a variance (σ^2) of the log of the distribution equal to 0.41 and a mean (μ) of the log of the distribution equal to -8.0 (corresponding to a hydraulic conductivity of 10^{-8} m/sec or a permeability of 1 millidarcy).

The random fields have been generated over a 50 $\times$ 50 two-dimensional horizontal mesh, with each grid block being 100 m $\times$ 100 m square. The total grid is therefore 5 km along each axis, which corresponds to the horizontal distance from the edge of the disturbed zone to the accessible environment. The random fields were generated by means of the turning bands method using the program TUBA (Mantoglou and Wilson [*4*]). The statistical properties of the fields were subsequently checked as indicated in the tabulation on the next page.

It is clear that the 2000-m correlation length random field yielded a log K variance over the 2500 grid blocks a bit less than the input value of 0.41. This is as expected as the variance over a specified spatial domain (in this case the 5000-m distance to the accessible environment) will be reduced as the ratio of λ to the domain

Random Field λ	μlog K	σ^2log K
650 m	−8.0	0.41
2000 m	−8.0	0.32

size increases. If one goes to the extreme of a very large λ (several times the size of the domain of interest), then there will be very little variance in the random property within this domain. In order to reduce possible numerical inaccuracies associated with generating a field which is correlated over approximately half of its domain, a 10 km $\times$ 10 km permeability field was initially generated with λ of 2000 m. The 5 km $\times$ 5 km permeability field used in this study was subsequently taken from the center of the generated 100 km^2 field, thus yielding the lower variance.

Following the generation of the two-dimensional random fields and the definition of appropriate hydrologic boundary conditions along the edges of these fields, the ground-water flow regime across the field has been simulated. In the present study, we have specified constant pressures along upgradient and downgradient boundaries such that the average horizontal gradient is 0.004, which is analogous to the observed gradient in the LSA Unit 4 interbed in Deaf Smith County (see Andrews et al.[*1*]). The ground-water flow and transport code SWENT (INTERA [*5*]) has been used to generate the simulated flow regime.

Once the flow regime has been simulated, a particle-tracking routine is used to determine the ground-water travel paths and corresponding travel times from the upgradient to the downgradient grid boundaries (see Figure 1). As in all analyses of the travel time performance measure, we do not consider molecular diffusion, matrix diffusion, dispersion, adsorption, or decay, but only advective transport. A total of 48 particle trajectories are tracked. Each particle originates from the edge of the model grid, which corresponds to the edge of the disturbed zone. We have modeled the flow in the four coordinate directions (i.e., W to E, S to N, E to W, and N to S) and tracked 12 particles for each flow direction.

We have used the particle tracking program STLINE embedded within the program GRASP (INTERA [*6*]) to evaluate the ground-water travel time. We have considered the porosity of the entire domain to be a constant value of 0.04, similar to observed porosities in the LSA Unit 4 interbed (Andrews et al. [*1*]). While porosity/permeability correlations could be incorporated in the analysis, we feel that this level of sophistication is not warranted given the intended purposes of our study. Following the simulation of the travel times, routine statistical analyses have been performed on the results.

In addition to the above, we have investigated the effect that limited observations made within the "real" field have on the predicted travel time uncertainty. This phase of the study essentially mimics the approach one might actually take to quantify the travel time uncertainty, and as such allows us the opportunity to examine the effect of various sampling schemes, as well as the robustness of the sampling

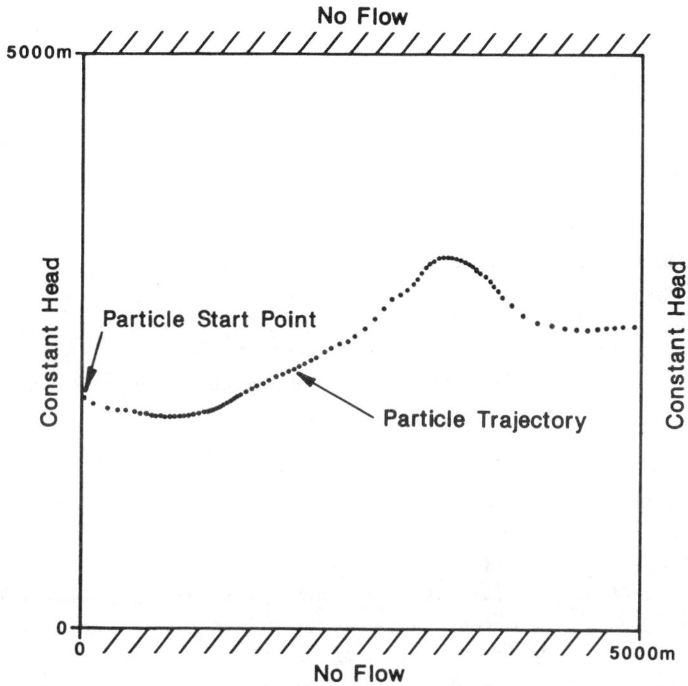

Figure 1. **Representation of particle tracking used to determine ground-water travel path and travel time.**

distribution. The approach taken to evaluate the effect of limited observations consists of the following steps:

1. Randomly locate 15 sampling points within each of the random fields generated previously. While we acknowledge that 15 observations is a very small number for normal geostatistical analysis, this is a relatively large number of observations for the problem at hand given the cost associated with acquiring punctual hydrogeologic data at such depths below the surface.
2. At these 15 sampling points, assume that the "observed" hydraulic conductivity is exactly equal to the conductivity of the 100 m × 100 m block in which the sampling point resides. This is analogous to assuming that not only is there *no* measurement error, but also that the measured conductivity exactly corresponds to the average property over the entire block. These assumptions are, in reality, never realized, and future studies will examine the effect of this measurement uncertainty and averaging scale uncertainty.
3. Using the 15 "observations" of hydraulic conductivity, perform semivariogram analyses to determine the range and sill of the variogram representing the conductivity field.
4. Based on the theoretical semivariogram generated in Step 3, generate a kriged hydraulic conductivity field conditioning the field with the 15 "observations."

During this kriging step, the estimation error of the kriged log hydraulic conductivities is also produced.

5. Simulate the ground-water flow regime through these "kriged" fields using the same gradients as used in the "real" field analyses.
6. Determine the ground-water travel paths and the travel times along these paths for the same 48 starting points used in the "real" field analyses. The same porosity field is also used in the analysis of ground-water travel time within the "kriged" field.
7. Using a representative particle trajectory through the "kriged" field, analyze the sensitivity derivatives of the travel time along this path to the hydraulic conductivity field.
8. Combine the sensitivity derivatives determined in Step 7 and the estimation error (variance) of the kriged log hydraulic conductivities determined in Step 4 to evaluate the uncertainty in the travel time along the representative path.
9. Combine the uncertainty (variance) in travel time due to path uncertainty (determined in Step 6) with the variance in the travel time along a path (determined in Step 8) to quantify the expected total travel time uncertainty within the "kriged" conductivity field.

In the present study we have used a modified version of the program K603 developed by Skrivan and Karlinger (*7*) to perform the semivariogram analysis, to estimate the "kriged" log conductivity fields, and to calculate the corresponding estimation errors. We have performed additional semivariogram analyses using different sets of randomly located "observation" points to test the robustness of the semivariograms used in the analysis. In both cases, we have performed consistency checks to assure minimum bias.

The sensitivity derivatives of the travel time along the representative trajectory to the hydraulic conductivity field have been determined by the adjoint technique. This technique is described in Sykes et al. (*8*) and is incorporated in the program GRASP (INTERA [*6*]), which was used in this study. GRASP calculates marginal and normalized sensitivities of specified performance measures (e.g., ground-water travel time) to specified system parameters (e.g., hydraulic conductivity in a particular grid block). In the present application, we utilize only the normalized sensitivities, which represent the percent change in the travel time caused by a 1% change in the hydraulic conductivity, or

$$S_i = \frac{dTT/TT}{dK_i/K_i} \equiv \frac{d(\ell nTT)}{d(\ell nK_i)} \qquad (1)$$

where S_i is the normalized sensitivity derivative, TT is the travel time, and K_i is the hydraulic conductivity in block i. The normalized sensitivities are more relevant when the performance measure and system parameters are lognormally distributed as they are in this study.

The uncertainty in the travel time along the representative particle trajectory due to the uncertainty in the hydraulic conductivity field has been estimated using the first order, second moment technique. The first order approximation of the second moment or variance of the performance measure is

$$\mathrm{Var(TT)} = \sigma^2\,\mathrm{TT} = \sum_{i=1}^{p}\sum_{j=1}^{p}\left[\mathrm{cov}(K_i, K_j)\cdot\frac{d\mathrm{TT}}{dK_i}\cdot\frac{d\mathrm{TT}}{dK_j}\right] \tag{2}$$

where p is the total number of blocks (i.e., 2500) and $\mathrm{cov}(K_i, K_j)$ is the covariance of the hydraulic conductivity field, defined by

$$\mathrm{cov}(K_i, K_j) = \rho_{ij}\sigma_{K_i}\sigma_{K_j} \tag{3}$$

in which ρ_{ij} is the correlation between K_i and K_j. In the present application all sensitivity derivatives, standard deviations, correlations, variances, and covariances are with respect to the log of the respective distributions (whether travel time or hydraulic conductivity). Therefore, we may rewrite Equation (2) as follows:

$$\sigma^2_{\log \mathrm{TT}} = \sum_{i=1}^{p}\sum_{j=1}^{p}\left[\rho_{ij}\sigma_{\log K_i}\sigma_{\log K_j}\frac{d(\log \mathrm{TT})}{d(\log K_i)}\cdot\frac{d(\log \mathrm{TT})}{d(\log K_j)}\right] \tag{4}$$

The covariance matrix is derived directly from the kriging analysis, while the sensitivity derivatives are generated from the adjoint analysis.

Step 9 requires combining two statistical distributions, the first being the travel time uncertainty due to the path uncertainty (i.e., each of the 48 paths is equally likely) and the second being the travel time uncertainty along the representative path caused by parameter uncertainty. There is no unique way to combine two independent distributions. To generate the combined or total uncertainty we essentially performed a Monte Carlo-type simulation, sampling first from the distribution representing the travel time uncertainty due to path uncertainty and then sampling from the second distribution derived from the calculated uncertainty along the path. For each trial the second distribution's mean is equivalent to the sampled value from the first distribution with a variance based on the calculated variance or uncertainty along the representative trajectory. This method of course assumes that the variance of the travel time along the representative trajectory is similar for each travel path. While we did not check this assumption due to time constraints, we feel it is appropriate because most of the particle travel time occurs through areas away from observation points. This sampling procedure was repeated 1000 times, resulting in the combination of the two distributions.

Effect of Spatial Correlation on Travel Time Uncertainty

The random log K fields generated with λ equals 650 m and λ equals 2000 m are illustrated in Figures 2a and 2b, respectively. As is clearly observed, the short-scale variability of the log K field is much less using the higher correlation length.

Forty-eight particle trajectories and travel times from the edge of the disturbed zone to the accessible environment were simulated for each random field. Twelve

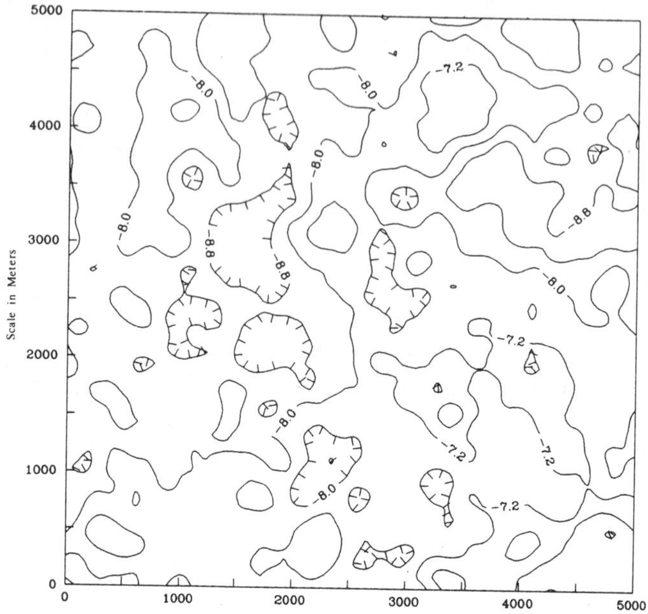

Figure 2a. **Random conductivity field with correlation scale (e^{-1}) of 650 m.**

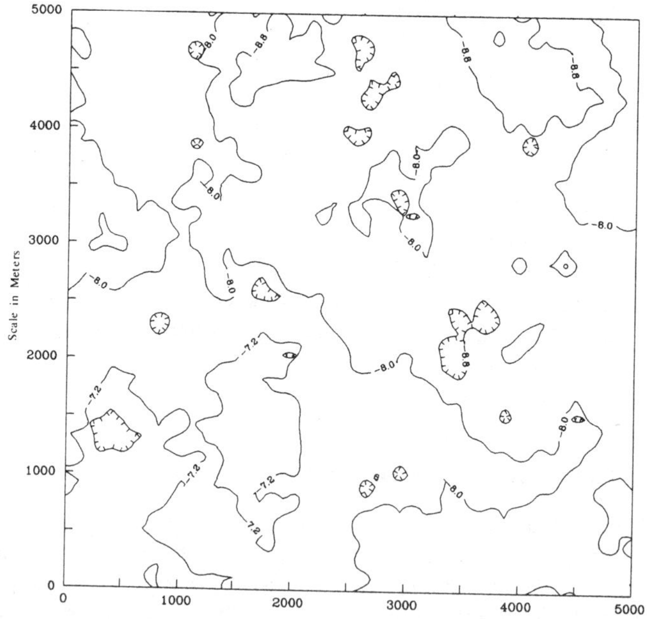

Figure 2b. **Random conductivity field with correlation scale (e^{-1}) of 2000 m.**

representative trajectories are depicted in Figures 3a and 3b, for correlation lengths of 650 m and 2000 m, respectively. The spacing between individual dots on these figures corresponds to the distance the particles traveled in 10^{11} sec (3200 years). Therefore, closely spaced dots indicate low velocities and large spacings between dots indicate high ground-water velocities. It is interesting to note that the simulated trajectories are highly irregular due to the particle attempting to find the path of least resistance. As expected, the particles move faster in the high conductivity zones and slower in the low conductivity regions, even though the gradient is also being adjusted. It is interesting to observe that a certain degree of "channeling" appears to occur through the higher conductivity zones. While it may appear in Figure 3b that the particles in the upper right-hand corner would have exited the right side boundary had this been a specified head boundary instead of a no-flow boundary, when the boundary condition was changed, the results remained essentially unchanged.

It bears noting in those figures that the actual path the particle travels from the starting point ("edge of the disturbed zone") is always greater than the straight-line distance of 5000 m between these two points. Because the distance of interest is indeed the straight-line distance, we have not adjusted our results to reflect the slightly increased travel distance.

The statistical results of the simulated travel times along the 48 independent particle travel paths are tabulated below:

Random Field		**Travel Times (in log years)**		
λ	μ**log TT**	σ^2**log TT**	$\mu + 2\sigma$	$\mu - 2\sigma$
650 m	5.83	0.065	6.34	5.32
2000 m	5.85	0.031	6.20	5.50

The means of the travel time distributions are essentially identical and are slightly larger than the deterministic value (μlog TT = 5.2) one would expect by dividing the 5000 m distance to the accessible environment by the product of the mean hydraulic conductivity and the mean gradient divided by the mean porosity. This increase is a result of the increased travel length described above as well as the reduced hydraulic gradient in higher conductivity regions.

The simulated travel time variances illustrate the effect one might expect; incorporating the random, albeit correlated, process in the travel time assessment greatly reduces the variance that would be predicted if one considered only the variance in log K. If one considered only the log K variability, the travel time variance would be expected to be the same as the log K variance, i.e., 0.41. This observation was made earlier in a fairly simplistic one-dimensional analysis using a Monte Carlo-type approach (Andrews et al. [*1*]). The small reduction in predicted variance as the correlation scale is increased from 650 m to 2000 m may be explained as follows: (1) the variance of log K is slightly lower in the 2000-m correlation-scale random field due to the increased ratio of λ to the 5000-m length of the domain of interest, and (2) over an e^{-1} correlation distance (i.e., 2000 m and 650 m), a particle must

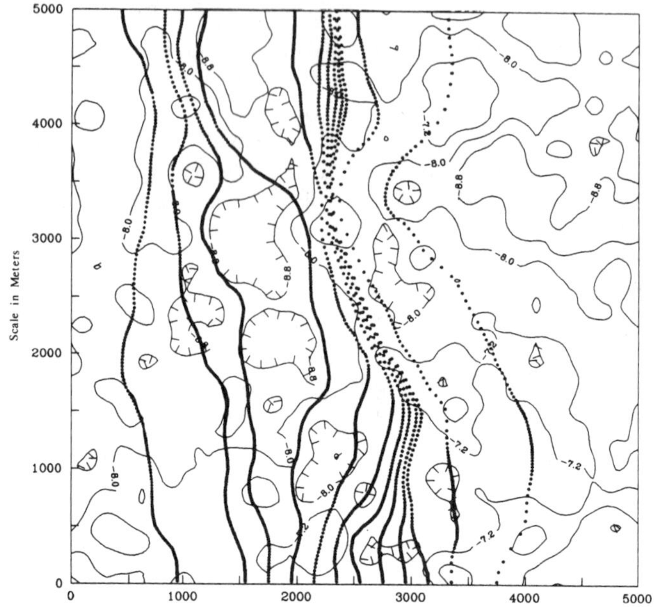

Figure 3a. **Ground-water particle trajectories through random conductivity field (λ = 650 m).**

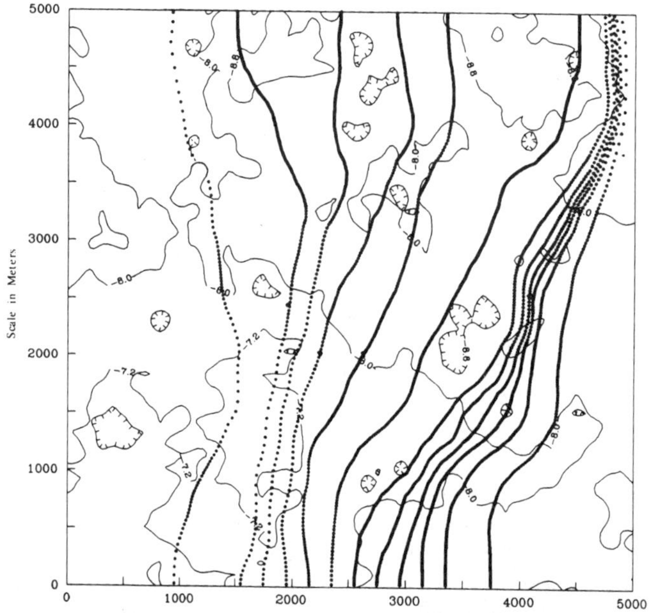

Figure 3b. **Ground-water particle trajectories through random conductivity field (λ = 2000 m).**

pass through more blocks (20 versus 6.5) for the higher correlation length. It is interesting to note that these small variances yield relatively tight travel time distributions, with a total range from $\mu + 2\sigma$ to $\mu - 2\sigma$ of only one order of magnitude. This is, however, what hydrogeologic intuition would have let us conclude anyway.

Effect of Limited Observations on Travel Time Uncertainty

In the real world, we of course do not have sufficient observations to exactly define the permeability field. Therefore, it is impossible for us to obtain as much data as would be required to completely specify the permeability distribution in space as was done in the analyses presented above. For all practical purposes, we may consider the results of the previous section to represent "reality," while the results we next describe illustrate the effects of our limited ability to "observe" this reality. This section examines the effect a limited number of observations have on our ability to quantify the travel time uncertainty and evaluates the different effects caused by varying correlation lengths.

Fifteen observation points have been randomly located within the 5 km × 5 km "real" fields used previously. At each of these 15 points, the "observed" hydraulic conductivity is considered to be exactly the grid block conductivity at that location from the "real" field, i.e., there is no measurement error and the observation point is assumed to be able to measure the average conductivity over the 100 m × 100 m block.

Using these 15 spatially distributed values we perform standard semivariogram analyses to determine the range and sill of the theoretical distribution that is consistent with these data. The observed and theoretical semivariograms for the cases of λ equal to 650 m and 2000 m are illustrated in Figures 4a and 4b, respectively. It bears noting that the correlation distance generated by the semivariogram analysis using the 15 values from the λ equal to 650 m random field is only 600 m. From here on we will therefore refer to the kriged field as having λ equal to 600 m. "Jackknife tests" were performed to check the consistency of the determined geostatistical parameters. In order to test the robustness of our geostatistical parameter estimates, we evaluated nine different sets of the 15 randomly located points. The results of these semivariogram analyses are presented below:

Random Field		**Range (m)**		**Sill**	
λ (m)	**σ^2log K**	**Trial 1**	**Range**	**Trial 1**	**Range**
650	0.41	600	600–1400*	0.55	0.22–0.55
2000	0.32	2000	1800–2200	0.38	0.14–0.70

*Note: Two of nine realizations could not be fit due to an apparent drift in the data.

It is clear that as the correlation scale increases, the robustness of the geostatistical parameters estimated from a particular sampled subset of the entire domain increases. In addition to the above, we tested the effect an increased number of observations has on our ability to estimate the geostatistical properties of the random stationary

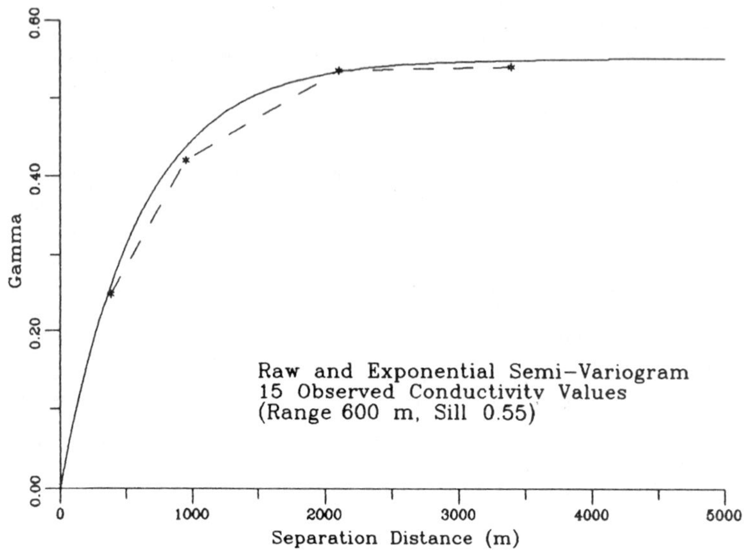

Figure 4a. **Raw and theoretical semivariograms for 15 sample points from random conductivity field (λ = 650 m).**

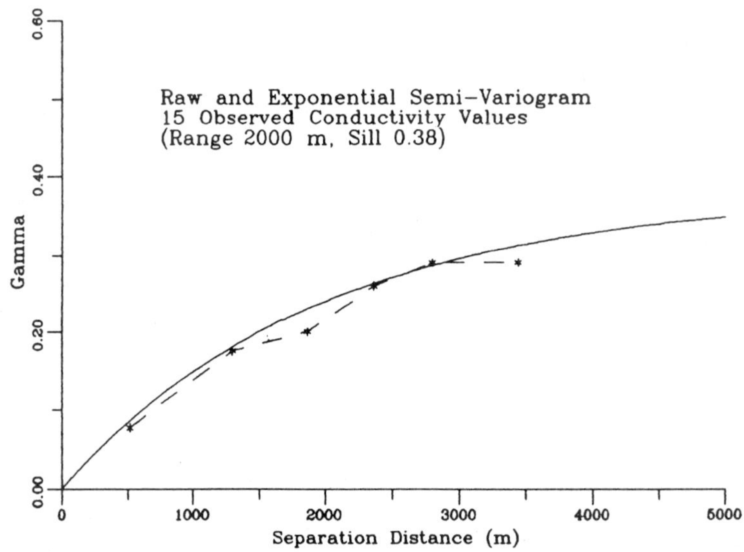

Figure 4b. **Raw and theoretical semivariograms for 15 sample points from random conductivity field (λ = 2000 m).**

field. As expected, increasing the number of observations increases the confidence in the estimated geostatistical properties. These observations were also made by Russo and Jury (*9*), wherein they state "...the uncertainty about variogram values or about the associated integral scales decreases as the number of sample points increases and when the size of the correlation scale of the underlying process increases relative to the field area."

Using the theoretical semivariograms illustrated in Figures 4a and 4b, kriging was used to generate kriged log K fields for λ equal to 600 and 2000 m (Figures 5a and 5b, respectively). These fields have been conditioned based on the 15 observations. Corresponding maps of the kriged estimation errors are illustrated in Figures 6a and 6b for λ equal to 600 and 2000 m, respectively. As expected, the kriged fields are considerably smoother than the underlying "real" fields illustrated in Figures 2a and 2b. (Note: Different contour intervals are used in Figure 2 compared to Figure 5.)

As in the previous analyses, the ground-water flow travel paths and travel times have been simulated for 48 trajectories using the kriged conductivity fields. Twelve particle trajectories are illustrated in Figure 7a for λ equal to 600 m and Figure 7b for λ equal to 2000 m. As expected, the particle trajectories are considerably smoother than those generated using the "real" random conductivity fields. The statistical summary of the travel time uncertainty due to path uncertainty is tabulated below:

Kriged Field	**Travel Time (in log years)**	
λ (m)	**μlog TT**	**σ^2log TT**
600	5.60	0.031
2000	5.77	0.020

While the mean travel times do not change significantly from the "real" field results, the variance of the travel time distribution becomes even smaller than before. This could have been anticipated in that the variance of the log K of the kriged field is also reduced, as indicated below:

Kriged Field		
λ (m)	**μlog K**	**σ^2log K**
600	−7.81	0.098
2000	−7.96	0.080

The above travel time variance refers to the variance due to the uncertainty in the path. That is, each of the 48 paths modeled must be considered equally likely. In the case of the travel path/travel time analyses corresponding to the kriged fields, we

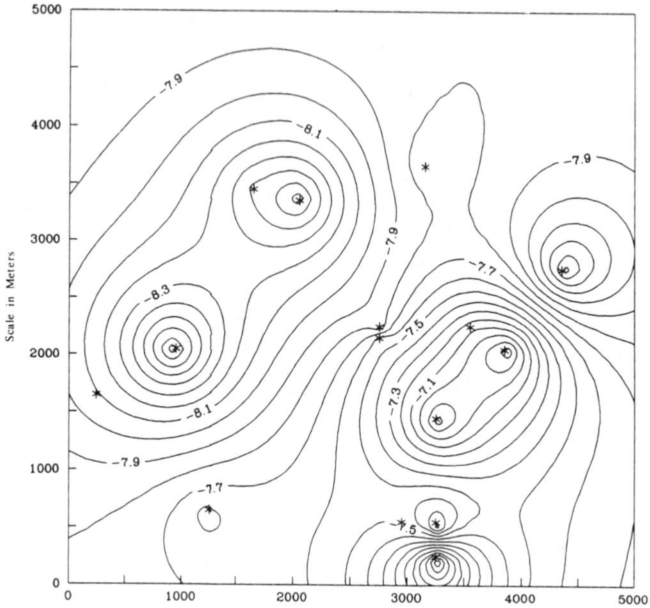

Figure 5a. **Log kriged conductivity field 1 (range = 600 m, sill = 0.55).**

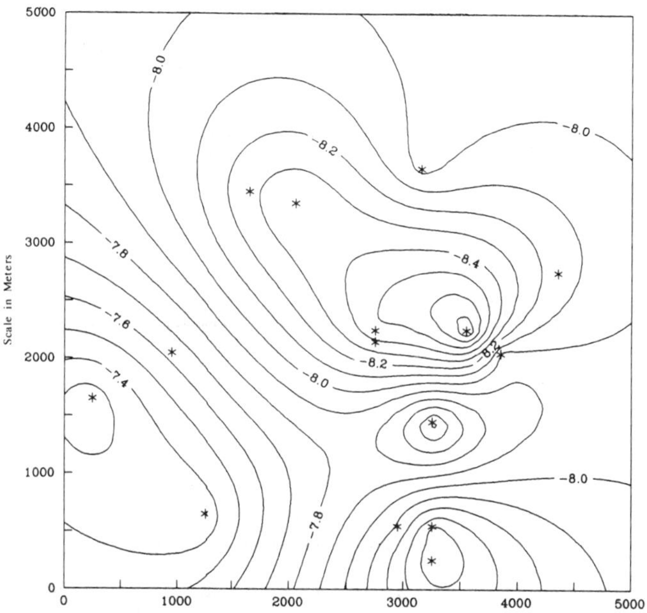

Figure 5b. **Log kriged conductivity field 2 (range = 2000 m, sill = 0.38).**

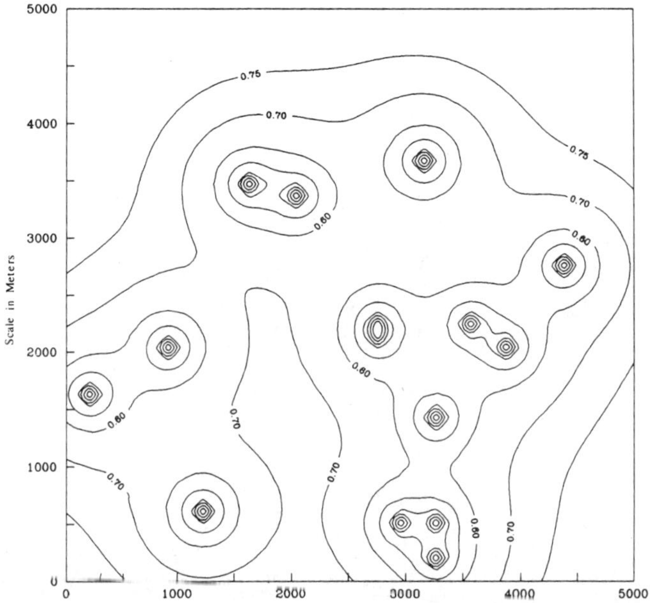

Figure 6a. **Estimation errors of kriged conductivity field (λ = 600 m).**

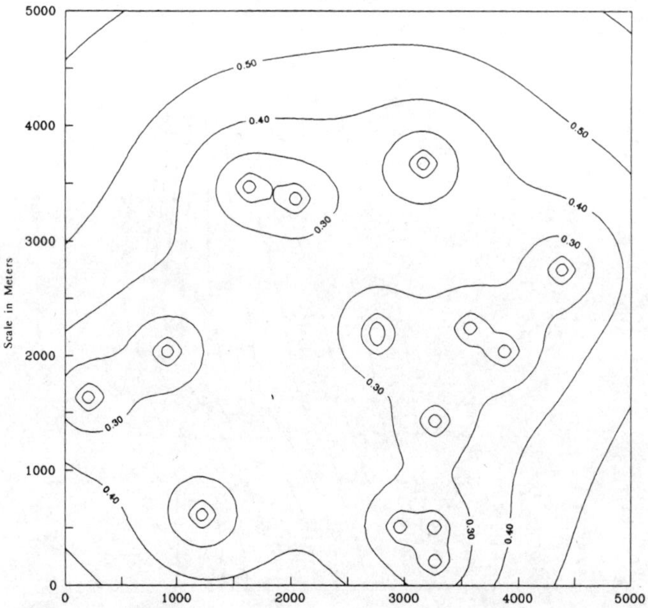

Figure 6b. **Estimation errors of kriged conductivity field (λ = 2000 m).**

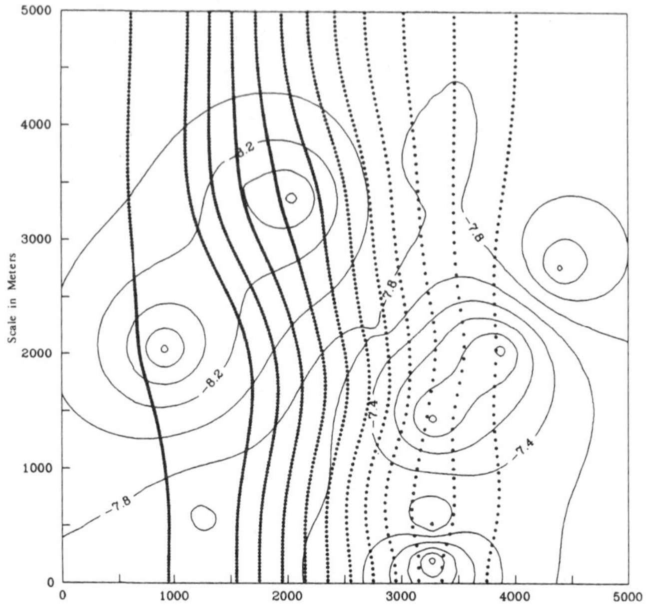

Figure 7a. **Ground-water particle trajectories through kriged conductivity field (λ = 600 m).**

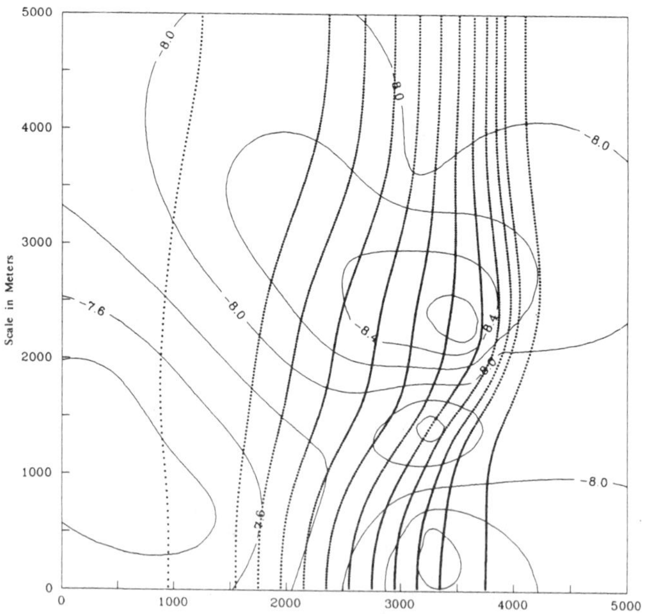

Figure 7b. **Ground-water particle trajectories through kriged conductivity field (λ = 2000 m).**

must also consider the uncertainty of the travel time along each of the 48 paths. This step was not required in the "real" field analysis presented previously because in that case there is by definition no parameter uncertainty. The simulated travel paths and travel times, in that case, are exactly what one would expect given the permeability field used.

In order to evaluate the travel time uncertainty along the paths, we have selected one representative path for performing a first order, second moment (FOSM) estimate of travel time variance due to parameter uncertainty. We have limited ourselves to one path due to the computational burden of performing FOSM for a large number of paths. In this analysis, the first step is to evaluate the normalized sensitivity of the travel time along this path to the permeability field. The particle paths selected in each field and the logs of the normalized sensitivity coefficients are presented in Figures 8 and 9. All of the normalized sensitivity coefficients have an absolute value less than one; therefore we have taken the absolute value of the logs and added the sign corresponding to the sign of the sensitivity coefficient. Hence a value of -3.8 indicates that 1% increase in hydraulic conductivity in this region would cause a $10^{-3.8}$% decrease in particle travel time. Although not shown in Figures 8 and 9, each block along the particle travel path has relatively high, negative normalized sensitivity value on the order of 10^{-2}. This reflects the direct effect an increase in hydraulic conductivity along the travel path would have in decreasing the travel time along this path. The values are less than 1.0 along the path due to the indirect effect that an increase in hydraulic conductivity would decrease the hydraulic gradient as well as the heterogeneous nature of the hydraulic conductivity field.

Several interesting features are apparent upon close comparison of Figures 8 and 9. For both kriged fields, transitions between positive and negative normalized sensitivity derivatives occur. The positive values were expected and result from the indirect effect that decreasing the conductivity in grid blocks away from the travel path decreases the hydraulic gradient along the particle trajectory, hence increasing the travel time. The negative values were somewhat unexpected; however, they can be explained by the fact that they tend to occur in low conductivity regions of the modeled domain, and increasing the conductivity in these regions would tend to increase the total flux through this domain (given that specified pressure boundary conditions are utilized) and thus increase the gradient and decrease the travel time.

Using the normalized sensitivity derivatives (Figures 8 and 9) and the covariance matrix generated during kriging, the travel time variance along the representative paths is estimated using Equation (4). These values are then combined with the earlier travel time uncertainty due to path variability to yield the total uncertainty for the kriged field. These results are tabulated below:

Kriged Field	**Log Travel Time Variance**		
λ (m)	**Path**	**Along Path**	**Total**
600	0.031	0.134	0.29
2000	0.020	0.190	0.33

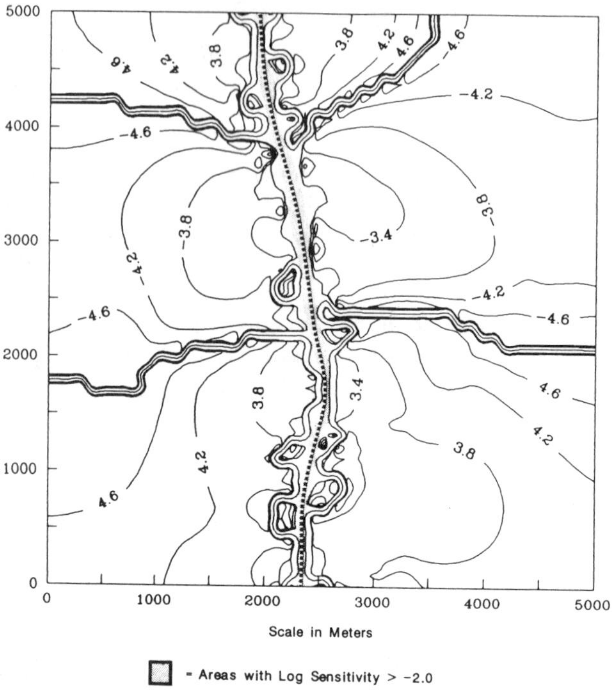

Figure 8. **Log sensitivity derivatives of travel time with respect to kriged conductivities (λ = 600 m).**

It is interesting to note that the travel time variance along the representative path is greater for the large correlation distance even though the variance of log K is less for the larger λ (see Figure 6b). This result is probably due to the larger role indirect effects play on the sensitivity derivatives and hence travel time uncertainties in the case of increased λ. However, it must also be added that this difference may possibly be attributed solely to the expected variability between different representative paths. It should be noted that the travel time uncertainty along the paths is much greater than the travel time uncertainty due to different paths, and that the uncertainty along the path apparently controls the total variability in that increasing λ causes an increase in the variance of the log of the travel time.

Summary and Conclusions

Three general conclusions were determined from the results of this investigation:

1. As the correlation length of the permeability field increases, the uncertainties associated with the semivariogram decrease as do the number of observation points needed to determine the representative geostatistical parameters.
2. If the permeability field is perfectly defined, the uncertainty of the groundwater travel time is based only on travel path uncertainty. Furthermore, the uncertainty decreases as the permeability correlation length increases.

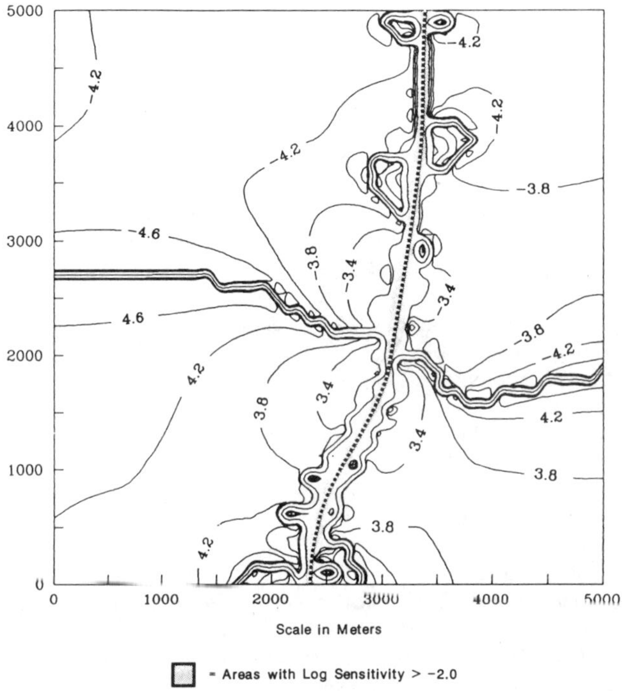

Figure 9. **Log sensitivity derivatives of travel time with respect to kriged conductivities (λ = 2000 m).**

3. For permeability fields with limited observations or sampling points, increasing the correlation length causes an increase in the uncertainty of the ground-water travel time. The increased uncertainty is due to an increase in the magnitude of the covariance values and an increase in the magnitude of the sensitivity derivatives.

In this study, we have employed a relatively straightforward framework in which pieces of associated methods (e.g., geostatistics, FOSM, Monte Carlo) are coupled to investigate the effects of sampling and permeability correlation length on ground-water travel time uncertainties. This is the first attempt of our investigation to determine the effects of limited observations and spatial correlation on predictions of ground-water travel time uncertainty. The classical Monte Carlo approach, where many random fields with the same geostatistical parameters are used to determine ground-water travel time distributions, could provide the user with more accurate quantitative results; however, we believe the general responses of the ground-water travel time uncertainty to changes in the permeability correlation length determined in this study should remain the same. We do recommend, however, that additional means of evaluating travel time uncertainty using various realizations of kriged conductivity fields should be investigated to quantify the effect of different sampling schemes. Such analyses may include performing FOSM analysis on all particle trajectories to eliminate the possible biasedness of using only representative paths.

Literature Cited

1. Andrews, R. W., V. A. Kelley, J. A. McNeish, A. M. LaVenue, and J. E. Campbell, 1986. *Travel Path/Travel Time Uncertainties at Salt Sites Proposed for High Level Waste Repositories,* prepared by INTERA Technologies, Inc. for Office of Nuclear Waste Isolation, Battelle Memorial Institute, Columbus, Ohio.
2. Andrews, R. W., A. M. LaVenue, B. M. Thompson, and James E. Campbell, 1987. ''Application of Uncertainty Analysis Techniques to Address NRC Travel Time Performance Measures at a Potential HLW Disposal Site in Deaf Smith County, Texas,'' National Water Well Association Proceedings of Solving Ground Water Problems with Models, Denver, Colorado.
3. LaVenue, A. M. and R. W. Andrews, 1987. ''Ground-Water Travel Time Uncertainty Analysis: A Comparison Between the Differential and Probabilistic Techniques,'' paper presented at American Geophysical Union, Baltimore Meeting, *EOS,* Vol. 68, No. 16, April 21, 1987.
4. Mantoglou, A. and J. L. Wilson, 1981. *Simulation of Random Fields with the Turning Bands Method,* Report No. 264, Ralph M. Parsons Laboratory, Massachusetts Institute of Technology, Cambridge, Massachusetts.
5. INTERA Environmental Consultants, 1983. *SWENT: Model Documentation and User's Manual,* ONWI-457, prepared for Office of Nuclear Waste Isolation, Battelle Memorial Institute, Columbus, Ohio.
6. INTERA Technologies, Inc., 1986. *GRASP (Ground-Water Adjoint Sensitivity Program): A Computer Code to Perform Post-SWENT Adjoint Sensitivity Analysis of Steady-State Ground-Water Flow,* BMI/ONWI-625, prepared for Office of Nuclear Waste Isolation, Battelle Memorial Institute, Columbus, Ohio.
7. Skrivan, James A. and Michael R. Karlinger, 1980. *Semi-Variogram Estimation and Universal Kriging Program,* USGS Water Resources Division, Tacoma, Washington, PB81-120560.
8. Sykes, J. F., J. L. Wilson, and R. W. Andrews, 1985. ''Sensitivity Analysis for Steady State Groundwater Flow Using Adjoint Operators,'' *Water Resources Research,* Vol. 21, No. 3, pp. 359–371.
9. Russo, David and William A. Jury, 1987. ''A Theoretical Study of the Estimation of the Correlation Scale in Spatially Variable Fields 1. Stationary Fields,'' *Water Resources Research,* Vol. 23, No. 7, pp. 1257–1268.

Stochastic Parameter Estimation for Nuclear Repository Site Characterization at Hanford, Washington

A. M. DJERRARI and V. V. NGUYEN
EWA, Incorporated
Minneapolis, Minnesota

Abstract

Pre waste emplacement groundwater travel-times are being used by the U.S. Department of Energy as one measure for ultimately selecting a suitable site for a high-level nuclear waste repository. To compute groundwater travel times, the groundwater flow pattern in the vicinity of the proposed repository location must be delineated and pertinent hydrologic parameters must be identified from the baseline data. In this paper, transmissivity and hydraulic head fields are estimated using a geostatistical solution to the inverse problem. An application to the Hanford Site is presented. The sensitivity of the estimated hydrologic fields to the number of available data points is analyzed through simulations. The predicted fields can be significantly different from one another depending on the number of data points used in the identification of the hydraulic head and transmissivity field geostatistical parameters.

Introduction

Under the Nuclear Waste Policy Act of 1982, the U.S. Department of Energy (DOE) is responsible for locating the first nuclear waste repository site. The Hanford Site, located in Washington, is one of the three sites recommended by DOE and approved by the President of the United States for site characterization. Federal regulation requires that exceedance of the 1000-year groundwater travel time of radionuclides be one of the disqualifying criteria. To assess the groundwater travel time (GWTT) of a released radionuclide through a hydrogeologic system, the groundwater flow pattern in the vicinity of the proposed repository must be delineated and pertinent parameters must be identified from the baseline data.

In this work, the problem of estimating hydrogeologic properties, specifically the transmissivity and hydraulic head, using field measurements is examined in a geostatistical framework. Unknown properties are represented as the realizations of a random field, and the estimation procedure consists of two main steps. First, the stochastic structure of the transmissivity field is identified using all available information, including head and transmissivity measurements. Second, linear estimation theory (cokriging) is applied to provide minimum variance and unbiased point estimates of hydrogeologic parameters. Structure identification is achieved iteratively in three substeps: structure selection, maximum likelihood estimation, and model validation.

This methodology is applied to the data collected in the Priest Rapids Member of the Wanapum Basalt Formation at the Hanford Site. Because of the scarcity of the available data, the sensitivity of the parameter identification procedure to the number of measurements is analyzed. This analysis sheds some light on whether the available hydrologic data at the Hanford Site are adequate to make reliable inferences about the hydrologic fields.

Geostatistical Solution to the Inverse Problem

The inverse problem of groundwater modeling is usually defined as a boundary value problem in which input and corresponding response function are considered known and in which the spatially varying transmissivity field is treated as a dependent variable to be estimated. The transmissivity field is estimated based on available information, including (1) point observations of the piezometric head and (2) measurements of transmissivity. The most recent effort in the solution of the inverse problem represents the piezometric head and transmissivity fields as realizations of random functions. In this paper, the geostatistical approach of Kitanidis and Vomvoris (*1*) is applied to estimate the spatially variable transmissivity and hydraulic head fields for a basalt layer at Hanford. A two-dimensional model of a confined aquifer under steady flow conditions with no leakage is assumed for this layer.

The geostatistical approach to the inverse problem is described in detail by Kitanidis and Vomvoris (*1*). Applications of this geostatistical method are presented by Hoeksema and Kitanidis (*2*,*3*) and Dagan (*4*). The geostatistical solution is viewed as a five-step procedure:

1. A model for the statistical spatial variability of the transmissivity field is proposed.
2. The governing differential equation of groundwater flow is used to relate the spatial variability of the piezometric head to that of the transmissivity.
3. The observed values of head and transmissivity are used to estimate the unknown parameters in the transmissivity spatial variability model.
4. The validity of the spatial variability model is tested. If the model passes the validity test to confirm its soundness,
5. Linear estimation procedures are applied to estimate the spatially distributed transmissivity and head field. The variance of estimation error is also computed. This latter is a measure of the accuracy of the transmissivity and hydraulic head field predictions.

Head Elevation and Transmissivity Estimation Following the Solution to the Inverse Problem: Application to the Hanford Site

The deep basalt flows in the Pasco Basin beneath the Hanford Site of south-central Washington are being investigated for the permanent storage of high-level nuclear waste. In general, the geology of this area is composed of thick basalt flows of low hydraulic conductivity interlayered with several thinner flow top layers of higher hydraulic conductivity. Lateral groundwater movement beneath the Pasco Basin is known to occur within the deep confined to semiconfined aquifers consisting of basalt flow tops, flow bottoms, and sedimentary interbeds. The multiaquifer system within the Pasco Basin has been broadly conceptualized as consisting of four geohydrologic units. One of them, the Wanapum geohydrologic unit, is the focus of the present analysis. Interested readers may refer to the final environmental assessment of the Hanford Site [DOE (*5*)] for additional information.

The hydraulic head and transmissivity data analyzed in this study were gathered in the Priest Rapids Member which is one of the three geologic units that comprise the Wanapum Formation. The selection of the Priest Rapids Member was based on (1) the potential for this geologic unit to act as a discharge zone for the groundwater system under the operating conditions of the repository, and (2) the greater number of transmissivity and head measurements that have been collected in this layer. The overall methodology described previously has been used (1) to identify the geostatistical parameters of the head and transmissivity field and (2) to estimate, using cokriging, the hydraulic head and transmissivity at each node of a grid that overlays the southern part of the Hanford Site boundaries (Figure 1).

Description of the data used

Hydraulic head data. Rockwell Hanford Operations has been monitoring water levels at three piezometer cluster sites at the Reference Repository Location (RRL) and at 35 additional boreholes at the Hanford Site (Figure 1). The water-level data are being used to evaluate time-variant hydraulic head behavior and to establish a head baseline for selected hydrogeologic units. The water-level data used in this study were provided by the DOE (*6*).

The monitoring boreholes are screened in several stratigraphic units in the Columbia River basalts. Fourteen boreholes are monitored in the Priest Rapids Member (Table 1). The water levels used in this study are averages of water levels observed from 1982 to 1986. Among the fourteen observations, the water levels observed at the Ford, Enyeart, O'Brian, DB-11, and McGee wells are 500 feet higher than those in the rest of the boreholes. The hydraulic heads in the upper Wanapum Basalt of the Cold Creek Valley are generally higher than the head elevations in the same stratigraphic horizon within the RRL east of the Cold Creek "Barrier" (Figure 1). This anomaly is interpreted as the effect of the aforementioned barrier [DOE (*5*)], i.e., as a no-flow or low-flow lateral boundary, although this interpretation has not yet been substantiated by sufficient evidence. Since the water levels at the Ford, Enyeart, O'Brian, DB-11, and McGee wells behave differently than those at the other wells, and since such an anomaly cannot be accounted for by

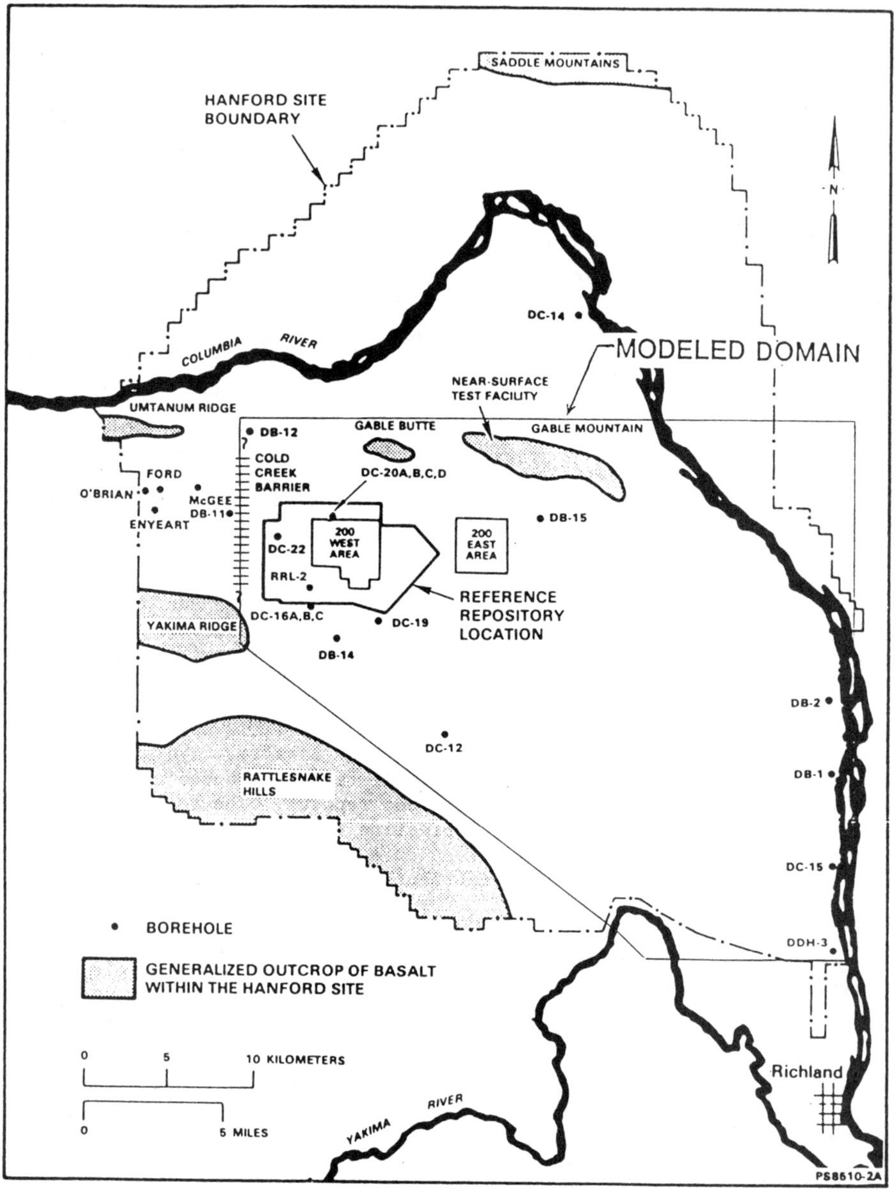

Figure 1. **Location map for selected boreholes.**

Table 1. **Water-level measurements in the Priest Rapids Member (average, 1982–1986).**

Borehole #	Location North	Location East	Water level (ft)
DB-01	406,971	2,308,893	392.5
DB-02	420,657	2,308,000	394.0
DB-12	468,067	2,200,144	399.0*
DB-14	430,190	2,215,764	400.2
DC-16C	436,377	2,211,009	405**
DC-19C	433,933	2,225,012	400.0
DC-20C	451,884	2,215,288	401.8
DC-22C	448,600	2,204,188	400.7
DDH-3	374,957	2,304,900	391.1
Enyeart	454,397	2,183,844	----- *
Ford	458,009	2,183,788	----- *
O'Brian	457,656	2,181,139	----- *
DB-11	454,471	2,194,850	----- *
McGee	457,773	2,191,775	----- *

*Important fluctuation during the year due to recharge from irrigation, pumpage or effect of the variations of the Columbia River stages.

**Important unexplained fluctuations.

a covariance function derived from a limited number of observations, these observations were dropped in the analysis.

Transmissivity data. Transmissivity values for the Priest Rapids Member have been compiled from published DOE documents in a Yakima Indian Nation (YIN) technical report [YIN (7)]. Table 2 lists the aquifer test data available for the Priest Rapids Member. Transmissivities of the Priest Rapids Member have been reported for eight boreholes, generally in terms of intervals. The transmissivity values considered in this study correspond to the median values of the intervals. Only six transmissivity measurements are considered in the following analysis for two reasons. First, the median transmissivity value observed at the McGee well is two orders of magnitude larger than the rest of transmissivity values considered in this study (Table 3). This, along with the fact that the McGee well is located west of the Cold Creek anomaly (a region that is not studied here), is the reason this observation was dropped in the analysis. Second, little is known about the hydraulic head configuration north of Gable Mountain. For this reason, the modeled domain considered in this study (Figure 1) does not extend north of Gable Mountain, and the transmissivity observed at well DC-14 is excluded from the analysis. Therefore, the predictions for the hydraulic head and transmissivity fields do not extend west of the Cold Creek anomaly, or north of Gable Mountain.

In order to characterize the spatially distributed transmissivity field, one transmissivity value must be assigned to each location. As shown in Table 2, several transmissivity values have been measured at the same borehole, depending on the

Table 2. **Transmissivity values for the Priest Rapids Member.**

Well No.	Source of Data (Document No.)	Transmissivity Range (ft²/day)	Median Transmissivity (ft²/day)
DB-12	RHO-BWI-ST-5		NR*
DB-15	RSO-BWI-TI-06	9.99 E02	9.99 E02
DB-15	RSD-BWI-TI-062	9.99 E02-9.99 E03	5.49 E03
DC-12	RSD-BWI-TI-064	9.99 E01	9.99 E01
DC-14	RHO-BWI-80-100 3Q	1.08 E02-1.08 E03	5.94 E02
DC-14	RSD-BWI-TI-065	9.99 E01-9.99 E02	5.49 E02
DC-14	RSD-BWI-TI-065	9.99 E02-9.99 E03	5.49 E03
DC-14	RSD-BWI-80-100 3Q	1.08 E03	1.08 E03
DC-14	RSD-BWI-TI-065	9.99 E01-9.99 E02	5.49 E02
DC-14	RHO-BWI-80-100 3Q	1.08 E02-1.08 E03	5.94 E02
DC-15	RHO-BWI-80-100 3Q	1.08 E−02	1.08 E−02
DC-15	RSD-BWI-TI-066	9.99 E−02	9.99 E−02
DC-15	RSD-BWI-TI-066	9.99 E+02	9.99 E02
DC-16A	SD-BWI-TI-135	9.99 E01-9.99 E02	5.49 E02
DC-16A	SD-BWI-TI-135 RHO-BW-SR-82-2	9.99 E−01-9.99 E00	5.49 E00
DC-16A	SD-BWI-TI-135 RHO-BW-SR-82-2 1QP	8.64 E−01-8.64 E01	4.36 E01
DC-19C	RHO-BW-SR-83-1 4QR	8.64 E01-8.64 E02	4.75 E02
Enyeart	RHO-BWI-ST-5	NR	
Ford	RHO-BWI-ST-5	NR	
McGee	SD-BWI-TI-227	9.99 E03-9.99 E04	5.49 E04
O'Brian	RHO-BWI-ST-5	NR	
RRL-2A	SD-BWI-TI-113 RHO-BW-SR-82-2 1QP	9.99 E01-9.99 E02 8.64 E00-8.64 E01	5.49 E02 4.75 E01

*NR: Not reported.

depth at which tests were performed. In the following analysis, each location has been assigned the maximum median transmissivity value observed at the monitoring borehole. The analysis of the maximum median transmissivity values may yield results that err on the side of conservatism (i.e., shorter travel times). However, even though conservative estimates of absolute transmissivity values are obtained, a conservative spatial configuration (in terms of yielding shorter groundwater travel times) of the transmissivity field may not be guaranteed.

Transmissivity covariance function parameter estimation

The five-step procedure is applied to identify the log-transmissivity field from the remaining six transmissivity and nine hydraulic head measurements as discussed

Table 3. **Maximum values of median transmissivity for the Priest Rapids Member.**

Well No.	Maximum Value of Transmissivities (ft^2/day)	Natural Log Maximum Transmissivities
DB-15	5.49 E03	8.6
DC-12	9.99 E01	4.6
DC-14*	5.49 E03	8.6
DC-15	9.99 E02	6.9
DC-16A	5.49 E02	6.3
DC-19C	4.75 E02	6.2
McGee*	5.49 E04	10.9
RRL-2A	5.49 E02	6.3

*Borehole out of the study area.

above. As it will be demonstrated later, this number of measurements is too small to yield accurate parameter estimates.

Step 1: Select a statistical spatial variability model for the transmissivity field. It is assumed throughout this study that the transmissivity follows a log-normal probability distribution. Some evidence in support of this assumption is provided by Freeze (*8*) and by Hoeksema and Kitanidis (*9*). Instead of dealing with the transmissivity, the natural logarithm of the transmissivity (Ln T) is used.

Since the log-transmissivity is assumed to be normally distributed, all that is needed is a model for its mean and covariance. The model consists of a constant Ln T mean and stationary zero mean random terms. The model used for the application is

$$E[Y_i] = F \tag{1.a}$$

$$Cov(Y_i, Y_j) = \theta_1 \delta_{ij} + \theta_2 \exp(-d_{ij}/I) \tag{1.b}$$

where Y_i is the Ln T value at point i of cartesian coordinates (x_i, y_i); E[ ] represents the mean or expected value; δ_{ij} is Kronecker's delta function ($\delta_{ij}=1$ if $i=j$ and $\delta_{ij}=0$ if $i \neq j$); d_{ij} is the scalar distance between points i and j, and I is the correlation length. The covariance parameters θ_1 and θ_2 are considered to be unknowns that will be estimated from the data.

The first term in Equation (1.b) represents small-scale Ln T variability. This variability is due to measurement error and to property variations at a scale smaller than the typical distance between neighboring measurements. The second term in Equation (1.b) represents the structured, or large-scale, Ln T variability. In Equation (1.b), I is the Ln T integral scale or correlation length.

Step 2: Relate the spatial variability of the hydraulic head to that of the transmissivity. The two geostatistical parameters θ_1 and θ_2 can be estimated by using only Ln T measurements. However, due to the paucity of Ln T measurements, a more general approach is followed. Measurements of hydraulic head as well as measurements of log-transmissivity are used to estimate the parameters. The complete use of information is accomplished by constructing the measurement covariance matrix (for both head and log-transmissivity measurements) as a function of θ_1 and θ_2. The small perturbation approach is applied to the governing partial differential equation. This latter is simplified by neglecting the second-order terms (first-order approximation). The simplified stochastic differential equation is used to express the autocovariance function of the hydraulic head measurements, as well as the cross-covariance function between hydraulic head and log-transmissivity measurements. A detailed description of the method is given in Hoeksema and Kitanidis (*2*).

The prior expected value of the piezometer head is calculated using a numerical model with constant transmissivity and prescribed head boundary conditions. The values of the hydraulic heads at the boundaries were obtained from a previous study [YIN (*10*)] in which the hydraulic head measurements were used to obtain a kriged hydraulic head field. In the geostatistical solution of the inverse problem described in Hoeksema and Kitanidis (*2*), one can account for the uncertainty of the hydraulic head values prescribed on the model boundaries by assigning a variance for these values. In this analysis, a large value (e.g., 1000 ft^2) was assigned to this variance in order to reduce the impact of the prescribed heads on the estimates of θ_1 and θ_2. Finally, the hydraulic head measurements were considered to be contaminated by measurement errors. A variance of 10 ft^2 was assumed for the head measurement error. This value of the variance has been chosen to account for potential errors associated with water-level measurements (e.g., uncertainty in measurement point elevation and error due to lack of vertical plumbness as a result of borehole deviation).

Step 3: Estimate the unknown geostatistical parameters. The measured values of head and log-transmissivity are used to estimate the unknown parameters of the model θ_1 and θ_2. The maximum likelihood parameter estimation procedure is used. The numerical procedure implemented is the Gauss-Newton method described in detail in Kitanidis and Lane (*11*).

Four cases were considered in the parameter estimation:

Case 1: Parameter estimation using all nine head and six transmissivity measurements.

Case 2: Parameter estimation using eight head and six transmissivity measurements (head at DC-16C has been omitted from the analysis).

Case 3: Parameter estimation using eight head and five transmissivity measurements (head at DC-16C and transmissivity at DC-12 have been omitted from the analysis).

Case 4: Parameter estimation using eight head and five transmissivity measurements (head at DC-16C and transmissivity at DB-15 have been omitted from the analysis).

In all four cases, the correlation length I of the transmissivity field is assumed to be 5 km. Several different values for the correlation length are investigated in a later phase of the analysis. This first estimate of the transmissivity integral scale is the one assumed in the DOE groundwater travel time analysis [Clifton (*12*)].

Step 4: Check validity of fitted spatial variability model. Model validation is accomplished by tests performed on a set of uncorrelated residuals obtained from the complete measurement vector Z, the measurement expectation vector E[Z], and the fitted measurement covariance matrix Q. The estimated parameters and the measurement vector Z are used to generate two normalized measurement vectors y_1 and y_2. The first normalized vector y_1 is a vector of normalized measurement errors. A large component of this vector indicates that the corresponding measurement may be an outlier. The second vector y_2 has (under the model) independently normally distributed components. The tests on this vector are designed to test the validity of the assumed structure. It should be pointed out that since the analysis is based on a small perturbation solution to the partial differential equation, when $\theta_2 > 1$, the validation tests may be somewhat suspect [Kitanidis, personal communication]. The reason is that by using head measurements, the estimate for θ_2 may be smaller than it should be, and transmissivity may appear as outliers unjustly. However, for I=5 km, this problem is not encountered since θ_2 is smaller than one (Table 4), a value for which small perturbation analysis gives reasonably accurate results [Dagan (*13*)].

In Case 1, three outliers were detected in the S_1 vector: the head measurement at DC-16C, and the transmissivity measurements at DB-15 and at DC-12.

In Case 2, two outliers were detected in the y_1 vector: the transmissivity measurements at DB-15 and at DC-12.

In Case 3, one outlier was detected in the y_1 vector: the transmissivity measurement at DB-15.

In Case 4, one outlier was detected in the y_1 vector: the transmissivity measurement at DC-12. The normalized residual observed for this measurement, however, is smaller than the ones observed for the outliers in the other three cases.

The tests on the y_2 vector held for all four cases.

In the first three cases considered, no large-scale structure was identified. All variability observed in the transmissivity values was recognized as small-scale variability or measurement error (i.e., $\theta_1 \neq 0$ and $\theta_2 = 0$).

In light of these results, only eight hydraulic heads and five transmissivity measurements (as in Case 4) were considered in all of the following analyses. It should be noted that by dropping the transmissivity observed at DB-15, which presented the highest value among the maximum transmissivity values observed, the cokriging procedure using only the thirteen remaining measurements is bound to yield transmissivity estimates smaller than what would have been obtained by using the transmissivity at DB-15.

The parameter estimate procedure used assumes that the correlation length I of the transmissivity field is known. The parameters θ_1 and θ_2 that describe small- and large-scale transmissivity variability are estimated by maximizing their likelihood (or minimizing their negative likelihood function). If one applies this procedure for different values of I and plots the minimum negative likelihood of θ_1 and θ_2 versus I, the best set of parameters θ_1, θ_2 and I corresponds to the one that gives the minimum negative likelihood value. This procedure has been applied to estimate θ_1, θ_2 and I.

First, it should be noted that the estimate for θ_2 is smaller than one for I varying up to 10 km (Table 4). This justifies the applicability of the first-order approximation, at least for I< 10 km. As will be seen in the sensitivity analysis of the cokriging

Table 4. **Values of the parameters θ_1, θ_2, and θ_2/I at various correlation lengths.**

I (km)	θ_1	θ_2	θ_2/I
1	0	0.710	0.7100
2	0	0.733	0.3665
4	0	0.741	0.1775
5	0	0.785	0.1570
6	0	0.789	0.1315
10	0	0.953	0.0953
15	0	1.216	0.0811
20	0	1.500	0.0750
30	0	2.103	0.0701
35	0	3.430	0.0980

results to the integral scale, the estimates of the parameter θ_2, for I between 10 km and 30 km, behave in a predictable manner. This may indicate that the first-order approximation may still be valid for I ranging from 10 km to 30 km.

Second, the variation of the negative log-likelihood of the parameters θ_1 and θ_2 with the correlation range is shown in Figure 2. As illustrated in Figure 2, this procedure could not distinguish a single value for the correlation length, i.e., any value for I between 10 km and 30 km is an equally acceptable value. This estimate, however, does not strongly affect the cokriging results. It was shown in Kitanidis and Lane (*11*), that when the correlation length is larger than the typical distance between measurement locations, the covariance function of the exponential model is approximately linear, with a slope θ_2/I. The parameter θ_2/I, in this case, is the effective parameter in kriging. Values of the ratio θ_2/I for different correlation length between 10 km and 30 km are presented in Table 4. It is obvious that the degree of variability of the effective parameter (i.e., the one that really matters) is much less than that of θ_2 and I. Therefore, an accurate estimate of the integral scale is not necessary if the purpose is only to interpolate hydrologic fields; however, it becomes crucial to have a reasonable estimate of I if the parameters of the spatially variable fields are to be used for GWTT probability determinations.

To illustrate the nonsensitivity of the cokriged log-transmissivity and hydraulic head fields to the value of the correlation length I, Ln T and hydraulic head fields were predicted for the two extreme cases (i.e., I=10 km and I=30 km). Figure 3 displays superimposed contours of kriged log-transmissivities using correlation ranges of 10 km and 30 km, respectively. The contours match closely near the measurement locations. However, in the central and northern parts of the modeled domain, the match is not as close. As for the hydraulic heads, a perfect match was observed between head contour maps constructed from kriged hydraulic heads using the correlation length of 10 km and 30 km, respectively.

Step 5: Minimum variance unbiased linear estimation of hydrogeologic parameter (cokriging). Once the statistical parameters of the log-transmissivity

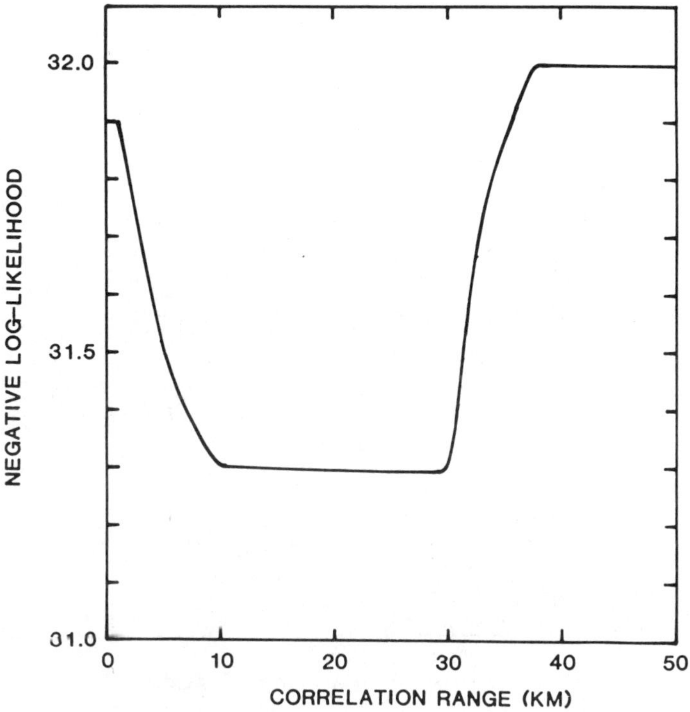

Figure 2. **Negative log-likelihood of parameters θ_1 and θ_2 as a function of the transmissivity correlation range.**

covariance function have been estimated, the value of log-transmissivity is inferred at any point in the model domain. Linear minimum variance, unbiased estimation theory, or cokriging, was employed.

The correlation range of the transmissivity field was assumed to be equal to 30 km. Figure 4 displays log-transmissivity contours, while Figure 6 displays transmissivity contours. The identified log-transmissivity field has some interesting characteristics. Along the Cold Creek syncline, the transmissivity decreases from the RRL, in the northwestern portion of the modeled domain, to the central part and then increases toward the southeast (i.e., Columbia River). This spatial configuration of the transmissivity field along this trend is favorable because it increases radionuclide GWTT from the RRL to the accessible environment (i.e., 5 km as defined in DOE [5]). Uncertainty in the delineation of radionuclide pathlines as well as in the GWTT calculations, however, may also increase. This will be discussed later.

Figure 5 displays the variance of log-transmissivity estimation errors. It can be seen that the variance of estimation is at a minimum along the Cold Creek syncline and increases to the northeast. This result reflects the site characterization strategy that focuses all efforts toward the characterization of the hydrologic fields in the RRL and along the Cold Creek syncline.

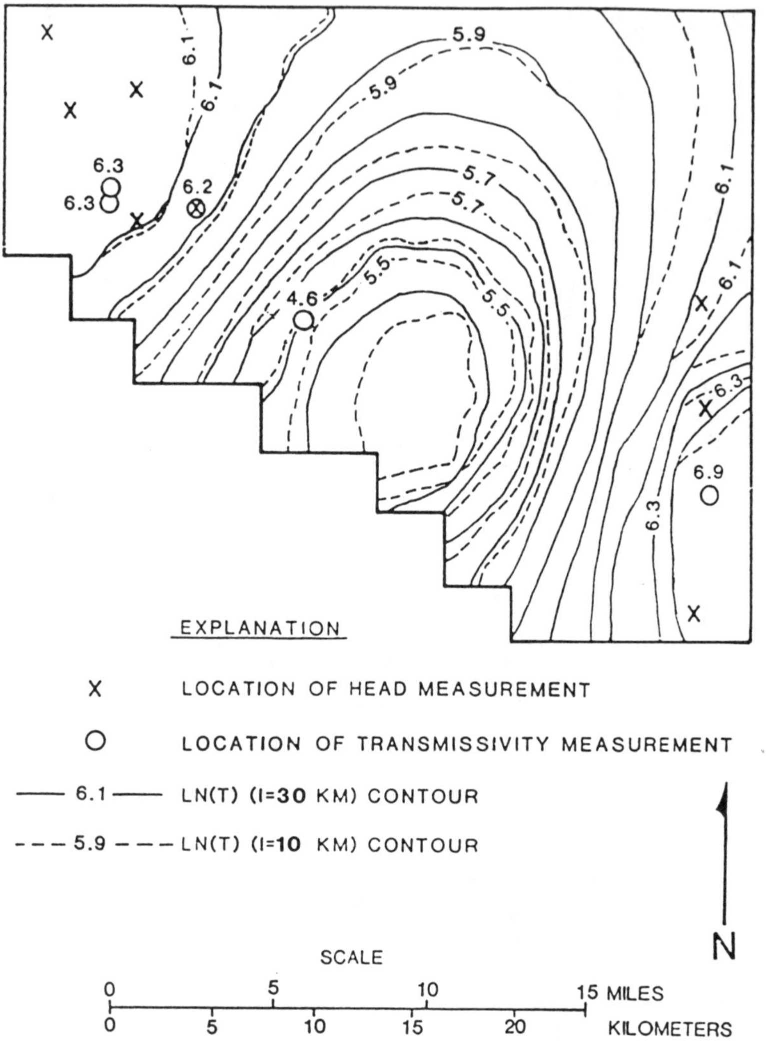

Figure 3. **Prediction of log-transmissivity field (correlation range 30 km and 10 km).**

Figures 7 and 8 display the transmissivity contours for the upper and lower 95% confidence limit, respectively. As shown, all five transmissivity measurements considered are contained between these two limits.

Figure 9 displays the hydraulic head contour map. The expected general trend of the groundwater flow is southeasterly toward the Yakima River. The groundwater flow in the RRL, however, is still not well defined. A northwestward groundwater flow is again predicted in the northern portion of the RRL [see YIN (*10*) for a discussion of the groundwater flow in the northern portion of the RRL].

Figure 10 displays the variance of hydraulic head estimation error. The minimum value of the variance of estimation is very high (approximately 40 ft^2). It should be noted, however, that this high value is partially due to very high variance assumed for the heads at the boundaries (i.e., 1000 ft^2), which is equivalent to assuming complete ignorance about the boundary heads. Therefore, this contour map of variance should be interpreted in a relative sense rather than in an absolute sense. Once again the minimum variance is found along the Cold Creek syncline, while the high variances are found in the northern and southern portions of the modeled domain.

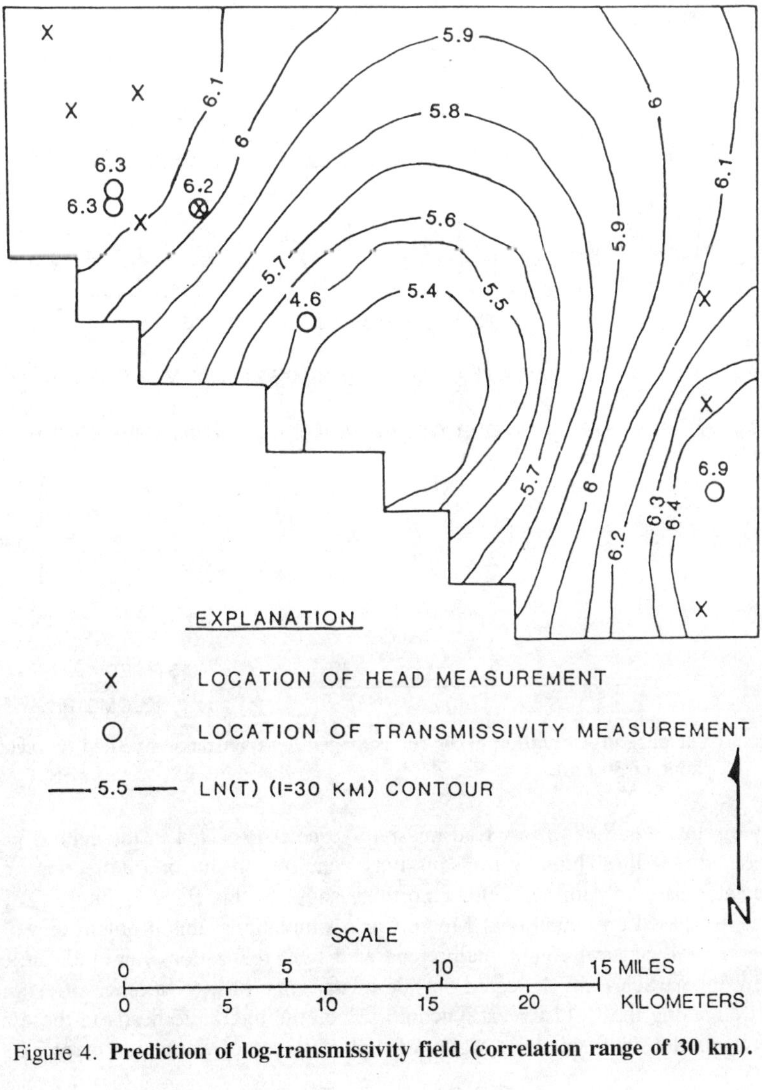

Figure 4. **Prediction of log-transmissivity field (correlation range of 30 km).**

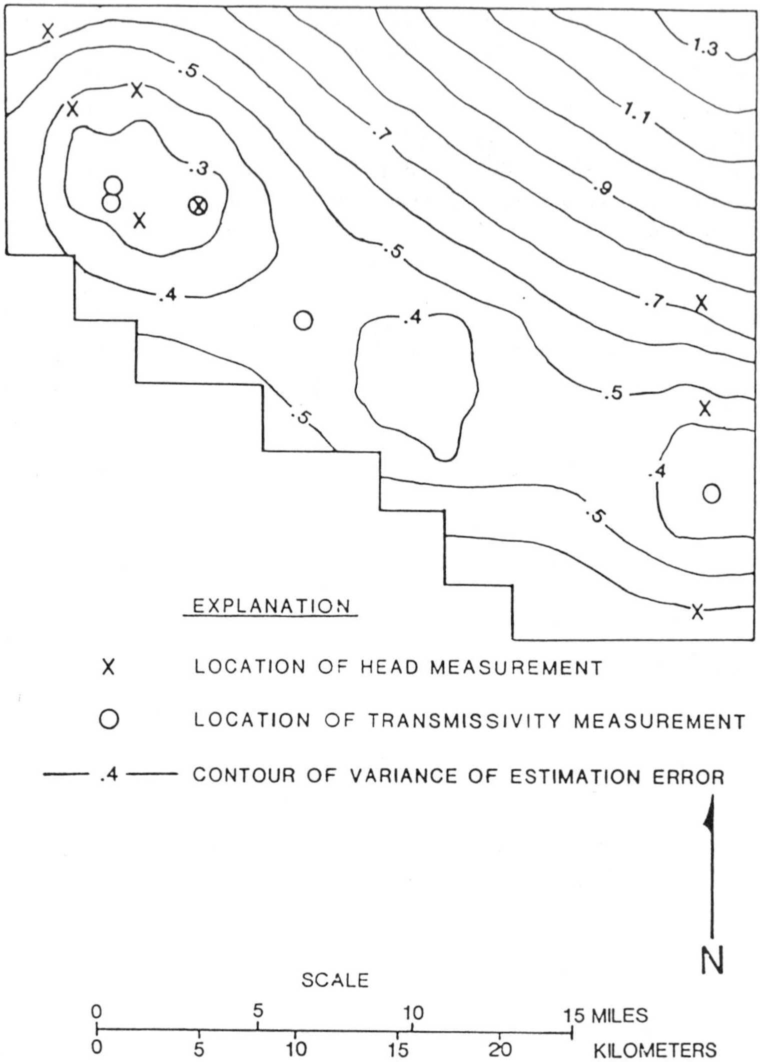

Figure 5. **Variance of estimation error for cokriged log-transmissivity field (correlation range of 30 km).**

As discussed earlier, a low transmissivity zone is expected in the central part of the modeled domain. This low transmissivity zone may result in a deflection of the groundwater particle pathlines, either north or south of this zone. If future GWTT analyses are based on conditional Monte Carlo simulations, this problem may become crucial. Several conditionally generated field realizations may indicate that groundwater particles are deflected outside of the zone of low variance of estimation error (i.e., along the Cold Creek syncline). Since the hydraulic heads in these areas are not accurately defined (i.e., region of high variance of estimation error as

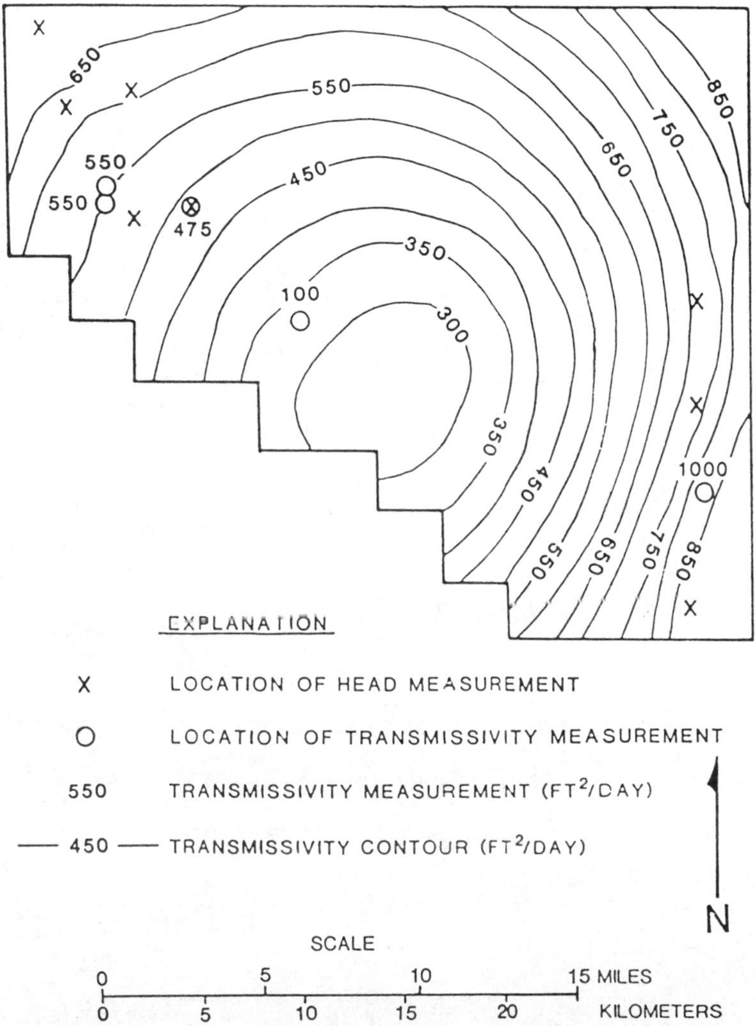

Figure 6. **Expected transmissivity field (correlation range 30 km).**

depicted in Figure 10), velocities and pathlines (and subsequently travel times) of the deflected particles may be associated with a high degree of uncertainty. This, once again, illustrates the need for better characterization of the hydrologic fields outside of the band that overlies the Cold Creek syncline.

Sensitivity analysis to the number of measurement points

A common concern in physical modeling is whether the number of available data points is adequate to produce a viable model. This section studies the sensitivity of

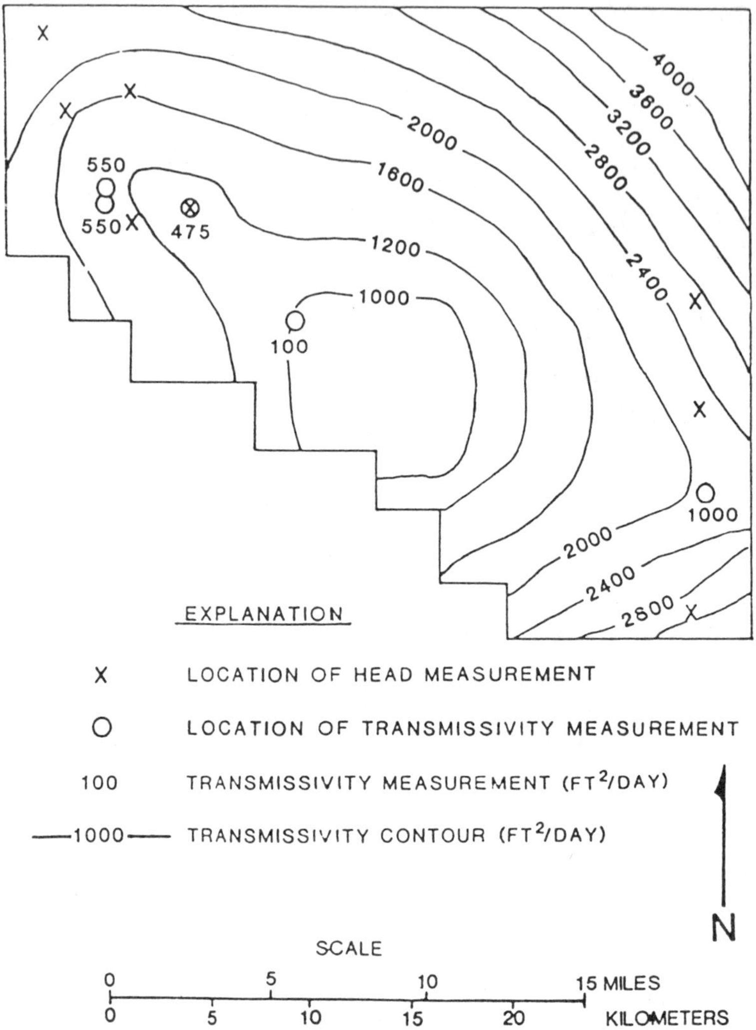

Figure 7. **Upper 95% transmissivity confidence limit (correlation range of 30 km).**

the parameter identification procedure to the number of measurements utilized in this process. The sensitivity of parameter estimates obtained from data sets of limited size was investigated by generating an ideal set of transmissivity and head values and randomly selecting fewer and fewer data from this set to produce models with increasing uncertainty and decreasing reliability. This exercise also sheds some light on whether the available hydrogeologic data at the Hanford Site are adequate to make reliable inferences about the hydrologic fields.

The initial step of this process was to generate a set of head and transmissivity data at each node of a grid that overlays the Hanford Site boundaries. The generated fields have the same spatial variability characteristics as the field identified for I=10 km (i.e., $\theta_1=0$, and $\theta_2=0.953$). This setting represents the ideal scenario in which the spatially dependent head and transmissivity fields are known. Once this ideal setting was generated, decreasing numbers of randomly chosen values of head and transmissivity (chosen from this complete data set) were used to assess the capability of reproducing the known hydrogeologic configuration. The randomly chosen values of head and transmissivity were used to estimate the parameters of the spatial structure. These parameters were then used to estimate, using cokriging, head values

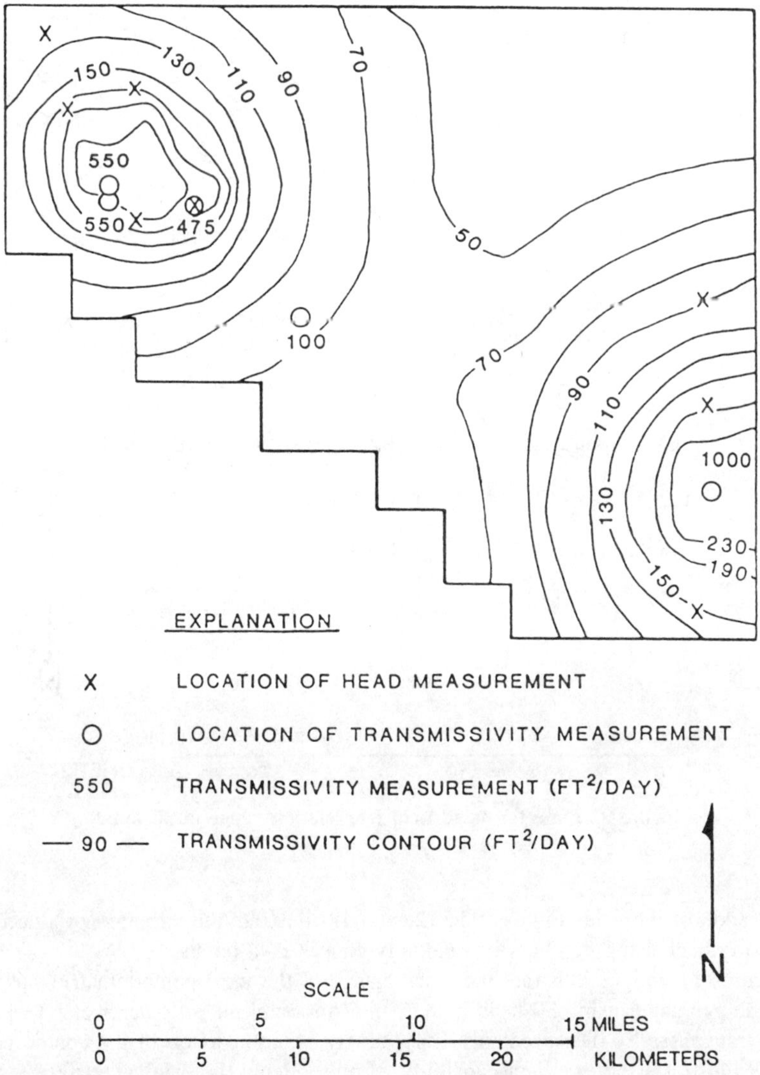

Figure 8. **Lower 95% transmissivity confidence limit (correlation range of 30 km).**

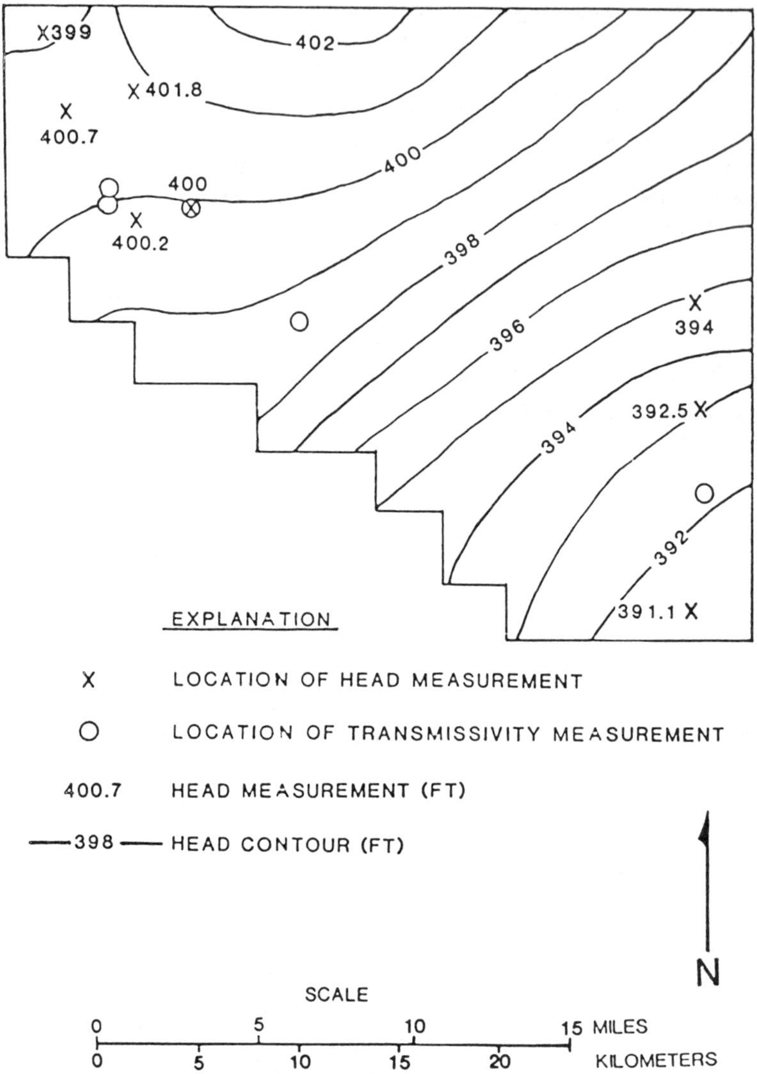

Figure 9. **Expected head field (correlation range of 30 km).**

at each node of the grid. Figures 11, 12, and 13 illustrate the variation produced from using decreasing numbers of randomly chosen data points.

Figures 11 and 12 illustrate the resemblance of the ideal potentiometric surface and those generated using 30 head and 20 transmissivity measurements and 18 head and 12 transmissivity measurements, respectively. A comparison of the contoured surfaces illustrates the excellent capability of reproducing the original surface used as the source of these data points.

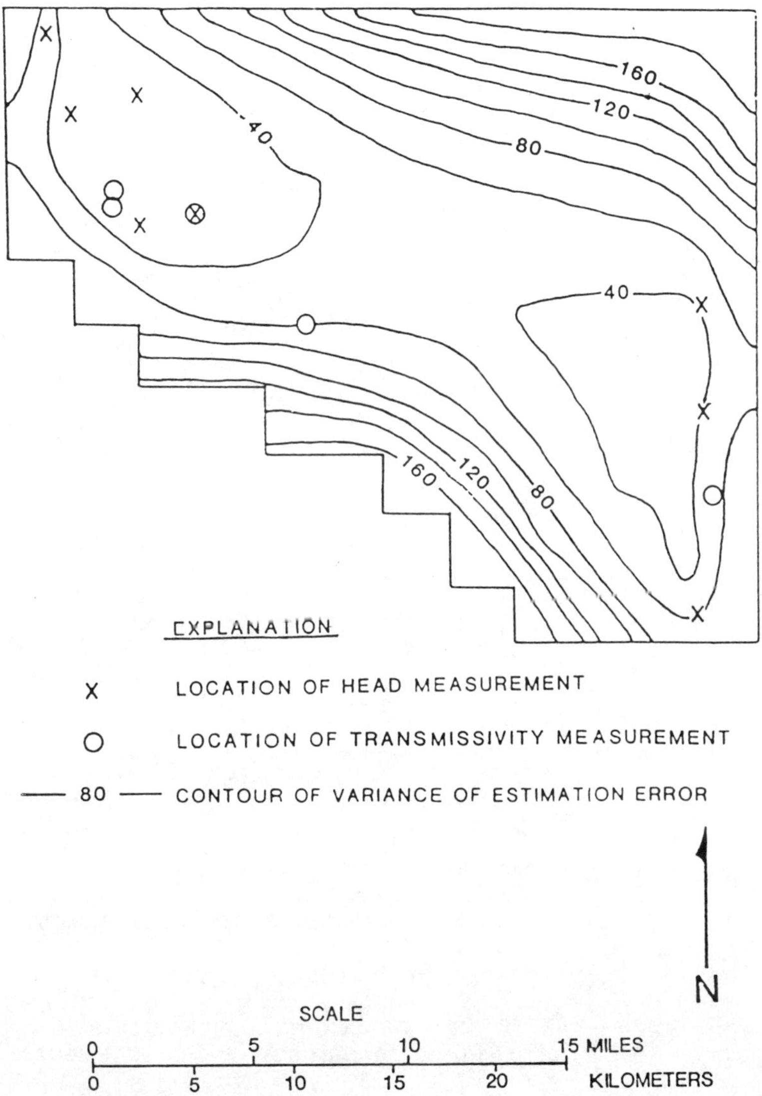

Figure 10. **Variance of estimation error for cokriged head field (correlation range of 30 km).**

The number of points used to generate the solid line contours in Figure 13 represents the more realistic situation of having few data points (8 head and 5 transmissivity measurements). As we can see, the surface generated from these few, randomly chosen points does not resemble the original surface from which these data were chosen. This exercise illustrates the extreme deviance from the actual situation that can be produced when using only a few data points and no additional information.

A similar situation may occur at the Hanford Site where a limited data base may not be able to provide reliable information on groundwater flow directions. The previous example demonstrates the need for a better delineation of the groundwater flow movement at the Hanford Site. The locations of additional monitoring wells should be selected using a process combining conceptual models and geostatistics.

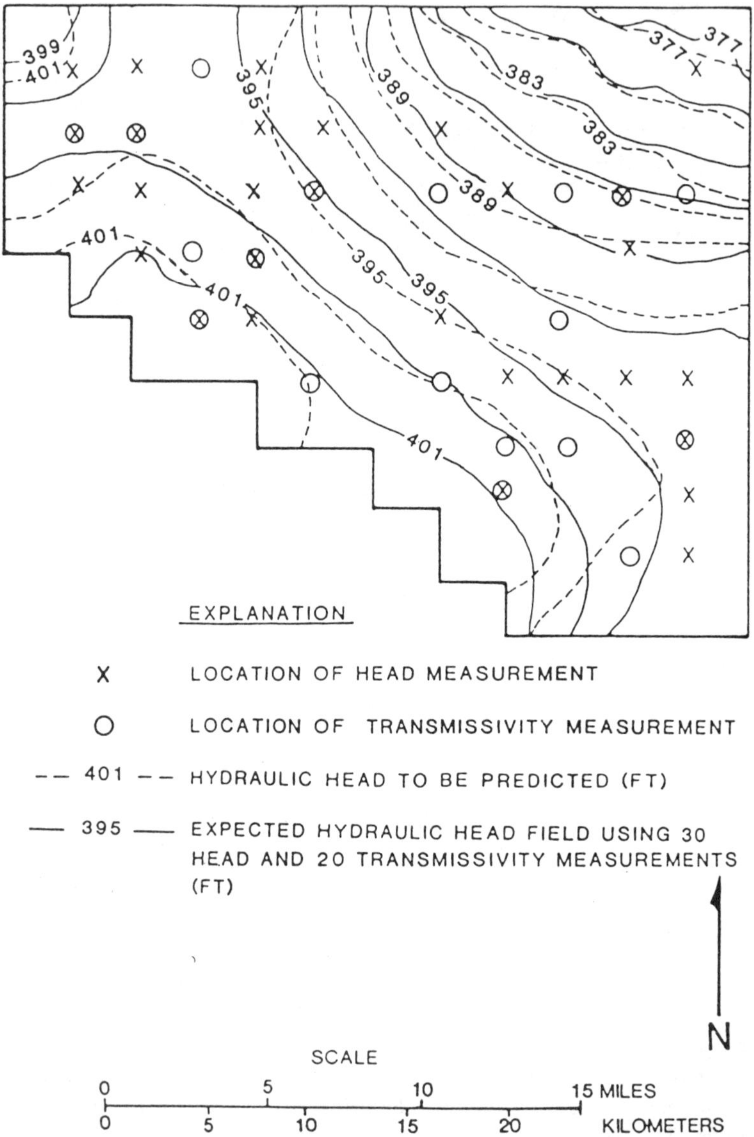

Figure 11. **Prediction of hydraulic head field using 30 head and 20 transmissivity measurements.**

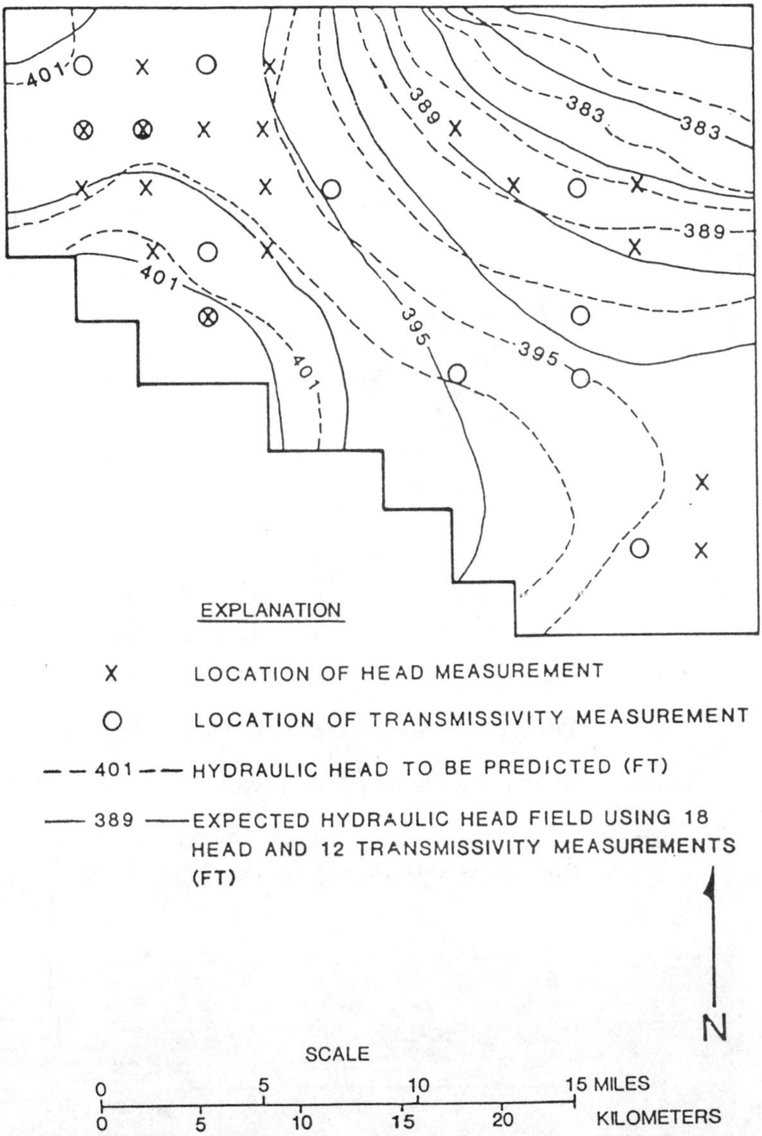

Figure 12. **Prediction of hydraulic head field using 18 head and 12 transmissivity measurements.**

Conclusions

The problem of estimating hydrogeologic transmissivity and hydraulic head fields using field measurements has been examined in a geostatistical framework. A geostatistical approach has been applied to the hydrologic data collected in the Priest Rapids Member of the Wanapum Basalt Formation at the Hanford Site. In this

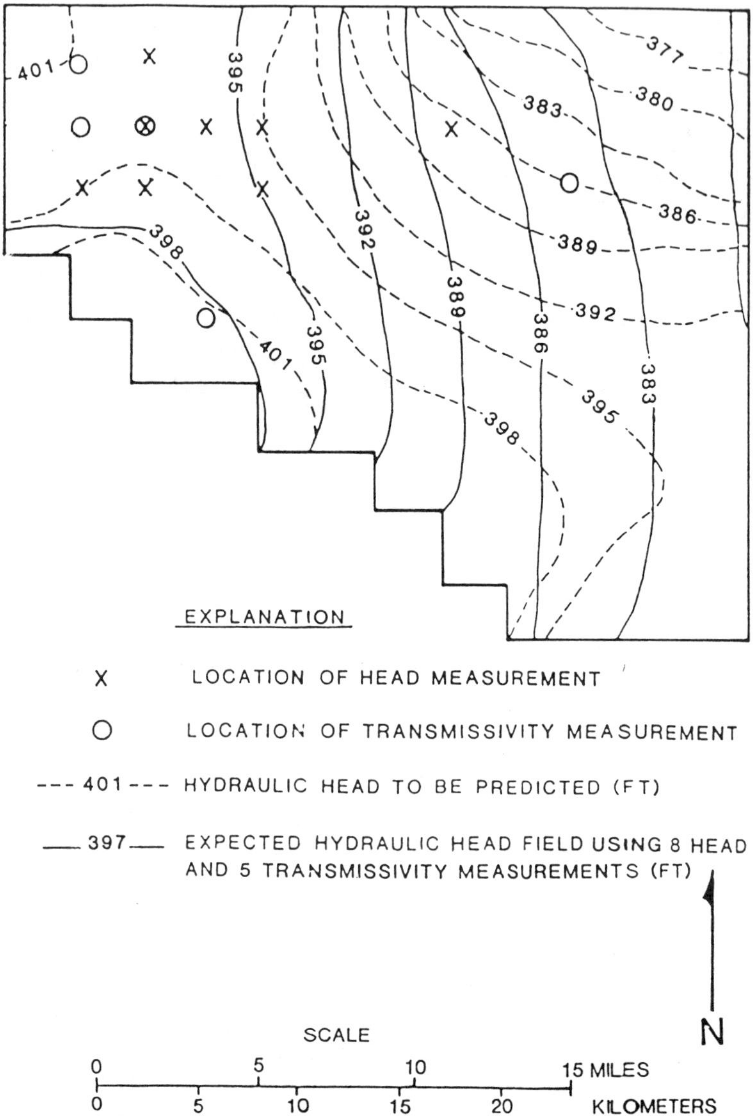

Figure 13. **Prediction of hydraulic head field using 8 head and 5 transmissivity measurements.**

approach, transmissivities and hydraulic heads have been regarded as realizations of random fields.

A maximum likelihood parameter estimation procedure has been applied to identify the stochastic structure of the transmissivity field using head and transmissivity measurements. Because of the limited data base, a proper identification of the transmissivity field was not possible.

A single value for the correlation length I of the transmissivity field could not be identified with any reasonable degree of accuracy. This result might not be of such importance if the purpose were to interpolate hydrologic fields. As shown in this paper, these estimates are rather insensitive to the value of I. But it is crucial to have a reasonable estimate of I if the geostatistical parameters are used to generate conditional realizations of transmissivity for the purpose of deriving the GWTT probability distribution. It has been shown in YIN (*14*) that the GWTT probability distribution is very sensitive to the correlation length.

When the data are scarce, the parameter estimates are shown through simulations to be very sensitive to the number of available head and transmissivity measurements. The estimated hydrologic fields can be dramatically different from the original fields in the case of a small number of measurements. This situation might be encountered at the Hanford Site.

The hydrologic investigations at the Hanford Site have been carried out for only a thin band situated along the Cold Creek syncline. The site characterization strategy has depended heavily on conceptual modeling based on preconceived ideas concerning the hydrologic fields. As a result, a considerable area of the domain in which released radionuclides may travel is not sufficiently characterized at the present time. This study has shown that it is essential that additional field tests be performed to better delineate the groundwater flow movement in the areas where groundwater particle pathlines (involved in travel-time calculations) might occur. The locations of future monitoring wells should be selected by relying on both conceptual models and geostatistics. Furthermore, methods that explicitly account for the effect on predictions of uncertain parameters [e.g., Kitanidis (*15*)] need to be adopted.

Literature Cited

1. Kitanidis, P. K., and E. G. Vomvoris, "A Geostatistical Approach to the Inverse Problem in the Groundwater Modeling (Steady State) and One Dimensional Simulations," *Water Resour. Res.*, *19*(3), 677 (1983).
2. Hoeksema, R. J., and P. K. Kitanidis, "An Application of the Geostatistical Approach to the Inverse Problem in Two-Dimensional Groundwater Modeling," *Water Resour. Res.*, *20*(7), 1003 (1984).
3. Hoeksema, R. J., and P. K. Kitanidis, "Comparison of Gaussian Conditional Mean and Kriging Estimation in the Geostatistical Solution of the Inverse Problem," *Water Resour. Res.*, *21*(6), 825 (1985).
4. Dagan, G., "Stochastic Modeling of Groundwater Flow by Unconditional and Conditional Probabilities: The Inverse Problem," *Water Resour. Res.*, *21*(1), 65 (1985).
5. DOE, *Environmental Assessment, Reference Repository Location, Hanford Site, Washington*, DOE/RW-0070, 3 Vols., U.S. Department of Energy, Washington, D.C. (1986).
6. DOE, "Water Level Hydrographs of BWIP Monitoring Boreholes," presented at the NRC/DOE Data Review Meeting, Richland, Washington (1986).
7. YIN, *Evaluation of Hydraulic Conductivity Data of Selected Stratigraphic Units at the BWIP Hanford Site, Washington*, Yakima Indian Nation Nuclear Waste Program, Toppenish, Washington (1986).
8. Freeze, R. A., "A Stochastic-Conceptual Analysis of One-Dimensional Groundwater Flow in Nonuniform Homogeneous Media," *Water Resour. Res.*, *11*(5), 725 (1975).

9. Hoeksema, R. J., and P. K. Kitanidis, "Analysis of the Spatial Structure of Properties of Selected Aquifers," *Water Resour. Res., 21*(4), 563 (1985).
10. YIN, *Evaluation of Hydraulic Head Data of Selected Hydrogeologic Units at the Hanford Site, Washington*, Yakima Indian Nation Nuclear Waste Program, Toppenish, Washington (1987).
11. Kitanidis, P. K., and R. W. Lane, "Maximum Likelihood Parameter Estimation of Hydrologic Spatial Processes by the Gauss-Newton Method," *Journal of Hydrology, 79*, 53 (1985).
12. Clifton, P. M., *Groundwater Travel Time Uncertainty Analysis—Sensitivity of Results to Model Geometry and Correlations and Cross-Correlations Among Input Parameters,* SD-BWI-TI-256, Basalt Waste Isolation Project, Rockwell Hanford Operations, Richland, Washington (1984).
13. Dagan, G., "A Note on Higher-Order Corrections of the Head Covariances in Steady Aquifer Flow," *Water Resour. Res., 21*(4), 573 (1985).
14. YIN, *Evaluation of DOE Analysis of Groundwater Travel Time, Hanford Site*, Yakima Indian Nation Nuclear Waste Program, Toppenish, Washington (1986).
15. Kitanidis, P. K., "Parameter Uncertainty in Estimation of Spatial Functions: Bayesian Analysis," *Water Resour. Res., 22*(4), 499 (1986).

6 The Use of Subjective Information to Supplement Hard Data

Imaging of Spatial Uncertainty: A Non-Gaussian Approach

A. G. JOURNEL
Applied Earth Sciences Department
Stanford University

Abstract

Gaussian-related models for stochastic simulation (imaging) of spatial uncertainty do not offer the flexibility to reproduce patterns of spatial connectivity specific to extreme values (e.g., fractures). The nonparametric indicator approach allows reproduction of several spatial connectivity functions, and the indicator data coding allows conditioning by both hard (e.g., cores) and soft (e.g., interpretative) data. The resulting alternative, equiprobable images of the spatial distribution of the attribute considered present low entropy features that may be critical to repository characterization. The paper is methodological in nature.

Introduction

The characterization of a nuclear repository, an aquifer, or an oil reservoir differs in many respects from the more static problem of reserves evaluation for a mineral deposit, and would thus call for an approach different from those proposed in traditional mining geostatistics.

i. First and utmost, the most critical parameters for the design of a repository, or the management of an aquifer, are flow-related parameters such as porosity, transmissivity, fracture density, and geometry. When mapping such parameters, the criterion of local accuracy is not as important as the reproduction of patterns of connectivity in space, particularly those of extreme values.

ii. Second, the hard data (e.g., cores) available within a repository are typically scarce, and thus any characterization exercise must rely upon "soft" information of interpretative nature or taken outside the site. Consequently such characterization is uncertain; an appropriate measure of that uncertainty must be provided.

Most traditional mapping algorithms, including kriging, aim towards local accuracy at the cost of failing to reproduce patterns of spatial continuity and do not provide any measure of joint spatial uncertainty. For example, regression-type algorithms consider as a criterion the minimization of a model-dependent local error variance that entails a smoothing of the spatial distribution of the estimates: the regression-type estimated values are less variant in space than their counterpart true values. The resulting error variance, also called "kriging variance" is generally just a ranking index of data configuration, which is a poor measure of local accuracy, the case corresponding to a Gaussian model being a major exception [see Journel (*1*)]. In any case, a series of error variances, say $\sigma_E^2(u_1)$, $\sigma_E^2(u_2)$, $\sigma_E^2(u_3)$ at different locations u_1, u_2, u_3 in space does not inform about joint spatial uncertainty, e.g., about the probability that jointly the attribute value at u_1, u_2, u_3 exceeds a given threshold value. Indication about such joint probability is often needed, e.g., to evaluate whether a pattern of faults or connected fractures may be present or not.

Geostatistical algorithms designed for the mapping of metal grades in a mineral deposit need not be appropriate for the characterization of spatial distributions of flow-related parameters in a repository. In fact, the very concept of a unique map is itself not appropriate. Rather, several alternative yet equiprobable maps will deliver an image of spatial uncertainty, allowing that uncertainty to be accounted for in the engineering design. Also, all such alternative maps must reproduce an a priori model of spatial continuity (connectivity) of the attributes of interest, still honoring data deemed relevant at their locations.

Simulation vs. Estimation

The process of site characterization may be viewed as that of providing the correct input to a transfer function modeling some aspect of the site design (see Figure 1). For example, the transfer function may be a flow simulator requiring for input an unsmoothed spatial distribution of permeability values, the output being leakage time and rates.

Short of a unique, perfect, input map of permeabilities that would require exhaustive sampling, and thus destruction, of the site, the idea is to propose equiprobable alternative maps as input. All such maps should carry in common whatever information, hard or soft, numerical and interpretative, is deemed relevant for the site design. Their differences should provide a measure of spatial uncertainty about the input. That measure of input uncertainty can then be processed through the flow simulator (transfer function Ψ in Figure 1) to yield the corresponding measure of uncertainty on the response function, and answers to such critical questions as "What is the probability of an early breakthrough?" The L alternative images of Figure 1 are called "simulations" or "simulated"

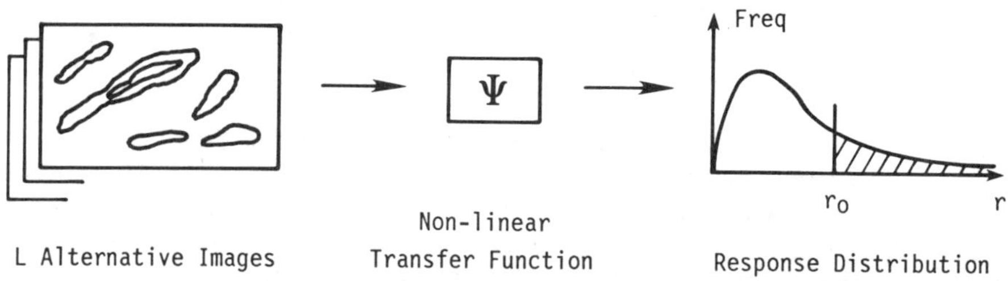

Figure 1. **Processing uncertain spatial information. The L alternative input images provide a measure of spatial uncertainty which is then processed into a distribution of response values.**

images as opposed to the single "estimated" image provided by interpolation algorithms [Journel (*2*)].

Denote by $\{k(u), u \epsilon S\}$ the set of attribute values, say horizontal permeability, distributed over the site or area S. In an estimation approach, the previous unknown set is replaced by a unique set of estimated values $\{k^*(u), u \epsilon S\}$. In a simulation approach, L alternative sets $\{k^{(\ell)}(u), u \epsilon S\}$, $\ell = 1, \ldots, L$ are proposed.

In the first case, a unique response with no measure of its uncertainty is obtained by application of the transfer function Ψ:

$$R^* = \Psi\,(k^*(u), u \epsilon S) \tag{1}$$

In the second case, a distribution of responses is obtained:

$$R_\ell = \Psi\,(k^{(\ell)}(u), u \epsilon S),\ \ell = 1, \ldots, L \tag{2}$$

This distribution of responses provides a measure of uncertainty and an assessment of such probabilities as that of exceedence of a threshold value r_0, estimated by the proportion of observed responses $R_\ell > r_0$. Knowledge of the estimation variance $\sigma_E^2(u)$ attached to each estimated value $k^*(u)$, $u \epsilon S$, as provided, e.g., by a kriging algorithm, provides little, if any, insight into the consequent error on the response function. Indeed the transfer function Ψ cannot usually be approximated by a sum of independent operators $\Psi^{(j)}$ acting on each attribute value $k(u_j)$, one at a time, as in expression (3).

If the following approximation were to hold,

$$R = \Psi\,(k(u), u \epsilon S) \cong \sum_{i=1}^{N} \Psi^{(j)}\,(k(u_j)),\ u_j \epsilon S, \tag{3}$$

with the N operators $\Psi^{(j)}$ known and independent from each other, then the variance of each stochastic term $\Psi^{(j)}\,(k(u_j))$ could be evaluated by randomizing the unknown $k(u_j)$ according to a certain distribution with the estimation variance $\sigma_E^2\,(u_j)$ used as a measure of distributional spread. Then,

$$\mathrm{Var}(R) = \sum_{i=1}^{N} \mathrm{Var}\ \{\Psi^{(j)}\,(k(u_j))\}$$

Still, knowledge of the variance of R does not yield a distribution for the response R unless some convenient central limit theorems can be called upon to justify a Gaussian-related distribution for R.

Another and better approach consists of relaxing the independence restriction in relation (3) and evaluating Var(R) from knowledge of the bivariate distribution of any pair of unknown $k(u_j)$, $k(u_j + h)$ separated by a distance vector h. This is the key assumption underlying most of the stochastic approaches to flow uncertainty proposed by hydrogeologists [Gelhar et al. (*3*), Gelhar and Axness (*4*)]. Such an approach is still not fully satisfactory [Desbarats (*5*)], for two reasons:

i. Typical flow-related transfer functions are highly multivariate in nature and cannot be approximated as the sum of operators acting on one or even two unknowns at a time.

ii. A covariance model $C(h)$ between two attributes $k(u_j)$, $k(u_j+h)$ falls short of characterizing the important (for Ψ) spatial features of the distribution $\{k(u), u\epsilon S\}$, as described in the later discussion on the limitation of the Gaussian model.

The approach corresponding to relation (2) and proposed in this paper requires running the transfer function Ψ on each alternative image $\{k^{(\ell)}(u), u\epsilon S\}$ and is tedious. The iteration can be somewhat reduced by selecting a few "extreme" configurations $\{k^{(\ell_0)}(u), u\epsilon S\}$, $\ell_0 = 1, \ldots, L_0$, with $L_0 << L$, and performing a worst case-type study. Such worst cases would provide indications about the tail of the response distribution. With presently available computers, repetitive iterations of the type (2) are now less of a concern, provided of course that both the transfer function Ψ and the random function underlying the alternatives $\{k^{(\ell)}(u), u\epsilon S\}$ are adequately modeled.

Selecting a mapping criterion

Whether a single set or L sets of input values $\{k_\ell(u), u\epsilon S\}$, $\ell = 1, \ldots, L$, are required, the critical question is which criterion should be considered when deriving the values $k_\ell(u)$ replacing the unknown true values $k(u)$. The answer depends on the particular transfer function Ψ under consideration. The values $k_\ell(u)$ should reproduce those features and statistics of the true values $k(u)$ that most severely impact the response function $R = \Psi(k(u), u\epsilon S)$.

For most flow-related problems the function Ψ is highly multivariate and nonlinear, thus an approximation of type (3) would not be appropriate. Flow in a fractured medium is governed by the connectivity within S of the high permeability (transmissivity) values, $k(u) > k_0$, not so much on whether at any particular location u the permeability is high or low, nor even on the proportion within S of these high values. In such a case, the $k_\ell(u)$'s should reproduce the patterns of connectivity of the extreme values even if this means partially trading off such properties as unbiasedness or reproduction of the overall histogram of the $k(u)$'s over S.

A set of input values, whether $\{k^*(u), u\epsilon S\}$ or $\{k^{(\ell)}(u), u\epsilon S\}$ is good inasmuch as it is drawn from a model that identifies the relevant spatial characteristics of the unknown true values $\{k(u), u\epsilon S\}$. The next section presents the random function model and insists on some limitations of the most commonly adopted Gaussian model.

The Random Function Model

A random function of the coordinate(s) u, $K(u)$, denoted with the capital letter K, can be seen as a set of spatially dependent random variables $\{K(u), u\epsilon S\}$, one at each location u within the site S [Journel (*1*)]. Stochastic simulation amounts to drawing joint outcomes from all random variables constituting the random function model. The essence of a random function model, i.e., of any stochastic simulation, is the imposed model for spatial dependence between the various component random variables $K(u)$, $u\epsilon S$.

Spatial dependence between any two random variables (RV's) $K(u)$, $K(u+h)$, separated by the vector h, is fully characterized by a bivariate (two points) distribution function:

$$P\{K(u) \le k, K(u+h) \le k'\} = F(h;k,k'). \tag{4}$$

Expression (4) does not depend on the two coordinate vectors u and $u' = u+h$, but only on their separation distance h. This corresponds to a stationary random function (RF) model within S, and in practice to a decision of averaging statistics over S.

The bivariate cumulative distribution function (cdf) can be seen as an elementary noncentered indicator variance:

$$F(h;k,k') = E\{I(u;k) \cdot I(u+h;k')\} = C_I(h;k,k') \tag{5}$$

with $$I(u;k) = \begin{cases} 0, \text{ if } k < K(u) \\ 1, \text{ if } k \ge K(u) \end{cases}$$

Thus, a full characterization of the spatial dependence between any two RV's $K(u)$, $K(u+h)$ is given by the set of all indicator covariances $C_I(h;k,k')$, for all k,k′, which can be inferred from the experimental covariances run from indicator data found over the averaging area S. For example, the particular two-points probability $F(h;k_0,k_0)$ could be inferred from the proportion of data approximately separated by h and jointly valued below or equal to k_0.

An indicator (noncentered) covariance is a two-points measure of spatial connectivity that quantifies the probability that two values $K(u)$, $K(u+h)$ exceed jointly a given threshold:

$$P\{K(u) > k, K(u+h) > k\} = C_J(h;k,k) \tag{6}$$

with $J(u;k) = 1 - I(u;k)$.

Short of enough indicator data available within S, the connectivity characteristic $C_J(h;k,k)$ may be synthesized from geological interpretations about the continuity (connectivity) of, e.g., high permeability values in any particular direction. The measure of connectivity $C_J(h;k,k)$ need not be the same for all threshold values k. For example, very high permeability values (fractures) may be connected in a given direction and not in others, resulting in a strongly anisotropic covariance $C_J(h;k,k)$, whereas low to median permeability values may be more isotropic in their spatial variability. A function $C(h)$ of a vector $h = (h_1,h_2)$ is said to be "anisotropic" if it actually depends on the direction of the vector h; if it depends only on the modulus of the vector h, $C(h) = C(\sqrt{h_1^2 + h_2^2})$, it is said to be "isotropic."

The characterization of a random function model $K(u)$ by a series of indicator covariance functions $C_I(h;k,k)$ is, in general, much richer than its characterization by the sole traditional covariance: $C_K(h) = E\{K(u), K(u+h)\} - [E\{K(u)\}]^2$. It has been shown [Alabert (*6*)] that the traditional covariance $C_K(h)$ is an average of the various centered indicator covariances $C_I'(h;k,k')$, and thus an average of possibly different patterns of connectivity corresponding to different threshold values k, k′:

$$C_K(h) = \text{Cov}\{K(u), K(u+h)\}$$

$$= \int_{-\infty}^{+\infty} \int_{-\infty}^{+\infty} C_I'(h;z,z')dzdz', \tag{7}$$

with $$C_I'(h;z,z') = \text{Cov}\{I(u;z), I(u+h;z')\}.$$

Just like the variance of a univariate cdf does not fully characterize that cdf, the covariance $C_K(h)$ does not fully characterize the bivariate cdf $F(h;k,k')$ as defined by relation (5). In particular, a simulation $\{k^{(\ell)}(u), u \epsilon S\}$ that reproduces only the covariance $C_K(h)$ does not necessarily identify the various connectivity functions $C_J(h;k,k)$. There are applications where identification of one or more critical connectivity functions $C_J(h;k,k)$ is more relevant than reproduction of the covariance $C_K(h)$. For example, flow in a fractured medium is influenced mainly by the pattern of connectivity of very high permeability values, i.e., by the covariance $C_J(h;k,k)$ for very high threshold values k.

The Gaussian model

The Gaussian random function model [Anderson (*7*)] is remarkable in the sense that it is fully determined by means and a single covariance function $C_K(h)$ of type (7). Conversely, the Gaussian model has no flexibility for the reproduction of a specific indicator covariance model, i.e., spatial connectivity function.

Consider a Gaussian RF model $Y(u)$, with zero mean and correlogram $\boldsymbol{\rho}_Y(h) = C_Y(h)/C_Y(0)$. Any indicator covariance of type (5) or (6) is fully determined from $\boldsymbol{\rho}_Y(h)$:

$$P\{Y(u) > y_p, Y(u+h) > y_p\} =$$

$$(1-p)^2 + \frac{1}{2\Pi} \int_0^{\arcsin \rho_Y(h)} \exp\left(\frac{-y_p^2}{1+\sin\Theta}\right) d\Theta \tag{8}$$

with y_p being the p-quantile of the standard normal cdf. For example at the median cutoff, $y_{.5} = 0$, and:

$$P\{Y(u) > 0, Y(u+h) > 0\} = \frac{1}{4} + \frac{1}{2\Pi} \arcsin \boldsymbol{\rho}_Y(h) \tag{9}$$

[see Abramovitz and Stegun (*8*)].

In practice when the histogram (univariate distribution) of the initial variable $K(u)$ is not normal, the Gaussian model would apply to its normal score transform: $Y(u) = T(K(u))$, e.g., $Y(u) = \text{Ln}\, K(u)$ if the histogram of $K(u)$ is lognormal.

For any monotone transform function T, such as the normal score transform T,

$$P\{K(u) > k_p, K(u+h) > k_p\} = P\{Y(u) > y_p, Y(u+h) > y_p\}, \tag{10}$$

for all values of $p \epsilon [0, 1]$. The k_p and y_p are the p-quantiles of the univariate cdf's of $K(u)$ and $Y(u)$, respectively.

Thus, the connectivity function $P\{K(u) > k_p, K(u+h) > k_p\}$ is determined by relation (8) as soon as Gaussian model for either $K(u)$ or its normal score transform $Y(u)$ is adopted. The normal score transform, including the popular log-transform, does not introduce any additional degree of freedom to allow for modeling the specific indicator covariances, i.e., spatial connectivity functions of type (6).

Maximum entropy

Not only do Gaussian-related models lack the flexibility to identify indicator covariances, but they also impart to the outcomes $\{k^{(\ell)}(u), u\epsilon S\}$ a severe maximum entropy character. It can be shown [Jones (*9*)] that, among all possible random function models with an imposed covariance $C_Y(h)$, the Gaussian model maximizes entropy, i.e., spatial disorder.

Maximum entropy means minimum continuity (connectivity) between high values or between low values, i.e., the minimization of the correlation ranges of indicator covariances of type (6) at high (or low) threshold values k. Indeed, in expression (8) for extreme threshold values y_p (p close to either 0 or 1), y_p^2 is large leaving the integrand close to zero, and

$$P\{Y(u) > y_p, Y(u+h) > y_p\} \cong P\{Y(u) > y_p\} \cdot P\{Y(u+h) > y_p\} = (1 - p)^2. \quad (11)$$

A Gaussian model does not properly allow for connectivity of extreme values in space: the events $\{I(u;y_p) = 0 \leftrightarrow Y(u) > y_p\}$ and $\{I(u+h;y_p) = 0 \leftrightarrow Y(u+h) > y_p\}$ are considered independent for extreme p's. Thus, if high permeability values might connect each other in space, generating potential flow paths, a Gaussian-related model (e.g., multivariate lognormal) is inadequate, if not dangerously biased. Maximum entropy, i.e., minimum ordering of extreme values, does *not* represent a conservative (i.e., less favorable) model for flow simulations and the testing of a repository. It behooves the modeler to develop non-Gaussian alternatives allowing reproduction of unfavorable patterns of spatial connectivity of high/low values, if only for worst-case sensitivity analyses.

Features of a Simulation

Stochastic simulation provides alternative, equiprobable images $\{k^{(\ell)}(u), u\epsilon S\}$, $\ell = 1, \ldots, L$ reproducing some spatial variability model assumed shared by the unknown true values $\{k(u), u\epsilon S\}$. The vast majority of simulation algorithms are Gaussian related. They yield outcomes $\{k^{(\ell)}(u), u\epsilon S\}$, $\ell = 1, \ldots, L$ that feature:

i. Reproduction of a common cdf, usually some histogram assumed representative of the distribution of $k(u)$ within S.
ii. Honoring the data values at data locations, i.e., $k^{(\ell)}(u_i) = k(u_i)$ at any datum location u_i, for any outcome ℓ.
iii. Reproduction of the attribute covariance $C_K(h)$ as defined in relation (7).
iv. A multivariate Gaussian distribution for the normal score transforms $\{y^{(\ell)}(u) = T(k^{(\ell)}(u)), u\epsilon S\}$ [see Journel (*1*), Mantoglou and Wilson (*10*), Luster (*11*)].

Because of the previously discussed limitation of the Gaussian model, the last two features (items iii and iv) may be traded for:

iii-a. Reproduction of any number Ω of indicator covariances $C_I(h;k_\omega,k_\omega)$, $\omega = 1, \ldots, \Omega$, of type (6). (13)

iv-a. Honoring not only the hard data values $k(u_i)$ but also any number of soft data, such as constraint intervals $k(u_\alpha) \in]a_\alpha,b_\alpha]$ and/or prior probability distributions for the unknown value $k(u_\alpha)$, as shown in Figures 2b and 2c.

Sequential Indicator Simulation

Consider Ω threshold values k_ω, $\omega = 1, \ldots, \Omega$ discretizing the range of variability of the attribute $K(u)$. Ω can be made as large as needed; however, in practice a small number from 3 to 9 would often be sufficient, defining from 4 to 10 classes for $k(u)$. Indeed, in many applications the exact value of $k(u)$ with 2 or 3 digits is not as important as the fact that it is low, median, or high *and* whether those lows, medians, or highs are spatially connected or not. Making Ω large would require a large and complex inference, which the available data may not support.

Indicator coding

Any particularly hard datum (more precisely assumed so) $k(u_1)$ can be coded into an indicator data column $\{i(u_1, k_\omega), \omega = 1, \ldots, \Omega\}$ with the indicator data $i(u_1; k_\omega)$ defined as 1 if the datum $k(u_1)$ is less than or equal to the threshold value k_ω, and 0 otherwise [see definition (5) and Figure 2a].

At the location u_2 of a soft datum known only to be within an interval, $k(u_2) \in]a_2,b_2]$, the indicator data column is incomplete: the values $i(u_2, k_\omega)$, for $k_\omega \in]a_2,b_2[$, are unknown, see Figure 2b.

At some other locations u_3 the soft datum $k(u_3)$ is known to lie within an interval $]a_3,b_3]$ with some prior probability distribution. At such a location,

$$i(u_3; k_\omega) = \begin{cases} 0, \text{ for all } k_\omega \leq a_3 \\ P\{K(u_3) \leq k_\omega\}, \text{ for all } k_\omega \in]a_3,b_3[\\ 1, \text{ for all } k_\omega \geq b_3, \end{cases} \tag{14}$$

[see Figure 2c and Journel (*12*)].

At any unsampled location u, there exists a minimum piece of information corresponding to the physical minimum and maximum value of the attribute k (see Figure 2d). For example a porosity value is necessarily within [0, 100%], and a permeability value cannot be negative.

Single threshold simulation

Consider any one of the Ω threshold values k_ω. The corresponding indicator random function $I(u; k_\omega)$ defined in (5) has

i. mean: $E\{I(u;k_\omega)\} = P\{K(u) \leq k_\omega\} = p_\omega$ (15)

ii. centered covariance:

$$C_I'(h;k_\omega,k_\omega) = E\{[I(u; k_\omega) - p_\omega][I(u+h; k_\omega) - p_\omega]\}.$$

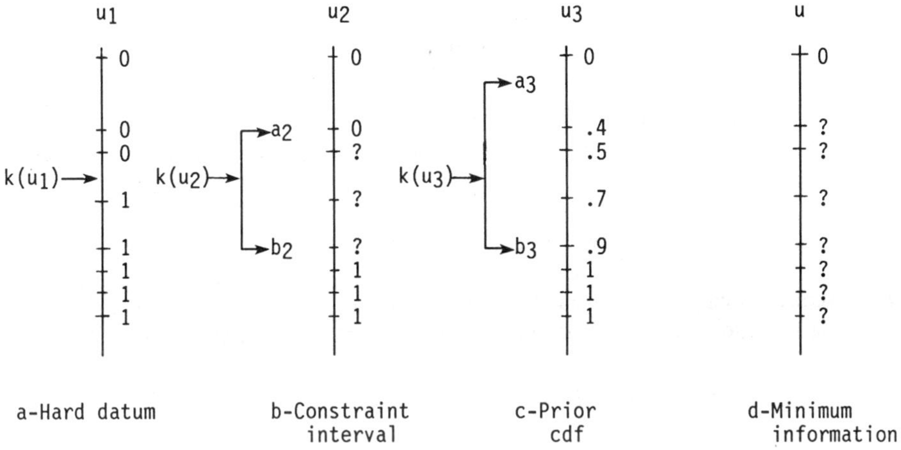

Figure 2. **Indicator coding of prior information.**

In the following the shorter notations $I(u)$, p, $C_I'(h)$ will be used, dropping the reference to the particular threshold value k_ω.

A set of N indicator values $i(u_j')$, $j = 1, \ldots, N$, $u_j' \epsilon S$, is to be generated over a grid covering the site S. This set should be an outcome of a binary RF $I(u)$ featuring:

i. a mean: $E\{I(u)\} = p$, actually, $p = P\{K(u) \le k_\omega\}$

ii. an indicator covariance $C_I'(h)$, actually $C_I'(h;k_\omega,k_\omega)$ (16)

iii. honoring the indicator data of all types, see Figures 2a,b,c,d, at data locations.

Note that the cross-indicator covariances $C_I'(h;k_\omega,k_{\omega'})$ with $k_\omega \neq k_{\omega'}$ will not be reproduced.

First step. Define a random path u_j', $j = 1, \ldots, N$ through the N nodes to be simulated. Start at the first nodal location u_1' which need not be at a corner of the site S.

Perform a linear regression of the unknown indicator value $i(u_1')$, an algorithm called "indicator kriging" (in short IK), using the n surrounding data with usually $n < N$:

$$i^*(u_1') - p = \sum_{\alpha=1}^{n} \lambda_\alpha \cdot [i(u_\alpha) - p] \tag{17}$$

The weights λ_α are given by the traditional normal equations system, or IK system:

$$\sum_{\beta=1}^{n} \lambda_\beta \cdot C_I'(u_\beta - u_\alpha) = C_I(u_1' - u_\alpha) \tag{18}$$

$$\text{for all } \alpha = 1, \ldots, n.$$

The resulting IK estimate $i^*(u_1')$ is shown to be an estimate of the conditional probability $P\{I(u_1') = 1 \mid \text{Information (n)}\}$ and is used as such [*not* as an estimate of $I(u_1')$]; see

Journel (*12*), Solow (*13*). An indicator value such as $i(u_1')$ is valued either zero or unity, whereas both the IK estimate and the conditional probability are valued between zero and unity.

Second step. Draw a simulated indicator value $i^{(\ell)}(u_1')$ from the binary distribution $B(i^*(u_1'))$ with mean equal to the estimated probability value $i^*(u_1')$. By definition of such a drawing:

$$E\{I^{(\ell)}(u_1') \mid i^*(u_1')\} = i^*(u_1').$$

By "deconditioning," i.e., after randomization of the n indicator data $i(u_\alpha)$ found in expression (17) for $i^*(u_1')$, the first requisite of (16) is seen to be verified:

$$E\{I^{(\ell)}(u_1')\} = E\{I^*(u_1')\} = p + \sum_{\alpha=1}^{n} \lambda_\alpha \cdot E\{I(u_\alpha) - p\} = p. \tag{19}$$

Similarly, consider the conditional covariance of $I^{(\ell)}(u)$ with any one of the n indicator data RV's:

$$\begin{aligned} &E\{I^{(\ell)}(u_1') \cdot I(u_\alpha) \mid \text{Information (n)}\} = \\ &E\{I^{(\ell)}(u_1') \cdot i(u_\alpha) \mid \text{Information (n)}\} = \\ &i(u_\alpha) \cdot E\{I^{(\ell)}(u_1') \mid \text{Information (n)}\} = i(u_\alpha) \cdot i^*(u_1'). \end{aligned}$$

After deconditioning:

$$\begin{aligned} &\text{Cov}\{I^{(\ell)}(u_1'), I(u_\alpha)\} = \text{Cov}\{I(u_\alpha), I^*(u_1')\} \\ &= \sum_{\beta=1}^{n} \lambda_\beta \text{Cov}\{I(u_\alpha), I(u_\beta)\}, \text{ according to relation (17)} \\ &= \text{Cov}\{I(u_1), I(u_\alpha)\}, \text{ according to the system (18).} \end{aligned}$$

Finally:

$$\text{Cov}\{I^{(\ell)}(u_1'), I(u_\alpha)\} = C_I'(u_1' - u_\alpha), \text{ for all } \alpha = 1, \ldots, n. \tag{20}$$

The n normal equations (18) are read as an identification of the n covariances Cov $\{I^{(\ell)}(u_1'), I(u_\alpha)\}$ to the corresponding model values $C_I'(u_1' - u_\alpha)$. Therefore, and with regard to the n data locations, the second requisite (16) is verified.

If a nodal point u_1' identifies an indicator datum location u_α, the uniqueness of the solution to the normal system (18) entails the indicator datum identification:

$$i^*(u_1') \equiv i(u_\alpha) = \begin{cases} 0 \text{ or } 1, \text{ in the case of Figure 2a,b} \\ \in [0,1], \text{ in the case of Figure 2c} \end{cases} \tag{21}$$

at $u_1' = u_\alpha$.

This is the well-known "exactitude" property of kriging [Journel (*12*)]. Since the simulated values $i^{(\ell)}(u_1')$ are drawn from the binary distribution $B(i^*(u_1'))$, with $i^*(u_1') = i(u_1')$, they will honor the prior indicator information at the datum location $u_1' = u_\alpha$. Thus, the last requisite of (16) is verified.

Third step. Proceed to the next node u_2', more generally u_{j+1}', along the random path u_j', j=1, . . . , N. Perform the IK of the unknown indicator value $i(u_2')$ using the corresponding n surrounding data, including the previously simulated value $i^{(\ell)}(u_1')$ if its location u_1' falls into the data search neighborhood centered on u_2':

$$i^*(u_2') - p = \sum_{\alpha=1}^{n} \lambda_\alpha [i(u_\alpha) - p] \tag{22}$$

If the previously simulated value $i^{(\ell)}(u_1')$ is included in the n data the IK system corresponding to the probability estimate (22) is written:

$$\lambda_1 \operatorname{Cov}\{I^{(\ell)}(u_1'), I^{(\ell)}(u_1')\} + \sum_{\beta=2}^{n} \lambda_\beta C_I'(u_\beta - u_1') = C_I'(u_2' - u_1') \tag{23}$$

$$\lambda_1 \operatorname{Cov}\{I^{(\ell)}(u_1'), I(u_\alpha)\} + \sum_{\beta=2}^{n} \lambda_\beta C_I'(u_\beta - u_\alpha) = C_I'(u_2' - u_\alpha)$$

$$\text{for all } \alpha = 2, \ldots, n, \text{ and } u_\alpha \neq u_1'.$$

Draw the simulated indicator value $i^{(\ell)}(u_2')$ from the binary distribution $B(i^*(u_2'))$. It can be shown, as was done for the node u_1', that the resulting simulated indicator RV $I^{(\ell)}(u_2')$ verifies the three requisites (16).

The first equation of the IK system (23) can be read as an identification of the covariance of $I^{(\ell)}(u_2')$ with the previously simulated indicator RV $I^{(\ell)}(u_1')$, indeed:

$$\begin{aligned} &\operatorname{Cov}\{I^{(\ell)}(u_2'), I^{(\ell)}(u_1')\} = \\ &\lambda_1 \operatorname{Cov}\{I^{(\ell)}(u_1'), I^{(\ell)}(u_1')\} + \sum_{\beta=2}^{n} \lambda_\beta \operatorname{Cov}\{I(u_\beta), I^{(\ell)}(u_1')\} \\ &= C_I'(u_2' - u_1') \text{ according to (23).} \end{aligned}$$

Loop over all nodes u_j', j = 1, . . . ,N. At each node u_{j+1}' an indicator kriging is performed using *all* neighboring original data and previously simulated indicator values.

The final output of a single threshold indicator simulation is one alternative indicator map $\{i^{(\ell)}(u_j'), j = 1, \ldots, N\}$ or correspondingly $\{j^{(\ell)}(u_j') = 1 - i^{(\ell)}(u_j'), j = 1, \ldots, N\}$. Such a map gives an indication of any joint exceedence in space of the threshold considered, and an answer to such questions as "is there a high permeability flow path crossing the whole site?" The consideration of several such indicator maps $\{j^{(\ell)}(u_j'), j = 1, \ldots, N\}$ and $\ell = 1, \ldots, L$ would provide the probability of occurrence of such a flow path, and other important statistics such as length distributions of flow paths and their geometry. By varying the threshold value considered, one could evaluate the critical threshold limit for which a fair proportion (say greater than 5%) of simulated indicator maps show connected paths—string of values $j^{(\ell)}(u_j') = 1$—crossing the entire site.

Possible uses of such simulated indicative maps are multiple and not yet fully appreciated.

Remarks. At each nodal point u_j' the indicator kriging weights λ_α do not depend on the actual indicator values $i(u_\alpha)$ retained. Thus, for a given random path u_j', $j = 1, \ldots, N$ and a given configuration of prior information, one can store the N sets of weights $\{\lambda_\alpha(u_j'), \alpha = 1, \ldots, n\}$ one for each node u_j', and generate at low CPU cost any number L of simulations $\{i^{(\ell)}(u_j), j = 1, \ldots, N\}, \ell = 1, \ldots, L$.

Also following the common practice of kriging, the simple kriging estimate (17) and system (18) are best replaced by an ordinary kriging estimate and system:

$$i^*(u_1') = \sum_{\alpha=1}^{n} \lambda_\alpha \, i(u_\alpha) \,, \tag{24}$$

with the constraint: $\sum_{\alpha=1}^{n} \lambda_\alpha = 1$

The constraint $\sum_\alpha \lambda_\alpha = 1$ allows the derivation of a probability $i^*(u_1')$ solely conditioned by the local data $i(u_\alpha)$, and not by the global parameter p. The price paid is the reproduction of the indicator variogram rather than the indicator covariance [Journel and Huijbregts (*14*)].

Multiple thresholds simulation

Consider now the case of Ω threshold values k_ω, $\omega = 1, \ldots, \Omega$, where a full indicator column $\{i^{(\ell)}(u_j'; k_\omega), \omega = 1, \ldots, \Omega\}$ is to be simulated at each nodal location u_j', $j = 1, \ldots, N$. From now on the full notations $I(u; k_\omega)$ and $C_I(h; k_\omega)$ for the indicator covariance (7) will be used.

Procedure. Start with any threshold value, say k_1, which need not be the smallest (see Remarks hereafter). Then start with the first nodal point u_1' along the random path u_j', $j = 1, \ldots, N$.

Perform a simple (alternatively ordinary) kriging to determine the estimate $i^*(u_1'; k_1)$ of the conditional probability $P\{I(u_1'; k_1) = 1 \mid \text{Information (n)}\}$, then draw $i^{(\ell)}(u_1'; k_1)$ from the binary distribution $B(i^*(u_1'; k_1))$.

Since an indicator column is ordered, the following setting of indicator values is done:

i. If $i^{(\ell)}(u_1'; k_1) = 0$, set $i^{(\ell)}(u_1'; k_\omega) = 0$, for all $k_\omega < k_1$ (25)

ii. If $i^{(\ell)}(u_1'; k_1) = 1$, set $i^{(\ell)}(u_1'; k_\omega) = 1$, for all $k_\omega > k_1$.

All such set indicator values, together with the indicator value $i^{(\ell)}(u_1'; k_1)$ just drawn, will be considered as indicator data for *all* subsequent indicator krigings.

Loop over all nodal points u_j', $j = 1, \ldots, N$ for the threshold k_1 then loop over all remaining $(\Omega-1)$ threshold values k_ω.

Simulation of the continuous attribute. One realization of the multiple thresholds indicator simulation $\{i^{(\ell)}(u_j'; k_\omega), j = 1, \ldots, N; \omega = 1, \ldots, \Omega\}$ is considered completed when all N indicator columns are filled, each one with Ω simulated indicators valued either 0 or 1.

Simulation of a continuous attribute value, say $k^{(\ell)}(u_j')$, is done by drawing from the single transition class lower-bounded by a 0 indicator value and upper-bounded by a 1, as for example Figure 2a. This drawing is done by Monte Carlo from the experimental distribution of hard data $k(u_\alpha)$ falling into that class, or from some model fitting that distribution.

A continuous attribute value, $k^{(\ell)}(u_j')$, is drawn at all nodal points generating one simulated realization $\{k^{(\ell)}(u_j'), j = 1, \ldots, N\}$. Another realization would call for a complete rerun of the process starting at (u_1', k_1).

Remarks. First, the path taken through the N nodal points is arbitrary. Each new path would generate a different outcome of the N values $\{k^{(\ell)}(u_j'), j = 1, \ldots, N\}$. It is preferable that two sequential nodes u_j' and u_{j+1}' not be systematically contiguous (i.e., very close) in space; otherwise the simulated value $i^{(\ell)}(u_j'; k_\omega)$ may overcondition the outcome $i^{(\ell)}(u_{j+1}'; k_\omega)$ beyond reproduction of the indicator covariance.

Second, the resetting of simulated indicator values, as described in relation (25), allows speeding the simulation process by reducing the number of indicator kriging systems needed. As a counterpart the indicator covariance reproduction (20) becomes approximate. For example, if the first threshold value k_1 is the smallest, all indicator values $i^{(\ell)}(u_j'; k_\omega)$ reset to 1 (there are, however, few of them) meet the covariance $C_I'(h; k_1)$ rather than $C_I'(h; k_\omega)$.

To minimize the covariance mismatch, the number of resettings should be minimized. This is achieved by starting with extreme threshold values, alternatively high and low, progressing towards the median value. For example, if 5 threshold values, $k_1 < k_2 < k_3 < k_4 < k_5$, are considered, the sequence to be used for indicator simulation is k_5, k_1, k_4, k_2, k_3 or k_1, k_5, k_2, k_4, k_3.

If one particular indicator covariance $C_I'(h; k_\omega)$ is critical, i.e., if the pattern of spatial connectivity at one particular threshold value is critical, it should be reproduced as well as possible, and thus considered first in the sequence.

Third, when kriging at node u_j' for the threshold value k_ω, only the nearest indicator data $i(u_\alpha; k_\omega)$ for that same threshold value are retained. The influence of data farther away is screened by the nearest data. Thus, the dimension of the indicator kriging systems need not increase considerably as the sequence progresses, even though the data set increases with all previously simulated indicator values.

Returning to the four features (13) requested from a simulation, the sequential indicator simulation (SIS) algorithm allows:

i. Reproduction of a histogram model over the site S, or more precisely approximate (due to the resetting process) reproduction of the series of Ω cdf values,

$$p_\omega = P\{K(u) \leq k_\omega\}, \omega = 1, \ldots, \Omega.$$

ii. Exact honoring of all prior indicator information whether originating from hard or soft data. The resolution of the indicator coding does depend on the number of threshold values k_ω retained.

iii. Exact reproduction of the indicator covariance model for the first threshold value considered, and approximate reproduction of the $(\Omega-1)$ other indicator covariance models.

Concerning the last point (iii), our argument is that it is preferable to reproduce approximately a series of Ω indicator covariances deemed critical to the response function under study (see Figure 1), rather than reproduce *none* (or an average) through a traditional Gaussian-related simulation algorithm, if these indicator covariance models happen not to fit the corresponding Gaussian models (8) and (9).

There is no place in this methodological paper to present in sufficient detail the complete case study of the sequential indicator simulation (SIS) performed by Alabert (*6*) in his M.Sc. thesis. The reader is urged to refer to this excellent work. Alabert used as his reference data base a large (75,000) elevation-related data set. For the simulation exercises he used only 30 hard data and 250 soft data under the format of constraint intervals of the type of Figure 2b. The results include several simulations (alternative equiprobable reconstitutions) of the reference data base using either only the 30 hard data or both 30 hard data and 250 soft constraint intervals. The improvement brought by the latter appears dramatic.

The case is made for the practicality of the SIS approach and the necessity to include soft information in the presence of sparse hard data.

Conclusions

Evaluation of the multiple risks involved in the engineering of a repository site requires an appreciation of the impact of spatial uncertainty, that uncertainty associated with the distribution in space of such attributes as porosity, transmissivity, and density of fractures. A measure of spatial uncertainty is not given by a series of local estimation variances or a map of error variances; it requires an evaluation of joint uncertainty, e.g., the probability that a string of estimated values is jointly in error by having failed to reveal a connected path of high permeabilities.

The notion of stochastic simulation, providing alternative equiprobable images of the site and its heterogeneities, is proposed in replacement to the traditional single estimated map. These alternative images can be processed by the transfer function(s) modeling the site design, and they yield a distribution of responses instead of a single one. Such a response distribution measures the impact of spatial uncertainty on the particular design aspect being studied.

Stochastic simulations are no better than the model of spatial continuity they image. Most simulation algorithms rely on Gaussian-related models characterized by a single covariance model mixing together the spatial correlation of high-valued attributes with that of low-valued attributes. It is argued that, notwithstanding the decision of stationarity within a site or part thereof, the pattern of spatial connectivity of high values can be separated from that of low or median values. In many flow-related issues, modeling the spatial connectivity of extremes (e.g., fractures or flow barriers) is more important than modeling the overall histogram or covariance. Gaussian-related models offer little flexibility for modeling spatial connectivity of extremes, and worse, they entail a little understood property of maximum entropy whereby extreme values are minimally connected. The indicator covariance(s) concept is introduced and shown to be a measure of connectivity, a probability of joint exceedence of any given threshold value. It is suggested that reproduction of a series of indicator covariance models is a more relevant requisite for stochastic simulations.

Beyond the notion of indicator covariances the indicator formalism allows a common coding of hard data (e.g., core data) and soft information (e.g., constraint interval or interpretative data). Both hard and soft information is then used for conditioning the alternative spatial images. The possibility of using soft information, accounting for its uncertainty, should prove essential in site characterization where paucity of hard data is common.

The sequential indicator simulation (SIS) algorithm is proposed to generate alternative images of the spatial distribution of a given attribute, say permeability. All such images reproduce prior models for the attribute histogram and indicator covariances, and honor at data locations prior information whether hard or soft. These simulations need not carry the maximum entropy character of traditional Gaussian-related algorithms and should provide more flexible tools for the important task of imaging spatial uncertainty in site characterization.

Literature Cited

1. Journel, A. G., *Math Geology, 18* (1), 119–140 (1986).
2. Journel, A. G., *Economic Geology, 69*, 673–687 (1984).
3. Gelhar, L. W., Gutjahr, A. L., and Naff, R. L., *Water Resour. Res., 15* (6), 1387–1397 (1979).
4. Gelhar, L. W., and Axness, C. L., *Water Resour. Res., 19* (1), 161–180 (1983).
5. Desbarats, A. J., *Water Resour. Res., 23* (2), 273–286 (1987).
6. Alabert, F. G., "Stochastic imaging of spatial distribution using hard and soft information," M.Sc. Thesis, Stanford University, Stanford, CA, 197 pp. (1987).
7. Anderson, T. W., *An introduction to multivariate statistical analysis,* Wiley & Sons, pp. 5–100 (1958).
8. Abramovitz, M., and Stegun, I., *Handbook of mathematical functions,* Dover, pp. 936–940 (1964).
9. Jones, D. S., *Elementary information theory,* Clarendon Press, Oxford p. 137 (1979).
10. Mantoglou, A., and Wilson, T., "Simulation of random fields with the turning band method," Rept. No. 264, Dept. of Civil Engr., Massachusetts Institute of Technology, Boston, MA (1981).
11. Luster, G. R., "Raw materials for Portland cement: Applications of conditional simulations of coregionalization," Ph.D. Thesis, Stanford University, Stanford, CA, 531 pp. (1985).
12. Journel, A. G., *Math Geology, 18* (3), 269–286 (1986).
13. Solow, A. R., *Math Geology, 18* (3), 335–352 (1986).
14. Journel, A. G., and Huijbregts, Ch., *Mining Geostatistics,* Academic Press, pp. 497–498 (1978).

An Alternative Approach to Uncertainty in Risk Analysis and an Exactly Soluble Model for a Nuclear Waste Repository

J. O. CORBETT
Berkeley Nuclear Laboratories
Berkeley
Gloucestershire
United Kingdom

D. P. HODGKINSON and P. C. ROBINSON
Harwell Laboratory
Oxfordshire
United Kingdom

Abstract

Probabilistic risk analysis needs careful interpretation if the spread of predicted consequences is dominated by uncertainty due to lack of data. This paper treats the situation where the risk appears to be dominated by the possibility of a number of uncertain quantities all lying in a narrowly defined region that gives rise to high consequences. The problem is demonstrated with a simple model of radionuclide dispersal from an underground repository. It is shown that the convergence of Monte Carlo calculations may be very erratic, and the results may strongly reflect any arbitrariness in assigning the original probability distributions. An iterative approach is suggested, in which a standard optimization program is used to locate and measure the high consequence peaks. The aim is to ensure that the final analysis correctly represents expert judgement in relation to the high consequence regions.

Introduction

An early stage in probabilistic risk analysis (PRA) is to assign distributions to all of the parameters that are treated as random variables. In some cases this task is quite straightforward, for example when treating the meteorological variables in a release to the atmosphere; however, probability distributions are often used to represent uncertainty that is due to lack of data rather than genuine chance. This is particularly so in analyzing the release of radioactivity from a waste repository and its transfer by groundwater transport to the biosphere.

Cook and Unwin (*1*) have approached this problem by using information theory. In the simplest case, they propose a neutral distribution that is uniform between an upper and a lower limit, outside of which it is zero; but if the range spans many orders of magnitude of a variable that is essentially positive, the distribution is taken

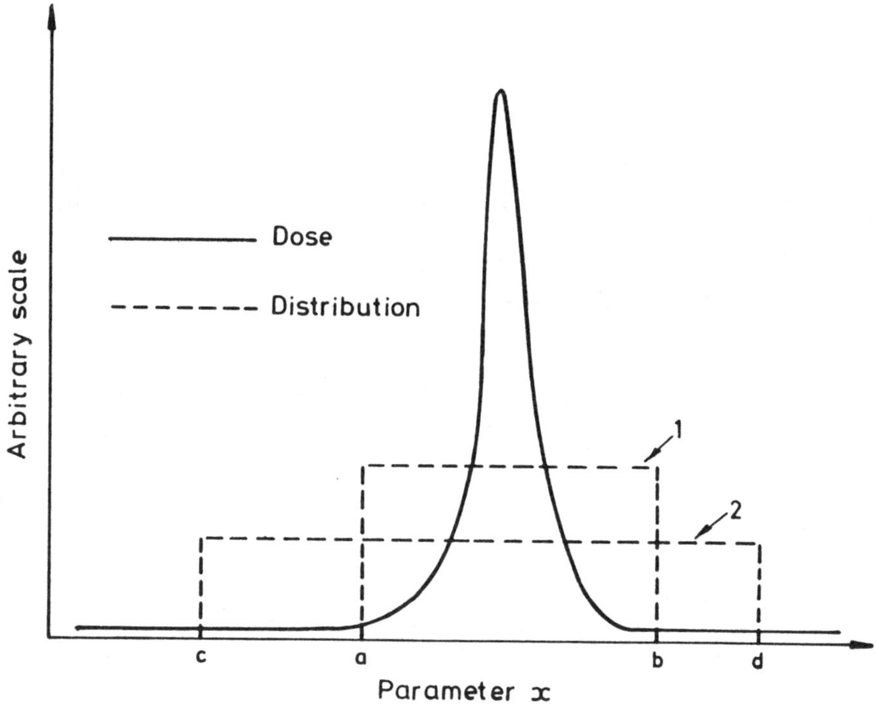

Figure 1. **Isolated peak.**

to be log-uniform. We are concerned here with some problems that can arise when estimates of adverse consequences, such as the radiation dose to a critical group, are highly sensitive to one or more of the parameters.

Although the problem is multidimensional in any practical case, it is most easily illustrated in just one dimension. In Figure 1, the dose is a sharply peaked function of the parameter x. The expectation value of the dose, or mean dose estimate, is therefore proportional to $1/(b-a)$ according to Distribution 1, or $1/(d-c)$ according to Distribution 2. If the risk is considered to be proportional to the mean dose estimate, then by admitting to greater uncertainty in Distribution 2 we have actually reduced our estimate of the risk. This anomalous situation, which has come to be known as 'risk dilution' (*2*), appears to stem from the requirement that an expert judgement of the range of values a parameter can take must be expressed as a probability distribution normalized to unity.

A different, but related situation is shown in Figure 2, where the consequence increases sharply towards the upper end of the parameter's range. Clearly, the mean dose estimate is highly sensitive to the precise choice of an upper limit, which in reality is likely to be somewhat arbitrary. Furthermore, if there is no obvious reason why, say, the uniform Distribution 3 should be preferred to the log-uniform Distribution 4, this arbitrariness is also reflected directly in the PRA results. The

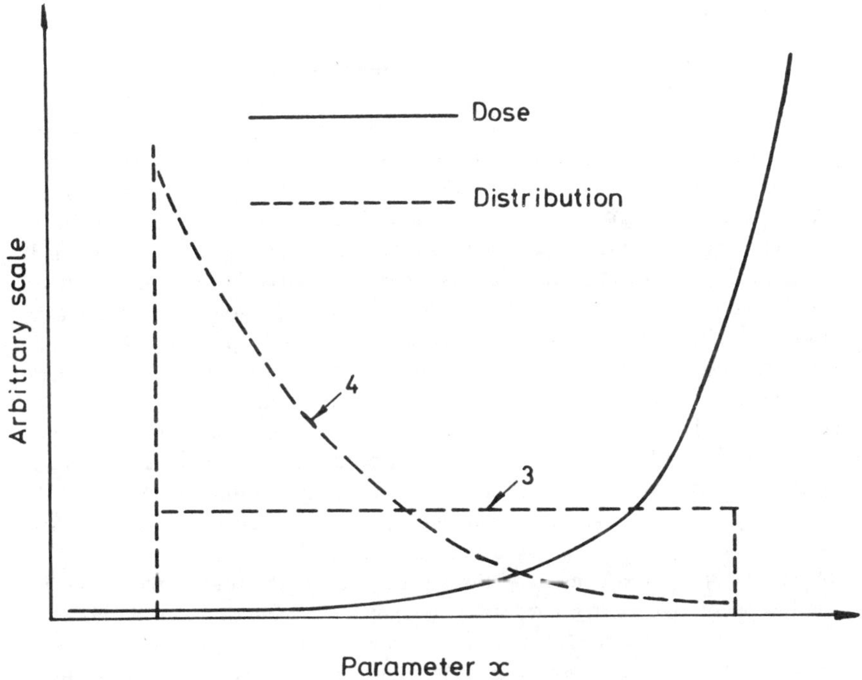

Figure 2. **Sharply increasing function.**

situation with a rapidly decreasing function at the lower end of the parameter's range is very similar.

The Role of Pessimization

Where there is strong parameter dependence, dose estimates are likely to exceed acceptable limits only in a very restricted region of parameter space, while throughout most of the remaining space the dose is very low. In these circumstances, PRA is essentially concerned only with the peak region. The important quantities then are the peak value of the dose estimate, its sensitivity to each of the parameters at the peak, and the likelihood that all the parameters lie in this region. We shall return to the likelihood question later.

Although the traditional approach to PRA is by Monte Carlo methods, random sampling is a very inefficient way of finding isolated peaks. On the other hand, efficient computer codes are commercially available for solving optimization problems, that is, minimizing a given function of several variables subject to various constraints on the variables. Such codes are equally suitable for finding maxima, which in the present context we call pessimization. We now consider both the Monte Carlo and the pessimization approach in terms of a simple model of groundwater transport.

A Simple Model

Despite the availability of comprehensive numerical models of groundwater transport, there is a continuing need for simpler models to provide insight and understanding, perform sensitivity and uncertainty analyses, and verify the accuracy of general-purpose numerical codes. The most appropriate approximations are to consider a one-dimensional flow field and assume constant physical parameters. Hodgkinson and Maul (*3*) showed that a wide range of phenomena, including one-dimensional advection, linear equilibrium sorption, hydrogeological dispersion and radioactive decay chains in permeable rock, and also matrix diffusion in fractured rock, could be accommodated in a model that is analytically soluble when Laplace-transformed in the time dimension. A numerical method due to Talbot (*4*) was then used to invert the Laplace transform with remarkable accuracy and efficiency. Our present model is simpler, being restricted to one radionuclide in a single stratum of permeable rock. The solution of the model has sufficient detail to provide a serious test of the Monte Carlo approach, but its Laplace transform is so simple that it can be integrated analytically over up to five uniform or log-uniform parameter distributions at a time. Thus the mean dose estimate can be calculated with just one Talbot inversion in place of several thousand random samples.

Physically, the model consists of three parts, namely the source, the geosphere, and the biosphere. The source term has two components. First there is an initial containment time, T, before which no radionuclide escapes from the repository. During this time only radioactive decay, with decay constant λ, occurs. After time T the repository begins to release radionuclides at a rate assumed to be proportional to the current inventory. This may represent diffusive or convective transport away from the repository. The release rate is determined by a single parameter, k, with units of y^{-1}. The initial inventory, M_0, is simply the total activity of a given nuclide present at time zero, in units of Becquerels.

The geosphere term includes advection, longitudinal dispersion, equilibrium sorption, and radioactive decay. Advection is characterized by the flow velocity, v. Dispersion is treated as a Fickian process, governed by a dispersion length, d. Equilibrium sorption is represented by a retardation factor, R. The length of the path through the geosphere is ℓ.

Finally, the biosphere path to man is taken as a stream which is used as the major source of drinking water for the critical group of individuals. Radionuclides leaving the geosphere are assumed to enter a stream which has flow rate W. An individual is assumed to drink an amount, w, of stream water each year, and the activity he consumes is converted to a dose by a factor q. The overall effect of this model is a simple multiplicative factor, B = wq/W.

The solution of this model for fixed parameters gives an annual dose, C(t), for a member of the critical group, which is zero for times less than T, after which

$$
\begin{aligned}
C(T + t) = {} & 0.5\ BkM_0 \exp\left(-\lambda(T + t) + \ell/2d - R\ell^2/4dvt - vt/4dR\right) \\
& [\Phi((R\ell^2/4dvt)^{1/2} - (vt/4dR - kt)^{1/2}) \\
& + \Phi((R\ell^2/4dvt)^{1/2} - (vt/4dR - kt)^{1/2})], \qquad (1)
\end{aligned}
$$

where

$$\Phi(z) = e^{z^2} \operatorname{erfc}(z),$$

z may be complex and Φ is related to the function W(z) defined by Abramowitz and Stegun (*5*) so that $\Phi(z) = W(iz)$. We shall refer to Equation (1) as the dose function.

For details of some of the different functions obtained by integrating over various parameter distributions in Laplace-transform space, and the use of the Talbot inversion algorithm, reference must be made to the original paper by Hodgkinson and Robinson (*6*). It is worth noting, however, that some of these expressions involve the exponential integral, $E_1(z)$, which must be evaluated in the complex z plane.

Results

The use of the simple model can be illustrated by an example in which up to five of the parameters are varied. Table 1 summarizes the data for a central case and five cases in which variability is introduced progressively into the parameters T, k, v, R, and ℓ, so that the final case represents an integration over five parameter distributions. This example is not intended to represent any realistic situation and in fact the choice of distributions is intentionally arbitrary. Figure 3 shows how this process affects the mean dose estimate.

Table 1. **Parameter values and distributions used for the illustrative example.**

Parameter	Central Case Value	Distribution Type	Range	Cases When Varied
M_0 (Bq)	1.0×10^{12}	Fixed	—	—
$k(y^{-1})$	3.0×10^{-5}	Log-uniform	10^{-6} to 10^{-3}	2 to 5
T(y)	100.0	Log-uniform	10 to 10^4	1 to 5
$\lambda(y^{-1})$	3.254×10^{-6}	Fixed	—	—
d(m)	20.0	Fixed	—	—
R(−)	10.0	Uniform	1 to 100	4 and 5
v(m/y)	0.2	Log-uniform	10^{-2} to 1	3 to 5
ℓ(m)	200.0	Uniform	100 to 1000	5 only
B(Sv/Bq)	1.0×10^{-16}	Fixed	—	—

Key:
- M_0 Initial quantity of nuclide in repository
- k Fractional release rate
- T Time delay before release
- λ Radioactive decay constant
- d Dispersion length
- R Retardation factor
- v Groundwater velocity
- ℓ Length of path through geosphere
- B Dose per unit of radionuclide entering biosphere

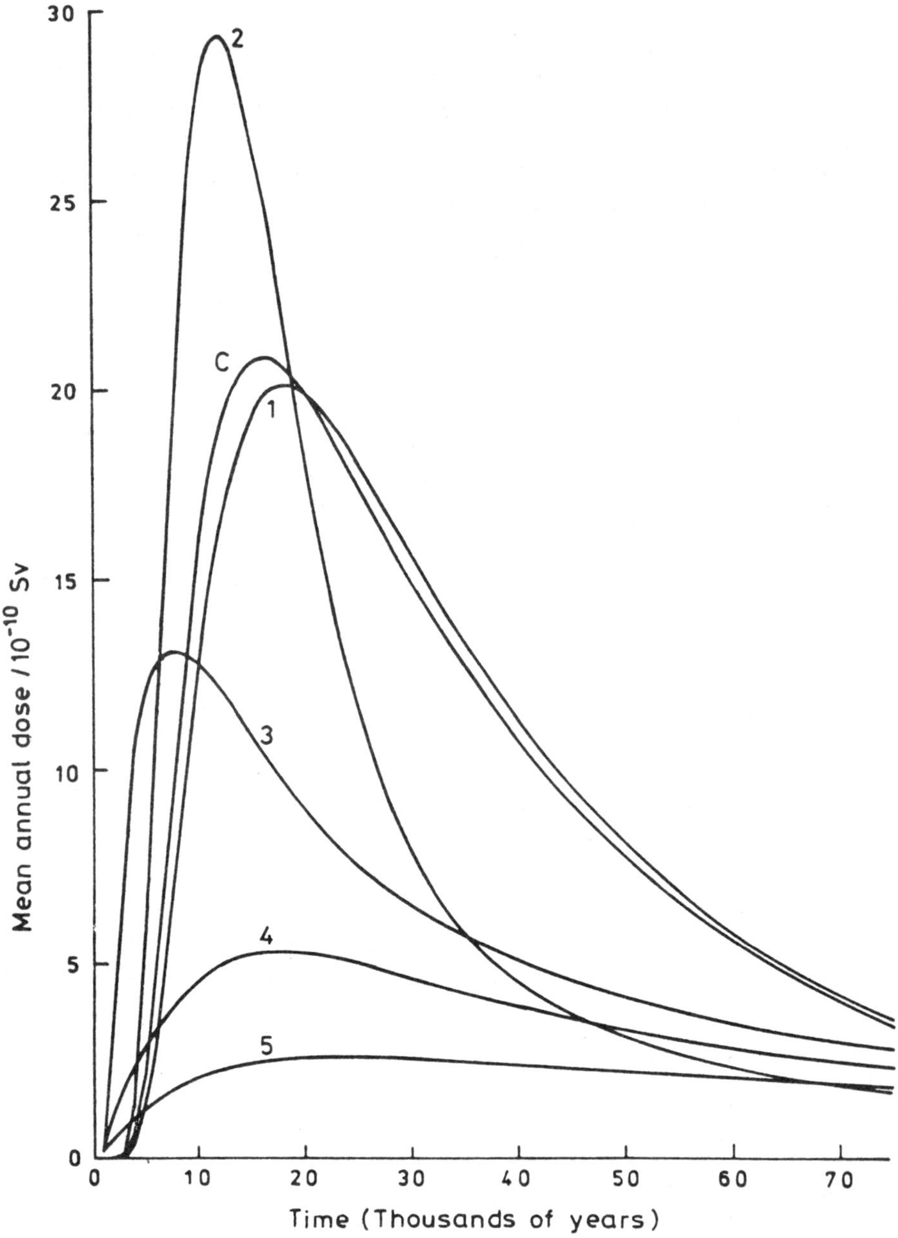

Figure 3. **Exact solution for the central case and the five variations.**

We now consider one case in greater detail, namely the annual dose at 10,000 years after closure with all five parameters varied. Figure 4 shows the results of a Monte Carlo calculation as the number of samples is increased. Convergence is erratic, with the exact solution being approached only after several thousand samples, which indicates a peak region occupying less than 0.1 percent of the available parameter space. Note also the false plateau after a few hundred samples, which in the absence of an exact solution might have been interpreted as the converged result.

In the pessimization approach, we define the sensitivity to a parameter x as

$$S = \text{Range }(x) \left| \frac{\partial}{\partial x} \ell n \, C \right|_{\text{Peak}}, \tag{3}$$

if x is at the limit of its range, or

$$S = \frac{\text{Range }(x)}{\sqrt{2\pi}} \left| \frac{\partial^2}{\partial x^2} \ell n \, C \right|^{1/2}_{\text{Peak}}, \tag{4}$$

if the peak occurs within the range of x. Both these expressions are dimensionless and, to a first approximation, the proportion of parameter space occupied by the peak can be estimated as

$$P = 1/\prod_i S_i, \quad S_i > 1, \tag{5}$$

where the index i runs through all the parameters for which $S > 1$. Equation (5) rests on the assumption that the distribution is uniform over all the parameters; for the present analysis, therefore, we consider ℓn k, ℓn T, R, ℓn v, and ℓ.

A readily available optimization program, based on a quasi-Newton algorithm, was applied to the dose function, Equation (1), subject to the constraints given in Table 1. The results are shown in Table 2. While the position of the peak merely corresponds to the fastest possible passage of radioactivity to the biosphere, some interesting conclusions can be drawn from the sensitivity estimates. The peak lies within the range of T, and the high sensitivity to ℓn T shows that it is very sharp. The situation is therefore similar to Figure 1; the peak's contribution to the mean dose estimate is proportional to the probability density in the region of T = 9780 years and thus inversely proportional to the range of uncertainty in ℓn T.

The sensitivity to R resembles that of Figure 2, although here the dose estimate falls steeply away from the lower end of the parameter's range. The probability density in R here is 1/99, since it is taken to be uniform over the range 1 to 100. If a log-uniform distribution is substituted over the same range, the probability density in R becomes $1/(\ell n \, 100)R$, which is about 20 times higher in the region of the peak. This is a further example of the arbitrariness discussed in relation to Figure 2.

The value of P in Table 2 implies that the peak region, with a maximum annual dose of about 8×10^{-8} Sv, occupies little more than 0.01 percent of the parameter space, whereas Figure 4 shows that this proportion is nearer 0.1 percent. This discrepancy arises because the peak is highly asymmetric in the ℓn T dimension; Equation (4) is based on the assumption that the peak is gaussian. To explain the

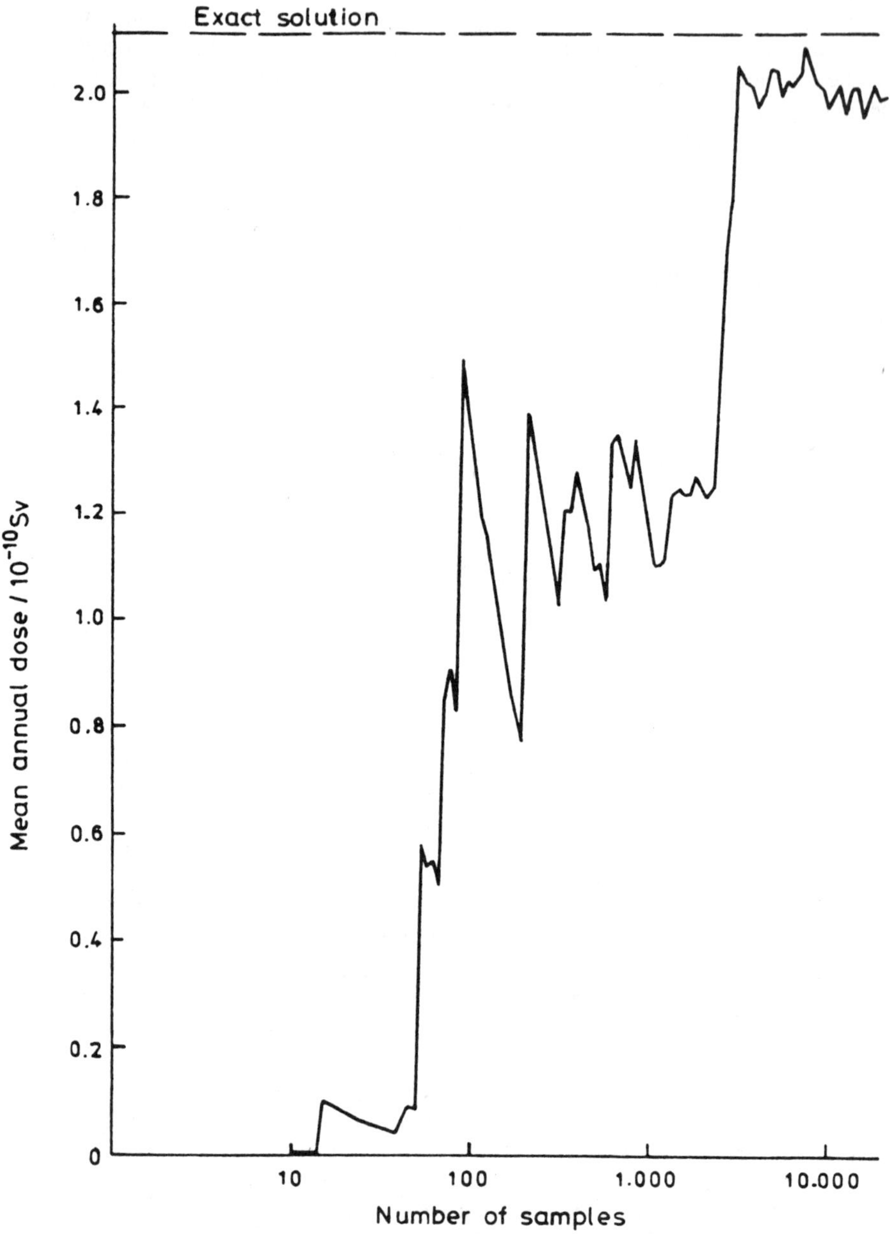

Figure 4. **Mean dose estimate against sample size at 10,000 years.**

Table 2. **Pessimization results for 10,000 years, case 5.**

Variable	Value at Peak	Sensitivity
k	10^{-3}	6.0
T	9780	110
R	1	13
v	1	0.6
ℓ	100	0.6

$C = 8.0 \times 10^{-8}$; $P = 1/8580$.

plateau in Figure 4 we now consider the sensitivity to R in more detail. Table 3 shows the pessimization results when R is fixed at 80, near the opposite end of its range from the peak. This extra constraint reduces the maximum estimated dose by only a factor of eight and the sensitivity to R at this point is low. Thus the risk function's initially rapid decrease with increasing R is not continued throughout much of the range. The peak is also less sharply defined with respect to ℓn T and the value of P rises to 1/120. Thus the mean dose estimate consists of almost equal contributions from a peak occupying about 0.1 percent of the parameter space and a more extended region at a lower level occupying about 1 percent of the space. The plateau is essentially an artifact of the probability distributions chosen for R and T.

Several other cases are analyzed in a paper by Corbett (*7*), which also provides further technical details of the optimization program and its application to this type of problem. The model used here is so simple that in many cases, including the one treated above, it is possible to locate the peak approximately by purely physical reasoning. In more realistic systems, however, physical reasoning may be less helpful. There may also be more than one peak within the parameter space. In these circumstances, it may be necessary to run the program several times with starting points chosen at random within the parameter space, in order to ensure that all the contributing peaks are found. We have not yet had the opportunity to apply the pessimization approach to more complicated models.

Implications

We have shown how problems can arise when the estimated risk from a waste repository depends strongly on a number of uncertain parameters. The convergence of a Monte Carlo calculation is liable to be very erratic and the results of a PRA are highly sensitive to the distributions assigned to some of the parameter values in just the regions where these distributions are likely to be most arbitrary. Although

Table 3. **Pessimization results for 10,000 years, case 5 with R fixed at 80.**

Variable	Value at Peak	Sensitivity
k	10^{-3}	0.9
T	4280	4.0
R	80	1.1
v	1	4.0
ℓ	100	6.8

$C = 1.10 \times 10^{-8}$; $P = 1/120$.

uniform distributions have been used above, these remarks apply equally well to normal distributions or any others that are chosen to represent uncertainty due to lack of data rather than genuine randomness. The solution to this problem is far from obvious, but we suggest that pessimization should be regarded as the first stage of an iterative process.

In this approach, the initial pessimization is based on a cautiously wide range of parameter values. One or more peaks are identified, and the sensitivity to each of the parameters at the peaks is obtained. The input data are then re-examined. The first possibility is that the combination of parameter values at a peak may actually be impossible; in other words, a correlation between some of the parameters has been overlooked in setting up the analysis. If this is so, the PRA should be restructured so as to take account of the correlation, possibly through a transformation of the variables.

Second, the question of arbitrariness must be faced. We realize that there are strongly differing opinions as to how this should be done. Our own view is that the experts originally consulted about the various parameter values should be approached again with much more specific questions relating to the probability of parameter values being inside or outside the important ranges identified in the pessimization stage. This does not imply that an expert is being asked to change his mind in view of the predicted consequences, for in many cases the original distributions are derived prescriptively by the analyst and not dictated directly by the expert. Furthermore, a highly specific question can normally be answered more completely and satisfactorily than one that is by necessity general and wide-ranging. Our purpose here is to ensure that the analysis correctly represents expert opinion in the most important regions.

A third possibility is that the pessimization study may lead to a decision to make further laboratory or field measurements on a limited number of parameters.

After this refinement of the input data, the whole process may be repeated. Ideally, we may thus be able to eliminate all the regions of parameter space that lead to dose estimates above an acceptable limit. Otherwise we can proceed with PRA using the final set of probability distributions derived from expert judgement and with the sensitivity data as a guide towards the most efficient sampling procedure in a Monte Carlo calculation. A major possible advantage of this approach lies in the final presentation and explanation of PRA results to any nonexpert body. We can now argue that, although many of the parameters that enter into our calculations are highly uncertain, the vast majority of possible values that they might take lead to estimated consequences that are totally innocuous. The probabilistic side of the analysis represents an aggregate of expert opinion concerning the likelihood, however small, that the actual parameter values lie in one or more specific regions that give rise to the higher estimated consequences.

Acknowledgment

This work is published by permission of the UK Central Electricity Generating Board, the UK Atomic Energy Authority, and UK Nirex Limited.

Literature Cited

1. Cook, I. and S. D. Unwin, *Controlling Principles for Prior Probability Assignments in Nuclear Risk Assessment*, Sandia Report No. SAND85-1325 (NUREG/CR-4514) (1986).
2. Corbett, J. O., *Risk Assessment Methods for Radioactive Waste Disposal: Problems Associated with the Annual Risk Approach and How to Avoid Them*, CEGB Report No. TPRD/B/0896/R87 (1987).
3. Hodgkinson, D. P. and P. R. Maul, *One-Dimensional Modelling of Radionuclide Migration Through Permeable and Fractured Rock for Arbitrary Length Decay Chains Using Numerical Inversion of Laplace Transforms*, UKAEA Report No. AERE R 11880 (1985).
4. Talbot, A., *J. Inst. Math. Appl.* (London) *23*, 97 (1979).
5. Abramowitz, M. and I. A. Stegun, *Handbook of Mathematical Functions*, Dover, New York (1965).
6. Robinson, P. C. and D. P. Hodgkinson, *Exact Solutions for Radionuclide Transport in the Presence of Uncertainty*, UKAEA Report No. AERE R 12125 (1986).
7. Corbett, J. O., "Uncertainty in Risk Analysis: An Alternative Approach Through Pessimization," *J. Radiol. Prot.* *8*(2), 107 (1987).

Use of Bayesian Analysis for Incorporating Subjective Information

PAUL W. ESLINGER[1]
BCS Richland, Inc.
Richland, Washington

BUDHI SAGAR[1]
Westinghouse Hanford Company
Richland, Washington

Abstract

Large uncertainties are inherent in characterizing geologic media and in defining the future conditions under which geologic repositories must perform. Therefore, incorporating subjective information is an important aspect of quantitative assessment of repository performance.

One method of incorporating subjective information in a quantitative analysis is demonstrated in this paper. Lacking objective data from laboratory and field tests, subjective information is coded to form prior distributions of the unknown parameters of the distribution of a variable. However, this technique presents some complications, as will be discussed further. As objective data are obtained, Bayes' theorem is used to derive estimates of the parameters. The use of conjugate prior distributions for performing this analysis is also discussed.

Two applications for the Basalt Waste Isolation Project (BWIP) are presented. The first pertains to subjective information obtained through probability encoding of the two parameters of the distribution of effective porosity at the macro scale. Only one measured value of effective porosity is currently available. Estimates for the two parameters are obtained by incorporating the single data point. The example is extended further to illustrate the effects that additional data may have on estimating the posterior distributions. The second application is for hydraulic transmissivity in a multivariate setting. No multivariate data are currently available.

Introduction

During the decision process leading to the licensing of a high-level nuclear waste repository, subjective information is expected to play a significant role. This is likely to be true of not only determination of "reasonable assurance," a regulatory descrip-

---

[1]Now with Battelle Pacific Northwest Laboratory, Richland, Washington.

tive term that occurs in 10 CFR Part 60 (*1*), but also of quantitative probability estimates as required by 40 CFR Part 191 (*2*). This paper explores the second aspect (i.e., a method to introduce subjective information into quantitative probability estimates).

Subjective information referred to here includes expert opinion, data obtained without direct testing, data from other similar sites, and information from related fields of study—in short, anything that helps form a judgment about the variable under study. There are many reasons why such information will have a substantial role in the geologic repository program. Primary among them are the difficulties in fully characterizing the geologic formations and the long time spans (10,000 years or longer) for which system performance must be estimated. Properties of geologic formations are known to vary in space, some more than others. Characterization of this variation in the deterministic sense would require an inordinately large number of borings, which would be destructive of the site itself. In practice, only a limited amount of such objective data can be obtained, which then must be augmented by related subjective information. Similarly, the conditions far into the future, under which a repository is to perform, can be estimated only through subjective information.

In the regulatory environment of a geologic repository, ad hoc subjectivity probably will not be acceptable. Structured auditable methods that can produce repeatable results have been developed for quantifying subjective information. The two better known such methods are the Delphi (*3*), and the Probability Encoding (*4*) methods. Information obtained through the use of these methods is used in a numerical example included in this paper. However, detailed discussion of these methods has been omitted because the same would be available in another paper (Merkhofer and Runchal) that is scheduled for presentation in this meeting.

The primary aim of this paper is to present the use of Bayes' theorem for incorporating prior information (which could be either entirely subjective or partially so) to arrive at probability estimates. In an earlier paper (*5*), we discussed the use of Bayes' theorem for a univariate case. Some of the material in that paper has been repeated here; however, the primary contribution in this paper is to extend the use of Bayes' theorem to a multivariate case. The incentive for this extension is the realization that many of the variables of interest would be either autocorrelated in space or crosscorrelated among themselves.

Mathematical Framework

Subjective probability

The concept of subjective probability is important for the use of subjective information. Definition of probability in terms of frequency of occurrence of an event is quite common. This definition is useful in analyzing events that occur repeatedly either in nature or by design through the performance of an experiment. Subjective probability, on the other hand, expresses a "degree of belief" in a hypothesis. This belief may be based on the frequency of observations, but it may also be formed in some other way. Thus, subjective probability is particularly suited for situations in which repeated observations are not possible. This could include probability of events that have not happened yet or happen very infrequently (e.g., core meltdown). In general, the prior subjective information would often be expressed in terms of subjective probability. Several standard textbooks contain a discussion of the use of subjective probability (*6*).

Bayes' theorem

In the following, capital letters represent random variables, their lowercase counterparts represent realizations of the random variable, one underscore indicates a vector, and two underscores indicate a matrix.

Bayes' theorem may be stated in the following form:

$$g(\underline{y}|\underline{x}) = f(\underline{x}|\underline{y})g(\underline{y}) \,/\, f(\underline{x}) \tag{1}$$

in which f(.|.) represents a conditional probability distribution function (pdf), and f(.) and g(.) are the marginal pdf's of the joint distribution of $\underline{X}$ and $\underline{Y}$. In general, $\underline{X}$ and $\underline{Y}$ are vectors. One may think of $\underline{X}$ as observations from direct field measurements of a variable of interest (e.g., porosity). Each measurement is, of course, a realization from the underlying distribution of porosity. The underlying distribution of the variable is assumed to come from some specified family of pdf's and is parametrized in terms of the unknown vector $\underline{Y}$. The parameter vector $\underline{Y}$, for example, may represent the unknown mean and variance. These parameters themselves are considered random variables. We are interested in obtaining point and/or interval estimates for the elements of $\underline{Y}$. The pdf of these parameters, $g(\underline{y})$, is called the prior distribution of the unknown parameters. Information from field observations is not yet included in formulating $g(\underline{y})$. The marginal distribution, $f(\underline{x})$, of the observations (or data) can be obtained from the following:

$$f(\underline{x}) = \int f(\underline{x}|\underline{y})\, g(\underline{y})\, d\underline{y} \tag{2}$$

The conditional distribution $f(\underline{x}|\underline{y})$ is also known as the likelihood function. The likelihood function quantifies the probability of obtaining the sample, $\underline{x}$, given a specified value, $\underline{y}$, of the parameter vector. The objective in applying the Bayes theorem is to obtain the posterior distribution of the vector of parameters, $g(\underline{y}|\underline{x})$. We emphasize that $g(\underline{y}|\underline{x})$ is the conditional pdf of the parameter vector $\underline{Y}$ given the set of collected data. It is not the pdf of the variable of interest (i.e., porosity).

Conjugate distributions

In order to write an expression for $f(\underline{x}|\underline{y})$, one has to assume some functional form for the pdf underlying the sample. In the absence of field data, the choice of the form for $f(\underline{x}|\underline{y})$ is purely subjective. An application of Bayes' theorem provides no information about the correctness of the form for $f(\underline{x}|\underline{y})$.

We shall assume that each sample value is statistically independent of all the other observed values when writing an expression for $f(\underline{x}|\underline{y})$. For example, assume that the variable of interest is normally distributed with unknown mean, m, and variance, s^2. Then, the likelihood function for a sample of size k would be as follows:

$$f_k(x_1, \ldots x_k; m,s) = (2\pi s^2)^{-k/2}\exp[-(1/2s^2) \sum_i (x_i - m)^2] \tag{3}$$

While the statement of Bayes' theorem puts no restrictions on the mathematical form of the likelihood function, it is useful to assume some standard form. Some parametric families of standard distributions possess the special property of "closure under sampling." The families of distributions possessing this property are also called "conjugate families" (*7*). Conjugate families of distributions have the following property: If the prior distribution for $\underline{Y}$, $g(\underline{y})$, belongs to this family of pdf's, then for any sample size and any values of observations in the sample, the posterior distribution of $\underline{Y}$, $g(\underline{y}|\underline{x})$, also belongs to the same family. The main advantage of choosing conjugate families of distributions is that the posterior distribution can be easily evaluated.

Applicable theorems

Many variables of interest in performance assessment are either normally distributed or can be transformed to normality (e.g., lognormal). The following two theorems (*7*) provide solutions to the estimation of the posterior pdf for univariate and multivariate settings under a normality assumption for the likelihood function. In the following, the term precision is defined as the reciprocal of the variance in a univariate setting or the inverse of the covariance matrix in a multivariate setting.

Theorem 1. Univariate Normal-Gamma Model. Suppose that $X_1, \ldots, X_k$ is a random sample from a normal distribution with unknown values of the mean, M, and the precision, R. Suppose also that the prior joint distribution of M and R is as follows: The conditional distribution of M when $R = r$ $(r > 0)$ is a normal distribution with mean m and precision vr $(v > 0)$, and the marginal distribution of R is a gamma distribution with parameters a and b $(a > 0,\ b > 0)$. Then, the posterior distribution of M given $R = r$ is normal with mean m' and precision $(v + k)r$, where:

$$m' = (vm + k\bar{x}) / (v + k) \tag{4}$$

and the marginal distribution of R is a gamma distribution with parameters $a + (k/2)$ and b', where:

$$b' = b + 1/2 \sum (x_i - \bar{x})^2 + 0.5vk(\bar{x} - m)^2/(v + k) \tag{5}$$

and $\bar{x}$ is the sample mean.

Theorem 2. Multivariate Normal-Wishart Model. Suppose that $\underline{X}_1, \ldots, \underline{X}_k$ is a random sample from a p-dimensional multivariate normal distribution with unknown values of the mean vector, $\underline{M}$, and unknown symmetric positive definite precision matrix, $\underline{\underline{R}}$, of size $p \times p$. Suppose also that the prior joint distribution of $\underline{M}$ and $\underline{\underline{R}}$ is as follows: The conditional distribution of $\underline{M}$ when $\underline{\underline{R}} = \underline{\underline{r}}$ is a multivariate normal distribution with mean vector, $\underline{m}$, and precision, $v\underline{\underline{r}}$, $(v > 0)$, and the marginal distribution of $\underline{\underline{R}}$ is a Wishart distribution with a degrees of freedom $(a > p - 1)$ and precision matrix $\underline{\underline{t}}$. Then, the posterior distribution of $\underline{M}$ given $\underline{\underline{R}} = \underline{\underline{r}}$ is a multivariate normal distribution with mean $\underline{m}'$ and precision matrix $(v + k)\underline{\underline{r}}$, where:

$$\underline{m}' = (v\underline{m} + k\underline{\bar{x}}) / (v + k) \tag{6}$$

and the marginal distribution of $\underline{\underline{R}}$ is a Wishart distribution with a + k degrees of freedom and precision matrix $\underline{\underline{t}}'$, where:

$$\underline{\underline{t}}' = \underline{\underline{t}} + \underline{\underline{c}} + vk(\underline{m} - \underline{\bar{x}}) \ (\underline{m} - \underline{\bar{x}})^T/(v + k) \tag{7}$$

the superscript T denotes matrix transpose, $\underline{\bar{x}}$ is the sample mean vector, and $\underline{\underline{c}}$ is the matrix formed from:

$$\underline{\underline{c}} = \sum (\underline{x}_i - \underline{\bar{x}}) \ (\underline{x}_i - \underline{\bar{x}})^T. \tag{8}$$

The Wishart distribution for the p × p matrix of random variables $\underline{\underline{R}}$ with degrees of freedom a and precision matrix $\underline{\underline{t}}$ can be written as follows:

$$f(\underline{\underline{R}}) = d|\underline{\underline{t}}|^{-a/2} \ |\underline{\underline{R}}|^{(a-p-1)/2} \exp[-1/2 \mathrm{tr}(\underline{\underline{t}}^{-1}\underline{\underline{R}})] \tag{9}$$

where tr denotes the trace of a matrix and d is the following constant

$$d = \left[2^{ap/2} \ \pi^{p(p-1)/4} \prod_{j=1}^{p} \Gamma((a + 1 - j)/2) \right]^{-1}. \tag{10}$$

Application Example for a Univariate Model

Prior information

A probability encoding method was used at the BWIP to obtain subjective information on the hydrologic parameter porosity (*8*). The variable of interest, "effective porosity at the macro scale," will be simply called porosity. Porosity at the macro scale (i.e., 1 to 10 cubic meters of rock volume) was encoded by eight experts. The encoding resulted in the generation of a cumulative probability distribution from each expert. These eight distributions are shown in Figure 1. Some percentiles from the distributions are given in Table 1 (*9*).

Field data

Field data on porosity are scarce. Data are difficult to obtain on the scale of 1 to 10 cubic meters or larger, because the only experimental method is that of tracer tests, which are costly and take considerable time to perform. Moreover, interpretation of test results in heterogeneous anisotropic media is not easy or unique. To date only one tracer test has been carried out at the BWIP site (*10*), yielding a value of 4.3×10^{-4}.

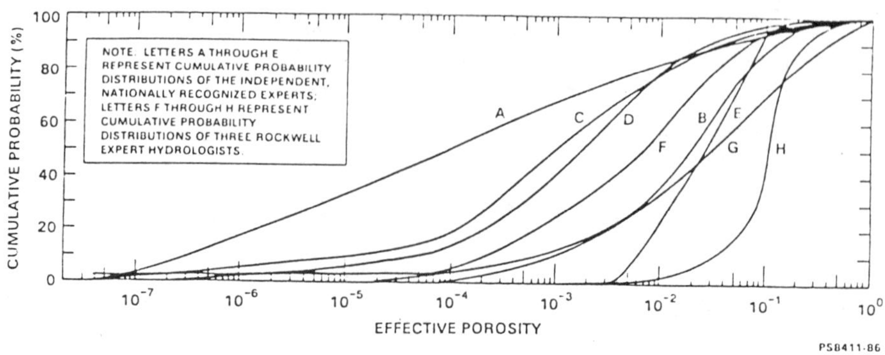

Figure 1. **Probability distributions of experts.**

Likelihood function

The form of the pdf (likelihood function) for porosity was guessed from the encoded data and therefore is based entirely on subjective information. The log of porosity was assumed to be normally distributed, so the likelihood function for porosity is given by equation 3. We chose the conjugate prior given in theorem 1, such that the mean, M, of the log of porosity is distributed as a normal random variable, and its precision, R, as a gamma variable. The parameters of the normal and gamma distributions were calculated from the information in Table 1. Using natural logarithms, the prior distributions are completely characterized by choosing $E(M) = -5.13$, $Var(M) = 5.04$, $E(R) = 0.1925$, and $Var(R) = 0.01984$.

Table 1. **Expert opinion on the cumulative probability distribution of Cohassett flow top average effective porosity.**

Expert	10th Percentile	Median Value	90th Percentile
A	3.3 E-07	1.0 E-04	5.0 E-02
B	1.0 E-03	1.9 E-02	1.2 E-01
C	1.0 E-05	9.5 E-04	3.7 E-02
D	4.5 E-05	1.8 E-03	3.0 E-02
E	6.0 E-03	2.5 E-02	8.5 E-02
F	1.9 E-04	6.8 E-03	6.2 E-02
G	6.0 E-04	2.8 E-02	2.2 E-01
H	3.0 E-02	1.0 E-01	3.5 E-01

Posterior estimates

Using Theorem 1 and the single data value, 4.3×10^{-4}, the posterior point estimates of the mean and variance of log porosity were calculated. The point estimates were -5.94 for the mean (median of 2.63×10^{-3}) and 5.098 for the variance. The prior values were -5.13 (median of 5.92×10^{-3}) and 5.195, respectively. A comparison of the likelihood function for porosity using the prior and posterior parameter estimates is provided in Figure 2.

Extension of results

Additional computations were performed to explore the rate of convergence of the estimated distribution to the true distribution, assuming that it is known. A "true" distribution of porosity that was lognormal with mean and variance of log porosity being -7.6009 and 0.29233, respectively, was chosen. These parameters were obtained by specifying 5.0×10^{-4} as the median of the true distribution and 1.0×10^{-3} as the upper 90th percentile; 1000 values were generated from the assumed true distribution, and Bayes' estimates for the parameters were computed using each data value. The prior distribution used was the one obtained from the eight experts. Point estimates and 95% confidence intervals for the mean and the precision are given in Figures 3 and 4. The estimate of the mean is reasonable for sample sizes of the order of 5 or 6 and quite accurate for sample sizes on the order of 50; however, much larger sample sizes are required to obtain a reliable estimate for the precision. Note that the confidence interval for the precision does not come close to the true value for the first 200 or so values.

Next, a "true" data distribution was chosen that agreed more closely with the experts' prior. The mean and the variance of the natural logarithm of porosity for

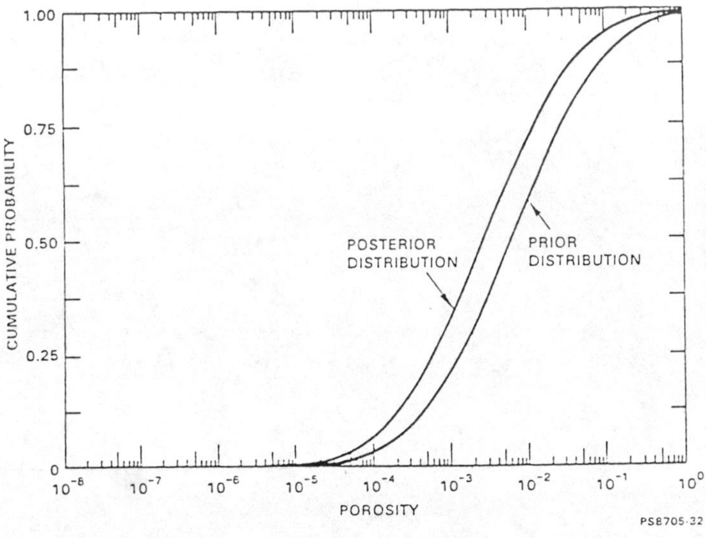

Figure 2. **Prior and posterior distributions for porosity.**

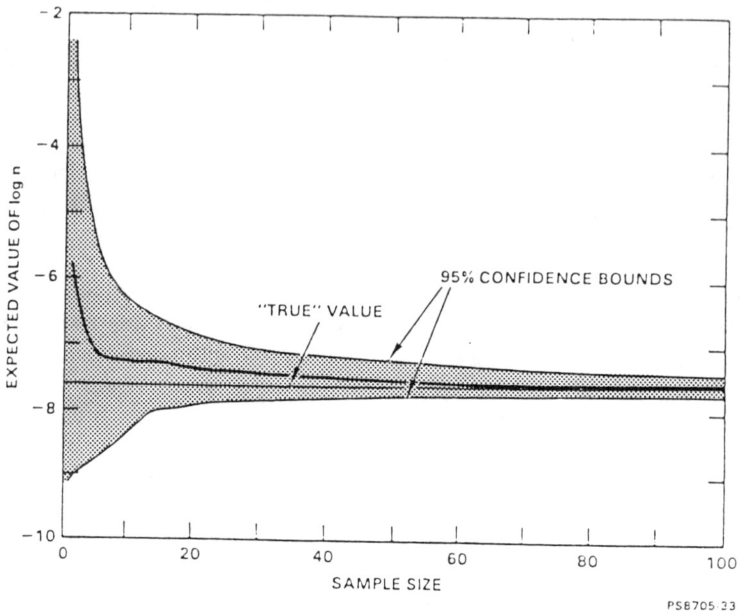

Figure 3. **Mean estimates using an inaccurate prior.**

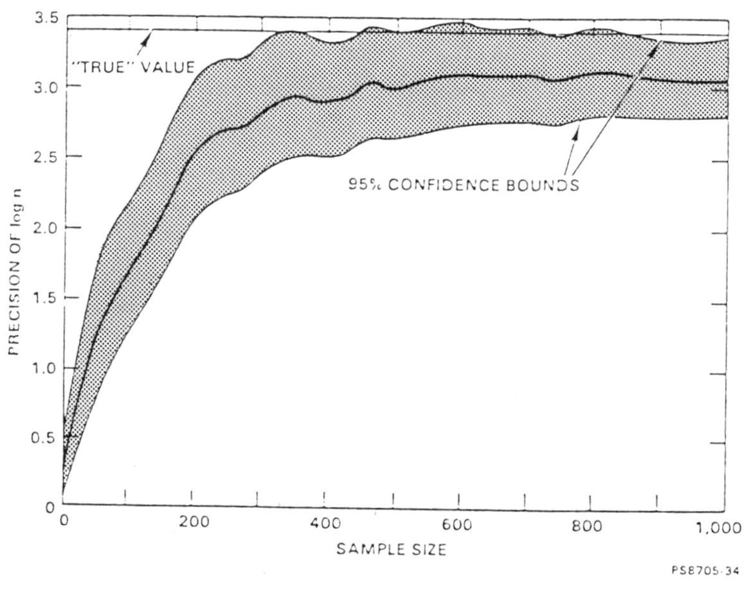

Figure 4. **Precision estimates using an inaccurate prior.**

this distribution were −6.91 and 6.25, respectively. Point estimates and 95% confidence intervals for the mean and precision, using this modified data distribution, are given in Figures 5 and 6. The estimates of the mean and precision are reasonable for very small sample sizes (on the order of 5).

Example for a Multivariate Model

Likelihood function

An example in two spatial dimensions was constructed to illustrate the use of the multivariate model given in Theorem 2. The likelihood function was assumed to be multivariate normal, and the variable of interest is hydraulic transmissivity of the tops of the layered sequence of basalt flows at the BWIP site in south-central Washington State.

Data distribution

Some 40 values of transmissivity have been collected at various locations within the BWIP site. The mean and variance of the common logarithm of the individual values are −7.147 and 3.409, respectively (original units are meters squared per year).

For this investigation, it was assumed that the underlying true distribution of hydraulic transmissivity was known so that any number of hypothetical "field" data could be generated. The data distribution was assumed to be multivariate normal; hence, it is completely specified when the mean vector and covariance matrix are defined. It was assumed that there were 10 locations, all at a common depth, at which measurements of transmissivity could be obtained; thus, the mean vector has

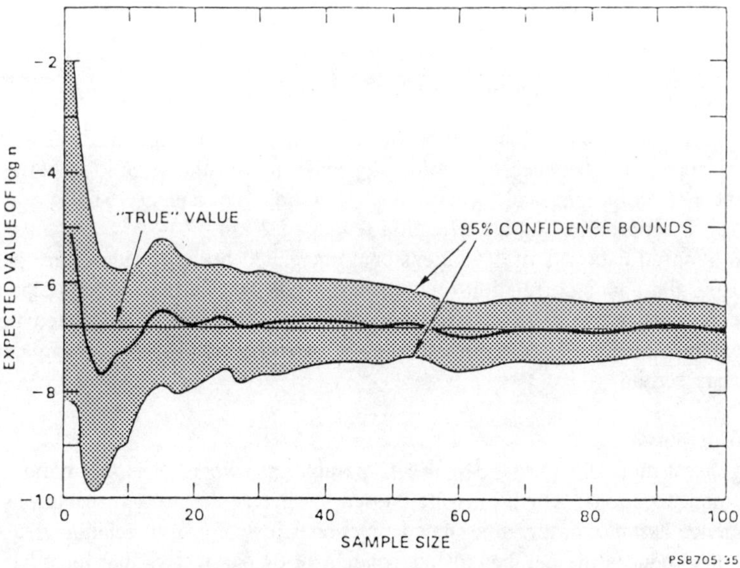

Figure 5. **Mean estimates using an accurate prior.**

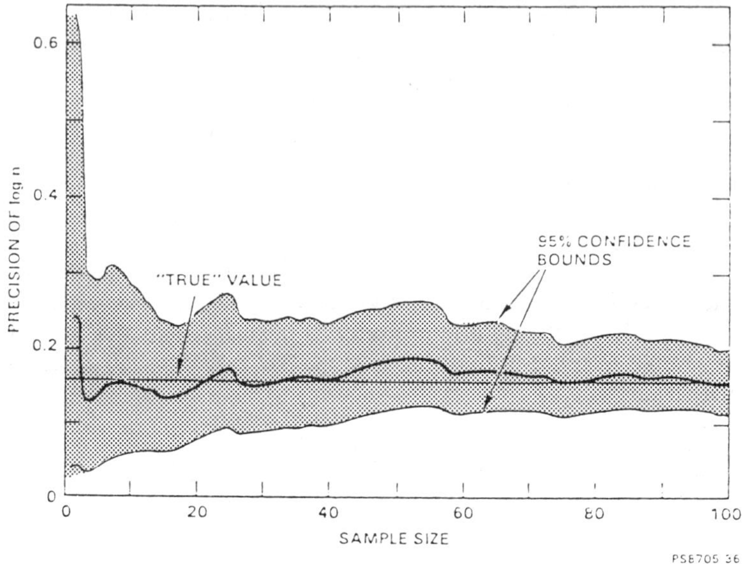

Figure 6. **Precision estimates using an accurate prior.**

10 components, and the covariance matrix is 10 × 10. Although not required in the mathematical formulation, it was also assumed that the stochastic process describing transmissivity was homogeneous. The homogeneous assumption implies that all 10 components of the mean vector are equal and also that the 10 variance values (on the diagonal of the covariance matrix) are also equal. A spatial correlation function of the form:

$$E[(k_1 - \bar{k}_1)\ (k_2 - \bar{k}_2)] = e^{-wu}, \tag{11}$$

where u is the horizontal separation between two measurement points, was assumed. Thus, the correlation between two points depends only on the separation between the points and not on the physical measurement location. The value 1.1513×10^{-3} was chosen for w, approximating a correlation length of 2 km.

One hundred data sets of 10-dimensional correlated normal values were generated using the above correlation relationship, a constant variance of 3.4225, and a constant mean vector of -7.0. These data sets played the role of correlated field measurements and were used to investigate the convergence of Bayes' estimates in a multivariate setting.

Prior information

In general there are $(N^2 + 3N + 4)/2$ parameters, where N is the number of measurement locations, which must be chosen to specify the prior distributions. An independence assumption or an assumed functional form for a correlation structure significantly reduces the number of independent prior parameters that must be specified.

Three priors were used in examining the performance of the Bayes estimates for the mean vector. The first prior, called the G prior, was chosen to closely approximate both the mean and covariance structure of the data distribution and place relatively high reliance on measured values. One interpretation for this choice of a prior is that the expert, in generating the prior, specified correct point (expected) values for the parameters of the distribution of transmissivity, yet reported a relatively large variance about the point estimates. A second prior, the B1 prior, was chosen by modifying the mean vector in the G prior to have the constant value -1.0. The covariance matrix was unchanged from the G prior. A third prior, the B2 prior, was chosen by reducing the covariance values of the distribution of the mean vector by an order of magnitude from those in the B1 prior, i.e., $v = 10$ rather than $v = 1$.

Two priors in addition to the G prior were used to examine the performance of the Bayes estimates for the covariance matrix. The V1 prior was chosen by defining the covariance matrix to be diagonal with constant value 0.04. The variance in this prior was in error by about 2 orders of magnitude, and the correlation structure was entirely missing. The mean vector was unchanged from the G prior. The third prior, called the V2 prior, was chosen by increasing the degrees of freedom of the Wishart distribution of the precision matrix from 10 to 50. The higher number of degrees of freedom for the prior puts more weight on the prior information and less weight on the sample values than does the V1 prior.

Posterior estimates

Estimates of the mean value at a single location for the G, B1, and B2 priors are shown in Figure 7. The values are representative of all locations because of the stationarity assumption. Note that the expression for the posterior mean estimate given in Theorem 2 is a linear combination of the prior mean and the sample mean. Estimates of the mean appear to be reliable when the sample size reaches 20 for the G and B1 priors. Both of these priors put a relatively high weight on the sample mean. Use of the B2 prior requires substantially larger sample sizes to obtain reliable estimates. An interpretation of the B2 prior is that the expert generating the prior was wrong yet believed he was quite accurate.

The estimated covariance matrix contains 55 distinct elements. Five of the 55 were examined in detail for the G, V1, and V2 priors. The elements are described in Table 2. Estimated values for the five selected elements of the estimated covariance matrix when the G prior was used are plotted in Figure 8. The horizontal lines on the plot represent the theoretical values from the data distribution. Deviations from the theoretical values are due to sampling variation. Sample sizes up to 1000 were examined (but are not presented here) to ensure that the data distribution was generated correctly and that the Bayes estimates converged to the corresponding parameters in the data distribution. At least 200 samples are required before the estimates are consistently close to the theoretical values.

Estimates for the same five elements using the V1 prior are given in Figure 9. This figure emphasizes again the sensitivity of estimates of the second moment of a distribution to the shape of the prior. The estimate of the (1,1) element is incorrect by more than a factor of 2 for the first 80 sample values, and sample sizes on the order of several hundred are required to ensure convergence to the parameters of the

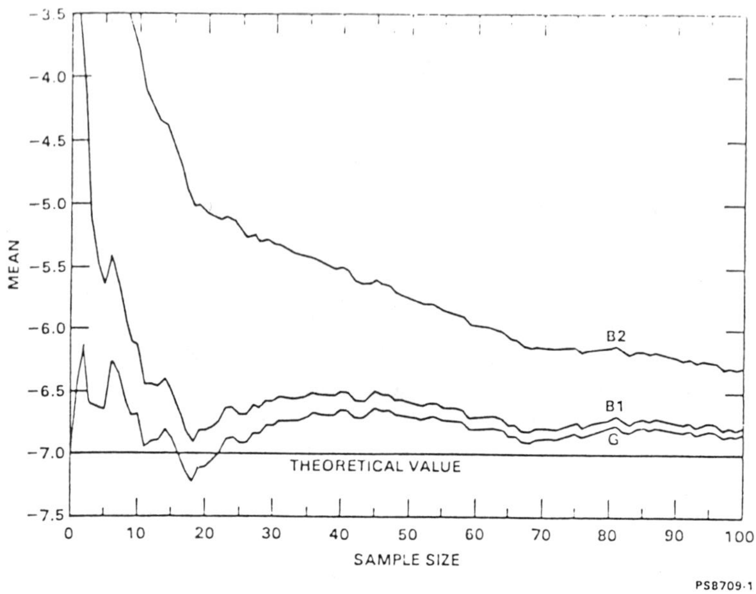

Figure 7. **Estimates of the mean vector using the G, B1, and B2 priors.**

data distribution. Estimates for the same elements using the V2 prior are presented in Figure 10. This prior places less weight on the data than the V1 prior, leading to slower convergence to the data distribution.

Discussion and Conclusions

The nature of the problem of investigating a site for a high-level nuclear waste repository is such that conclusions will have to be drawn from a limited data base. It

Table 2. **Selected elements of the data covariance matrix.**

Element Number (I,J)	Correlation	Covariance (m)	Separation Distance (m)
(1,1)	1.000	3.442	0.0
(1,2)	0.794	2.719	200.0
(2,3)	0.501	1.715	600.0
(2,8)	0.251	0.860	1200.0
(1,10)	0.076	0.261	2236.1

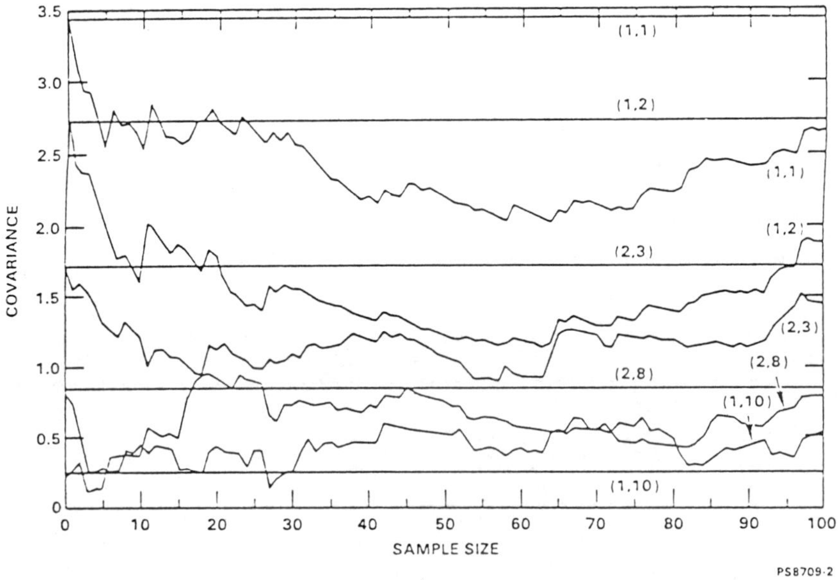

Figure 8. **Covariance matrix estimates using the G prior.**

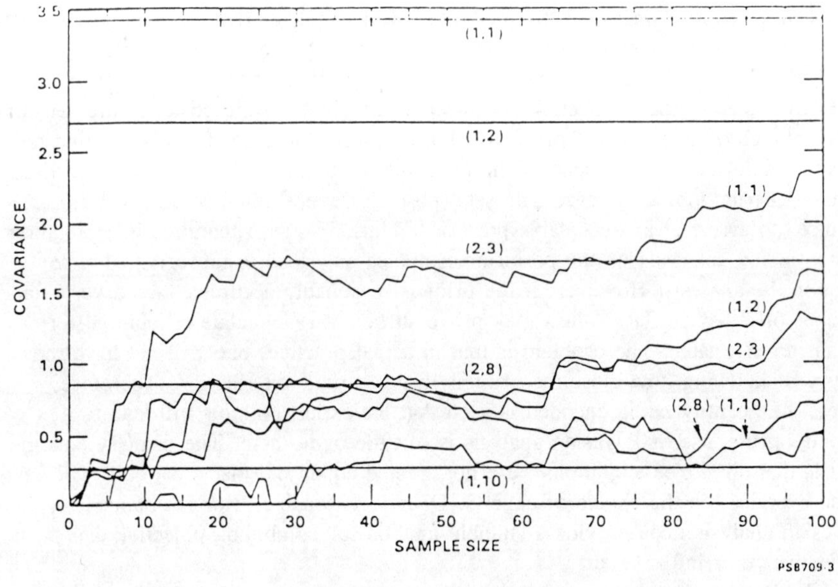

Figure 9. **Covariance matrix estimates using the V1 prior.**

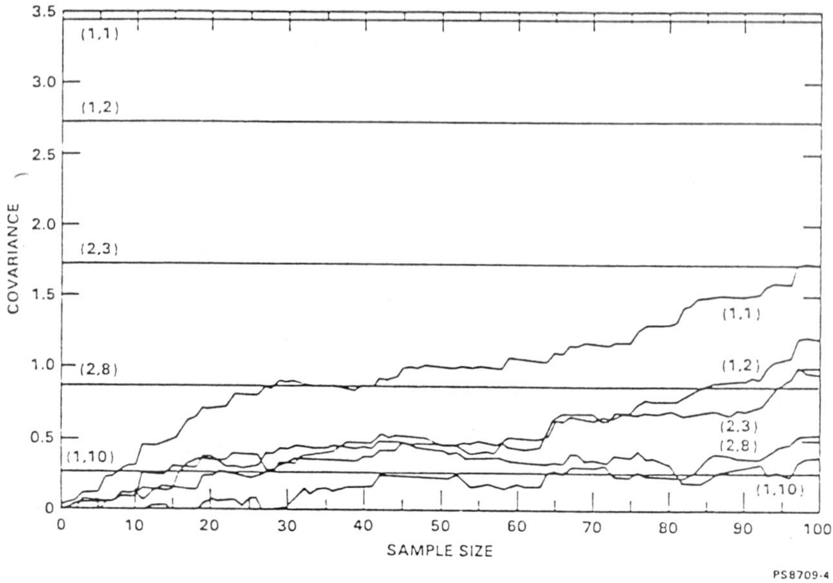

Figure 10. **Covariance matrix estimates using the V2 prior.**

is also apparent that "subjective judgments" and "expert opinions" about missing, inadequate, or impossible to obtain data will have to be used in decision making. Bayesian analysis is one available method by which quantitative analysis may be performed in situations where both objective and subjective data are available.

From the investigation of the convergence properties of the posterior as a function of the size of the data set, it appears both moments of the posterior are sensitive to the choice of shape of the prior distributions, with the second moment being more sensitive than the first moment. If the prior distribution is quite different from the "true" distribution, very large data sets (possibly several hundred data values) are required to arrive at an acceptable posterior. Thus, in such situations, interpretation of confidence intervals on the posterior estimates may lead to unwarranted conclusions in the analysis. However, if the prior is reasonably accurate, then a very small number of objective data values may prove sufficient to calculate reliable posterior parameter estimates. The problem is that in actual practice, one will not have the luxury of knowing how close the prior is to the "true" distribution. Therefore, unless the quantity being encoded is such that the expert opinion will lead to a "good" prior, use of Bayesian analysis is of little or no help. The primary conclusion is that all "expert opinions" are not created equal and that great care will have to be exercised in the choice of experts. Once a reasonable prior is obtained, Bayesian analysis does provide a suitable method for combining objective data with subjective prior information.

The Bayesian approach outlined in this paper enables a researcher to choose the parameters of a statistical distribution describing some physical quantity of interest (e.g., porosity). Note, however, that the researcher has to choose a parametric family of distributions for the data likelihood function, e.g., lognormal, before this method can be applied. The Bayesian approach then selects one member of the chosen family of distributions after some data has been collected. This approach does *not* guarantee that the data likelihood function chosen is a better description of the physical process than any other likelihood function.

Literature Cited

1. NRC, *Disposal of High-Level Radioactive Wastes in Geologic Repositories: Licensing Procedures,* Title 10, Code of Federal Regulations, Part 60, U.S. Nuclear Regulatory Commission, Washington, D.C. (1983).
2. EPA, *Environmental Radiation Protection Standards for Management and Disposal of Spent Nuclear Fuel, High-Level and Transuranic Radioactive Wastes,* Title 40, Code of Federal Regulations, Part 191, U.S. Environmental Protection Agency, Washington, D.C. (1986).
3. Linstone, H. A. and M. Truoff, eds, *The Delphi Method: Techniques and Applications,* Addison-Wesley, Reading, Massachusetts (1975).
4. Staël von Holstein, C.A.S. and J. E. Matheson, *A Manual for Encoding Probability Distributions,* Stanford Research Institute, Stanford Research Institute International, Menlo Park, California (1979).
5. Sagar, B. and P. W. Eslinger, 1987, "Use of Subjective Information in Probabilistic System Analysis," Presented at the Fifth Meeting of Probabilistic System Analysis Code Users' Group, Nuclear Energy Agency, Stockholm, Sweden, June 2–5. (Also RHO-BW-SA-704 P, Rockwell Hanford Operations, Richland, Washington.)
6. Martz, H. F. and R. A. Waller, *Bayesian Reliability Analysis,* John Wiley & Sons, New York, New York (1982).
7. DeGroot, Morris H., *Optimal Statistical Decisions,* McGraw-Hill Book Company, New York, New York (1970).
8. Runchal, A. K., M. W. Merkhofer, E. Olmsted, and J. D. Davis, *Probability Encoding of Hydraulic Parameters for Basalt,* RHO-BW-CR-145P, Rockwell Hanford Operations, Richland, Washington (1984).
9. Loo, W. W., R. C. Arnett, L. S. Leonhart, S. P. Luttrell, I-Seng Wang, and W. R. McSpadden, *Effective Porosities of Basalt: A Technical Basis for Values and Probability Distributions Used in Preliminary Performance Assessments,* SD-BWI-TI-254, Rockwell Hanford Operations, Richland, Washington (1984).
10. Gelhar, L. W., *Analysis of Two-Well Tracer Tests with a Pulse Input,* RHO-BW-CR-131 P, Rockwell Hanford Operations, Richland, Washington (1982).

Probability Encoding: Quantifying Judgmental Uncertainty Over Hydrologic Parameters for Basalt

MILEY W. MERKHOFER
Applied Decision Analysis, Inc. (ADA)
Menlo Park, California

AKSHAI K. RUNCHAL
Analytic and Computational Research, Inc. (ACRI)
West Los Angeles, California

Abstract

Judgmental probabilities (together with Bayesian analysis) provide a means for overcoming the lack-of-data problems associated with classical methods for uncertainty analysis. Judgmental probabilities may be obtained through a formal process called probability encoding. This paper describes probability encoding methodology and presents an application that quantified expert uncertainty regarding key hydrologic parameters at the potential repository site at Hanford, Washington.

The need for probability encoding arises from experimental research demonstrating that accurate probabilities *cannot* be obtained by simply asking someone for a probability number. Apparently, common modes of thinking introduce large and systematic errors into the estimates produced. Recent research has focused on developing methods for eliminating or reducing these biases. In particular, SRI International has developed a detailed, multistep process in which probability assessments are generated in a structured interview between a probability assessor and a subject who is an expert on the uncertain quantity in question.

The SRI probability encoding process was applied to quantify expert uncertainty regarding (1) the effective porosity of the Cohassett basalt flow top and flow interior and (2) the ratio of vertical to horizontal conductivity (anisotropy ratio) of the Cohassett basalt flow interior. Each variable was assessed on two scales, a megascale (100 to 1,000 meters) and a macroscale (1 to 10 meters). Background material, which included measured site-specific, test-based values of the pertinent parameters, was prepared and distributed to participants prior to the probability encoding sessions.

Two distinct groups of experts participated. The first group consisted of five individuals selected on the basis of a reputational survey from among nationally and internationally known hydrologists. Probability distributions were obtained as a two-phase Delphi process. In the first phase, each expert was individually interviewed to obtain initial distributions for each variable. In the second phase, the initial distributions were distributed to all panel members, along with explanations, and each expert provided revised distributions. The second group of experts consisted of three hydrologists selected by the prime contractor studying the Hanford site. A similar process was used for this group, except that the initial distributions were not revised. In the case of both groups, probability encoding interviews lasted between six and eight hours each.

The results of probability encoding indicate considerable judgmental uncertainty, areas of agreement and disagreement among experts within each group, and systematic differences among the two groups. In addition, the results suggest the potential for resolving differences of opinion through clarifying and sharing the information and the reasoning processes used by the experts.

Introduction

The appropriate method of accounting for uncertainty in quantitative analysis depends on the level of available information and understanding. A fairly common situation is the following. The functional relationships among variables are sufficiently well known that models can be constructed. The values of the model variables and associated coefficients, however, are unknown. In this situation, it is common to characterize uncertainties with probability distributions. Uncertainty analysis consists, then, of propagating probability distributions (or their moments) through analytical, numerical, or stochastic models.

To obtain probability distributions for model parameters, analysts have two options. One approach is to derive classical probability distributions from empirical data. The other is to rely on experts to provide judgmental probability distributions. Because available data are often insufficient to permit meaningful statistical analysis, judgmental probabilities may be the only viable alternative.

The selection of possible sites for a nuclear waste repository provides an example of a problem for which judgmental probabilities may be important. Explicit measures of uncertainty are needed to set priorities for the collection of site-specific information and to determine the overall level of resources that should be allocated to reducing uncertainty. At present, expert judgment appears to be the only feasible means for quantifying many important uncertainties.

Using judgmental probability distributions to analyze potential repository performance presents two serious problems. First, there is substantial empirical evidence that the probabilities provided by individuals who are untrained in probability assessment are biased. How can such biases be avoided or at least minimized? Second, different experts are likely to provide different distributions. How can such differences be resolved or at least reconciled with the desire to produce objective analysis?

This paper has three objectives. These are (1) to clarify the difficulties associated with obtaining and using judgmental probability distributions, (2) to describe a recommended approach, and (3) to present an application that quantified expert uncertainty over important hydrologic parameters at a potential repository site at Hanford, Washington.

Biases in Judgmental Probabilities

A probability distribution is biased if it inaccurately represents an individual's underlying beliefs and knowledge. Figure 1 illustrates the two effects that biases may have on a judgmental probability distribution. First, the whole probability distribution might be shifted to the right or left from that which would represent the subject's underlying knowledge and beliefs (displacement bias). Second, the shape of the distribution might be altered (variability bias) (*2*).

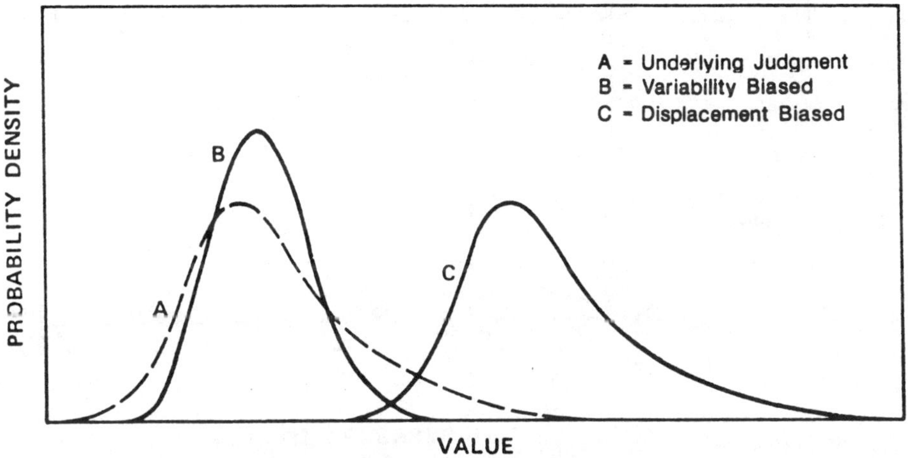

Figure 1. **Examples of variability and displacement biases. Source: Adapted from Reference (*1*).**

Calibration curves

Biases have been investigated primarily through research on calibration. Calibration is a measure of the degree to which a set of probability estimates matches the frequency of actual outcomes. The degree can be measured empirically by asking subjects to provide probability distributions for knowable but obscure quantities. For example, the quantities might be the populations of countries, dates of historical events, heights of mountains, or other quantities that can be looked up in an almanac. The fractiles of the assessed distributions (i.e., the values of the uncertain variable corresponding to the 1 percent, 25 percent, etc., points of the assessed cumulative probability distributions) are identified and the fraction of actual values that lies below each fractile is plotted, as shown in Figure 2. The result is termed a calibration curve.

A perfectly calibrated subject would have a calibration curve that is a diagonal line on which 1 percent of the actual values fall below the 0.01 fractile, 25 percent fall below the 0.25 fractile, etc. Overconfident subjects assign probability distributions that are too narrow. More than 25 percent of the time actual values are less than their 0.25 fractiles and greater than their 0.75 fractiles. Conversely, underconfident subjects assign probability distributions that are too broad. Less than 25 percent of the time actual values are less than or greater than their 0.25 and 0.75 fractiles, respectively.

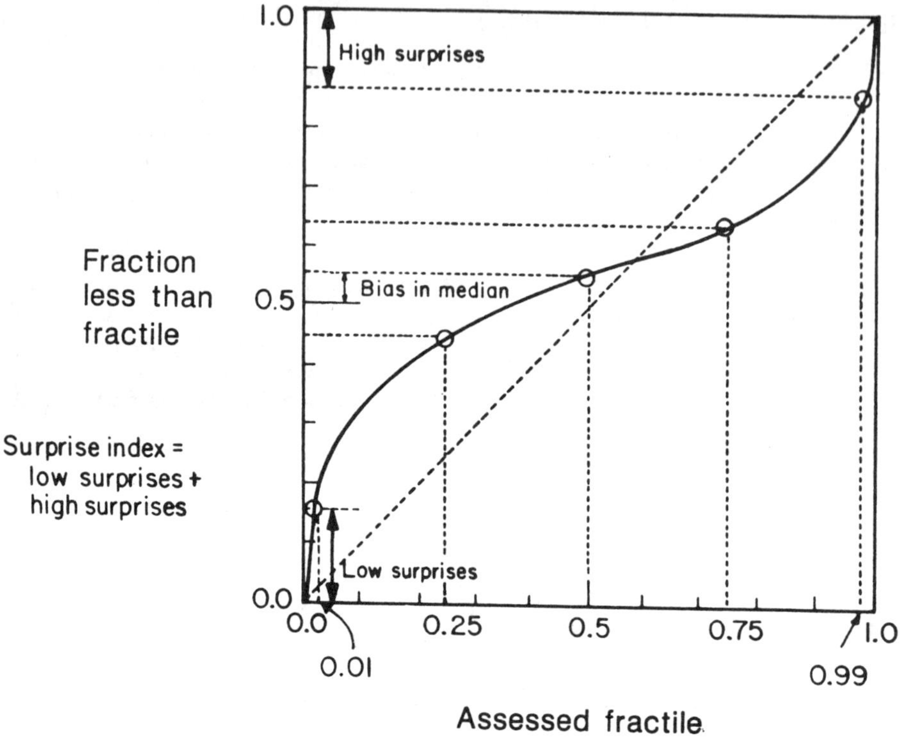

Figure 2. **Calibration curve for an overconfident subject. Source: Adapted from Reference (*3*).**

Empirical evidence indicates that subjects are almost always overconfident. This bias has been shown to occur for both lay people and technical experts, including experts knowledgeable about probability and statistics (*4*). The extent to which overconfidence distorts judgmental probability distributions may be illustrated by the "surprise index"—the fraction of times that an actual value is found to lie beyond the 1 percent or 99 fractiles of the subject's assessed probability distribution. Well-calibrated subjects should be surprised only 2 percent of the time. As indicated by Table 1, studies show that subjects untrained in probability assessment are typically surprised between 5 and 50 percent of the time.

Heuristics

Psychologists believe overconfidence and other biases important to probability assessment are caused by the simplified procedures, called heuristics, that people use to process information. There are particularly strong and prevalent heuristics:

- **Anchoring and adjustment**: People tend to estimate the numerical value of an uncertain quantity by starting with some initial value (e.g., suggested by the problem formulation or some partial calculation) and then adjusting this value in light of other factors (*5, 6*). Generally, the adjustment is insufficient, and the probabil-

Table 1. **Summary of calibration indices for continuous distributions.***

	Number of Assessments N	Surprise Index (ideal 2%)
Alpert and Raiffa (1969)		
Group 1-A	880	46
Groups 2 and 3	1670	39
Group 4	600	21
Hession and McCarthy (1974)		
Fractiles	2035	47
Selvidge (1975)		
Five Fractiles	400	10
Seven Fractiles	520	7
Schaefer and Borcherding (1973)		
Fractiles	396	39
Hypothetical Sample	396	50
Pickhardt and Wallace (1974)		
Group 1	?	32
Group 2	?	46
Seaver, von Winterfeldt, and Edwards (1975)		
Fractiles	160	34
Odds-Fractiles	160	24
Probabilities	180	5
Odds	180	5
Log Odds	140	20
Staël von Holstein (1971)		
Fixed Intervals	1269	30
Schaefer (1976)		
Fixed Interval	660	25
Lichtenstein and Fischhoff (1978)		
Fractiles	924	41
Seaver (1978b)		
Parameters of Beta Distribution	3200	25

*These results are for subjects without feedback training.
Source: Adapted from Reference (*3*).

ity distribution is too narrow. This heuristic is a principal cause of overconfidence in probability distributions.

- **Availability**: People tend to assess the probability of an event by the ease with which they can think of previous occurrences (*5, 7*). The problem is that many characteristics other than actual frequency influence the ease of retrieval from memory, such as recency and emotional impact. Thus, highly publicized causes of death, such as homicide or botulism, tend to be assigned incorrectly high probabilities while underpublicized causes, such as death by stroke, are assigned probabilities that are too low.

- **Representativeness**: People seem to judge the likelihood of an event in part by the degree to which it seems consistent with the population of possibilities (*5, 8*). For example, studies show that most people will incorrectly judge the sequence of coin flips head-head-head-tail-tail-tail to be less likely than the sequence head-tail-head-tail-tail-head, because the latter seems more random.
- **Base rates and the law of small numbers:** People seem to pay considerable attention to information specific to the particular quantity being assessed, even if that information is unreliable. At the same time, relevant base-rate information for that class of quantities is often ignored (*9, 10, 11*). For example, a scientist may place too much faith in a single measurement of a geohydrologic parameter, even though he admits that the parameter is hard to estimate, and may place too little attention to analogous data collected for similar situations of the same general category.

The strength and prevalence of these biases have been demonstrated in a wide variety of judgment tasks (see, e.g., *12*, for more references). They have been compared to optical illusions (*13*): Even knowing that they're there doesn't make them go away.

Formal Assessment Methods

Two approaches have been proposed for countering the problem of biases in judgmental probabilities. One approach is to attempt to correct for biases by using calibration curves in reverse. A calibration curve for a subject might be developed using test questions. The resulting curve might then be used as a function to map assessed distributions into well-calibrated ones.

This sort of recalibration has several significant problems. First, developing a calibration curve is very time consuming. Calibration curves differ from individual to individual (*14*). Furthermore, for a given individual, calibration curves seem to depend on the subject matter and the difficulty of the question (*15, 4*). Thus, finding a suitable set of test questions is likely to be very difficult. Recalibration can also cause a set of probabilities to become incoherent; that is, to violate axioms of probability theory such as Bayes' theorem. The most significant problem for recalibration is its likely effect on the relationship between those who provide probabilities and those who recalibrate them. The former are likely to resent the fact that their opinions are not being taken at face value and may change their calibration, producing a vicious circle of mutual mistrust.

A better approach for dealing with biases is to use formal methods for probability assessment. Formal methods may be used to reduce biases in three ways. First, methods are available for educating a subject about biases and training him to estimate probabilities. Some experienced probability assessors have been found to be excellently calibrated (*16, 17*). Second, methods have been designed to eliminate or minimize the subject's use of counterproductive heuristics that introduce biases. Third, methods are available for reducing biases by aggregating experts' opinions.

In the early 1970s, a detailed probability assessment process was recommended by the Decision Analysis Department of SRI International (*2*, *18*). The SRI process is conducted as a structured interview between a trained interviewer and the subject. It consists of five distinguishable stages: motivating, structuring, conditioning, encoding, and verifying. The original SRI process has been expanded to include methods for dealing with expert differences (*19*). One such method is the Delphi method, originally developed by Rand Corporation to promote reliable consensus of opinion (*20*). With the Delphi method, judgments and associated rationales are obtained independently from each member of an expert panel, and the results are summarized and fed back to panel members in a controlled way designed to eliminate the destructive pressures associated with group meetings. After one or more repetitions or "rounds," the information held by participants is exchanged and opinions often converge.

An Application of Formal Probability Encoding Methods

The remaining sections of this paper describe an application of a formal probability encoding methodology (*21*, *22*). The methodology combines the SRI process with elements of the Delphi method.

Problem background

At the time of this study, the Columbia River basalts underlying the Hanford Site in Washington State were being considered as a possible location for a geologic repository for high-level radioactive waste (*23*). In particular, the Cohassett basalt flow was considered a prime candidate horizon for repository studies. Hydrologic parameters, such as hydraulic conductivity and effective porosity, provide a basis for estimating the degree to which the site would contain and isolate buried wastes. The ratio of vertical-to-horizontal conductivity, the anisotropy ratio, determines the geometry and length of the primary radionuclide transport pathways and thus affects the travel time of dissolved radionuclides. The effective porosity determines the velocity with which fluid particles move through the groundwater system and thus also affects the time required for dissolved radionuclides to reach the biosphere.

The appropriate numeric values for these key hydraulic parameters at the Hanford site are uncertain. Effective porosity data are available from only one measurement location. Hydraulic conductivity data, particularly for vertical conductivity, are few and of limited quality. Additional field studies are being conducted to obtain more site-specific data; however, ongoing assessments ought to be based on best available knowledge and should fairly reflect existing uncertainties. For these reasons, judgmental probability distributions were desired for key hydraulic parameters.

Selection of experts

Two panels of experts were selected to provide the judgmental probability distributions. The first panel consisted of five nationally and internationally known hydrologists. These individuals were selected using a process recommended in the Delphi literature (*24*). Candidates were ranked through a reputational survey and then screened according to their availability and potential conflicts of interest. Table 2 identifies the selected participants and their affiliations at the time of the study. This group is collectively referred to as "independent experts." The second panel consisted of three hydrologists employed by Rockwell Hanford Operations (the Department of Energy contractor responsible for studying the Hanford site). The selection criteria for Rockwell participants were knowledge about the site and availability. Table 3 identifies the Rockwell participants, collectively referred to as the "Rockwell experts."

To prepare for the assessment exercise, three information packages were developed. The first summarized the background of the study, its purpose, and its scope (Appendix C of *21*). The second information package summarized data and information thought to be directly relevant to the assessment process (Appendix D of *21*). This included available data pertaining to the specified hydrologic parameters, such as measurements of effective porosity, transmissivity, and hydraulic conductivity from about 40 boreholes in and around the Hanford site. Also included were summaries of other published estimates and relevant information derived from previous reports. These first two packages, which were prepared in accordance with general principles of the Delphi method, were mailed to the participating experts and reviewed by them prior to attempting the encoding exercise.

Table 2. **Independent experts.**

Name	Position (at time of study)
Stanley Davis	Professor Department of Hydrology and Water Resources University of Arizona Tucson, Arizona
Paul Fenske	Professor Water Resources Center Desert Research Institute University of Nevada Reno, Nevada
Lynn Gelhar	Professor Civil Engineering Massachusetts Institute of Technology Cambridge, Massachusetts
Shlomo Neuman	Professor Department of Hydrology and Water Resources University of Arizona Tucson, Arizona
Charles Wilson	Partner Hydrotechnique Associates Berkeley, California

Table 3. **Rockwell experts.**

Name	Position (at time of study)
Roy Gephart	Staff Hydrologist Basalt Waste Isolation Project Site Analysis Group Hanford Operations Richland, Washington
Leo Leonhart	Staff Hydrologist Basalt Waste Isolation Project Site Analysis Group Rockwell Hanford Operations Richland, Washington
Frank Spane	Staff Hydrologist Basalt Waste Isolation Project Drilling and Testing Group Rockwell Hanford Operations Richland, Washington

The third information package consisted of a written protocol for probability assessment based in the SRI probability encoding process. The purpose of the protocol was to facilitate the implementation of the SRI process by outlining the detailed steps and providing a convenient framework for recording the panelists' responses. The usual Delphi approach of obtaining judgments solely through questionnaires was judged unworkable for probability assessment. Therefore, the protocol was not distributed to the panelists. Rather, it was used by the interviewers to guide and document the process and, to the extent practicable, promote uniformity across the assessment sessions.

Variables encoded

The hydrologic parameters whose uncertainties were encoded were those judged to be most important for the preliminary assessments of repository performance. These were (1) the average effective porosity of the Cohassett flow top, (2) the average effective porosity of the Cohassett flow interior, and (3) the anisotropy ratio in the Cohasset flow interior. Each of these variables was assessed both on a megascale (on the order of 1,000 meters) and a macroscale (on the order of 10 meters). Thus, each expert was asked to provide six probability distributions.

To minimize ambiguity and misunderstanding, precise definitions were specified for all terms used to describe the variables. Table 4 summarizes these definitions.

Probability encoding interviews

Probability distributions were first obtained from the independent experts. Probability assessment interviews were conducted at the experts' home institutions under circum-

Table 4. **Definitions of variables to be encoded.**

Variable	Definition
Effective porosity	The in situ volume proportion of rock that contributes to solute transport if a hydraulic gradient is applied across the volume. The volume size is specified as megascale or macroscale as defined below.
Average effective porosity	The average is defined so that an accurate gross experiment performed on this entire volume would yield this value.
Anisotropic ratio	The vertical conductivity divided by horizontal conductivity (K_v/K_h), where conductivity is defined as the flow in square meters per second that comes out of a volume cross section (specified at a megascale or macroscale) for a unit hydraulic gradient that is applied under in situ conditions.
Flow	The vesicular and/or brecciated upper portion of a basalt flow.
Flow interior	The relatively dense portion of the basalt flow that has a characteristic cooling joint pattern and typically contains no vesicularity.
Megascale	The volumes mentioned above are specified to be 1,000 m per side and the depth is such that the volume lies *entirely within* the flow top or flow interior (as specified).
Macroscale	The volumes mentioned above are specified to be 10 m per side and the depth is such that the volume lies *entirely within* the flow top or flow interior (as specified).

Note: All definitions are applied to the Cohassett basalt flow beneath the Hanford site.

stances that in most cases allowed full access to files and data. A decision analyst experienced in probability encoding directed each interview, and a second analyst was present for note-taking and recording responses. Each interview took between six and eight hours.

The protocol included each of the five stages of the SRI process. In the initial (motivating) stage, the analysts discussed the objectives of the study and answered questions regarding the background information packages. In two cases, panelists interrupted the analysts' introductory comments by protesting that the lack of data made it impossible for them to predict the value of desired hydrologic parameters. However, when it was clarified that the intent of the study was to measure the individual's knowledge and uncertainty, not to predict a value, each expert felt more comfortable with the process. The motivating stage also included approximately a one-half-hour briefing on the psychological theory and data regarding heuristics and biases. This information helped the experts to understand the reason for the very detailed process that was to be applied. By admitting the difficulty of probability assessment and demonstrating that we were trying to do the best job possible, rapport with the experts was established, and they seemed to become more inclined to think hard when answering elicitation questions.

The second stage was devoted to defining and structuring the assessment tasks. The analysts reviewed the definitions of the variables whose probability distributions were to be assessed. Understanding was checked by applying a "clairvoyance test" (*2*, *19*). Each aspect of the definition was carefully reviewed until the expert agreed that a clairvoyant could theoretically provide a single, well-defined numeric value for each variable. This helped to assure that vagueness of understanding would not contribute to the expert's uncertainty.

The third stage consisted of preparing and conditioning the expert to promote the full use of relevant knowledge. An extended technical discussion lasting more than an hour was conducted. During this stage the expert was encouraged to describe his personal view of the status of information relevant to the variables. The quality and applicability of the Hanford data were discussed. The experts described various theories and models and their opinions as to which were most compelling. In addition, the experts were asked to describe more general information, such as data collected from other basalt flows, which might be relevant to their considerations.

Throughout the conditioning stage, the analysts asked probing questions to encourage the expert to identify the full range of considerations relevant to the exercise. However, the ideas and opinions of previously interviewed experts were not suggested by the analysts. The goal was to draw out into the expert's immediate consciousness all relevant information pertaining to the variables but not to influence those judgments by revealing what other experts had said.

The fourth stage consisted of actually encoding probability distributions. Each of the variables was addressed in turn with estimates using the megascale conducted before the estimates using the macroscale. Each megascale estimate required about an hour to complete. Macroscale estimates were typically obtained by modifying the corresponding megascale estimates and required roughly one-half hour each.

To help counter biases caused by the anchoring and availability heuristic, the analyst began each assessment by asking the expert to provide maximum and minimum credible values. The expert was then asked to imagine the possibility that the actual value might be found to lie outside his specified limits and to offer plausible scenarios that would explain such a finding. This technique has been found useful for reducing overconfidence bias (*25*).

Cumulative probabilities for various values of each variable were elicited using an SRI probability encoding wheel. The wheel consists of a disk constructed so that two complementary orange and blue sectors can be varied in angle to occupy any fraction of the disc's area. The disc may be spun in relation to a fixed pointer and will come to rest with the pointer falling in one sector or the other at random. Thus, the probability wheel is similar to a gambling wheel, except that with the former the analyst can alter the chance of "winning."

Initially, probabilities were obtained through the standard decision analysis approach of asking the expert to express his preferences over gambles (*26*, *19*). For example, the expert was asked, "Consider a volume lying in the Cohassett flow directly below the

Hanford site. This volume is 1,000 meters wide, 1,000 meters deep, and has a height just small enough to fit entirely within the flow interior. Would you rather bet that the average value of the effective porosity within this volume is less than 10^{-2}, or would you prefer to bet that when I spin this wheel the pointer will end up in the orange part of the disc?" The relative areas of the blue and orange sectors on the probability wheel were then adjusted until a setting was found where the expert was indifferent between the two bets. The probability of the defined event (i.e., the cumulative probability of an effective porosity value of 10^{-2}) was then assumed to be equal to the proportion of area covered by orange on the probability wheel. Additional points on the cumulative probability distributions were subsequently obtained in a similar fashion. However, to prevent the expert from anchoring on previous judgments, the gambles were varied to involve bets that actual values would be greater, as well as less than, specified values, and wheel bets were defined in terms of the blue, as well as the orange, sectors.

As experts gained familiarity with the process, they typically shifted to a more straightforward approach of directly providing probability estimates. Whenever an expert directly estimated a probability, the analyst asked him to confirm that value by comparing it with the probability wheel set at that probability or, if the probability was less than 0.05, by a comparison with some other reference event. For very low probabilities the concept of an urn filled with colored balls was used as a reference event. For example, if the expert estimated that the probability of a value of effective porosity greater than 10^{-3} was only 1 percent, he was asked, "In other words, if I had an urn filled with 99 red balls and one white ball, are you saying that you think it is just as likely that the first ball drawn from that urn would be white as you do that the effective porosity is actually greater than 10^{-3}?"

After a sufficient number of points on the distribution were established using the above methods, a smooth curve was sketched in. This curve was interpreted to represent a preliminary estimate of the expert's cumulative probability distribution for the variable in question.

As each preliminary probability distribution was obtained, it was subjected to verification in the fifth stage of the assessment process. The purpose of the verification stage was to test the judgments implied by the preliminary distribution to see if the expert really believed in them. To investigate whether the expert felt comfortable with the curve, the cumulative distribution was converted to a probability density function (using graphical techniques). Skewness and any sharp extremes were noted and discussed. Then, based on the revised cumulative distribution, the range for the variable was split into three regions of equal probability and the expert was asked to express any preference over which interval he would prefer to bet the variable would fall within. The distributions for megascales and macroscales were also compared. For example, experts were asked whether the expected value for the macroscale of a variable ought to be equal to that for the megascale. Most experts believed that the expected values for the two scales ought to be equal or nearly equal, but two advanced arguments for why the expected values for the two scales might differ.

During the verification stage, the experts often took considerable care to make small revisions to their distributions. Evidently, they were sensitive to discrepancies in their curves of a few percentage points. The process was not concluded until the expert stated his belief that the curves provided an accurate representation of his level of uncertainty.

Resolution of expert differences

A comparison of the probability curves obtained from the various independent experts indicated considerable diversity of opinion. In an effort to reduce differences of opinions, the complete set of probability curves was distributed among the independent experts together with comments that they had made during the encoding exercise to justify high or low judgments. Following the Delphi method, the anonymity of the panelists was maintained by using the letter code rather than identifying the experts' names. The experts were also provided with a workbook describing a step-by-step procedure by which they might systematically revise their distributions.

In each case, the revised curves provided by the experts indicated that no or only minor revisions were made. Although it was apparent that the various independent experts often relied on markedly different conceptual models and information to generate their curves, the mere knowledge that other experts disagreed with them was evidently not sufficient to sway any of the experts from their original curves. When returning their workbooks, most experts stated that they would need to know who believed the conflicting opinions and have a clearer understanding of why before they would feel any compulsion to change their own opinions.

Probability distributions

Figures 3 through 8 show the final probability curves provided by the independent experts. Figure 3 shows the experts' cumulative probability distributions of flow top average effective porosity at the megascale. In this figure and those that follow, the curves belonging to the different experts are distinguished through a letter code. The median values estimated by the experts range from 10^{-4} to 3.5×10^{-2}. The highest values deemed possible are clustered just above 10^{-1}. The lowest values range over four orders of magnitude, from 10^{-7} to 2×10^{-3}. The corresponding curves for the macroscale are very similar, as shown in Figure 4. However, the highest estimated values are approximately one order of magnitude higher and the lowest values are one order of magnitude lower than the comparable values for the megascale.

Figures 5 and 6 show the probability distributions for effective porosity in the flow interior at the mega- and macroscales, respectively. As shown, the estimated values tend to be about an order of magnitude lower than the analogous flow top values. This disparity is more pronounced near the upper end of the range of the distributions than near the lower end.

Figures 7 and 8 show the probability distributions for anisotropy ratio of hydraulic conductivity. With respect to anisotropy ratio, there is somewhat less disagreement among the experts. The estimated median values range from just over 1 to about 20. Four of the five probability distributions are roughly within one order of magnitude of each other throughout their ranges.

Aggregation and comparison of independent and Rockwell probability distributions

Evidence suggests that aggregating judgmental estimates from multiple experts can improve assessment quality. In addition, aggregating may be necessary to simplify the comparison of the views of different groups or to provide a single probability distribu-

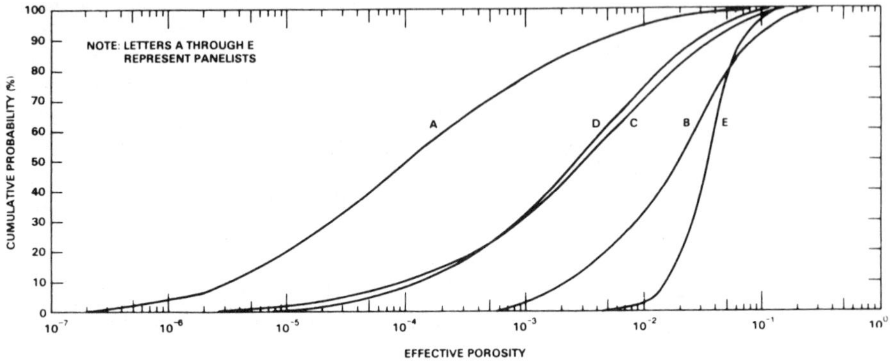

Figure 3. **Cumulative probability distributions for estimated values of Cohassett basalt flow top average effective porosity at megascale.**

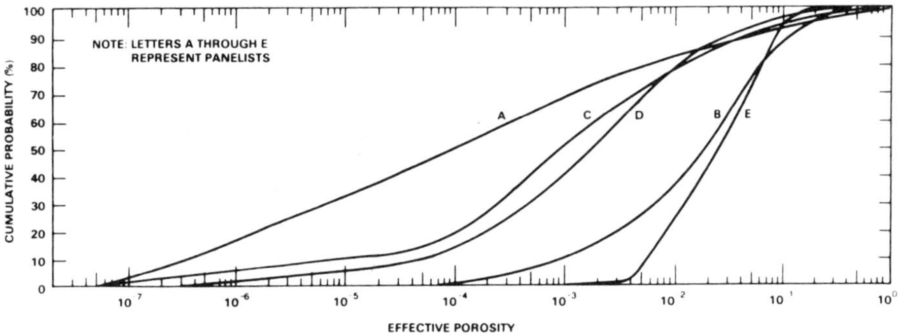

Figure 4. **Cumulative probability distributions for estimated values of Cohassett basalt flow top average effective porosity at macroscale.**

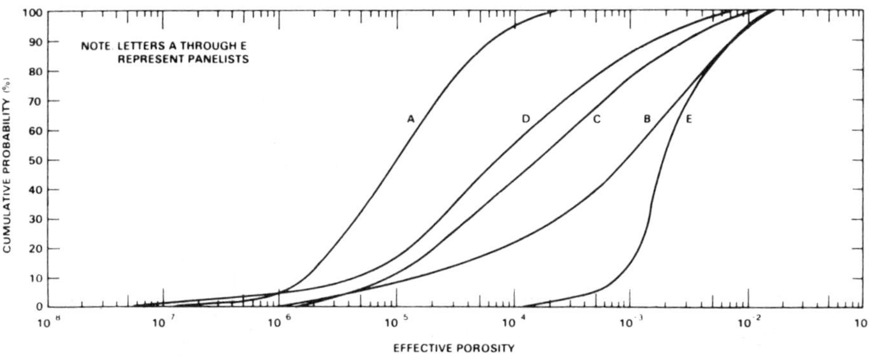

Figure 5. **Cumulative probability distributions for estimated values of Cohassett basalt flow interior average effective porosity at megascale.**

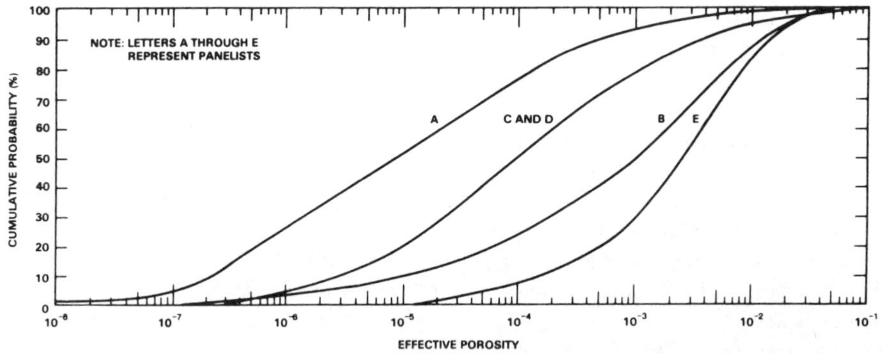

Figure 6. **Cumulative probability distributions for estimated values of Cohassett basalt flow interior for average effective porosity at macroscale.**

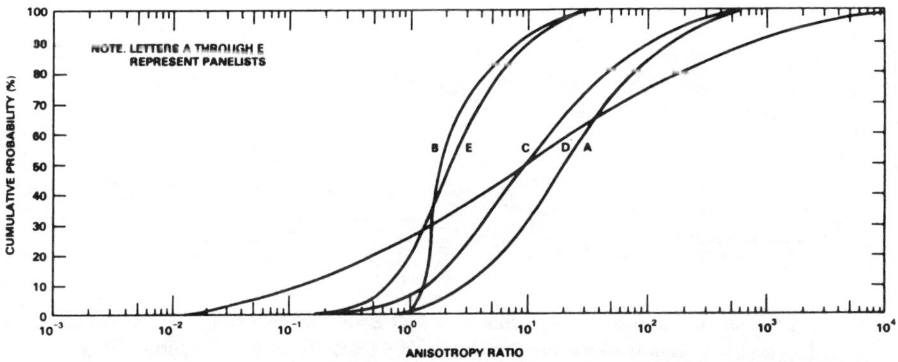

Figure 7. **Cumulative probability distributions for estimated values of Cohassett basalt flow interior anisotropy ratio at megascale.**

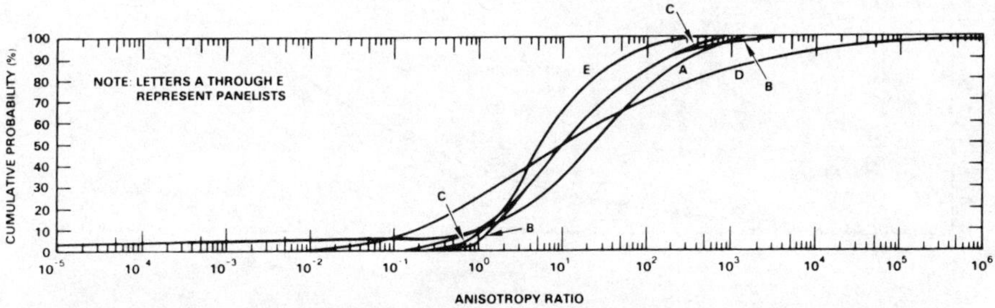

Figure 8. **Cumulative probability distributions for estimated values of Cohassett basalt flow interior anisotropy ratio at macroscale.**

tion for probabilistic analysis. Various methods of aggregation have been proposed (e.g., *27*). In situations where expert differences may be traced to conflicting underlying assumptions, face-to-face exchanges of information according to a structured format are generally recommended (*19*). Mechanical aggregations, such as averaging, provide a last recourse, but such methods have been criticized for requiring strong and difficult-to-support assumptions (e.g., *28*).

Figures 9 through 14 show the results of averaging the five experts' probability distributions together with averages of probability distributions similarly obtained from the three Rockwell experts. As with the independent experts, there was considerable disagreement among the Rockwell experts. The trimodal nature of the Rockwell averages in Figures 11 and 12 reflects the three distinct distributions of the three experts. The other curves appear smooth because of the larger number of distributions used in the averages or because the individual distributions were more in agreement.

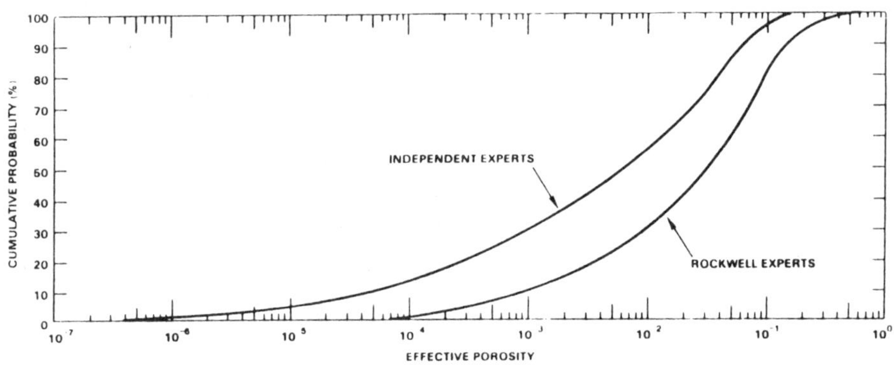

Figure 9. **Comparison of averages of cumulative probability distributions of two groups of expert hydrologists for estimated values of Cohassett basalt flow top average effective porosity at megascale.**

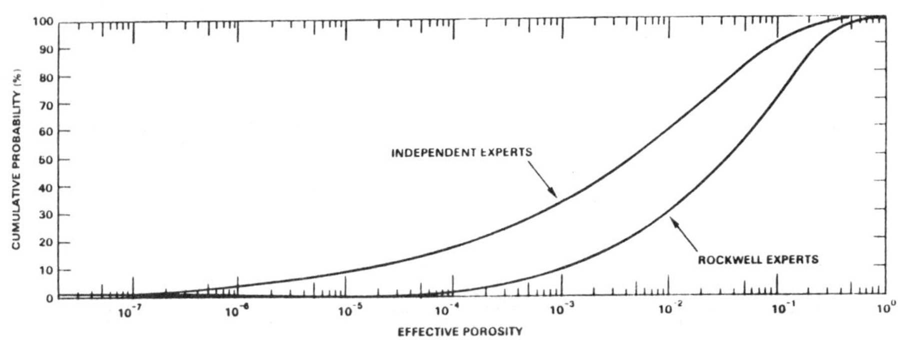

Figure 10. **Comparison of averages of cumulative probability distributions of two groups of expert hydrologists for estimated values of Cohassett basalt flow top average effective porosity at macroscale.**

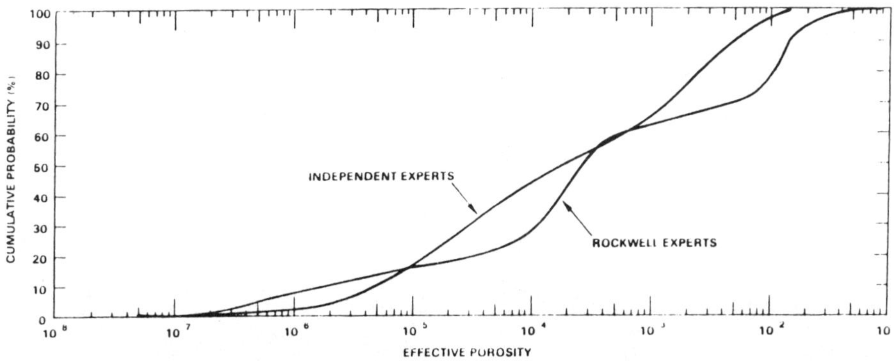

Figure 11. **Comparison of averages of cumulative probability distributions of two groups of expert hydrologists for estimated values of Cohassett basalt flow interior average effective porosity at megascale.**

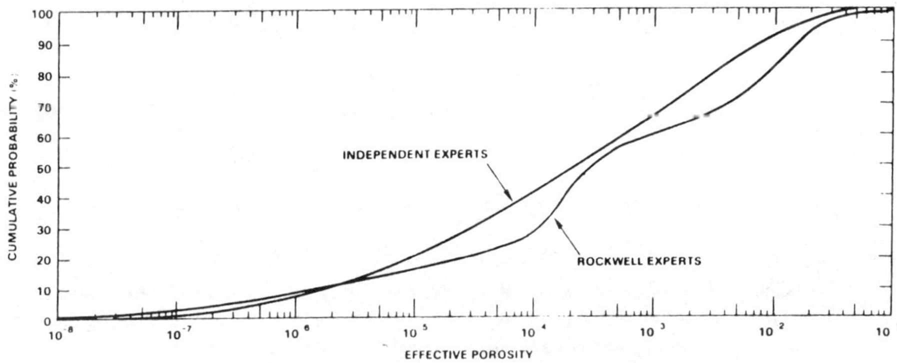

Figure 12. **Comparison of averages of cumulative probability distributions of two groups of expert hydrologists for estimated values of Cohassett basalt flow interior average effective porosity at macroscale.**

In the case of both the Rockwell and independent experts, the average curves were obtained by averaging the probabilities provided by the experts for each value of the uncertain variable. Averaging probabilities rather than values was regarded to be a more appropriate approach since it is less prone to allowing one expert's high estimate to dominate the estimates of other experts and because it preserves the range of values estimated by the various experts.

The figures show that, on average, the Rockwell hydrologists estimated the effective porosities of the Cohassett basalt flow to be generally higher and the anistropic ratios to be generally lower than did the independent experts. With the exception of the flow interior (Figures 11 and 12), the Rockwell experts felt somewhat more certain about the values of effective porosity (as indicated by the steeper curves) than did the independent experts. Conversely, the Rockwell experts felt slightly more uncertainty over the values of anisotropy ratio (Figures 13 and 14).

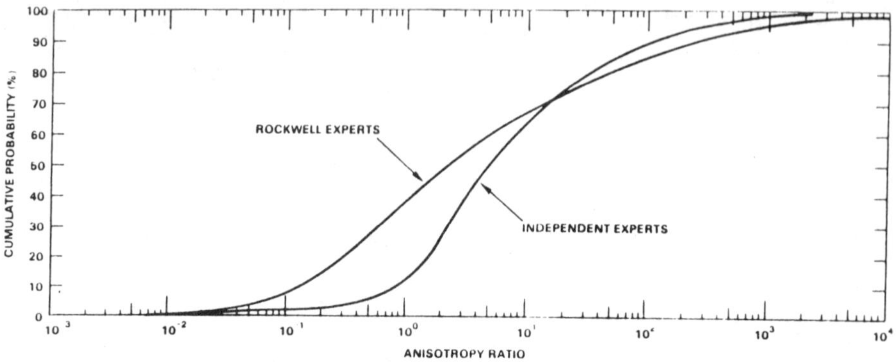

Figure 13. **Comparison of averages of cumulative probability distributions of two groups of expert hydrologists for estimated values of Cohassett basalt flow interior hydraulic conductivity anisotropy ratio at megascale.**

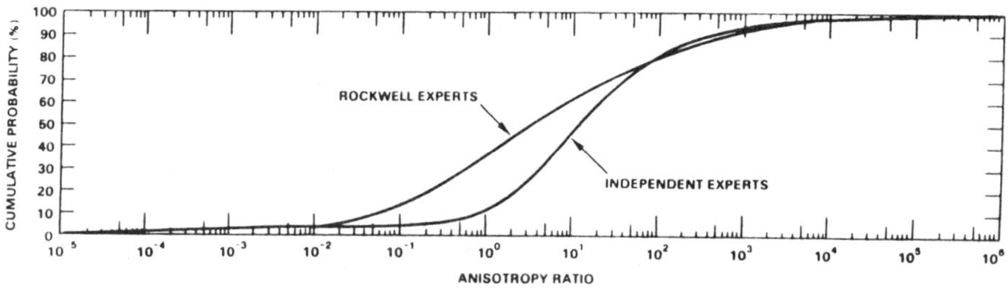

Figure 14. **Comparison of averages of cumulative probability distributions of two groups of expert hydrologists for estimated values of Cohassett basalt flow interior hydraulic conductivity anisotropy ratio at macroscale.**

Although averaging facilitates comparisons, it should be noted that this method of aggregating the experts' opinions is necessarily ad hoc. The average curves should not be allowed to obscure the substantial differences of opinion held by the experts.

Conclusions

Probability encoding of expert judgments regarding the six hydrologic parameters was successfully accomplished in the sense that the experts regarded their cumulative distributions as accurate representations of their professional judgments. Although it is unclear whether biases might nevertheless contaminate these distributions, a rigorous process was used in an attempt to minimize bias.

The results of the study clearly demonstrate a lack of scientific consensus. Uncertainty over key hydrologic parameters ranges over as many as seven orders of magnitude. Rockwell and independent experts are approximately an order of magnitude apart in their estimates. Providing each expert with the complete set of probability curves assessed from other experts produced virtually no movement toward consensus.

It is unclear whether more powerful consensus-building techniques involving face-to-face information exchange would be more effective at resolving expert differences, since such techniques were not applied in this study. However, the recent results from a pair of related studies are encouraging (*29*, *30*). These studies obtained probability distributions from multiple experts for 39 geologic and hydrologic parameters at the Hanford site, excluding effective porosity and anisotropy ratio. Experts who initially participated in individual probability-encoding sessions were found to have small to large differences of opinion (*29*). However, when most of the original experts were subsequently convened for a group probability-encoding process, consensus curves were obtained in all cases.

Whether or not probability encoding leads to consensus probability curves, it must be acknowledged that the methods can be criticized on several grounds. First, probability encoding is clearly more art than science. Second, qualified interviewers are scarce. Third, conducting probability encoding exercises is expensive and time-consuming. Even so, it is often faster and cheaper to ask a group of experts what they think than to sponsor the research needed to learn the true answers, and this in itself may be a problem. As Morgan et al. (*31*) point out, there is always the danger that the methods will be used as a substitute for scientific research. Finally, some people, including some experts, may feel that the methods may give individual scientists too strong a role in the policymaking process.

Nevertheless, probability encoding provides a practical means for providing a more complete and accurate understanding of the nature and magnitude of the uncertainties surrounding a complex problem. Clarifying such uncertainties can assist in setting priorities for information-gathering activities. Being explicit about uncertainties also helps to separate issues of science from policy and value issues. By expressing uncertainties quantitatively in the form of probability distributions, probability encoding permits the application of mathematical logic to analyze risks.

Literature Cited

1. Zamora, R. M., "ONR Probability Encoding Task," SRI Working Paper, SRI Project 4030, prepared for Office of Naval Research, Operational Decision Aids Project, Stanford Research Institute, Menlo Park, CA (April 25, 1975).
2. Spetzler, C. S. and C.-A.S. Staël von Holstein, *Management Science 22* (November), 340–358 (1975).
3. Morgan, M. G., M. Henrion, and S. C. Morris, "Expert Judgments for Policy Analysis," Report of an invitational workshop held at Brookhaven National Laboratory July 8–11, sponsored by the Dept. of Engineering and Public Policy of Carnegie-Mellon Univ. and the Biomedical and Environmental Assessment Div. of Brookhaven National Laboratory, BNL 51358 UC-13, NTIS, Springfield, VA (1979).
4. Lichtenstein, S. and B. Fischhoff, *Organizational Behavior and Human Performance 20*, 159–183 (1977).
5. Tversky, A. and D. Kahneman, *Science* (26 September), 1124–1131 (1974).
6. Slovic, P. and S. Lichtenstein, *Organizational Behavior and Human Performance 6*, 649–744 (1971).
7. Tversky, A. and D. Kahneman, *Science 211* (30 January), 453–458 (1981).
8. Tversky, A., *Cognitive Psychology 5*, 207–232 (1973).
9. Makridakis, S. and S. C. Wheelwright (eds.), *Forecasting*, North-Holland, New York (1979).

10. Bar-Hillel, M., *Acta Psychologica 44*, 211–233 (1980).
11. Tversky, A. and S. Sattagh, *Evidential Impact of Base Rates*, Technical Report 4, Stanford University, Dept. of Psychology, Stanford, CA. (May 15, 1981).
12. Kahneman, D., P. Slovic, and A. Tversky, *Judgment Under Uncertainty: Heuristics and Biases*, Cambridge University Press, New York (1982).
13. Kahneman, D. and A. Tversky, *Econometrica 47* (2), 263–292 (1979).
14. Lichtenstein, S., B. Fischhoff, and L. D. Phillips, "Calibration of probabilities: The state of the art to 1980," in *Judgment Under Uncertainty: Heuristics and Biases*, D. Kahneman, P. Slovic, and A. Tversky (Eds.), Cambridge University Press, New York (1982).
15. Pitz, G. F., "Subjective Probability Distributions for Imperfectly Known Quantities," in *Knowledge and Cognition*, L. W. Gregg (Ed.), Wiley, New York (1974).
16. Murphy, A. H. and R. L. Winkler, *Bull. American Meteorological Society 55*, 1449–1453 (1974).
17. Murphy, A. H. and R. L. Winkler, *Applied Statistics 26* (1), 41–47 (1977).
18. Staël von Holstein, C.-A. S. and J. E. Matheson, *A Manual for Encoding Probability Distributions*, SRI Project 770778, SRI International, Menlo Park, CA (1979).
19. Merkhofer, M. W., *IEEE Transactions on Systems, Man, and Cybernetics 17* (5) (1987).
20. Dalkey, N. and O. Helmer, *Management Science 9* (3), 458–467 (1963).
21. Runchal, A. K., M. W. Merkhofer, Elizabeth Olmsted, and J. D. Davis, *Probability Encoding of Hydrologic Parameters for Basalt; Elicitation of Expert Opinions from a Panel of Five Consulting Hydrologists*, RHO-BW-CR145P, prepared under Contract DE-AC06-77RL01030, Rockwell International, Rockwell Hanford Operations, Richland, WA (1984).
22. Runchal, A. K., M. W. Merkhofer, Elizabeth Olmsted, and J.D. Davis, *Probability Encoding of Hydrologic Parameters for Basalt; Elicitation of Expert Opinions from a Panel of Three Basalt Waste Isolation Project Staff Hydrologists*, RHO-BW-CR-146P, prepared under Contract DE-AC06-77RL01030, Rockwell International, Rockwell Hanford Operations, Richland, WA (1984).
23. U.S. Department of Energy, *Recommendation by the Secretary of Energy of Candidate Sites for Site Characterization for the First Radioactive Waste Repository*, DOE/S-0048, U.S. Department of Energy, Office of Civilian Radioactive Waste Management, Washington, DC (May 1986).
24. Davis, J. D., A. K. Runchal, and N. A. Bauman, *Delphi Analysis of Radionuclide Release Scenarios for a Nuclear Waste Repository at the Hanford Site, Washington State: An Initial Iteration*, RHO-BW-ST-42P, Rockwell Hanford Operations, Richland, WA (1983).
25. Koriat, A., S. Lichtenstein, and B. Fischhoff, *Journal of Experimental Psychology: Human Learning, Memory 6*, 107–118 (1980).
26. Raiffa, H., *Decision-Analysis: Introductory Lectures on Choices Under Uncertainty*, Addison-Wesley, Reading, MA (1970).
27. Ashton, H. A. and R. H. Ashton, *Management Science 31* (December), 1499–1508 (1985).
28. Stewart, T. R. and M. H. Glantz, *Climatic Change 7*, 159–183 (1985).
29. Golder Associates, Inc., *Probability Encoding of Variables Affecting Inflow into the Exploratory Shaft Test Facility at the Basalt Waste Isolation Project, Richland, Washington*, Technical Report No. 863-1032, report to Rockwell Hanford Operations, Redmond (Seattle), WA (1986).
30. Golder Associates, Inc., in conjunction with Applied Decision Analysis, Inc., *Groundwater and Methane Inflow Study for the BWIP ESF at Hanford, Washington*, SD-BWI-DER-001, Rev. O, Final Report to Basalt Waste Isolation Project, Westinghouse Hanford Company of Richland, Washington, No. 873-1400.009, Redmond (Seattle), WA (1987).
31. Morgan, M. G., S. C. Morris, Max Henrion, D.A.L. Amaral, and W. R. Rish, *Risk Analysis 4* (3), 201–216 (1984).

General Discussion

At the conclusion of the six technical sessions, an informal discussion session was conducted. The discussion session was moderated by Bruce Buxton of Battelle, and questions were generally directed toward a panel made up of several conference speakers and technical organizers:

Albin Brandstetter (Battelle-OWTD)
Tin Chan (Atomic Energy of Canada Limited)
Jean-Paul Chilès (Bureau de Recherches Géologiques et Minières)
Isobel Clark (Geostokos Limited)
Peter Clifton (Rockwell Hanford Operations-BWIP)
Bill Harper (Battelle-ONWI)
André Journel (Stanford University)
Jane Long (Lawrence Berkeley Laboratory)
Ivars Neretnieks (Royal Institute of Technology)
Brian Worley (Oak Ridge National Laboratory)

Bruce Buxton (Battelle): Where should we go from here? Does anyone have an idea?

Bill Harper (Battelle-ONWI): I was talking to Dae Young last night and he brought up an ongoing problem, and I think it has at various times surfaced here in the conference. The people that are gathering the data in many cases are not working that closely with the modelers. There is a real question on what data we really need to feed these big complex models and how is it really collected out in the field? I think Dick Codell was talking about some of the permeability or transmissivity measurements and alluding to that particular problem. I think one of the things we need to do, one of many, is to take a serious look at how that data is being gathered out in the field, what sort of errors can happen there, and what do we really need for these complex models. What's the relationship between the field data and the method that it's gathered and what's assumed in terms of volumes, or what have you, in our large, finite-element and finite-difference codes, whatever the particular performance assessment model is?

Peter Hufschmied (NAGRA): I would like to address a question to Professor Journel that might also, to some extent, be a comment on what kind of field data we might be able to collect. Being partly responsible for the design of programs to collect hydrogeologic data in the field, what I would like to say is that we might be in a position to provide you with these 50 hard numbers. But I see some problems in providing you also with the 250 spatially located soft data. However, rather than being able to give you this spatially located information, it might be possible to give you some information about the spatial extent of these highs, say high-conductivity features, for example. Also we might have some information about the orientation in which these features extend. From this I would like to derive my question to you. Would that also help you to come up with models that are conditioned on this information?

André Journel (Stanford): Absolutely; soft information can be soft in the value, but can be soft also in the location. In other words, that spatial information need not be at an exact location as in this exercise. The soft information may be located at random within a certain volume. As for the main direction, you're talking about soft information that is used for a synthetic variogram. For example, you don't have any data on the highs, but from some geological interpretation you know that if there are highs they would be oriented north-south rather than east-west. This is very critical information that could be put to use. So my answer is absolutely yes!

Bruce Buxton (Battelle): I'd like to ask a followup that occurred to me with Lee Merkhofer's talk about eliciting PDFs from experts. It occurred to me that if we have a hard time getting a probability distribution from an expert, can we hope to get spatially located data, for example the permeability values for finite-element grids, from experts, or are we asking too much in that case? That's sort of what you are talking about here, spatially located soft information.

André Journel (Stanford): I think it's very difficult to answer; it depends on each case. Our role is to provide the tools and it is your role to provide information, whether hard or soft. I imagine there are some cases where you could provide soft spatially located information. There are other cases where you can't. But then you may have inexpensive information that will provide you with that soft information. And the reason why is that inexpensive information is usually seismic; and the seismic data, believe me, are pretty soft. But they are inexpensive.

Albin Brandstetter (Battelle): I'd like to add to that. Really when we are talking about ground-water flow and radionuclide transport, there are a number of variables or characteristics that we have to know. So we need a number of experts and a number of data or different data. The soft information, for instance points like the geologic structure, have one set of information you need, while for the conductivities you may have other people. We need to get these groups together to get the entire system characterized before we actually start our analysis. And I think it's really happening. It hadn't for a long time, but because of the regulatory pressure on showing total system performance, people are starting to talk to each other and they are realizing that they cannot by themselves solve the entire problem.

Bill Harper (Battelle-ONWI): Do we really need all that additional information? This ties into sensitivity analysis. Take for example the case of Barfield flow. If your performance measure, say a global one such as travel time, is relatively insensitive to the particular assumed parameter values of, say permeability or head at certain spatial locations, don't worry about quizzing the experts on that. There are a variety of techniques to address the sensitivity of the spatial locations; some are the deterministic approaches, adjoint or direct, that Brian Worley was talking about earlier. So that might be treated as a screening process to say, "Do we really need it?" If it doesn't impact your prediction, who cares? That is, of course, assuming you believe the underlying conceptual model and believe that Darcy flow is appropriate. But with all those caveats, that might be one way to make the first cut.

Ram Kulkarni (Woodward-Clyde): It seems clear to me, listening to some of the papers for the last two or three days, that we'll continue to face the problem of lack of data—adequate data, field data. One of the solutions that has been proposed, particularly in the afternoon sessions, is to use soft information, to use expert opinions. But then also some of the papers pointed out the pitfalls associated with that: biases, conflict of interest, the variability among experts. It seems to me we need to give some serious thought to how can we find an objective professional and an unemotional way of calibrating experts. Let me share some of my thoughts on that, and we'll see if anyone else has comments on it. One way might be some kind of a national test program where we collect data, field data, on some of the critical parameters, and use this data set to calibrate experts whose opinions would be included in the modeling. Nonetheless, we need to get some serious thought to that problem because the variability seems to be so large that one can easily ban the results. If you use this expert with some other group of experts, you can drive the results in any direction you want to. There's got to be some way of calibrating these experts.

André Journel (Stanford): I would like to make a comment on Ram Kulkarni's comment. He said it's very difficult to calibrate human expert opinions because they are highly nonstationary. You can calibrate them on one data set; that doesn't mean that the calibration will be the same in another data set. This doesn't mean that we shouldn't try, at least on several data sets. The other point I would like to make is that it has always struck me that people are nervous when using expert opinion. But they are not nervous when making hidden, very consequential judgmental decisions like going systematically with a Gaussian distribution. That is perfectly accepted; and it may have much more consequence on the project than a well-documented expert opinion. So it seems that when you are honest and you document your expert decision, you are shot at. If you are dishonest and you just hide it under a mountain of equations, most of the time Gaussian-diseased, then you are safe. So, what's wrong with expert opinion? We're doing that every single time. Any time we pick one algorithm rather than another we are making an expert opinion; but most of the time we just don't say it.

Bruce Buxton (Battelle): I think that's a good point. All the modeling assumptions that we make are judgmental decisions at some point, and those should be subject to a stringent peer review just like expert opinions are.

Lee Merkhofer (Applied Decision Analysis): Just one more comment on the recalibration which has been proposed several times as a way of getting around the problem with biases. In addition to the fact that calibration curves tend to depend on the type, the subject matter, I think that it's also important to recognize that having the full support of an expert in the sense of saying "Yes, you use the distribution as an input for your analysis that I truly believe" is very important. If we get into the habit of recalibrating or altering the judgments we get from experts, I think we risk the danger of a mutual circle spiraling down to distrust and recalibration and altering. Because of that I think an approach which strives to get the best distributions we can, the least biased distributions we can from an expert, is a better one in the long run than worrying about recalibration.

Jim Channell (New Mexico Environmental Evaluation Group): We're very concerned of course with the WIPP Project and with the performance assessment on that project for the EPA standard. Now Sandia's timetable is to have that completed in the fall of 1992, five years from now. I'm concerned from a lot of what I've heard this week. Are we converging on the methodology that would be acceptable as a group and the field data that would be acceptable?

Albin Brandstetter (Battelle-OWTD): This conference was organized to concentrate on uncertainties. I think there was an overemphasis on uncertainties. Maybe we need another conference on certainty analysis. I'm quite encouraged. I've been to similar conferences in the past, and my impression became that the relative importance of uncertainty, of different types of uncertainty, like model data and so on, depends on the state of the art for evaluating that type of uncertainty, or on the expertise of an individual who is talking about a specific type of uncertainty. At this conference I think we had a wide spectrum of coverage of different types of uncertainties which maybe gave the impression that we don't know much. But I don't think that's the case. I think we actually know more and we are in much better shape than I thought in evaluating what these uncertainties are. As a result I think we should be better able to provide that reasonable assurance that the regulators expect from us. I think it will still be difficult to determine or guess what that reasonable assurance is and what's involved in convincing the public and the regulators that we do have reasonable assurance. So it'll be a tortuous path to get to that point, but I think we are well on the way. Of course the uncertainties vary greatly depending on the type of hydrologic or geologic settings we are evaluating. It's better for some than it is for others, but I think we're well on the way.

Bill Harper (Battelle-ONWI): I would like to jump back to an earlier question on how do we show reasonable assurance, how do we satisfy the public and the NRC. Several people at previous conferences have referred to reasonable assurance as providing that warm, fuzzy feeling. I think one way of doing that is to look at different models, such as a very simple one that predicts ground-water travel time, look at conditional simulation, look at the detailed models such as a finite-element, finite-difference code and see if you end up in the same ballpark. If you are within the same order of magnitude on travel times, you are probably OK, and that should provide part of that warm, fuzzy feeling. Bob Andrews talked a little about some of the work that INTERA has done. They've looked at the relatively simple P-TRACK code, and they've done other things with more detailed codes such as SWENT, and the results compared very well. That is one way of helping to address that warm fuzzy feeling. It's not the be-all and end-all, but it's a start.

Thomas Nicholson (U.S. NRC): The point I was going to try to make is that I don't think just going back and trying other codes is good enough. With regard to the study I described yesterday, we thought about this long and hard and the first thing we had to ask ourselves is, "Do we understand the geologic system?" Even though we put in a trench, the trench was actually the most direct method of understanding the flow system by visually seeing it. Then you go with the tensiometers, the neutron probes, different arrangements. The idea was to have multiple means of identifying what's going on in the flow system—both the geometries involved and the hydraulics involved. Then we had a tremendous wealth of data. I said before we had 607 separate data points, representing 9 flights at a 6-meter interval. We also took cores, over 450 cores, and are in the process of analyzing those in the laboratory. Then we did an experiment, and we're not satisfied with that experiment, obviously, because we are going to do another one in the same vicinity. So the point I'm trying to make is that I think the way you go about getting reasonable assurance is you don't stop with just one site investigation; you have multiple means. And, as Dr. Neretnieks has said, the whole concept here is radionuclide transport. So we're not done by any stretch of the imagination; we haven't gotten to that point yet. You have to do numerous testing in the field, understanding the objective of what you're doing, and the most important part of the study I was talking about yesterday is that we force people ahead of time to actually model the system based upon very preliminary models. That's how we did a design of the experiment. What are the important processes? How do you question yourself, go back into the field, collect more data rather than just doing more modeling?

Bill Harper (Battelle-ONWI): I totally agree with that, and I might add that other people have mentioned looking at analog studies also to supplement that.

Harald Ahagen (SKB): I think we have in the past three days focused on one barrier out of four. The ultimate licensing of a repository will be based on the performance of the total system. Each portion, or each component of the system, will be based on the knowledge of that system at that date, and more emphasis will be put on another barrier to compensate for the lack of knowledge. I think we need to keep the broad picture in mind here. We do not really have to solve all the data needs and all the little question marks that we might have here.

Mark Larson (Lehman and Associates): I was wondering if any of the experts think it's appropriate to use matrix diffusion arguments in calculating ground-water travel time. If you plan on using them, what do you suggest as field tests?

Ivars Neretnieks (Royal Institute of Technology): This meeting was on water flow and transport modeling. Well, where are the transport models? We have heard precious little about that. But back to the matrix diffusion. What should we do with water travel time? And how do we measure porosity? I'd like to point out that in practically all of the models which I use, we are interested in radionuclide travel time. Radionuclide travel time simplistically is determined from water velocity and a retardation factor. The water velocity we will obtain indirectly from the hydrologists who will tell us what the Darcy velocity is, then some estimate of porosity with two independent estimates which we need. Then we'll have the retardation factor. We take the water velocity and we divide that by the retardation factor and we get the nuclide velocity. But the nuclide velocity also contains information called porosity, flow porosity, either if we model it as volume sorption, that is we assume that all the matrix will be labeled for sorption, or if we say only surface sorption will be active. Now we need to know how much surface the water volume will have. Or we may say, "How much surface do we have per rock volume?" and divide that by the porosity which as yet is unknown.

Now for matrix diffusion, the retardation factor varies with time. So it's a nonsensical parameter. But now combine these equations with that equation and we see after looking at the fact that for dissolving nuclides, and most of them are, then the porosity will be very, very small in this entity compared to the sorption parameter and we can neglect that, both for volume sorption and for surface sorption. And, I'm mostly used to fractured rock, where the porosity is very small so that we can effectively neglect it. What we end up with is for the volume sorption case, we need the Darcy velocity and sorption coefficient; we need no porosity for the nuclide velocity. The same happens if we have surface sorption. We need no porosity and we need no velocity of the water; we need only the Darcy velocity, that's the flow per square meter which the hydrogeologists give us with their models. And we need the sorption coefficient on how much surface we have per rock volume. If we go to the matrix diffusion case, we have partial utilization of the rock. We don't utilize it all to sorb the nuclides, but we utilize as much as is accessed by the diffusion from the surfaces. So again, we need information on how much surface we have per rock volume from which the nuclides can diffuse into the matrix and sorb there. I'm asking myself,"Who is looking for this entity?" And I find precious few in the world who are doing it. Yes, I feel there are some entities which we don't have a good handle on. I feel that if in the matrix diffusion case we have sorption or long contact times, we really are not primarily interested at all in the water travel time, because the capacity for the nuclide retardation is not in the water which is flowing but in the stagnant water and sorption of the surfaces.

Berge Gureghian (Battelle-OWTD): As far as the diffusion process is concerned, you wrote up justifiably the retardation factor. The retardation factor does play a very important role as far as the diffusion process is concerned. Now, if you want to be on the conservative side, you shouldn't assume any kind of matrix diffusion process. So the other problem is how to characterize the diffusion process in a flow region where fractures prevail. Yesterday I brought out the point about the case of the single fracture which assumes infinite diffusion. In that case, of course, we are not on the conservative side. And when I brought up the point of why should we be concerned so much about a dispersion process in a fracture system, I was trying to suggest that if we want to be on the conservative side again, we shouldn't really bother too much about this dispersion process. So, the two things go hand in hand, and obviously since we always come about with very conservative estimates, I think in some cases perhaps the diffusion process should be neglected, although it's very important.

U Sun Park (Science Applications International): Many talks were centered around how to minimize the uncertainties given the scarce data. There is data scatter and sometimes lack of data. My question is, to begin with, how do we get those data? Is there any better way of getting data to minimize the uncertainties? The NNWSI project just finished their SCP draft, and site characterization hasn't started yet. We have drilling programs that have some drillholes planned. But is there any way, any statistically justified method, of reducing the uncertainties by preselecting a certain pattern of drilling holes?

Bill Harper (Battelle-ONWI): I think we've seen several different ways. Al Freeze talked about a Bayesian approach. Within the salt program, one approach that we've looked at for large codes is to combine some of the deterministic sensitivity methods such as adjoint or direct methods, along with some of the geostatistical methods, to help determine spatially where the important data are. What we are doing is melding together the sensitivities of some key performance measures such as ground-water travel time, along with the uncertainties that go into that calculation. There are different ways to do it. That particular one I mentioned is using a complex model.

Jane Long (Lawrence Berkeley Laboratory): I believe someone asked,"How can we get better data?" I think that we need a lot more work on conceptualizing the system that we're testing. If you go out and make a test you have to do a small, inverse problem to back out the data that you're then going to use in your model. By using incorrect conceptual models in doing that, you're getting parameters which don't make any sense. I think that we need to spend more time looking at how porosity is connected in the rock in order to have models that make any sense. So if you're going to ask about how we can get better data, I think we need better conceptual models, more work on interpreting well tests, for instance, in fractured rock rather than a porous medium.

Rainer Senger (Bureau of Economic Geology): I wanted to follow up on conceptual modeling. Most of the conceptualization came from numerical flow models. In the Palo Duro Basin where I'm working, we still have problems combining hydrologic modeling with geochemical interpretation of the flow system. I was curious if that can be somehow addressed in the uncertainty terms.

Peter Clifton (Rockwell Hanford Operations-BWIP): I'd just like to raise a question or comment following on from what Jane was saying with regard to the conceptual models. With the various conceptual models, porous media, fracture flow, or as with some of the points that Dr. Journel was making today with regard to degree of connectivity on a larger scale, are we not looking at a similar phenomenon manifest on a different scale? In a sense in all these conceptual models we have a thread of commonality. Recognizing that, how can we then design tests in order to sufficiently interrogate the flow regime so that we can use the information from those tests in the predictive models? I think, if you look at the tests in hydrology that we commonly use to interrogate a formation, many of these tests have been designed for the purpose of determining ground-water yields for the design of pumping wells, and for the exploitation of ground water. Are these tests the right sort of tests that we should use in order to determine data that we can subsequently make parameter estimates for use in performance prediction models? Or should we be looking at designing new types of tests? I think the same comment could be made with regard to tracer tests where we use those tests in order to determine transport properties used in those predictive models.

K. Karasaki (Lawrence Berkeley Laboratory): I'd like to make a followup comment on well test methods in collecting flow parameters in highly heterogeneous rocks. I've seen many field people going out there doing pulse tests and slug tests or swabbing, and I think through my experience that pulse testing is not a very reliable test to get flow parameters or fracture characteristics. I think pulse tests and slug tests have too small a region of influence. Also they get affected by the well bore conditions and all sorts of problems associated with it.

Peter Flavelle (Atomic Energy Control Board): I just want to follow along and ask a question on some of the conceptual models and expert opinion. How can we evaluate expert opinion on what conceptual models, research strategies, and test designs to use as opposed to expert opinion on just developing data distributions? That's going to be a much more subjective question to ask an expert.

Albin Brandstetter (Battelle-OWTD): If I understand your question right it is, "How do we treat uncertainties in conceptual models?" Is that it?

Peter Flavelle (Atomic Energy Control Board): No, how do we evaluate a suite of expert opinions on what conceptual models to adopt? Do we take a porous flow, a dual porosity approach, a fracture flow approach? Which one is going to give the correct answer? A lot of it depends on how you define the problem and subsequently, what conceptual model you will adopt to interpret a site.

Albin Brandstetter (Battelle-OWTD): The NRC, in fact, has suggested that if there are questions on what the correct conceptual model is for a particular site, then we do indeed analyze different conceptual models, and not just make a guess what the best is, so that we do get a range of analyses on a range of conceptual models. Now, of course it's still conjecture which is the right one, but you may come up then with a probability distribution in your results that not only includes the uncertainties in the data but also the uncertainties in the conceptual models.

Peter Flavelle (Atomic Energy Control Board): But that should require evaluation of different types of parameters for each conceptual model. Does that mean that you go out with a shotgun approach and measure everything you can, so that it all can be interpreted in different methods?

Albin Brandstetter (Battelle-OWTD): No, I think you still have to use professional judgment to determine what is a reasonable conceptual model and what is important for each of these conceptual models.

Peter Flavelle (Atomic Energy Control Board): That's essentially what I'm asking you. If you call a series of experts on which ones to use for a site, how do you evaluate the different opinions? Because it strikes me it would be more difficult than what's indicated in the last couple of talks we've heard, where you actually can ask what those specific numbers are.

Jane Long (Lawrence Berkeley Laboratory): I think there is a lot to be learned from the process by which geologists go about developing their science. A physicist can go into the lab and design an experiment that can prove or disprove a point. But a geologist is forced to collect evidence, and make a hypothesis and defend it. I think that we're in the same position. You have to go in, look at the evidence that you have, make a hypothesis, and then go back and test that hypothesis, and collect more evidence. I think, for instance, Ivars Neretnieks is in that position in Stripa, where he has hypothesized that channeling within the fractures is causing bizarre, or somewhat bizarre, results that he observes. It now is incumbent on him to go back and say, "I need to design an experiment, a further experiment, that will show that my hypothesis of channeling has more validity." I think it is an iterative process. You're not going to be able to choose that. And it points out the second great need, I was sort of amused to hear someone get up before and say that they want to have some test sites so that they can calibrate the experts. How about some test sites so we can get some data for the models?

Mark Larson (Lehman Associates): I want to make a comment on just what Jane was saying. We are discovering all this uncertainty, and a lot of it has to do with the fact that hydrology is based on porous flow, and we decide to put our nuclear waste in a fractured environment or salt. I'm not making a comment on that one way or the other, but I don't think there is any doubt that, between now and when we come up with some acceptable answers, the best way to get data is going to be using graduate students to figure out ways with masking tape and plastic to shove things in fractures to monitor. That's how it's going to happen. I'm fairly convinced that the efforts put forth in those directions are going to be a lot better than doing a lot more math. I think the math is excellent so far, and they know what to do once they get the data, but the data is not there. I think that it's going to go back to the gadget people. That's just an opinion.

Jean Paul Chilès (BRGM): I want to comment on the model. Models are generally relative to a discipline and also to a scale. What seems to be important is to assess the models. In particular, global models should be understood and related to models which are of smaller scale. It seems to me that we have now a certain puzzle of models. In coming years it will be important to go to a bigger imbrication of the various models in order to assess the problem model.

Dwayne Chesnut (Science Applications International): I'd like to return to the data question for just a moment. Something that's been nagging at me since I got involved in this business is the lack of something we can actually measure or observe, if and when we get a repository in operation, that tells us anything about what's going to happen in the kind of time span that we're having to worry about. Let's take 1000 years, sort of the minimum end of what we're looking at. We're going to have an opportunity for a period of 5% of that, 50 years before closure, to be putting waste in a repository. We have an opportunity to measure some things at that point. Now the question is, "What can we measure during that period of time that's going to help us distinguish among all the different kinds of models we've got and all the different predictions that we are making about a time period 20 times that long?" Can we focus a little bit on what kind of monitoring system, what sort of thing should we be looking at, and how much can that help us reduce the uncertainty, while we still have some opportunity to correct the situation if it's wrong?

Berge Gureghian (Battelle-OWTD): The deterministic models should complement each other. This consensus was reached by nonstatisticians or geostatisticians. From what we have seen over the last few days, there was tremendous emphasis, particularly with the paper given by Allan Freeze, on the importance of stochastic models and yet living on the side of deterministic models. I suppose that they must play an important role, if for nothing else, minimizing the cost of the investigations, for example, if we take the case of the Monte Carlo approach. I'd like to have your opinion as to the role that deterministic models should play in such investigations.

Harald Åhagen (SKB): I would like to make a comment on the previous question. I don't think we can solve your problem with one single model or one single test or one set of data. I think we need to build a set of evidence that starts in the laboratory on small scale, goes through data gathering, in situ testing, and natural analogs to get the whole time span on all the scales. I think that's the sequence we have to go to build the base for our final conclusions.

Brian Worley (Oak Ridge): I'd like to comment on the last three speakers, especially the question of the gentlemen in the front [Dwayne Chesnut]. It seems from my viewpoint, if we're trying to assess what is going to happen on a time scale of 1000 years, then at some point we have to rely on a deterministic approach, or some model for that period of time. And if we are interested in the next 50 years in deciding which experiments we should perform, then we should keep in mind that our ultimate objective is to decide what's going to happen in a thousand years. If we are going to use a deterministic model to do that, and if we are going to rely on either Monte Carlo or deterministic methods to determine what the cumulative distribution functions are for those performance measures from that model, then we should look at the model distributions, decide where those tails of the uncertainties are coming from, and address those questions in the data that we want to gather in the next 50 years.

Bill Harper (Battelle-ONWI): The approach that one might take is to do a very comprehensive sensitivity analysis that at least does a first-order screening on all the parameters in a given model, and then reduce the scope of the problem to one that you could do a more in-depth uncertainty analysis on. You would then worry about what are the PDFs, the correlations, etc. on that reduced model, rather than starting out with a full-blown uncertainty program.

Albin Brandstetter (Battelle-OWTD): I'd like to point out that in spite of all the computations and the fancy terms we use in mathematics and statistics, a lot of it will depend on just how good we are in convincing, on the basis of just qualitative information we have on the past stability of a site, the past geologic evolution of a site, whether that site is suitable and safe. So I think we shouldn't forget that our arguments will be not be just based on mathematics.

Tin Chan (AECL): Since this is a part of the U.S. DOE and AECL cooperative agreement, I would comment on some parts of our agreement which actually address some of the comments we have heard in the last 30 minutes or so. One thing is the interaction between experimenters and modelers. Under the cooperative agreement we are designing a suite of field experiments and lab experiments. We always want to make sure to involve the modelers in the planning of experiments right from the start. Also, we don't let the modelers get away without telling us what data they need in order to crank out their model. What input data do they need and is there any chance of getting that type of data? The second point is the synthesis of data. Part of the cooperative agreement is also to develop site characterization technology or methodology. We emphasize the synthesis of all types of available data including hydrology data, geophysical data, survey mapping type of data, and also expert judgment. I'm very, very glad that both Al Freeze and André Journel could talk about some methods which can put expert judgment and also soft data in a kind of rational basis. The third comment is about getting the parameters that are necessary for transport models. We have lab experiments on cores containing natural fractures, cores 200 millimeters in diameter and about a meter or so long, where we actually inject epoxy to make casts of the fracture and then open it up and look at the epoxy to see how the pores are connected or the channel is connected. We have a large block experiment with a 1-meter cube containing a natural fracture. We take it back to a lab, do flow testing, do tracer testing with radionuclides. Then we open it up and do autoradiography, which is similar to taking an X-ray picture, and try to map out the channels. In addition, we open up the fracture after the experiment, core into the fracture, and slice up the core to see how far the radionuclide has penetrated.

Bruce Buxton (Battelle): I would just like to thank all of you, on behalf of the organizing committee, DOE, and AECL, for all the excellent participation we had, especially from those of you who came from out of the country. I also want to make one note on behalf of the organizing committee again; we had a lot of excellent abstracts, and we had a hard time deciding among them. We tried to go for breadth, and I hope you all enjoyed the meeting as much as we did.

Rich Schassburger (U.S. DOE): I too would like to thank all of the speakers that have spoken here this week as well as the attendees for participating in the conference and making it a successful one. I hope that this conference has opened some new lines of communication between researchers in different countries and disciplines. We have heard about a broad range of statistical problems and statistical tools in different areas concerning performance assessment and uncertainty analysis, such as policy and decision making, methods for complex computer codes, stochastic hydrology, fracture flow, geostatistics, and the use of subjective information. We have also heard, from Ralph Stein, on the U.S. Department of Energy's commitment to performance assessment as an important area in the licensing arena for any future repository in the U.S. I hope that you are leaving this conference feeling as I do, more informed and anxious to pursue new ideas and new directions. Again, thank you for coming, and please have a safe trip home.

List of Attendees

Pradeep Aggarwal
Battelle—OWTD
7000 S. Adams Street
Willowbrook, IL 60521

Harald Åhagen
SKB—Sweden
c/o DOE/RTP
9800 S. Cass Avenue
Building 201
Argonne, IL 60439

Kaj Ahlbom
Swedish Geological Co.
P.O. Box 1424
S-75144 Uppsala, Sweden

Robert E. Allan
Ontario Hydro
700 University Avenue
H18D16 Toronto, Ontario
M5G 1X6 Canada

Robert W. Andrews
INTERA Technologies, Inc.
Suite 300
6850 Austin Center Blvd.
Austin, TX 78731

Stefan Bachu
Alberta Research Council
P.O. Box 8330, Postal Station F
Edmonton, AB, TGH5X2, Canada

David Back
Roy F. Weston, Inc.
955 L'Enfant Plaza, SW
8th Floor
Washington, D.C. 20024

Richard C. Baker
U.S. Department of Energy
Chicago Operations Office
9800 S. Cass Avenue
Argonne, IL 60439

C. B. Bennett
E. I. du Pont de Nemours & Co., Inc.
Savannah River Plant
Aiken, SC 29808-0001

Daniel Billaux
Lawrence Berkeley Laboratory
1 Cyclotron Road
Berkeley, CA 94720

E. J. Bonano
Sandia National Laboratories
Division 6416
Albuquerque, NM 87185

Kenneth R. Bradbury
Wisconsin Geological &
Natural History Survey
University of Wisconsin
3817 Mineral Point Road
Madison, WI 53705

Albin Brandstetter
Battelle—OWTD
7000 S. Adams Street
Willowbrook, IL 60521

Kenneth Brog
Battelle—OWTD
7000 S. Adams Street
Willowbrook, IL 60521

Bruce E. Buxton
Battelle Columbus Division
505 King Avenue
Columbus, OH 43201-2693

Katherine Campbell
Los Alamos National Laboratory
MS F600
Los Alamos, NM 87545

Tin Chan
AECL—Waste Management Division
Whiteshell Nuclear Research Establishment
Pinawa, Manitoba ROE 1L0, Canada

James K. Channell
New Mexico Environmental Evaluation Group
P.O. Box 968
Santa Fe, NM 87503

Dwayne Chesnut
Science Applications International Corporation
101 Convention Center Drive, Suite 407
Las Vegas, NV 89109

Jean-Paul Chilès
Bureau de Recherches Géologiques et Minières
Orleans, France

Isobel Clark
Geostokos Limited
36 Baker Street
London W1M 1DG, England

Tom Clemo
EG&G Idaho
2151 North Blvd.
Mail Stop 2212
Idaho Falls, ID 83401

Peter M. Clifton
Basalt Waste Isolation Project
Westinghouse Hanford Co.
P.O. Box 1970
Richland, WA 99352

Kevin Coakley
Lawrence Berkeley Laboratory
Earth Sciences Division
Building 50-E
1 Cyclotron Road
Berkeley, CA 94720

Richard Codell
U.S. Nuclear Regulatory Commission
Mail Stop 623-SS
Washington, D.C. 20555

J. O. Corbett
Central Electricity Generating Board
Berkeley, Gloucestershire GL13 9PB
United Kingdom

Jim Danna
Science Applications International Corporation
101 Convention Center Dr., Suite 407
Las Vegas, NV 89109

Paul A. Davis
Sandia National Laboratories
Division 6416, P.O. Box 5800
Albuquerque, NM 87185

Alexander Desbarats
Geological Survey of Canada
601 Booth Street
Ottawa, Ontario K1A 0E8, Canada

A. M. Djerrari
EWA, Inc.
133 First Avenue, North
Minneapolis, MN 55401

James M. Doesburg
Battelle—OWTD
7000 S. Adams Street
Willowbrook, IL 60521

Kenneth W. Dormuth
AECL—Whiteshell Nuclear Research Establishment
Pinawa, Manitoba, ROE 1L0, Canada

Christine Eddy
Argonne National Laboratory
9800 S. Cass Avenue
Argonne, IL 60439

Norman Eisenberg
U.S. Department of Energy
Office of Civilian Radioactive
Waste Management
Mail Stop RW-23-2
1000 Independence Avenue, SW
Washington, D.C. 20585

Aly I. El-Kadi
Holcomb Research Institute
Butler University
4600 Sunset Avenue
Indianapolis, IN 46208

Paul Eslinger
Battelle Pacific Northwest Laboratories
3160 George Washington Way
Richland, WA 99352

Elaine Ferguson
Battelle—OWTD
7000 S. Adams Street
Willowbrook, IL 60521

Irene T. Findikakis
Bechtel Civil Inc.
45/31/18C
45 Freemont Street
San Francisco, CA 94105

Robert A. Fjeld
Environmental Systems Engineering
Department
Clemson University
501 Rhodes Building
Clemson, SC 29634-0919

Peter A. Flavelle
Atomic Energy Control Board
P.O. Box 1046
Ottawa, Ontario K1P 5S9, Canada

Charles Fox
Department of Petroleum Engineering
School of Earth Sciences
Stanford University
Stanford, CA 94305

R. A. Freeze
Department of Geological Sciences
University of British Columbia
Vancouver, BC V6T 2B4, Canada

Daniel A. Galson
U.S. Nuclear Regulatory Commission
Mail Stop 623-SS
Washington, D.C. 20555

Michael Grant
Dames & Moore
6459 W. Quaker Street
Orchard Park, NY 14127

Berge Gureghian
Battelle—OWTD
7000 S. Adams Street
Willowbrook, IL 60521

V. Guvanasen
Battelle Project Management Division
ONWI
P.O. Box 16594
Columbus, OH 43201-6594

Amado Guzman
Hydrology & Water Resources Dept.
University of Arizona
Building 11, Room 308
Tucson, AZ 85721

Susan Hall
Battelle—OWTD
7000 S. Adams Street
Willowbrook, IL 60521

William V. Harper
Battelle—ONWI
P.O. Box 16594
Columbus, OH 43201-6594

Margaret Hart
Texas Water Commission
P.O. Box 13087
Austin, TX 78711

Andreas Haug
INTERA Technologies, Inc.
6850 Austin Center Blvd.
Austin, TX 78731

Kevin Hestir
Lawrence Berkeley Laboratory
1 Cyclotron Road
Berkeley, CA 94720

Robert Hiergesell
Westinghouse Hanford Co.
P.O. Box 1970
Richland, WA 99352

Susanne Hogan
Battelle—OWTD
7000 S. Adams Street
Willowbrook, IL 60521

Deborah Hopkins
90/3147 Lawrence Berkeley Laboratory
University of California
Berkeley, CA 94720

Roland N. Horne
Department of Petroleum Engineering
School of Earth Sciences
Stanford University
Stanford, CA 94305

Richard Horner
University of Washington
FX-10
Seattle, WA 98195

Lowell Howlett
IT Corporation
2790 Masside Blvd.
Monroeville, PA 15146

Peter Hufschmied
NAGRA
Parkstrasse 23
CH-5401 Baden, Switzerland

Satoshi Imamura
Taisei Corporation
c/o Lawrence Berkeley Laboratory
851 Richmond Street
El Cerrito, CA 94530

Elizabeth Jacobson
Battelle—Pacific Northwest Laboratory
P.O. Box 999, Sigma 5 Building
Richland, WA 99352

Bob Johns
Department of Petroleum Engineering
School of Earth Sciences
Stanford University
Stanford, CA 94305

Dr. André G. Journel
Department of Applied Earth Sciences
Mitchell Building
Stanford University
Stanford, CA 94305

Paul G. Kaplan
Sandia National Laboratories
P.O. Box 5800/Division 6315
Albuquerque, NM 87185

K. Karasaki
Lawrence Berkeley Laboratory
1 Cyclotron Road
Berkeley, CA 94720

John D. Kasprowicz
U.S. Department of Energy
9800 South Cass Avenue
Argonne, IL 60439

Catherine Kearney
Battelle—OWTD
7000 S. Adams Street
Willowbrook, IL 60521

M. Kehnemuyi
Battelle—OWTD
7000 S. Adams Street
Willowbrook, IL 60521

Nils A. Kjelbert
Swedish Nuclear Fuel & Waste Management Co. (SKB)
P.O. Box 5864
S-1028 Stockholm, Sweden

A. J. Knepp
Federal Building
U.S. Department of Energy
Richland Operations Office
Richland, WA 99352

Ram B. Kulkarni
Woodward-Clyde Consultants
100 Pringle Avenue, Suite 300
Walnut Creek, CA 94596-3564

Mark Larson
L. Lehman & Associates, Inc.
1103 West Burnsville Parkway
Suite 209
Minneapolis, MN 55337

A. M. LaVenue
INTERA Technologies, Inc.
6850 Austin Center Blvd.
Suite 300
Austin, TX 78731

Bill Lee
Department of Nuclear Engineering
University of California
Berkeley, CA 94720

Linda L. Lehman
L. Lehman & Associates, Inc.
1103 W. Burnsville Parkway
Suite 209
Minneapolis, MN 55337

Lutz Liedtke
Federal Institute for Geosciences & Natural Resources
Stillweg 2
D-3000 Hannover 51, West Germany

Don Livingston
U.S. Department of Energy
Waste Management Project Office
101 Convention Center Drive
Phase 2, Suite 200
Las Vegas, NV 89109

Jane C. S. Long
Lawrence Berkeley Laboratory
1 Cyclotron Road
Berkeley, CA 94720

Gordon R. Luster
Woodward-Clyde Consultants
P.O. Box 1374
Richland, WA 99352

Nai-hsien Mao
Lawrence Livermore National Laboratory
P.O. Box 808, L-208
Livermore, CA 94550

Walter W. Matyskiela
TRW Inc.
1 Federal Systems Park Drive
Fairfax, VA 22033

David McNelis
TRW
R2/2068
1 Space Park
Redondo Beach, CA 90278

Dave McTigue
Sandia National Laboratories
Division 1511
Albuquerque, NM 87185

Joyce Meints
Westinghouse/Hanford
P.O. Box 800
PBB/1100/325
Richland, WA 99352

L. Merkhofer
Applied Decision Analysis, Inc.
Suite 4-255
3000 Sand Hill Road
Menlo Park, CA 94025

Ian Miller
Golder Associates
4104 148th Avenue, NE
Redmond, WA 98052

Donald Myers
University of Arizona
Department of Mathematics
Building 89
Tucson, AZ 85721

Tommy Nedbal
Kemakta Consultants Co.
Pipersg. 27
S-112 28 Stockholm, Sweden

Ivars Neretnieks
Royal Institute of Technology
Department of Chemical Engineering
S10044 Stockholm, Sweden

Walter E. Newcomb
Battelle—OWTD
7000 S. Adams Street
Willowbrook, IL 60521

V. V. Nguyen
EWA, Inc.
133 First Avenue, North
Minneapolis, MN 55401

Thomas J. Nicholson
U.S. Nuclear Regulatory Commission
Waste Management Branch
Washington, D.C. 20555

Stan Niehoff
Argonne National Laboratory
9700 S. Cass Avenue
Argonne, IL 60439

Karla J. Olson
U.S. Department of Energy
Chicago Operations Office
9800 S. Cass Avenue
Argonne, IL 60439

Yuzo Onishi
Kyoto University
c/o Professor H. H. Einstein
Department of Civil Engineering
Massachusetts Institute of Technology
Cambridge, MA 02139

John Osnes
RE/SPEC Inc.
P.O. Box 725
Rapid City, SD 57709

U Sun Park
Science Applications International Corporation
101 Convention Center Drive, Suite 407
Las Vegas, NV 89109

Ralph Patt
State of Oregon
Water Resources Dept.
3850 Portland Road, NE
Salem, OR 97310

Esko Peltonen
Industrial Power Co.
Fredrikinkatu 51-53B
SF-00100 Helsinki, Finland

Harry A. Perry
Science Applications International Corporation
101 Convention Center Dr., Suite 407
Las Vegas, NV 89109

Edwin H. Price
CER Corporation
9800 S. Cass Avenue
Argonne, IL 60439

Robert W. Prindle
Sandia National Laboratories
P.O. Box 5800, Division 6312
Albuquerque, NM 87185

Todd C. Rasmussen
University of Arizona
College of Engineering & Mines
Dept. of Hydrology & Water Resources
Building 11, Room 246
Tucson, AZ 85721

C. A. Rautman
Sandia National Laboratories
P.O. Box 5800
Albuquerque, NM 87185

Dr. J. A. Keith Reid
Waste Management Division
AECL—Whiteshell Nuclear Research Establishment
Pinawa, Manitoba ROE 1L0, Canada

Andrew Reisenauer
Woodward-Clyde Consultants
P.O. Box 1374
Richland, WA 99352

Brian Rutherford
Sandia National Laboratories
Department 7223
Albuquerque, NM 87185

Jamshid Sadeghipour
Battelle—OWTD
7000 S. Adams Street
Willowbrook, IL 60521

Budhi Sagar
Basalt Waste Isolation Project
Westinghouse Hanford Co.
P.O. Box 1970
Richland, WA 99352

Atul M. Salhotra
Woodward-Clyde Consultants
100 Pringle Avenue, -300
Walnut Creek, CA 94596-3564

Rich Schassburger
U.S. Department of Energy
Chicago Operations Office
9800 S. Cass Avenue
Argonne, IL 60439

Todd Schrauf, Senior Hydrogeologist
Golder Associates
4104-148th Avenue, NE
Redmond, WA 98052

Robert F. Schurtz, Consultant
1200 California Street
San Francisco, CA 94109

Rainer K. Senger
Bureau of Economic Geology
University of Texas
Austin, TX 78713

Michael Shea
Battelle—OWTD
7000 S. Adams Street
Willowbrook, IL 60521

Grant Sheng
McMaster University
Room GS216
Hamilton, Ontario, Canada
L8S 4K1

Malcolm Siegel
Sandia National Laboratories
Division 6331
Albuquerque, NM 87185

Leslie Smith
University of British Columbia
6339 Stores Road
Vancouver, British Columbia
V6T 2B4, Canada

Bruce T. Stanley
PB-KBB, Inc.
11767 Katy Freeway, Suite 700
Houston, TX 77079

Ralph Stein
U.S. Department of Energy
Office of Geologic Repositories
1000 Independence Avenue
RW-23
Washington, D.C. 20585

Al Stephens
CAE Support
7 Cyrus Place
San Francisco, CA 94109

Benny Sundstrom
SKI
P.O. Box 27106, Sehistedtsgatan 11
S-10252
Stockholm, Sweden

Edward C. Taylor
TRW, Inc.
1 Federal Systems Park Drive
Fairfax, VA 22033

Shi-Ping Teng
Department of Nuclear Engineering
National Tsing Hua University
Hsinchu, Taiwan 30043 R.O.C.

Tom R. Thomas
Westinghouse Idaho Nuclear
P.O. Box 4000
Idaho Falls, ID 83403

Brian Thompson
H. M. Inspectorate of Pollution
Department of the Environment
Romney House, 43 Marsham Street
London SWI, England

Fonda L. Thompson
Stone & Webster Engineering Corp.
P.O. Box 2325
Boston, MA 02107

Dr. David Tomasko
Energy and Environmental Systems Division
Building 362, Mail Stop 3D
Argonne National Laboratory
9700 S. Cass Avenue
Argonne, IL 60439

Chin-Fu Tsang
Lawrence Berkeley Laboratory
1 Cyclotron Road
Berkeley, CA 94720

Yvonne Tsang
Lawrence Berkeley Laboratory
1 Cyclotron Road
Berkeley, CA 94720

Scott Tyler
Water Resources Center
Desert Research Institute
P.O. Box 60220
Reno, NV 89506

Danielle Veneziano
Massachusetts Institute of Technology
77 Massachusetts Avenue
Room 1-382
Cambridge, MA 02139

Scott Versluis
Battelle—OWTD
Whiteshell Nuclear Research Establishment
Pinawa, Manitoba ROE 1L0, Canada

A. Von Kanel
Federal Institute of Technology
VAW-ETH Zentrum
Ch-8092 Zurich, Switzerland

Charles F. Voss
Battelle—Pacific Northwest Laboratory
Richland, WA 99352

Joseph S. Y. Wang
Earth Sciences Division
Lawrence Berkeley Laboratory
1 Cyclotron Road
Berkeley, CA 94720

Dr. Stephen Wheatcraft
Water Resources Center
Desert Research Center
P.O. Box 60220
Reno, NV 89506

Paul Witherspoon
Lawrence Berkeley Laboratory
1 Cyclotron Road
Berkeley, CA 94720

Pat Wilkey
CER Corporation
9800 S. Cass Avenue
Argonne, IL 60439

Caroline Wittwer
NAGRA
Parkstrasse 23
CH 5401 Baden, Switzerland

Brian Worley
Oak Ridge National Laboratory
P.O. Box X
Oak Ridge, TN 37831-6364

Arto Ylinen
Technical Research Center of Finland
VTT/GEO
Betonimichenkuja 1
02150 ESPOO
Finland

Dae S. Young
Mining Engineering Department
Michigan Technological University
Houghton, MI 49931

Author Index